Lecture Notes in Computer Science 11813

FoLLI Publications on Logic, Language and Information

Subline of Lectures Notes in Computer Science

More information about this series at http://www.springer.com/series/7407

Patrick Blackburn · Emiliano Lorini ·
Meiyun Guo (Eds.)

Logic, Rationality, and Interaction

7th International Workshop, LORI 2019
Chongqing, China, October 18–21, 2019
Proceedings

Springer

Editors
Patrick Blackburn
Roskilde University
Roskilde, Denmark

Emiliano Lorini
Université Paul Sabatier, IRIT-CNRS
Toulouse, France

Meiyun Guo
Southwest University
Chongqing, China

ISSN 0302-9743 ISSN 1611-3349 (electronic)
Lecture Notes in Computer Science
ISBN 978-3-662-60291-1 ISBN 978-3-662-60292-8 (eBook)
https://doi.org/10.1007/978-3-662-60292-8

LNCS Sublibrary: SL1 – Theoretical Computer Science and General Issues

This Springer imprint is published by the registered company Springer-Verlag GmbH, DE part of Springer Nature.
The registered company address is: Heidelberger Platz 3, 14197 Berlin, Germany

Preface

This book contains the papers presented at the 7th International Workshop on Logic, Rationality, and Interaction (LORI-VII 2019), held during October 18–21, 2019, in Chongqing, China, and hosted by the Institute of Logic and Intelligence (ILI) of Southwest University.

As with previous LORI conferences, the focus of the workshop was on the following topics: Agency, Argumentation, and Agreement; Belief Revision and Belief Merging; Belief Representation, Cooperation, Decision Making, and Planning; Natural Language, Philosophy, and Philosophical Logic; and Strategic Reasoning. We received 56 full paper submissions and ended up selecting 33 of them (the authors of two accepted papers chose not to have them published here). The papers were selected on the basis of at least two blind reviews. We decided not to impose a long-paper/short-paper distinction, and to allow all authors 30 minutes of presentation time; this meant that we had to have a number of parallel sessions at the workshop.

In addition, there were presentations by six keynote speakers:

Leila Amgoud	IRIT-CNRS, Toulouse University, France
Kevin Kelly	Carnegie Mellon University, USA
Rineke Verbrugge	University of Groningen, The Netherlands
Michael Wooldridge	University of Oxford, UK
Yanjing Wang	Peking University, China
Thomas Ägotnes	University of Bergen, Norway, and Southwest University, China

The LORI series dates back to August 2007, when the first meeting was held at Beijing Normal University. Since then, a meeting has been held every two years: for further information about the previous conferences, see www.golori.org. Here we will simply remark that over the past 12 years the LORI series has acted as a focus point for research in East Asia on the topics related to logic, rationality, and interaction, and has simultaneously succeeded in drawing scholars from out-side the region to interact and collaborate in this area. The "LORI-community" which has been built up in this way seems destined to become bigger, broader in scope, and to give rise to new and exciting research in the coming years.

As Organizing and Program Committee chairs we would like to thank all the members of the Program Committee for their hard work in a short space of time; the reviewing called forth intense and generous efforts, for which we are deeply grateful.

We are also grateful to Zuojun Xiong who handled the website practicalities and responded fast to our emails, and to Fenrong Liu, Johan van Benthem, and Jeremy Seligman whose timely advice, based on their experience with all the previous LORI conferences, helped us to keep things on track.

Finally, special thanks must also go to the School of Political Science and Public Administration at Southwest University and Institute of Logic and Intelligence (ILI) of Southwest University, China, for their sponsorship for the conference and for financially supporting the proceedings of LORI-VII 2019.

October 2019

Patrick Blackburn
Emiliano Lorini
Meiyun Guo

Organization

Organization Committee

Meiyun Guo (Chair)	Southwest University, China
Zhanglv Li	Southwest University, China
Junli Jiang	Southwest University, China
Zuojun Xiong	Southwest University, China

Program Committee Chairs

Patrick Blackburn	Roskilde University, Denmark
Emiliano Lorini	IRIT-CNRS, Université Paul Sabatier, France

Steering Committee

Johan van Benthem	University of Amsterdam, The Netherlands, Tsinghua University, China, and Stanford University, USA
Shier Ju	Sun Yat-sen University, China
Frank Veltman	University of Amsterdam, The Netherlands
Jialong Zhang	Chinese Academy of Social Sciences, China
Xiangdong He	Southwest University, China
Huaxin Huang	Zhejiang University, China
Wen-fang Wang	Yang Ming University, China
Tomoyuki Yamada	Hokkaido University, China

Scientific Secretaries

Fenrong Liu	Tsinghua University, China
Minghui Xiong	Sun Yat-sen University, China

Program Committee

Natasha Alechina	University of Nottingham, UK
Luciana Benotti	Universidad Nacional de Cordoba, Argentina
Thomas Bolander	Technical University of Denmark, Denmark
Nina Gierasimczuk	Technical University of Denmark, Denmark
Valentin Goranko	Stockholm University, Sweden
Davide Grossi	University of Groningen, The Netherlands
Andreas Herzig	CNRS, IRIT, University of Toulouse, France
Wesley Holliday	University of California Berkeley, USA
Jeff Horty	University of Maryland, USA

Thomas Icard	Stanford University, USA
Guifei Jiang	Nankai University, China
Fengkui Ju	Beijing Normal University, China
Kohei Kishida	Dalhousie University, Canada
Dominik Klein	University of Bamberg, Germany
Jérôme Lang	CNRS, LAMSADE, Université Paris-Dauphine, France
Hu Liu	Sun Yat-sen University, China
Guo Meiyun	South-West University, China
Sara Negri	University of Helsinki, Finland
Hiroakira Ono	JAIST, Japan
Eric Pacuit	University of Maryland, USA
Laurent Perrussel	IRIT, University of Toulouse, France
Gabriella Pigozzi	Université Paris-Dauphine, France
Soroush Rafiee Rad	University of Amsterdam, The Netherlands
Ram Ramanujam	The Institute of Mathematical Sciences, India
Olivier Roy	Universität Bayreuth, Germany
Katsuhiko Sano	Hokkaido University, Japan
François Schwarzentruber	École normale supérieure de Rennes, France
Jeremy Seligman	The University of Auckland, New Zealand
Sonja Smets	University of Amsterdam, The Netherlands
Kaile Su	Griffith University, Australia
Paolo Turrini	Imperial College London, UK
Hans van Ditmarsch	LORIA, CNRS, University of Lorraine, France
Robert van Rooij	University of Amsterdam, The Netherlands
Fernando R. Velázquez-Quesada	University of Amsterdam, The Netherlands
Gregory Wheeler	Frankfurt School of Finance & Management, Germany
Yì N. Wáng	Zhejiang University, China
Audrey Yap	University of Victoria, Australia
Junhua Yu	Tsinghua University, China
Dongmo Zhang	Western Sydney University, Australia
Thomas Ågotnes	University of Bergen, Norway

Additional Reviewers

Carlos Areces
Ryuta Arisaka
Fausto Barbero
Tristan Charrier
Zoé Christoff
Tiziano Dalmonte
Liangda Fang
Rustam Galimullin
Malvin Gattinger
Sujata Ghosh
Marianna Girlando
Umberto Grandi
Zhe Lin
Andrés Occhipinti Liberman
Alessandra Marra
Karl Nygren
Anantha Padmanabha
Carlo Proietti
Arthur Queffelec
Rasmus K. Rendsvig
Chenwei Shi
Xin Sun
Apostolos Tzimoulis
Jonni Virtema
Ren-June Wang
Yanjing Wang
Kun Xing
Zuojun Xiong
Aybüke Özgün

Abstract of Invited Talks

The Dynamics of Group Knowledge and Belief

Thomas Ågotnes

University of Bergen/Southwest University
Thomas.Agotnes@uib.no

Principles of reasoning about group knowledge and belief have received attention over the past decade, in particular in the context of reasoning about the dynamics of interaction. In the talk I will review some of this work, hopefully provide some new insights, and pose some open problems. I will focus on formalisations in modal logic.

What we mean when we say that a group knows something can be radically different depending on context. Well-known notions of group knowledge that have been proposed in the literature include general knowledge (everybody-knows), distributed knowledge, common knowledge, relativised common knowledge. What group belief is, however, is murkier. Applying the same definitions to belief, group belief is not actually always belief. The existence of group belief depends on the particular properties one assumes of belief, and I will map out different possible notions of group belief under different notions of belief. I will also discuss intermediate notions of group belief between distributed and common belief.

Moving to dynamics, we first look at the consequences of adding new group knowledge operators to dynamic epistemic logics such as public announcement logic and action model logic. The relationship between distributed and common knowledge has been of special interest in the dynamic setting, an intuitive idea being that distributed knowledge is potential common knowledge. However this idea is clearly false: it is possible to have distributed knowledge of a Moore-like sentence, which can never even become individual knowledge. I will discuss a dynamic operator that exactly captures what is true after the group have shared all their information with each other; this is what we call resolving distributed knowledge. Intuitions about group knowledge, such as the one just mentioned, are often related to group ability; which states of knowledge a group can make come about. I will thus discuss group knowledge first in the context of general group ability operators such as those found in Alternating-time Temporal Logic and Coalition Logic, and then circle back to dynamic epistemic logics again and discuss cases where ability means ability to achieve some state of knowledge by using public announcements. I will have something to say about how all these different static and dynamic takes on group knowledge and belief are tied together.

Argument-Based Paraconsistent Logics

Leila Amgoud

CNRS
Leila.Amgoud@irit.fr

Handling inconsistency in propositional knowledge bases (KBs) has been studied in AI for a long time. Several two-level logics have been defined: They start with classical propositional logic and define on top of it a non-monotonic logic that infers non-trivial conclusions from an inconsistent KB. There are at least two families of such logics: coherence-based and argument-based logics. The former compute the set of all maximal (for set inclusion) consistent subbases (MCSs) of a KB, and then they apply an inference mechanism for drawing consequences from the MCSs. Argument-based logics follow another process. They justify every candidate consequence of a KB by arguments, generated using the classical consequence relation, then they identify possible conflicts between arguments, evaluate arguments using a formal method, called semantics, and finally keep among the candidate consequences those that are supported by "strong" arguments.

In this talk, I present three families of argument-based logics that use respectively on extension semantics, ranking semantics, and gradual semantics in the evaluation step. I discuss the properties of those logics, and compare them with coherence-based logics.

Realism, Simplicity, and Topology

Kevin Kelly

Carnegie Mellon University
kk3n@andrew.cmu.edu

This is joint work with Hanti Lin, University of California, Davis and Konstantin Genin, University of Toronto.

Scientific realists assure us that simpler theories are better-confirmed by simple data and are, therefore, more worthy of belief. Scientific anti-realists respond that the data might look simple for eternity if the complex theory is true, in which case the realist's assurance would lead to eternal error. We show that the realist's position follows from a learning-theoretic argument for Ockham's razor. The argument is based on the interaction of two topologies on possible worlds: the realist topology of arbitrary similarity in reality, and the empirical topology of arbitrary empirical similarity. The former pertains to the ends of inquiry, and the latter characterizes the means.

Reasoning in Dynamic Games: From Rationality to Rationalization

Rineke Verbrugge

Department of Artificial Intelligence, Bernoulli Institute,
University of Groningen
L.C.Verbrugge@rug.nl

Game theorists have proposed backward induction as the reasoning procedure that rational players follow in dynamic games, on the basis of their collective belief that all participating players are rational. An alternative reasoning procedure is forward induction, in which a player rationalizes any previous apparently irrational move by the opponent. Do people's choices in centipede-like dynamic games fit better with backward or forward induction?

In our experiments (with Sujata Ghosh, Aviad Heifetz, and Harmen de Weerd), participants played a centipede-like game called Marble Drop. The computer opponent was programmed to surprise the participant by deviating often from its backward induction strategy at the beginning of the game. Participants had been told that the computer was optimizing against some belief about the participant's future strategy.

In the aggregate, participants tended to favor the forward induction choice. However, their verbalized strategies usually depended on other features, such as risk aversion, trust or cooperativeness. In a follow-up experiment we compared participants from India, Israel and The Netherlands in the Marble Drop game, with surprising results.

Understanding Equilibrium Properties of Multi-agent Systems

Michael Wooldridge

University of Oxford
michael.wooldridge@cs.ox.ac.uk

Over a twenty minute period on the afternoon 6 May 2010, the Dow Jones industrial average collapsed, at one point wiping a trillion dollars off the value of the US markets. Remarkably, the market recovered in a similarly short period of time, to nearly its position before the collapse. While the precise causes of the so-called "Flash Crash" are complex and controversial, the Flash Crash was only possible because modern international markets are multi-agent systems, in which high frequency trading agents autonomously buy and sell on timescales that are so small that they are far beyond human comprehension or control. There is no reason to believe that the 2010 Flash Crash was an isolated event: and the next one could be even bigger, with potentially devastating global consequences. The 2010 Flash Crash provides a stark illustration of something we have long known: that systems composed of large numbers of multiple interacting components can be subject to rapid, unpredictable swings in behaviour. We urgently need to develop the theory and tools to understand such multi-agent system dynamics.

In this talk, I will present two very different approaches to this problem.

The first views a multi-agent system as a game, in the sense of game theory, with decision-makers interacting strategically in pursuit of their goals. I describe a model we have developed in which players in such a game act in pursuit of temporal logic goals. In such a setting, the key decision problems relate to the properties of a system that hold under the assumption that players choose strategies in (Nash) equilibrium. I conclude by describing a tool, developed by DPhil student Muhammed Najib, through which we can automatically analyse the properties of such equilibria.

The second approach takes a very different approach, in which we use agent-based financial models, involving very large numbers of agents, to understand specifically the factors that can contribute to Flash Crash events, and in particular the phenomenon of "contagion", where stress on one asset leads to other assets being stressed.

This talk will report joint work with Ani Calinescu, Julian Gutierrez, Paul Harrenstein, Muhammed Najib, James Paulin, and Giuseppe Perelli.

Beyond Knowing that: A New Generation of Epistemic Logics

Yanjing Wang

Peking University
wangyanjing@gmail.com

Epistemic logic is a major field of philosophical logic studying reasoning patterns about knowledge and belief. Despite its various applications in epistemology, theoretical computer science, AI, and game theory, the technical developments in the field have been mainly focusing on the propositional part, i.e., the propositional modal logics of "knowing that". However, knowledge is also expressed in everyday life by "knowing whether", "knowing what", "knowing how", "knowing why" and so on (know-wh hereafter). Recent years witnessed a growing interest in new epistemic logics of know-wh motivated by questions in philosophy, AI and linguistics. The new epistemic modalities introduced in those logics usually share, in their semantics, the general schema of 'exists x [] phi' (where [] is a box-modality), e.g., knowing how to achieve phi roughly means that there exists a way such that you know that it is a way to make sure that phi. Therefore they are natural fragments of first-order modal logic. The new axioms of those logics intuitively capture the essential interactions of know-that and other know-wh operators, and the resulting logics are non-normal but decidable.

In this talk, I will first explain the core ideas behind this new research program with some examples, and then propose a more general framework inspired by the concrete know-wh logics, which leads to the discovery of new decidable fragments of first-order modal logic.

Contents

On the Right Path: A Modal Logic for Supervised Learning

Alexandru Baltag[1], Dazhu Li[1,2](✉), and Mina Young Pedersen[1]

[1] ILLC, University of Amsterdam, Amsterdam, The Netherlands
thealexandrubaltag@gmail.com, minaypedersen@gmail.com
[2] Department of Philosophy, Tsinghua University, Beijing, China
lidazhu91@163.com

Abstract. Formal learning theory formalizes the process of inferring a general result from examples, as in the case of inferring grammars from sentences when learning a language. Although empirical evidence suggests that children can learn a language without responding to the correction of linguistic mistakes, the importance of Teacher in many other paradigms is significant. Instead of focusing only on learner(s), this work develops a general framework—the *supervised learning game (SLG)*—to investigate the interaction between *Teacher* and *Learner*. In particular, our proposal highlights several interesting features of the agents: on the one hand, Learner may make mistakes in the learning process, and she may also ignore the potential relation between different hypotheses; on the other hand, Teacher is able to correct Learner's mistakes, eliminate potential mistakes and point out the facts ignored by Learner. To reason about strategies in this game, we develop a *modal logic of supervised learning (SLL)*. Broadly, this work takes a small step towards studying the interaction between graph games, logics and formal learning theory.

Keywords: Formal learning theory · Modal logic · Dynamic logic · Undecidability · Graph games

1 Introduction

Formal learning theory formalizes the process of inferring a general result from examples, as in the case of inferring grammars from sentences when learning a language. A good way of understanding this general process is by treating it as a game played by *Learner* and *Teacher*. It starts with a class of possible worlds, where one of them is the actual one chosen by Teacher. Learner's aim is to get to know which one it is. Teacher inductively provides information about the world, and whenever receiving a piece of information Learner picks a conjecture from the class, indicating which one she thinks is the case. Different success conditions for Learner can be defined. In this article we require that at some finite stage of the procedure Learner decides on a correct hypothesis. This kind of learnability is known as *finite identification* [14].

P. Blackburn et al. (Eds.): LORI 2019, LNCS 11813, pp. 1–14, 2019.
https://doi.org/10.1007/978-3-662-60292-8_1

Although empirical evidence suggests that children can learn a language without responding to the correction of linguistic mistakes [11], the importance of teachers in many other paradigms is significant. For instance, in the paradigm of *learning from queries and counterexamples* [1], Teacher has a strong influence on whether the process is successful. Moreover, results in [10] suggest that a helpful Teacher may make learning easier. In this work, instead of focusing only on Learner, we highlight the interactive natural of learning.

As noted in [10], a concise model for characterizing the interaction between Learner and Teacher is the *sabotage game (SG)*. A SG is played on a graph with a starting node and a goal node, and it goes in rounds: Teacher first cuts an edge in the graph, then Learner makes a step along one of the edges still available. Both of them win iff Learner arrives at the goal node [16]. From the perspective of formal learning theory, this step-by-step game depicts a guided learning situation. Say, a natural interpretation is the situation of theorem proving. In this case, the starting node is given by axioms, the goal node stands for the theorem to be proved, other nodes represent lemmas conjectured by Learner, and edges capture Learner's possible inferences between them. Inferring is represented by moving along those edges. The information provided by Teacher can be treated as his feedback, i.e., removing edges to eliminate wrong inferences. The success condition is given by the winning condition: the learning process has been successful if Learner reaches the goal node, i.e., proving the theorem. For the general correspondence between SG and learning models, we refer to [10].

However, we would argue that this application of SG gives a highly restricted model of learning. For instance,

- Intuitively, all links in the graph are inferences conjectured by Learner, which may include mistakes. From the perspective of Learner, the wrong inferences cannot be distinguished from the correct ones. Although it is reasonable to assume that Teacher is able to do so, SG does not highlight that Learner lacks perfect information. Besides, Teacher in SG has to remove a link in each round, which is overly restrictive.
- Links removed represent wrong inferences between lemmas. So, whether or not a link deleted occurs in Learner's current proof (i.e., the current process) is important. If the proof includes a mistake, any inference after the mistake should not make sense. However, if a potential transition having not occurred in the proof is wrong, Learner can continue with her current proof. Clearly, SG cannot distinguish between these two cases.
- The game does not distinguish between all the various ways Learner can reach the goal. That is, as long as Learner has come to the right conclusion, the game cannot tell us whether Learner has come to this conclusion in a coherent way. Reaching the correct hypothesis by wrong transitions is not reliable. The well-known Gettier cases [9] where one has justified true belief, but not knowledge are also examples of situations in which one wrongly reaches the right conclusion. Thus, the theory developed in [10] is subject to the Gettier problems.

- Teacher can only *delete* links to decide what Learner will not learn, and thus he only teaches what Learner has already conjectured. However, during the process of learning, 'possibilities may also be ignored due to the more questionable practice if assuming that one of the theories under consideration must be true. And complexity can come to be ignored through convention or habit' ([13], p. 260). Hence, it is natural to assume that Learner may ignore the correct relation between different hypotheses.

In this paper, we therefore propose a new game, called the *supervised learning game (SLG)*. This game differs from the SG on several accounts, motivated by the mentioned restrictions. Before introducing its definition, we first define some auxiliary notions.

Let $S = \langle w_0, w_1, ..., w_n \rangle$ be a non-empty, finite sequence. We use $e(S)$ to denote its last element. Define $Set(S) := \{\langle w_0, w_1\rangle, \langle w_1, w_2\rangle, ..., \langle w_{n-1}, w_n\rangle\}$. For the particular case when S is a singleton, $Set(S) := \emptyset$. Besides, for any $\langle w_i, w_{i+1}\rangle \in Set(S)$, define $S|_{\langle w_i, w_{i+1}\rangle} := \langle w_0, w_1, ..., w_u\rangle$, where $\langle w_u, w_{u+1}\rangle = \langle w_i, w_{i+1}\rangle$ and $\langle w_u, w_{u+1}\rangle \neq \langle w_j, w_{j+1}\rangle$ for any $j < i$. Intuitively, $S|_{\langle w_i, w_{i+1}\rangle}$ is obtained by deleting all elements occurring after w_u from S, where $\langle w_u, w_{u+1}\rangle$ is the first occurrence of $\langle w_i, w_{i+1}\rangle$ in S. Say, when $S = \langle a, b, c, a, b\rangle$, we have $S|_{\langle a,b\rangle} = \langle a\rangle$. Now let us introduce SLG.

Definition 1 (SLG). *A SLG $\langle W, R_1, R_2, \langle s\rangle, g\rangle$ is given by a graph $\langle W, R_1, R_2\rangle$, the starting node s and the goal node g. A position of the game is a tuple $\langle R_1^i, S^i\rangle$. The initial position $\langle R_1^0, S^0\rangle$ is given by $\langle R_1, \langle s\rangle\rangle$. Round $n+1$ from position $\langle R_1^n, S^n\rangle$ is as follows: first, Learner moves from $e(S^n)$ to any of its R_1-successors s'; then Teacher does nothing or acts out one of the following three choices:*

(1) Extend R_1^n with some $\langle v, v'\rangle \in R_2$;
(2) Transfer $\langle S^n, s'\rangle$ to $\langle S^n, s'\rangle|_{\langle v,v'\rangle}$ by cutting $\langle v, v'\rangle$ from $Set(\langle S^n, s'\rangle) \setminus R_2$;
(3) Delete some $\langle v, v'\rangle \in (R_1 \setminus R_2) \setminus Set(\langle S^n, s'\rangle)$ from R_1.

The new position, denoted $\langle R_1^{n+1}, S^{n+1}\rangle$, is $\langle R_1^n, S^n\rangle$ (when Teacher does nothing), $\langle R_1^n \cup \{\langle v, v'\rangle\}, \langle S^n, s'\rangle\rangle$ (when he chooses (1)), $\langle R_1^n \setminus \{\langle v, v'\rangle\}, \langle S^n, s'\rangle|_{\langle v,v'\rangle}\rangle$ (if he acts as (2)), or $\langle R_1^n \setminus \{\langle v, v'\rangle\}, \langle S^n, s'\rangle\rangle$ (if he chooses (3)). It ends if Learner arrives at g through an R_2-path $\langle s, ..., g\rangle$ or cannot make a move, with them winning in the former case and losing in the latter.

Intuitively, the clause for Learner illustrates that she cannot distinguish the links starting from the current position. The sequence S^i is her current learning process, which may include mistakes; R_1 represents Learner's possible inferences between conjectures; and R_2 is the correct inferences. For any position $\langle R_1^n, S^n\rangle$ we have $Set(S^n) \subseteq R_1^n$. Besides, both (2) and (3) above are concerned with the case that Teacher eliminates wrong transitions, but there is an important difference. The former one concerns the case that Teacher gives Learner a counterexample to show that she has gone wrong somewhere in her current process, so Learner should move back to the conjecture right before the wrong transition. In contrast, (3) illustrates that Teacher eliminates a wrong transition conjectured that has not occurred in Learner's process yet, therefore this action does not modify Learner's current process.

From the winning condition, we know that both the players cooperate with each other. It is important to recognize that Learner's action does not conflict with her cooperative nature: she makes an effort to achieve the goal in each round. Moreover, it is not hard to see that players cannot win when there exists no R_2-path from the starting node to the goal node. This is reasonable, since their interaction makes sense only when the goal is learnable. The correlation between the situation of theorem proving and SLG is shown in Table 1.

Table 1. Correspondence between theorem proving and supervised learning games.

Theorem proving	Supervised learning games
Axioms	Starting node
Theorem	Goal node
Lemmas conjectured by Learner	Other states except the starting state and the goal state
Learner's possible inference from a to b	R_1-edge from a to b
Correct inference from a to b	R_2-edge from a to b
Inferring b from a	Transition from a to b
Proof for a	R_1-sequence from the starting node to a
Correct proof for a	R_1-sequence S from the starting node to a and $Set(S) \subseteq R_2$
Giving a counterexample to the inference from a to b in the proof S	Modifying S to $S\|_{\langle a,b \rangle}$ ($\langle a,b \rangle \in Set(S)$)
Giving a counterexample to the conjectured inference from a to b not in the proof S	Deleting $\langle a,b \rangle$ from R_1 ($\langle a,b \rangle \notin Set(S)$)
Pointing out a potential inference from a to b not conjectured by Learner before	Extending R_1 with $\langle a,b \rangle$

Remark 1. The interpretation of SLG presented in Table 1 can be easily adapted to characterize other paradigms in formal learning theory, such as language learning and scientific inquiry. More generally, any single-agent games, such as solitaire and computer games, can be converted into SLG. Say, the player (Learner) does not know the correct moves well, but she knows the starting position and the goal position, and has some conjectures about the moves of the game. Besides, she can be taught by Teacher: she just attempts to play it, while Teacher instructs her positively (by revealing more correct moves) or negatively (by pointing out incorrect moves, in which case Learner may have to be moved back to the moment previous to the first incorrect move, if she made any).

Example 1. Let us consider a simple example of SLG, as depicted in Fig. 1. The starting node is a and the goal node is G. We show that players have a winning strategy by depicting the game to play out as follows. Learner begins with moving along the only available edge to node b. Teacher in his turn can make $\langle e, f\rangle$ 'visible' to Learner by adding it to R_1. Then, Learner proceeds to move along $\langle b, c\rangle$, and Teacher extends $\langle b, e\rangle$ to R_1. Afterwards, Learner continues on the only option $\langle c, G\rangle$. Although she now has already arrived at the goal node, her path $\langle a, b, c, G\rangle$ is not an R_2-sequence. So, Teacher can remove $\langle b, c\rangle$ moving Learner back to node b. Next, Learner has to move to e, and Teacher can delete $\langle c, G\rangle$ from R_1. Finally, Learner can arrives at G in 2 steps with Teacher doing nothing. Now we have $Set(\langle a, b, e, f, G\rangle) \subseteq R_2$, so they win.

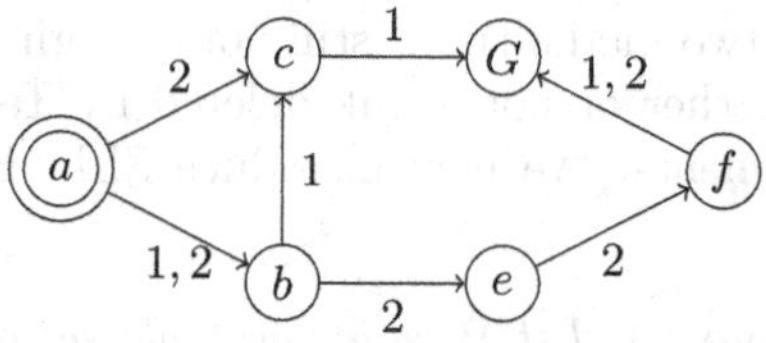

Fig. 1. A SLG game (R_1 is labelled with '1' and R_2 with '2').

To reason about players' strategies in SLG, in what follows we will study SLG from a modal perspective. *Sabotage modal logic (SML)* [5] is known to be a suitable tool to characterize SG, which extends basic modal logic with a sabotage modality $\langle -\rangle\varphi$ stating that there is an edge such that, φ is true at the evaluation node after deleting the edge from the model. However, given the differences between SG and SLG, we will develop a novel *modal logic of supervised learning (SLL)* to capture SLG.

Outline. Section 2 introduces SLL along with its application to SLG and some preliminary observations. Section 3 studies the expressivity of SLL. Section 4 investigates the model checking problem and satisfiability problem for SLL. We end this paper by Sect. 5 on conclusion and future work.[1]

2 Modal Logic of Supervised Learning (SLL)

To be an ideal tool, the logic SLL should at least be able to precisely express players' actions and depict their winning strategies. In this section, we first introduce its language and semantics. Then we analyze its applications to SLG. Finally, we make various observations, including some logical validities and relations between SLL and other logics.

---

[1] Due to page-limit constraints, in this Proceedings version all the proofs are omitted, as well as some auxiliary definitions. The reader interested in these details is encouraged to communicate with the authors.

2.1 Language and Semantics

We begin by considering the action of Learner. In SML, the standard modality $\Diamond$ characterizes the transition from a node to its successors and corresponds well to Learner's actions in SG. However, operator $\Diamond$ is not any longer sufficient in our case. Note that after Teacher cuts a link $\langle w, v\rangle$ from Learner's current process S, Learner should start from w with the new path $S|_{\langle w,v\rangle}$ in the next round. Therefore, the desired operator should remember the history of Learner's movements.

To capture Teacher's action, a natural place to start is by defining operators that correspond to link addition and deletion. There is already a body of literature on logics of these modalities, such as the sabotage operator $\langle -\rangle$ and the bridge operator $\langle +\rangle$ [3]. As mentioned, each occurrence of $\langle -\rangle$ in a formula deletes exactly one link whereas the bridge operator *adds* links stepwise to models. Yet, including these two modalities is still not enough. For instance, we need to take into account whether or not a link deleted by Teacher is a part of the path of Learner's movements. We now introduce SLL. First, let us define its language.

Definition 2 (Language $\mathcal{L}$). *Let* $\mathbf{P}$ *be a countable set of propositional atoms. The formulas of* $\mathcal{L}$ *are recursively defined in the following way:*

$$\varphi ::= p \mid \neg\varphi \mid (\varphi \wedge \varphi) \mid \blacklozenge\varphi \mid \langle -\rangle_1\varphi \mid \langle -\rangle_2\varphi \mid \langle +\rangle\varphi$$

where $p \in \mathbf{P}$. *Notions* $\top$, $\bot$, $\vee$ *and* $\rightarrow$ *are as usual. Besides, we use* $\blacksquare, [-]_1, [-]_2$ *and* $[+]$ *to denote the dual operators of* $\blacklozenge$, $\langle -\rangle_1$, $\langle -\rangle_2$ *and* $\langle +\rangle$ *respectively.*

Several fragments of $\mathcal{L}$ will be studied in the following of the article. For brevity, we use a notational convention listing in subscript all modalities of the corresponding language. For instance, $\mathcal{L}_\blacklozenge$ is the fragment of $\mathcal{L}$ that has only the operator $\blacklozenge$ (besides Boolean connectives $\neg$ and $\wedge$); $\mathcal{L}_{\langle -\rangle_2}$ has only the modality $\langle -\rangle_2$; $\mathcal{L}_{\blacklozenge\langle -\rangle_1}$ has only the modality $\blacklozenge$ and $\langle -\rangle_1$, etc. We now proceed to define the models of SLL.

Definition 3 (Models, Pointed Models and Frames). *A model of SLL is a tuple* $\mathcal{M} = \langle W, R_1, R_2, V\rangle$, *where* W *is a non-empty set of possible worlds,* $R_{i\in\{1,2\}} \subseteq W^2$ *are two binary relations and* $V : \mathbf{P} \rightarrow 2^W$ *is a valuation function.* $\mathcal{F} = \langle W, R_1, R_2\rangle$ *is a frame. Let* S *be an* R_1*-sequence, i.e.,* $Set(S) \subseteq R_1$. *We name* $\langle \mathcal{M}, S\rangle$ *a pointed model, and* S *an evaluation sequence. Usually we write* $\mathcal{M}, S$ *instead of* $\langle \mathcal{M}, S\rangle$.

For brevity, we call R_1 the *conjectured relation* and R_2 the *correct relation.* Besides, we use $\mathfrak{M}$ to denote the class of pointed models and $\mathfrak{M}^\bullet$ the class of pointed models whose sequence S is a singleton. Before introducing the semantics, let us define some preliminary notations.

Assume that $\mathcal{M} = \langle W, R_1, R_2, V\rangle$ is a model, $w \in W$ and $i \in \{1, 2\}$. We use $R_i(w) := \{v \in W | R_i wv\}$ to denote the set of R_i-successors of w in $\mathcal{M}$. Besides, for a sequence S, define $R_i(S) := R_i(e(S))$, i.e., the R_i-successors of a sequence S are exactly the R_i-successors of its last element. For brevity, we also use $S;v$ to denote the sequence extending S with node v. Moreover, $\mathcal{M}\ominus\langle u,v\rangle := \langle W, R_1 \setminus \{\langle u,v\rangle\}, R_2, V\rangle$ is the model obtained by removing $\langle u,v\rangle$ from R_1, and $\mathcal{M}\oplus\langle u,v\rangle := \langle W, R_1 \cup \{\langle u,v\rangle\}, R_2, V\rangle$ is obtained by extending R_1 in $\mathcal{M}$ with $\langle u,v\rangle$. We now have enough background to introduce the semantics of SLL.

Definition 4 (Semantics). *Let $\langle\mathcal{M}, S\rangle$ be a pointed model and $\varphi \in \mathcal{L}$. The semantics of SLL is defined as follows:*

$$
\begin{aligned}
&\mathcal{M}, S \vDash p \Leftrightarrow e(S) \in V(p)\\
&\mathcal{M}, S \vDash \neg\varphi \Leftrightarrow \mathcal{M}, S \nvDash \varphi\\
&\mathcal{M}, S \vDash \varphi\wedge\psi \Leftrightarrow \mathcal{M}, S \vDash \varphi \textit{ and } \mathcal{M}, S \vDash \psi\\
&\mathcal{M}, S \vDash \blacklozenge\varphi \Leftrightarrow \exists v \in W \textit{ s.t. } R_1 e(S) v \textit{ and } \mathcal{M}, S; v \vDash \varphi\\
&\mathcal{M}, S \vDash \langle -\rangle_1\varphi \Leftrightarrow \exists\langle v,v'\rangle \in \mathit{Set}(S)\setminus R_2 \textit{ s.t. } \mathcal{M}\ominus\langle v,v'\rangle, S|_{\langle v,v'\rangle} \vDash \varphi\\
&\mathcal{M}, S \vDash \langle -\rangle_2\varphi \Leftrightarrow \exists\langle v,v'\rangle \in (R_1\setminus R_2)\setminus \mathit{Set}(S) \textit{ s.t. } \mathcal{M}\ominus\langle v,v'\rangle, S \vDash \varphi\\
&\mathcal{M}, S \vDash \langle +\rangle\varphi \Leftrightarrow \exists\langle v,v'\rangle \in R_2\setminus R_1 \textit{ s.t. } \mathcal{M}\oplus\langle v,v'\rangle, S \vDash \varphi
\end{aligned}
$$

By the semantics, a propositional atom p is true at a sequence S if and only if p is true at the last element of S. The cases for $\neg$ and $\wedge$ are as usual. Formula $\blacklozenge\varphi$ states that $e(S)$ has an R_1-successor v such that φ is true at sequence $S;v$. Additionally, $\langle -\rangle_1\varphi$ means that after deleting a link $\langle v,v'\rangle$ from $\mathit{Set}(S)\setminus R_2$, φ is true at $S|_{\langle v,v'\rangle}$. Moreover, $\langle -\rangle_2\varphi$ states that φ holds at S after cutting a link $\langle v,v'\rangle$ belonging to $(R_1\setminus R_2)\setminus\mathit{Set}(S)$. Both $\langle -\rangle_1$ and $\langle -\rangle_2$ require that the link deleted cannot be an R_2-edge. Intuitively, whereas $\langle -\rangle_1$ depicts the case when Teacher deletes a link from Learner's path S, $\langle -\rangle_2$ captures the situation that the link deleted is not a part of S. Finally, $\langle +\rangle\varphi$ means that after extending R_1 with a new link belonging to R_2, φ holds at the current sequence.

A formula φ is *satisfiable* if there is a pointed model $\langle\mathcal{M}, S\rangle \in \mathfrak{M}$ with $\mathcal{M}, S \vDash \varphi$. *Validity* in a model and in a frame is defined in the usual way. Note that the relevant class of models to specify SLL is $\mathfrak{M}^\bullet$, that is, models where the evaluation sequence S starts with a singleton. Hence SLL is the set of $\mathcal{L}$-formulas that are valid in the class $\mathfrak{M}^\bullet$.

For any $\langle\mathcal{M}, S\rangle$ and $\langle\mathcal{M}', S'\rangle$, we say that they are *learning modal equivalent* (notation: $\langle\mathcal{M}, S\rangle \leftrightsquigarrow_l \langle\mathcal{M}', S'\rangle$) iff $\mathcal{M}, S \vDash \varphi \Leftrightarrow \mathcal{M}', S' \vDash \varphi$ for any $\varphi \in \mathcal{L}$. Besides, we define a relation $\mathbf{U} \subseteq \mathfrak{M}\times\mathfrak{M}$ with $\langle\langle\mathcal{M}, S\rangle, \langle\mathcal{M}', S'\rangle\rangle \in \mathbf{U}$ iff $\langle\mathcal{M}', S'\rangle$ is $\langle\mathcal{M}\ominus\langle v,v'\rangle, S|_{\langle v,v'\rangle}\rangle$ for some $\langle v,v'\rangle \in \mathit{Set}(S)\setminus R_2$, $\langle\mathcal{M}\ominus\langle v,v'\rangle, S\rangle$ for some $\langle v,v'\rangle \in (R_1\setminus R_2)\setminus\mathit{Set}(S)$, or $\langle\mathcal{M}\oplus\langle v,v'\rangle, S\rangle$ for some $\langle v,v'\rangle \in R_2\setminus R_1$. We can also iterate this order, to talk about models reachable in finitely many $\mathbf{U}$-steps, obtaining the relation $\mathbf{U}^*$.

2.2 Application: Winning Strategies in SLG

By Definition 4, $\blacklozenge$ captures the actions of Learner, and operators $\langle +\rangle$, $\langle -\rangle_1$ and $\langle -\rangle_2$ characterize those of Teacher. Besides, our logic is expressive enough to describe the winning strategy (if there is one) for players in finite graphs.[2]

[2] Generally speaking, to define the existence of winning strategies for players, we need to extend SLG with some fixpoint operators. We leave this for future inquiry.

Given a finite SLG, let p be a distinguished atom holding only at the goal node. Generally, the winning strategy of Learner and Teacher will be of the following form:

$$\blacksquare \bigcirc_0 \blacksquare \bigcirc_1 \blacksquare \cdots \bigcirc_n \blacksquare (p \wedge [-]_1 \bot) \tag{1}$$

where $\bigcirc_i$ is blank or one of $\langle - \rangle_1$, $\langle - \rangle_2$ and $\langle + \rangle$ for each $i \leq n$. In this formula, the recurring $\blacksquare$ operator depicts Learner's actions and $\bigcirc_i$ Teacher's response. The proposition p signalizes Learner's arrival at the goal, and $[-]_1 \bot$ states that there are no edges in Learner's path that Teacher can cut. Hence, we can conclude that Learner has reached the goal through a sequence of correct edges. It is worth noting that in formula (1) we use $\blacksquare$, other than $\blacklozenge$, to represent Learner's action, although SLG is a cooperative game. Recall the graph in Fig. 1. We observe that $\blacksquare \langle + \rangle \blacksquare \langle + \rangle \blacksquare \langle - \rangle_1 \blacksquare \langle - \rangle_2 \blacksquare \blacksquare (p \wedge [-]_1 \bot)$ holds at the starting node a, so there exists a winning strategy in this specific SLG.

Remark 2. In SG we know that links cut by Teacher represent wrong inferences. However, SG does not tell us anything about the links that remain in the graph. Therefore, winning strategies of the players in SG cannot guarantee against situations like Gettier cases. In contrast, the formula $[-]_1 \bot$ in (1) ensures that Teacher is not allowed to remove any more links from Learner's path. In SLG, a Gettier-style case is that Learner arrives at the goal node with some $\langle u, v \rangle \in R_2 \setminus R_1$ occurring in her path, so Teacher now would be allowed to cut those links. Therefore Gettier cases cannot be winning strategies in SLG.

2.3 Preliminary Observations

As observed, the semantics of SLL is not simple. In this section, we make some preliminary observations on SLL. In particular, we will discuss the relations between SLL and other related logics, and present some logical validities.

First of all, we have the following result on the relation between $\mathcal{L}_\blacklozenge$ and standard modal logic:

Proposition 1. *Let $\mathcal{M} = \langle W, R_1, R_2, V \rangle$ be a model. For any $\langle \mathcal{M}, S \rangle \in \mathfrak{M}$ and $\varphi \in \mathcal{L}_\blacklozenge$, we have $\mathcal{M}, S \vDash \varphi \Leftrightarrow \langle W, R_1, V \rangle, e(S) \vDash \varphi^*$, where φ^* is a standard modal formula obtained by replacing every occurrence of $\blacklozenge$ in φ with $\Diamond$.*

Therefore, essentially the fragment $\mathcal{L}_\blacklozenge$ of $\mathcal{L}$ is the standard modal logic. Moreover, the operator $\langle - \rangle_2$ is much similar to the sabotage operator $\langle - \rangle$:

Proposition 2. *Let $\mathcal{M} = \langle W, R_1, R_2, V \rangle$ be a model, and $R = R_1 \setminus R_2$. For any $\langle \mathcal{M}, w \rangle \in \mathfrak{M}^\bullet$ and $\varphi \in \mathcal{L}_{\langle - \rangle_2}$, we have $\mathcal{M}, w \vDash \varphi \Leftrightarrow \langle W, R, V \rangle, w \vDash \varphi'$, where φ' is a SML formula obtained by replacing every occurrence of $\langle - \rangle_2$ in φ with $\langle - \rangle$.*

Next, we have the following result on the relation between $\mathcal{L}_{\blacklozenge\langle + \rangle}$ and the '*bridge modal logic (BML)*' (i.e., the logic expanding the standard modal logic with the bridge operator):

Proposition 3. *Let $\mathcal{M} = \langle W, R_1, W^2, V\rangle$ be a model. For any $\langle \mathcal{M}, S\rangle \in \mathfrak{M}$ and $\varphi \in \mathcal{L}_{\blacklozenge\langle + \rangle}$, we have $\mathcal{M}, S \vDash \varphi \Leftrightarrow \langle W, R_1, V\rangle, e(S) \vDash \varphi^\star$, where $\varphi^\star$ is a bridge modal formula obtained by replacing every occurrence of $\blacklozenge$ in φ with $\Diamond$.*[3]

From Propositions 1–3, we know that several fragments of SLL are similar to other logics that have been studied. However, as a whole, SLL is not a loose aggregation of these fragments: different operators interact with each other. A typical example is that, for any $\langle \mathcal{M}, w\rangle \in \mathfrak{M}^\bullet$, the formula

$$[-]_1\varphi \tag{2}$$

is valid, as $Set(w) = \emptyset$. However, formula $\blacklozenge\neg[-]_1\varphi$ is satisfiable. This presents a drastic difference between SLL and other logics mentioned so far: in those logics, it is impossible that the evaluation point has access to a node satisfying a contradiction. In order to understand how operators in SLL work, we present some other validities of SLL.

Proposition 4. *Let $p \in \mathbf{P}$ and $\varphi, \psi \in \mathcal{L}$. The following formulas are validities of SLL (w.r.t. $\mathfrak{M}^\bullet$):*

$$p \to \blacksquare[-]_1 p \tag{3}$$

$$p \wedge \blacklozenge\top \to \blacklozenge[-]_1 p \tag{4}$$

$$p \to \bigcirc p \qquad \bigcirc \in \{[-]_2, [+]\} \tag{5}$$

$$\bigcirc(\varphi \to \psi) \to (\bigcirc\varphi \to \bigcirc\psi) \qquad \bigcirc \in \{[-]_2, [+]\} \tag{6}$$

$$\blacksquare^n[-]_1(\varphi \to \psi) \to (\blacksquare^n[-]_1\varphi \to \blacksquare^n[-]_1\psi) \qquad n \in N \tag{7}$$

$$\blacksquare^n\langle -\rangle_1\varphi \to \blacksquare^{n+m}\langle -\rangle_1\varphi \qquad n, m \in N \tag{8}$$

$$\blacklozenge^n\langle -\rangle_1\varphi \to \bigvee_{m<n} \blacklozenge^m\langle -\rangle_2\varphi \qquad 1 \le n \in N \tag{9}$$

Note that formulas (3)–(5) above are not schemata. Although they will still be valid if we replace each propositional atom occurring in them with any Boolean formula, substitution fails in the general case.

Proposition 5. *Validities of SLL are not closed under substitution.*

Interestingly, SLL also has other features that are very different from standard modal logic. For instance, it lacks the tree model property, which holds directly by the following result:

Proposition 6. *$\mathcal{L}_{\blacklozenge\langle - \rangle_1}$ does not have the tree model property.*

As observed, many instances of validities in our logic are not straightforward, and SLL has some distinguishing features. The results that we have so far are not sufficient enough to understand SLL. In the sections to come we will make a deeper investigation into our logic.

---

[3] By abuse of notation, for any $\varphi \in \mathcal{L}_{\blacklozenge\langle + \rangle}$, $\varphi^\star$ is a formula of the bridge modal logic.

3 Expressive Power of SLL

In this section, we study the expressivity of SLL. First, we will show that SLL is still a fragment of first-order logic even though it looks complicated. After this, a suitable notion of bisimulation for SLL is introduced. Finally, we provide a characterization theorem for the logic.

3.1 First-Order Translation

Given the complicated semantics, is SLL still a fragment of FOL? In this section we will provide a positive answer to this question. To do so, we will describe a translation from SLL to FOL. However, compared with that for standard modal logic [7], we now need some new devices.

Let $\mathcal{L}_1$ be the first-order language consisting of countable unary predicates $P_{i\in N}$, two binary relations $R_{i\in\{1,2\}}$, and equivalence $\equiv$. Take any finite, non-empty sequence E of variables. Let y and y' be two fresh variables not appearing in E. When there exists $\langle x,x'\rangle \in Set(E)$ with $x \equiv y$ and $x' \equiv y'$, we define $E|_{\langle y,y'\rangle} := E|_{\langle x,x'\rangle}$. Now let us define the first-order translation.

Definition 5 (First-Order Translation). *Let $E = \langle x_0, x_1, ..., x_n\rangle$ be a finite sequence of variables (non-empty), and $E^- = \{\langle v_0, v_0'\rangle, ..., \langle v_i, v_i'\rangle\}$ and $E^+ = \{\langle u_0, u_0'\rangle, ..., \langle u_j, u_j'\rangle\}$ two finite sets of links. The translation $\mathcal{T}(\varphi, E, E^+, E^-)$ from $\mathcal{L}$-formulas φ to first-order formulas is defined recursively as follows:*

$$\mathcal{T}(p, E, E^+, E^-) = Pe(E)$$

$$\mathcal{T}(\neg\varphi, E, E^+, E^-) = \neg\mathcal{T}(\varphi, E, E^+, E^-)$$

$$\mathcal{T}(\varphi\wedge\psi, E, E^+, E^-) = \mathcal{T}(\varphi, E, E^+, E^-) \wedge \mathcal{T}(\psi, E, E^+, E^-)$$

$$\mathcal{T}(\blacklozenge\varphi, E, E^+, E^-) = \exists y((\bigvee_{\langle x,x'\rangle\in E^+} (e(E)\equiv x \wedge y\equiv x') \vee (R_1 e(E) y \wedge \neg \bigvee_{\langle v,v'\rangle\in E^-} (e(E)\equiv v\wedge y\equiv v'))) \wedge \mathcal{T}(\varphi, E; y, E^+, E^-))$$

$$\mathcal{T}(\langle -\rangle_1\varphi, E, E^+, E^-) = \exists y\exists y'(\bigvee_{\langle x,x'\rangle\in Set(E)\setminus(E^-\cup E^+)} (y\equiv x\wedge y'\equiv x')\wedge R_1 yy' \wedge \neg R_2 yy' \wedge \mathcal{T}(\varphi, E|_{\langle y,y'\rangle}, E^+, E^- \cup \{\langle y,y'\rangle\}))$$

$$\mathcal{T}(\langle -\rangle_2\varphi, E, E^+, E^-) = \exists y\exists y'(R_1 yy' \wedge \neg \bigvee_{\langle x,x'\rangle\in Set(E)\cup E^-\cup E^+} (y\equiv x\wedge y'\equiv x')\wedge \neg R_2 yy' \wedge \mathcal{T}(\varphi, E, E^+, E^- \cup \{\langle y,y'\rangle\}))$$

$$\mathcal{T}(\langle +\rangle\varphi, E, E^+, E^-) = \exists y\exists y'(\neg \bigvee_{\langle x,x'\rangle\in E^-\cup E^+} (y\equiv x\wedge y'\equiv x') \wedge \neg R_1 yy' \wedge R_2 yy' \wedge \mathcal{T}(\varphi, E, E^+ \cup \{\langle y,y'\rangle\}, E^-))$$

From the perspective of SLG, the sequence E denotes Learner's process, and sets E^+ and E^- represent the links that have already been added and deleted respectively. In any translation $\tau(\varphi, E, E^+, E^-)$, each of E^+ and E^- may be extended. For any their extensions $E^+ \cup X$ and $E^- \cup Y$, we have $X \cap Y = \emptyset$. Intuitively, this fact is in line with our semantics: for any $\langle\langle W, R_1, R_2, V\rangle, S\rangle$, we always have $Set(S) \subseteq R_1$ and $(R_1 \setminus R_2) \cap (R_2 \setminus R_1) = \emptyset$, therefore links deleted are different from those added. Another point worth mentioning is that, unlike the case of standard modal logic, generally the translation does not yield a first-order formula with only one free variable. However, it does so when we set E, E^+ and E^- to be a sequence consisting of a singleton, $\emptyset$ and $\emptyset$ respectively. By Definition 5, we have the following result:

Lemma 1. *Let $\mathcal{M}$ be a model and $\tau(\varphi, E, E^+, E^-)$ a translation s.t. $E^+ \cap E^- = \emptyset$. Assume that y and y' are two fresh variables. For any assignment σ, we have $\mathcal{M} \ominus \langle v, v'\rangle \vDash \mathcal{T}(\varphi, E, E^+, E^-)[\sigma]$ iff $\mathcal{M} \vDash \mathcal{T}(\varphi, E, E^+, E^- \cup \{\langle y, y'\rangle\})[\sigma_{y^{(\prime)}:=v^{(\prime)}}]$, for any $\langle v, v'\rangle \in R_1 \setminus R_2$; and $\mathcal{M} \oplus \langle v, v'\rangle \vDash \mathcal{T}(\varphi, E, E^+, E^-)[\sigma]$ iff $\mathcal{M} \vDash \mathcal{T}(\varphi, E, E^+ \cup \{\langle y, y'\rangle\}, E^-)[\sigma_{y^{(\prime)}:=v^{(\prime)}}]$, for any $\langle v, v'\rangle \in R_2 \setminus R_1$.*

With Lemma 1, we now can show the correctness of the translation:

Theorem 1. *Let $\langle \mathcal{M}, S\rangle$ be a pointed model and E an R_1-sequence of variables with the same size as S. For any $\varphi \in \mathcal{L}$, $\mathcal{M}, S \vDash \varphi$ iff $\mathcal{M} \vDash \mathcal{T}\varphi, E, \emptyset, \emptyset)[\sigma_{E:=S}]$.*

Note that the translation in Theorem 1 has an extra requirement on the sequence E, i.e., $Set(E) \subseteq R_1$. Intuitively, this restriction corresponds to the definition of pointed models. For each $\langle \mathcal{M}, w\rangle \in \mathfrak{M}^\bullet$, any extension E' of w fulfils the requirement naturally by Definition 5.

3.2 Bisimulation and Characterization for SLL

The notion of bisimulation serves as a useful tool for establishing the expressive power of modal logics. However, it is not hard to see that SLL is not closed under the standard bisimulation [7]. In this section we introduce a novel notion of 'learning bisimulation (l-bisimulation)' tailored to our logic, which leads to a characterization theorem for SLL as a fragment of first-order logic.

Definition 6 (l-Bisimulation). *For any two models $\mathcal{M} = \langle W, R_1, R_2, V\rangle$ and $\mathcal{M}' = \langle W', R_1', R_2', V'\rangle$, a non-empty relation $Z_l \subseteq \mathbf{U}^*(\langle \mathcal{M}, \mathbf{S}\rangle) \times \mathbf{U}^*(\langle \mathcal{M}', \mathbf{S}'\rangle)$ is an l-bisimulation between the two pointed models $\langle \mathcal{M}, S\rangle$ and $\langle \mathcal{M}', S'\rangle$ (notation: $\langle \mathcal{M}, S\rangle Z_l \langle \mathcal{M}', S'\rangle$) if the following conditions are satisfied:*

Atom: *$\mathcal{M}, S \vDash p$ iff $\mathcal{M}', S' \vDash p$, for each $p \in \mathbf{P}$.*
Zig$_\blacklozenge$: *If there exists $v \in W_1$ s.t. $R_1 wv$, then there exists $v' \in W_1$ s.t. $R_1' w'v'$ and $\langle \mathcal{M}, S; v\rangle Z_l \langle \mathcal{M}', S'; v'\rangle$.*
Zig$_{\langle - \rangle_1}$: *If there is $\langle u, v\rangle \in Set(S) \setminus R_2$, then there is $\langle u', v'\rangle \in Set(S') \setminus R_2'$ with $\langle \mathcal{M} \ominus \langle u, v\rangle, S|_{\langle u,v\rangle}\rangle Z_l \langle \mathcal{M}' \ominus \langle u', v'\rangle, S'|_{\langle u',v'\rangle}\rangle$.*
Zig$_{\langle - \rangle_2}$: *If there exists $\langle u, v\rangle \in (R_1 \setminus R_2) \setminus Set(S)$, then there exists $\langle u', v'\rangle \in (R_1' \setminus R_2') \setminus Set(S')$ with $\langle \mathcal{M} \ominus \langle u, v\rangle, S\rangle Z_l \langle \mathcal{M}' \ominus \langle u', v'\rangle, S'\rangle$.*
Zig$_{\langle + \rangle}$: *If there exists $\langle u, v\rangle \in R_2 \setminus R_1$, then there exists $\langle u', v'\rangle \in R_2' \setminus R_1'$ with $\langle \mathcal{M} \oplus \langle u, v\rangle, S\rangle Z_l \langle \mathcal{M}' \oplus \langle u', v'\rangle, S'\rangle$.*

Zag$_\blacklozenge$, **Zag**$_{\langle - \rangle_1}$, **Zag**$_{\langle - \rangle_2}$ *and* **Zag**$_{\langle + \rangle}$: *the analogous clauses in the converse direction of* **Zig**$_\blacklozenge$, **Zig**$_{\langle - \rangle_1}$, **Zig**$_{\langle - \rangle_2}$ *and* **Zig**$_{\langle + \rangle}$ *respectively.*

For brevity, we write $\langle \mathcal{M}_1, w\rangle \underline{\leftrightarrow}_l \langle \mathcal{M}_2, v\rangle$ *if there is an l-bisimulation* Z_l *with* $\langle \mathcal{M}_1, w\rangle Z_l \langle \mathcal{M}_2, v\rangle$.

The clauses for $\blacklozenge$ is similar to those for the basic modality in the standard bisimulation: they keep the model fixed and extend the evaluation sequence with some of its R_1-successors. In contrast, all of the conditions for $\langle - \rangle_1$, $\langle - \rangle_2$ and $\langle + \rangle$ change the model. In particular, clauses for $\langle - \rangle_2$ and $\langle + \rangle$ do not modify the evaluation sequence, while those for $\langle - \rangle_1$ change both the model and the current sequence. Now we can show the following result:

Theorem 2 ($\underline{\leftrightarrow}_l \subseteq \leftrightsquigarrow_l$). *For any pointed models* $\langle \mathcal{M}, S\rangle$ *and* $\langle \mathcal{M}', S'\rangle$, *it holds that:* $\langle \mathcal{M}, S\rangle \underline{\leftrightarrow}_l \langle \mathcal{M}', S'\rangle \Rightarrow \langle \mathcal{M}, S\rangle \leftrightsquigarrow_l \langle \mathcal{M}', S'\rangle$.

Moreover, the converse direction of Theorem 2 holds for the models that are ω-saturated. To introduce its definition, we need some auxiliary notations. For each finite set Y, we denote the expansion of $\mathcal{L}_1$ with a set Y of constants with $\mathcal{L}_1^Y$, and denote the expansion of $\mathcal{M}$ to $\mathcal{L}_1^Y$ with $\mathcal{M}^Y$. Let $\mathbf{x}$ be a finite tuple of variables. A model $\mathcal{M} = \langle W, R_1, R_2, V\rangle$ is *ω-saturated* if, for every finite subset Y of W, the expansion $\mathcal{M}^Y$ realizes every set $\Gamma(\mathbf{x})$ of $\mathcal{L}_1^Y$-formulas whose finite subsets $\Gamma'(\mathbf{x})$ are all realized in $\mathcal{M}^Y$.

Theorem 3 ($\leftrightsquigarrow_l \subseteq \underline{\leftrightarrow}_l$). *For any ω-saturated* $\langle \mathcal{M}, S\rangle$ *and* $\langle \mathcal{M}', S'\rangle$, *it holds that:* $\langle \mathcal{M}, S\rangle \leftrightsquigarrow_l \langle \mathcal{M}', S'\rangle \Rightarrow \langle \mathcal{M}, S\rangle \underline{\leftrightarrow}_l \langle \mathcal{M}', S'\rangle$.

Thus we have established a match between learning modal equivalence and learning bisimulation for the ω-saturated models. Now, by a simple adaptation of standard arguments (cf. [5,7]), we can show the following result:

Theorem 4. *For any* $\alpha(x) \in \mathcal{L}_1$ *with only one free variable,* $\alpha(x)$ *is equivalent to the translation of some* $\mathcal{L}$*-formula* φ *iff* $\alpha(x)$ *is invariant under l-bisimulation.*

Therefore, in terms of the expressivity, SLL is as powerful as the one free variable fragment of first-order logic that is invariant for l-bisimulation.

4 Model Checking and Satisfiability for SLL

In this section, we consider the model checking problem and satisfiability problem for SLL. Fortunately, the results that we have already shown are quite helpful to establish the complexity result for its model checking problem.

Theorem 5. *Model checking for SLL is PSPACE-complete.*

Note that Theorem 5 also establishes an upper bound for the complexity of SLG. Now we move to considering the satisfiability problem for SLL. In particular, we have the following result:

Theorem 6. *$\mathcal{L}_{\blacklozenge\langle - \rangle_1}$ does not enjoy the finite model property, and its satisfiability problem is undecidable.*

Therefore, it holds directly that:

Theorem 7. *SLL lacks the finite modal property, and its satisfiability problem is undecidable.*

5 Conclusion and Future Work

Summary. Motivated by restrictions on learning in SG, we have extended the game to SLG by naming right and wrong paths of learning, and let Teacher not only delete but also add links. Afterwards, logic SLL was presented, which enables us to reason about players' strategies in SLG. Besides, to understand the new device, we provided some interesting observations and logical validities. Next, we studied basics of its expressivity, including its first-order translation, a novel notion of bisimulation and a characterization theorem for SLL as a fragment of FOL that is invariant under the bisimulation introduced. Finally, it was proved that model checking for SLL is PSPACE-complete, and via the research on $\mathcal{L}_{\blacklozenge\langle - \rangle_1}$ we shown that SLL does not enjoy the finite model property and its satisfiability problem is undecidable.

Relevant and Future Research. Broadly, this work takes a small step towards studying the interaction between graph games, logics and formal learning theory. We are inspired by the work on SG [16], SML [5] and their application to formal learning theory [10]. This article is also relevant to other work studying graph games with modal logics, such as [8,12,15,17]. Technically, the logic SLL has resemblances to several recent logics with model modifiers, such as [2–4]. Besides, instead of updating links, [18] considers a logic of stepwise point deletion, which sheds light on the long-standing open problem of how to axiomatize the sabotage-style modal logics. Moreover, [18] is also helpful to understand the complexity jumps between dynamic epistemic logics of model transformations and logics of freely chosen graph changes recorded in current memory. Another relevant line of research for this paper is epistemic logics. As mentioned already, one goal of our work is to avoid the Gettier problem. Similar to this, [6] uses the topological semantics to study the full belief.

Except what have been studied in this article, there are still various open problems deserving to be studied. From the logic point of view, Sect. 2.2 shows that logic SLL is able to express the winning positions for players in finite games, but to capture those for infinite games, can SLL be expanded with some least-fixpoint operators? From the translation described in Definition 5 we know that SLL are effectively axiomatizable. However, is it possible to axiomatize the logic via a Hilbert-style calculus? In terms of games, we do not know the complexity of SLG, although Theorem 5 provides us with an upper bound. Besides, SLG includes exactly two players, and it is also meaningful to study the cases that are more general.

Acknowledgments. We thank Johan van Benthem and Fenrong Liu for their inspiring suggestions. We also wish to thank three anonymous LORI-VII referees for improvement comments. Dazhu Li is supported by China Scholarship Council and the Major Program of the National Social Science Foundations of China [17ZDA026].

References

1. Angluin, D.: Learning regular sets from queries and counterexamples. Inf. Comput. **75**, 87–106 (1987)
2. Areces, C., Fervari, R., Hoffmann, G.: Moving arrows and four model checking results. In: Ong, L., de Queiroz, R. (eds.) WoLLIC 2012. LNCS, vol. 7456, pp. 142–153. Springer, Heidelberg (2012). https://doi.org/10.1007/978-3-642-32621-9_11
3. Areces, C., Fervari, R., Hoffmann, G.: Relation-changing modal operators. J. IGPL **23**(4), 601–627 (2015)
4. Areces, C., Fervari, R., Hoffmann, G., Martel, M.: Satisfiability for relation-changing logics. J. Logic Comput. **28**, 1143–1470 (2018)
5. Aucher, G., van Benthem, J., Grossi, D.: Modal logics of sabotage revisited. J. Logic Comput. **28**(2), 269–303 (2018)
6. Baltag, A., Bezhanishvili, N., Özgün, A., Smets, S.: A topological approach to full belief. J. Philos. Logic **48**(2), 205–244 (2019)
7. Blackburn, P., de Rijke, M., Venema, Y.: Modal Logic. Cambridge University Press, New York (2001)
8. Blando, F.Z., Mierzewski, K., Areces, C.: The modal logics of the poison game. In: Liu, F., Ono, H., Yu, J. (eds.) Knowledge, Proof and Dynamics. Logic in Asia: Studia Logica Library. Springer (2019, to appear)
9. Gettier, E.: Is justified true belief knowledge? Analysis **23**(6), 121–123 (1963)
10. Gierasimczuk, N., Kurzen, L., Velázquez-Quesada, F.R.: Learning and teaching as a game: a sabotage approach. In: He, X., Horty, J., Pacuit, E. (eds.) LORI 2009. LNCS (LNAI), vol. 5834, pp. 119–132. Springer, Heidelberg (2009). https://doi.org/10.1007/978-3-642-04893-7_10
11. Gold, E.: Language identification in the limit. Inf. Control **10**, 447–474 (1967)
12. Grossi, D., Rey, S.: Credulous acceptability, poison games and modal logic. In: Syntax Meets Semantics (SYSMICS) (2019)
13. Kelly, K.T., Schulte, O., Juhl, C.: Learning theory and the philosophy of science. Philos. Sci. **64**(2), 245–267 (1997)
14. Mukouchi, Y.: Characterization of finite identification. In: Jantke, K.P. (ed.) AII 1992. LNCS, vol. 642, pp. 260–267. Springer, Heidelberg (1992). https://doi.org/10.1007/3-540-56004-1_18
15. Thompson, D.: Local fact change logic. In: Liu, F., Ono, H., Yu, J. (eds.) Knowledge, Proof and Dynamics. Logic in Asia: Studia Logica Library. Springer (2019, to appear)
16. van Benthem, J.: Logic in Games. The MIT Press, Cambridge (2013)
17. van Benthem, J., Liu, F.: Graph games and logic design. Tsinghua Univ. J. **64**(1), 131–139 (2018)
18. van Benthem, J., Mierzewski, K., Blando, F.Z.: The modal logic of stepwise removal, Working manuscript (2019)

Elementary Iterated Revision and the Levi Identity

Jake Chandler[1] and Richard Booth[2](✉)

[1] La Trobe University, Melbourne, Australia
[2] Cardiff University, Cardiff, UK
boothr2@cardiff.ac.uk

Abstract. Recent work has considered the problem of extending to the case of iterated belief change the so-called 'Harper Identity' (HI), which defines single-shot contraction in terms of single-shot revision. The present paper considers the prospects of providing a similar extension of the Levi Identity (LI), in which the direction of definition runs the other way. We restrict our attention here to the three classic iterated revision operators–natural, restrained and lexicographic, for which we provide here the first collective characterisation in the literature, under the appellation of 'elementary' operators. We consider two prima facie plausible ways of extending (LI). The first proposal involves the use of the rational closure operator to offer a 'reductive' account of iterated revision in terms of iterated contraction. The second, which doesn't commit to reductionism, was put forward some years ago by Nayak *et al.*. We establish that, for elementary revision operators and under mild assumptions regarding contraction, Nayak's proposal is equivalent to a new set of postulates formalising the claim that contraction by $\neg A$ should be considered to be a kind of 'mild' revision by A. We then show that these, in turn, under slightly weaker assumptions, jointly amount to the conjunction of a pair of constraints on the extension of (HI) that were recently proposed in the literature. Finally, we consider the consequences of endorsing both suggestions and show that this would yield an identification of rational revision with natural revision. We close the paper by discussing the general prospects for defining iterated revision in terms of iterated contraction.

Keywords: Belief revision · Iterated belief change · Levi identity

1 Introduction

The crucial question of iterated belief change–that is, the question of the rationality constraints that govern the beliefs resulting from a sequence of changes in view–remains very much a live one.

In recent work [3], we have studied in some detail the problem of extending, to the iterated case, a principle of single-step change known as the 'Harper Identity' (henceforth '(HI)') [15]. This principle connects single-step contraction and revision, the two main types of change found in the literature, in a manner that allows one to define the former in terms of the latter. We presented a family of extensions of (HI) characterised by the

This research was partially supported by the Australian Government through an ARC Future Fellowship (project number FT160100092) awarded to Jake Chandler.

P. Blackburn et al. (Eds.): LORI 2019, LNCS 11813, pp. 15–28, 2019.
https://doi.org/10.1007/978-3-662-60292-8_2

satisfaction of an intuitive pair of principles and showed how these postulates could be used to translate principles of iterated revision into principles of iterated contraction.

But (HI) also has a well known companion principle which reverses the direction of definition, allowing one to define single-step revision in terms of single-step contraction: the Levi Identity (henceforth '(LI)') [20]. To date, furthermore, the issue of extending (LI) to the iterated case has barely been discussed. Two noteworthy exceptions are the short papers of Nayak et al. [21] and of Konieczny and Pino Pérez [17]. The second paper argues that no reasonable extension of (LI) will enable us to reduce iterated revision to iterated contraction. The first paper introduces a non-reductionist extension of (LI) consonant with this claim.

The present contribution aims to provide a more comprehensive discussion of the issue, carried out against the backdrop of the aforementioned recent work on (HI). The plan of the paper is as follows. After a preliminary introduction of the formal framework in Sect. 2, we provide, in Sect. 3, a novel result that is of general interest in itself. We collectively characterise the three classic belief revision operators that are the focus of the paper (natural, restrained and lexicographic) under the appellation of 'elementary' operators, showing that they are in fact the *only* operators satisfying a particular set of properties. Section 4 turns to the issue of extending (LI) to the iterated case. We present, in Sect. 4.1, an extension of (LI) based on the concept of rational closure, which would result in a reduction of two-step revision to two-step contraction. Section 4.2 then discusses the non-reductive proposal of [21]. We first establish that, for elementary revision operators and under mild assumptions regarding contraction, it is in fact equivalent to a new set of postulates formalising the claim that contraction by $\neg A$ should be considered to be a kind of 'mild' revision by A. These, in turn, under slightly weaker assumptions, are proven to jointly amount to the conjunction of the aforementioned constraints on the extension of (HI) that were proposed in [3]. In Sect. 4.3, we consider the consequences of endorsing both suggestions and show that this would yield an identification of rational revision with natural revision. In Sect. 5, we briefly discuss the general prospects for defining iterated revision from iterated contraction, critically assessing the central argument of [17]. We conclude, in Sect. 6, with some remaining open questions.

Due to space limitations, only sketches of the more important proofs have been provided. A version of the paper containing all proofs can be accessed online at https://arxiv.org/abs/1907.01224.

2 Preliminaries

The beliefs of an agent are represented by a *belief state* Ψ. The latter determines a *belief set* $[\Psi]$, a deductively closed set of sentences, drawn from a finitely generated, propositional, truth-functional language L. The set of classical logical consequences of $\Gamma \subseteq L$ will be denoted by $\mathrm{Cn}(\Gamma)$. The set of propositional worlds or valuations will be denoted by W, and the set of models of a given sentence A by $[\![A]\!]$.

We consider the three classic belief change operations mapping a prior state Ψ and input sentence A in L onto a posterior state. The operation of *revision* $*$ returns the posterior state $\Psi * A$ that results from an adjustment of Ψ to accommodate the inclusion of A, in such a way as to maintain consistency of the resulting belief set when $\neg A \in [\Psi]$.

The operation of *expansion* $+$ is similar, save that consistency of the resulting beliefs needn't be ensured. Finally, the operation of *contraction* $\div$ returns the posterior state $\Psi \div A$ that results from an adjustment of Ψ to accommodate the retraction of A.

2.1 Single-Step Change

In terms of single-step change, revision and contraction are assumed to satisfy the postulates of Alchourrón, Gärdenfors and Makinson [1] (henceforth 'AGM'), while the behaviour of expansion is constrained by $[\Psi + A] = \mathrm{Cn}([\Psi] \cup \{A\})$. AGM ensures a useful order-theoretic representability of the single-shot revision or contraction dispositions of an agent, such that each Ψ is associated with a total preorder (henceforth '*TPO*') $\preccurlyeq_\Psi$ over W, such that $[\![[\Psi * A]]\!] = \min(\preccurlyeq_\Psi, [\![A]\!])$ ([14, 16]). In this context, the AGM postulate of *Success* ($A \in [\Psi * A]$) corresponds to the requirement that $\min(\preccurlyeq_{\Psi * A}, W) \subseteq [\![A]\!]$. We denote by TPO($W$) the set of all TPOs over W and shall assume that, for every $\preccurlyeq \in \mathrm{TPO}(W)$, there is a state Ψ such that $\preccurlyeq = \preccurlyeq_\Psi$.

Equivalently, these revision dispositions can be represented by a '*conditional belief set*' $[\Psi]_c$. This set extends the belief set $[\Psi]$ by further including various 'conditional beliefs', expressed by sentences of the form $A \Rightarrow B$, where $\Rightarrow$ is a non-truth-functional conditional connective and $A, B \in L$ (we shall call L_c the language that extends L to include such conditionals). This is achieved by means of the so-called *Ramsey Test*, according to which $A \Rightarrow B \in [\Psi]_c$ iff $B \in [\Psi * A]$. In terms of constraints on $[\Psi]_c$, AGM notably ensures that its conditional subset corresponds to a *rational consequence relation*, in the sense of [19] (we shall say, in this case, that $[\Psi]_c$ is rational).

Following convention, we shall call principles couched in terms of belief sets 'syntactic', and call 'semantic' those principles couched in terms of TPOs, denoting the latter by subscripting the corresponding syntactic principle with '$\preccurlyeq$'.

The operations $*$ and $\div$ are assumed to be related in the single-shot case by the Levi and Harper identities, namely

(LI) $\quad [\Psi * A] = \mathrm{Cn}([\Psi \div \neg A] \cup \{A\})$

(HI) $\quad [\Psi \div A] = [\Psi] \cap [\Psi * \neg A]$

with single-shot revision determining single-shot expansion via a third identity:

(TI) $\quad [\Psi + A] = [\Psi * A]$, if $\neg A \notin [\Psi]$

$\quad = L$, otherwise

(LI) can of course alternatively be presented as $[\Psi * A] = [(\Psi \div \neg A) + A]$. Note that, given (HI) and (LI), the constraint $[\![[\Psi * A]]\!] = \min(\preccurlyeq_\Psi, [\![A]\!])$ is equivalent to $[\![[\Psi \div \neg A]]\!] = \min(\preccurlyeq_\Psi, W) \cup \min(\preccurlyeq_\Psi, [\![A]\!])$, so that $\preccurlyeq_\Psi$ equally represents both revision and contraction dispositions.

The motivation for (LI) is the following: The most parsimonious way of modifying $[\Psi]$ so as to include A is to simply add the joint logical consequences of $[\Psi]$ and A. However, $\mathrm{Cn}([\Psi] \cup \{A\})$ needn't be consistent. Hence we first 'make room' for A by considering instead the belief set $[\Psi \div \neg A]$ that results from making the relevant minimal change necessary to achieve consistency.

2.2 Iterated Change

In terms of iterated revision, we shall considerably simplify the discussion by restricting our attention to the three principal operators found in the literature. These are natural revision [8]:

$x \preccurlyeq_{\Psi *_{\mathrm{N}} A} y$ iff (1) $x \in \min(\preccurlyeq_\Psi, [\![A]\!])$, or (2) $x, y \notin \min(\preccurlyeq_\Psi, [\![A]\!])$ and $x \preccurlyeq_\Psi y$

restrained revision [6]:

$x \preccurlyeq_{\Psi *_{\mathrm{R}} A} y$ iff (1) $x \in \min(\preccurlyeq_\Psi, [\![A]\!])$, or (2) $x, y \notin \min(\preccurlyeq_\Psi, [\![A]\!])$ and either (a) $x \prec_\Psi y$ or (b) $x \sim_\Psi y$ and ($x \in [\![A]\!]$ or $y \in [\![\neg A]\!]$)

and lexicographic revision [22]:

$x \preccurlyeq_{\Psi *_{\mathrm{L}} A} y$ iff (1) $x \in [\![A]\!]$ and $y \in [\![\neg A]\!]$, or (2) ($x \in [\![A]\!]$ iff $y \in [\![A]\!]$) and $x \preccurlyeq_\Psi y$.[1]

See Fig. 1.

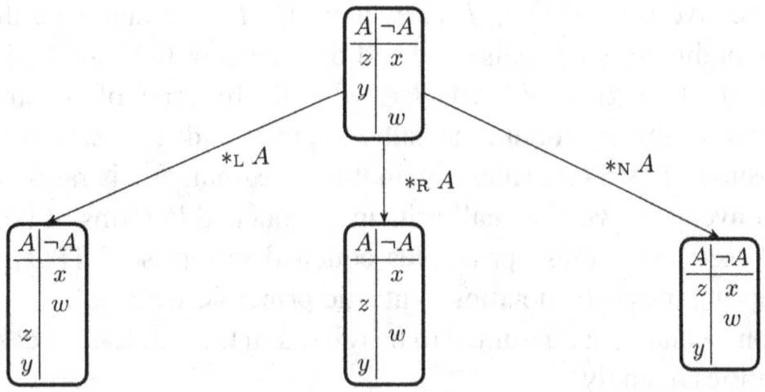

Fig. 1. Elementary revision by A. The boxes represent states and associated TPOs. The lower case letters, which represent worlds, are arranged in such a way that the lower the letter, the lower the corresponding world in the relevant ordering. The columns group worlds according to the sentences that they validate. So, for example, in the initial ordering, we have $w \prec y \prec x \sim z$, with $y, z \in [\![A]\!]$ and $x, w \in [\![\neg A]\!]$ and then, after lexicographic revision by A, $y \prec z \prec w \prec x$.

All three suggestions operate on the assumption that a state Ψ is to be identified with its corresponding TPO $\preccurlyeq_\Psi$ and that belief change functions map pairs of TPOs and sentences onto TPOs, in other words, they entail:

(Red) If $\preccurlyeq_\Psi = \preccurlyeq_{\Psi'}$, then, for any A, $\preccurlyeq_{\Psi * A} = \preccurlyeq_{\Psi' * A}$

[1] These are three of the four iterated revision operators mentioned in Rott's influential survey [23]. The remaining operator the irrevocable revision operator of [24], which has the unusual characteristic of ensuring that prior inputs to revision are retained in the belief set after *any* subsequent revision.

The proposals ensure that $*$ satisfies the postulates of Darwiche and Pearl [10]. In their semantic forms, these are:

(C1$^*_{\preccurlyeq}$) If $x, y \in [\![A]\!]$ then $x \preccurlyeq_{\Psi * A} y$ iff $x \preccurlyeq_{\Psi} y$

(C2$^*_{\preccurlyeq}$) If $x, y \in [\![\neg A]\!]$ then $x \preccurlyeq_{\Psi * A} y$ iff $x \preccurlyeq_{\Psi} y$

(C3$^*_{\preccurlyeq}$) If $x \in [\![A]\!]$, $y \in [\![\neg A]\!]$ and $x \prec_{\Psi} y$, then $x \prec_{\Psi * A} y$

(C4$^*_{\preccurlyeq}$) If $x \in [\![A]\!]$, $y \in [\![\neg A]\!]$ and $x \preccurlyeq_{\Psi} y$, then $x \preccurlyeq_{\Psi * A} y$

Regarding $\div$, we assume that it satisfies the postulates of Chopra et al. [9], given semantically by:

(C1$^{\div}_{\preccurlyeq}$) If $x, y \in [\![\neg A]\!]$ then $x \preccurlyeq_{\Psi \div A} y$ iff $x \preccurlyeq_{\Psi} y$

(C2$^{\div}_{\preccurlyeq}$) If $x, y \in [\![A]\!]$ then $x \preccurlyeq_{\Psi \div A} y$ iff $x \preccurlyeq_{\Psi} y$

(C3$^{\div}_{\preccurlyeq}$) If $x \in [\![\neg A]\!]$, $y \in [\![A]\!]$ and $x \prec_{\Psi} y$ then $x \prec_{\Psi \div A} y$

(C4$^{\div}_{\preccurlyeq}$) If $x \in [\![\neg A]\!]$, $y \in [\![A]\!]$ and $x \preccurlyeq_{\Psi} y$ then $x \preccurlyeq_{\Psi \div A} y$

Concerning the relations between the belief change operators in the iterated case, we will be discussing the extension of (LI), as well as that of (TI), later in the paper. Regarding (HI), a proposal for extending the principle to the two-step case was recently floated in [3]. Semantically speaking, this involved the characterisation of a binary TPO combination operator $\oplus$, such that $\preccurlyeq_{\Psi \div A} = \preccurlyeq_{\Psi} \oplus \preccurlyeq_{\Psi * \neg A}$. Among the baseline constraints on $\oplus$, were a pair of conditions that were shown to be respectively equivalent, in the presence of (C1$^*_{\preccurlyeq}$) and (C2$^*_{\preccurlyeq}$), to the following joint constraints on $\preccurlyeq_{\Psi \div A}$, $\preccurlyeq_{\Psi}$ and $\preccurlyeq_{\Psi * \neg A}$:

(SPU$_{\preccurlyeq}$) If $x \prec_{\Psi} y$ and $x \prec_{\Psi * \neg A} y$ then $x \prec_{\Psi \div A} y$

(WPU$_{\preccurlyeq}$) If $x \preccurlyeq_{\Psi} y$ and $x \preccurlyeq_{\Psi * \neg A} y$ then $x \preccurlyeq_{\Psi \div A} y$

We called operators satisfying such postulates, in addition to (HI), 'TeamQueue combinators'.

3 Elementary Revision Operators

In this section, we demonstrate the relative generality of the results that follow by providing a characterisation result according to which natural, restrained and lexicographic revision operators are the *only* operators satisfying a small set of potentially appealing properties. We shall call operators that satisfy these properties *elementary* revision operators. We define elementary revision operators semantically by:

Definition 1. $*$ *is an* elementary *revision operator iff it satisfies* (C1$^*_{\preccurlyeq}$)–(C4$^*_{\preccurlyeq}$), (IIAP$^*_{\preccurlyeq}$), (IIAI$^*_{\preccurlyeq}$) *and* (Neut$^*_{\preccurlyeq}$).

We have already introduced (C1$^*_{\preccurlyeq}$)-(C4$^*_{\preccurlyeq}$). The remaining principles are new. We call the first of these 'Independence of Irrelevant Alternatives with respect to the prior TPO', after an analogous precept in social choice. For this, we first define the notion of 'agreement' between TPO's on a pair of worlds:

Definition 2. *Where* $\preccurlyeq_\Psi, \preccurlyeq_{\Psi'} \in \mathrm{TPO}(W)$, $\preccurlyeq_\Psi$ *and* $\preccurlyeq_{\Psi'}$ agree *on* $\{x, y\}$ *iff* $\preccurlyeq_\Psi \cap \{x, y\}^2 = \preccurlyeq_{\Psi'} \cap \{x, y\}^2$.

then offer:

($\mathrm{IIAP}^*_{\preccurlyeq}$) If $x, y \notin \min(\preccurlyeq_\Psi, [\![A]\!]) \cup \min(\preccurlyeq_{\Psi'}, [\![A]\!])$, then, if $\preccurlyeq_\Psi$ and $\preccurlyeq_{\Psi'}$ agree on $\{x, y\}$, so do $\preccurlyeq_{\Psi * A}$ and $\preccurlyeq_{\Psi' * A}$

The second new principle–'Independence of Irrelevant Alternatives with respect to the input'–is formally similar to the first. For this we first introduce some helpful notation:

Definition 3. *(i)* $x \preccurlyeq^A y$ *iff* $x \in [\![A]\!]$ *or* $y \in [\![\neg A]\!]$, *(ii)* $x \sim^A y$ *when* $x \preccurlyeq^A y$ *and* $y \preccurlyeq^A x$, *and (iii)* $x \prec^A y$ *when* $x \preccurlyeq^A y$ *but not* $y \preccurlyeq^A x$.

The principle is then given by:

($\mathrm{IIAI}^*_{\preccurlyeq}$) If $x, y \notin \min(\preccurlyeq_\Psi, [\![A]\!]) \cup \min(\preccurlyeq_\Psi, [\![B]\!])$, then, if $\preccurlyeq^A$ and $\preccurlyeq^B$ agree on $\{x, y\}$, so do $\preccurlyeq_{\Psi * A}$ and $\preccurlyeq_{\Psi * B}$

The final principle is a principle of 'Neutrality', again named after an analogous condition in social choice. To the best of our knowledge, it appears here for the first time in the context of belief revision. Its presentation makes use of the following concept:

Definition 4. *Where* $A \in L$, π *is an* A-preserving order isomorphism *from* $\langle W, \preccurlyeq_\Psi, \preccurlyeq^A \rangle$ *to* $\langle W, \preccurlyeq_{\Psi'}, \preccurlyeq^A \rangle$ *iff it is a 1:1 mapping from* W *onto itself such that*

(i) $x \preccurlyeq_\Psi y$ *iff* $\pi(x) \preccurlyeq_{\Psi'} \pi(y)$, *and*
(ii) $x \preccurlyeq^A y$ *iff* $\pi(x) \preccurlyeq^A \pi(y)$

and proceeds as follows:

($\mathrm{Neut}^*_{\preccurlyeq}$) $x \preccurlyeq_{\Psi * A} y$ iff $\pi(x) \preccurlyeq_{\Psi' * A} \pi(y)$, for any A-preserving order isomorphism π from $\langle W, \preccurlyeq_\Psi, \preccurlyeq^A \rangle$ to $\langle W, \preccurlyeq_{\Psi'}, \preccurlyeq^A \rangle$

($\mathrm{IIAP}^*_{\preccurlyeq}$) and ($\mathrm{IIAI}^*_{\preccurlyeq}$) say that the relative ordering of x and y after revising by A depends on only (i) their relative order prior to revision (from ($\mathrm{IIAP}^*_{\preccurlyeq}$)) and (ii) their relative positioning with respect to A (i.e., whether or not they satisfy A) *unless* one of x or y is a minimal A-world, in which case this requirement acquiesces to the Success postulate (from ($\mathrm{IIAI}^*_{\preccurlyeq}$)). ($\mathrm{Neut}^*_{\preccurlyeq}$) is a form of language-independence property, stating that the labels (in terms of valuations) of worlds are irrelevant in determining the posterior TPO. The prima facie appeal of these principles is similar to that of their analogues in social choice, substituting a doxastic interpretation of the ordering for a preferential one.

With this in hand, we can now report that:

Theorem 1. *The only elementary revision operators are lexicographic, restrained and natural revision.*

Proof Sketch: The claim that lexicographic, restrained and natural revision operators are elementary operators is straightforward to establish and we shall not provide it here.

Regarding the other direction of the result: (IIAP$^*_{\preccurlyeq}$) and (Neut$^*_{\preccurlyeq}$) jointly allows us to represent revision by a given sentence A as a quadruple of functions from prior to posterior relations between two arbitrary worlds x and y, such that $x, y \notin \min(\preccurlyeq_\Psi, [\![A]\!])$, one for each of the three following possibilities: (1) $x \in [\![A]\!]$, $y \in [\![\neg A]\!]$, (2) $x, y \in [\![A]\!]$, (3) $x, y \in [\![\neg A]\!]$ (the case in which $x \in [\![\neg A]\!]$, $y \in [\![A]\!]$ is determined by (1), by virtue of (Neut$^*_{\preccurlyeq}$)). These functions can be represented by state diagrams in which the set of states is $\{x \prec y, x \sim y, y \prec x\}$ and the edges represent revisions by A. The postulates (C1$^*_{\preccurlyeq}$) and (C2$^*_{\preccurlyeq}$) entail that, whatever degrees of freedom there are, they are associated with (1). Furthermore, (C3$^*_{\preccurlyeq}$) and (C4$^*_{\preccurlyeq}$) then leave us with at most *six* possible diagrams for (1) (see Fig. 2).

Diagrams (a), (b) and (c) respectively correspond to $*_R A$, $*_L A$ and $*_N A$. However, (d) and (e) are inconsistent with (C2$^*_{\preccurlyeq}$), on pains of triviality. Indeed assume that there exist two worlds $y, z \in [\![\neg A]\!]$ and a world $x \in [\![A]\!]$, such that $z \prec_\Psi y \prec_\Psi x$. Then $z \sim_{\Psi * A} y$, in violation of (C2$^*_{\preccurlyeq}$). (f) exhibits a similar inconsistency. Consider this time the prior TPO given by $y \prec_\Psi \{x, w\}$. We have $z \prec_{\Psi * A} y$. Given $y \prec_\Psi z$, this is again inconsistent with (C2$^*_{\preccurlyeq}$).

So we have established that (IIAP$^*_{\preccurlyeq}$) and (Neut$^*_{\preccurlyeq}$) collectively entail that, for any A, $\Psi * A$ is equal to one of either $\Psi *_R A$, $\Psi *_L A$ or $\Psi *_N A$. But it still remains the case that $*$ coincides with one elementary operator for one input but with another elementary operator for another, so that, for example, $\Psi * A = \Psi *_R A$ while $\Psi * A = \Psi *_L A$. This is ruled out by the final condition (IIAI$^*_{\preccurlyeq}$). □

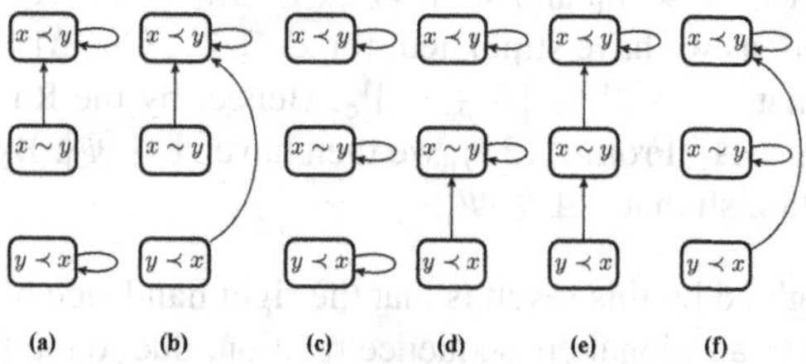

Fig. 2. State diagrams for all $x \in [\![A]\!]$ and all $y \in [\![\neg A]\!]$.

(IIAP$^*_{\preccurlyeq}$) significantly weakens a principle introduced under the name of '(IIA)' in [12], which simply corresponds to the embedded conditional: If $\preccurlyeq_\Psi$ and $\preccurlyeq_{\Psi'}$ agree on $\{x, y\}$, so do $\preccurlyeq_{\Psi * A}$ and $\preccurlyeq_{\Psi' * A}$. (IIAI$^*_{\preccurlyeq}$) amounts to a similar weakening of a condition found in [7]. An interesting question, therefore, arises as to why the stronger principles do not figure in our characterisation.

The unqualified version of (IIAP$^*_{\preccurlyeq}$) is only satisfied by $*_L$, assuming (C1$^*_{\preccurlyeq}$) and (C2$^*_{\preccurlyeq}$) and that the domain of the revision function is TPO(W). Indeed, let $x \in [\![A]\!]$ and $y \in [\![\neg A]\!]$. Then, for any $\preccurlyeq_\Psi$, there will exist $\preccurlyeq_{\Psi'}$ in TPO(W) that agrees with $\preccurlyeq_\Psi$ on $\{x, y\}$ and is such that $x \in \min(\preccurlyeq_{\Psi'}, [\![A]\!])$ (and, since $y \in [\![\neg A]\!]$, $y \notin \min(\preccurlyeq_{\Psi'}, [\![A]\!])$). But by AGM, if $x \in \min(\preccurlyeq_{\Psi'}, [\![A]\!])$ but $y \notin \min(\preccurlyeq_{\Psi'}, [\![A]\!])$, then $x \prec_{\Psi' * A} y$. So, by the unqualified version of ($\mathrm{IIAP}^*_{\preccurlyeq}$), $x \prec_{\Psi * A} y$. Hence, if $x \in [\![A]\!]$ and $y \in [\![\neg A]\!]$, then $x \prec_{\Psi * A} y$, a condition only satisfied by $*_{\mathrm{L}}$, assuming ($\mathrm{C1}^*_{\preccurlyeq}$) and ($\mathrm{C2}^*_{\preccurlyeq}$).[2]

Similarly, Booth and Meyer's strong version of ($\mathrm{IIAI}^*_{\preccurlyeq}$), in conjunction with ($\mathrm{C1}^*_{\preccurlyeq}$)–($\mathrm{C4}^*_{\preccurlyeq}$), can be shown to entail a principle that we have called '($\beta 1+^*_{\preccurlyeq}$)' in previous work [5], where we showed (see Corollary 1 there) to characterise lexicographic revision, given AGM and ($\mathrm{C1}^*_{\preccurlyeq}$)-($\mathrm{C2}^*_{\preccurlyeq}$).

4 Extending the Levi Identity

4.1 A Proposal Involving Rational Closure

The most straightforward syntactic extension of (LI) would involve replacing all belief sets by conditional belief sets, leaving all else unchanged. This would require extending the domain of Cn to subsets of the conditional language L_c, which can be naturally achieved by setting, for $\Delta \subseteq L_c$, $\mathrm{Cn}(\Delta) = \Delta \cup \mathrm{Cn}(\Delta \cap L)$. So we would be considering the claim that $[\Psi * A]_c = \mathrm{Cn}([\Psi \div \neg A]_c \cup \{A\})$. This, however, is a bad idea, since it is easy to show that:

Proposition 1. *If* $[\Psi * A]_c = \mathrm{Cn}([\Psi \div \neg A]_c \cup \{A\})$, *then there are no consistent belief sets, given the two following AGM postulates:*

(K2*) $A \in [\Psi * A]$

(K2÷) $[\Psi \div A] \subseteq [\Psi]$

Proof. Assume $[\Psi * A]_c = \mathrm{Cn}([\Psi \div \neg A]_c \cup \{A\})$. By Success, $A \in [\Psi * A]$. By the Ramsey Test, $\top \Rightarrow A \in [\Psi * A]_c$ and so $\top \Rightarrow A \in \mathrm{Cn}([\Psi \div \neg A]_c \cup \{A\})$. But then, since $\top \Rightarrow A \notin L$ and, as we have stipulated, for $\Delta \subseteq L_c$, $\mathrm{Cn}(\Delta) = \Delta \cup \mathrm{Cn}(\Delta \cap L)$, it must be the case that $\top \Rightarrow A \in [\Psi \div \neg A]_c$. Hence, by the Ramsey Test again, it follows that $A \in [\Psi \div \neg A]$. From (K2÷), we then have $A \in [\Psi]$. By a similar chain of reasoning, we can establish that $\neg A \in [\Psi]$. □

The core issue highlighted by this result is that the right hand side of the equality won't generally correspond to a rational consequence relation, due to the fact that Cn simply yields too small a set of consequences. So a natural suggestion here would be to make use of the rational closure operator $\mathrm{C_{rat}}$ of [19] instead of Cn. Indeed, $\mathrm{C_{rat}}$ has been touted as offering the appropriately conservative way of extending a set of conditionals to something that corresponds to a rational consequence relation (see [19]). This gives us the 'iterated Levi Identity using Rational Closure' (or '(iLIRC)' for short):

(iLIRC) $[\Psi * A]_c = \mathrm{C_{rat}}([\Psi \div \neg A]_c \cup \{A\})$[3]

---

[2] We note that [12] offers a rather different characterisation of lexicographic revision that also involves the unqualified version of ($\mathrm{IIAP}^*_{\preccurlyeq}$).

[3] Strictly speaking, $\mathrm{C_{rat}}$ is an operation on *purely* conditional belief sets. However, it can be obviously generalised to the case in which the set includes non-conditionals, since for any $A \in L$, $A \in [\Psi]_c$ iff $\top \Rightarrow A \in [\Psi]_c$.

4.2 Nayak *et al.*'s 'New Levi Identity'

An alternative extension of (LI) can be obtained by using an iterable expansion operator $+$. This is the 'New Levi Identity' of Nayak *et al.*, which is briefly presented in [21]. Semantically, it is given by:

$$(\mathrm{NLI}_{\preccurlyeq}) \quad \preccurlyeq_{\Psi * A} = \preccurlyeq_{(\Psi \div \neg A)+A}$$

Syntactically, in terms of conditional belief sets, we then would have: $[\Psi * A]_c = [(\Psi \div \neg A) + A]_c$.

It is easily verified that (LI) follows from $(\mathrm{NLI}_{\preccurlyeq})$, if one assumes, for instance, that $\div$ satisfies $(\mathrm{C1}^{\div}_{\preccurlyeq})$. Indeed, (LI) amounts to $\min(\preccurlyeq_{\Psi}, [\![A]\!]) = \min(\preccurlyeq_{\Psi\div\neg A}, W) \cap [\![A]\!] = \min(\preccurlyeq_{\Psi\div\neg A}, [\![A]\!])$, which immediately follows from $(\mathrm{C1}^{\div}_{\preccurlyeq})$. $(\mathrm{NLI}_{\preccurlyeq})$ also has some other interesting general properties. For example, one can show, rather trivially, that:

Proposition 2. *If* $*$ *and* $\div$ *satisfy* $(\mathrm{NLI}_{\preccurlyeq})$*, then, for* $i \in \{1,2,3,4\}$, $(\mathrm{C}i^{\div}_{\preccurlyeq})$ *entails* $(\mathrm{C}i^{*}_{\preccurlyeq})$*, if* $+$ *also satisfies* $(\mathrm{C}i^{*}_{\preccurlyeq})$*.*

This result mirrors a result in [3], in which it was shown that TeamQueue combination allows one to move from each $(\mathrm{C}i^{*}_{\preccurlyeq})$ to the corresponding $(\mathrm{C}i^{\div}_{\preccurlyeq})$.

Assuming, as Nayak *et al.* do, the following natural semantic iterated version of (TI):

$$(\mathrm{iTI}_{\preccurlyeq}) \quad \preccurlyeq_{\Psi+A} = \begin{cases} \preccurlyeq_{\Psi*A}, & \text{if } \min(\preccurlyeq, W) \not\subseteq [\![\neg A]\!] \\ \preccurlyeq_{\Psi_\bot}, & \text{otherwise} \end{cases}$$

where $\Psi_\bot$ is an 'absurd' epistemic state such that $[\Psi_\bot] = L$,[4] $(\mathrm{NLI}_{\preccurlyeq})$ is equivalent to:

$$(\mathrm{iLI}{*}_{\preccurlyeq}) \quad \preccurlyeq_{\Psi * A} = \preccurlyeq_{(\Psi \div \neg A) * A}$$

In what follows, then, we shall use $(\mathrm{NLI}_{\preccurlyeq})$ and $(\mathrm{iLI}{*}_{\preccurlyeq})$ interchangeably. Importantly, while the proposal considered in the previous section was reductive, in the sense that the operator $*$ on the left-hand side of the identity did not appear on the right, $(\mathrm{iLI}{*}_{\preccurlyeq})$ features $*$ on both sides.

To date, however, the implications of this principle have not been studied in any kind of detail. In what follows, we offer some new results of interest. We first note:

Theorem 2. *If* $*$ *is an elementary revision operator and* $\div$ *satisfies* $(\mathrm{C1}^{\div}_{\preccurlyeq})$-$(\mathrm{C4}^{\div}_{\preccurlyeq})$*, then* $*$ *and* $\div$ *satisfy* $(\mathrm{NLI}_{\preccurlyeq})$ *iff they satisfy the following:*

$(\mathrm{C1}^{\div/*}_{\preccurlyeq})$ *If* $x, y \in [\![A]\!]$*, then* $x \preccurlyeq_{\Psi\div\neg A} y$ *iff* $x \preccurlyeq_{\Psi*A} y$

$(\mathrm{C2}^{\div/*}_{\preccurlyeq})$ *If* $x, y \in [\![\neg A]\!]$*, then* $x \preccurlyeq_{\Psi\div\neg A} y$ *iff* $x \preccurlyeq_{\Psi*A} y$

$(\mathrm{C3}^{\div/*}_{\preccurlyeq})$ *If* $x \in [\![A]\!]$, $y \in [\![\neg A]\!]$ *and* $x \prec_{\Psi\div\neg A} y$*, then* $x \prec_{\Psi*A} y$*.*

$(\mathrm{C4}^{\div/*}_{\preccurlyeq})$ *If* $x \in [\![A]\!]$, $y \in [\![\neg A]\!]$ *and* $x \preccurlyeq_{\Psi\div\neg A} y$*, then* $x \preccurlyeq_{\Psi*A} y$*.*

---

[4] Nayak *et al.* have little to say about $\Psi_\bot$, aside from its being the case that $\preccurlyeq_{\Psi_\bot \div A}$ is such that $x \sim_{\Psi_\bot \div A} y$ for all $x, y \in W$. More recently, [11] have suggested that the state resulting from expansion into inconsistency be defined in a more fine-grained manner, in a proposal that involves introducing an 'impossible' world such that $w_\bot \models A$ for all $A \in L$. We refer the reader to their paper for further details, since nothing here hinges on the distinction between these views.

Proof Sketch: Each direction is proven separately. Regarding the right-to-left direction:

Lemma 1. *If* $*$ *is an elementary revision operator,* $\div$ *satisfies* $(\mathrm{C1}^{\div}_{\preccurlyeq})$-$(\mathrm{C4}^{\div}_{\preccurlyeq})$*, and* $\div$ *and* $*$ *satisfy* $(\mathrm{C1}^{\div/*}_{\preccurlyeq})$-$(\mathrm{C4}^{\div/*}_{\preccurlyeq})$*, then* $*$ *and* $\div$ *satisfy* $(\mathrm{NLI}_{\preccurlyeq})$.

Regarding the left-to-right direction, we actually prove the following stronger claim:

Lemma 2. *If* $*$ *satisfies* $(\mathrm{C1}^{*}_{\preccurlyeq})$–$(\mathrm{C4}^{*}_{\preccurlyeq})$*, then there exists* $\div$ *such that* $*$ *and* $\div$ *satisfy* $(\mathrm{NLI}_{\preccurlyeq})$ *only if* $*$ *and* $\div$ *satisfy* $(\mathrm{C1}^{\div/*}_{\preccurlyeq})$–$(\mathrm{C4}^{\div/*}_{\preccurlyeq})$. □

The principles $(\mathrm{C1}^{\div/*}_{\preccurlyeq})$-$(\mathrm{C4}^{\div/*}_{\preccurlyeq})$ are new to the literature and bear an obvious formal resemblance to the postulates of Darwiche and Pearl and of Chopra *et al.*. Taken together, they require contraction by $\neg A$ to be a kind of '*mild revision*' by A, since they tell us that the position of any A-world with respect to any $\neg A$-world is at least as good after revision by A as it is after contraction by $\neg A$.

Somewhat surprisingly (to us), it turns out that these principles are *also* closely connected to the semantic 'TeamQueue combinator' approach to extending the Harper Identity to the iterated case that was proposed in [3]. Indeed, one can show that:

Theorem 3. *If* $*$ *satisfies* $(\mathrm{C1}^{*}_{\preccurlyeq})$-$(\mathrm{C4}^{*}_{\preccurlyeq})$ *and* $\div$ *satisfies* $(\mathrm{C1}^{\div}_{\preccurlyeq})$-$(\mathrm{C4}^{\div}_{\preccurlyeq})$*, then* $*$ *and* $\div$ *satisfy* $(\mathrm{C1}^{\div/*}_{\preccurlyeq})$-$(\mathrm{C4}^{\div/*}_{\preccurlyeq})$ *iff they satisfy* $(\mathrm{SPU}_{\preccurlyeq})$ *and* $(\mathrm{WPU}_{\preccurlyeq})$.

Proof: We prove the result in two parts. Firstly we establish the following strengthening of the right-to-left direction of the claim:

Lemma 3. *Given* $(\mathrm{SPU}_{\preccurlyeq})$ *and* $(\mathrm{WPU}_{\preccurlyeq})$*, for all* $1 \leq i \leq 4$, $(\mathrm{C}i^{*}_{\preccurlyeq})$ *entails* $(\mathrm{C}i^{\div/*}_{\preccurlyeq})$.

Regarding $i = 1, 2$: We provide the proof for $i = 1$, since the case in which $i = 2$ is handled analogously. Assume $x, y \in [\![A]\!]$. From left to right: Assume $x \preccurlyeq_{\Psi \div \neg A} y$. By the contrapositve of $(\mathrm{SPU}_{\preccurlyeq})$, either $x \preccurlyeq_{\Psi} y$ or $x \preccurlyeq_{\Psi * A} y$. If the latter holds, we are done. So assume that $x \preccurlyeq_{\Psi} y$. Then the required result follows by $(\mathrm{C1}^{*}_{\preccurlyeq})$. From right to left: Assume $x \preccurlyeq_{\Psi * A} y$. By $(\mathrm{C1}^{*}_{\preccurlyeq})$, $x \preccurlyeq_{\Psi} y$. By $(\mathrm{WPU}_{\preccurlyeq})$, $x \preccurlyeq_{\Psi \div \neg A} y$, as required.

Regarding $i = 3, 4$: We provide the proof for $i = 3$, since the case in which $i = 4$ is handled analogously (using $(\mathrm{SPU}_{\preccurlyeq})$ rather than $(\mathrm{WPU}_{\preccurlyeq})$). We derive the contrapositive. Assume $x \in [\![A]\!]$, $y \in [\![\neg A]\!]$ and $y \preccurlyeq_{\Psi * A} x$. If $y \preccurlyeq_{\Psi} x$, then, from $y \preccurlyeq_{\Psi * A} x$, we have $y \preccurlyeq_{\Psi \div \neg A} x$, by $(\mathrm{WPU}_{\preccurlyeq})$, as required. So assume $x \prec_{\Psi} y$. By $(\mathrm{C3}^{*}_{\preccurlyeq})$, $x \prec_{\Psi * A} y$. Contradiction. This completes the proof of Lemma 3.

Concerning the left-to-right direction of our principal claim, we show:

Lemma 4. $(\mathrm{C3}^{\div}_{\preccurlyeq})$, $(\mathrm{C4}^{\div}_{\preccurlyeq})$*, and* $(\mathrm{C1}^{\div/*}_{\preccurlyeq})$–$(\mathrm{C4}^{\div/*}_{\preccurlyeq})$ *entail* $(\mathrm{SPU}_{\preccurlyeq})$ *and* $(\mathrm{WPU}_{\preccurlyeq})$.

We just prove this in relation to $(\mathrm{WPU}_{\preccurlyeq})$, using $(\mathrm{C1}^{\div/*}_{\preccurlyeq})$, $(\mathrm{C2}^{\div/*}_{\preccurlyeq})$, $(\mathrm{C3}^{\div/*}_{\preccurlyeq})$ and $(\mathrm{C4}^{\div}_{\preccurlyeq})$. The proof in relation to $(\mathrm{SPU}_{\preccurlyeq})$ is analogous but uses $(\mathrm{C1}^{\div/*}_{\preccurlyeq})$, $(\mathrm{C2}^{\div/*}_{\preccurlyeq})$, $(\mathrm{C4}^{\div/*}_{\preccurlyeq})$ and $(\mathrm{C3}^{\div}_{\preccurlyeq})$ instead. Assume that $x \preccurlyeq_{\Psi} y$ and $x \preccurlyeq_{\Psi * A} y$. We want to show $x \preccurlyeq_{\Psi \div \neg A} y$. If (a) $x, y \in [\![A]\!]$, (b) $x, y \in [\![\neg A]\!]$, or (c) $x \in [\![\neg A]\!]$ and $y \in [\![A]\!]$, this follows from $x \preccurlyeq_{\Psi * A} y$, by $(\mathrm{C1}^{\div/*}_{\preccurlyeq})$, $(\mathrm{C2}^{\div/*}_{\preccurlyeq})$ or $(\mathrm{C3}^{\div/*}_{\preccurlyeq})$, respectively. If (d) $x \in [\![A]\!]$ and $y \in [\![\neg A]\!]$, then it follows from $x \preccurlyeq_{\Psi} y$, by $(\mathrm{C4}^{\div}_{\preccurlyeq})$. □

In conjunction with Theorem 2, Theorem 3 entails:

Corollary 1. *If* $*$ *is an elementary revision operator and* $\div$ *satisfies* $(\mathrm{C1}^{\div}_{\preccurlyeq})$-$(\mathrm{C4}^{\div}_{\preccurlyeq})$, *then* $*$ *and* $\div$ *satisfy* $(\mathrm{NLI}_{\preccurlyeq})$ *iff they satisfy* $(\mathrm{SPU}_{\preccurlyeq})$ *and* $(\mathrm{WPU}_{\preccurlyeq})$.

In this particular context, then, $(\mathrm{NLI}_{\preccurlyeq})$ simply amounts to the conjunction of a pair of constraints proposed in the context of extending (HI) to the iterated case.

4.3 Rational Closure and the New Levi Identity

At this stage, we have considered both a potentially promising reductive proposal and a promising non-reductive one. A natural question, then, is: How would these two suggestions fare in conjunction with one another? To answer this question, we provide the semantic counterpart for our first principle, which was formulated only syntactically:

Theorem 4. *Given AGM,* (iLIRC) *is equivalent to :*

$(\mathrm{iLIRC}_{\preccurlyeq})$ $\preccurlyeq_{\Psi * A} = \preccurlyeq_{(\Psi \div \neg A) *_{\mathrm{N}} A}$[5]

Proof Sketch: We prove the claim by establishing that (iLIRC) ensures that $\preccurlyeq_{\Psi * A}$ is the 'flattest' TPO such that the following *lower bound principle* is satisfied: $[\Psi \div \neg A]_{\mathrm{c}} \cup \{A\} \subseteq [\Psi * A]_{\mathrm{c}}$. In view of Definitions 20 and 21 of [19], the upshot of this is then that $\preccurlyeq_{(\Psi \div \neg A) *_{\mathrm{N}} A}$ corresponds to the rational closure of $[\Psi \div \neg A]_{\mathrm{c}} \cup \{A\}$.

We first note that, given AGM, the lower bound principle can be semantically expressed as follows: (a) If $x \prec_{\Psi \div \neg A} y$, then $x \prec_{\Psi * A} y$ and (b) $\min(\preccurlyeq_{\Psi * A}, W) \subseteq [\![A]\!]$. With this in hand, we prove two lemmas, which are individually easy to establish. First:

Lemma 5. *If* $*$ *and* $\div$ *satisfy* $(\mathrm{iLIRC}_{\preccurlyeq})$, *then they satisfy the lower bound principle.*

For the second lemma, we make use of the convenient representation of TPOs by their corresponding *ordered partitions* of W. The ordered partition $\langle S_1, S_2, \ldots S_m \rangle$ of W corresponding to a TPO $\preccurlyeq$ is such that $x \preccurlyeq y$ iff $r(x, \preccurlyeq) \leq r(y, \preccurlyeq)$, where $r(x, \preccurlyeq)$ denotes the 'rank' of x with respect to $\preccurlyeq$ and is defined by taking $S_{r(x,\preccurlyeq)}$ to be the cell in the partition that contains x. The lemma is given as follows:

Lemma 6. $\preccurlyeq_{(\Psi \div \neg A) *_{\mathrm{N}} A} \sqsupseteq \preccurlyeq$, *for any TPO* $\preccurlyeq$ *satisfying the lower bound principle.*

where:

Definition 5. $\sqsupseteq$ *is a binary relation on the set of TPOs over* W *such such that, for any TPOs* $\preccurlyeq_1$ *and* $\preccurlyeq_2$, *whose corresponding ordered partitions are given by* $\langle S_1, S_2, \ldots, S_m \rangle$ *and* $\langle T_1, T_2, \ldots, T_m \rangle$ *respectively,* $\preccurlyeq_1 \sqsupseteq \preccurlyeq_2$ *iff either (i)* $S_i = T_i$ *for all* $i = 1, \ldots, m$, *or (ii)* $S_i \supset T_i$ *for the first* i *such that* $S_i \neq T_i$.

$\sqsupseteq$ partially orders TPO(W) according to comparative 'flatness', with the flatter TPOs appearing 'greater' in the ordering, so that $\preccurlyeq_1 \sqsupseteq \preccurlyeq_2$ iff $\preccurlyeq_1$ is at least as as flat as $\preccurlyeq_2$.

[5] Note that [17] explicitly mention $(\mathrm{iLIRC}_{\preccurlyeq})$ and flag it out as a potentially desirable principle.

Let $\langle T_1, \ldots, T_m \rangle$ be the ordered partition corresponding to the TPO $\preccurlyeq_{(\Psi \div \neg A) *_{\mathrm{N}} A}$, which we will denote by $\preccurlyeq_{\mathrm{N}}$. Let $\preccurlyeq$ be any TPO satisfying the lower bound condition: (a) If $x \prec_{\Psi \div \neg A} y$, then $x \prec y$ and (b) $\min(\preccurlyeq, W) \subseteq [\![A]\!]$. Let $\langle S_1, \ldots, S_n \rangle$ be its corresponding ordered partition. The proof of the lemma then amounts to showing that $\preccurlyeq_{\mathrm{N}} \sqsupseteq \preccurlyeq$. □

With this in hand, the consequences of endorsing (iLIRC) on the heels of ($\mathrm{NLI}_{\preccurlyeq}$) should be obvious: rational iterated revision would have to coincide with natural revision.

This raises an interesting question: For each remaining elementary operator $*$, does there exist a suitable alternative closure operator C, such that $[\Psi * A]_{\mathrm{c}} = \mathrm{C}([\Psi \div \neg A]_{\mathrm{c}} \cup \{A\})$ iff $\preccurlyeq_{\Psi * A} = \preccurlyeq_{(\Psi \div \neg A) * A}$?[6] Indeed, although rational closure is by far the most popular closure operator in the literature, alternative closure operators have been proposed, including, for instance the lexicographic closure operator of [18] or again the maximum entropy closure operator of [13]. Furthermore, there has been some limited work on potential connections between closure operators and revision operators (namely [2]). However, this work has only focussed on the relation between lexicographic closure and lexicographic revision and its pertinence to the current problem remains unclear.

Although we do not currently have an answer to our question, we can report that the existence of suitable relevant closure operators will very much depend on the manner in which one extends (HI) to the iterated case. To illustrate, in a previous discussion of the issue [3], we considered a particular TeamQueue combinator, $\oplus_{\mathrm{STQ}}$. We showed, in Sect. 6 of that paper, that for $* = *_{\mathrm{L}}$ or $* = *_{\mathrm{R}}$, the equality $\preccurlyeq_{\Psi \div A} = \preccurlyeq_{\Psi} \oplus_{\mathrm{STQ}} \preccurlyeq_{\Psi * \neg A}$ entails that $\div = \div_{\mathrm{STQL}}$, where $\div_{\mathrm{STQL}}$ is an iterated contraction operator that we call 'STQ-Lex'. We can, however, show the following:

Proposition 3. *If* $* = *_{\mathrm{L}}$ *or* $* = *_{\mathrm{R}}$ *and* $\div = \div_{\mathrm{STQL}}$, *then there exists no closure operator* C, *satisfying the property of Rational Identity:*

($\mathrm{RID^c}$) *If* Δ *is rational, then* $\mathrm{C}(\Delta) = \Delta$.

such that both ($\mathrm{NLI}_{\preccurlyeq}$) *and* $[\Psi * A]_{\mathrm{c}} = \mathrm{C}([\Psi \div \neg A]_{\mathrm{c}} \cup \{A\})$ *are true.*

($\mathrm{RID^c}$) seems a desirable property of closure operators, which aim to extend a set of conditionals Δ to that rational set of conditionals whose endorsement is mandated by that of Δ. The standard postulate of Inclusion ($\Delta \subseteq \mathrm{C}(\Delta)$) tell us that C must extend Δ to a rational superset of Δ. ($\mathrm{RID^c}$) adds to this the notion that if Δ 'ain't broke', it needn't be 'fixed'.

Interestingly, the proof of this impossibility result fails to go through when $* = *_{\mathrm{L}}$ and $\div = \div_{\mathrm{P}}$, where $\div_{\mathrm{P}}$ is the priority contraction operator of [21]. In [3] we note that priority contraction can be recovered from lexicographic revision via a particular TeamQueue combinator. Furthermore, the same combinator can be used to define a contraction operator from restrained revision (call it $\div_{\mathrm{R}}$). Again, the proof of the above result breaks down when $* = *_{\mathrm{R}}$ and $\div = \div_{\mathrm{R}}$.

[6] Note the importance of (Red) in making this kind of correspondence even *prima facie* possible. Indeed, if (Red) fails, then $\preccurlyeq_{\Psi \div \neg A}$ and A will fail to jointly determine $\preccurlyeq_{\Psi * A}$. In syntactic terms, $[\Psi \div \neg A]_{\mathrm{c}}$ and A will fail to jointly determine $[\Psi * A]_{\mathrm{c}}$.

5 Is Iterated Revision Reducible to Iterated Contraction?

Konieczny and Pino Pérez [17, Theorem 5] plausibly claim that, for a finitely generated language, the cardinality of (i) the set of revision operators that satisfy both the AGM postulates for revision and $(C1^*_{\preccurlyeq})$–$(C4^*_{\preccurlyeq})$ is strictly greater than the cardinality of (ii) the set of contraction operators that satisfy both the AGM postulates for contraction and $(C1^{\div}_{\preccurlyeq})$–$(C4^{\div}_{\preccurlyeq})$. From this, they conclude that there is no bijection between rational iterated revision and contraction operators and hence no reduction of iterated revision to iterated contraction.

But this conclusion is not warranted without a further argument to the effect that every member of (i) is rational. In other words, it could be the case that $(C1^*_{\preccurlyeq})$–$(C4^*_{\preccurlyeq})$ need supplementing. This has certainly been the belief of the proponents of the various elementary revision operators that we have discussed in the present paper. And indeed, the proponent of $*_N$ could claim, endorsing our $\oplus_{STQ}$-based extension of (HI), that rational contraction goes by natural contraction. By the same principle, proponents of $*_R$ or $*_L$ could respectively claim that rational contraction goes by natural contraction or STQ-Lex contraction, respectively (see [3, Sect. 6]). Those are three candidate bijections that are all consistent, furthermore, with $(NLI_{\preccurlyeq})$.

One could nevertheless run an arguably plausible argument to Konieczny and Pino Pérez's desired conclusion based on the observation that natural and restrained revision are both mapped onto natural contraction by the $\oplus_{STQ}$ method. Even if one thinks that it is implausible to claim that iterated change *must* comply with one of either restrained or natural revision, it is not implausible to claim that it sometimes *may* comply with either. In other words: There plausibly exists at least one prior TPO that is rationally consistent with two distinct potential posterior TPOs, respectively obtained via natural and restrained revision by a given sentence A. Given the $\oplus_{STQ}$-based extension of (HI), only one posterior TPO can be obtained by contraction by $\neg A$, namely the one obtained by natural contraction by $\neg A$. But if this is true, iterated revision dispositions cannot be recovered from iterated contraction dispositions.

6 Conclusions and Further Work

We have considered two possible extensions of (LI) to the iterated case: a reductive proposal (iLIRC) based on the rational closure operator, and a non-reductive proposal $(NLI_{\preccurlyeq})$ that involves a contraction step, followed by an expansion. We have shown that, when restricted to a popular class of 'elementary' revision operators, $(NLI_{\preccurlyeq})$ is in fact equivalent, under weak assumptions, to both (i) a new set of postulates $(C1^{\div/*}_{\preccurlyeq})$-$(C4^{\div/*}_{\preccurlyeq})$ and (ii) a pair of principles recently defended in the literature on (HI).

However, it has also been noted that (iLIRC) has strong consequences when conjoined with $(NLI_{\preccurlyeq})$. This suggests the need for (1) a future consideration of various alternatives to the former that make use of surrogate closure operators.

Furthermore, the revision operators of the class that we have focussed on have been criticised for their equation of belief states with TPOs (the principle (Red); see [4]). One obvious extension of our work would be (2) an exploration of the extent to which the results reported in Sect. 4 carry over to operators that avoid this identification, such as the POI operators of [5].

References

1. Alchourrón, C.E., Gärdenfors, P., Makinson, D.: On the logic of theory change: partial meet contraction and revision functions. J. Symb. Log. **50**(2), 510–530 (1985)
2. Booth, R.: The lexicographic closure as a revision process. J. Appl. Non-Class. Log. **11**(1–2), 35–58 (2001)
3. Booth, R., Chandler, J.: Extending the Harper identity to iterated belief change. In: Proceedings of IJCAI 2016, pp. 987–993 (2016)
4. Booth, R., Chandler, J.: The irreducibility of iterated to single revision. J. Philos. Log. **46**(4), 405–418 (2017)
5. Booth, R., Chandler, J.: On strengthening the logic of iterated belief revision: proper ordinal interval operators. In: Proceedings of KR 2018, pp. 210–219 (2018)
6. Booth, R., Meyer, T.: Admissible and restrained revision. J. Artif. Intell. Res. **26**(1), 127–151 (2006)
7. Booth, R., Meyer, T.: How to revise a total preorder. J. Philos. Log. **40**, 1–46 (2011)
8. Boutilier, C.: Iterated revision and minimal change of conditional beliefs. J. Philos. Log. **25**(3), 263–305 (1996)
9. Chopra, S., Ghose, A., Meyer, T., Wong, K.S.: Iterated belief change and the recovery axiom. J. Philos. Log. **37**(5), 501–520 (2008)
10. Darwiche, A., Pearl, J.: On the logic of iterated belief revision. Artif. Intell. **89**(1), 1–29 (1997)
11. Fermé, E., Wassermann, R.: On the logic of theory change: iteration of expansion. J. Braz. Comput. Soc. **24**(1), 8 (2018)
12. Glaister, S.M.: Symmetry and belief revision. Erkenntnis **49**(1), 21–56 (1998)
13. Goldszmidt, M., Morris, P.H., Pearl, J.: A maximum entropy approach to nonmonotonic reasoning. In: Proceedings of AAAI 1990, pp. 646–652 (1990)
14. Grove, A.: Two modellings for theory change. J. Philos. Log. **17**(2), 157–170 (1988)
15. Harper, W.L.: Rational conceptual change. In: PSA: Proceedings of the Biennial Meeting of the Philosophy of Science Association, pp. 462–494. JSTOR (1976)
16. Katsuno, H., Mendelzon, A.O.: Propositional knowledge base revision and minimal change. Artif. Intell. **52**(3), 263–294 (1991)
17. Konieczny, S., Pino Pérez, R.: On iterated contraction: syntactic characterization, representation theorem and limitations of the levi identity. In: Moral, S., Pivert, O., Sánchez, D., Marín, N. (eds.) SUM 2017. LNCS (LNAI), vol. 10564, pp. 348–362. Springer, Cham (2017). https://doi.org/10.1007/978-3-319-67582-4_25
18. Lehmann, D.: Another perspective on default reasoning. Ann. Math. Artif. Intell **15**, 61–82 (1992)
19. Lehmann, D., Magidor, M.: What does a conditional knowledge base entail? Artif. intell. **55**(1), 1–60 (1992)
20. Levi, I.: Subjunctives, dispositions and chances. Synthese **34**(4), 423–455 (1977)
21. Nayak, A.C., Goebel, R., Orgun, M.A., Pham, T.: Iterated belief change and the levi identity. In: Belief Change in Rational Agents: Perspectives from Artificial Intelligence, Philosophy, and Economics, pp. 7–12, August 2005
22. Nayak, A.C., Pagnucco, M., Peppas, P.: Dynamic belief revision operators. Artif. Intell. **146**(2), 193–228 (2003)
23. Rott, H.: Shifting priorities: simple representations for twenty-seven iterated theory change operators. In: Makinson, D., Malinowski, J., Wansing, H. (eds.) Towards Mathematical Philosophy. TL, vol. 28, pp. 269–296. Springer, Dordrecht (2009). https://doi.org/10.1007/978-1-4020-9084-4_14
24. Segerberg, K.: Irrevocable belief revision in dynamic doxastic logic. Notre Dame J. Formal Log. **39**(3), 287–306 (1998)

Undefinability in Inquisitive Logic with Tensor

Ivano Ciardelli[1](✉) and Fausto Barbero[2]

[1] Munich Center for Mathematical Philosophy, LMU Munich, Munich, Germany
ivano.ciardelli@lrz.uni-muenchen.de

[2] Department of Philosophy, University of Helsinki, Helsinki, Finland
fausto.barbero@helsinki.fi

Abstract. Logics based on team semantics, such as inquisitive logic and dependence logic, are not closed under uniform substitution. This leads to an interesting separation between expressive power and definability: it may be that an operator O can be added to a language without a gain in expressive power, yet O is not definable in that language. For instance, even though propositional inquisitive logic and propositional dependence logic have the same expressive power, inquisitive disjunction and implication are not definable in propositional dependence logic. A question that has been open for some time in this area is whether the tensor disjunction used in propositional dependence logic is definable in inquisitive logic. We settle this question in the negative. In fact, we show that extending the logical repertoire of inquisitive logic by means of tensor disjunction leads to an independent set of connectives; that is, no connective in the resulting logic is definable in terms of the others.

Keywords: Definability · Inquisitive logic · Tensor disjunction · Dependence logic · Team semantics · Conjunction

1 Introduction

The notion of definability of a connective in terms of a set of other connectives is one of the basic notions of propositional logic. In classical logic every truth-functional connective is definable from, say, the set $\{\neg, \wedge\}$; for instance, disjunction can be defined as $\varphi \vee \psi := \neg(\neg\varphi \wedge \neg\psi)$ and implication as $\varphi \to \psi := \neg(\varphi \wedge \neg\psi)$. By contrast, it was shown by McKinsey [20] that in intuitionistic logic none of the primitive connectives $\neg, \wedge, \to, \vee$ whose semantics is characterized by the BHK interpretation is definable in terms of the other. In the setting of classical and intuitionistic logic, questions of definability are tighly connected to questions about the expressive power of a certain repertoire of connectives. If a connective $\circ$ is definable from other connectives, then each occurrence of this connective can be paraphrased away according to its definition; therefore, the connective can be omitted from the language without loss of expressive power. Conversely, if $\circ$ can be eliminated from the language without

P. Blackburn et al. (Eds.): LORI 2019, LNCS 11813, pp. 29–42, 2019.
https://doi.org/10.1007/978-3-662-60292-8_3

loss of expressive power, then $\circ$ can also be defined. To see why, suppose for concreteness that $\circ$ is binary. If $\circ$ can be omitted from the language, then in particular the formula $p \circ q$, where p and q are atomic formulas, is equivalent to some formula $\varphi(p, q)$ which does not contain $\circ$. In both classical and intuitionistic logic, logical equivalences remain valid when we replace atoms by arbitrary formulas. Therefore, we also have that for all ψ and χ, $\psi \circ \chi$ is equivalent to $\varphi(\psi, \chi)$. This means that the formula $\varphi(p, q)$, when viewed as a scheme where p and q are placeholders for the two arguments, provides a definition of the connective $\circ$.

This connection between definability and expressive power holds because of an important property of classical and intuitionistic logic, namely, *closure under uniform substitution*: logical equivalences remain valid when we replace atomic formulas by arbitrary formulas. In the recent literature, a number of logics have been studied which lack this property: in these logics, atomic formulas are not viewed as placeholders for *arbitrary* sentences, but rather as placeholders for sentences of a particular kind, which may have special logical properties. Two families of non substitution-closed logics which have been investigated thoroughly in the last few years are inquisitive logic (see, e.g., [6–8,15,22,23]) and dependence logic (see, e.g., [1,11,12,14,18,19,24,28,30]).[1] These approaches arose independently, and from different enterprises: inquisitive logic is concerned with extending the scope of logic to questions, while dependence logic is concerned with enriching classical logic with formulas that talk about dependencies and independencies between variables. However, these two approaches turned out to be tightly connected from a mathematical perspective [5,28,30]; and the convergence is not accidental: as discussed in [5,7] the relation of dependency can be seen as a special case of the relation of entailment, once the latter is generalized to questions. In the propositional setting, the standard incarnation of these two approaches is given by the systems of propositional inquisitive logic, InqB [8], and propositional dependence logic, PD [30]. These systems are interpreted in the same semantic setting, namely, in terms of a relation $\models$ which is assessed relative to sets of propositional valuations. However, these systems differ in their repertoire of logical operation. The set of primitives in InqB comprises the constant $\bot$ and the binary connectives $\wedge$ (conjunction), $\to$ (inquisitive implication) and $\mathbin{\rotatebox[origin=c]{-90}{$\geqslant$}}$ (inquisitive disjunction); from these operators negation is defined as $\neg\varphi := \varphi \to \bot$. The logical repertoire of PD comprises the connectives $\wedge$ (conjunction, the same operator as in InqB) and $\otimes$ (tensor disjunction) as well as two operators that can be applied only to atomic formulas: negation $\neg$, and the dependence atom $=(\, ; \,)$ which is an operator of variable arity, taking an arbitrary sequence of atoms in the first coordinate and an atom in the second.

InqB, PD, and their extensions are not closed under uniform substitution. As a consequence, in these logics expressive power and definability come apart. It is quite possible that a connective $\circ$ can be dropped from the language without loss

[1] Other examples of non-substitution closed logics are data logic [25], several dynamic epistemic logics [2,10,21], as well as logics arising from dynamic semantics [9,16,26] and from expressivist approaches to modals and conditionals [3,17,27].

of expressive power, yet this connective is not definable from the remaining connectives. This happens when each particular occurrence of $\circ$ can be paraphrased away, but the required paraphrase depends on the specific arguments to which $\circ$ applies. For instance, it was shown in [4] that every formula of InqB is equivalent to one in the language $\{\neg, \mathbin{\vee\!\!\!\vee}\}$, and also to one in the language $\{\bot, \to, \mathbin{\vee\!\!\!\vee}\}$; yet, implication is not definable in terms of $\{\neg, \mathbin{\vee\!\!\!\vee}\}$; and conjunction is not definable in terms of either $\{\neg, \mathbin{\vee\!\!\!\vee}\}$ or $\{\bot, \to, \wedge\}$.

More recently, [30] showed that PD has the same expressive power as InqB. Adding the inquisitive connectives $\mathbin{\vee\!\!\!\vee}$ and $\to$ to the logical repertoire of PD would not increase the expressive power of the language. However, one may still ask whether these operators are definable in PD. As shown in [29], the answer is negative: both $\to$ and $\mathbin{\vee\!\!\!\vee}$ are *not* definable in PD.[2] Similarly, adding the dependence logic operators $=(\,;\,)$ and $\otimes$ to the inquisitive repertoire does not increase the expressive power of InqB. But are these operators definable in InqB? For the dependence atom, the answer is easy: it is definable in InqB by means of the inquisitive connectives by the following scheme:

$$=(p_1, \ldots, p_n; q) \;:=\; ?p_1 \wedge \cdots \wedge ?p_n \to ?q$$

where $?\varphi$ is used as an abbreviation for $\varphi \mathbin{\vee\!\!\!\vee} \neg\varphi$. For tensor disjunction, however, the question has so far remained open. It was conjectured in [29] and [6] that $\otimes$ is not definable from the inquisitive connectives, but a proof has so far been missing. Our main aim in this paper is to prove this conjecture. In fact, we will take this opportunity to investigate more thoroughly matters of expressive power and definability in the context of inquisitive logic extended with the tensor disjunction connective. We will show that adding $\otimes$ to InqB leads to an independent set of primitives: that is, not only is $\otimes$ not definable from the inquisitive connectives, but none of the inquisitive connectives becomes definable from the other inquisitive connectives in the presence of tensor disjunction.

The paper is structured as follows. In Sect. 2 we specify the notions of definability and eliminability of a connective and discuss the relations between the two. In Sect. 3 we define $\mathsf{InqB}^{\otimes}$, inquisitive logic with tensor, and mention some key properties of this logic. In Sect. 4 we look into expressive completeness: we show that, starting from the repertoire $\{\bot, \wedge, \to, \mathbin{\vee\!\!\!\vee}, \otimes\}$, there is a unique minimal set of connectives which is expressively complete, namely, $\{\bot, \to, \mathbin{\vee\!\!\!\vee}\}$; thus, $\wedge$ and $\otimes$ can be removed from the language without loss of expressive power. In Sect. 5 we look into definability, showing that no connective in $\{\bot, \wedge, \to, \mathbin{\vee\!\!\!\vee}, \otimes\}$ is definable from the others; the novel results are the undefinability of tensor disjunction from the inquisitive connectives, and the undefinability of conjunction from $\{\bot, \to, \mathbin{\vee\!\!\!\vee}, \otimes\}$. Section 6 concludes.

[2] For other undefinability results in the setting of dependence logic, see also [12,13]. It is worth noting that, in the dependence logic literature, the standard notion of definability is called *uniform definability*; since there seems to be no special reason to add the qualification *uniform* (the notion of definability is intrinsically "uniform" in the relevant sense) we prefer to stick with the standard terminology.

2 Definability and Eliminability

Throughout this section, let L be an arbitrary propositional logic with language $\mathcal{L} = \mathcal{L}[C]$ generated by a set C of connectives, giving rise to a relation of logical equivalence $\equiv_L \subseteq \mathcal{L} \times \mathcal{L}$. We assume $\equiv_L$ to be an equivalence relation and a congruence with respect to the connectives: that is, we suppose that for every n-ary connective $\circ$, if $\varphi_i \equiv_L \psi_i$ for $i \leq n$ then $\circ(\varphi_1, \ldots, \varphi_n) \equiv_L \circ(\psi_1, \ldots, \psi_n)$.

Definition 1 (Context). *A* propositional context $\varphi(p_1, \ldots, p_n)$ *consists of a formula* $\varphi \in \mathcal{L}$ *together with a sequence of designated atomic formulas* $p_1, \ldots, p_n$*. Note that* $\varphi(p_1, \ldots, p_n)$ *is allowed to contain other atoms besides* $p_1, \ldots, p_n$*. If* $\varphi(p_1, \ldots, p_n)$ *is a context and* $\chi_1, \ldots, \chi_n \in \mathcal{L}$*, we write* $\varphi(\chi_1, \ldots, \chi_n)$ *for the result of replacing* $p_1, \ldots, p_n$ *by* $\chi_1, \ldots, \chi_n$ *throughout* φ*.*

Definition 2 (Definability). *We say that an n-ary connective* $\circ \in C$ *is defined by a context* $\xi(p_1, \ldots, p_n)$ *in case for all* $\chi_1, \ldots, \chi_n \in \mathcal{L}[C]$*:*

$$\circ(\chi_1, \ldots, \chi_n) \equiv_L \xi(\chi_1, \ldots, \chi_n)$$

We say that $\circ$ *is definable from a set* $C' \subseteq C$ *of connectives in case there is a context* $\xi(p_1, \ldots, p_n)$*, with* $\xi \in \mathcal{L}[C']$ *which defines* $\circ$*. If we just say that* $\circ$ *is definable then we mean that it is definable from* $C - \{\circ\}$*.*

In terms of definability we define the notion of an independent set of connectives.

Definition 3 (Independence). *We say that a set* C' *of connectives is* independent *if no connective* $\circ \in C'$ *is definable from* $C' - \{\circ\}$*.*

In addition to the notion of definability, we also introduce a notion of *eliminability* of a connective, which means that the connective can be omitted from the language without a loss in expressive power.

Definition 4 (Eliminability). *Let* $C' \subseteq C$ *be a set of connectives. We say that the set of connectives* C' *is* eliminable *if for each formula* $\varphi \in \mathcal{L}[C]$ *there is a formula* $\varphi^* \in \mathcal{L}[C - C']$ *such that* $\varphi \equiv_L \varphi^*$*. We say that a connective* $\circ$ *is eliminable if* $\{\circ\}$ *is eliminable.*

Notice that definability implies eliminability.

Proposition 1. *If a connective* $\circ$ *is definable, then it is eliminable.*

Proof. Suppose $\circ$ is defined by $\xi(p_1, \ldots, p_n) \in \mathcal{L}[C - \{\circ\}]$. We show by induction that every $\varphi \in \mathcal{L}[C]$ is equivalent to some $\varphi^* \in \mathcal{L}[C - \{\circ\}]$. The only non-trivial case is the one for $\varphi = \circ(\psi_1, \ldots, \psi_n)$. By induction hypothesis there are $\psi_1^*, \ldots, \psi_n^* \in \mathcal{L}[C - \{\circ\}]$ s.t. $\psi_i \equiv_L \psi_i^*$ for $i \leq n$. Then $\varphi \equiv_L \circ(\psi_1^*, \ldots, \psi_n^*)$. Since $\circ$ is defined by ξ we have $\circ(\psi_1^*, \ldots, \psi_n^*) \equiv_L \xi(\psi_1^*, \ldots, \psi_n^*)$. Since $\xi, \psi_1^*, \ldots, \psi_n^* \in \mathcal{L}[C - \{\circ\}]$, we have that $\xi(\psi_1^*, \ldots, \psi_n^*) \in \mathcal{L}[C - \{\circ\}]$. So φ is L-equivalent to some formula in $\mathcal{L}[C - \{\circ\}]$, which completes the inductive step. □

Another important notion in this area is the notion of completeness of a set C' of connectives, which holds when every formula in the language is L-equivalent to one built up using only connectives from C'.

Definition 5 (Completeness). *We say that a set of connectives $C' \subseteq C$ is complete for L if for all $\varphi \in \mathcal{L}[C]$ there exists some $\varphi^* \in \mathcal{L}[C']$ s.t. $\varphi \equiv_L \varphi^*$. We say that a set C' is a* minimal complete *set of connectives for L if C' is complete for L, and no proper subset $C'' \subset C'$ is complete for L.*

The notions of definability and eliminability often go hand in hand. This is because the logics L one typically considers are closed under uniform substitution. Closure under uniform substitution is usually formulated in terms of preservation of validity under substitutions. For our purposes, a characterization in terms of preservation of equivalence is more suitable. The two coincide if L validates a deduction theorem, and more generally if $\varphi \equiv_L \psi$ amounts to the validity of a formula $\chi(\varphi, \psi)$ in L (if the deduction theorem holds, $\chi(\varphi, \psi)$ will be $\varphi \leftrightarrow \psi$).

Definition 6 (Closure under uniform substitution). *L is* closed under uniform substitution *if for any two contexts $\varphi(p_1, \ldots, p_n)$ and $\psi(p_1, \ldots, p_n)$, if $\varphi \equiv_L \psi$ then for all $\chi_1, \ldots, \chi_n \in \mathcal{L}$ we have $\varphi(\chi_1, \ldots, \chi_n) \equiv_L \psi(\chi_1, \ldots, \chi_n)$.*

In a logic which is closed under uniform substitution, the notions of definability and eliminability for a connective coincide.

Proposition 2. *If L is closed under uniform substitution and $\circ$ is eliminable, then $\circ$ is definable.*

Proof. Let $p_1, \ldots, p_n$ be n distinct atomic formulas. Suppose L is closed under uniform substitution and $\circ$ is eliminable. Then the formula $\circ(p_1, \ldots, p_n)$ is L-equivalent to some $\xi \in \mathcal{L}[C - \{\circ\}]$. Consider the context $\xi(p_1, \ldots, p_n)$: by closure under uniform substitution, for all $\chi_1, \ldots, \chi_n \in \mathcal{L}$ we have $\circ(\chi_1, \ldots, \chi_n) \equiv_L \xi(\chi_1, \ldots, \chi_n)$. Thus, $\circ$ is defined by $\xi(p_1, \ldots, p_n)$. □

As we discussed in the introduction, systems of inquisitive and dependence logic are typically not closed under uniform substitution. As a consequence, in these logics we find an interesting gap between the eliminability of a connective and its definability. We will examine this gap carefully for one particular logic, the system $\mathsf{InqB}^{\otimes}$ introduced in the next section.

3 Inquisitive Logic with Tensor Disjunction

In this section we introduce the system $\mathsf{InqB}^{\otimes}$ that we will be concerned with. This system extends the standard system of propositional inquisitive logic, InqB [4,6,8] with the tensor disjunction used in dependence logic [28,30]. We will present this logic from a purely mathematical point of view. For an introduction to the motivations of inquisitive logic and to the conceptual interpretation of formulas in this system, see [6,7].

Formally, the language $\mathcal{L}$ of $\mathsf{InqB}^{\otimes}$ is generated by the set of connectives $\{\bot, \wedge, \mathbin{\rotatebox[origin=c]{-90}{$\geqslant$}}, \to, \otimes\}$, where all connectives are binary except for $\bot$, which has arity 0:

$$\varphi ::= p \mid \bot \mid \varphi \wedge \varphi \mid \varphi \vee\!\!\!\vee \varphi \mid \varphi \to \varphi \mid \varphi \otimes \varphi$$

In addition to the primitive connectives, some defined connectives are used:

$$\neg\varphi := \varphi \to \bot \qquad ?\varphi := \varphi \vee\!\!\!\vee \neg\varphi \qquad \top := \neg\bot$$

In classical propositional logic, the semantics of a formula is given in terms of truth-conditions relative to a valuation function $w : \mathcal{P} \to \{0, 1\}$, which represents a complete state of affairs. By contrast, in inquisitive logic, formulas are interpreted relative to a set of such valuation functions. Informally, such a set s is taken to represent an *information state*, where the idea is that s embodies the information that the actual state of affairs corresponds to one of the valuations $w \in s$; thus, the smaller the set s, the stronger the information it encodes.

Definition 7 (Information states)
An information state is a set s of valuations for $\mathcal{P}$, i.e., a subset $s \subseteq \{0, 1\}^{\mathcal{P}}$.

The semantics is given in terms of a relation of *support* relative to information states, which is defined inductively as follows.

Definition 8 (Support)

- $s \models p \iff w(p) = 1$ *for all* $w \in s$
- $s \models \bot \iff s = \emptyset$
- $s \models \psi \wedge \chi \iff s \models \psi$ *and* $s \models \chi$
- $s \models \psi \vee\!\!\!\vee \chi \iff s \models \psi$ *or* $s \models \chi$
- $s \models \psi \to \chi \iff \forall t \subseteq s : t \models \psi$ *implies* $t \models \chi$
- $s \models \psi \otimes \chi \iff \exists t_1, t_2$ *s.t.* $t_1 \models \psi, t_2 \models \chi$ *and* $s = t_1 \cup t_2$.

It is immediate to see that the defined operators have the following semantics:

- $s \models \neg\psi \iff \forall t \subseteq s : t \models \psi$ implies $t = \emptyset$
- $s \models ?\psi \iff s \models \psi$ or $s \models \neg\psi$
- $s \models \top$ for all information states s.

The support relation has the following properties.

- Persistency: if $s \models \varphi$ and $t \subseteq s$, then $t \models \varphi$
- Empty state property: $\emptyset \models \varphi$ for all φ.

From the notion of support at a state, a notion of truth relative to a valuation w is recovered by defining truth at w as support with respect to $\{w\}$.

Definition 9 (Truth). *We say that φ is true at w, notation $w \models \varphi$, if and only if $\{w\} \models \varphi$.*

It is easy to see that the truth-conditions that this delivers are simply the ones familiar from classical propositional logic, when both $\mathbin{\rotatebox[origin=c]{-90}{$\geqslant$}}$ and $\otimes$ are identified with classical disjunction $\vee$. Notice that, due to persistency, if a formula φ is supported at s, then it is true at all valuations $w \in s$. The converse is not true in general, but it may hold for particular formulas. The formulas for which this is the case are called *truth-conditional*, since the semantics of these formulas is completely determined at the level of truth-conditions.[3]

Definition 10 (Truth-conditionality). *We say that* $\varphi \in \mathcal{L}$ *is* truth-conditional *if for all information states s we have:* $s \models \varphi \iff \forall w \in s : w \models \varphi$.

There is an important fragment of our language which consists entirely of truth-conditional formulas. We refer to the formulas in this fragment as *declaratives.*

Definition 11 (Declaratives). *A formula* $\varphi \in \mathcal{L}$ *is a* declarative *if it is* $\mathbin{\rotatebox[origin=c]{-90}{$\geqslant$}}$*-free. In other words, the set of declaratives is the set* $\mathcal{L}_! := \mathcal{L}[\bot, \wedge, \rightarrow, \otimes]$.

Proposition 3 (cf. [5], Prop. 3). *Every* $\alpha \in \mathcal{L}_!$ *is truth-conditional.*

Formulas containing $\mathbin{\rotatebox[origin=c]{-90}{$\geqslant$}}$ are not in general truth-conditional. For instance, consider $?p$ (which abbreviates $p \mathbin{\rotatebox[origin=c]{-90}{$\geqslant$}} \neg p$). Since truth-conditions coincide with truth-conditions in classical propositional logic, $?p$ will be true with respect to all valuations w whatsoever. However, the support-conditions for this formula are:

$$s \models ?p \iff s \models p \text{ or } s \models \neg p \iff (\forall w \in s : w(p) = 1) \text{ or } (\forall w \in s : w(p) = 0)$$

where the last equivalence uses the fact that p and $\neg p$ are declaratives, and thus truth-conditional. Thus, $?p$ is supported at a state s only if all the valuations in s agree about the truth-value of p. Thus, if s is a state containing some valuations that make p true and some that make p false, then $s \not\models ?p$, even though $?p$ is true at all $w \in s$. This shows that $?p$ is not truth-conditional.

Logical entailment and equivalence are defined naturally in terms of support.

Definition 12 (Logical entailment and equivalence). *We say that:*

- φ *entails* ψ, $\varphi \models \psi$, *if for all states s:* $s \models \varphi$ *implies* $s \models \psi$;
- φ *and* ψ *are equivalent,* $\varphi \equiv \psi$, *if for all states s:* $s \models \varphi$ *iff* $s \models \psi$.

In addition to these purely logical notions, it will also be useful to have notions of entailment and equivalence relativized to an information state s. The idea is that, when looking at entailment and equivalence relative to s, only valuations in s are taken into account (for discussion of the significance of logical and contextual entailment in inquisitive logic, see [6]§1 and [7]).

Definition 13 (Relativized entailment and equivalence). *Let* $\varphi, \psi \in \mathcal{L}$ *and let s be an information state. We say that:*

- φ *entails* ψ *in s,* $\varphi \models_s \psi$, *if for all states* $t \subseteq s$*:* $t \models \varphi$ *implies* $t \models \psi$;
- φ *and* ψ *are equivalent in s,* $\varphi \equiv_s \psi$, *if for all states* $t \subseteq s$*:* $t \models \varphi$ *iff* $t \models \psi$.

[3] Truth-conditional formulas are called *flat* formulas in the dependence logic literature.

Notice that logical entailment implies entailment relative to any information state; similarly, logical equivalence implies equivalence relative to any state.

For formulas in the declarative fragment of the language, logical entailment and equivalence coincide with entailment and equivalence in classical propositional logic, when the operator $\otimes$ is identified with classical disjunction. We state the case of equivalence as a proposition, since we will make use of it below.

Proposition 4 (cf. [5], Prop. 6)
For all $\alpha, \beta \in \mathcal{L}_!$: $\alpha \equiv \beta \iff \alpha$ and β are equivalent in classical propositional logic, when $\otimes$ is replaced by the classical disjunction symbol $\vee$.

An important feature of propositional inquisitive logic, which extends to $\mathsf{InqB}^{\otimes}$ [5], is that every formula is equivalent to an inquisitive disjunction of declaratives.

Theorem 1 (cf. [5], Prop. 11). *For every $\varphi \in \mathcal{L}$ there are declarative formulas $\alpha_1, \dots, \alpha_n \in \mathcal{L}_!$, called the* resolutions *of φ, such that $\varphi \equiv \alpha_1 \mathbin{\rotatebox[origin=c]{-90}{$\geqslant$}} \dots \mathbin{\rotatebox[origin=c]{-90}{$\geqslant$}} \alpha_n$.*

Having reviewed the main definitions and facts about $\mathsf{InqB}^{\otimes}$, we can now delve into the issue of eliminability and definability of connectives in this system.

4 Eliminability

In this section we discuss which connectives from our language $\mathcal{L}$ are eliminable, and show that there is a unique minimal complete set of connectives for $\mathsf{InqB}^{\otimes}$. These results are straightforward adaptations to our language of results from [4], although we will lay out the consequences of these results more systematically. First, let us show that the connectives $\bot$, $\mathbin{\rotatebox[origin=c]{-90}{$\geqslant$}}$, and $\to$ are not eliminable: omitting any of these connectives from our language results in a loss of expressive power.

Proposition 5. *$\bot$ is not eliminable.*

Proof. We have to show that no $\bot$-free formula of $\mathcal{L}$ is equivalent to $\bot$. Let w_t be the valuation function mapping all atomic formulas to 1. A straightforward induction shows that every $\bot$-free formula is supported at the state $\{w_t\}$. Since $\{w_t\} \not\models \bot$, it follows that no $\bot$-free formula is equivalent to $\bot$. □

Proposition 6. *$\mathbin{\rotatebox[origin=c]{-90}{$\geqslant$}}$ is not eliminable.*

Proof. We must show that some formula of $\mathcal{L}$ is not equivalent to any $\mathbin{\rotatebox[origin=c]{-90}{$\geqslant$}}$-free formula. Consider the formula $?p := p \mathbin{\rotatebox[origin=c]{-90}{$\geqslant$}} \neg p$. Proposition 3 ensures that every $\mathbin{\rotatebox[origin=c]{-90}{$\geqslant$}}$-free formula is truth-conditional. Since we saw above that $?p$ is not truth-conditional, it follows that it is not equivalent to any $\mathbin{\rotatebox[origin=c]{-90}{$\geqslant$}}$-free formula. □

Proposition 7. *$\to$ is not eliminable.*

Proof. We show that $\top$ (defined as $\neg\bot$, i.e., $\bot \to \bot$) is not equivalent to any $\to$-free formula. Let w_f be the valuation mapping all atomic formulas to 0. A straightforward induction shows that no $\to$-free formula is supported at $\{w_f\}$. Since $\top$ is supported at $\{w_f\}$, no $\to$-free formula is equivalent to $\top$. □

This shows that, starting from the repertoire of connectives $\{\bot, \wedge, \vee\!\!\!\vee, \to, \otimes\}$, none of $\bot$, $\vee\!\!\!\vee$, and $\to$ can be dropped without a loss in expressive power. We are now going to see that these three connectives together are sufficient to express anything that is expressible in $\mathsf{InqB}^{\otimes}$.

Proposition 8. $\{\bot, \to, \vee\!\!\!\vee\}$ *is a complete set of connectives for* $\mathsf{InqB}^{\otimes}$.

Proof. We need to show that for all $\varphi \in \mathcal{L}$ there is a $\varphi^* \in \mathcal{L}[\bot, \to, \vee\!\!\!\vee]$ such that $\varphi \equiv \varphi^*$. First, it follows from Proposition 4 and the fact that $\{\bot, \to\}$ is complete for classical propositional logic that any any declarative formula $\alpha \in \mathcal{L}[\bot, \wedge, \to, \otimes]$ is equivalent to a formula $\alpha^* \in \mathcal{L}[\bot, \to]$.

Now take any $\varphi \in \mathcal{L}$. By Theorem 1 we have $\varphi \equiv \alpha_1 \vee\!\!\!\vee \ldots \vee\!\!\!\vee \alpha_n$ for some formulas $\alpha_1, \ldots, \alpha_n \in \mathcal{L}[\bot, \wedge, \to, \otimes]$. Now let $\varphi^* := \alpha_1^* \vee\!\!\!\vee \ldots \vee\!\!\!\vee \alpha_n^*$, where $\alpha_1^*, \ldots, \alpha_n^*$ are defined as above. Since $\alpha_i \equiv \alpha_i^*$ for $i \leq n$, we have $\varphi \equiv \varphi^*$. And since $\alpha_i^* \in \mathcal{L}[\bot, \to]$ for $i \leq n$, we have $\varphi^* \in \mathcal{L}[\bot, \to, \vee\!\!\!\vee]$. □

Corollary 1. *The set of connectives* $\{\wedge, \otimes\}$ *is eliminable in* $\mathsf{InqB}^{\otimes}$. *In particular, both* $\wedge$ *and* $\otimes$ *are eliminable.*

Proposition 8 together with the non-eliminability of $\bot$, $\to$, and $\vee\!\!\!\vee$ implies that $\mathsf{InqB}^{\otimes}$ admits only one minimal complete set of connectives, namely, $\{\bot, \to, \vee\!\!\!\vee\}$

Theorem 2. $\{\bot, \to, \vee\!\!\!\vee\}$ *is the only minimal complete set for* $\mathsf{InqB}^{\otimes}$.

Proof. It follows from Propositions 5, 6, and 7 that any complete set of connectives C for $\mathsf{InqB}^{\otimes}$ must include $\{\bot, \to, \vee\!\!\!\vee\}$. Since $\{\bot, \to, \vee\!\!\!\vee\}$ is itself a complete set, it is the only minimal one. □

We have thus achieved a complete characterization of:

- which connectives are eliminable in $\mathsf{InqB}^{\otimes}$ ($\wedge$ and $\otimes$, but not $\bot, \to$ and $\vee\!\!\!\vee$);
- which sets of connectives are complete (those which include $\{\bot, \to, \vee\!\!\!\vee\}$);
- which sets of connectives are minimal complete (only $\{\bot, \to, \vee\!\!\!\vee\}$).

5 Independence of the Connectives

In this section, we turn to the issue of definability. The main contribution of the paper is to answer the following question, which is open in the literature: which connectives of $\mathsf{InqB}^{\otimes}$ are definable in terms of the remaining connectives? We will see that the answer is *none*: thus, although $\{\bot, \wedge, \vee\!\!\!\vee, \to, \otimes\}$ is not a minimal set of connectives for $\mathsf{InqB}^{\otimes}$, it is an independent set of connectives.

For a start, notice that the question of definability only arises for $\wedge$ and $\otimes$: as we saw, the connectives $\bot$, $\vee\!\!\!\vee$ and $\to$ are not even eliminable in $\mathsf{InqB}^{\otimes}$, which *a fortiori* implies that they are not definable. We will examine first the case of tensor disjunction, and then the case of conjunction.

5.1 Undefinability of $\otimes$

Fix three propositional letters q_1, q_2, q_3 and consider:

- $\psi := q_1 \vee\!\!\!\vee q_2 \vee\!\!\!\vee q_3$
- $s := \{w_1, w_2, w_3\}$ where $w_i(q_i) = 1$ and $w_i(r) = 0$ for all $r \neq q_i$

Lemma 1. *For any context $\varphi(p)$, with $\varphi \in \mathcal{L}[\bot, \wedge, \to, \vee\!\!\!\vee]$ not containing q_1, q_2, q_3, one of the following holds:*

$$\varphi(\psi) \equiv_s \top \qquad \varphi(\psi) \equiv_s \bot \qquad \varphi(\psi) \equiv_s \psi$$

Proof. We proceed by induction on φ. To ease notation, given a formula $\chi(p)$, let us write $\chi^\star$ for $\chi(\psi)$.

- $\varphi = p$ or $\varphi = \bot$. Then $\varphi^\star$ is either ψ or $\bot$, and we are done.
- $\varphi = r$ for an atomic formula r different from p, q_1, q_2, q_3. Then $\varphi^\star = r$. Since r is false at all worlds in s we have $r \equiv_s \bot$.
- $\varphi = \eta \wedge \theta$. Then $\varphi^\star = \eta^\star \wedge \theta^\star$. By induction hypothesis, each of $\eta^\star$ and $\theta^\star$ is equivalent modulo s to either one of $\top, \bot, \psi$. Since the conjunction of any two formulas from $\{\top, \bot, \psi\}$ is logically equivalent to a formula from this set, we can conclude that $\varphi^\star$ is equivalent modulo s either to $\top, \bot$ or ψ.
- $\varphi = \eta \vee\!\!\!\vee \theta$. Then $\varphi^\star = \eta^\star \vee\!\!\!\vee \theta^\star$. This case is similar to the previous one, since the inquisitive disjunction of any two formulas from $\{\top, \bot, \psi\}$ is logically equivalent to a formula from this set.
- $\varphi = \eta \to \theta$. Then $\varphi^\star = \eta^\star \to \theta^\star$. We consider three cases:
  - $\eta^\star \equiv_s \top$. In this case, by persistency we have $\varphi^\star \equiv_s \top \to \theta^\star \equiv \theta^\star$. By induction hypothesis, $\theta^\star$ is equivalent in s to $\top$, $\bot$, or ψ, so we are done.
  - $\eta^\star \equiv_s \bot$. In this case, by the empty state property $\varphi^\star \equiv_s \bot \to \theta^\star \equiv \top$.
  - $\eta^\star \equiv_s \psi$. In this case, we need to distinguish three sub-cases:
    - $\theta^\star \equiv_s \top$. In this case, $\varphi^\star \equiv_s \psi \to \top \equiv \top$.
    - $\theta^\star \equiv_s \psi$. In this case, $\varphi^\star \equiv_s \psi \to \psi \equiv \top$.
    - $\theta^\star \equiv_s \bot$. In this case $\varphi^\star \equiv_s \psi \to \bot$. We claim that $\psi \to \bot \equiv_s \bot$. We need to show that the only $t \subseteq s$ that supports $\psi \to \bot$ is $\emptyset$. Notice that, by the definitions of ψ and negation, $\psi \to \bot = \neg(q_1 \vee\!\!\!\vee q_2 \vee\!\!\!\vee q_3)$. Suppose for a contradiction that $\emptyset \neq t \subseteq s$ and $t \models \neg(q_1 \vee\!\!\!\vee q_2 \vee\!\!\!\vee q_3)$. Take a valuation $w_i \in t$. By persistency, $\{w_i\} \models \neg(q_1 \vee\!\!\!\vee q_2 \vee\!\!\!\vee q_3)$. But this is impossible, since by construction $\{w_i\} \models q_i$. □

Now just observe the support conditions for the formulas ψ and $\psi \otimes \psi$ relativized to the given state s. Letting $\#t$ be the cardinality of t, for all $t \subseteq s$ we have:

- $t \models \psi \iff \#t \leq 1$
- $t \models \psi \otimes \psi \iff \#t \leq 2$

This means that we have:

$$\psi \otimes \psi \not\equiv_s \top \qquad \psi \otimes \psi \not\equiv_s \bot \qquad \psi \otimes \psi \not\equiv_s \psi$$

From this observation and Lemma 1, we get the following proposition.

Proposition 9. *For all* $\varphi(p) \in \mathcal{L}[\bot, \wedge, \rightarrow, \mathbin{\vee\!\!\!\vee}]$ *which do not contain* q_1, q_2, q_3*:* $(q_1 \mathbin{\vee\!\!\!\vee} q_2 \mathbin{\vee\!\!\!\vee} q_3) \otimes (q_1 \mathbin{\vee\!\!\!\vee} q_2 \mathbin{\vee\!\!\!\vee} q_3) \not\equiv \varphi(q_1 \mathbin{\vee\!\!\!\vee} q_2 \mathbin{\vee\!\!\!\vee} q_3)$

It is now easy to get the desired undefinability result.

Theorem 3. $\otimes$ *is not definable from* $\{\bot, \wedge, \rightarrow, \mathbin{\vee\!\!\!\vee}\}$.

Proof. Consider a context $\varphi(p_1, p_2)$ with $\varphi \in \mathcal{L}[\bot, \wedge, \rightarrow, \mathbin{\vee\!\!\!\vee}]$ (possibly containing other atoms besides p_1, p_2). Let q_1, q_2, q_3 be three atoms which do not occur in φ and let $\psi = q_1 \mathbin{\vee\!\!\!\vee} q_2 \mathbin{\vee\!\!\!\vee} q_3$. By Proposition 9 we have:

$$\psi \otimes \psi \not\equiv \varphi(\psi, \psi)$$

which implies that $\varphi(p_1, p_2)$ does not define $\otimes$. □

5.2 Undefinability of $\wedge$

We will follow a strategy similar to that we used for $\otimes$. Fix two atomic formulas q_1, q_2, and let $s = \{w_{12}, w_1, w_2\}$ where w_{12} makes both q_1 and q_2 true, w_1 makes only q_1 true, and w_2 makes only q_2 true. Moreover, suppose that all atoms different from q_1, q_2 are false at all three worlds. We will show the following.

Lemma 2. *Take a context* $\varphi(p_1, p_2)$*, with* $\varphi \in \mathcal{L}[\bot, \rightarrow, \mathbin{\vee\!\!\!\vee}, \otimes]$ *not containing* q_1, q_2*. Then one of the following claims holds:*

1. $\varphi(?q_1, ?q_2) \equiv_s \bot$
2. $\varphi(?q_1, ?q_2)$ *is supported by all subsets of* s *of cardinality 1, and by at least one subset of* s *of cardinality 2.*

Proof. We proceed by induction on φ. To ease notation, given a context $\chi(p_1, p_2)$, let us write $\chi^\star$ for $\chi(?q_1, ?q_2)$.

- $\varphi = p_1$. Then $\varphi^\star = ?q_1$ is supported by all singleton states, and also by the state $\{w_{12}, w_1\} \subseteq s$, which has cardinality 2. So claim 2 holds.
- $\varphi = p_2$. Analogous.
- $\varphi = \bot$. Then $\varphi^\star = \bot$, so claim 1 holds.
- $\varphi = r$ for $r \neq p_1, p_2, q_1, q_2$. Then $\varphi^\star = r$. Since r is false in all worlds in s we have $r \equiv_s \bot$, so claim 1 holds.
- $\varphi = \eta \rightarrow \theta$. We distinguish three cases:
  - $\eta^\star \equiv_s \bot \equiv_s \theta^\star$. Then $\varphi^\star \equiv_s \bot \rightarrow \bot = \top$, therefore claim 2 holds.
  - $\eta^\star \not\equiv_s \bot \equiv_s \theta^\star$. Then $\varphi^\star \equiv_s \eta^\star \rightarrow \bot$. We will show that $\varphi^\star \equiv_s \bot$. Take any non-empty $t \subseteq s$, and let $\{w\} \subseteq t$. By the induction hypothesis on η we have $\{w\} \models \eta^\star$, but $\{w\} \not\models \bot$. Therefore, $t \not\models \eta^\star \rightarrow \bot$. Thus, $\varphi^\star$ is not supported at any non-empty subset of s, which means that $\varphi^\star \equiv_s \bot$.
  - $\theta^\star \not\equiv_s \bot$. Then by induction hypothesis $\theta^\star$ is supported by all singleton substates of s, and also by a substate of s of cardinality 2. Since any state that supports $\theta^\star$ also supports $\varphi^\star = \eta^\star \rightarrow \theta^\star$, by persistency, claim 2 holds for $\varphi^\star$.

- $\varphi = \eta \mathbin{\mathbb{V}} \theta$ or $\varphi = \eta \otimes \theta$. If $\eta^\star \equiv_s \theta^\star \equiv_s \bot$ then $\varphi^\star \equiv_s \bot$ and case 1 applies. Otherwise, at least one of $\eta^\star$ and $\theta^\star$ is not s-equivalent to $\bot$. Suppose it is $\eta^\star$. Then by induction hypothesis $\eta^\star$ is supported by all singleton substates of s, and by a substate of s of cardinality 2. Since any state that supports $\eta^\star$ also supports $\eta^\star \mathbin{\mathbb{V}} \theta^\star$ as well as $\eta^\star \otimes \theta^\star$, claim 2 applies to $\varphi^\star$. □

Now consider the formula $?q_1 \wedge ?q_2$. For all $t \subseteq s$ we have:

- $t \models ?q_1 \wedge ?q_2 \iff \#t \leq 1$

Thus, $?q_1 \wedge ?q_2 \not\equiv_s \bot$, and moreover $?q_1 \wedge ?q_2$ is not supported by any substate of s of cardinality 2. It follows from Lemma 2 that $?q_1 \wedge ?q_2 \not\equiv_s \varphi(?q_1, ?q_2)$ for any context $\varphi(p_1, p_2)$ which does not contain the atoms q_1, q_2. Since logical equivalence implies equivalence in every state, we obtain the following proposition.

Proposition 10. *For all* $\varphi(p_1, p_2) \in \mathcal{L}[\bot, \to, \mathbin{\mathbb{V}}, \otimes]$ *which do not contain* q_1, q_2*:*

$$?q_1 \wedge ?q_2 \not\equiv \varphi(?q_1, ?q_2)$$

This yields as a corollary the undefinability of $\wedge$.

Theorem 4. $\wedge$ *is not definable from* $\{\bot, \to, \mathbin{\mathbb{V}}, \otimes\}$.

Proof. Take a context $\varphi(p_1, p_2)$, where $\varphi \in \mathcal{L}[\bot, \to, \mathbin{\mathbb{V}}, \otimes]$ (possibly containing other atoms besides p_1, p_2). Let q_1, q_2 be two atoms that do not occur in φ. By Proposition 10 we have $?q_1 \wedge ?q_2 \not\equiv \varphi(?q_1, ?q_2)$, which implies that $\varphi(p_1, p_2)$ does not define $\wedge$. □

This completes our study of definability in $\mathsf{InqB}^\otimes$: no connective in $\mathsf{InqB}^\otimes$ is definable in terms of the others; although the set of connectives $\{\bot, \wedge, \mathbin{\mathbb{V}}, \to, \otimes\}$ is not a minimal complete set for $\mathsf{InqB}^\otimes$, it is an independent set.

6 Conclusion

We have studied connectives in inquisitive logic enriched with tensor disjunction. We have shown that, starting from the set of primitives $\{\bot, \wedge, \to, \mathbin{\mathbb{V}}, \otimes\}$, the only minimal complete subset is $\{\bot, \to, \vee\}$. Thus, $\wedge$ and $\otimes$ do not add to the expressive power of the language. However, the set $\{\bot, \wedge, \to, \mathbin{\mathbb{V}}, \otimes\}$ is independent, i.e., no connective is definable in terms of the others. The undefinability results for $\wedge$ and $\otimes$ use nontrivial combinatorial arguments. The result for $\otimes$ answers a question raised in [6] and [29]. It is worth pointing out that our proof establishes something slightly stronger than undefinability. Undefinability amounts to:

> For every putative definition $\varphi(p, q)$ of connective $\circ$, there are formulas $\psi_\varphi, \chi_\varphi$ such that $\varphi(\psi_\varphi, \chi_\varphi) \not\equiv \psi_\varphi \circ \chi_\varphi$.

However, our proofs for $\otimes$ and $\wedge$ establish results of the following form:

> There are ψ, χ such that, for every putative definition $\varphi(p, q)$ of $\circ$, $\varphi(\psi', \chi') \not\equiv \psi' \circ \chi'$, where ψ', χ' differ from ψ, χ by a renaming of atoms.

In future work, it would be interesting to extend our study of expressiveness and definability to a setting where the operators $\neg$ and $?$ are taken as primitives.

References

1. Abramsky, S., Väänänen, J.: From IF to BI. Synthese **167**(2), 207–230 (2009)
2. Baltag, A., Moss, L.S., Solecki, S.: The logic of public announcements, common knowledge, and private suspicions. In: Arló-Costa, H., Hendricks, V.F., van Benthem, J. (eds.) Readings in Formal Epistemology. SGTP, vol. 1, pp. 773–812. Springer, Cham (2016). https://doi.org/10.1007/978-3-319-20451-2_38
3. Bledin, J.: Logic informed. Mind **123**(490), 277–316 (2014)
4. Ciardelli, I.: Inquisitive semantics and intermediate logics. MSc Thesis, University of Amsterdam (2009)
5. Ciardelli, I.: Dependency as question entailment. In: Abramsky, S., Kontinen, J., Väänänen, J., Vollmer, H. (eds.) Dependence Logic, pp. 129–181. Springer, Cham (2016). https://doi.org/10.1007/978-3-319-31803-5_8
6. Ciardelli, I.: Questions in logic. Ph.D. thesis, Institute for Logic, Language and Computation, University of Amsterdam (2016)
7. Ciardelli, I.: Questions as information types. Synthese **195**, 321–365 (2018). https://doi.org/10.1007/s11229-016-1221-y
8. Ciardelli, I., Roelofsen, F.: Inquisitive logic. J. Philos. Logic **40**(1), 55–94 (2011)
9. Dekker, P.: Transsentential meditations. ups and downs in dynamic semantics. Ph.D. thesis, ILLC, University of Amsterdam (1993)
10. van Ditmarsch, H., van der Hoek, W., Kooi, B.: Dynamic Epistemic Logic. SYLI, vol. 337. Springer, Dordrecht (2007). https://doi.org/10.1007/978-1-4020-5839-4
11. Galliani, P.: Inclusion and exclusion dependencies in team semantics - on some logics of imperfect information. Ann. Pure Appl. Logic **163**(1), 68–84 (2012)
12. Galliani, P.: Epistemic operators in dependence logic. Stud. Logica **101**(2), 367–397 (2013)
13. Goranko, V., Kuusisto, A.: Logics for propositional determinacy and independence. Rev. Symb. Logic **11**(3), 470–506 (2018)
14. Grädel, E., Väänänen, J.: Dependence and independence. Stud. Logica **101**(2), 399–410 (2013)
15. Grilletti, G.: Disjunction and existence properties in inquisitive first-order logic. Stud. Logica, 1–36 (2018). https://doi.org/10.1007/s11225-018-9835-3
16. Groenendijk, J., Stokhof, M., Veltman, F.: Coreference and modality in the context of multi-speaker discourse. In: Kamp, H., Partee, B.H. (eds.) Context Dependence in the Analysis of Linguistic Meaning, pp. 195–216. IMS, Stuttgart (1997)
17. Kolodny, N., MacFarlane, J.: Ifs and oughts. J. Philos. **107**(3), 115–143 (2010)
18. Kontinen, J.: Coherence and computational complexity of quantifier-free dependence logic formulas. Stud. Logica **101**(2), 267–291 (2013)
19. Kontinen, J., Väänänen, J.: On definability in dependence logic. J. Logic Lang. Inf. **18**(3), 317–332 (2009)
20. McKinsey, J.C.C.: Proof of the independence of the primitive symbols of Heyting's calculus of propositions. J. Symb. Logic **4**(4), 155–158 (1939). http://www.jstor.org/stable/2268715
21. Plaza, J.: Logics of public communications. In: Emrich, M., Pfeifer, M., Hadzikadic, M., Ras, Z. (eds.) Proceedings of the Fourth International Symposium on Methodologies for Intelligent Systems. pp. 201–216. Oak Ridge National Laboratory (1989)
22. Punčochář, V.: Weak negation in inquisitive semantics. J. Logic Lang. Inf. **24**(3), 323–355 (2015)
23. Roelofsen, F.: Algebraic foundations for the semantic treatment of inquisitivecontent. Synthese **190**(1), 79–102 (2013). https://doi.org/10.1007/s11229-013-0282-4

24. Väänänen, J.: Dependence Logic: A New Approach to Independence Friendly Logic. Cambridge University Press, Cambridge (2007)
25. Veltman, F.: Data semantics. In: Groenendijk, J., Janssen, T., Stokhof, M. (eds.) Formal Methods in the Study of Language. Mathematical Centre, Amsterdam (1981)
26. Veltman, F.: Defaults in update semantics. J. Philos. Logic **25**(3), 221–261 (1996)
27. Yalcin, S.: Epistemic modals. Mind **116**(464), 983–1026 (2007)
28. Yang, F.: On extensions and variants of dependence logic: a study of intuitionistic connectives in the team semantics setting. Ph.D. thesis, University of Helsinki (2014)
29. Yang, F.: Uniform definability in propositional dependence logic. Rev. Symb. Logic **10**(1), 65–79 (2017)
30. Yang, F., Väänänen, J.: Propositional logics of dependence. Ann. Pure Appl. Logic **167**(7), 557–589 (2016)

Minimal-Change Counterfactuals in Intuitionistic Logic

Ivano Ciardelli[1](✉) and Xinghan Liu[2](✉)

[1] Munich Center for Mathematical Philosophy, LMU, Munich, Germany
ivano.ciardelli@lrz.uni-muenchen.de

[2] Department of Philosophy, LMU, Munich, Germany
kennarsliu@gmail.com

Abstract. In this paper we study the logic IVC obtained by adding Lewis-style counterfactual conditionals to intuitionistic propositional logic. Building on recent work by Weiss [21], we first show how to introduce a Lewisian counterfactual operator into intuitionistic Kripke semantics. We then establish a complete axiomatization of the resulting logic.

Keywords: Counterfactuals · Intuitionistic logic · Variably strict conditionals · Minimal-change semantics · Non-monotonic reasoning

1 Introduction

Lewis [14] proposed an analysis of counterfactuals within modal logic which remains the most influential logical approach to counterfactuals to date. In this semantics, a counterfactual $\varphi > \psi$ is seen as a kind of box modality, which asserts that the consequent holds throughout a certain set $f(w, \varphi)$ of possible worlds, depending on the world of evaluation and, crucially, on the antecedent. Lewis' intuition is that the worlds in $f(w, \varphi)$ are those worlds in which φ is true, and which are otherwise minimally different from w.[1] This intuitive characterization yields certain constraints on the behavior of the selection function f, which give rise to a specific logic of counterfactuals, called VC. This logic, though not without its issues,[2] does a remarkable job at accounting for the ways in which the logical behavior of counterfactuals differs from that of the implication connective

[1] This description of Lewis's view presupposes the *limit assumption*, i.e., the assumption that for any w and any entertainable φ there be worlds where φ is true and which differ minimally from w in the relevant sense. In our study we will take this assumption for granted, for two reasons. First, this assumption allows for a nice characterization of the semantics in terms of selection functions, and does not affect the resulting propositional logic. Second, there are in fact good conceptual reasons to make the limit assumption: as [12] showed, this assumption is needed to guarantee that an entertainable antecedent has a consistent set of counterfactual consequences.

[2] For criticism of different aspects of this logic, see, e.g., [2,6,8,9,16].

P. Blackburn et al. (Eds.): LORI 2019, LNCS 11813, pp. 43–56, 2019.
https://doi.org/10.1007/978-3-662-60292-8_4

in classical (and intuitionistic) logic: for counterfactuals, VC invalidates strengthening of the antecedent, transitivity, and contraposition. And for a good reason, since both Lewis [14] and Stalnaker [19] give examples where these principles seem to be fallacious.

In the decades following Lewis' work, VC and its relatives have been thoroughly investigated from the perspective of modal logic (see [15] for an overview), and connected to a number of other topics in logic such as probabilistic reasoning [1], belief revision [3,5,11] and default reasoning [13,20]. This body of work focuses on adding counterfactual-like operators against the background of classical propositional logic. However, there seems to be no special reason why the study of such operators should be restricted to a classical setting. The central ideas of Lewis's account are fully modular with respect to the specific semantics used to interpret the constituents of a counterfactual, as long as an intensional setting is available. Thus, it seems interesting to look at how Lewisian counterfactuals could be added to various non-classical logics. This is particularly natural in the case of intuitionistic propositional logic, since the most commonly used semantics for this logic, intuitionistic Kripke semantics, is already intensional in nature, and therefore provides an ideal environment to implement Lewis' idea.

Somewhat surprisingly, a study of Lewisian counterfactuals in intuitionistic Kripke semantics is missing in the literature. Recently, Weiss [21] took a first step in this direction, showing how to extend intuitionistic Kripke semantics with the structure needed to interpret a non-monotonic conditional operator $>$, and studying the intuitionistic counterparts of some very weak conditional logics, including the minimal logic CK of Chellas [7]. These weak logics are interesting, as they bring out most clearly the connection between counterfactuals and modal operators. However, they are generally regarded as too weak to capture many interesting logical principles about counterfactuals. For instance, they *never* allow strengthening of the antecedent, which seems too restrictive. In this paper, we follow up on Weiss' work by studying the intuitionistic counterpart of Lewis's logic VC. In order to achieve this goal, we will propose a semantics that departs slightly from the one given by Weiss. The modifications do not affect the generality of the semantics: it can be shown that a Weiss model can be turned into one of our models without affecting the satisfaction of formulas, and vice versa. However, the modified notion of models will facilitate a simple and elegant correspondence between semantic conditions and syntactic axioms, which seems hard to obtain using Weiss' original semantics.

Before delving into the technical material, let us mention two reasons why adding counterfactuals in an intuitionistic setting is an interesting enterprise. First, intuitionistic logic is already equipped with its own conditional operator $\rightarrow$. Unlike the material conditional of classical logic, which is truth-functional, the conditional of intuitionistic logic is an intensional operator, and its semantics is very similar to the one we will use for the operator $>$: both check whether the consequent is satisfied everywhere within a certain set of possible worlds determined by the antecedent. The difference is which set of worlds is picked out by each of them. Crucially, $\rightarrow$ is constrained to quantifying over states of affairs which are *possible* from the perspective of the evaluation world, while $>$ is allowed to quantify over counterfactual states of affairs as well. A natural interpretation is to view $p \rightarrow q$ as standing for an indicative conditional like (1-a), and $p > q$ as standing for a counterfactual conditional like (1-b).

(1) a. If the butler didn't do it, the gardener did.
 b. If the butler had not done it, the gardener would have.

Second, studying the principles of conditional logic from the perspective of intuitionistic logic allows us to ask which of these principles stem only from assumptions about the semantics of counterfactuals, and which stem partly from the classicality of the logic. As an example, consider a principle central to Lewis' logic, the *rational monotonicity* principle. Informally, this principle says that, if in making a counterfactual assumption φ one leaves open the possibility that ψ, then one is justified in strengthening the antecedent from φ to $\varphi \wedge \psi$. Classically, this principle could be equivalently formulated in either of the following ways:

- $((\varphi > \chi) \wedge \neg(\varphi > \neg\psi)) \rightarrow (\varphi \wedge \psi > \chi)$
- $(\varphi > \chi) \rightarrow ((\varphi > \neg\psi) \vee (\varphi \wedge \psi > \chi))$

It turns out that, in the intuitionistic setting, the latter, and not the former, is the appropriate way to capture the rational monotonicity constraint. Thus, the former can be seen as a consequence of rational monotonicity plus classical logic.

The paper is structured as follows: in Sect. 2 we describe how to extend intuitionistic Kripke semantics with the structure needed to interpret a counterfactual conditional operator; in Sect. 3 we give intuitionistic versions of the assumptions of minimal change semantics, which lead to IVC, an intuitionistic counterpart of Lewis' logic VC; in Sect. 4 we describe an axiomatization of this logic and show that it is sound; in Sect. 5 we describe how to construct a canonical model for IVC, and use this construction to prove completeness; Sect. 6 summarizes our findings and outlines some directions for further work.

2 Counterfactuals in Intuitionistic Kripke Semantics

In this section we describe how to extend intuitionistic Kripke semantics with the structure needed to interpret a counterfactual conditional operator $>$. The idea is to enrich a Kripke structure with a selection function, which picks for each world and each antecedent a set of "relevant antecedent worlds".

Definition 1 (Intuitionistic selection models). *An intuitionistic selection model is a tuple* $M = \langle W, \leq, \mathcal{A}, f, V\rangle$ *where:*

- W *is a set, whose elements are called* worlds.
- $\leq$ *is a partial order on* W*, the* refinement *ordering; the set of* $\leq$*-successors of a world* w *is denoted* $w^{\uparrow}$*; in symbols:* $w^{\uparrow} := \{v \in W \mid w \leq v\}$.
- $Up_{\leq}(W)$ *denotes the set of up-sets of* W*, i.e.:*

$$Up_{\leq}(W) = \{X \subseteq W \mid \forall w, v : w \in X \text{ and } w \leq v \text{ implies } v \in X\}$$

- $\mathcal{A} \subseteq Up_{\leq}(W)$ *is a set of up-sets called* propositions, *which contains* $\emptyset$ *and is closed under union, intersection, and the following operations:*

$$X, Y \mapsto \{w \in W \mid X \cap w^{\uparrow} \subseteq Y\} \qquad X, Y \mapsto \{w \in W \mid f(w, X) \subseteq Y\}$$

- $f : W \times \mathcal{A} \to \wp(W)$, *the* selection function, *is a map assigning to each world* w *and proposition* X *a subset* $f(w, X) \subseteq X$ *(the* relevant X-worlds at w*).*
- $V : \mathcal{P} \to \mathcal{A}$ *is a* valuation function, *assigning to each atom a proposition.*

The refinement ordering and the selection function are required to be linked by the following conditions:

- *Upwards-closure:* $f(w, X) \in Up_{\leq}(W)$ *for any* $w \in W$ *and* $X \in \mathcal{A}$.
- *Monotonicity of* f *in the first coordinate: if* $w \leq v$ *then* $f(w, X) \supseteq f(v, X)$.

One may think of worlds as partial stages in a process of inquiry. At each world, a sentence φ may or may not have been established. The relation $w \leq v$ means that v is a refinement of w: if this holds, then v establishes everything that w establishes, and possibly more. The closure conditions on $\mathcal{A}$ are needed to ensure that the object $|\varphi|$ expressed by a sentence φ in a model is always a proposition, and thus that the hypothetical context $f(w, |\varphi|)$ needed to interpret counterfactuals with antecedent φ is defined. The intended interpretation of f is that the elements $v \in f(w, X)$ are those worlds which, from the standpoint of w, may have obtained if X had been the case. We refer to $f(w, X)$ as the *hypothetical context* generated at w by the making the counterfactual assumption that X.[3]

The condition that $f(w, X)$ be upwards-closed can be motivated as follows: if $v \in f(w, X)$, then at w we think that, had X been the case, v may have obtained. Since v may evolve into any of its successors, any such point may have obtained if X had been the case. Thus, each successor of v should be in $f(w, X)$.

Finally, the monotonicity condition says that, if $w \leq v$, then the hypothetical context $f(v, X)$ is at least as strong as the context $f(w, X)$. This is a natural constraint: $w \leq v$ means that all the information available at w is also available at v; this includes counterfactual information about how things would be if X were the case; so, any counterfactual possibility u which can be ruled out at w ($u \notin f(w, X)$) can also be ruled out at v ($u \notin f(v, X)$). Thus, $f(v, X) \subseteq f(w, X)$.[4]

[3] In the work of Lewis, the selection function takes formulas, rather than propositions, as its second argument. It would in principle be possible to do the same here. However, the presentation would become more complicated: some conditions in the definition of a model (in particular, the requirement that f should yield the same result when applied to intensionally equivalent formulas) appeal to the semantics of sentences, which in turn is defined with reference to the notion of a model. Letting selection functions take propositions allows us to avoid such seeming circularities.

[4] Our semantics departs from the one recently proposed by Weiss [21] in two ways: first, Weiss does not require $f(w, X)$ to be upwards-closed; second, he requires $f(w, X)$ to be defined for all subsets $X \subseteq W$, not just for a designated set of such subsets. Both differences are important for our completeness result. At the same time, however, a Weiss model can be translated to one of our models, and vice versa, without affecting the satisfaction of formulas. A detailed comparison must be left for another occasion.

The addition of the selection function component to intuitionistic Kripke models allows us to interpret a propositional language extended with a counterfactual conditional connective $>$. More precisely, the language $\mathcal{L}^{>}$ that we will work with is given by the following BNF definition:

$$\varphi ::= p \mid \bot \mid \varphi \wedge \varphi \mid \varphi \vee \varphi \mid \varphi \rightarrow \varphi \mid \varphi > \varphi$$

As usual in intuitionistic logic, negation and the biconditional are defined as:

$$\neg\varphi := \varphi \rightarrow \bot \qquad \varphi \leftrightarrow \psi := (\varphi \rightarrow \psi) \wedge (\psi \rightarrow \varphi)$$

Satisfaction relative to a model M and a world w is defined as follows.

Definition 2 (Semantics)

1. $M, w \models p \iff w \in V(p)$
2. $M, w \not\models \bot$
3. $M, w \models \varphi \wedge \psi \iff M, w \models \varphi$ *and* $M, w \models \psi$
4. $M, w \models \varphi \vee \psi \iff M, w \models \varphi$ *or* $M, w \models \psi$
5. $M, w \models \varphi \rightarrow \psi \iff \forall v \geq w : M, v \models \varphi$ *implies* $M, v \models \psi$
6. $M, w \models \varphi > \psi \iff \forall v \in f(w, |\varphi|) : M, v \models \psi$

where the set $|\varphi|$*, called the* proposition expressed *by of* φ *in* M*, is defined as:*

- $|\varphi| := \{w \in W \mid M, w \models \varphi\}$

To lighten notation, in the following we will write $f(w, \varphi)$ *instead of* $f(w, |\varphi|)$*.*

Clauses 1–5 are just the standard clauses of intuitionistic Kripke semantics. Clause 6 says that $\varphi > \psi$ is satisfied at w if ψ is satisfied at all the relevant φ-worlds at w; that is, $\varphi > \psi$ is satisfied at w iff ψ holds throughout the hypothetical context generated from making the counterfactual assumption φ at w.

It is interesting to note that the semantics for the two conditionals can be made more parallel than it looks at first. To see this, notice that we can think of those $\leq$-successors of w which satisfy φ as those states which may become actual if φ is established. We can then think that supposing φ as an indicative (as opposed to counterfactual) assumption amounts to imagining that we are in one of these worlds. More technically, let us define a selection function g as $g(w, X) := \{v \in X \mid w \leq v\}$; then the semantics of the intuitionistic conditional $\rightarrow$ can also be presented in the selection function format:

- $M, w \models \varphi \rightarrow \psi \iff \forall v \in g_{\leq}(w, |\varphi|) : M, v \models \psi$

Thus, the difference between the two conditionals lies not in the mathematical workings of their semantics, but rather in the different selection functions that they invoke, corresponding to the difference between supposing φ an indicative assumption and as a counterfactual assumption.

As usual in intuitionistic Kripke semantics, we have a *persistency* property: whatever is established at a world remains established at any refinement of it.

Proposition 1 (Persistency). *For every ICM M, if $w \leq v$ then $M, w \models \varphi$ implies $M, v \models \varphi$.*

Proof. By induction on φ. We only give the inductive step for $\varphi = \psi > \chi$. Suppose $w \leq v$ and $M, w \models \psi > \psi$. This means that $f(w, \psi) \subseteq |\chi|$. By the monotonicity condition we have $f(v, \psi) \subseteq f(w, \psi)$. Therefore also $f(v, \psi) \subseteq |\chi|$, which means that $M, v \models \psi > \chi$. □

It is worth pointing out that classical selection function semantics (e.g., [6,17]) can be retrieved as a special case: classical selection models can be identified with intuitionistic selection models where the refinement relation $\leq$ is the identity; restricted to these models, our semantic clauses boil down to the classical ones.

3 Minimal Change Conditions

Lewis's minimal change semantics can be seen as obtained from selection function semantics by imposing some constraints on how the selection function works (see [14] §2.7, [6] §1.2). In this section we propose analogues of these constraints in the intuitionistic setting, and discuss some repercussions of these constraints for the logical behavior of conditionals.

Definition 3 (Intuitionistic minimal change models). *An intuitionistic selection model M is called an* intuitionistic minimal change model *if it satisfies the following conditions:*

1. *if $w \in X$ then $w \in f(w, X)$*
2. *if $w \in X$ then $f(w, X) \subseteq w^{\uparrow}$*
3. *if $f(w, X) = \emptyset$ and $Y \subseteq X$ then $f(w, Y) = \emptyset$*
4. *if $Y \subseteq X$ and $f(w, X) \cap Y \neq \emptyset$ then $f(w, Y) \subseteq f(w, X)$*
5. *if $Y \subseteq X$ then $f(w, X) \cap Y \subseteq f(w, Y)$*

Condition 1 is known as the *weak centering* condition: it says that if X is true at w, then w is one of the worlds which might be the case if X were the case. That is, if the antecedent is true, then the consequent must be true in order for the counterfactual to be true. In our intuitionistic setting, this constraint implies another interesting property: at every world w, any refinement of w which satisfies φ is relevant to the truth of a conditional $\varphi > \psi$. To state this precisely, recall that we used the notation $g(w, X)$ for the set of refinements of w which are in X, that is, $g(w, X) = \{v \in X \mid v \geq w\}$. We have the following.

Proposition 2. *If 1 holds in M, then for any w and any X: $g(w, X) \subseteq f(w, X)$.*

Proof. Suppose $v \in g(w, X)$, i.e., $v \geq w$ and $v \in X$. By condition 1, $v \in f(v, X)$. By the monotonicity condition, $f(v, X) \subseteq f(w, X)$. Therefore, $v \in f(w, X)$. □

Since $g(w, \varphi)$ and $f(w, \varphi)$ provide, respectively, the domains of quantifications used to assess $\varphi \rightarrow \psi$ and $\varphi > \psi$, this proposition implies the following corollary.

Proposition 3. *If 1 holds in M, then $M, w \models \varphi > \psi$ implies $M, w \models \varphi \to \psi$.*

Condition 2 says that, if φ is true at w, then no *counterfactual* world—i.e., no world which is not a refinement of w—is relevant to the truth of $\varphi > \psi$ at w. In combination with Condition 1, this gives the *strong centering* condition, which in our setting is formulated as follows.

Proposition 4. *Suppose 1 and 2 hold in M. If $w \in X$, then $f(w, X) = w^{\uparrow}$.*

This condition looks a bit different than the classical strong centering condition, which requires that if $w \in X$ then $f(w, X) = \{w\}$. Note, however, that the classical formulation would not be compatible with the upwards-closure requirement on $f(w, X)$, since $\{w\}$ is not upwards-closed if w is not an endpoint (i.e., if there are proper extensions $v > w$). If $f(w, X)$ includes w, then it must contain the whole set $w^{\uparrow}$; thus, $w^{\uparrow}$ is the smallest hypothetical context which includes w.

Nevertheless, if we look at the special case of classical selection models, that is, models where the relation $\leq$ is the identity, then we retrieve the classical formulation of strong centering: for then $w^{\uparrow} = \{w\}$. Furthermore, in our setting the condition $f(w, \varphi) = w^{\uparrow}$ captures exactly the central idea of strong centering, namely: if φ is true at w, then the only world which is relevant to assessing the truth of a conditional $\varphi > \psi$ is w itself.

Proposition 5. *Suppose conditions 1 and 2 hold in M. If $w \in |\varphi|$, for every ψ we have $M, w \models \varphi > \psi \iff M, w \models \psi$.*

Proof. If 1 and 2 hold and $w \in |\varphi|$, by the previous proposition we have $f(w, \varphi) = w^{\uparrow}$. Suppose $M, w \models \varphi > \psi$. Then $f(w, \varphi) \subseteq |\psi|$, and since $w \in w^{\uparrow} = f(w, \varphi)$ we have $M, w \models \psi$. Conversely, suppose $M, w \models \psi$. By persistency, every $v \geq w$ satisfies ψ, so $w^{\uparrow} \subseteq |\psi|$. Since $f(w, \varphi) = w^{\uparrow}$, it follows that $M, w \models \varphi > \psi$. □

The third condition says that if X cannot be consistently supposed, then any proposition stronger than X cannot be consistently supposed either. This gives:

Proposition 6. *If 3 holds in M, $M, w \models \varphi > \bot$ implies $M, w \models \varphi \wedge \psi > \bot$.*

The fourth condition is a restricted monotonicity constraint. It says that, when we strengthen an antecedent from X to $Y \subseteq X$, we must get a stronger hypothetical context $f(w, Y) \subseteq f(w, X)$, as long as the stronger antecedent is consistent with the hypothetical context determined by the weaker antecedent.

The fifth condition is also about the effect of strengthening an antecedent. It says that if v is one of the relevant X-worlds at w, and if v also satisfies a stronger proposition $Y \subseteq X$, then v is also one of the relevant Y-worlds at w.

Notice that conditions 4 and 5 jointly determine the effect of strengthening an antecedent in those cases in which the stronger antecedent is consistent with the hypothetical context for the weak one.

Proposition 7. *Let M obey 4 and 5. If $Y \subseteq X$ and $f(w, X) \cap Y \neq \emptyset$, then $f(w, Y) = f(w, X) \cap Y$.*

Proof. Suppose $Y \subseteq X$ and $f(w, X) \cap Y \neq \emptyset$. By definition of selection function, $f(w, Y) \subseteq Y$, and by condition 4, $f(w, Y) \subseteq f(w, X)$. So, $f(w, Y) \subseteq f(w, X) \cap Y$. The converse inclusion is given by Condition 5.

Notice that, when we restrict ourselves to classical selection models, where $\leq$ is the identity, the above conditions pick out exactly the class of classical selection models which characterize Lewis's logic VC (see [14], §2.7). The logic of all intuitionistic minimal change models can thus be naturally regarded as an intuitionistic counterpart of VC. We will denote this logic as IVC.

Definition 4 (Logic IVC). *For $\Phi \cup \{\psi\} \subseteq \mathcal{L}^{>}$, we write $\Phi \models_{IVC} \psi$ iff for any intuitionistic minimal change model M and any world w, if $M, w \models \varphi$ for all $\varphi \in \Phi$, then $M, w \models \psi$.*

4 Axiomatization

In this section we describe a Hilbert-style system for the logic IVC, and show that it is sound. In the following sections we will show that it is also complete. The system has three groups of axioms: axioms for intuitionistic propositional logic; axioms that pertain to selection function semantics in general; and axioms that correspond to the minimal change conditions.

- Intuitionistic schemata:
  - $\varphi \to (\psi \to \varphi)$
  - $(\varphi \to (\psi \to \chi)) \to ((\varphi \to \psi) \to (\varphi \to \chi))$
  - $\varphi \to (\psi \to \varphi \wedge \psi)$
  - $\varphi \wedge \psi \to \varphi, \ \varphi \wedge \psi \to \psi$
  - $\varphi \to \varphi \vee \psi, \ \psi \to \varphi \vee \psi$
  - $(\varphi \to \chi) \to ((\psi \to \chi) \to (\varphi \vee \psi \to \chi))$
  - $\bot \to \varphi$
- Selection function schemata:
  - $\varphi > \varphi$
  - $(\varphi > \psi \wedge \chi) \leftrightarrow (\varphi > \psi) \wedge (\varphi > \chi)$
- Minimal change schemata:
  - $(\varphi > \psi) \to (\varphi \to \psi)$
  - $(\varphi \wedge \psi) \to (\varphi > \psi)$
  - $(\varphi > \bot) \to (\varphi \wedge \psi > \bot)$
  - $(\varphi > \chi) \to (\varphi > \neg\psi) \vee ((\varphi \wedge \psi) > \chi)$
  - $(\varphi \wedge \psi > \chi) \to (\varphi > (\psi \to \chi))$

The system has three inference rules: *modus ponens*, *replacement of equivalent antecedents*, and *replacement of equivalent consequents*:

$$\frac{\varphi \quad \varphi \to \psi}{\psi}\ (\mathsf{MP}) \qquad \frac{\varphi \leftrightarrow \psi}{(\varphi > \chi) \leftrightarrow (\psi > \chi)}\ (\mathsf{RCEA}) \qquad \frac{\varphi \leftrightarrow \psi}{(\chi > \varphi) \leftrightarrow (\chi > \psi)}\ (\mathsf{RCEC})$$

As usual in modal logic, some care is needed when defining derivability from a set of assumptions in the system. We define this as follows.

Definition 5. *For $\Phi \cup \{\psi\} \subseteq \mathcal{L}^{>}$, we write $\Phi \vdash_{\mathsf{IVC}} \psi$ to mean that there exist $\varphi_1, \ldots, \varphi_n \in \Phi$ such that $\varphi_1 \wedge \cdots \wedge \varphi_n \to \psi$ is derivable in the system above.*

A fact that will be useful below is that, in IVC, conditionals are right monotonic.

Proposition 8. *If $\psi \vdash_{\mathsf{IVC}} \chi$ then $\varphi > \psi \vdash_{\mathsf{IVC}} \varphi > \chi$.*

Proof. If $\psi \vdash_{\mathsf{IVC}} \chi$ then $\vdash_{\mathsf{IVC}} \psi \to \chi$, and therefore by intuitionistic reasoning, $\vdash_{\mathsf{IVC}} \psi \leftrightarrow \psi \wedge \chi$. By closure under replacement of equivalent consequents, we have $\vdash_{\mathsf{IVC}} (\varphi > \psi) \leftrightarrow (\varphi > \psi \wedge \chi)$. But we also have the axiom $(\varphi > \psi \wedge \chi) \leftrightarrow (\varphi > \psi) \wedge (\varphi > \chi)$. By intuitionistic reasoning it follows that $\vdash_{\mathsf{IVC}} (\varphi > \psi) \to (\varphi > \chi)$. Thus, $\varphi > \psi \vdash_{\mathsf{IVC}} \varphi > \chi$. □

Proposition 9 (Soundness). *If $\Phi \models_{\mathsf{IVC}} \psi$ then $\Phi \vdash_{\mathsf{IVC}} \psi$.*

Proof. As usual, the proof amounts to checking that the axioms of the system are valid, and that the inference rules preserve validity. We focus on the soundness of the minimal change schemata, since the other cases are straightforward. The soundness of the first three schemata is an immediate consequence of Propositions 3, 5, and 6. Consider the fourth minimal change schema. Suppose that $M, w \models \varphi > \chi$, i.e., $f(w, \varphi) \subseteq |\chi|$. We want to show that $M, w \models (\varphi \to \neg\psi) \vee (\varphi \wedge \psi > \chi)$. We distinguish two cases:

- Case 1: $f(w, \varphi) \cap |\psi| = \emptyset$. Take any $v \in f(w, \varphi)$. Since $f(w, \varphi)$ is upwards closed, for any successor $u \geq v$ we have $u \in f(w, \varphi)$, and therefore $u \notin |\psi|$. This means that $M, v \models \neg\psi$. Therefore, $M, w \models \varphi > \neg\psi$.
- Case 2: $f(w, \varphi) \cap |\psi| \neq \emptyset$. Then we have $|\varphi \wedge \psi| \subseteq |\varphi|$ and $f(w, \varphi) \cap |\varphi \wedge \psi| \neq \emptyset$. Therefore, Condition 4 implies $f(w, \varphi \wedge \psi) \subseteq f(w, \varphi)$, and since $f(w, \varphi) \subseteq |\chi|$ we have $M, w \models \varphi \wedge \psi > \chi$.

In both cases, $M, w \models (\varphi \to \neg\psi) \vee (\varphi \wedge \psi > \chi)$.

Finally, consider an instance of the schema $(\varphi \wedge \psi > \chi) \to (\varphi > (\psi \to \chi))$. Suppose $M, w \models \varphi \wedge \psi > \chi$. Then $f(w, \varphi \wedge \psi) \subseteq |\chi|$. We need to show that $w \models \varphi > (\psi \to \chi)$. So, take any $v \in f(w, \varphi)$. We want to show $v \models \psi \to \chi$. Consider any $u \geq v$ with $M, u \models \psi$. Since $f(w, \varphi)$ is upwards closed, $u \in f(w, \varphi)$. Thus, $u \in f(w, \varphi) \cap |\psi| = f(w, \varphi) \cap |\varphi \wedge \psi|$. By Condition 5, $f(w, \varphi) \cap |\varphi \wedge \psi| \subseteq f(w, \varphi \wedge \psi) \subseteq |\chi|$. Therefore, $u \in |\chi|$, which means that $M, u \models \chi$, as we wanted.

Notice that the proof makes crucial use of the upwards closure condition of $f(w, \varphi)$, which distinguishes our semantics from the one of Weiss [21]. It is not hard to show that, if $f(w, \varphi)$ is not required to be upwards closed, then the last two minimal change schemata are not sound in general.

5 Canonical Model Construction

In this section we define a canonical model for IVC, which will allow us to show the completeness of our proof system. As usual in intuitionistic logic, the model is based on consistent theories with the disjunction property.

Definition 6. *Let $\Gamma \subseteq \mathcal{L}^{>}$. We say that:*

- *Γ is an IVC*-theory *if for all $\varphi \in \mathcal{L}^{>}$: $\Gamma \vdash_{\mathsf{IVC}} \varphi$ implies $\varphi \in \Gamma$;*
- *Γ is a* consistent *IVC*-theory *if Γ is an IVC-theory and $\bot \notin \Gamma$;*
- *Γ has the* disjunction property *if $\varphi \vee \psi \in \Gamma$ implies $\varphi \in \Gamma$ or $\psi \in \Gamma$.*

The next lemma, familiar in intuitionistic logic, says that there are enough consistent IVC-theories with the disjunction property to witness all non-entailments in IVC. The standard proof of the lemma is omitted (see, e.g., Lemma 11 in [4]).

Lemma 1. *If $\Phi \nvdash_{\mathsf{IVC}} \psi$ then there exists a consistent IVC-theory with the disjunction property Γ such that $\Phi \subseteq \Gamma$ and $\psi \notin \Gamma$.*

For an IVC-theory Γ, the set of counterfactual consequences of φ in Γ is:

- $\mathrm{Cn}_\varphi(\Gamma) := \{\psi \in \mathcal{L}^{>} \mid \varphi > \psi \in \Gamma\}$

The following feature of $\mathrm{Cn}_\varphi(\Gamma)$ will play an important role below.

Lemma 2. *If Γ is an IVC-theory, then so is $Cn_\varphi(\Gamma)$.*

Proof. Straightforward, using the right monotonicity of $>$ (Proposition 8).

Definition 7 (Canonical model). *The canonical model for IVC is the model $M^c_{\mathsf{IVC}} = \langle W^c, \leq^c, \mathcal{A}^c, f^c, V^c \rangle$, where:*

- *W^c is the set of consistent IVC-theories with the disjunction property;*
- *$\Gamma \leq^c \Gamma' \iff \Gamma \subseteq \Gamma'$*
- *$\mathcal{A}^c = \{\widehat{\varphi} \mid \varphi \in \mathcal{L}^{>}\}$ where $\widehat{\varphi} := \{\Gamma \in W^c \mid \varphi \in \Gamma\}$*
- *$f^c(\Gamma, \widehat{\varphi}) = \{\Gamma' \in W^c \mid Cn_\varphi(\Gamma) \subseteq \Gamma'\}$*
- *$V^c(p) = \widehat{p}$*

We need to make sure that f^c is well-defined, i.e., that if $\widehat{\varphi} = \widehat{\chi}$ then $f^c(\Gamma, \widehat{\varphi}) = f^c(\Gamma, \widehat{\chi})$. This is guaranteed by the following proposition.

Proposition 10. *If $\widehat{\varphi} = \widehat{\chi}$, then for all $\Gamma \in W^c$: $Cn_\varphi(\Gamma) = Cn_\chi(\Gamma)$.*

Proof. Suppose $\widehat{\varphi} = \widehat{\chi}$. First, note that this implies $\vdash_{\mathsf{IVC}} \varphi \leftrightarrow \chi$. For suppose not: then $\varphi \nvdash_{\mathsf{IVC}} \chi$ or $\chi \nvdash_{\mathsf{IVC}} \varphi$. Without loss of generality, suppose the former. By Lemma 1 there exists a theory $\Gamma \in W^c$ with $\varphi \in \Gamma$ and $\psi \notin \Gamma$, which means that $\Gamma \in \widehat{\varphi}$ but $\Gamma \notin \widehat{\chi}$, contrary to $\widehat{\varphi} = \widehat{\chi}$. So, $\vdash_{\mathsf{IVC}} \varphi \leftrightarrow \chi$. By replacement of equivalent antecedents, it follows that for any ψ, $\vdash_{\mathsf{IVC}} (\varphi > \psi) \leftrightarrow (\chi > \psi)$. Now take any $\Gamma \in W^c$. Since Γ is an IVC-theory, for any ψ we have $(\varphi > \psi) \in \Gamma \iff (\chi > \psi) \in \Gamma$. As this holds for all ψ, it follows that $\mathrm{Cn}_\varphi(\Gamma) = \mathrm{Cn}_\varphi(\Gamma)$.

The next proposition ensures that M^c_{IVC} satisfies all the conditions required by Definition 1.

Proposition 11. *M^c_{IVC} is an intuitionistic selection model.*

Proof. We need to check that all the conditions in Definition 1 are satisfied. Clearly, the relation $\subseteq$ is a partial order on W^c. Every element of $\mathcal{A}^c$ is upwards closed, and $V^c(p) \in \mathcal{A}^c$. Four conditions remains to be checked:

- Closure of $\mathcal{A}^c$ under logic. We show in detail the most interesting case, namely, closure of $\mathcal{A}^c$ under the operation corresponding to $>$. So, suppose $\widehat{\varphi}, \widehat{\chi} \in \mathcal{A}$; we need to show that $\{\Gamma \mid f^c(\Gamma, \widehat{\varphi}) \subseteq \widehat{\chi}\} \in \mathcal{A}^c$. This will follow if we can show that
$$\{\Gamma \mid f^c(\Gamma, \widehat{\varphi}) \subseteq \widehat{\chi}\} = \widehat{\varphi > \chi}$$
This amounts to the claim that, for $\Gamma \in W^c$: $f^c(\Gamma, \widehat{\varphi}) \subseteq \widehat{\chi} \iff \varphi > \chi \in \Gamma$. In one direction, suppose $\varphi > \chi \in \Gamma$. Then $\chi \in \mathrm{Cn}_\varphi(\Gamma)$. Therefore any $\Gamma' \in f^c(\Gamma, \widehat{\varphi})$ must contain χ, which means that $f^c(\Gamma, \widehat{\varphi}) \subseteq \widehat{\chi}$. Conversely, suppose $\varphi > \chi \notin \Gamma$. Then $\chi \notin \mathrm{Cn}_\varphi(\Gamma)$. By Lemma 2, $\mathrm{Cn}_\varphi(\Gamma) \not\vdash_{\mathsf{IVC}} \chi$. Therefore, by Lemma 1 there is $\Gamma' \in W^c$ with $\mathrm{Cn}_\varphi(\Gamma) \subseteq \Gamma'$ and $\chi \notin \Gamma'$. Thus, $\Gamma' \in f^c(\Gamma, \widehat{\varphi})$ but $\Gamma' \notin \widehat{\chi}$, witnessing that $f^c(\Gamma, \widehat{\varphi}) \nsubseteq \widehat{\chi}$.
In a similar fashion, it is easy to prove that $\mathcal{A}^c$ is closed under intersection, union, and the operation corresponding to $\rightarrow$, since $\widehat{\varphi} \cup \widehat{\chi} = \widehat{\varphi \vee \chi}$, $\widehat{\varphi} \cap \widehat{\chi} = \widehat{\varphi \wedge \chi}$, and $\{\Gamma \in W^c \mid \widehat{\varphi} \cap \Gamma^\uparrow \subseteq \widehat{\psi}\} = \widehat{\varphi \rightarrow \chi}$.
- $f^c(\Gamma, \widehat{\varphi}) \subseteq \widehat{\varphi}$. Take $\Gamma' \in f^c(\Gamma, \widehat{\varphi})$. This means that $\mathrm{Cn}_\varphi(\Gamma) \subseteq \Gamma'$. By the axiom $\varphi > \varphi$ we have $\varphi \in \mathrm{Cn}_\varphi(\Gamma)$. Thus, $\varphi \in \Gamma'$, which shows that $\Gamma' \in \widehat{\varphi}$.
- $f^c(\Gamma, \widehat{\varphi})$ is upwards closed. This is clear since, if $\Gamma' \subseteq \Gamma''$ we have $\Gamma' \in f^c(\Gamma, \widehat{\varphi}) \iff \mathrm{Cn}_\varphi(\Gamma) \subseteq \Gamma' \Longrightarrow \mathrm{Cn}_\varphi(\Gamma) \subseteq \Gamma'' \iff \Gamma'' \in f^c(\Gamma, \widehat{\varphi})$.
- f^c is monotonic in the first coordinate. This is also clear: if $\Gamma \subseteq \Gamma'$ then $\mathrm{Cn}_\varphi(\Gamma) \subseteq \mathrm{Cn}_\varphi(\Gamma')$, and therefore $\Gamma'' \in f^c(\Gamma', \widehat{\varphi}) \iff \mathrm{Cn}_\varphi(\Gamma') \subseteq \Gamma'' \Longrightarrow \mathrm{Cn}_\varphi(\Gamma) \subseteq \Gamma'' \iff \Gamma'' \in f^c(\Gamma, \widehat{\varphi})$. □

Moreover, we can prove that M^c_L behaves in the way expected of a canonical model: satisfaction at a theory Γ amounts to membership in Γ.

Lemma 3 (Truth Lemma). *For any $\Gamma \in W^c$ and any $\varphi \in \mathcal{L}^>$: $M^c_{\mathsf{IVC}}, \Gamma \models \varphi \iff \varphi \in \Gamma$.*

Proof. As usual, the proof is by induction on φ. We only give the inductive step for $\varphi = \chi > \psi$, since the other steps are the same as in the case of intuitionistic propositional logic.

For the right-to-left direction, suppose $\chi > \psi \in \Gamma$. Then $\psi \in \mathrm{Cn}_\chi(\Gamma)$, so $\psi \in \Gamma'$ for every $\Gamma' \in f^c(\Gamma, \widehat{\chi})$, by definition of $f^c(\Gamma, \widehat{\chi})$. By the induction hypothesis on ψ, this means that for every $\Gamma' \in f^c(\Gamma, \widehat{\chi})$ we have $M^c_{\mathsf{IVC}}, \Gamma' \models \psi$. Moreover, by the induction hypothesis on χ we have $\widehat{\chi} = |\chi|$. Thus, $M^c_{\mathsf{IVC}}, \Gamma \models \chi > \psi$.

For the converse direction, suppose $\chi > \psi \notin \Gamma$. Then $\psi \notin \mathrm{Cn}_\chi(\Gamma)$, which by Lemma 2 means that $\mathrm{Cn}_\chi(\Gamma) \not\vdash_{\mathsf{IVC}} \psi$. By Lemma 1 there exists a theory $\Gamma' \in W^c$ such that (i) $\mathrm{Cn}_\chi(\Gamma) \subseteq \Gamma'$ and (ii) $\psi \notin \Gamma'$. By (i) we have $\Gamma' \in f^c(\Gamma, \widehat{\chi})$, and by (ii) and the induction hypothesis on ψ, $M^c_{\mathsf{IVC}}, \Gamma' \not\models \psi$. Moreover, by the induction hypothesis on χ we have $\widehat{\chi} = |\chi|$. Therefore, it is not the case that all worlds in $f^c(\Gamma, |\chi|)$ satisfy ψ, which means that $M^c_{\mathsf{IVC}}, \Gamma \not\models \chi > \psi$. □

This lemma implies that any IVC-invalid entailment can be falsified in M^c_{IVC}.

Proposition 12. *If $\Phi \not\vdash_{\mathsf{IVC}} \psi$, then there exists a world $\Gamma \in W^c$ such that $M^c_{\mathsf{IVC}}, \Gamma \models \varphi$ for all $\varphi \in \Phi$ but $M^c_{\mathsf{IVC}}, \Gamma \not\models \psi$.*

Proof. If $\Phi \not\vdash_{\mathsf{IVC}} \psi$, by Lemma 1 there is $\Gamma \in W^c$ s.t. $\Phi \subseteq \Gamma$ but $\psi \notin \Gamma$. By the truth-lemma, world Γ in M^c_{IVC} satisfies all formulas in Φ but not ψ. □

To show that our system is complete with respect to the logic of intuitionistic minimal change models, all that remains to be shown is the following.

Proposition 13. *M^c_{IVC} is an intuitionistic minimal change model.*

Proof. We will show that each of the five minimal-change axioms of the logic IVC yields one of the corresponding minimal change properties for M^c_{IVC}.

- Condition 1. Suppose $\Gamma \in \widehat{\varphi}$, which means that $\varphi \in \Gamma$. We want to show that $\Gamma \in f^c(\Gamma, \widehat{\varphi})$, which amounts to $\text{Cn}_\varphi(\Gamma) \subseteq \Gamma$. Consider any $\psi \in \text{Cn}_\varphi(\Gamma)$. This means that $\varphi > \psi \in \Gamma$. By the axiom $(\varphi > \psi) \to (\varphi \to \psi)$, also $\varphi \to \psi \in \Gamma$. Since $\varphi \in \Gamma$, it follows $\psi \in \Gamma$.
- Condition 2. Take a point $\Gamma \in \widehat{\varphi}$, which means that $\varphi \in \Gamma$. We want to show that $f^c(\Gamma, \widehat{\varphi}) \subseteq \Gamma^{\uparrow}$. So, take any $\Gamma' \in f^c(\Gamma, \widehat{\varphi})$. This means that $\text{Cn}_\varphi(\Gamma) \subseteq \Gamma'$. We need to show that $\Gamma' \in \Gamma^{\uparrow}$, which amounts to $\Gamma \subseteq \Gamma'$. This will follow if we can show that $\Gamma \subseteq \text{Cn}_\varphi(\Gamma)$. So, take any $\psi \in \Gamma$. Since $\varphi, \psi \in \Gamma$, also $\varphi \wedge \psi \in \Gamma$. By the axiom $(\varphi \wedge \psi) \to (\varphi > \psi)$, it follows that $\varphi > \psi \in \Gamma$, so $\psi \in \text{Cn}_\varphi(\Gamma)$.
- Condition 3. Suppose $\widehat{\psi} \subseteq \widehat{\varphi}$. This implies $\psi \vdash_{\mathsf{IVC}} \varphi$: for otherwise, by Lemma 1 there would be a theory $\Gamma \in W^c$ such that $\psi \in \Gamma$ and $\varphi \notin \Gamma$; and then $\Gamma \in \widehat{\psi} - \widehat{\varphi}$, contrary to the inclusion $\widehat{\psi} \subseteq \widehat{\varphi}$.
  Now suppose that $f^c(\Gamma, \widehat{\varphi}) = \emptyset$. We want to show that also $f^c(\Gamma, \widehat{\psi}) = \emptyset$. First, notice that $f^c(\Gamma, \widehat{\varphi}) = \emptyset$ implies that $\text{Cn}_\varphi(\Gamma) \vdash_{\mathsf{IVC}} \bot$. For otherwise, by Lemma 1, $\text{Cn}_\varphi(\Gamma)$ could be extended to a world $\Gamma' \in W^c$, and then we would have $\Gamma' \in f^c(\Gamma, \widehat{\varphi})$.
  By Lemma 2, $\text{Cn}_\varphi(\Gamma) \vdash_{\mathsf{IVC}} \bot$ implies $\bot \in \text{Cn}_\varphi(\Gamma)$, that is, $\varphi > \bot \in \Gamma$. By the axiom $(\varphi > \bot) \to (\varphi \wedge \psi > \bot)$ we also have $\varphi \wedge \psi > \bot \in \Gamma$. Since $\psi \vdash_{\mathsf{IVC}} \varphi$, the counterfactual $\varphi \wedge \psi > \bot$ is inter-derivable with $\psi > \bot$. Therefore, $\psi > \bot \in \Gamma$, which implies that $\bot \in \text{Cn}_\psi(\Gamma)$.
  Now for every $\Gamma' \in W^c$ we have $\bot \notin \Gamma'$, and therefore $\text{Cn}_\psi(\Gamma) \not\subseteq \Gamma'$. This shows that $f^c(\Gamma, \varphi \wedge \psi) = \emptyset$.
- Condition 4. Suppose $\widehat{\psi} \subseteq \widehat{\varphi}$. As discussed in the previous point, this implies $\psi \vdash_{\mathsf{IVC}} \varphi$. Suppose moreover that $f^c(\Gamma, \widehat{\varphi}) \cap \widehat{\psi} \neq \emptyset$. This means that there is $\Gamma' \in W^c$ with $\text{Cn}_\varphi(\Gamma) \cup \{\psi\} \subseteq \Gamma'$. This implies that $\neg\psi \notin \text{Cn}_\varphi(\Gamma)$, since otherwise Γ' would not be consistent. Hence, $\varphi > \neg\psi \notin \Gamma$.
  We want to show that $f^c(\Gamma, \widehat{\psi}) \subseteq f^c(\Gamma, \widehat{\varphi})$. Given the definition of f^c, this will follow if we can show that $\text{Cn}_\psi(\Gamma) \supseteq \text{Cn}_\varphi(\Gamma)$. So, take $\chi \in \text{Cn}_\varphi(\Gamma)$. This means that $\varphi > \chi \in \Gamma$. Since Γ is an IVC-theory, by the axiom $(\varphi > \chi) \to ((\varphi > \neg\psi) \vee (\varphi \wedge \psi > \chi))$, it follows that $(\varphi > \neg\psi) \vee (\varphi \wedge \psi > \chi) \in \Gamma$. Since Γ has the disjunction property, one of the disjuncts is in Γ. Since we already know that $\varphi > \neg\psi \notin \Gamma$, it follows that $\varphi \wedge \psi > \chi \in \Gamma$. Since

$\psi \vdash_{\mathsf{IVC}} \varphi$, the counterfactual $\varphi \wedge \psi > \chi$ is IVC-equivalent to $\psi > \chi$. Therefore, $\psi > \chi \in \Gamma$, which implies that $\chi \in \mathrm{Cn}_\psi(\Gamma)$. This proves the desired inclusion $\mathrm{Cn}_\psi(\Gamma) \supseteq \mathrm{Cn}_\varphi(\Gamma)$.

– Condition 5. Suppose $\widehat{\psi} \subseteq \widehat{\varphi}$, which implies $\psi \vdash_{\mathsf{IVC}} \varphi$. We want to show that $f^c(\Gamma, \widehat{\varphi}) \cap \widehat{\psi} \subseteq f^c(\Gamma, \widehat{\psi})$. So, take any $\Gamma' \in f(\Gamma, \widehat{\varphi}) \cap \widehat{\psi}$: this means that $\mathrm{Cn}_\varphi(\Gamma) \subseteq \Gamma'$ and $\psi \in \Gamma'$. We need to prove that $\Gamma' \subseteq f^c(\Gamma, \widehat{\psi})$, which amounts to showing that $\mathrm{Cn}_\psi(\Gamma) \subseteq \Gamma'$.
So, suppose $\chi \in \mathrm{Cn}_\psi(\Gamma)$. This means that $\psi > \chi \in \Gamma$. Since $\psi \vdash_{\mathsf{IVC}} \varphi$, the counterfactual $\psi > \chi$ is inter-derivable with $\varphi \wedge \psi > \chi$, therefore also $\varphi \wedge \psi > \chi \in \Gamma$. By the axiom $(\varphi \wedge \psi > \chi) \to (\varphi > (\psi \to \chi))$, it follows that also $\varphi > (\psi \to \chi) \in \Gamma$. Thus, $\psi \to \chi \in \mathrm{Cn}_\varphi(\Gamma) \subseteq \Gamma'$. Since also $\psi \in \Gamma'$, it follows that $\chi \in \Gamma'$. This shows the required inclusion $\mathrm{Cn}_{\varphi \wedge \psi}(\Gamma) \subseteq \Gamma'$. □

Our main result is now a corollary of Propositions 12 and 13.

Theorem 1 (Completeness). *If* $\Phi \models_{\mathsf{IVC}} \psi$, *then* $\Phi \vdash_{\mathsf{IVC}} \psi$.

6 Conclusion and Outlook

In this paper we saw that it is possible to extend intuitionistic Kripke semantics in a natural way with a Lewisian counterfactual conditional. The resulting logic, IVC, is an intuitionistic counterpart of Lewis' VC: indeed, modulo the replacement of classical propositional logic by intuitionistic logic, the axioms for IVC are the same as the axioms for VC, provided the formulation of the latter is chosen in a suitable way. In particular, the rational monotonicity axiom should be formulated in a constructive form as $(\varphi > \chi) \to (\varphi > \neg\psi) \vee (\varphi \wedge \psi > \chi)$ and not, as more common in the literature, as $(\varphi > \chi) \wedge \neg(\varphi > \neg\chi) \to (\varphi \wedge \psi > \chi)$.

This work may be taken further in several directions. First, while we focused here on a specific set of constraints, determining a specific intuitionistic counterfactual logic IVC, it would be interesting to take a broader perspective, and study the correspondence between constraints on f and conditional axioms in a more general way. This may yield intuitionistic counterparts of other notable logics of counterfactuals, such as Stalnaker's logic C2. Second, we may study how counterfactuals can be added not just to intuitionistic logic, but to intermediate logics more generally. Technically, this could be done by placing constraints not just on the selection function f, but also on the intuitionistic accessibility relation $\leq$. Third, it would be interesting to study not only the operator $>$, which Lewis denotes by $\Box\!\!\rightarrow$, but also its dual, which Lewis denotes by $\Diamond\!\!\rightarrow$. Whereas in the classical case the two are inter-definable via negation, in the intuitionistic case they must be treated as two independent operators, just like $\Box$ and $\Diamond$ need to be treated both as primitives in intuitionistic modal logic [10,18]. Finally, it would be interesting to look at the relevance of IVC for the analysis of counterfactuals in natural language. For instance, recent work [8] provided experimental evidence that antecedents of the form $\neg p \vee \neg q$ and $\neg(p \wedge q)$ do not make the same contribution to a counterfactual: sentences of the form $(\neg p \vee \neg q) > r$ and $\neg(p \wedge q) > r$

do not in general have the same truth value. This runs against the predictions of any intensional account based on classical logic, since any such account renders $(\neg p \vee \neg q) > r$ and $\neg(p \wedge q) > r$ equivalent. By contrast, since the relevant de Morgan law is invalid intuitionistically, this is perfectly compatible with IVC.

References

1. Adams, E.: The Logic of Conditionals: An Application of Probability to Deductive Logic. Synthese Library, vol. 86. Springer, Dordrecht (1975). https://doi.org/10.1007/978-94-015-7622-2
2. Alonso-Ovalle, L.: Counterfactuals, correlatives, and disjunction. Linguist. Philos. **32**, 207–244 (2009)
3. Baltag, A., Smets, S.: Dynamic belief revision over multi-agent plausibility models. In: Proceedings of LOFT (2006)
4. Bezhanishvili, N., de Jongh, D.: Intuitionistic logic, lecture Notes. Institute for Logic, Language and Computation (ILLC), University of Amsterdam (2006)
5. Board, O.: Dynamic interactive epistemology. Games Econ. Behav. **49**(1), 49–80 (2004)
6. Briggs, R.: Interventionist counterfactuals. Philos. Stud. **160**(1), 139–166 (2012)
7. Chellas, B.: Basic conditional logic. J. Philos. Logic **4**(2), 133–153 (1975)
8. Ciardelli, I., Zhang, L., Champollion, L.: Two switches in the theory of counterfactuals. Linguist. Philos. **41**(6), 577–621 (2018). https://doi.org/10.1007/s10988-018-9232-4
9. Fine, K.: Critical notice on Counterfactuals by D. Lewis. Mind **84**(1), 451–458 (1975)
10. Servi, G.F.: Semantics for a class of intuitionistic modal calculi. In: Dalla Chiara, M.L. (ed.) Italian Studies in the Philosophy of Science. BSPS, vol. 47, pp. 59–72. Springer, Dordrecht (1980). https://doi.org/10.1007/978-94-009-8937-5_5
11. Grove, A.: Two modellings for theory change. J. Philos. Logic **17**(2), 157–170 (1988)
12. Herzberger, H.G.: Counterfactuals and consistency. J. Philos. **76**(2), 83–88 (1979)
13. Kraus, S., Lehmann, D., Magidor, M.: Nonmonotonic reasoning, preferential models and cumulative logics. Artif. Intell. **44**(1–2), 167–207 (1990)
14. Lewis, D.: Counterfactuals. Blackwell, Oxford (1973)
15. Nute, D.: Conditional logic. In: Gabbay, D., Guenthner, F. (eds.) Handbook of Philosophical Logic Synthese Library. SYLI, vol. 165. Springer, Dordrecht (1984). https://doi.org/10.1007/978-94-009-6259-0_8
16. Santorio, P.: Interventions in premise semantics. Philosophers' Imprint **19**(1), 1–27 (2019)
17. Segerberg, K.: Notes on conditional logic. Studia Logica **48**(2), 157–168 (1989)
18. Simpson, A.: The proof theory and semantics of intuitionistic modal logic. Ph.D. thesis, University of Edinburgh (1994)
19. Stalnaker, R.: A theory of conditionals. In: Rescher, N. (ed.) Studies in Logical Theory. Blackwell, Oxford (1968)
20. Veltman, F.: Defaults in update semantics. J. Philos. Logic **25**(3), 221–261 (1996)
21. Weiss, Y.: Basic intuitionistic conditional logic. J. Philos. Logic **48**, 447–469 (2018)

Consolidation of Belief in Two Logics of Evidence

Yuri David Santos(✉)

University of Groningen, Groningen, The Netherlands
y.david.santos@rug.nl

Abstract. Recently, several logics have emerged with the goal of modelling evidence in a more relaxed sense than that of justifications. Here, we explore two of these logics, one based on neighborhood models and the other being a four-valued modal logic. We establish grounds for comparing these logics, finding, for any model, a counterpart in the other logic which represents roughly the same evidential situation. Then we propose operations for *consolidation*, answering our central question: *What should the doxastic state of a rational agent be in a given evidential situation?* These operations map evidence models to Kripke models. We then compare the consolidations in the two logics, finding conditions under which they are isomorphic. By taking this dynamic perspective on belief formation we pave the way for, among other things, a study of the complexity, and an AGM-style analysis of rationality of these belief-forming processes.

Keywords: Evidence logics · Epistemic logic · Many-valued logic

1 Introduction

Epistemic and doxastic logics have been used for decades to model the knowledge and beliefs of agents [16,22]. Intelligent agents, especially in real-world settings, however, build up their beliefs from inputs that might be incomplete or even inconsistent. We think of these inputs as *evidence*, broadening of the concept of justification featured in justification logics [4–6,19,23]. Real agents normally have access to raw, imperfect data, which they process into a (preferably consistent) set of beliefs, which only then can be used to make sensible decisions and to act.

Like [12–15,20,24,30], the paper [29] presents a multi-agent *four-valued epistemic logic* (FVEL) to model evidence. But differently from those, it does not feature a belief modality. Our initial goal here is to add beliefs to that framework. It is of little use to model evidence and not derive any beliefs from it. In the spirit of [12], we assume that rational belief can be determined from evidence. However, we do not do that by extending FVEL models, similarly to the strategy in [12]. Instead, we extract a doxastic Kripke model representing the agents' beliefs from the FVEL model, which represents their evidence. With that, we not only accomplish the first goal of adding beliefs to the FVEL framework, but also introduce a dynamic perspective on forming beliefs from evidence. This new perspective, compared to the static one in [14], where evidence and belief

P. Blackburn et al. (Eds.): LORI 2019, LNCS 11813, pp. 57–70, 2019.
https://doi.org/10.1007/978-3-662-60292-8_5

coexist, is akin to public announcement logic [16,25,26] compared to epistemic logic: it adds a model-changing aspect. Rational beliefs, although pre-encoded in evidence, are not obtained for free, but require "computation". This process of forming beliefs from evidence, which we call *consolidation*, is represented by transformations from evidence models to Kripke models. This idea generalises the static approach, because we can represent the "consolidation" of models where belief and evidence coexist as an automorphism from these models to themselves.

This paper is structured as follows. In Sect. 2 we introduce FVEL, a logic that models evidence but no beliefs. In Sect. 3, we present the main idea of this paper, the so-called *cautious consolidation*, a transformation from FVEL evidence models to doxastic Kripke models. We also discuss some of its properties. The remainder of the paper is concerned with comparing our work with another approach in the literature: the work started by Van Benthem and Pacuit [14] and extended together with Fernández-Duque [12,30]. Baltag et al. [8] also built upon those logics, offering more general topological semantics, but for the purpose of this paper the models of [14] will suffice. We cannot compare our consolidations with the ones from Van Benthem et al. if we cannot compare those evidence models in the first place, so that is what is done in Sect. 4. Then in Sect. 5 we finally compare the consolidations per se. We lay out our conclusions and ideas left for future work in Sect. 6. Proofs were omitted, but are available online[1].

2 A Multi-agent Logic of Evidence

Now we concisely describe the *four-valued epistemic logic* (FVEL, in short) [29], the logic of evidence to which we apply our idea of consolidations.

Definition 1. *[29] Let At be a countable set of atomic propositions and A a finite set of agents. A formula φ in the language $\mathscr{L}^n_{\Box\sim}$ is defined as follows:*

$$\varphi ::= p \mid {\sim}\varphi \mid \neg\varphi \mid (\varphi \wedge \varphi) \mid \Box_i \varphi$$

with $p \in At$ and $i \in A$. Let $(\varphi \vee \psi) \stackrel{\text{def}}{=} \neg(\neg\varphi \wedge \neg\psi)$.

The intended readings of literals such as p and $\neg p$ are *there is evidence for* p and *there is evidence against* p, respectively. We read ~ as classical negation: ${\sim}\varphi$ means that *it is not the case that* φ. Formulas with the modal operator such as $\Box_i\varphi$ and $\Box_i\neg\varphi$, finally, have the intended meaning of *agent* i *knows that there is evidence for* φ and *agent* i *knows that there is evidence against* φ, respectively.

Definition 2. *[29] Given a set $A = \{1, 2, ..., n\}$ of agents, an FVEL model is a tuple $\mathscr{M} = (S, R, \mathscr{V})$, where $S \neq \emptyset$ is a set of states, $R = (R_1, R_2, ..., R_n)$ is an n-tuple of binary relations on S and $\mathscr{V} : At \times S \rightarrow \mathscr{P}(\{0, 1\})$ is a valuation*

1 https://github.com/ydsantos/appendix_cons/blob/master/proofs.pdf.

function that assigns to each proposition at each state one of four truth values[2]. *With* $p \in At$, $s \in S$, $i \in A$ *and* $\varphi, \psi \in \mathscr{L}^n_{\Box\sim}$, *the relation* $\models$ *is defined as follows:*

$$\mathscr{M}, s \models p \text{ iff } 1 \in \mathscr{V}(p,s) \qquad \mathscr{M}, s \models \neg p \text{ iff } 0 \in \mathscr{V}(p,s)$$
$$\mathscr{M}, s \models (\varphi \wedge \psi) \quad \text{iff} \quad \mathscr{M}, s \models \varphi \text{ and } \mathscr{M}, s \models \psi$$
$$\mathscr{M}, s \models \neg(\varphi \wedge \psi) \quad \text{iff} \quad \mathscr{M}, s \models \neg\varphi \text{ or } \mathscr{M}, s \models \neg\psi$$
$$\mathscr{M}, s \models \Box_i \varphi \quad \text{iff} \quad \forall t \in S \text{ s.t. } sR_i t : \mathscr{M}, t \models \varphi$$
$$\mathscr{M}, s \models \neg\Box_i \varphi \quad \text{iff} \quad \exists t \in S \text{ s.t. } sR_i t \text{ and } \mathscr{M}, t \models \neg\varphi$$
$$\mathscr{M}, s \models {\sim}\varphi \quad \text{iff} \quad \mathscr{M}, s \not\models \varphi$$
$$\mathscr{M}, s \models \neg{\sim}\varphi \text{ iff } \mathscr{M}, s \models \varphi \qquad \mathscr{M}, s \models \neg\neg\varphi \text{ iff } \mathscr{M}, s \models \varphi$$

Definition 3. *[29] The* extended valuation function $\overline{\mathscr{V}} : \mathscr{L}^n_{\Box\sim} \times S \to \mathscr{P}(\{0,1\})$ *is defined as follows:* $1 \in \overline{\mathscr{V}}(\varphi, s)$ *iff* $\mathscr{M}, s \models \varphi$; $0 \in \overline{\mathscr{V}}(\varphi, s)$ *iff* $\mathscr{M}, s \models \neg\varphi$.

Using Definition 3, we say that φ has value *both* at s, for example, iff $\overline{\mathscr{V}}(\varphi, s) = \{0,1\}$, which is the case when both $\mathscr{M}, s \models \varphi$ and $\mathscr{M}, s \models \neg\varphi$. Semantic conditions for negated and non-negated formulas are defined separately, due to the independence of positive and negative atoms. Based on this semantics, it will be handy to define formulas discriminating which of the four truth values a formula φ has:

Definition 4. *[29]* $\varphi^n \stackrel{\text{def}}{=} ({\sim}\varphi \wedge {\sim}\neg\varphi)$; $\varphi^f \stackrel{\text{def}}{=} {\sim}{\sim}({\sim}\varphi \wedge \neg\varphi)$; $\varphi^t \stackrel{\text{def}}{=} {\sim}{\sim}(\varphi \wedge {\sim}\neg\varphi)$; $\varphi^b \stackrel{\text{def}}{=} {\sim}{\sim}(\varphi \wedge \neg\varphi)$.

Now we can read $\Box_i \varphi^x$ as *Agent i knows that the status of evidence for* φ *is* x (where $x \in \{t, f, b, n\}$).

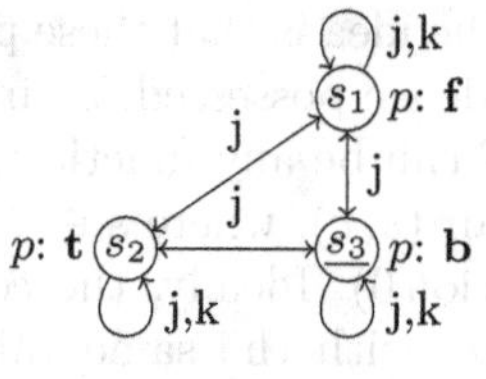

Fig. 1. Some evidence about p.

Example 1. John (j) knows that there are studies about health effects of coffee. However, he never read those articles, so he is sure that there is evidence *for* or *against* (or even *both for and against*) *coffee being beneficial for health* (p), but he does not know what the status of the evidence about p is, only that there is *some* information. Looking at Fig. 1, one can see that $\Box_j((p \wedge {\sim}\neg p) \vee (\neg p \wedge {\sim} p) \vee (p \wedge \neg p))$, which is equivalent to $\Box_j(p \vee \neg p)$, holds in the "actual" world (s_3).

[2] Abbrev.: $\{0\}$ is *false* or f, $\{1\}$ is *true* or t, $\{\}$ is *none* or n, and $\{0,1\}$ is *both* or b.

Kate (k), on the other hand, is a researcher on the effects of coffee on health, and for this reason she knows exactly what evidence is available (R_k has only reflexive arrows). Notice that $\mathscr{M}, s_3 \models \Box_k(p \wedge \neg p)$, that is, in the actual state, Kate knows that there is evidence both for and against the benefits of coffee. Moreover, John knows Kate and her job, so he also knows that she knows about p, whatever its status is: $\Box_j(\Box_k p^f \vee \Box_k p^t \vee \Box_k p^b)$. Likewise, Kate knows that John simply knows that there is some information about p: $\Box_k(\Box_j(p \vee \neg p) \wedge {\sim}\Box_j(p \wedge \neg p))$.

Thus, FVEL expresses two types of facts: whether there is evidence for and/or against propositions (in a public sense); and first and higher-order knowledge of agents about these evidential facts.

3 A Consolidation Operation

Now that we have seen how FVEL works, we want to be able to extract a Kripke model from an FVEL model, representing the beliefs obtained from the evidence in the latter, constituting a so-called *consolidation* operation.

3.1 Definitions

To define this operation we will need some essential notions:

Definition 5 *(Selection Function and Accepted Valuations). Let $Val = \{v : At \to \{0,1\}\}$ be the set of all* binary valuations. *Given an FVEL model $\mathscr{M} = (S, R, \mathscr{V})$ and the set of agents $A = \{1, 2, ..., n\}$, we define $\mathcal{V} = (\mathcal{V}_1, \mathcal{V}_2, ..., \mathcal{V}_n)$, where $\mathcal{V}_i(s) \subseteq Val$ and $\mathcal{V}_i(s) \neq \emptyset$, for all $i \in A$ and $s \in S$. $\mathcal{V}$ is called a* (valuation) selection function *for $\mathscr{M}$, and $\mathcal{V}_i(s)$ is the set of binary valuations that agent i* accepts *at s. $U_s = \bigcup_{i \in A} \mathcal{V}_i(s)$ are the valuations accepted by* some *agent at s.*

Intuitively, the selection function $\mathcal{V}$ gives the set of valuations that each agent finds plausible at each state. The idea is that these plausible valuations will bear a strong connection to the evidence possessed, by means of constraints imposed on $\mathcal{V}$. In principle, however, $\mathcal{V}$ can be any function conforming to Definition 5.

We use s_v to denote the pair (s, v), where $s \in S$ and $v \in Val$. Now we define *cluster consolidations* (Definition 6). Ideally, the consolidation would generate one state for each state in $\mathscr{M}$, with the same valuation. If FVEL were two-valued, that would be possible, but since it is four-valued, we generate a cluster of states for each state s, with one state s_v for each valuation v accepted at s according to $\mathcal{V}$.

Definition 6 *(Cluster Consolidation). Let $\mathscr{M} = (S, R, \mathscr{V})$ be an FVEL model, $\mathcal{V}$ be a selection function for $\mathscr{M}$. The* cluster consolidation *of $\mathscr{M}$ (based on $\mathcal{V}$) is the Kripke model $\mathscr{M}! = (S', R', V)$, where: (i) $S' = \{s_v \mid s \in S, v \in U_s\}$; (ii) if $s_v, t_u \in S'$ then: $s_v R'_i t_u$ iff sR_it and $u \in \mathcal{V}_i(t)$; and (iii) $V(p, s_v) = v(p)$.*[3]

[3] Since the number of states in $\mathscr{M}!$ can be exponential in the number of elements of At, if At is countably infinite, S' may be uncountable (by Cantor's Theorem).

Definition 6 hopefully covers most reasonable consolidations, modulo some notion of equivalence. It covers a lot of unreasonable ones too. It does not reflect, however, any specific "consolidating policy": it only defines a technically convenient class of consolidations, due to their modular nature (each state generating a cluster of states) and the way they link accepted valuations and evidence.

Now we define a type of cluster consolidation reflecting an actual policy: *cautious consolidation.* It is based on the following consolidating principle: *If there is only positive evidence for a proposition, then the agent believes it; if there is only negative evidence, then the agent believes its negation; otherwise, the agent has no opinion about it.* Consider the set of functions $H = \{h : \mathscr{P}(\{0,1\}) \to \{-1,0,1\}\}$, mapping status of evidence to doxastic attitudes (1 standing for belief, 0 for disbelief and -1 for abstention of judgement). This principle, then, can be codified in a function h_1 such that $h_1(n) = h_1(b) = -1$, $h_1(t) = 1$, $h_1(f) = 0$.[4]

Definition 7[5] *(Compatibility). Let $h \in H$ and $Val_s^h = \{v \in Val \mid$ for all $p \in At$, if $h(\mathscr{V}(p,s)) \neq -1$ then $v(p) = h(\mathscr{V}(p,s))\}$ be the set of binary valuations* h-compatible *with $\mathscr{V}$ at s.*

Definition 8 *(Implementation). If $\mathcal{V}_i(s) = Val_s^h$ for all $s \in S$ and some $i \in A$, we say that $\mathcal{V}$* implements *h for agent i.*

Definition 9 *(h-consolidation). Let $h \in H$. $\mathscr{M}!$ is called an h-consolidation of $\mathscr{M}$ for agent i iff $\mathscr{M}!$ is the cluster consolidation of $\mathscr{M}$ based on $\mathcal{V}$, and $\mathcal{V}$ implements h for agent i.*

Let *cautious consolidation* be synonymous with h_1-consolidation. A consolidation is characterised in Definition 9 relative to an agent. This allows consolidations to implement different belief formation policies for each agent.

3.2 Examples

Figure 2 (left) shows a simple cautious consolidation, with one agent and one proposition with value *true*. The selection function is cautious, so the set of valuations accepted by the agent has to be h_1-compatible with $\mathscr{V}$ at s_1. This is the case for a valuation v only if $v(p) = 1$. Then, according to Definition 6, there is only one state in the consolidated model (s_1'), which conforms to v (that is, p holds) and has a reflexive arrow, because the original state s_1 has one as well. In Fig. 2 (right), the value *both* for p admits two h_1-compatible valuations: one in which p holds, and one in which p does not hold. Then, by Definition 6, two states

---

[4] Out of 81 functions in H, only h_1 and h_0 ($h_0(x) = -1$, $x \in \{t,f,b,n\}$) respect some permissive postulates. They are: if evidence is only positive (negative) then you should not disbelieve (believe); if only positive (negative) evidence is not enough to generate belief (disbelief), nothing is; $h(n) = h(b) = -1$, justified by the fact that $\varphi^b = (\neg\varphi)^b$ (similarly for n), so only abstention can avoid inconsistency; and $h(t) = 1$ iff $h(f) = 0$, justified by the fact that $\varphi^t = (\neg\varphi)^f$ and $\varphi^f = (\neg\varphi)^t$ in FVEL.

[5] For this and coming definitions, keep in mind that whenever $\mathscr{V}$, S or $\mathcal{V}$ are mentioned, they are always relative to an underlying FVEL model $\mathscr{M} = (S, R, \mathscr{V})$.

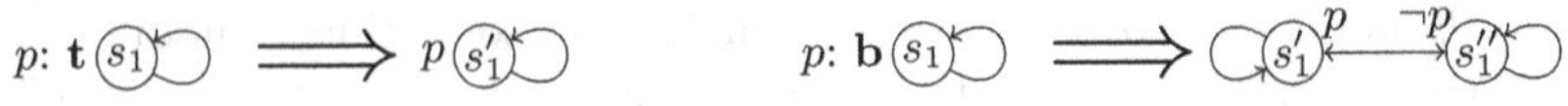

Fig. 2. Cautious consolidations on positive (left) and conflicting evidence (right).

must exist in the consolidation, and they should contain all possible arrows, because the original state has a reflexive arrow. The consolidation would be identical if p had value *none*: cautious consolidations do not distinguish between *none* and *both* (due to h_1). Figure 3 illustrates cautious consolidation applied to Example 1.

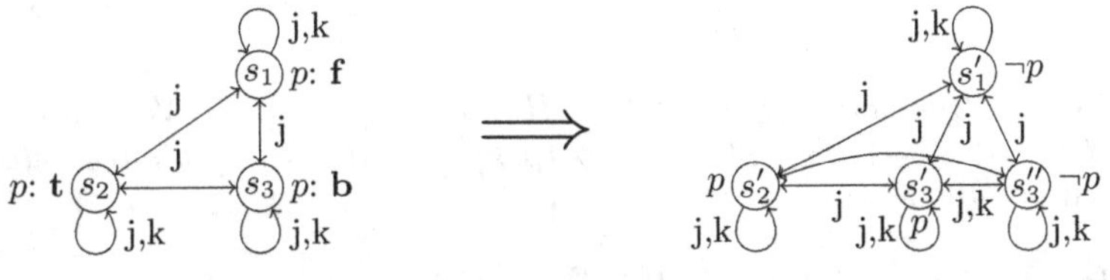

Fig. 3. Cautious consolidation of Example 1.

3.3 Properties

In this section we explore formal properties of the consolidations. Proposition 1 represents a desideratum for cluster consolidations: that they "respect" the function h upon which they are based. In a cautious consolidation, for example, we want that if an agent a knows that the status of evidence for p is t in state s, that is, $\mathscr{M}, s \models \Box_a p^t$, then in the corresponding state of $\mathscr{M}!$ a will believe p. Now if $\Box_a p^f$ holds, a will believe $\neg p$, and otherwise a will believe neither p nor $\neg p$. Proposition 1 generalises this result for any function $h \in H$, for any number of "stacked boxes", and for disjunctions of truth values of p. For example, with h_1, if $\Box_a(p^b \vee p^n)$ holds, then the agent will not form beliefs about p. Let $h^{-1}(y)$ be the preimage of y by h: $h^{-1}(y) = \{x \in \mathscr{P}(\{0,1\}) \mid h(x) = y\}$.

Proposition 1. *Given any FVEL model $\mathscr{M} = (S, R, \mathscr{V})$ and a function $h \in H$, consider an h-consolidation $\mathscr{M}! = (S', R', V)$ of $\mathscr{M}$ for agent i_0. For any such consolidation, for all $p \in At$ and $s \in S$: $\mathscr{M}, s \models \Box_{i_n} ... \Box_{i_0}(p^{x_1} \vee ... \vee p^{x_m}) \Rightarrow$*

$$\begin{cases} \mathscr{M}!, f(s) \models B_{i_n} ... B_{i_0} p & \text{if } \{x_1, ..., x_m\} \subseteq h^{-1}(1) \\ \mathscr{M}!, f(s) \models B_{i_n} ... B_{i_0} \neg p & \text{if } \{x_1, ..., x_m\} \subseteq h^{-1}(0) \\ \mathscr{M}!, f(s) \not\models B_{i_n} ... B_{i_0} p & \text{if } \{x_1, ..., x_m\} \cap h^{-1}(1) = \emptyset \\ \mathscr{M}!, f(s) \not\models B_{i_n} ... B_{i_0} \neg p & \text{if } \{x_1, ..., x_m\} \cap h^{-1}(0) = \emptyset \end{cases}$$

where for all $s \in S$, $f(s) = s_v$ for some $s_v \in S'$, and B_a is the belief modality associated with R'_a.

Function h is respected in a weak way, namely, only for atoms. Now consider the following translation function for formulas.

Definition 10. *Let* $\mathsf{t} : \mathscr{L}^n_{\Box\sim} \to \mathscr{L}^n_B$ *be a function that translates FVEL formulas into a standard multimodal language with modal operators* B_a *for each* $a \in A$ *such that* $\sim$ *is replaced by* $\neg$, $\Box_a$ *is replaced by* B_a, *and the rest remains the same.*

The following result, as Proposition 1, establishes a correspondence between formulas in an FVEL model and in its consolidation. The result is limited to formulas with "classically-valued" atoms, but encompasses all formulas instead of only atoms.

Proposition 2. *Let* $\mathscr{M} = (S, R, \mathscr{V})$ *be an FVEL model and* $\mathscr{M}! = (S', R', V)$ *its cautious consolidation, and let* φ *be an FVEL formula such that for all atoms* p *occurring in* φ, $\mathscr{V}(p, s) \in \{\{0\}, \{1\}\}$ *for all* $s \in S$. *Then, for all* $s \in S$, $\mathscr{M}, s \models \varphi$ *iff* $\mathscr{M}!, s_v \models \mathsf{t}(\varphi)$, *for any* $s_v \in S'$.

Now let us check the preservation of frame properties under consolidations. Seriality, transitivity and Euclideanicity are preserved in general. Reflexivity and symmetry, however, are only preserved if there is a certain similarity among the selection functions $\mathcal{V}_i$. Notice that for all R'_i to be reflexive, all functions $\mathcal{V}_i$ have to be equal. The following propositions are all relative to an FVEL model $\mathscr{M} = (S, R, \mathscr{V})$ and a cluster consolidation $\mathscr{M}! = (S', R', V)$ of $\mathscr{M}$, where $R = (R_1, ..., R_n)$ and $R' = (R'_1, ..., R'_n)$.

Proposition 3. *If* R_i *is serial (transitive, Euclidean), then* R'_i *is serial (transitive, Euclidean).*

Proposition 4. *If* R_i *is reflexive, then* R'_i *is reflexive iff for all* $j \in A$ *and all* $s \in S$ *it holds that* $\mathcal{V}_j(s) \subseteq \mathcal{V}_i(s)$.

Proposition 5. *If* R_i *is symmetric, then* R'_i *is symmetric iff for all* $s, t \in S$ *such that* sR_itR_is *it holds that* $\mathcal{V}_j(s) \subseteq \mathcal{V}_i(s)$ *for all* $j \in A$.

In the case where all the agents consolidate in the same manner (for example, through cautious consolidation), reflexivity, symmetry, transitivity, seriality and Euclideanicity are all preserved. Since we want the consolidated model to be a *doxastic* model, it is desirable that its relation be Euclidean, serial and transitive (KD45 models). These results provide sufficient conditions for that.

3.4 A Unified Language for Evidence and Beliefs

A detailed study of an extension of the language and logic of FVEL with beliefs is beyond the scope of this paper, but we will suggest here how this can be done.

First, we have to recall that propositional formulas in FVEL are not about facts, but about evidence. For this reason, it is better to define belief over formulas of $\mathscr{L}_B$, the doxastic language of the consolidated model. We can define belief in FVEL model as follows:

$$\mathscr{M}, s \models B_a\mathsf{t}(\varphi) \quad \text{iff} \quad \mathscr{M}!, s_v \models B_a\mathsf{t}(\varphi)$$

where $\mathscr{M}! = (S', R', V)$ is the cautious consolidation of $\mathscr{M}$, and $s_v \in S'$.

In this language it is now possible to talk about formulas such as $\Box_a p^t \tilde{\leftrightarrow} B_a p$ or $\Box_a p^f \tilde{\leftrightarrow} B_a \neg p$, i.e., only positive (negative) evidence equals belief (disbelief), where $\varphi \tilde{\leftrightarrow} \psi \overset{\text{def}}{=} {\sim}(\varphi \wedge {\sim}\psi) \wedge {\sim}(\psi \wedge {\sim}\varphi)$. These formulas are valid, but if we employ another type of consolidation in the semantic definition above, they may not be.

Notice also that if $\mathscr{M}!$ is a KD45 model, for example, the behaviour of this new B_a operator in FVEL will be governed by that logic. But since the consolidation is completely determined by the original FVEL model, it should be possible to define semantics for B_a in FVEL without mentioning $\mathscr{M}!$.

4 Equivalence Between Evidence Models

Now we introduce Van Benthem and Pacuit's (hereafter, B&P) models [14].

Definition 11. *[14] A* B&P model *is a tuple $M = (S, E, V)$ with $S \neq \emptyset$ a set of states, $E \subseteq S \times \mathscr{P}(S)$ an evidence relation, and $V : At \rightarrow \mathscr{P}(S)$ a valuation function. We write $E(w)$ for the set $\{X \mid wEX\}$. We impose two constraints on E: for all $w \in S$, $\emptyset \notin E(w)$ and $S \in E(w)$.*

In B&P models, propositional formulas are about facts (not evidence), as usual.

Definition 12. *[12] A* w-scenario *is a maximal $\mathcal{X} \subseteq E(w)$ such that for any finite $\mathcal{X}' \subseteq \mathcal{X}$, $\bigcap \mathcal{X}' \neq \emptyset$. Let $Sce_E(w)$ be the collection of w-scenarios of E.*

Definition 13. *[14] A standard bimodal language $\mathscr{L}_{\Box B}$ (with $\Box$ for evidence and B for belief) is interpreted over a B&P model $M = (S, E, V)$ in a standard way, except for B and $\Box$:*

$$M, w \models \Box\varphi \textit{ iff } \exists X \textit{ with } wEX \textit{ and } \forall v \in X : M, v \models \varphi$$

$$M, w \models B\varphi \textit{ iff } \forall \mathcal{X} \in Sce_E(w) \textit{ and } \forall v \in \bigcap \mathcal{X}, M, v \models \varphi$$

Formulas such as $\Box\varphi$ mean that *the agent has evidence for* φ. Notice that an agent can have evidence for φ and $\neg\varphi$ at the same time, or have no evidence about φ whatsoever. This makes the status of evidence (in any given state) four-valued, just as in FVEL. Note also that the conditions for the satisfaction of $B\varphi$ tell us how the consolidation in B&P logic is done: *You believe what is supported by all pieces of evidence in all maximal consistent subsets of your evidence* (w-scenarios).

Now we want to be able to compare consolidations of B&P models to consolidations of FVEL models. For this, first, we need a way of establishing that an FVEL model and a B&P model are "equivalent" with respect to how evidence is represented. It only makes sense to compare consolidations if they depart from (roughly) the same evidential situation.

The "logics of evidence" in B&P logic and FVEL differ, the former being non-normal (so, for example, $\Box\varphi \wedge \Box\psi$ does not imply $\Box(\varphi \wedge \psi)$ in B&P logic, while in FVEL it does), and the latter being *First Degree Entailment* (FDE)

[17,27][6]. Note, however, that this difference is more about how evidence is *manipulated* in these logics, than about how it is *represented*. For this reason, our equivalence in evidence is, fittingly, limited to literals.

Definition 14 *(ev-equivalence). Let $M = (S, E, V)$ be a B&P model and let $\mathscr{M} = (S', R, \mathscr{V})$ be an FVEL model. A relation $\doteq \subseteq S \times S'$ is an ev-equivalence between M and $\mathscr{M}$ iff:*

1. *$\doteq$ is a bijection;*
2. *If $s \doteq s'$, where $s \in S$ and $s' \in S'$, then, for all $p \in At$: $M, s \models \Box p$ iff $\mathscr{M}, s' \models \Box p$; and $M, s \models \Box\neg p$ iff $\mathscr{M}, s' \models \Box\neg p$.*

We write $M \doteq \mathscr{M}$ if there exists an ev-equivalence between M and $\mathscr{M}$. $M \doteq M'$, $\mathscr{M} \doteq M$ and $\mathscr{M} \doteq \mathscr{M}'$ are defined analogously.

Now our job is to find, for each B&P or FVEL model, a model of the other type which is ev-equivalent to it, that is, that *represents* the same evidence[7]. Since B&P models are single-agent, we assume from now on that all models are single-agent. Much of the conversions between models that follow will be about removing aspects of evidence that are not represented in the other type of model.

4.1 From B&P to FVEL Models

Consider the following *conversion* from B&P to FVEL models:

Definition 15. *Let $M = (S, E, V)$ be a B&P model. Define the FVEL model $\mathsf{FV}(M) = (S, R, \mathscr{V})$, where $R = \{(s, s) \mid s \in S\}$ and for all $p \in At$ and states $s \in S$: $1 \in \mathscr{V}(p, s)$ iff $M, s \models \Box p$; and $0 \in \mathscr{V}(p, s)$ iff $M, s \models \Box\neg p$.*

We cannot expect a complete correspondence between M and $\mathsf{FV}(M)$ in terms of satisfaction of formulas (in the vein of Proposition 11), for while propositional formulas in B&P models represent facts and $\Box$ formulas represent the agent's evidence, in FVEL propositional formulas represent generally available evidence, while $\Box$ formulas represent agents' knowledge of such evidence. This public/personal distinction for evidence in FVEL would be superfluous in B&P models, since they are not multi-agent. Nevertheless, we have the following correspondence:

Proposition 6. *For any B&P model $M = (S, E, V)$ and its FVEL counterpart $\mathsf{FV}(M)$, for all states $s \in S$ and all literals $l \in \{p, \neg p\}$, with $p \in At$, we have:*

$$M, s \models \Box l \textit{ iff } \mathsf{FV}(M), s \models l \textit{ iff } \mathsf{FV}(M), s \models \Box l$$

Corollary 1. *For any B&P model M, $M \doteq \mathsf{FV}(M)$.*

---

[6] In other words: if there is evidence for Σ and $\Sigma \vdash_{\mathsf{FDE}} \varphi$, then there is evidence for φ.

[7] I opted for Definition 14 instead of an equivalence between $\Box p$ in B&P and p in FVEL models, because even though we do restrict FVEL models to the single-agent case, these models are still multi-agent in nature. So, while $\mathscr{M}, s \models p$ indicates that *there is evidence for* p (at s), it is only when $\mathscr{M}, s \models \Box_a p$ holds that we should think that an agent a *has (knowledge of)* this evidence. On the other hand, in single-agent B&P models there is no semantic difference between *there is evidence for* p and *the agent has evidence for* p.

4.2 From FVEL to B&P Models

This direction is less straightforward than the conversion discussed above. Again we run into the problem of representing a four-valued model as a two-valued one.

Definition 16. *Let $\mathscr{M} = (S, R, \mathscr{V})$ be an FVEL model. We build a B&P model* $\mathsf{BP}(\mathscr{M}) = (S', E, V)$ *where* $S' = \{s_v \mid s \in S \text{ and } v \in Val_s^{h_1}\}$ *and* $s_v \in V(p)$ *iff* $v(p) = 1$. *Let* $C(s) = \{t_v \in S' \mid sRt\}$. *E is defined as follows:* $E(s_v) = \{S'\} \cup \{X_p \subseteq C(s) \mid X_p \neq \emptyset, p \in At; t_u \in X_p \text{ iff } \mathscr{M}, s \models \Box p \text{ and } t_u \in V(p)\} \cup \{X_{\neg p} \subseteq C(s) \mid X_{\neg p} \neq \emptyset, p \in At; t_u \in X_{\neg p} \text{ iff } \mathscr{M}, s \models \Box\neg p \text{ and } t_u \notin V(p)\}$.

Definition 16 creates clusters of states for each original state in $\mathscr{M}$ (similarly to the technique for cluster consolidations). Then, all clusters accessible from a state s_v are grouped together and "filtered" to form the "pieces of evidence" in $E(s_v)$, one for each literal that is known to be evidence in the corresponding state of the FVEL model. E.g. if in a state s only evidence for the literal $\neg p$ is known (that is, $\mathscr{M}, s \models \Box\neg p$), then $E(s_v)$ will be $\{S', X_{\neg p}\}$, where $X_{\neg p}$ is a piece of evidence made up of all states accessible from s_v where $\neg p$ holds. See Fig. 4.

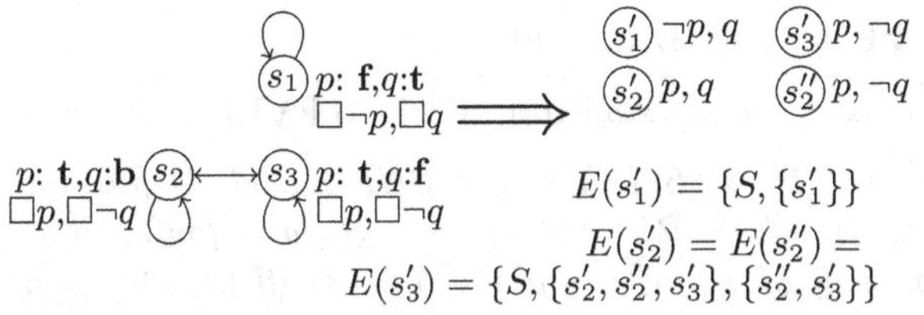

Fig. 4. An example of BP being applied to an FVEL model.

Proposition 7. *Let $\mathscr{M} = (S, R, \mathscr{V})$ be a serial FVEL model with* $\mathsf{BP}(\mathscr{M}) = (S', E, V)$. *Then, for all $s \in S$, all v such that $s_v \in S'$ and all $l \in \{p, \neg p\}$, with $p \in At$: $\mathscr{M}, s \models \Box l$ iff $\mathsf{BP}(\mathscr{M}), s_v \models \Box l$.*

Corollary 2. *For all serial FVEL models $\mathscr{M}$, $\mathsf{BP}(\mathscr{M}) \mathrel{\mathring{=}} \mathscr{M}$.*

4.3 Evaluating the Conversions

Our conversions are satisfactory enough to produce ev-equivalent models, but unfortunately the following proposition can be easily verified:

Proposition 8. *Let M be a B&P model and $\mathscr{M}$ be an FVEL model. Then, neither* $\mathsf{BP}(\mathsf{FV}(M)) \cong M$ *nor* $\mathsf{FV}(\mathsf{BP}(\mathscr{M})) \cong \mathscr{M}$ *are guaranteed to hold; where $M \cong M'$ denote that M is isomorphic to M', and similarly for $\mathscr{M} \cong \mathscr{M}'$.*

One reason why the above do not hold in general is simple: $\mathsf{BP}(\mathscr{M})$ has more states than $\mathscr{M}$ if the latter has any state where some atom has value b or n.

Definition 17. *Let $M = (S, E, V)$ be a B&P model. We define the following conditions on M:*

- ***Consistent Evidence (CONS)*** $\forall s \in S \forall X, Y \in E(s)$: *if* $\forall x \in X, M, x \models l$ *then* $\exists y \in Y, M, y \models l$, *for all literals* $l \in \{p, \neg p\}$, $p \in At$;
- ***Complete Evidence (COMP)*** $\forall s \in S \forall p \in At \exists X \in E(s)$ *s.t.* $\forall x \in X, M, x \models p$ *or* $\forall x \in X, M, x \models \neg p$;
- ***Good Evidence (GOOD)*** $s \in V(p)$ *iff* $\exists X \in E(s)$ *s.t.* $\forall x \in X, M, x \models p$
- ***Simple Evidence (SIMP)*** $\forall s \in S, E(s) = \{\{s\}, S\}$.

Proposition 9. *SIMP entails CONS, COMP and GOOD. CONS and COMP are sufficient and necessary for the preservation of S. CONS, COMP and GOOD are sufficient (but GOOD is not necessary) for preservation of V. SIMP is sufficient and necessary for preservation of E.*

Corollary 3. $\mathsf{BP}(\mathsf{FV}(M)) \cong M$ *iff SIMP holds.*

Definition 18. *Let $\mathscr{M} = (S, R, \mathscr{V})$ be an FVEL model. We define the following conditions on $\mathscr{M}$:*

- ***Classicality (CLAS)*** $\forall p \in At, \forall s \in S : \mathscr{V}(p, s) \in \{t, f\}$;
- ***Knowledge of Evidence (KNOW)*** $\mathscr{M}, s \models p$ *iff* $\mathscr{M}, s \models \Box p$; $\mathscr{M}, s \models \neg p$ *iff* $\mathscr{M}, s \models \Box \neg p$;
- ***Only-Reflexivity (REFL)*** $R = \{(s, s) \mid s \in S\}$

Proposition 10. *REFL entails KNOW. CLAS is necessary and sufficient for preservation of S. CLAS and KNOW are sufficient (but KNOW is not necessary) for preservation of $\mathscr{V}$. CLAS and REFL are the necessary and sufficient conditions for preservation of R.*

Corollary 4. $\mathsf{FV}(\mathsf{BP}(\mathscr{M})) \cong \mathscr{M}$ *iff CLAS and REFL hold.*

The desired correspondences only hold under fairly strong conditions. These conditions are not arbitrary restrictions, but idealising conditions[8]. This means that B&P and FVEL models have perfectly (ev-)equivalent counterparts under idealised scenarios, where evidence is factive, always present, complete and consistent, and where agents have perfect knowledge of what evidence is available. This correspondence breaks when we deviate from these assumptions to cover situations of imperfect evidence and imperfect knowledge. Now we can compare the two consolidations.

5 Comparing Consolidations

In [12], a method for obtaining a relation from B&P models is provided:

[8] S is added in SIMP and in the evidence sets generated by BP just to comply with the last condition of Definition 11. If we remove it from both places, Proposition 9 still holds.

Definition 19. *[12] Given a B&P model* $M = (S, E, V)$*, define* $B_E \subseteq S \times S$ *by* sB_Et *if* $t \in \bigcap \mathcal{X}$ *for some* $\mathcal{X} \in Sce_E(s)$.

Consider a monomodal language $\mathscr{L}_B$ with B as its modality.

Proposition 11. *Let* $M = (S, E, V)$ *be a B&P model and* $M! = (S, B_E, V)$ *its relational counterpart. Then, for all* $\varphi \in \mathscr{L}_B$ *and* $s \in S$: $M, s \models \varphi$ *iff* $M!, s \models \varphi$.

This effectively proves that $M!$ is the consolidation for M found "implicitly" in [12]. Now given two models M (B&P) and $\mathscr{M}$ (FVEL) such that $M \triangleq \mathscr{M}$, how does $M!$ compare to $\mathscr{M}!$ ($\mathscr{M}$'s cautious consolidation)?

Definition 20. *Given* $M \triangleq \mathscr{M}$ *under bijection* f*, we say that* V *matches* $\mathscr{V}$ *iff: for all* $p \in At$ *and all* $s' \in S'$, $\mathscr{V}(p, s') \in \{t, f\}$*; and* $s \in V(p)$ *iff* $\mathscr{V}(p, f(s)) = t$.

Proposition 12. *Let* $M \triangleq \mathscr{M}$ *under bijection* f. $M! \cong \mathscr{M}!$ *iff:* V *matches* $\mathscr{V}$*, and* $f(s)Rf(t)$ *iff* $t \in \bigcap \mathcal{X}$ *for some* $\mathcal{X} \in Sce_E(s)$.

So the conditions for consolidations of ev-equivalent B&P and FVEL models to be isomorphic are rather strong: they must have matching valuations and $\mathscr{M}$'s relation has to mirror B_E.

6 Conclusion

We introduced *consolidation* as the process of forming beliefs from a given evidential state, formally represented by transformations from evidential (FVEL and B&P) models into doxastic Kripke models. We established the grounds for comparison between these different models, and then found the conditions under which their consolidations are isomorphic. Future work can use bisimilarity instead of isomorphism, and extend this methodology to other evidence logics. Would it be possible to define belief without resorting to two-valued Kripke models? Certainly, as all information used in the consolidation is already in the initial evidential models. The rationale here is that, since Kripke models are standard and widely-accepted formal representations of belief, we should be able to represent the beliefs that implicitly exist in evidential models using this tool. We also wanted to highlight the *process* of transforming evidence into beliefs.

The dynamic perspective on consolidations allows us to study, for example, the complexity of these operations, which is important if we are concerned with real agents forming beliefs from imperfect data. It is clear that consolidations of FVEL models tend to be much larger than those of B&P models, but, on the other hand, might be much easier to compute, given that B&P consolidations rely on the hard-to-compute concept of maximally consistent sets. FVEL models can also deal with multiple agents, and accept a function from status of evidence to doxastic attitude as a parameter (in this case, function $h_1 \in H$), allowing for some flexibility in consolidation policies. It would also be interesting to see if a consolidation like B&P's, where maximal consistent evidence sets are taken into account, would be possible in the context of FVEL. Is the converse possible: to apply the idea of H functions in B&P models?

A future extension of this work taking computational costs of consolidations into account would be in line with other work that tries to fight "logical omniscience" or to model realistic resource-bounded agents [1–3,7,18]. Other *aspects* of evidence can also be considered, such as the *amount* of evidence for or against a certain proposition, the *reliability* of a source or a piece of evidence, etc.

Agents form different beliefs in ev-equivalent situations when departing from an FVEL or a B&P model. Part of this is explained by the fact that these logics do not represent exactly the same class of evidence situations. But clearly the consolidation policies also differ. Is one better than the other? At first glance, both seem to be reasonable, but more investigation could be done in this direction.

Moreover, how are changes in an FVEL (or other) evidence model reflected in its consolidation? *Evidence dynamics* for B&P logic are explored in [14], in line with other dynamic logics of knowledge update and belief revision [9–11,16,21,26,28,31].

Acknowledgement. Thanks to Rineke Verbrugge and Barteld Kooi, and also to Malvin Gattinger, Stipe Pandžić, Davide Grossi, Thiago Dias Simão, the anonymous reviewers and the RUG MAS group, for very useful suggestions. Research supported by Ammodo KNAW project "Rational Dynamics and Reasoning".

References

1. Ågotnes, T., Alechina, N.: The dynamics of syntactic knowledge. J. Logic Comput. **17**(1), 83–116 (2007)
2. Alechina, N., Logan, B.: Ascribing beliefs to resource bounded agents. In: Proceedings of the First International Joint Conference on Autonomous Agents and Multiagent Systems (AAMAS 2002), pp. 881–888. ACM (2002)
3. Alechina, N., Logan, B., Whitsey, M.: A complete and decidable logic for resource-bounded agents. In: Proceedings of the Third International Joint Conference on Autonomous Agents and Multiagent Systems (AAMAS 2004), pp. 606–613. IEEE Computer Society (2004)
4. Artemov, S.: Logic of proofs. Ann. Pure Appl. Logic **67**(1–3), 29–59 (1994)
5. Artemov, S.: Operational modal logic. Technical report, MSI 95–29, Cornell University, December 1995
6. Artemov, S.: Explicit provability and constructive semantics. Bull. Symbolic Logic **7**(1), 1–36 (2001)
7. Balbiani, P., Fernández-Duque, D., Lorini, E.: A logical theory of belief dynamics for resource-bounded agents. In: Proceedings of the 2016 International Conference on Autonomous Agents and Multiagent Systems (AAMAS 2016), pp. 644–652 (2016)
8. Baltag, A., Bezhanishvili, N., Özgün, A., Smets, S.: Justified belief and the topology of evidence. In: Väänänen, J., Hirvonen, Å., de Queiroz, R. (eds.) WoLLIC 2016. LNCS, vol. 9803, pp. 83–103. Springer, Heidelberg (2016). https://doi.org/10.1007/978-3-662-52921-8_6
9. Baltag, A., Smets, S.: Conditional doxastic models: a qualitative approach to dynamic belief revision. Electron. Notes Theor. Comput. Sci. **165**, 5–21 (2006)

10. van Benthem, J.: Dynamic logic for belief revision. J. Appl. Non-Classical Logics **17**(2), 129–155 (2007)
11. van Benthem, J.: Logical Dynamics of Information Flow. Cambridge University Press, Cambridge (2011)
12. van Benthem, J., Fernández-Duque, D., Pacuit, E.: Evidence and plausibility in neighborhood structures. Ann. Pure Appl. Logic **165**(1), 106–133 (2014)
13. van Benthem, J., Pacuit, E.: Dynamic logics of evidence-based belief. Technical report, University of Amsterdam, ILLC (2011)
14. van Benthem, J., Pacuit, E.: Dynamic logics of evidence-based beliefs. Stud. Logica **99**(1), 61–92 (2011)
15. Carnielli, W., Rodrigues, A.: An epistemic approach to paraconsistency: a logic of evidence and truth. Synthese **196**, 3789–3813 (2017)
16. van Ditmarsch, H., van der Hoek, W., Kooi, B.: Dynamic Epistemic Logic. Synthese Library, vol. 337. Springer, Dordrecht (2007). https://doi.org/10.1007/978-1-4020-5839-4
17. Dunn, J.M.: Intuitive semantics for first-degree entailments and 'coupled trees'. Philos. Stud. **29**(3), 149–168 (1976)
18. Fagin, R., Halpern, J.: Belief, awareness, and limited reasoning. Artif. Intell. **34**(1), 39–76 (1987)
19. Fitting, M.: The logic of proofs, semantically. Ann. Pure Appl. Logic **132**(1), 1–25 (2005)
20. Fitting, M.: Paraconsistent logic, evidence and justification. Stud. Logica **105**(6), 1149–1166 (2017)
21. Gerbrandy, J.: Bisimulations on Planet Kripke. Ph.D. thesis, Institute for Logic, Language and Computation, University of Amsterdam (1999)
22. Meyer, J.J., van der Hoek, W.: Epistemic Logic for AI and Computer Science. Cambridge University Press, Cambridge (1995)
23. Mkrtychev, A.: Models for the logic of proofs. In: Adian, S., Nerode, A. (eds.) LFCS 1997. LNCS, vol. 1234, pp. 266–275. Springer, Heidelberg (1997). https://doi.org/10.1007/3-540-63045-7_27
24. Özgün, A.: Evidence in epistemic logic: a topological perspective. Ph.D. thesis, University of Amsterdam (2017)
25. Plaza, J.: Logics of public communications. In: Emrich, M.L., Pfeifer, M.S., Hadzikadic, M., Ras, Z.W. (eds.) Proceedings of the 4th International Symposium on Methodologies for Intelligent Systems (ISMIS 1989): Poster Session Program, pp. 201–216. Oak Ridge National Laboratory ORNL/DSRD-24, Charlotte (1989)
26. Plaza, J.: Logics of public communications. Synthese **158**(2), 165–179 (2007)
27. Priest, G.: An Introduction to Non-classical Logic: From If to Is, 2nd edn. Cambridge University Press, Cambridge (2008)
28. Rott, H.: Shifting priorities: simple representations for twenty-seven iterated theory change operators. In: Makinson, D., Malinowski, J., Wansing, H. (eds.) Towards Mathematical Philosophy. TL, vol. 28, pp. 269–296. Springer, Dordrecht (2009). https://doi.org/10.1007/978-1-4020-9084-4_14
29. David Santos, Y.: A dynamic informational-epistemic logic. In: Madeira, A., Benevides, M. (eds.) DALI 2017. LNCS, vol. 10669, pp. 64–81. Springer, Cham (2018). https://doi.org/10.1007/978-3-319-73579-5_5
30. Van Benthem, J., Fernández-Duque, D., Pacuit, E.: Evidence logic: a new look at neighborhood structures. In: Bolander, T., Braäuner, T., Ghilardi, S., Moss, L. (eds.) Advances in Modal Logic, vol. 9, pp. 97–118. College Publications (2012)
31. Velázquez-Quesada, F.R.: Inference and update. Synthese **169**(2), 283–300 (2009)

From Classical to Non-monotonic Deontic Logic Using ASPIC$^+$

Huimin Dong[1(✉)], Beishui Liao[1], Réka Markovich[2], and Leendert van der Torre[1,2]

[1] Department of Philosophy, Zhejiang University, Hangzhou, China
huimin.dong@xixilogic.org, baiseliao@zju.edu.cn

[2] Department of Computer Science, University of Luxembourg, Esch-sur-Alzette, Luxembourg
{reka.markovich,leon.vandertorre}@uni.lu

Abstract. In this paper we use formal argumentation to design non-monotonic deontic logics, based on two monotonic deontic logics. In particular, we use the structured argumentation theory ASPIC$^+$ to define non-monotonic variants of well-understood modal logics. We illustrate the approach using argumentation about free-choice permission.

1 Using ASPIC$^+$ to Design Non-monotonic Deontic Logics

Deontic logic is the logic of obligation, prohibition and permission [7,18]. Many axioms of deontic logic have been criticised, and non-monotonic techniques have been applied widely [3,11,17,20,21,23]. In this paper we consider the use of so-called ASPIC$^+$ to design deontic argumentation systems and non-monotonic deontic logics and, in particular, to study strong and free-choice permission [9].

Modgil and Prakken [15] observe that "in ASPIC$^+$ and its predecessors, going back to the seminal work of John Pollock, arguments can be formed by combining strict and defeasible inference rules and conflicts between arguments can be resolved in terms of a preference relation on arguments. This results in abstract argumentation frameworks (a set of arguments with a binary relation of defeat), so that arguments can be evaluated with the theory of abstract argumentation."

In this paper, we use argumentation systems to define non-monotonic logics. Our ASPIC$^+$-based methodology consists of three steps.

Arguments We take literally Modgil and Prakken's idea that "Rule-based approaches in general do not adopt a single base logic but two base logics, one for the strict and one for the defeasible rules" [15]. We use monotonic modal logics as our base logics with Hilbert-style proof theory.

Strict arguments use only strict rules defined in terms of a "lower bound" logic, in the sense that it defines the minimal inferences which must be made. We use a variant of Von Wright's standard deontic logic [24].

P. Blackburn et al. (Eds.): LORI 2019, LNCS 11813, pp. 71–85, 2019.
https://doi.org/10.1007/978-3-662-60292-8_6

Defeasible arguments use also defeasible rules defined in terms of an "upper bound" logic in the sense that it defines all possible inferences that can be made. We use a variant of Van Benthem's logic of strong permission [4].

Preferences among arguments can be generic or depend on the logical languages used to build the arguments. We focus on **Argument types** defined in ASPIC$^+$ which distinguish between defeasible and plausible arguments.

Nonmonotonic inference relations can be based on skeptical or credulous relation, and on one of the argumentation semantics.

The layout of this paper is as follows. We first introduce the running example of this paper. Then we introduce the monotonic deontic logics, and we use the logics to define ASPIC$^+$ argumentation systems. Finally we define the non-monotonic deontic logics in terms of the argumentation systems.

2 Running Example: Free-Choice Permission

There are many ways in which the relation between obligation and permission has been defined. For example, in some papers permission is used to define exceptions to general obligations and prohibitions, and in such approaches, permission overrides obligation [13]. In other approaches, we can see examples where obligations and prohibitions override permissions [4,10]. For example, the general norm that product placement in TV programs is strongly permitted, is overridden by the particular case that product placement is forbidden in children programs. In this paper we work with an example where this latter is the case.

In particular, we take standard deontic logic without weak permission as our logic for strict rules, and for the defeasible rules we use an extension of this logic with strong permission, proposed by Van Benthem [4]. We consider three combinations of monotonic deontic logics and three ways to define the preferences, and we only consider stable semantics. So we define six non-monotonic deontic logics in this paper.

In Van Benthem's logic, what is obligated is the necessary condition of being ideal, while what is permitted is the sufficient condition for ideality. An intuitive example is the so-called "free-choice permission" [10]. If having a tea or having a coffee is permitted, then free-choice implies that both cases are permitted. Here we consider the following example in legal reasoning, and see that in what sense of non-monotonicity we say a free-choice permission holds or not.

1. It is permitted to freely use any of your property, for example, a knife.
2. It is forbidden to murder.

The question now is, is it permitted to use your knife to kill someone? The solution we adopt is that it is permitted to use the knife in normal situations, in sense of being non-defeated in ASPIC$^+$. So we can derive that (normally):

3. Knifes are not used to murder.

If we now add the information that "The knife is used to murder", or "The knife can be used to murder", and in addition we *prefer* this statement over the previous ones, then we would expect no longer to derive that (normally) knifes are not used for murder, and neither we would derive that knifed murder is permitted. However, we would still expect to predict that, for example, knifes can be used to cut the bread, because *no more preferable* argument for the contrary exists.

From a formal point of view, the problem of free-choice permission we focus on in this paper is the derivation of $P(\phi \wedge \psi)$ from $P\phi$. It has been observed by Glavaničová [8] that this is a strong rule which should not hold in case it leads to inconsistency. We adopt ASPIC$^+$ to explain it.

Example 1 (Knifed murder). Our aim is to define a logic such that the defeasible permission to use the knife, Pk, can infer the permission to cut the bread with the knife, $P(k \wedge b)$, but in some exceptional cases, for instance in case of murder, $\{Pk, O\neg m\}$ we cannot infer $P(k \wedge m)$. In the latter case, without the prohibition and in analogy with cutting the bread, the logic also derives $P(k \wedge m)$ from Pk only, and thus the logic is non-monotonic.

Each level of our approach can be analysed using the methods of that discipline, i.e. monotonic logic (e.g. possible world semantics), argumentation theory can be studied using rationality postulates [6], and non-monotonic inference can be analysed using, for example, the approach advocated by Kraus *et al.* [12].

3 Step 1: Arguments Based on Two Monotonic Logics

We use two monotonic logics to define the strict and defeasible rules of ASPIC$^+$, and use the crude approach to define arguments [15]: "A crude way is to simply put all valid propositional (or first-order) inferences over your language of choice in [the strict rules] R_s. So if a propositional language has been chosen, then R_s can be defined as follows (where $\vdash_{PL}$ denotes standard propositional-logic consequence). For any finite $S \subseteq \mathcal{L}$ and any $\phi \in \mathcal{L}$: $S \rightarrow \phi \in R_s$ if and only if $S \vdash_{PL} \phi$." This method can be applied to define defeasible rules, and this application, as stated in [15], is based on some cognitional or rational criteria. By using the crude method to define strict rules in the lower-bounded logic $\mathbf{S}^-$ and to defeasible rules in the upper-bounded logic $\mathbf{S}^+$, even when Hilbert style derivations are quite long, the arguments can be short.

Besides this way to define the defeasible rules, all the other definitions in this section like the arguments and the extensions are standard and taken from the handbook article of Modgil and Prakken. In particular, we consider three instantiations of ASPIC$^+$, by taking different monotonic logics ($\mathbf{D}_{-1}$ or $\mathbf{D}_{-2}$ defined later) as the basic logic and then treating either merely FCP or it together with OWP (in Table 1) as defeasible. In this section, we define the notion of argumentation theory. In the following section we use the argumentation theory to define non-monotonic logic as a combination of two selected monotonic logics $\mathbf{S}^-, \mathbf{S}^+$.

We first present a version of Van Benthem's deontic logic of obligation and permission [4]. This logic is different than Standard Deontic Logic [18], in the latter obligation and permission are a dual pair, while in the former they are the necessary and sufficient conditions of being ideal. The modal language contains the classic negation $\neg$, conjunction $\wedge$, universal modality $\Box$, as well as two additional deontic modalities, O for obligation and P for permission.

Definition 1 (Deontic Language). *Let p be any element of a given (countable) set Prop of atomic propositions. The deontic language $\mathcal{L}$ of modal formulas is defined as follows:*

$$\phi := p \mid \neg\phi \mid (\phi \wedge \phi) \mid \Box\phi \mid O\phi \mid P\phi$$

The disjunction $\vee$, the material condition $\rightarrow$ and the existential modality $\Diamond$ are defined as usual: $\phi \vee \psi := \neg(\neg\phi \wedge \neg\psi)$, $\phi \rightarrow \psi := \neg(\phi \wedge \neg\psi)$ and $\Diamond\phi := \neg\Box\neg\phi$.

The axiomatization presented in Table 1 is a variant of Van Benthem's logic [4]. We use **D** to denote it. **D** not only takes obligation and universal modality into account, but also considers free-choice permission and the connection between obligation and permission. In the logic **D**, except the essential $K_\Box$, $E_\Box$, $T_\Box$, $4_\Box$, $B_\Box$, and $NEC_\Box$, the axioms $\Box_O$ and $\Box_P$ are the core of the universal modality in normal modal logic. Moreover, $\Box_O$ claims that what is always the case is obligatory, but $\Box_P$ leaves the space for what is never the case to be permitted. The axiom D_O maintains obligation to be ideally consistent as usual. OWP states that "obligation as the weakest permission" [1,4]. RFC is one direction of free-choice permission, and FCP is the other. For further information about the logic and its motivations, see Van Benthem's paper.

Table 1. The logic **D** of obligation and permission.

- All instances of propositional tautologies	- $K_\triangle$: $\triangle(\phi \rightarrow \psi) \rightarrow (\triangle\phi \rightarrow \triangle\psi)$
- $E_\Box$: $\Box\phi \leftrightarrow \neg\Diamond\neg\phi$	- $T_\Box$: $\Box\phi \rightarrow \phi$
- $4_\Box$: $\Box\phi \rightarrow \Box\Box\phi$	- $B_\Box$: $\phi \rightarrow \Box\Diamond\phi$
- $\Box_O$: $\Box\phi \rightarrow O\phi$	- $\Box_P$: $\Box\neg\phi \rightarrow P\phi$
- D_O: $\neg(O\phi \wedge O\neg\phi)$	- OWP: $O\phi \wedge P\psi \rightarrow \Box(\psi \rightarrow \phi)$
- RFC: $P\phi \wedge P\psi \rightarrow P(\phi \quad \psi)$	- FCP: $P\psi \wedge \Box(\phi \rightarrow \psi) \rightarrow P\phi$
- MP: $\phi, \phi \rightarrow \psi / \psi$	- $NEC_\triangle$: $\phi / \triangle\phi$
where $\triangle \in \{\Box, O\}$	

In this paper we consider sub-systems of **D** that contain a strict subset of the axioms and inference rules of **D**. In particular, we define $\mathbf{D}_{-1}$ as the axiomatization which does not contain FCP, and we define $\mathbf{D}_{-2}$ as the axiomatization which does not contain FCP and OWP.

We define the notions of derivation based on modal logic $\mathbf{S} \in \{\mathbf{D}, \mathbf{D}_{-1}, \mathbf{D}_{-2}\}$ in the usual way, see e.g. [5]. Note that modal logic provides two related kinds of derivation according to the application of necessitation, i.e. necessitation can only be applied to theorems but not to an arbitrary set of formulas. We use both notions in the formal argumentation theory.

Definition 2 (Derivations without Premises). *Let* $\mathbf{S} \in \{\mathbf{D}, \mathbf{D}_{-1}, \mathbf{D}_{-2}\}$ *be a deontic logic. A derivation for* ϕ *in* $\mathbf{S}$ *is a finite sequence* $\phi_1, \ldots, \phi_{n-1}, \phi_n$ *such that* $\phi = \phi_n$ *and for every* $\phi_i (1 \leqslant i \leqslant n)$ *in this sequence is*

1. *either an instance of one of the axioms in* $\mathbf{S}$*;*
2. *or the result of the application of one of the rules in* $\mathbf{S}$ *to those formulas appearing before* ϕ_i*.*

We write $\vdash_{\mathbf{S}} \phi$ *if there is a derivation for* ϕ *in* $\mathbf{S}$*, or,* $\vdash \phi$ *when the context of* $\mathbf{S}$ *is clear. We say* ϕ *is a theorem of* $\mathbf{S}$ *or* $\mathbf{S}$ *proves* ϕ*. We write* $Cn(\mathbf{S})$ *as the set of all theorems of* $\mathbf{S}$*.*

Definition 3 (Derivations from Premises). *Let* $\mathbf{S} \in \{\mathbf{D}, \mathbf{D}_{-1}, \mathbf{D}_{-2}\}$ *be a deontic logic. Given a set* Γ *of formulas, a derivation for* ϕ *from* Γ *in* $\mathbf{S}$ *is a finite sequence* $\phi_1, \ldots, \phi_{n-1}, \phi_n$ *such that* $\phi = \phi_n$ *and for every* $\phi_i (1 \leqslant i \leqslant n)$ *in this sequence*

1. *either* $\phi_i \in Cn(\mathbf{S}) \cup \Gamma$*;*
2. *or the result of the application of one of the rules (which is neither* $NEC_\Box$ *nor* NEC_O*) to those formulas appearing before* ϕ_i*.*

We write $\Gamma \vdash_{\mathbf{S}} \phi$ *if there is a derivation from* Γ *for* ϕ *in* $\mathbf{S}$[1]*, or,* $\Gamma \vdash \phi$ *when the context of* $\mathbf{S}$ *is clear. We say this that* ϕ *is derivable in* $\mathbf{S}$ *from* Γ*. We write* $Cn_{\mathbf{S}}(\Gamma)$ *as the set of formulas derivable in* $\mathbf{S}$ *from* Γ*, or* $Cn(\Gamma)$ *if the context of* $\mathbf{S}$ *is clear.*

A system $\mathbf{S}$ is consistent iff $\bot \notin Cn(\mathbf{S})$; otherwise, inconsistent. A set Γ is consistent iff $\bot \notin Cn(\Gamma)$; otherwise, inconsistent. A set $\Gamma' \subseteq \Gamma$ is maximally consistent subset of Γ, denoted as $\Gamma' \in MC(\Gamma')$ iff there is no $\Gamma'' \supset \Gamma'$ such that Γ'' is consistent.

The following example explains in what sense in monotonic logics we can say that Pk and $O\neg m$ are in conflict. Notice that the set of Pk and $O\neg m$ is consistent even in $\mathbf{D}$. This matches our intuition. We say that it is not consistent, as shown below, when it is not normal that using a knife is not a murder, i.e. $\Diamond(k \wedge m)$ holds. The conditional will play an important role in the ASPIC$^+$-based analysis of the running example.

Example 2 (Knifed murder, continued). The following derivation shows that $\{Pk, O\neg m, \Diamond(k \wedge m)\}$ is inconsistent in $\mathbf{D}_{-1}$ or $\mathbf{D}$.

1. $O\neg m \wedge \Diamond(k \wedge m)$	assumption
2. $O\neg m \wedge Pk \to \Box(k \to \neg m)$	OWP
3. $\Diamond(k \wedge m) \leftrightarrow \neg\Box(k \to \neg m)$	$E_\Box$
4. $O\neg m \wedge \Diamond(k \wedge m) \to \neg Pk$	2, 3, MP
5. $\neg Pk$	1, 4, MP

[1] Alternatively, it can be seen as a theorem $\vdash_{\mathbf{S}} \bigwedge \Gamma \to \phi$ by the so-called deduction theorem.

Since we want to represent $\{Pk, O\neg m, \Diamond(k \wedge m)\}$ in a consistent way, we use **D** only to derive conclusions which are defeasible, and we use one of the subsystems of **D** to define the monotonic conclusions.

We involve one spirit of ASPIC$^+$ by considering the inference rules which are uncertain and fallible defeasible rules, while the ones which are unfallible are strict rules. This type of uncertainty or fallibility is represented by the distinction between lower-bounded and upper-bounded logics. However, to simplify the issue addressed, namely, how we can use ASPIC$^+$ to define non-monotonic logics, we are not necessary to fully adopt all methods in ASPIC$^+$ to define arguments. We only consider a general knowledge base here. To distinguish the types of knowledge we leave it to the future work.

Definition 4 (Argumentation Theory). *Let $\mathcal{L}$ be the deontic language and $(\mathbf{S}^-;\mathbf{S}^+) \in \{(\mathbf{D}_{-2};\mathbf{D}_{-1}), (\mathbf{D}_{-2};\mathbf{D}), (\mathbf{D}_{-1};\mathbf{D})\}$ be a Cartesian product of two monotonic logics. An argumentation theory AT based on $(\mathbf{S}^-;\mathbf{S}^+)$ is a tuple (AS, K) where AS is an argumentation system $(\mathcal{L}, R)$, $K \subseteq \mathcal{L}$ is a knowledge base, and $R = R_s \cup R_d$ is a set of rules, such that*

- $R_s = \{\phi_1, \ldots, \phi_n \mapsto \phi \mid \{\phi_1, \ldots, \phi_n\} \vdash_{\mathbf{S}^-} \phi\}$ *is the set of strict rules, and*
- $R_d = \{\phi_1, \ldots, \phi_n \Rightarrow \phi \mid \{\phi_1, \ldots, \phi_n\} \vdash_{\mathbf{S}^+} \phi \ \& \ \{\phi_1, \ldots, \phi_n\} \nvdash_{\mathbf{S}^-} \phi\}$ *is the set of defeasible rules.*

If the context of $(\mathbf{S}^-;\mathbf{S}^+)$ is clear, we mention AT without $(\mathbf{S}^-;\mathbf{S}^+)$.

So the requirement of $R_s \cap R_d = \varnothing$ holds.

In contrast to derivations, arguments are different structures. Although each argument corresponds to a derivation defined as a top rule, the former has to explicitly consider each step of this derivation as a finite sequence.

Definition 5 (Arguments). *Let AT be an argumentation theory with a knowledge base K and an argumentation system $(\mathcal{L}, R)$. Given each $n \in \mathbb{N}$, the set $\mathcal{A}_n$ where $n \in \mathbb{N}$ is defined by induction as follows:*

$$\mathcal{A}_0 = K$$
$$\mathcal{A}_{n+1} = \mathcal{A}_n \cup \{B_1, \ldots, B_m \rhd \psi \mid B_i \in \mathcal{A}_n \text{ for all } i \in \{1, \ldots, m\}\}$$

where for an element $B \in \mathcal{A}_i$ with $i \in \mathbb{N}$:

- *If $B \in K$, then $Prem(B) = \{\phi\}, Conc(B) = \phi, Sub(B) = \{\phi\}, Rules_d(B) = \varnothing, TopRule(B) = undefined$ where $\psi \in K$.*
- *If $B = B_1, \ldots, B_m \rhd \psi$ where $\rhd$ is $\mapsto$ then $\{Conc(B_1), \ldots, Conc(B_m)\} \mapsto \psi \in R_s$ with $Prem(B) = Prem(B_1) \cup \ldots \cup Prem(B_m), Conc(B) = \psi$, $Sub(B) = Sub(B_1) \cup \ldots \cup Sub(B_m) \cup \{B\}$, $Rules_d(B) = Rules_d(B_1) \cup \ldots \cup Rules_d(B_m)$, $TopRule(B) = Conc(B_1), \ldots, Conc(B_m) \mapsto \psi$.*
- *If $B = B_1, \ldots, B_m \rhd \psi$ where $\rhd$ is $\Rightarrow$, then each condition is similar to the previous item, except that the rule is defeasible and $Rules_d(B) = Rules_d(B_1) \cup \ldots \cup Rules_d(B_m) \cup \{Conc(B_1), \ldots, Conc(B_m) \Rightarrow \psi\}$.*

We define $\mathcal{A} = \bigcup_{n \in \mathbb{N}} \mathcal{A}_n$ *as the set of arguments on the basis of* AT*, and define* $Conc(E) = \{\varphi \subseteq Conc(A) \mid A \in E\}$ *where* $E \subseteq \mathcal{A}$.

The following example illustrates the arguments in the running example. We consider the defeats (arrows) in Fig. 1 in the following section.

Example 3 (Knifed murder, continued). We illustrate the argumentation theory in our running example shown in Fig. 1. Let $(\mathbf{S}^-; \mathbf{S}^+) \in \{(\mathbf{D}_{-2}; \mathbf{D}_{-1}), (\mathbf{D}_{-2}; \mathbf{D}), (\mathbf{D}_{-1}; \mathbf{D})\}$ be a pair of two monotonic logics, and AT be an argumentation theory based on $(\mathbf{S}^-; \mathbf{S}^+)$ that takes $K = \{\Diamond(k \wedge m), O\neg m, Pk\}$ where k (using knife) and m (murder) are atomic propositions. We know that $K \vdash_{\mathbf{S}} \bot$ for any consistent system $\mathbf{S} \in \{\mathbf{D}_{-1}, \mathbf{D}\}$. Three arguments refer to the knowledge that it is forbidden to kill (A), it is permitted to use the knife (B), and knifed murder is possible (C). In the logic $\mathbf{D}_{-2}$, we can derive, for example, that knifed murder is forbidden from the premise that murder is forbidden. This derivation is the strict rule to construct the argument A''' (in closed circle) from the knowledge A. Also, $TopRule(A''')$ is this derivation. In the stronger logic $\mathbf{D}_{-1}$, we can derive, for example, that knifed murders are not permitted from the premises that murder is forbidden and knifed murder is possible. This derivation as a defeasible rule together with the knowledges A and C construct the argument A'' (in densely dashed circle). In the strongest logic $\mathbf{D}$, we can derive, for example, that knifed murders are permitted (loosely dashed circles for B'', B'''').

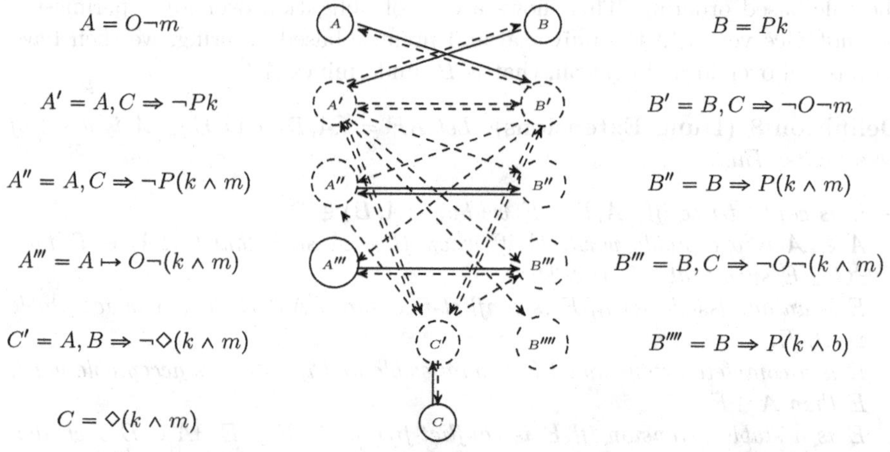

Fig. 1. It shows some of the arguments and defeats. Closed circles are arguments of $\mathbf{D}_{-2}$, densely dashed circles are arguments of $\mathbf{D}_{-1}$, and loosely dashed circles are arguments of $\mathbf{D}$. Straight arrows are defeats among these arguments in the rule-based ordering for $S^+ = \mathbf{D}_{-2}$, and the dashed arrows are defeats in all orderings.

4 Step 2: Preferences Among Arguments

In this paper we consider three orders: universal, rule-based, and premise-based. They emphasize different perspectives of selecting proper arguments for constructing non-monotonic inferences.

Definition 6 (Argument Properties). *Let A be an argument and E a set of arguments. Then A is strict if $Rules_d(A) = \varnothing$; defeasible if $Rules_d(A) \neq \varnothing$; firm if $Prem(A) \subseteq K$; plausible if $Prem(A) \cap K \neq \varnothing$. We define $Concs(E) = \{Conc(A) \mid A \in E\}$. The partial order $\leqslant$ rule-based iff we have $A \leqslant B$ iff A is defeasible; and premise-based iff $A \leqslant B$ iff A is plausible.*

We use $\leqslant^{\tau}$ to denote the τ-ordering with $\tau \in \{r, p\}$, where r for rule-based and p for premise-base. Next we introduce the notion of defeat. The first is a rebuttal while the second is a undermining [15]. In the next section this distinction will give different consequences in non-monotonic reasoning.

Definition 7 (Argumentation Frameworks). *An abstract argumentation framework AF corresponding to $\langle AT, \leqslant\rangle$ is a pair $(\mathcal{A}, \mathcal{D})$, where $\mathcal{D}$ is a set of pairs of arguments in which argument A defeats argument B is defined as:*

- *either $Conc(A) = \neg\phi$ for some $B' \in Sub(B)$ and $TopRule(B') \in R_d, Conc(B') = \phi$ and $A \nless B'$.*
- *or $Conc(A) = \neg\phi$ for knowledge $\phi \in Prem(B)$ of B and $A \nless \phi$.*

As shown in Fig. 1, A'' rebuts B'' but B' does not undermine A''' when it is in the rule-based ordering. This shows a case of obligation overriding permission but not vice versa. In the universal and premise-based ordering, we then have permission overrides obligation, that is B' undermines A'''.

Definition 8 (Dung Extensions). *Let $AF = (\mathcal{A}, \mathcal{D})$ and $E \subseteq \mathcal{A}$ is a set of arguments. Then*

- *E is conflict-free iff $\forall A, B \in E$ we have $(A, B) \notin \mathcal{D}$.*
- *$A \in \mathcal{A}$ is acceptable w.r.t. E iff when $B \in \mathcal{A}$ such that $(B, A) \in \mathcal{D}$ then $\exists C \in E$ such that $(C, B) \in \mathcal{D}$.*
- *E is an admissible set iff E is conflict-free and if $A \in E$ then A is acceptable w.r.t. E.*
- *E is a complete extension iff E is admissible and if $A \in \mathcal{A}$ is acceptable w.r.t. E then $A \in E$.*
- *E is a stable extension iff E is conflict-free and $\forall B \notin E\ \exists A \in E$ such that $(A, B) \in \mathcal{D}$.*

The following example illustrates a different sense of consistency in ASPIC$^+$ by using stable extensions, in order to explain, given the inconsistent knowledge base K, why $B \Rightarrow P(k \wedge m)$ is sometimes defeated and why $B \Rightarrow P(k \wedge b)$ is always non-defeated.

Example 4 (Knifed murder, continued). Consider the arrows in Fig. 1. The straight arrows represent defeat relations under the rule-based ordering, and the dashed arrows represent additional defeat relations under the premise-based or universal ordering. Under the rule-based ordering the arguments A, B and C will not be defeated and thus in every extension, whereas in the premise-based or universal ordering, they will not. For this reason, we prefer the rule-based ordering in this example. Furthermore, under the rule-based ordering, we at least have two stable extensions, one contains $B \Rightarrow P(k \wedge m)$ and another $A, C \Rightarrow \neg P(k \wedge m)$. As $B'''' = B \Rightarrow P(k \wedge b)$ will be non-defeated, we have B'''' in every stable extension. Similarly, arguments in form of $A_1, \ldots, A_n \mapsto Pk \vee O\neg m \vee \Diamond(k \wedge m)$ are contained in every stable extension.

Apart from comparing plausible and defeasible arguments in the preference ordering, factual statements can be preferred over deontic statements, prohibitions over permissions, or vice versa. We leave such further investigations for the journal extension of this paper.

5 Step 3: Designing Non-monotonic Logics

Our non-monotonic logics are designed by using the stable extensions regarding to different monotonic logics and to different orderings. In order to do so, the following proposition provides a guideline to search for these stable extensions. In the case of universal/premise-based ordering, strict rules are as equally preferable as defeasible rules. So a stable extension can be considered as a maximally consistent subset of the knowledge base K. We call this the *undermining* mechanism, see e.g. [2,22]. But this is not enough to capture the case of rule-based ordering, in which the defeasible argument is less preferable than the others. So the second item of this proposition provides a general method to construct the desired extensions, stable extensions. We construct each stable extension in the style of Lindenbaum's Lemma [5]. That is, we first consider the maximally consistent subset K' of the knowledge base w.r.t. the lower-bounded logic $\mathbf{S}^-$ for strict rules, and then a consistent subset of K' w.r.t. the upper-bounded logic $\mathbf{S}^+$ for defeasible rules, such that no argument w.r.t. $\mathbf{S}^+$ defeat that w.r.t. $\mathbf{S}^-$ and it is a maximal set satisfying these two conditions. This is called the *rebuttal* mechanism. See the following for details.

Proposition 1.[2] *Consider the deontic language $\mathcal{L}$ and a combination of two monotonic logics $(\mathbf{S}^-; \mathbf{S}^+) \in \{(\mathbf{D}_{-2}; \mathbf{D}_{-1}), (\mathbf{D}_{-2}; \mathbf{D}), (\mathbf{D}_{-1}; \mathbf{D})\}$. Let AF corresponding to $\langle AT, \leqslant^\tau \rangle$ be an abstract argumentation framework $(\mathcal{A}, \mathcal{D})$, such that AT is based on $(\mathbf{S}^-; \mathbf{S}^+)$, K is a knowledge base, and $\tau \in \{p, r\}$. We define $F(D) = Prem(D) \cup \{Conc(D)\}$ where $D \in \mathcal{A}$. Let $F(E) = \bigcup\{F(D) \mid D \in E \subseteq \mathcal{A}\}$.*

[2] For the proof please check: https://pan.zju.edu.cn/share/793a363c53083fbf2c00433b1b

1. *When* $\tau = p$ *then* $E = \{D \in \mathcal{A} \mid Conc(D) \in Cn_{\mathbf{S}^+}(\Gamma)\}$ *is a stable extension if and only if* Γ *is a maximally consistent subset of the knowledge base* K *in* AT *w.r.t.* $\mathbf{S}^+$.
2. *We define* $E_1 = \{D \in \mathcal{A} \mid F(D) \subseteq Cn_{\mathbf{S}^-}(\Gamma_1)\}$ *where* Γ_1 *is a maximally consistent subset of* K *w.r.t.* $\mathbf{S}^-$. *Let* $E_2 = \{D \in \mathcal{A} \mid F(D) \subseteq Cn_{\mathbf{S}^+}(\Gamma_1)\}$ *such that (i)* $F(D)$ *is consistent w.r.t.* $\mathbf{S}^+$*; (ii)* $F(D) \cup \{\varphi\}$ *is consistent w.r.t.* $\mathbf{S}^-$ *where* $\varphi \in Cn_{\mathbf{S}^-}(\Gamma_1)$*; (iii) there is no* $\Gamma \supset F(D)$ *such that* Γ *is consistent w.r.t.* $\mathbf{S}^+$, *and for any* $\varphi \in Cn_{\mathbf{S}^-}(\Gamma_1)$ *we have* $\Gamma \cup \{\varphi\}$ *be* $\mathbf{S}^-$*-consistent. If* $\tau \in \{p, r\}$ *then* $E = E_1 \cup E_2$ *is a stable extension.*

Given the knowledge base $K = \{Pk, O\neg m, \Diamond(k \wedge m)\}$ of the running example, ASPIC$^+$ provides a mechanism to decide whether the two arguments $A, C \Rightarrow \neg P(k \wedge m)$ and $A, C \Rightarrow \neg Pk$ can be accepted. In the case of premise-based, the undermining together with stability is a mechanism to ensure that in conflict like K the maximally consistent subsets form the stable extensions. In the case of rule-based ordering, we cannot use the undermining mechanism to ensure that we derive the first but not the second. Instead, we need to use the rebuttal mechanism. Rebuttal corresponds to closed world assumption. So the above two arguments hold, unless there is a proof to the contrary. That is how the two are then distinguished in the logics.

We now present the central definition of the paper, namely the definition of the non-monotonic logic in terms of the formal argumentation theory. This is well in line with current practice in ASPIC$^+$. We first take the desired conclusions in each stable extension (as shown in Proposition 1) and then the intersection of all the stable extensions.

Definition 9 (Non-Monotonic Inferences). *Let* $\Gamma \subseteq \mathcal{L}, \phi \in \mathcal{L}$, $(\mathbf{S}^-; \mathbf{S}^+) \in \{(\mathbf{D}_{-2}; \mathbf{D}_{-1}), (\mathbf{D}_{-2}; \mathbf{D}), (\mathbf{D}_{-1}; \mathbf{D})\}$ *be a Cartesian product of two monotonic logics, and* $\leqslant^\tau$ *be a* τ*-ordering such that* $\tau \in \{r, p\}$. *Let* AT *be the* Γ*-argumentation theory based on* $(\mathbf{S}^-; \mathbf{S}^+)$ *iff the argumentation theory* AT *obtains with* $K = \Gamma$, *and* $AF^\tau = \langle AT, \leqslant^\tau \rangle$. *The non-monotonic inference* $|\!\!\sim^\tau_{\mathbf{S}^-;\mathbf{S}^+}$ *is defined as follows:*

- $\Gamma |\!\!\sim^\tau_{\mathbf{S}^-;\mathbf{S}^+} \phi$ *iff every stable extension of the* Γ*-*AT *based on* $(\mathbf{S}^-; \mathbf{S}^+)$ *corresponded by* AF^τ *contains an argument* A *with* $Conc(A) = \phi$.

We define the closure operator corresponding to this inference relation as usual: $\mathcal{C}^\tau_{\mathbf{S}^-;\mathbf{S}^+}(\Gamma) = \{\phi \mid \Gamma |\!\!\sim^\tau_{\mathbf{S}^-;\mathbf{S}^+} \phi\}$. *Moreover, we write* $|\!\!\sim^\tau_{\mathbf{S}^-;\mathbf{S}^+} \phi$ *when* $\varnothing |\!\!\sim^\tau_{\mathbf{S}^-;\mathbf{S}^+} \phi$.

The resulting non-monotonic inference relations are standard relations among sets of formulas of the logical language, i.e. they no longer refer to ASPIC$^+$. An alternative way to define non-monotonic logics is to first consider the intersection of all stable extensions and then the conclusions. For instance, $Pk \vee O\neg m \vee \Diamond(k \wedge m)$ is an element in $\mathcal{C}^\tau_{\mathbf{D}_{-2};\mathbf{D}}(\{Pk, O\neg m, \Diamond(k \wedge m)\}$ where $\tau \in \{p, r\}$. If the proposed order is reversed, this cannot be inferred. Because it is possible to have many different arguments which contain the same conclusion but from different premises.

The following proposition offers a detailed explanation of the mechanisms we proposed. First, the undermining mechanism states that the non-monotonic consequences are the intersection of all maximally consistent subsets of the knowledge base under the universal or premise-based ordering. Second, and more generally, the rebuttal mechanism states that the non-monotonic consequences are encased by all unions of a maximally consistent subset of the knowledge base w.r.t. the lower-bounded logic and a consistent subset of it w.r.t. the upper-bounded logic in certain maximal behavior.

Proposition 2. *Let $\Gamma \subseteq \mathcal{L}$, $(\mathbf{S}^-;\mathbf{S}^+) \in \{(\mathbf{D}_{-2};\mathbf{D}_{-1}), (\mathbf{D}_{-2};\mathbf{D}), (\mathbf{D}_{-1};\mathbf{D})\}$ be a Cartesian product of two monotonic logics, $\leqslant^\tau$ be a τ-ordering such that $\tau \in \{r, p\}$, and K be a knowledge base of AT. Then*

1. *$\mathcal{C}^{\tau}_{\mathbf{S}^-;\mathbf{S}^+}(K) = \bigcap_{\Gamma \in MC(K)} Cn_{\mathbf{S}^+}(\Gamma)$, where $\tau = p$.*
2. *We define $\Gamma' \in M(\Gamma)$ as follows: (i) Γ' is a consistent subset of Γ w.r.t. $\mathbf{S}^+$, (ii) $Cn_{\mathbf{S}^+}(\Gamma')$ is consistent with φ w.r.t. $\mathbf{S}^-$ where $\varphi \in Cn_{\mathbf{S}^-}(\Gamma)$, and (iii) there is no $\Gamma'' \subseteq \Gamma'$ such that Γ'' is $\mathbf{S}^+$-consistent and for all $\varphi \in Cn_{\mathbf{S}^-}(\Gamma)$ we have $\Gamma'' \cup \{\varphi\}$ be $\mathbf{S}^-$-consistent. Then*

$$\mathcal{C}^{r}_{\mathbf{S}^-;\mathbf{S}^+}(K) = \bigcap_{\Gamma \in MC(K)} \bigcap_{\Gamma' \in M(\Gamma)} (Cn_{\mathbf{S}^-}(\Gamma) \cup Cn_{\mathbf{S}^+}(\Gamma')),$$

where $\tau \in \{r, p\}$.

To prove Proposition 2, as inspired by Proposition 1, we first consider the maximally consistent subset of the knowledge base w.r.t the lower-bounded logic $\mathbf{S}^-$, and then consider the consistent subset of the knowledge base w.r.t. the upper-bounded logic $\mathbf{S}^+$, such that this set is maximal in the sense that it is consistent with each element of the previous set w.r.t. the lower-bounded logic. Moreover, Proposition 2.2 illustrates a new understanding of maximality of consistency, which not only has to consider the consistency of the upper-bounded logic but also the consistency with each element in the lower-bounded logic.

A formal analysis of the non-monotonic inference relation is left to further research, as well as the development of alternative non-monotonic relations in terms of the formal argumentation theory.

Example 5 (Knifed murder, continued). Given the set $K = \{\Diamond(k \wedge m), O\neg m, Pk\}$ as the premises, we have different non-monotonic consequences shown in Table 2, depending on the combinations of monotonic logics and the orderings. They are non-monotonic, in the sense that, even given Pk as one premise, $P(k \wedge m)$ is excluded in every non-monotonic consequences, while $P(k \wedge b)$ is a non-monotonic consequence w.r.t. $(\mathbf{D}_{-2}; \mathbf{D})$ under the rule-based ordering.

Table 2. Examples of the non-monotonic inferences in the case of knifed murder. We have $T_u = \bigcap_{\Gamma \in MC(K)} Cn_{\mathbf{D}_{-1}}(\Gamma)$ and $T_r^1 = \bigcap_{\Gamma \in M(K)}(Cn_{\mathbf{D}_{-2}}(K) \cup Cn_{\mathbf{D}_{-1}}(\Gamma))$ and $T_r^2 = \bigcap_{\Gamma \in M(K)}(Cn_{\mathbf{D}_{-2}}(K) \cup Cn_{\mathbf{D}}(\Gamma))$.

	Order	K	Example of Consequences $\{\Diamond(k \wedge m), O\neg m, Pk\}$
$(\mathbf{D}_{-2}; \mathbf{D}_{-1})$	p	T_u	$\bigvee K$
$(\mathbf{D}_{-2}; \mathbf{D}_{-1})$	r	T_r^1	$\Diamond(k \wedge m), O\neg m, Pk, O\neg(k \wedge m), \bigvee K$
$(\mathbf{D}_{-2}; \mathbf{D})$	p	T_u	$\bigvee K$
$(\mathbf{D}_{-2}; \mathbf{D})$	r	T_r^2	$\Diamond(k \wedge m), O\neg m, Pk, O\neg(k \wedge m), P(k \wedge b), \bigvee K$
$(\mathbf{D}_{-1}; \mathbf{D})$	p, r	T_u	$\bigvee K$

6 Related Work

Concerning the formalization of non-monotonic reasoning about norms, obligations and permissions, there is a large amount of work. For instance, Horty [11] formalized the reasoning in the presence of conflicting obligations and reasoning with conditional obligations based on default logic and a model preference logic, Prakken [20] proposed a combination of standard deontic logic with an early-generation formal argumentation system to formalize defeasible deontic reasoning, and Prakken and Sartor [21] formulated arguments about norms as the application of argument schemes to knowledge bases of facts and norms, among others. Our work is in line with the existing methodology by using non-monotomic formalisms to deal with the conflicts between norms, obligations and permissions. Besides this point, our work focuses more on how to capture the intuition of reasoning about free-choice permission, by using different monotonic logics (lower bound and upper bound) to define strict rules and defeasible rules, and different types of arguments (rule-based, premise-based and universal) to define the preference relation between arguments.

Connecting formal argumentation and deontic logic is an increasingly active research topic in recent years [19]. In the direction of using argumentation to represent various non-monotonic logics, Young *et al.* [25] proposed an approach to represent prioritized default logic by using ASPIC$^+$. Liao et al. [14] represented three logics of prioritized norms by using argumentation. While existing works use argumentation to represent existing non-monotonic logics or non-monotonic reasoning, this paper uses argumentation to define new logics. A recent work that is close to our work is by Straßer and Arieli [22], which presented an argumentative approach to normative reasoning by using standard deontic logic as base logic. Similar to this paper, our logic $\mathbf{D}_{-2}$ is also a variant of standard deontic logic. The difference is that we use the extension $\mathbf{D}_{-1}$ and $\mathbf{D}$ to capture the permission by using FCP and OWP.

7 Summary and Concluding Remarks

In this paper, ASPIC$^+$ relates formal argumentation to non-monotonic logic. We believe this approach benefits both areas. For formal argumentation, the resulting non-monotonic logics can be studied to provide new insights in the adopted argumentation systems, for example in the effect of the adopted argumentation semantics. For the non-monotonic logics, the underlying argumentation theory can be used for explaining deontic conclusions. Our case-study with the logic of obligations and permissions provides first evidence for this.

Within this general ambitious setting, the contributions of this paper are as follows. First, concerning the definitions, in Definition 4 we show how to use two logics in ASPIC$^+$, and in Definition 9 we show how to build a non-monotonic logic on top of ASPIC$^+$. For the formal results, Propositions 1 and 2 characterise the consequences of the non-monotonic logic. Finally, the example illustrates how to apply this approach to formalise the analysis of Glavaničová [8] of strong and free-choice permission.

The relation between the argumentation system and the non-monotonic logic can be studied in more detail. Consider the possibility of post-rationalization in law. The models describing decision making as people deliberate and argue and then, at the end, a group decision is proposed might be considered as naive: we cannot identify the cases of post-rationalization, where a decision is made first, then arguments in favour of that decision are sought. The interaction between the argumentation system and the non-monotonic logic is not a trivial relation where one is the master and the other is the slave, but both the argumentation system and the non-monotonic logic should be seen as different conceptualizations with different concerns, which are related, but one cannot be reduced to the other.

The main tool for studying formal argumentation in the setting of ASPIC$^+$ is based on the use of rationality postulates [6]. It immediately follows from the two propositions of this paper, that all rationality postulates are satisfied. This can also be proven as a corollary of the more general theorems of Caminada, and of Modgil and Prakken.

Our study opens up many lines of further research. For example, as done by Beirlaen *et al.* [3], we can consider the alternatives of monotonic combinations, like **D** minus K_O and NEC_O as the logic for strict rules and SDL minus weak permission as the logic for defeasible rules. In this case, we can go for the approach of permission overriding obligation. Second, we can study the sophisticated method rather than the crude method by using the natural deduction proof theories [16], in which we take axioms as the knowledge base and the others as the rules for arguments. Also, we then can explore the challenge of obtaining a normalizing system of natural deduction for deontic logic with the sub-formula property. Further, we have discussed the distinction of strict/defeasible rules in this paper, and have checked the relation of the non-monotonic inferences with the monotonic one. We can distinguish the premises of arguments from strict to defeasible, and then study the relation with supra-classical logics. We believe that this future work will bring us an interesting insight of non-monotonicity.

Acknowledgement. Huimin Dong is supported by the China Postdoctoral Science Foundation funded project [No. 2018M632494], the National Science Centre of Poland [No. UMO-2017/26/M/HS1/01092], the National Social Science Fund of China [No. 18ZDA290, No. 17ZDA026, No. 16AZX017], the MOE Project of Key Research Institute of Humanities and Social Sciences in Universities [No. 17JJD720008], and the Fundamental Research Funds for the Central Universities of China. Beishui Liao and Huimin Dong are supported by the Convergence Research Project for Brain Research and Artificial Intelligence, Zhejiang University. Leendert van der Torre, Beishui Liao, and Réka Markovich have received funding from the European Union's Horizon 2020 research and innovation programme under the Marie Sklodowska-Curie grant agreement No 690974 for the project "MIREL: MIning and REasoning with Legal texts".

References

1. Anglberger, A.J., Gratzl, N., Roy, O.: Obligation, free choice, and the logic of weakest permissions. Rev. Symbolic Log. **8**, 807–827 (2015)
2. Arieli, O., Borg, A., Straßer, C.: Reasoning with maximal consistency by argumentative approaches. J. Log. Comput. **28**(7), 1523–1563 (2018)
3. Beirlaen, M., Heyninck, J., Straßer, C.: Structured argumentation with prioritized conditional obligations and permissions. J. Log. Comput. **29**(2), 187–214 (2018)
4. van Benthem, J.: Minimal deontic logics. Bull. Sect. Log. **8**(1), 36–42 (1979)
5. Blackburn, P., De Rijke, M., Venema, Y.: Modal Logic, vol. 53. Cambridge University Press, Cambridge (2002)
6. Caminada, M.: Rationality postulates: applying argumentation theory for non-monotonic reasoning. In: Baroni, P., Gabbay, D., Giacomin, M., van der Torre, L. (eds.) Handbook of Formal Argumentation. College Publication, London (2018)
7. Gabbay, D., Horty, J., Parent, X., van der Meyden, R., van der Torre, L. (eds.): Handbook of Deontic Logic and Normative Systems. College Publications, London (2013)
8. Glavaničová, D.: The free choice principle as a default rule. Organon F **25**(4), 495–516 (2018)
9. Governatori, G., Rotolo, A.: Is free choice permission admissible in classical deontic logic? arXiv preprint arXiv:1905.07696 (2019)
10. Hansson, S.O.: The varieties of permissions. In: Gabbay, D., Horty, J., Parent, X., van der Meyden, R., van der Torre, L. (eds.) Handbook of Deontic Logic and Normative Systems. College Publication, London (2013)
11. Horty, J.F.: Deontic logic as founded on nonmonotonic logic. Ann. Math. Artif. Intell. **9**(1–2), 69–91 (1993)
12. Kraus, S., Lehmann, D.J., Magidor, M.: Nonmonotonic reasoning, preferential models and cumulative logics. Artif. Intell. **44**(1–2), 167–207 (1990)
13. Lewis, D.: A problem about permission. In: Saarinen, E., Hilpinen, R., Niiniluoto, I., Hintikka, M.P. (eds.) Essays in Honour of Jaakko Hintikka. Synthese Library (Studies in Epistemology, Logic, Methodology, and Philosophy of Science), vol. 124, pp. 163–175. Springer, Dordrecht (1979). https://doi.org/10.1007/978-94-009-9860-5_11
14. Liao, B., Oren, N., van der Torre, L., Villata, S.: Prioritized norms in formal argumentation. J. Log. Comput. **29**(2), 215–240 (2019)
15. Modgil, S., Prakken, H.: Abstract rule-based argumentation. In: Baroni, P., Gabbay, D., Giacomin, M., van der Torre, L. (eds.) Handbook of Formal Argumentation. College Publication, London (2018)

16. Negri, S.: Proof theory for modal logic. Philos. Compass **6**(8), 523–538 (2011)
17. Nute, D. (ed.): Defeasible Deontic Logic (1997)
18. Parent, X., van der Torre, L.: Introduction to Deontic Logic and Normative Systems. College Publications, London (2018)
19. Pigozzi, G., van der Torre, L.: Arguing about constitutive and regulative norms. J. Appl. Non-Classical Log. **28**(2–3), 189–217 (2018)
20. Prakken, H.: Two approaches to the formalisation of defeasible deontic reasoning. Stud. Logica **57**(1), 73–90 (1996)
21. Prakken, H., Sartor, G.: Formalising arguments about norms. In: Legal Knowledge and Information Systems (JURIX 2013), pp. 121–130. IOS Press, Amsterdam (2013)
22. Straßer, C., Arieli, O.: Normative reasoning by sequent-based argumentation. J. Log. Comput. **29**(3), 387–415 (2019)
23. van der Torre, L.: Reasoning About Obligations: Defeasibility in Preference-based Deontic Logic (1007)
24. von Wright, G.H.: Deontic logic. Mind **60**, 1–15 (1951)
25. Young, A.P., Modgil, S., Rodrigues, O.: Prioritised default logic as rational argumentation. In: Proceedings of AAMAS 2016, pp. 626–634 (2016)

A Discrete Representation of Lattice Frames

Ivo Düntsch[1,3P] (✉) and Ewa Orłowska[2]

[1] School of Mathematics and Computer Science, Fujian Normal University, Fuzhou, Fujian, China
D.ivo@fjnu.edu.cn

[2] National Institute of Telecommunications, Szachowa 1, 04–894 Warsaw, Poland
orlowska@itl.waw.pl

[3] Department of Computer Science, Brock University, St. Catharines, ON, Canada
duentsch@brocku.ca

Abstract. We characterize those doubly ordered frames $\langle X, \leq_1, \leq_2 \rangle$ that are embeddable into the canonical frames of their complex algebras defined by Alasdair Urquhart in his representation theorem for bounded general lattices [31]. Our result together with the topology-free version of Urquhart's representation leads to a discrete (i.e. topology free) duality for bounded general lattices. We also show that doubly ordered frames are definable neither in a logic endowed with only a possibility operator nor a logic with only a sufficiency operator, but in a logic based on mixed algebras with both a possibility and a sufficiency operator.

1 Introduction

Let $\mathfrak{A}$ be a class of structures, and $\mathfrak{B}$ a subclass of $\mathfrak{A}$. A *representation of* $\mathfrak{A}$ *with respect to* $\mathfrak{B}$ says that every member of $\mathfrak{A}$ is isomorphic to a member of $\mathfrak{B}$. For algebras, famous examples include

- Every group is isomorphic to a permutation group [2].
- Every Boolean algebra is isomorphic to the algebra of clopen sets of a zero–dimensional compact topological space [29].
- Every distributive lattice is isomorphic to an algebra of increasing subsets of a compact totally order disconnected topological space [28].

Many of such theorems embed a class of algebras into a class of subsets of topological spaces, which, invariably, involves second order constructions. A different approach can be learned from the two kinds of semantics of propositional modal logic, namely, frame semantics and algebraic semantics. This leads to the concept of *duality via truth*, proposed in [25]: Here, a class of algebras and a class of relational structures (frames) are considered dual, if both are semantics for the same logic; in this sense, algebras and frames are considered on an equal footing. More concretely, a *discrete duality* relates a class Alg of algebras to a class Frm of frames and vice versa in the following way:

1. With each algebra A from Alg associate a relational structure $\mathfrak{Cf}(A)$ from Frm, called the *canonical frame of* A.

P. Blackburn et al. (Eds.): LORI 2019, LNCS 11813, pp. 86–97, 2019.
https://doi.org/10.1007/978-3-662-60292-8_7

2. With each frame $\mathcal{F} = \langle U, R, S, \ldots \rangle$ from Frm associate an algebra $\mathfrak{Cm}(\mathcal{F})$ from Alg, called the *complex algebra of F*.
3. Prove two representation theorems:
   (a) For each $A \in \mathsf{Alg}$ there is an embedding $h : A \hookrightarrow \mathfrak{Cm}\,\mathfrak{Cf}(A)$.
   (b) For each frame $\mathcal{F} \in \mathsf{Frm}$ there is an embedding $k : \mathcal{F} \hookrightarrow \mathfrak{Cf}\,\mathfrak{Cm}(\mathcal{F})$.

Many such discrete dualities have been established for structures in various areas relevant to the theme of the conference, among others in

- Information logics [3] ,
- Spatial reasoning [10],
- Preference relations [8],
- Apartness frames [7],
- Substructural logics [24].

In [21] applications of discrete duality to prove completeness theorems, correspondence theory, data analysis and reasoning with incomplete information are outlined. In [22] we presented duality via truth for some nonclassical logics, and outlined its extension to duality of maps. For a comprehensive treatment and many examples, the reader is invited to consult [23].

While dualities were found for many classes of algebras based on Boolean algebras and also on distributive lattices, the known representations of general lattice frames as canonical frames of their complex algebras all involved some kind of topology. The earliest duality for general lattices was found by Urquhart [31], followed by those of Hartung [19], Hartonas and Dunn [18], Ploščica [27] and others. A first order representation theorem for general lattice frames which would enable us to prove a discrete duality for bounded lattices, however, was still missing. In this paper we present such a representation based on Urquhart's work. Independently, Hartonas [17] presented a representation based on his construction using frames with two universes introduced in [18]. We note in passing that formal presentation of nonclassical logics that are considered in philosophical logic is often based on general lattices.

We also show that doubly ordered sets are not definable by a possibility operator, neither are they definable by a sufficiency operator, but require both.

2 Notation and First Definitions

Throughout, $\langle L, \vee, \wedge, 0, 1 \rangle$ is a bounded lattice. If no confusion can arise we shall identify algebras with their base set. If $a \in L$, then $\downarrow_{\leq} a \stackrel{\mathrm{df}}{=} \{b \in L : b \leq a\}$, and $\uparrow_{\leq} a \stackrel{\mathrm{df}}{=} \{b \in L : b \geq a\}$.

An operator f on a Boolean algebra B is called a *possibility operator*, if $f(0) = 0$, and $f(a \vee b) = f(a) \vee f(b)$ for all $a, b \in B$. In this case, $\langle B, f \rangle$ is called a *possibility algebra.* Similarly, g is a *sufficiency operator*, if $g(0) = 1$, and $g(a \vee b) = g(a) \wedge g(b)$ for all $a, b \in B$. The pair $\langle B, g \rangle$ is called a *sufficiency algebra.* A *PS algebra* is a structure $\langle B, f, g \rangle$, where B is a Boolean algebra, f is a possibility operator, and g a sufficiency operator.

If R, S are binary relations on a set X and $x \in X$, we set $R(x) \stackrel{\text{df}}{=} \{y : xRy\}$, and $R \,;\, S \stackrel{\text{df}}{=} \{\langle x, y\rangle : (\exists z)xRzSy\}$ is the relational composition. The converse of R is the relation $R^{\smile} \stackrel{\text{df}}{=} \{\langle y, x\rangle : xRy\}$.

$[R]$ is the necessity operator and $\langle R\rangle$ is the possibility operator induced by R; in other words, for $Y \subseteq X$,

$$[R](Y) \stackrel{\text{df}}{=} \{x : R(x) \subseteq Y\}, \langle R\rangle(Y) \stackrel{\text{df}}{=} \{x : R(x) \cap Y \neq \emptyset\}. \tag{2.1}$$

We note that $\langle R\rangle$ is completely additive and $[R]$ is completely multiplicative. We also define the *sufficiency operator induced by* R as

$$[[R]](Y) = \{x \in X : Y \subseteq R(x)\}. \tag{2.2}$$

The identity relation on X is denoted by $1'$. We follow the convention that embeddings of relational structures are strong. In other words, $f : \langle X, R\rangle \to \langle X', R'\rangle$ is an *embedding*, if it is injective, and $f(x)R'f(y)$ if and only if xRy.

3 Urquhart's Lattice Representation

If L is not distributive, then prime ideals and prime filters need not exist, and more involved constructions are required. Urquhart [31] uses filter – ideal pairs $\langle F, I\rangle$ instead of prime ideals (or prime filters), where F and I are mutually maximally disjoint.

The collection of proper filters of L is denoted by $\mathcal{F}$, and the collection of proper ideals of L is denoted by $\mathcal{I}$. A *filter – ideal pair* is a pair $\langle F, I\rangle$ where $F \in \mathcal{F}$, $I \in \mathcal{I}$, and $F \cap I = \emptyset$. A filter – ideal pair $\langle F, I\rangle$ is called *maximal*, if F is maximally disjoint to I and I is maximally disjoint to F. In other words, if $F' \in \mathcal{F}$ such that $F \subsetneq F'$, then $F' \cap I \neq \emptyset$, and if $I' \in \mathcal{I}$ such that $I \subsetneq I'$, then $F \cap I' \neq \emptyset$. The following result is decisive:

Lemma 1 *[31]. Each filter – ideal pair can be extended to a maximal pair.*

Let X_L be the set of all maximal filter – ideal pairs. To facilitate notation, if $x \in X_L$ with $x = \langle F, I\rangle$ we let $x_1 = F$ and $x_2 = I$. We define two relations $\leq_1, \leq_2$ on X_L by $x \leq_i y$ if and only if $x_i \subseteq y_i$. Clearly, $\leq_1$ and $\leq_2$ are quasiorders on X_L. For $i = 1, 2$ and $x \in X$, we set $\uparrow_i x \stackrel{\text{df}}{=} \{y : x \leq_i y\}$.

A *doubly ordered frame* is a structure $\langle X, \leq_1, \leq_2\rangle$ such that

DO$_1$. $\leq_1$ and $\leq_2$ are quasiorders on X.
DO$_2$. If $x \leq_1 y$ and $x \leq_2 y$, then $x = y$ for all $x, y \in X$.

The structure $\langle X_L, \leq_1, \leq_2\rangle$ is called the *(Urquhart) canonical frame of* L, denoted by $\mathfrak{Cf}(L)$. Note that for any $Y \subseteq X$ and $i = 1, 2$,

$$[\leq_i](Y) \subseteq Y \subseteq \langle\leq_i\rangle(Y), \tag{3.1}$$

since $\leq_i$ is reflexive.

For $Y \subseteq X$,

$$l(Y) \stackrel{\text{df}}{=} \{x : \uparrow_1 x \cap Y = \emptyset\} = [\leq_1](-Y), \tag{3.2}$$

$$r(Y) \stackrel{\text{df}}{=} \{x : \uparrow_2 x \cap Y = \emptyset\} = [\leq_2](-Y). \tag{3.3}$$

Y is called an *l–stable set*, if $Y = l(r(Y))$. The collection of l–stable sets is denoted by L_X. Y is called an *r–stable set*, if $Y = r(l(Y))$. Observe that

$$l(r(Y)) = l([\leq_2](-Y)) = [\leq_1](-[\leq_2](-Y)) = [\leq_1]\langle\leq_2\rangle(Y). \tag{3.4}$$

For an l–stable Y we have $[\leq_1](Y) = Y$.

Galois connections play an important role in the theory of ordered sets and data analysis, see e.g. [11] and [12]. If $\langle A, \leq\rangle$ and $\langle B, \preceq\rangle$ are partially ordered sets, a *Galois connection* is a pair of antitone (order reversing) functions $f : A \to B$, $g : B \to A$ such that for all $a \in A, b \in B$

$$b \preceq f(a) \Longleftrightarrow a \leq g(b). \tag{3.5}$$

Lemma 2 *[31]. Let $(X, \leq_1, \leq_2)$ be a doubly ordered frame. The mappings l and r form a Galois connection between the lattice of $\leq_1$–increasing subsets of X and the lattice of $\leq_2$–increasing subsets of X.*

Thus, if Y is $\leq_1$–increasing and Z is $\leq_2$–increasing, then $Y \subseteq l(Z)$ if and only if $Z \subseteq r(Y)$.

Corollary 1

1. *$l(Y)$ is $\leq_1$–increasing, and $r(Y)$ is $\leq_2$–increasing.*
2. *If Y is $\leq_2$ increasing, then $l(Y)$ is an l–stable set, and if Y is $\leq_1$ increasing, then $r(Y)$ is an r–stable set.*
3. *lr is a closure operator on the $\leq_1$–increasing sets and rl is a closure operator on the $\leq_2$–increasing sets.*

3. is not necessarily true if Y is not increasing: Suppose that $X = \{x, y\}$, $\leq_1 = 1' \cup \{\langle x, y\rangle\}$, and $\leq_2 = 1'$. Then, $[\leq_1]\langle\leq_2\rangle(\{x\}) = \emptyset$.

For $Y, Z \in L_X$ let

$$Y \vee_X Z \stackrel{\text{df}}{=} [\leq_1]\langle\leq_2\rangle(Y \cup Z), \tag{3.6}$$

$$Y \wedge_X Z \stackrel{\text{df}}{=} Y \cap Z. \tag{3.7}$$

Theorem 1 *[31]. The structure $\langle L_X, \vee_X, \wedge_X, \emptyset, X\rangle$ is a complete bounded lattice.*

We call this structure the *(Urquhart) complex algebra of X*, and denote it by $\mathfrak{Cm}(X)$. Urquhart's representation theorem now is as follows:

Theorem 2 *[31]. Define $h : L \to 2^{X_L}$ by $h(a) \stackrel{\text{df}}{=} \{x \in X_L : a \in x_1\}$. Then h is a lattice embedding into $\mathfrak{Cm}\,\mathfrak{Cf}(L)$.*

4 Modal Definability of Doubly Ordered Frames

A common form of a mathematical theorem states that the truth of some properties for some objects is a necessary and/or sufficient condition for other properties to hold for the same or other objects. The treatment of "necessity" has a reliable foundation in the logic K, but what about formalizing "sufficiency"? Reducing "sufficiency" grammatically to "necessity" by simply stating that

$$x \text{ is sufficient for } q \text{ if and only if } q \text{ is necessary for } x$$

is true, but not useful, as it does not enhance our knowledge. In 1985, Solomon Passy (under the name Sulejman Tehlikely) [30] presented a sufficiency counterpart $K^{\star}$ to K with the unary modal operator $\boxbar$ ("window") and additional axiom and rule

$$\vdash \boxbar\neg(\varphi \implies \psi) \implies (\boxbar\neg\varphi \implies \boxbar\neg\psi) \tag{4.1}$$

$$\text{If } \vdash \varphi, \text{ then } \vdash \boxbar\neg\varphi. \tag{4.2}$$

Let Fml^* be the set of formulas in the language of K^*. The frame semantics is given by relational structures $\langle X, S, v\rangle$ in such a way that for a valuation $v : \mathsf{Fml}^* \to 2^X$ which acts on the Boolean connectives in the usual way; its action with respect to $\boxbar$ is given by

$$x \in v(\boxbar\varphi) \iff v(\varphi) \subseteq S(x), \tag{4.3}$$

which may be interpreted as φ *is sufficient for accessibility from* x if and only if $v(\varphi) \subseteq S(x)$, in other words,

$$\text{Whenever } y \models_v \varphi, \text{ then } xSy.$$

Assuming the usual interpretation of $\Box$, it is easy to see that

$$\langle X, S, v\rangle, x \models \boxbar\varphi \iff \langle X, X^2 \setminus S, v\rangle, x \models \Box\neg\varphi, \tag{4.4}$$

so that axiomatization, completeness etc. of $K^{\star}$ are reducible to the corresponding properties of K. Thus, K^* has all the positive as well as the negative qualities of K. Both K and $K^{\star}$ are lacking in expressive power, and "necessity and sufficiency split the modal theory into two dual branches each of which spreads over less than a half of the Boolean realm" [13].

Independently of earlier work by Goldblatt [14], van Benthem [1], Humberstone [20] and others, members of the logic group at Sofia University presented a bimodal logic $K^{\sim}$ which unified the two approaches. Its modal operators are the normal modality $\Box$, the operator $\boxbar$ satisfying (4.1), (4.2) and the condition that the auxiliary operator $[U]\varphi \stackrel{\text{df}}{=} \Box\varphi \wedge \boxbar\neg\varphi$ is an S5 modality [13]. Its algebraic semantics were presented in [9], and a somewhat simplified version in [4].

While the possibility algebras are the algebraic models of the logic K and the sufficiency algebras are the algebraic models of its sufficiency counterpart K^*, both are limited in their powers of expression if considered separately. For example, $\langle 2^X, \langle R\rangle\rangle$ can express reflexivity by $Y \subseteq \langle R\rangle(Y)$ but it cannot express irreflexivity of R. On the other hand, $\langle 2^X, [[R]]\rangle$ can express irreflexivity by $[[R]](Y) \subseteq -Y$, but not reflexivity. Neither $\langle 2^X, \langle R\rangle\rangle$ nor $\langle 2^X, [[R]]\rangle$ can express antisymmetry on its own, but together they can [6]. Other results of definability in extended modal languages can be found in [16].

In this section we show that condition DO_2 in the definition of a doubly ordered set requires both a possibility and a sufficiency operator to be modally definable.

If $F = \langle X, R_1, \ldots, R_n\rangle$ and $F' = \langle X', R'_1 \ldots, R'_n\rangle$ are binary frames, a mapping $f : X \to X'$ is a *bounded morphism* if

BM$_1$. xR_iy implies $f(x)R'_if(y)$ for all $1 \leq i \leq n$ and $x, y \in X$.
BM$_2$. If $f(x)R'_iy'$, then there exists some $y \in X$ such that xR_iy and $f(y) = y'$.

$f : X \to X'$ is a *co–bounded morphism* if

cBM$_1$. $x(-R_i)y$ implies $f(x)(-R'_i)f(y)$ for all $1 \leq i \leq n$ and $x, y \in X$.
cBM$_2$. If $f(x)(-R'_i)y'$, then there is some $y \in X$ such that $x(-R_i)y$ and $f(y) = y'$.

In the proof of the main theorem, we shall use (part of) the famous Goldblatt–Thomason Theorem and its analogue for sufficiency structures:

Lemma 3

1. *[15, Theorem 3] If a class of frames is definable by a possibility operator, then it is closed under bounded morphisms.*
2. *[5, Proposition 9] If a class of frames is definable by a sufficiency operator, then it is closed under co–bounded morphisms.*

Theorem 3

1. *The class of doubly ordered frames is not definable by a possibility logic.*
2. *The class of doubly ordered frames is not definable by a sufficiency logic.*
3. *The class of doubly ordered frames is definable by a mixed logic.*

Proof 1. By Lemma 3(1) it is enough to show that the class is not closed under bounded morphisms. Let $\mathcal{F} = \langle X, R_1, R_2\rangle$ be a frame such that $X = \{x, y, z\}$, $R_1 = 1' \cup \{\langle x, y\rangle\}$, and $R_2 = 1' \cup \{\langle x, z\rangle\}$; then, $\mathcal{F}$ is a doubly ordered frame. Next, let $\mathcal{F}' = \langle Y, S_1, S_2\rangle$, where $Y = \{s, t\}$, $S_1 = 1' \cup \{\langle s, t\rangle\}$, and $S_2 \stackrel{\text{df}}{=} S_1$; observe that $\mathcal{F}'$ is not doubly ordered. Let $f : X \twoheadrightarrow Y$ be defined by $f(x) = s, f(y) = f(z) = t$ (Fig. 1). Clearly, f preserves R_1 and R_2, and thus, it satisfies BM_1.

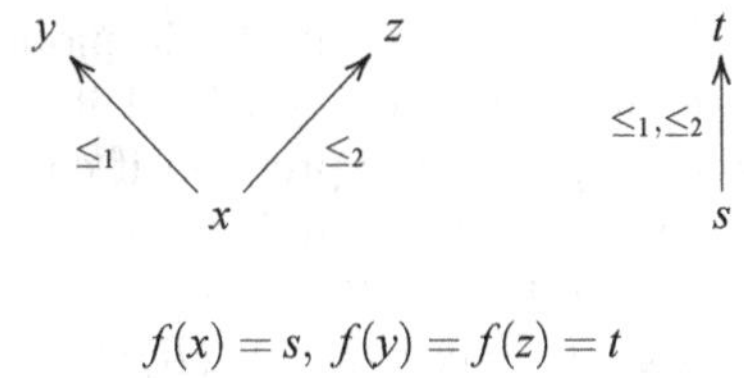

Fig. 1. Doubly ordered frames are not definable by possibility operators

For BM_2, let $f(u)S_1v$. We need to find some $w \in X$ such that uR_1w and $f(w) = v$. If $v = s$, then $x = u$. If $v = s$, then set $w = x$, if $v = t$, then set $w = y$. If $f(u) = t$, then $v = t$, and the reflexivity of R_1 gives the result. For R_2 the procedure is analogous, using z instead of y.

Thus $\mathcal{F}'$ is a bounded image of a doubly ordered frame. On the other hand, $s \neq t$ implies that $\mathcal{F}'$ is not doubly ordered.

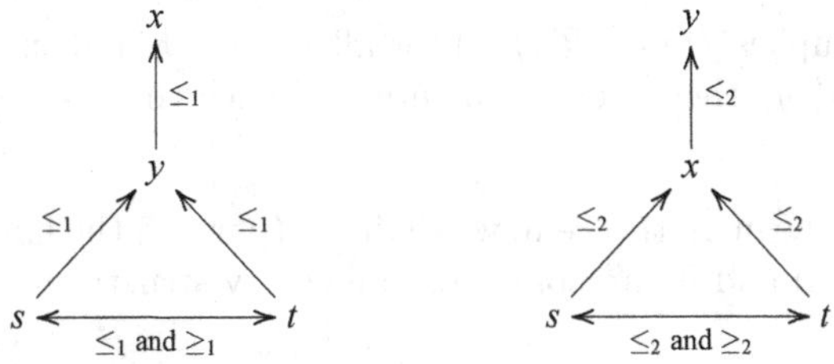

Fig. 2. Doubly ordered frames are not definable by sufficiency operators

2. By Lemma 3(2) it is enough to show that the class of doubly ordered frames is not closed under co–bounded morphisms. Let $X = \{x, y\}$, $X' = \{x, y, s, t\}$ with the quasiorders shown in Fig. 2. Then, $\langle X, \leq_1, \leq_2 \rangle$ is a doubly ordered frame, while $\langle X', \leq_1, \leq_2 \rangle$ is not. On the other hand, the mapping $f : X \to X'$ defined by $f(x) = x,\ f(y) = y$ is a co–bounded morphism.

3. As the properties of quasi–orders are definable in a logic with a possibility operator, all we need to express is DO_2. Let $\mathcal{F} = \langle X, \leq_1, \leq_2 \rangle$ be a frame, where $\leq_1, \leq_2$ are quasiorders. Consider the statement

$$(\forall Y \subseteq X)\langle \leq_1 \rangle([[\geq_2]](-Y) \cap Y) \subseteq Y. \tag{4.5}$$

$DO_2 \Rightarrow (4.5)$: Suppose that $\mathcal{F}$ is doubly ordered, i.e. $\mathcal{F}$ satisfies DO_2. Assume that there is some $Y \subseteq X$ such that $x \in \langle \leq_1 \rangle([[\geq_2]](-Y) \cap Y)$ and $x \notin Y$. Then, $\uparrow_1 x \cap [[\geq_2]](-Y) \cap Y \neq \emptyset$, so there is some $y \in X$ such that $x \leq_1 y$ and $y \in [[\geq_2]](-Y) \cap Y$. Now, $y \in [[\geq_2]](-Y)$ implies that $-Y \subseteq \downarrow_2 y$, and thus, $x \notin Y$ implies $x \leq_2 y$. DO_2 implies $x = y$ which contradicts $y \in Y, x \notin Y$.

(4.5) $\Rightarrow$ DO$_2$: Let $x, y \in X$, $x \leq_1 y$, $x \leq_2 y$, and assume that $x \neq y$. Set $Y \stackrel{\text{df}}{=} X \setminus \{x\}$. Substituting $X \setminus \{x\}$ for Y we obtain the following equivalent conditions:

$$\langle \leq_1 \rangle([[\geq_2]](\{x\}) \cap (X \setminus \{x\})) \subseteq X \setminus \{x\},$$
$$x \notin \langle \leq_1 \rangle([[\geq_2]](\{x\}) \cap (X \setminus \{x\}),$$
$$\uparrow_1 x \cap ([[\geq_2]](\{x\}) \cap (X \setminus \{x\}) = \emptyset.$$

Thus, $y \notin [[\geq_2]](\{x\}) \cap (X \setminus \{x\})$, and therefore, $y \notin [[\geq_2]](\{x\})$, since $x \neq y$ by our assumption. On the other hand,

$$y \notin [[\geq_2]](\{x\}) \iff x \notin (\geq_2)(y) \iff y \not\geq_2 x,$$

contradicting $x \leq_2 y$. □

5 Representability of Lattice Frames

Urquhart [31] proved that every doubly ordered frame endowed with a suitable topology can be embedded into the dual frame of its dual lattice. We show below on a first order level that his conditions suffice to prove that a suitably defined lattice frame can be embedded into the canonical frame of its complex algebra.

A *lattice frame* is a doubly ordered frame $\langle X, \leq_1, \leq_2 \rangle$ which satisfies the following conditions:

LF$_1$. Each element of X is below a $\leq_1$ maximal one and a $\leq_2$ maximal one,
LF$_2$. $x \not\leq_1 y \Rightarrow (\exists z)[y \leq_1 z \text{ and } (\forall w)(x \leq_1 w \Rightarrow z \not\leq_2 w)]$,
LF$_3$. $x \not\leq_2 y \Rightarrow (\exists z)[y \leq_2 z \text{ and } (\forall w)(x \leq_2 w \Rightarrow z \not\leq_1 w)]$.

LF$_2$ and LF$_3$ are the conditions given by Urquhart [31] for lattices of finite length. In such lattices, they guarantee embeddability of X into $\mathfrak{Cf}\mathfrak{Cm}(X)$. Indeed, they hold in all canonical frames:

Theorem 4. *If L is a lattice, then the structure $\langle X_L, \leq_1, \leq_2 \rangle$ is a lattice frame.*

Proof. LF$_1$: By Zorn's Lemma, each filter (ideal) is contained in a maximal one.

LF$_2$ Assume that LF$_2$ is not true. Then,

$$(\exists x, y)[x \not\leq_1 y \text{ and } (\forall z)(y \leq_1 z \Rightarrow (\exists w)(x \leq_1 w \text{ and } z \leq_2 w))]. \tag{5.1}$$

Let $x, y \in X$ witness (5.1). Since $x_1 \not\subseteq y_1$, there is some $a \in x_1, a \notin y_1$. Thus, $\downarrow_\leq a \cap y_1 = \emptyset$, and so there is a maximal pair z such that $y_1 \subseteq z_1$ and $a \in z_2$. The assumption (5.1) implies that there is a maximal pair w such that $x_1 \subseteq w_1$ and $z_2 \subseteq w_2$. Since w is a maximal pair, $w_1 \cap w_2 = \emptyset$ which contradicts $a \in x_1 \cap z_2$.

LF_3: This is shown similarly: Assume that LF_3 is not true. Then,

$$(\exists x, y)[x \not\leq_2 y \text{ and } (\forall z)(y \leq_2 z \Rightarrow (\exists w)(x \leq_2 w \text{ and } z \leq_1 w))]. \tag{5.2}$$

Since $x_2 \not\subseteq y_2$, there is some $a \in x_2, a \notin y_2$. Thus, $\uparrow_{\leq} a \cap y_2 = \emptyset$, and so there is a maximal pair z such that $y_2 \subseteq z_2$ and $a \in z_1$. The assumption (5.2) implies that there is a maximal pair w such that $x_2 \subseteq w_2$ and $z_1 \subseteq w_1$. Since w is a maximal pair, $w_1 \cap w_2 = \emptyset$ which contradicts $a \in x_2 \cap z_1$. □

We can now show the representation theorem for lattice frames. In preparation we prove a lemma:

Lemma 4. *Let L_X be the lattice of the l–stable sets of X. Define $k_1, k_2 : X \to 2^{L_X}$ by $k_1(x) \stackrel{\text{df}}{=} \{Y \in L_X : x \in Y\}$, and $k_2(x) \stackrel{\text{df}}{=} \{Y \in L_X : \uparrow_2 x \cap Y = \emptyset\}$. Then, $Y \in k_2(x)$ if and only if $x \in r(Y)$, and*

1. *$k_1(x)$ is the filter of 2^{L_X} generated by $\uparrow_1 x$.*
2. *$k_2(x)$ is the ideal of 2^{L_X} generated by $l(\uparrow_2 x)$.*

Proof. 1. Clearly, $k_1(x)$ is a filter of 2^{L_X}. All that remains to be shown is that $\uparrow_1 x \in L_X$, i.e. that $[\leq_1]\langle\leq_2\rangle(\uparrow_1 x) = \uparrow_1 x$.

"$\subseteq$": We show the contrapositive: Suppose that $x \not\leq_1 y$. By LF_2, there is some $z \in X$ such that $y \leq_1 z$ and $\uparrow_1 x \cap \uparrow_2 z = \emptyset$. This implies that $\uparrow_1 y \not\subseteq \langle\leq_2\rangle(\uparrow_1 x)$.

"$\supseteq$": This follows from the fact that $[\leq_1]\langle\leq_2\rangle$ is a closure operator on the $\leq_1$–increasing sets.

Observe that this implies that the smallest l–stable set containing x is $\uparrow_1 x$ for any $x \in X$.

2. Clearly, $k_2(x)$ is an ideal of 2^{L_X}. Since $\uparrow_2 x$ is $\leq_2$ – increasing, $l(\uparrow_2 x)$ is l–stable by [23, Lemma 2.8.3(b)]. Finally, suppose that $Y \in k_2(x)$; it is enough to show that $Y \subseteq l(\uparrow_2 x)$. Now, $Y \in k_2(x)$ implies that $Y \cap \uparrow_2 x = \emptyset$, i.e. $Y \subseteq -\uparrow_2 x$. Therefore, $[\leq_1](Y) \subseteq [\leq_1](-\uparrow_2 x) = l(\uparrow_2 x)$, and the facts that $Y = [\leq_1]\langle\leq_2\rangle(Y)$ and $\leq_1$ is transitive imply that $Y = [\leq_1](Y)$. □

Theorem 5. *Let X be a lattice frame. Then, X is embeddable into $\mathfrak{Cf}\,\mathfrak{Cm}(X)$.*

Proof. Let L_X be the lattice of the l–stable sets of X. Define k_1, k_2 as in Lemma 4, and set $k(x) \stackrel{\text{df}}{=} \langle k_1(x), k_2(x)\rangle$. We shall show that

1. For all $x, y \in X$, $x \leq_i y$ if and only if $k_i(x) \subseteq k_i(y)$, $i = 1, 2$.
2. k is injective.
3. $k(x)$ is a maximal pair of $\mathfrak{Cm}(X)$.

Preservation of $\leq_1$ and $\leq_2$ is immediate. Conversely, let $k_1(x) \subseteq k_1(y)$. Since $k_1(x)$ is the filter of L_X generated by $\uparrow_1 x$, and $k_1(y)$ is the filter of L_X generated by $\uparrow_1 y$, we have $\uparrow_1 y \subseteq \uparrow_1 x$ which implies $x \leq_1 y$. Next, suppose that $x \not\leq_2 y$. By LF_3, there is some $z \in X$ such that $y \leq_2 z$ and $\uparrow_2 x \cap \uparrow_1 z = \emptyset$. Then, $\uparrow_1 z \in k_2(x)$ and $\uparrow_1 z \notin k_2(y)$.

For injectivity, let $x \neq y$ and assume $k(x) = k(y)$, i.e. $k_1(x) = k_1(y)$ and $k_2(x) = k_2(y)$. Then, by Lemma 4(1), $k_1(x) = k_1(y)$ implies $\uparrow_1 x = \uparrow_1 y$, i.e.

$x \leq_1 y$ and $y \leq_1 x$. Since X is doubly ordered we may suppose w.l.o.g. that $x \not\leq_2 y$. By LF_3 there is some z such that $y \leq_2 z$ and $\uparrow_2 x \cap \uparrow_1 z = \emptyset$. Then, $x \in r(\uparrow_1 z)$ and $y \notin \uparrow_1 z$, contradicting $k_2(x) = k_2(y)$.

Clearly, $k_1(x) \cap k_2(x) = \emptyset$. All that is left to show is that $k(x)$ is a maximal pair. Assume that F is a filter of L_X strictly containing $k_1(x)$ and $F \cap k_2(x) = \emptyset$. Let $Y \in F \setminus k_1(x)$. Since $\uparrow_1 x \in k_1(x)$ and F is a filter, it follows that $Z \stackrel{\mathrm{df}}{=} Y \cap \uparrow_1 x \in F$ and $Z \subseteq \uparrow_1 x$. Then, $t \in Z$ implies $x \leq_1 t$, and thus, $x \not\leq_2 t$ for all $t \in Z$. By the assumption we have $Z \notin k_2(x)$, and thus, $x \notin r(Z)$. Hence, $x \notin [\leq_2](-Z)$, and there is some z such that $x \leq_2 z$ and $z \in Z$. This contradicts $x \not\leq_2 t$ for all $t \in Z$, and thus, $k_1(x)$ is maximally disjoint from $k_2(x)$.

Finally, we show that $k_2(x)$ is maximally disjoint from $k_1(x)$. Suppose that I is an ideal of L_X which strictly contains $k_2(x)$, and let $Y \in I \setminus k_2(x)$. Then, $Y \not\subseteq l(\uparrow_2 x)$ by Lemma 4(2), and it follows that there is some $y \in Y$ with $y \notin l(\uparrow_2 x)$; hence, $\uparrow_1 y \cap \uparrow_2 \neq \emptyset$, and there is some t such that $y \leq_1 t$ and $x \leq_2 t$. Since $y \in Y$, we have $\uparrow_1 t \subseteq \uparrow_1 y \subseteq Y$, and $Y \in I$ implies that $\uparrow_1 t \in I \setminus k_2(x)$; therefore, $l(\uparrow_2 x) \vee_X \uparrow_1 t \in I$. Now,

$$l(\uparrow_2 x) \vee_X \uparrow_1 t = [\leq_1](- \uparrow_2 x) \vee_X \uparrow_1 t = [\leq_1]\langle\leq_2\rangle[\leq_1]((- \uparrow_2 x) \cup \uparrow_1 t) \in I.$$

Therefore, $x \in [\leq_1]\langle\leq_2\rangle([\leq_1](- \uparrow_2 x) \cup \uparrow_1 t)$

$$\begin{aligned}
&\Longleftrightarrow \uparrow_1 x \subseteq \langle\leq_2\rangle([\leq_1](- \uparrow_2 x) \cup \uparrow_1 t) \\
&\Longleftrightarrow (\forall y)[x \leq_1 y \Rightarrow (\exists z)(y \leq_2 z \text{ and } z \in ([\leq_1](- \uparrow_2 x) \cup \uparrow_1 t))] \\
&\Longleftrightarrow (\forall y)[x \leq_1 y \Rightarrow (\exists z)(y \leq_2 z \text{ and } [(\forall u)(z \leq_1 u \Rightarrow x \not\leq_2 u) \text{ or } t \leq_1 z])], \\
&\Longleftrightarrow (\forall y)[x \leq_1 y \Rightarrow (\exists z)(y \leq_2 z \text{ and } [\uparrow_1 z \cap \uparrow_2 x = \emptyset \text{ or } t \leq_1 z])]
\end{aligned}$$

For the right hand side, we consider two cases:

1. $x = y$: Then, setting $z = t$, we have $x \leq_2 z$ by the property of t, and $t \leq_1 z$, so that x is in the right hand side.
2. $x \lneq_1 y$: Then, $x \not\leq_2 y$, and LF_3 implies that there is some z such that $\uparrow_2 x \cap \uparrow_1 z = \emptyset$.

Thus, the right hand side is fulfilled for all $x \leq_1 y$, and it follows that $x \in l(\uparrow_2 x) \vee_X \uparrow_1 t \in I$. Hence, $I \cap k_1(x) \neq \emptyset$. □

6 Summary and Outlook

In the present article we have filled a gap in the project of discrete dualities for various classes of general bounded lattices by defining a general lattice frame and showing that it is embeddable into the canonical frame of its Urquhart complex algebra. We have also shown that doubly ordered frames are not definable by possibility or sufficiency operators alone, but require both.

Future work will include extending the frames by relations corresponding to various modal logics, and show or disprove that a frame duality exists. In view of the representation theorems for lattice based modal (i.e., possibility, necessity, sufficiency, and dual sufficiency, respectively) algebras presented by Orlowska and Vakarelov [26], the existence of representation theorem for modal frames will lead to discrete duality and duality via truth for the corresponding logics.

The two major groups of logics considered in philosophical logic are substructural logics and paraconsistent logics. For several substructural logics we have a discrete duality, however, we have not yet considered dualities for paraconsistent logics. This will also be a task for future work.

Acknowledgements. We are grateful to Alasdair Urquhart for stimulating discussions. We also thank the anonymous referees for careful reviewing and helpful suggestions. I. Düntsch gratefully acknowledges support by the National Natural Science Foundation of China, Grant No. 61976053.

References

1. van Benthem, J.: Minimal deontic logics (Abstract). Bull. Sect. Logic **8**, 36–42 (1979)
2. Cayley, A.: On the theory of groups, as depending on the symbolic equation $\theta^n = 1$. Philos. Mag. Series **4**(7), 40–47 (1854)
3. Demri, S., Orłowska, E.: Incomplete Information: Stucture, Inference, Complexity. Monographs in Theoretical Computer Science, 1st edn. Springer, Heidelberg (2002)
4. Düntsch, I., Dzik, W., Orłowska, E.: On the semilattice of modal operators and decompositions of the discriminator. In: Madarasz, J., Szekely, G. (eds.) Hajnal Andréka and István Németi on Unity of Science: From Computing to Relativity Theory Through Algebraic Logic. Outstanding Contributions to Logic. Springer (2019, to appear)
5. Düntsch, I., Orlowska, E.: Beyond modalities: sufficiency and mixed algebras. In: Orlowska, E., Szalas, A. (eds.) Relational Methods for Computer Science Applications. Studies in Fuzziness and Soft Computing, vol. 65. Physica, Heidelberg (2001). https://doi.org/10.1007/978-3-7908-1828-4_16. MR1858531
6. Düntsch, I., Orłowska, E.: Boolean algebras arising from information systems. Ann. Pure Appl. Logic **127**, 77–98 (2004). MR2071169
7. Düntsch, I., Orłowska, E.: A discrete duality between apartness algebras and apartness frames. J. Appl. Non-Classical Logics **18**, 213–227 (2008). MR2462235
8. Düntsch, I., Orłowska, E.: An algebraic approach to preference relations. In: de Swart, H. (ed.) RAMICS 2011. LNCS, vol. 6663, pp. 141–147. Springer, Heidelberg (2011). https://doi.org/10.1007/978-3-642-21070-9_12
9. Düntsch, I., Orłowska, E., Tinchev, T.: Mixed algebras and their logics. J. Appl. Non-Classical Logics **27**(3–4), 304–320 (2018). MR3779213
10. Düntsch, I., Vakarelov, D.: Region-based theory of discrete spaces: a proximity approach. Ann. Math. Artif. Intell. **49**, 5–14 (2007). MR2348380
11. George, E., Erné, M., Koslowski, J., Melton, A., Strecker, G.: A primer on Galois connections. Ann. N.Y. Acad. Sci. **704**(1), 103–125 (1993)
12. Ganter, B., Wille, R.: Formal Concept Analysis: Mathematical Foundations. Springer, Heidelberg (1999). https://doi.org/10.1007/978-3-642-59830-2
13. Gargov, G., Passy, S., Tinchev, T.: Modal environment for Boolean speculations. In: Skordev, D. (ed.) Mathematical Logic and Applications, pp. 253–263. Plenum Press, New York (1987)

14. Goldblatt, R.I.: Semantic analysis of orthologic. J. Philos. Logic **3**, 19–35 (1974)
15. Goldblatt, R.I., Thomason, S.K.: Axiomatic classes in propositional modal logic. In: Crossley, J.N. (ed.) Algebra and Logic. LNM, vol. 450, pp. 163–173. Springer, Heidelberg (1975). https://doi.org/10.1007/BFb0062855
16. Goranko, V.: Modal definability in enriched languages. Notre Dame J. Formal Logic **31**(1), 81–105 (1990)
17. Hartonas, C.: Discrete duality for lattices with modal operators. J. Logic Comput. **29**(1), 71–89 (2019)
18. Hartonas, C., Dunn, J.M.: Duality theorems for partial orders, semilattices, Galois connections and lattices. In: Preprint IULG-93-26, Indiana University Logic Group (1993)
19. Hartung, G.: A topological representation of lattices. Algebra Univers. **29**, 273–299 (1992)
20. Humberstone, I.L.: Inaccessible worlds. Notre Dame J. Formal Logic **24**, 346–352 (1983)
21. Orłowska, E., Rewitzky, I.: Discrete duality and its applications to reasoning with incomplete information. In: Kryszkiewicz, M., Peters, J.F., Rybinski, H., Skowron, A. (eds.) RSEISP 2007. LNCS (LNAI), vol. 4585, pp. 51–56. Springer, Heidelberg (2007). https://doi.org/10.1007/978-3-540-73451-2_7
22. Orłowska, E., Rewitzky, I., Düntsch, I.: Relational semantics through duality. In: MacCaull, W., Winter, M., Düntsch, I. (eds.) RelMiCS 2005. LNCS, vol. 3929, pp. 17–32. Springer, Heidelberg (2006). https://doi.org/10.1007/11734673_2
23. Orłowska, E., Rewitzky, I., Radzikowska, A.: Dualities for Structures of Applied Logics, Studies in Logic, vol. 56. College Publications, London (2015)
24. Orłowska, E., Radzikowska, A.M.: Relational representability for algebras of substructural logics. In: MacCaull, W., Winter, M., Düntsch, I. (eds.) RelMiCS 2005. LNCS, vol. 3929, pp. 212–224. Springer, Heidelberg (2006). https://doi.org/10.1007/11734673_17
25. Orłowska, E., Rewitzky, I.: Duality via truth: semantic frameworks for lattice-based logics. Logic J. IGPL **13**(4), 467–490 (2005)
26. Orłowska, E., Vakarelov, D.: Lattices with modal operators and lattice-based modal logic. In: Hájek, P., Valdés-Villanueva, L., Westerståhl, D. (eds.) Proceedings of the 12th International Congress of Logic, Methodology and Philosophy of Science, pp. 147–170 (2005)
27. Ploščica, M.: A natural representation of bounded lattices. Tatra Mt. Math. Publ. **5**, 75–88 (1995)
28. Priestley, H.A.: Representation of distributive lattices by means of ordered Stone spaces. Bull. Lond. Math. Soc. **2**, 186–190 (1970)
29. Stone, M.: The theory of representations for Boolean algebras. Trans. Amer. Math. Soc. **40**, 37–111 (1936)
30. Tehlikeli, S.: An alternative modal logic, internal semantics and external syntax (A philosophical abstract of a mathematical essay) (1985). manuscript
31. Urquhart, A.: A topological representation theorem for lattices. Algebra Univers. **8**, 45–58 (1978)

Group Announcement Logic with Distributed Knowledge

Rustam Galimullin[1(✉)], Thomas Ågotnes[2,3], and Natasha Alechina[1]

[1] University of Nottingham, Nottingham, UK
{rustam.galimullin,natasha.alechina}@nottingham.ac.uk
[2] Southwest University, Chongqing, China
[3] University of Bergen, Bergen, Norway
thomas.agotnes@uib.no

Abstract. Public announcement logic (PAL) is an extension of epistemic logic with dynamic operators that model the effects of all agents simultaneously and publicly acquiring the same piece of information. One of the extensions of PAL, group announcement logic (GAL), allows quantification over (possibly joint) announcements made by agents. In GAL, it is possible to reason about what groups can achieve by making such announcements. It seems intuitive that this notion of coalitional ability should be closely related to the notion of distributed knowledge, the implicit knowledge of a group. Thus, we study the extension of GAL with distributed knowledge, and in particular possible interaction properties between GAL operators and distributed knowledge. The perhaps surprising result is that there in fact are no interaction properties, contrary to intuition. We make this claim precise by providing a sound and complete axiomatisation of GAL with distributed knowledge.

Keywords: Distributed knowledge · Resolved distributed knowledge · Group announcement logic · Dynamic epistemic logic

1 Introduction

The main contribution of this paper is extending Group Announcement Logic (GAL) [1] with distributed knowledge [11], and a sound and complete axiom system for the resulting logic GALD.

Our motivation for studying this combination of modalities is twofold. First, extending epistemic logics with quantifiers over information-changing actions [2,7,8,16], of which GAL is a representative, with group knowledge modalities is an open problem. Second, the quest for a better understanding of both types of logical operators and their interaction is interesting in its own right. Distributed knowledge is often intuitively understood as closely related to the knowledge the agents would arrive at if they could communicate their individual knowledge to each other. Deep analyses of this intuition [4,17,20] shows that it is not always accurate, but when we started investigating GAL with distributed knowledge operator D we nevertheless expected to find

P. Blackburn et al. (Eds.): LORI 2019, LNCS 11813, pp. 98–111, 2019.
https://doi.org/10.1007/978-3-662-60292-8_8

non-trivial interaction axioms. We consider several plausible candidates for such interaction axioms, and show that none of them are actually valid. Then we show that in fact *there are no interaction axioms at all*: the axiom system obtained by the independent combination of axioms for epistemic logic with distributed knowledge and GAL is *complete*. This is contrary to intuition, and therefore an interesting result.

We also consider the relationship between *resolved distributed knowledge* [4] and distributed knowledge in the context of announcement logics, and give some preliminary results on their relative expressive power.

The paper is organised as follows. In the next section we briefly review the technical background. In Sect. 3 we look at some potential interaction axioms relating group announcements and group knowledge. In Sect. 4 we present a Hilbert-style axiomatic system for group announcement logic with distributed knowledge, and show that it is sound and complete. Resolved distributed knowledge is discussed in Sect. 5, before we conclude in Sect. 6.

2 Background

In this section, we introduce the necessary background information on epistemic and group announcement logic.

2.1 Languages

The language of GALD is defined relative to a finite set of agents A and a countable set of propositional variables P. Below we also define a positive fragment of this language.

Definition 1. *The* language of group announcement logic with distributed knowledge *and its* positive fragment *are defined by the following BNFs:*

$$\mathcal{L}_{GALD} \quad \varphi ::= p \mid \neg\varphi \mid \varphi \wedge \varphi \mid K_a\varphi \mid D_G\varphi \mid [\varphi]\varphi \mid [G]\varphi$$

$$\mathcal{L}_{GALD^+} \quad \varphi^+ ::= p \mid \neg p \mid \varphi^+ \wedge \varphi^+ \mid \varphi^+ \vee \varphi^+ \mid K_a\varphi^+ \mid D_G\varphi^+ \mid [\neg\varphi^+]\varphi^+ \mid [G]\varphi^+$$

where $p \in P$, $a \in A$ *and* $G \subseteq A$.

The sublanguage only containing K_a modalities in addition to the propositional operators and variables is referred to as $\mathcal{L}_{EL}$ (epistemic logic) [11], with addition of D_G it becomes $\mathcal{L}_{ELD}$, $\mathcal{L}_{EL}$ with announcements $[\varphi]$ is referred to as $\mathcal{L}_{PAL}$ [19], with D_G as $\mathcal{L}_{PALD}$, and $\mathcal{L}_{GALD}$ without D_G is $\mathcal{L}_{GAL}$. The intuitive meaning of formulas is as follows: $K_a\varphi$ means that agent a knows that φ; $D_G\varphi$ means that G has distributed knowledge of φ (φ is true in the set of states that all agents in G consider possible); $[\varphi]\psi$ means that if φ is true, then after it is announced (and everyone's knowledge updated by removing states not satisfying φ), ψ is true; $[G]\varphi$ means that after any joint announcement by agents in G of formulas they know, φ is true. The quantification in the latter modality is over conjunctions of formulas of $\mathcal{L}_{ELD}$ prefixed by K_a for $a \in G$.

Duals are defined as $\widehat{K}_a\varphi := \neg K_a\neg\varphi$, $\langle\varphi\rangle\psi := \neg[\varphi]\neg\psi$, and $\langle G\rangle\varphi := \neg[G]\neg\varphi$.

2.2 Models and Bisimulation

Definition 2. *An* epistemic model *M is a triple $(S, \sim, V)$, where S is a non-empty set of states, $\sim: A \to 2^{S \times S}$ assigns to each agent an equivalence relation, and $V : P \to 2^S$ is a valuation. If necessary, we refer to the elements of the tuple as S^M, $\sim^M$, and V^M.*

A model M with a designated state $s \in S$ is called a pointed model *and denoted by M_s.*

Model M is called finite *if S is finite. Also, we write $M \subseteq N$ if $S^M \subseteq S^N$, $\sim^M$ and V^M are results of restricting $\sim^N$ and V^N to S^M, and call M a* submodel *of N.*

An updated model *M^φ is $(S^\varphi, \sim^\varphi, V^\varphi)$, where $S^\varphi = \{s \in S \mid M_s \models \varphi\}$, $\sim^\varphi_a = \sim_a \cap (S^\varphi \times S^\varphi)$ for all $a \in A$, and $V^\varphi(p) = V(p) \cap S^\varphi$ ($\models$ is given in Definition 5).*

For a group $G \subseteq A$, $\sim_G$ denotes $\bigcap_{a \in G} \sim_a$.

Definition 3 (Bisimulation). *Let $M = (S^M, \sim^M, V^M)$ and $N = (S^N, \sim^N, V^N)$ be two models. A non-empty binary relation $Z \subseteq S^M \times S^N$ is called a* bisimulation *if and only if for all $s \in S^M$ and $u \in S^N$ with $(s, u) \in Z$:*

- *for all $p \in P$, $s \in V^M(p)$ if and only if $u \in V^N(p)$;*
- *for all $a \in A$ and all $t \in S^M$: if $s \sim^M_a t$, then there is a $v \in S^N$ such that $u \sim^N_a v$ and $(t, v) \in Z$;*
- *for all $a \in A$ and all $v \in S^N$: if $u \sim^N_a v$, then there is a $t \in S^M$ such that $s \sim^M_a t$ and $(t, v) \in Z$.*

If there is a bisimulation between models M and N linking states s and u, we say that M_s and N_u are bisimilar, and write $M_s \rightleftarrows N_u$.

Definition 4 (Bisimulation contraction). *Let $M = (S, \sim, V)$ be an epistemic model. The* bisimulation contraction *of M is the model $\|M\| = (\|S\|, \|\sim\|, \|V\|)$, where $\|S\| = \{[s] \mid s \in S\}$ and $[s] = \{t \in S \mid M_s \leftrightarrows N_t\}$, $[s]\|\sim\|_a[t]$ if and only if $\exists s' \in [s]$, $\exists t' \in [t]$ such that $s' \sim_a t'$ in M, and $[s] \in \|V\|(p)$ if and only if $\exists s' \in [s]$ such that $s' \in V(p)$.*

It is a known result that $M_s \leftrightarrows \|M\|_{[s]}$ [15].

2.3 Semantics of GALD

Let us denote by ψ_G a formula of the form $\bigwedge_{a \in G} K_a \psi_a$ where $\psi_a \in \mathcal{L}_{EL}$. We refer to this fragment as $\mathcal{L}^G_{EL}$. We will also write $\top_G$ to denote $\bigwedge_{a \in G} K_a(p \vee \neg p)$.

Definition 5 (Semantics of GALD). *Let M_s be a pointed epistemic model. The* semantics *of GALD is defined as follows (boolean cases are as usual and we omit them):*

$$M_s \models K_a\varphi \text{ iff } \forall t \in S : s \sim_a t \text{ implies } M_t \models \varphi$$
$$M_s \models D_G\varphi \text{ iff } \forall t \in S : s \sim_G t \text{ implies } M_t \models \varphi$$
$$M_s \models [\psi]\varphi \text{ iff } M_s \models \psi \text{ implies } M^\psi_s \models \varphi$$
$$M_s \models [G]\varphi \text{ iff } \forall \psi_G : M_s \models [\psi_G]\varphi$$

Definition 6 (Validity and satisfiability). *φ is* valid *($\models \varphi$) if and only if for any pointed model M_s it holds that $M_s \models \varphi$. φ is* satisfiable *if and only if there is some M_s such that $M_s \models \varphi$.*

2.4 The Positive Fragment

Positive formulas can be considered as a particularly well behaved fragment of public announcement logic [9]. In particular, they remain true after an announcement.

Definition 7. *A formula φ is said to be* preserved *under submodels if and only if $M_s \models \varphi$ implies $N_s \models \varphi$ for any pointed models M_s and N_s such that $N_s \subseteq M_s$.*

Proposition 1. *Formulas of $\mathcal{L}_{GALD^+}$ are preserved under submodels.*

Proof. Let $M = (S^M, \sim^M, V^M)$ and $N = (S^N, \sim^N, V^N)$ be models such that $s \in S^M$, $s \in S^N$, and $N_s \subseteq M_s$. Boolean cases, case $K_a\varphi^+$, and case $[\neg\psi^+]\varphi^+$ are proved in [9, Proposition 8]. We show the remaining two cases $D_G\varphi^+$ and $[G]\varphi^+$.

Induction hypothesis. If $M_s \models \varphi^+$, then $N_s \models \varphi^+$.

Case $D_G\varphi^+$. Let $M_s \models D_G\varphi^+$. By the definition of semantics, this is equivalent to the fact that $\forall t \in S^M$: $s \sim_G t$ implies $M_t \models \varphi^+$. The latter implies that $\forall t \in S^N$: $s \sim_G t$ implies $M_t \models \varphi^+$. By the induction hypothesis, we have that $\forall t \in S^N$: $s \sim_G t$ implies $N_t \models \varphi^+$, which is equivalent to $N_s \models D_G\varphi^+$.

Case $[G]\varphi^+$. Assume towards a contradiction that $M_s \models [G]\varphi^+$ and $N_s \not\models [G]\varphi^+$. By the duality of group announcements, this is equivalent to $N_s \models \langle G\rangle\neg\varphi^+$, and by the definition of semantics, the latter is equivalent to $\exists\psi_G : N_s \models \langle\psi_G\rangle\neg\varphi^+$, which equals to $N_s \models \psi_G$ and $N_s^{\psi_G} \not\models \varphi^+$. Now observe that $N_s^{\psi_G} \subseteq N_s \subseteq M_s$. From that and the contraposition of the induction hypothesis, it follows that $M_s \not\models \varphi^+$. However, $M_s \models [G]\varphi^+$ implies that $M_s \models [\top_G]\varphi^+$. Finally, $M_s \models [\top_G]\varphi^+$ is equivalent to $M_s \models \varphi^+$, which contradicts $M_s \not\models \varphi^+$.

3 Ability, Announcements, and Group Knowledge

Distributed knowledge is often described as potential individual (or even common) knowledge, that the individual members of the group can establish "through communication" or by "pooling their knowledge together". However, this intuition is in fact not correct [4]. For example, a group can have distributed knowledge of a formula of the form $p \wedge \neg K_a p$ (sometimes called a *Moore sentence* [18]), which can never become individual knowledge in a group that contains agent a [4]. Nevertheless, that doesn't mean that there are no interaction properties between group announcements and group knowledge. Indeed, the natural intuition is that there is. In this section, we consider possible interaction axioms relating group announcements and group knowledge.

It is known that the following potential axioms are not valid [1]:

- $\langle G\rangle\varphi \to D_G\langle G\rangle\varphi$
- $D_G\langle G\rangle\varphi \to \langle G\rangle D_G\varphi$

It is also known that the following are valid:

- $\langle G\rangle D_G\varphi \to D_G\langle G\rangle\varphi$ (implied by Proposition 28 of [1] and the fact that knowledge de re implies knowledge de dicto)
- $D_G\langle G\rangle\varphi \to \langle G\rangle\varphi$ (distributed knowledge is veridical)

Consider weaker properties which involve ‘everybody knows’ operator E_G where $E_G\varphi := \bigwedge_{a\in G} K_a\varphi$. These properties encapsulate the intuition that distributed knowledge can be made explicit through public communication. It is known that the following is not valid:

- $D_G\varphi \to \langle G\rangle E_G\varphi$ (take $\varphi := p \wedge \neg K_a p$ where $a \in G$ [4])

The other direction also does not hold:

Fact 1. *$\langle G\rangle E_G\varphi \to D_G\varphi$ is not valid.*

Proof. Consider Fig. 1.

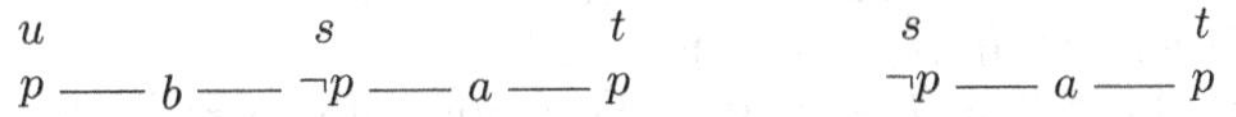

Fig. 1. Models M and $M^{\psi_{\{a,b\}}}$

Let $\varphi := K_b p \vee K_b\neg p$ and $\psi_{\{a,b\}} := K_a(p \to K_b p) \wedge K_b(p \vee \neg p)$. We have that $M_s \models \langle \psi_{\{a,b\}}\rangle E_{\{a,b\}}\varphi$, which is equivalent to $M_s \models \psi_{\{a,b\}}$ and $M_s^{\psi_{\{a,b\}}} \models E_{\{a,b\}}\varphi$. On the other hand, it is easy to verify that $M_s \not\models D_{\{a,b\}}\varphi$ as the only $\sim_{\{a,b\}}$-accessible state is s itself, and $M_s \not\models \varphi$.

In general, implicit knowledge in a group cannot be made explicit via public communication. However, there is an exception. Positive formulas can be made known on *bisimulation contracted models* (this restriction is not surprising given analysis in [17]).

Fact 2. *$D_G\varphi^+ \to \langle G\rangle E_G\varphi^+$ is valid on finite bisimulation contracted models.*

Proof. Let $M_s \models D_G\varphi^+$ for an arbitrary finite bisimulation contracted M_s. Since distributed knowledge is veridical, the latter implies $M_s \models \varphi^+$. Now let us a consider the maximally informative announcement by agents from G. Since M_s is finite and bisimulation contracted, each state in the model can be uniquely described by a characteristic formula. Moreover, disjunctions of these formulas correspond to sets of states. Agents from G can announce characteristic formulas that describe their equivalence classes and include s, i.e. $[s]_a \cap \ldots \cap [s]_b$ for $a, \ldots, b \in G$ (see [3,13] for details). In the resulting model $M_s^{\psi_G}$, relation $\sim_G$ on set of states S^{ψ_G} is universal. Moreover, since φ^+ is preserved under submodels, we have that $M_s^{\psi_G} \models E_G\varphi^+$, and, consequently, $M_s \models \langle G\rangle E_G\varphi^+$.

The restriction to finite bisimulation contracted models is essential in the previous proposition.

Fact 3. *$D_G\varphi^+ \to \langle G\rangle E_G\varphi^+$ is not valid.*

Proof. Consider the model in Fig. 2. It is easy to check that $M_s \models D_{\{a,b\}}p$ and $M_s \not\models \langle\{a,b\}\rangle E_{\{a,b\}}p$. Indeed, any announcement by a that preserves $\{s,t\}$ also preserves $\{u,v\}$. The same holds for agent b and sets $\{s,u\}$ and $\{t,v\}$.

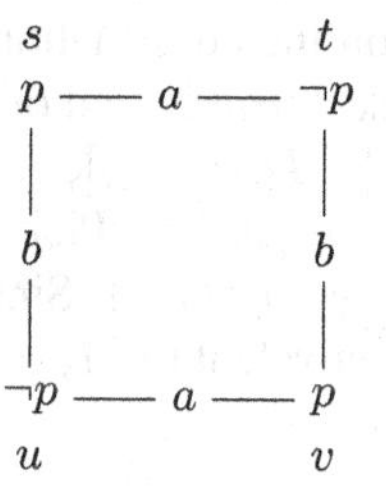

Fig. 2. Models M

4 Proof System for GALD

In this section, we provide a sound and complete axiomatisation for GALD. Our proofs follow general strategies of proofs from [1,5,22].

4.1 Axiomatisation of GALD

In order to provide the proof system, we first define necessity forms [14].

Definition 8. *Necessity forms are defined by the following grammar:*

$$\eta(\sharp) ::= \sharp \mid \varphi \rightarrow \eta(\sharp) \mid K_a\eta(\sharp) \mid D_G\eta(\sharp) \mid [\varphi]\eta(\sharp)$$

where $\varphi \in \mathcal{L}_{GALD}$, *and* $\sharp$ *has a unique occurrence in* $\eta(\sharp)$. *The result of substituting* φ *for* $\sharp$ *in* η *is denoted by* $\eta(\varphi)$.

Definition 9. *The axiomatisation of GALD is comprised of axiom systems for EL [11], PAL [10], GAL [1], and PALD [22].*

$(A0)$	*Propositional tautologies*	$(A11)$	$[\varphi]p \leftrightarrow (\varphi \rightarrow p)$
$(A1)$	$K_a(\varphi \rightarrow \psi) \rightarrow K_a\varphi \rightarrow K_a\psi$	$(A12)$	$[\varphi]\neg\psi \leftrightarrow (\varphi \rightarrow \neg[\varphi]\psi)$
$(A2)$	$K_a\varphi \rightarrow \varphi$	$(A13)$	$[\varphi](\psi \wedge \chi) \leftrightarrow ([\varphi]\psi \wedge [\varphi]\chi)$
$(A3)$	$K_a\varphi \rightarrow K_aK_a\varphi$	$(A14)$	$[\varphi]K_a\psi \leftrightarrow (\varphi \rightarrow K_a[\varphi]\psi)$
$(A4)$	$\neg K_a\varphi \rightarrow K_a\neg K_a\varphi$	$(A15)$	$[\varphi]D_G\psi \leftrightarrow (\varphi \rightarrow D_G[\varphi]\psi)$
$(A5)$	$D_G(\varphi \rightarrow \psi) \rightarrow D_G\varphi \rightarrow D_G\psi$	$(A16)$	$[\varphi][\psi]\chi \leftrightarrow [\varphi \wedge [\varphi]\psi]\chi$
$(A6)$	$D_G\varphi \rightarrow \varphi$	$(A17)$	$[G]\varphi \rightarrow [\psi_G]\varphi$
$(A7)$	$D_G\varphi \rightarrow D_GD_G\varphi$	$(R0)$	$\vdash \varphi \rightarrow \psi, \vdash \varphi \Rightarrow \vdash \psi$
$(A8)$	$\neg D_G\varphi \rightarrow D_G\neg D_G\varphi$	$(R1)$	$\vdash \varphi \Rightarrow \vdash K_a\varphi$
$(A9)$	$D_a\varphi \leftrightarrow K_a\varphi$	$(R2)$	$\vdash \varphi \Rightarrow \vdash [G]\varphi$
$(A10)$	$D_G\varphi \rightarrow D_H\varphi$, *if* $G \subseteq H$	$(R3)$	$\forall\psi_G : \vdash \eta([\psi_G]\varphi) \Rightarrow \vdash \eta([G]\varphi)$

We denote by GALD the smallest set that contains all instances of $A0$–$A17$ and is closed under $R0$–$R3$. Elements of GALD are called *theorems*.

Lemma 1. *Rule* $R3$ *is truth-preserving.*

Proof. The proof is by induction on the construction of η. Let M_s be a pointed epistemic model. We show only the case $D_H\eta(\sharp)$, and other cases are similar.

Case $D_H\eta(\sharp)$. Let $\forall\psi_G : M_s \models D_H\eta([\psi_G]\varphi)$. By the semantics this means that $\forall\psi_G, \forall t : s \sim_H t$ implies $M_t \models \eta([\psi_G]\varphi)$. Pick any t such that $s \sim_H t$. By the induction hypothesis we have $M_t \models \eta([G]\varphi)$. Since t was arbitrary, $\forall t : s \sim_H t$ implies $M_t \models \eta([G]\varphi)$, which is equivalent to $M_s \models D_H\eta([G]\varphi)$.

Theorem 1. *GALD is sound.*

Proof. Soundness of $R2$ is easy to show. The rest follows from the soundness of PALD [22], GAL [1], and Lemma 1.

4.2 Completeness

Following the technique from [22,23], we prove the completeness of GALD by making a detour through pre- and pseudo models, where distributed knowledge operators are treated as classic knowledge modalities.

Definition 10. *An epistemic pre-model is a tuple* $\mathcal{M} = (S, \sim, V)$, *where* $\sim$ *maps every agent* a *and every subset* $G \subseteq A$ *to an element of* 2^{2^S}. *A pre-model is called a pseudo-model (and is written* $\mathfrak{M}$*) if for all* a *it holds that* $\sim_{\{a\}} = \sim_a$, *and for all* $G, H \subseteq A$: *if* $G \subseteq H$, *then* $\sim_H \subseteq \sim_G$.

Next, we define theories that will be used for the construction of the canonical model.

Definition 11. *A set* x *of formulas of* $\mathcal{L}_{GALD}$ *is called a* theory, *if it contains all theorems and is closed under* $R0$ *and* $R3$. *A theory is* consistent *if for all* φ, *either* $\varphi \notin x$ *or* $\neg\varphi \notin x$. *A theory is called* maximal *if for all* φ, *either* $\varphi \in x$ *or* $\neg\varphi \in x$.

Theories are not required to be closed under $R1$ and $R2$ since this rules of inference, unlike $R0$ and $R3$, preserve only validity and not truth.

Lemma 2. *Let* x *be a theory, and* $\varphi, \psi \in \mathcal{L}_{GALD}$. *The following are theories:* $x + \varphi = \{\psi \mid \varphi \to \psi \in x\}$, $K_a x = \{\varphi \mid K_a\varphi \in x\}$, $D_G x = \{\varphi \mid D_G\varphi \in x\}$, *and* $[\varphi]x = \{\psi \mid [\varphi]\psi \in x\}$.

Proof. Cases for $x + \varphi$, $K_a x$, $[\varphi]x$ are proved in [5, Lemma 4.11]. Here we argue that $D_G x$ is a theory.

We need to show that $D_G x$ contains GALD and is closed under $R0$ and $R3$. Let $\varphi \in$ GALD. Then we also have that $D_G\varphi \in$ GALD by the necessitation of D_G, which is derivable in PALD [22]. Since x is a theory, and hence GALD $\subseteq x$, we have that $D_G\varphi \in x$, and $\varphi \in D_G x$. This establishes that GALD $\subseteq D_G x$.

Assume that $\varphi \to \psi, \varphi \in D_G x$. By $A5$ and $R0$ this implies that $D_G\psi \in x$, or, equivalently, $\psi \in D_G x$.

Suppose that $\forall\psi_G$: $\eta([\psi_G]\varphi) \in D_G x$. This means that $\forall\psi_G$: $D_G\eta([\psi_G]\varphi) \in x$, and from the fact that $D_G\eta(\sharp)$ is a necessity form, we conclude by $R3$ that $D_G\eta([G]\varphi) \in x$. Finally, by the definition of $D_G x$ we yield $\eta([G]\varphi) \in D_G x$.

Lemma 3. *For all consistent theories x, $\neg\varphi \notin x$ if and only if $x + \varphi$ is consistent.*

Lemma 4 (Theorem 2.5.2 of [14]). *Every consistent theory can be extended to a maximal consistent theory.*

Definition 12. *The canonical pseudo model is $\mathfrak{M}^C = (S^C, \sim^C, V^C)$, where $S^C = \{x \mid x$ is maximal consistent theory$\}$, $x \sim^C_a y$ if and only if $K_a x \subseteq y$, $x \sim^C_G y$ if and only if $D_H x \subseteq y$ and $H \subseteq G$, and $V^C(p) = \{x \in S^C \mid p \in x\}$.*

For the rest of the section, we employ the following strategy. First, we prove the truth lemma for the canonical pseudo model. Next, we unravel $\mathfrak{M}^C$ into the tree-like pre-model $\mathcal{M}^C$, which satisfies the same GALD formulas as $\mathfrak{M}^C$. After that, we fold $\mathcal{M}^C$ into the model M^C. Folding preserves trans-bisimulation, and hence we will be able to conclude the completeness of GALD.

Definition 13. *The* size *and* $[]$-depth *of $\varphi \in \mathcal{L}_{GALD}$ are defined as follows:*

$$\begin{array}{ll}
Size(p) = 1 & d_{[]}(p) = 0 \\
Size(\neg\varphi) = Size(K_a\varphi) = & d_{[]}(\neg\varphi) = d_{[]}(K_a\varphi) = \\
\quad = Size(D_G\varphi) = Size([G]\varphi) = & \quad = d_{[]}(D_G\varphi) = d_{[]}(\varphi) \\
\quad = Size(\varphi) + 1 & d_{[]}(\varphi \wedge \psi) = \max\{d_{[]}(\varphi), d_{[]}(\psi)\} \\
Size(\varphi \wedge \psi) = \max\{Size(\varphi), Size(\psi)\} + 1 & d_{[]}([\psi]\varphi) = d_{[]}(\psi) + d_{[]}(\varphi) \\
Size([\psi]\varphi) = Size(\psi) + 3 \cdot Size(\varphi) & d_{[]}([G]\varphi) = d_{[]}(\varphi) + 1
\end{array}$$

Definition 14 (Size Relation). *The binary relation $<^{Size}_{[]}$ between $\varphi, \psi \in \mathcal{L}_{GALD}$ is defined as follows:*
$\varphi <^{Size}_{[]} \psi$ iff $d_{[]}(\varphi) < d_{[]}(\psi)$, or $d_{[]}(\varphi) = d_{[]}(\psi)$ and $Size(\varphi) < Size(\psi)$. The relation is a well-founded strict partial order between formulas. Note that for all epistemic formulas ψ we have that $d_{[]}(\psi) = 0$.

Lemma 5. *Let $\varphi, \chi \in \mathcal{L}_{GALD}$.*

1. $\varphi <^{Size}_{[]} \neg\varphi$,
2. $\varphi <^{Size}_{[]} \varphi \wedge \psi$,
3. $\varphi <^{Size}_{[]} K_a\varphi$,
4. $\varphi <^{Size}_{[]} D_G\varphi$,
5. $\varphi \to p <^{Size}_{[]} [\varphi]p$,
6. $\varphi \to \neg[\varphi]\psi <^{Size}_{[]} [\varphi]\neg\psi$,
7. $[\varphi]\psi \wedge [\varphi]\chi <^{Size}_{[]} [\varphi](\psi \wedge \chi)$,
8. $[\varphi \wedge [\varphi]\chi]\psi <^{Size}_{[]} [\varphi][\chi]\psi$,
9. $[\psi_G]\varphi <^{Size}_{[]} [G]\varphi$,
10. $[\chi][\psi_G]\varphi <^{Size}_{[]} [\chi][G]\varphi$.

Lemma 6. *Let x be a theory. If $D_G\varphi \notin x$, then there is a maximal consistent theory y such that $D_G x \subseteq y$ and $\varphi \notin y$.*

Proof. Assume that $D_G\varphi \notin x$. This means that $\varphi \notin D_G x$, and hence $D_G x + \neg\varphi$ is a consistent theory by Lemma 3. By Lemma 4, $D_G x + \neg\varphi$ can be extended to a maximal consistent theory y. Since $\neg\varphi \in y$, by consistency we have that $\varphi \notin y$.

Lemma 7. *Let x be a theory. If $K_a\varphi \notin x$, then there is a maximal consistent theory y such that $K_a x \subseteq y$ and $\varphi \notin y$.*

Proof. Similar to the proof of Lemma 6.

Lemma 8. *For all formulas* φ *and maximal consistent theories* x *it holds that* $\mathfrak{M}^C_x \models \varphi$ *if and only if* $\varphi \in x$.

Proof. The proof is by induction on the size of φ. Boolean cases are straightforward, and cases with public announcements are dealt with using $A11$–$A16$. Here we show only cases with distributed knowledge and group announcements.

Case $D_G\varphi$. ($\Rightarrow$): Let $\mathfrak{M}^C_x \models D_G\varphi$. By the semantics we have that for all $y \in S^C$: $x \sim_G y$ implies $\mathfrak{M}^C_y \models \varphi$. By the definition of the canonical pseudo model, Lemma 5, and the induction hypothesis, the latter is equivalent to the fact that for all $y \in S^C$ and all $H \subseteq G$: $D_H x \subseteq y$ implies $\varphi \in y$. In particular, for all $y \in S^C$: $D_G x \subseteq y$ implies $\varphi \in y$. By the contraposition of Lemma 6 this implies that $D_G\varphi \in x$.

($\Leftarrow$): Assume that $D_G\varphi \in x$ and $x \sim^C_G y$ for some maximal consistent theory y. By $A7$ and $R0$ it holds that $D_G D_G\varphi \in x$. By the definition of the canonical model, we have that $D_G\varphi \in y$. Since y is a maximal consistent theory and thus contains $D_G\varphi \rightarrow \varphi$, it holds that $\varphi \in y$. Next, by the induction hypothesis we have that $\mathfrak{M}^C_y \models \varphi$. Since y was arbitrary, we have that $\mathfrak{M}^C_y \models \varphi$ for all y such that $x \sim^C_G y$. The latter is equivalent to $\mathfrak{M}^C_x \models D_G\varphi$ by the semantics.

Case $[\varphi]D_G\psi$. $\mathfrak{M}^C_x \models [\varphi]D_G\psi$ if and only if $\mathfrak{M}^C_x \models \varphi \rightarrow D_G[\varphi]\psi$ by $A15$. By Lemma 5 and the induction hypothesis, $\mathfrak{M}^C_x \models \varphi \rightarrow D_G[\varphi]\psi$ if and only if $\varphi \rightarrow D_G[\varphi]\psi \in x$ if and only if $[\varphi]D_G\psi \in x$ by $A15$.

Case $[\varphi][G]\psi$. ($\Rightarrow$): Let $\mathfrak{M}^C_x \models [\varphi][G]\psi$. By the semantics, $\forall\psi_G$: $\mathfrak{M}^C_x \models [\varphi][\psi_G]\psi$. By Lemma 5 and the induction hypothesis, $[\varphi][\psi_G]\psi \in x$ for all ψ_G. Note that $[\varphi](\sharp)$ is a necessity form, hence, by $R3$, we have that $[\varphi][G]\psi \in x$.

($\Leftarrow$): Let $[\varphi][G]\psi \in x$. The distributivity rule for public announcements is derivable in PAL [10, Proposition 4.46]. Hence, by $A17$ and $R0$ it holds that $[\varphi][\psi_G]\psi \in x$. By Lemma 5 and the induction hypothesis, $\forall\psi_G$: $\mathfrak{M}^C_x \models [\varphi][\psi_G]\psi$. By the semantics, $\forall\psi_G$: $\mathfrak{M}^C_x \models \varphi$ implies $(\mathfrak{M}^C_x)^\varphi \models [\psi_G]\psi$. The latter is equivalent to $\mathfrak{M}^C_x \models \varphi$ implies $(\mathfrak{M}^C_x)^\varphi \models [G]\psi$, and thus $\mathfrak{M}^C_x \models [\varphi][G]\psi$.

Case $[G]\varphi$. ($\Rightarrow$): Let $\mathfrak{M}^C_x \models [G]\varphi$. By the semantics, $\forall\psi_G$: $\mathfrak{M}^C_x \models [\psi_G]\varphi$. By Lemma 5 and the induction hypothesis, $\forall\psi_G$: $[\psi_G]\varphi \in x$, and by $R3$, $[G]\varphi \in x$.

($\Leftarrow$): Let $[G]\varphi \in x$. By $A17$, $[\psi_G]\varphi \in x$ for all ψ_G. By Lemma 5 and the induction hypothesis, $\forall\psi_G$: $\mathfrak{M}^C_x \models [\psi_G]\varphi$, which is equivalent to $\mathfrak{M}^C_x \models [G]\varphi$ by the semantics.

Due to the lack of space, we briefly sketch the second part of the proof. It follows closely [22] and details can be found there.

Canonical pseudo model $\mathfrak{M}^C$ can be unravelled into the treelike canonical pre-model $\mathcal{M}^C$. Such an operation preserves bisimulation. After that, the pre-model can be folded into the canonical model. Folding preserves trans-bisimulation (denoted $\leftrightarrows^T$), which can be considered as a generalisation of standard bisimulation with a separate case for groups of agents. In such a way we can relate pre-models and models. The corresponding notion of equivalence between pre-models and models is trans-equivalence (denoted $\equiv^T$).

Before stating the completeness, we need one more result.

Lemma 9. *Given* M_s, $\mathcal{M}_t$, *and* $\mathfrak{M}_u$, *if* $M_s \leftrightarrows^T \mathcal{M}_t \leftrightarrows \mathfrak{M}_u$, *then* $M_s \equiv^T \mathcal{M}_t$.

Proof. The proof is by induction on φ. Boolean cases, cases for knowledge and distributed knowledge, and the case for public announcements are proved in [22, Lemma 26]. We show the case of $[G]\psi$.

Assume that $M_s \models [G]\psi$. By the semantics this is equivalent to $\forall \psi_G$: $M_s \models [\psi_G]\psi$. By the induction hypothesis we have that $\forall \psi_G$: $\mathcal{M}_t \models [\psi_G]\psi$, which is equivalent to $\mathcal{M}_t \models [G]\psi$ by the semantics.

Finally, we have everything we need to prove the completeness of GALD.

Theorem 2. *For all $\varphi \in \mathcal{L}_{GALD}$, if φ is valid, then $\varphi \in GALD$.*

Proof. Suppose towards a contradiction that φ is valid and $\varphi \notin$ GALD. Since GALD is a consistent theory, by Lemma 3 GALD $+\neg\varphi$ is a consistent theory. By Lemma 4, GALD $+\neg\varphi$ can be extended to a maximal consistent theory x such that GALD $+\neg\varphi \subseteq x$, and $\neg\varphi \in x$. By Lemma 8, the latter is equivalent to $\mathfrak{M}^C_x \not\models \varphi$. Next, the canonical pseudo model $\mathfrak{M}^C_x$ can be unravelled into bisimilar canonical pre-model $\mathcal{M}^C_y$, and the latter can be folded into the trans-bisimilar canonical model M^C_z. So, we have that $\mathfrak{M}^C_x \leftrightarrows \mathcal{M}^C_y$ and $\mathcal{M}^C_y \leftrightarrows^T M^C_z$. From $\mathfrak{M}^C_x \not\models \varphi$ by bisimilarity we have $\mathcal{M}^C_y \not\models \varphi$. Finally, by Lemma 9, $\mathfrak{M}^C_x \leftrightarrows \mathcal{M}^C_y$ and $\mathcal{M}^C_y \leftrightarrows^T M^C_z$ imply $M^C_z \equiv^T \mathcal{M}^C_y$, and from $\mathcal{M}^C_y \not\models \varphi$ we can infer that $M^C_z \not\models \varphi$, which contradicts φ being a validity.

5 Resolved Distributed Knowledge

Resolved distributed knowledge models private publicly observable communication within a group [4]. Distributed knowledge deals with agents' knowledge *before* any communication has taken place, and resolved distributed knowledge models the situation *after* all agents within a group have shared their knowledge. In a way, resolved distributed knowledge is a kind of a dynamic operator in disguise. In this section, we consider the relationship between group announcements, distributed, and resolved distributed knowledge.

Definition 15. *Let $M = (S, \sim, V)$ be an epistemic model.* A global *G*-resolved update *of M is the model $M^G = (S^G, \sim^G, V^G)$, where $S^G = S$, $V^G = V$, and*

$$\sim^G_a = \begin{cases} \bigcap_{b \in G} \sim_b & \text{if } a \in G, \\ \sim_a & \text{otherwise.} \end{cases}$$

The semantics for $R_G\varphi$ is

$$M_s \models R_G\varphi \text{ iff } M^G_s \models \varphi$$

The immediate result is that resolution and distributed knowledge are indeed different.

Fact 4. *$D_G\varphi \to R_G\varphi$ and $R_G\varphi \to D_G\varphi$ are not valid.*

Proof. For the first formula, consider a model M_s such that $M_s \models D_{\{a,b\}}(p \wedge \neg K_a p) \wedge K_b p$. Then obviously $M_s \not\models R_{\{a,b\}}(p \wedge \neg K_a p)$.

And for the second formula, $R_{\{a\}}p$ does not necessarily imply $D_{\{a\}}p$. For example, there is an M_s such that $M_s \models p \wedge \neg K_a p$, and hence $M_s \models R_{\{a\}}p$ (the model remains the same and $R_G p \leftrightarrow p$) and $M_s \not\models D_{\{a\}}p$ (since $D_a \varphi \leftrightarrow K_a \varphi$).

An interesting thing to note is that in our counterexample to $R_G\varphi \rightarrow D_G\varphi$, the target φ was a positive formula $K_a p$. Thus, $R_G\varphi^+ \rightarrow D_G\varphi^+$ is not valid as well.

Since resolved distributed knowledge models private communication, it does not coincide with group announcements.

Fact 5. *$\langle G\rangle\varphi \rightarrow R_G\varphi$ and $R_G\varphi \rightarrow \langle G\rangle\varphi$ are not valid.*

Proof. For the first formula, consider a two-state model M_s such that $M_s \models \langle\{b\}\rangle(K_a p \wedge K_b p) \wedge \neg K_a p$. The b-resolved update of the model, leaves the model intact. Hence, $M_s \not\models R_{\{b\}}K_a p$.

For the second formula, consider a two-state model M_s such that $M_s \models R_{\{a,c\}}(K_a p \wedge \neg K_b p) \wedge K_c p \wedge \neg K_a p \wedge \neg K_b p$. In such a model, no announcement by $\{a, c\}$ can both inform a that p is true, and leave b unaware of this fact. Hence, $M_s \not\models \langle\{a, c\}\rangle(K_a p \wedge \neg K_b p)$.

Even if we require the target formula to be positive, neither resolution implies ability, nor ability implies resolution. In the previous proposition, the counterexample for $\langle G\rangle\varphi \rightarrow R_G\varphi$ used positive formula $K_a p$.

Fact 6. *$R_G\varphi^+ \rightarrow \langle G\rangle\varphi^+$ is not valid.*

Proof. Consider model M from Fig. 2. We have that $M_s \models R_{\{a,b\}}K_a p$ if and only if $M_s^{\{a,b\}} \models K_a p$, and at the same time $M_s \not\models \langle\{a, b\}\rangle K_a p$. The rest of the argument is similar to the one in the proof of Fact 3.

The surprising corollary of this proposition is that semi-private communication between *all* agents does not imply the possibility of equivalent public communication between *all* agents. Formally, $R_A\varphi \rightarrow \langle A\rangle\varphi$ is not valid even for positive φ.

The special case when private communication between all agents implies the ability of equivalent public communication is considered in the next proposition.

Fact 7. *$R_A\varphi^+ \rightarrow \langle A\rangle\varphi^+$ is valid on finite bisimulation contracted models.*

Proof. On a finite bisimulation contracted model M_s, resolution for all agents results in a model with the universal relation for all agents. This corresponds to the maximal informative announcement by all agents (see Fact 2).

5.1 First Step Towards the Relative Expressivity of GALR and GALD

In the future, we would like to study the language of GAL extended with resolved distributed knowledge. Let us denote such a language GALR. In this section we make a first step towards comparing it to GALD.

Definition 16. *Let L_1 and L_2 be two languages. We say that L_1* is at least as expressive as *L_2 ($L_2 \leqslant L_1$) if and only if for all $\varphi \in L_2$ there is an equivalent $\psi \in L_1$. If L_1 is* not *at least as expressive as L_2, we write $L_2 \nleqslant L_1$.*

Some results on the expressivity of logics with distributed knowledge and resolution are presented in [4]. Relative expressivity of GALR and GALD is an open question. Here we present a partial result which establishes that a fragment GALD *without distributed knowledge operators within public announcements* (we call such a fragment GALD$^-$) is not at least as expressive as GALR.

Proposition 2. *GALR $\nleqslant$ GALD$^-$.*

Proof. Consider a GALR formula $R_{\{b,c\}}\langle\{a,b,c\}\rangle(\neg p \wedge \widehat{K}_a(K_b p \wedge K_c p) \wedge \widehat{K}_a(\widehat{K}_b(\neg p \wedge K_a \neg p) \wedge (\widehat{K}_c(\neg p \wedge K_a \neg p)))$. Assume towards a contradiction that there is an equivalent GALD$^-$ formula ψ, and $|\psi| = n$.

Consider models M and N (Figs. 3 and 4), where p holds in white states.

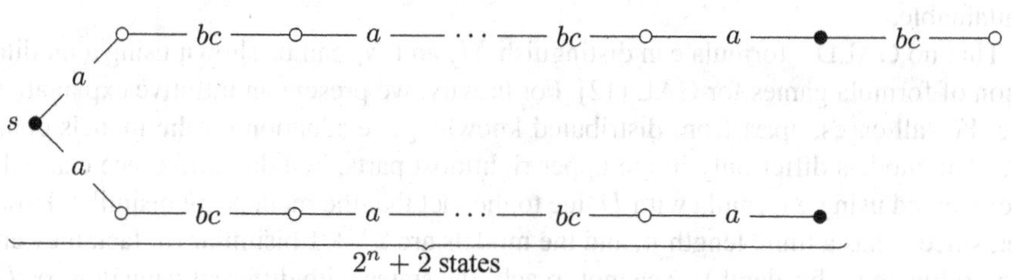

Fig. 3. Model M

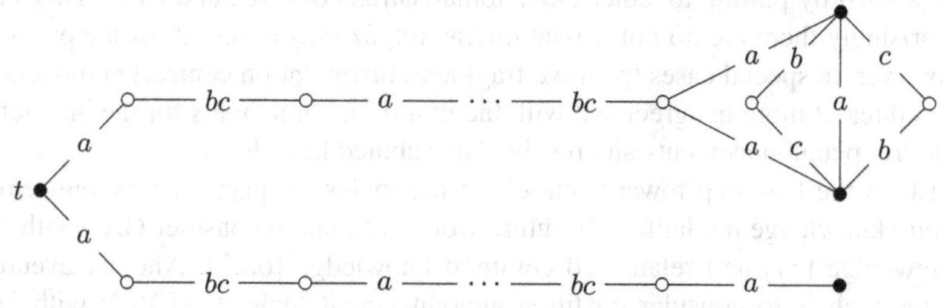

Fig. 4. Model N

These models are bisimilar, and hence they agree on formulas of GAL. Structurally, every model is almost symmetric, and the only difference are bits on the right.

For M it holds that $M_s \models R_{\{b,c\}}\langle\{a,b,c\}\rangle(\neg p \wedge \widehat{K}_a(K_b p \wedge K_c p) \wedge \widehat{K}_a(\widehat{K}_b(\neg p \wedge K_a \neg p) \wedge (\widehat{K}_c(\neg p \wedge K_a \neg p)))$. Indeed, resolution $R_{\{b,c\}}$ has no effect on the model, and the agents can make $\neg p \wedge \widehat{K}_a(K_b p \wedge K_c p) \wedge \widehat{K}_a(\widehat{K}_b(\neg p \wedge K_a \neg p) \wedge (\widehat{K}_c(\neg p \wedge K_a \neg p))$ true (note that intersection of agents' relations is the identity). This formula describes the configuration depicted in Fig. 5.

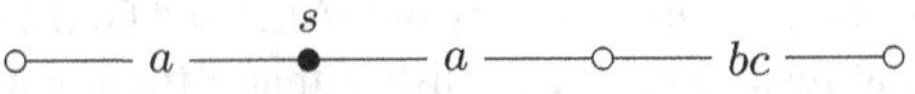

Fig. 5. An $\{a, b, c\}$-definable submodel of M

On the other hand, we have that $N_t \not\models R_{\{b,c\}}\langle\{a, b, c\}\rangle(\neg p \wedge \widehat{K}_a(K_b p \wedge K_c p) \wedge \widehat{K}_a(\widehat{K}_b(\neg p \wedge K_a \neg p) \wedge (\widehat{K}_c(\neg p \wedge K_a \neg p)))$. Update of N_t with $R_{\{b,c\}}$ results in model $N^{\{b,c\}}$, which is fully symmetric and bisimilar in both directions from state t ($R_{\{b,c\}}$ removes b and c relations in the upper right part). Thus, whatever the agents announce, if they preserve the $\{b, c\}$-equivalence class (on the right in Fig. 5), then they preserve the same equivalence class on the left. Hence, the configuration depicted in Fig. 5 is unattainable.

That no GALD$^-$ formula can distinguish M_s and N_t can be shown using a modification of formula games for GAL [12]. For brevity, we present an intuitive explanation here. For all cases, apart from distributed knowledge, evaluation on the models coincide. The models differ only in the upper rightmost parts, and this difference can only be expressed using a formula with D due to the fact that the models are bisimilar. However, since ψ has a finite length n, and the models are $2^n + 1$ bisimilar (in fact, they are isomorphic up to this depth), ψ cannot 'reach' the states with different valuations of D.

6 Conclusions and Future Work

In this paper, we have shown that GALD has a complete and sound axiomatisation that is obtained by putting together the axiomatisations of GAL and ELD. This shows that surprisingly there are no non-trivial interaction axioms required for the proof system. However, in special cases (positive fragment, bisimulation contracted models) the operators interact more in agreement with the intuition. Same holds for the interactions between group announcements and resolved distributed knowledge.

GALD is the first step towards enriching the logics of quantified announcements with group knowledge modalities. In future work, we plan to consider GAL with common knowledge [21] and relativised common knowledge [6,23]. Another avenue of further research is to consider coalition announcement logic (CAL) [2] with group knowledge. In CAL, as opposed to GAL, agents outside of a group make a simultaneous announcement as well, and thus they may preclude the group from reaching its epistemic goals. Finally, we would like to investigate GALR. In particular, we are interested in its axiomatisation and expressivity relative to GALD.

Acknowledgements. We would like to thank three anonymous reviewers for their helpful comments.

References

1. Ågotnes, T., Balbiani, P., van Ditmarsch, H., Seban, P.: Group announcement logic. J. Appl. Logic **8**(1), 62–81 (2010)
2. Ågotnes, T., van Ditmarsch, H.: Coalitions and announcements. In: Proceedings of AAMAS 2008, pp. 673–680 (2008)
3. Ågotnes, T., van Ditmarsch, H.: What will they say? - Public announcement games. Synthese **179**(1), 57–85 (2011)
4. Ågotnes, T., Wáng, Y.N.: Resolving distributed knowledge. Artif. Intell. **252**, 1–21 (2017)
5. Balbiani, P., Baltag, A., van Ditmarsch, H., Herzig, A., Hoshi, T., de Lima, T.: 'Knowable' as 'known' after an announcement. Rev. Symbolic Logic **1**(3), 305–334 (2008)
6. van Benthem, J., van Eijck, J., Kooi, B.P.: Logics of communication and change. Inf. Comput. **204**(11), 1620–1662 (2006)
7. Bozzelli, L., van Ditmarsch, H., French, T., Hales, J., Pinchinat, S.: Refinement modal logic. Inf. Comput. **239**, 303–339 (2014)
8. van Ditmarsch, H., van der Hoek, W., Kooi, B., Kuijer, L.B.: Arbitrary arrow update logic. Artif. Intell. **242**, 80–106 (2017)
9. van Ditmarsch, H., Kooi, B.: The secret of my success. Synthese **151**(2), 201–232 (2006)
10. van Ditmarsch, H., van der Hoek, W., Kooi, B.: Dynamic Epistemic Logic. Synthese Library, vol. 337. Springer, Heidelberg (2008)
11. Fagin, R., Halpern, J.Y., Moses, Y., Vardi, M.Y.: Reasoning About Knowledge. The MIT Press, Cambridge (1995)
12. French, T., Galimullin, R., van Ditmarsch, H., Alechina, N.: Groups versus coalitions: on the relative expressivity of GAL and CAL. In: Proceedings of AAMAS, vol. 2019, pp. 953–961 (2019)
13. Galimullin, R., Alechina, N., van Ditmarsch, H.: Model checking for coalition announcement logic. In: Trollmann, F., Turhan, A.-Y. (eds.) KI 2018: Advances in Artificial Intelligence. LNCS, vol. 11117, pp. 11–23. Springer, Cham (2018). https://doi.org/10.1007/978-3-030-00111-7_2
14. Goldblatt, R.: Axiomatising the Logic of Computer Programming. Lecture Notes in Computer Science, vol. 130. Springer, Heidelberg (1982)
15. Goranko, V., Otto, M.: Model theory of modal logic. In: Blackburn, P., van Benthem, J., Wolter, F. (eds.) Handbook of Modal Logic, Studies in Logic and Practical Reasoning, vol. 3, pp. 249–329. Elsevier, Amsterdam (2007)
16. Hales, J.: Arbitrary action model logic and action model synthesis. In: Proceedings of LICS 2013, pp. 253–262 (2013)
17. van der Hoek, W., Meyer, J.-J.Ch.: Making some issues of implicit knowledge explicit. Int. J. Found. Comput. Sci. **3**(2), 193–223 (1992)
18. Moore, R.C.: A formal theory of knowledge and action. In: Allen, J.F., Hendler, J., Tate, A. (eds.) Readings in Planning, pp. 480–519. Morgan Kaufmann Publishers, San Mateo (1990)
19. Plaza, J.: Logics of public communications. Synthese **158**(2), 165–179 (2007)
20. Roelofsen, F.: Distributed knowledge. J. Appl. Non-Classical Logics **17**(2), 255–273 (2007)
21. Vanderschraaf, P., Sillari, G.: Common knowledge. In: Zalta, E.N. (ed.) The Stanford Encyclopedia of Philosophy (2014)
22. Wáng, Y.N., Ågotnes, T.: Public announcement logic with distributed knowledge: expressivity, completeness and complexity. Synthese **190**(1), 135–162 (2013)
23. Wáng, Y.N., Ågotnes, T.: Relativized common knowledge for dynamic epistemic logic. J. Appl. Logic **13**(3), 370–393 (2015)

Towards a Logic for Conditional Local Strategic Reasoning

Valentin Goranko[1,2] and Fengkui Ju[3](✉)

[1] Stockholm University, Stockholm, Sweden
valentin.goranko@philosophy.su.se
[2] University of Johannesburg (visiting professorship), Johannesburg, South Africa
[3] School of Philosophy, Beijing Normal University, Beijing, China
fengkui.ju@bnu.edu.cn

Abstract. We consider systems of rational agents who act in pursuit of their individual and collective objectives and we study the reasoning of an agent or an external observer about the consequences from the expected choices of action of the other agents based on their objectives, in order to assess the reasoner's ability to achieve his own objective.

To formalize such reasoning we introduce new modal operators of conditional strategic reasoning and use them to extend Coalition Logic in order to capture variations of conditional strategic reasoning. We provide formal semantics for the new conditional strategic operators, introduce the matching notion of bisimulation for each of them and discuss and compare briefly their expressiveness.

Keywords: Concurrent game models · Conditional strategic reasoning · Coalition Logic · Expressiveness

1 Introduction

Consider the following scenario. Alice and Bob are students at DownTown University. Alice is coming to campus today, and has some agenda to complete. Bob wants to meet Alice somewhere on campus today. She does not know that (maybe, even does not know Bob) and they have no communication. Bob may, or may not, know what Alice is going to do on campus, or where and at what time she will go during the day.

Using his knowledge of what, where, and when Alice intends to do today, Bob wants to come up with a plan of how (where and when) to meet her.

The work of Valentin Goranko was supported by a research grant 2015-04388 of the Swedish Research Council. The work of Fengkui Ju was supported by the Major Program of the National Social Science Foundation of China (NO. 17ZDA026). We thank the reviewers for some helpful remarks.

P. Blackburn et al. (Eds.): LORI 2019, LNCS 11813, pp. 112–125, 2019.
https://doi.org/10.1007/978-3-662-60292-8_9

Put in a more general perspective, we consider a system of agents acting independently, and possibly concurrently, in pursuit of their individual and collective goals and we analyse the reasoning of an agent (or, just an observer) about the possible local actions (at the current state only) of the other agents and their possible effect for realising the outcome of interest for the observer or, respectively, for enabling the reasoner to act towards achieving his own goal.

Our Contributions. In this paper we identify several distinct cases of conditional strategic reasoning of an observer or an active agent, depending on his knowledge about the objectives and possible actions of the other agents. To formalize such reasoning we introduce new modal operators of conditional strategic reasoning and use them to extend Coalition Logic to capture variations of conditional strategic reasoning. We provide formal semantics for the new conditional strategic operators, introduce the matching notion of bisimulation for each of them and discuss and compare briefly their expressiveness.

Related Work. The kind of strategic reasoning discussed here is within the conceptual thrust motivating the research on logic-based strategic reasoning over the past two decades, starting with Coalition Logic ([10,11]) and its temporal extension ATL ([4]), and evolving towards increasingly expressive formalisms, such as Strategy Logic (cf. [9]) (cf. [5] and [3] for overviews of the area). Still, we are aware of very few works that deal more explicitly with *conditional* strategic reasoning in the sense of the present paper, with perhaps the closest being the recent [8], to which the present work relates both conceptually and technically. In the literature there has been some work on reasoning about agents' goals (cf. [6]).

Structure of the Paper. Section 2 provides an informal discussion on conditional strategic reasoning, motivating the further technical work. Section 3 introduces several modal operators formalising patterns conditional strategic reasoning and uses them to introduce the new logic ConStR as an extension of Coalition Logic with these operators. Section 4 introduces the matching notion of bisimulation for that logic and discuss briefly it expressiveness. We end with brief concluding remarks in Sect. 5.

2 Conditional Strategic Reasoning: An Informal Discussion

Suppose that Alice has an objective α to achieve (say, pick a book from a friend). Suppose also that Alice has several possible choices of an action (or strategy)[1] that would possibly, or certainly, guarantee the achievement of her objective.

[1] In this paper we focus on local reasoning, about once-off actions, but in this section the word 'action' can be conceived in a wider sense, and may mean either a once-off action, or a global strategy guiding the long term behaviour of the agent.

2.1 Conditional Reasoning of an Observer About an Agent's Actions

Let us first consider the case where Bob is just an observer who is not acting, but only reasoning about the consequences from Alice's possible actions with respect to the occurrence of another – intended or not – outcome β. More generally, we also assume that there are other agents, besides Alice, also acting in pursuit of their own goals, and Bob is reasoning about their individual and collective choices of action and the consequences from these choices. This leads to a *passive observer's conditional strategic reasoning* about statements of the type:

"*Some/every action of Alice that guarantees achievement of α also guarantees/enables occurrence of the (desired or expected) outcome β*".

Depending on Bob's knowledge about Alice's objective and of her expected choices of action there can be several possible cases for Bob's reasoning about the expected occurrence of the outcome β.

Bob's Reasoning, Case 1: Bob Knows Nothing About Alice. Suppose that *Bob does not know Alice's objective*, and therefore has no a priori expectations about her choice of action. E.g., if Alice is coming to the university and Bob is standing by the only entrance of the campus, he will know for sure that he is going to meet Alice, no matter what she will do there.

Then, Bob can *only* claim for sure that the outcome β will occur if β is inevitable, regardless of how Alice (and all others) will act. This can be expressed in Coalition Logic CL (cf. [11]) simply as $[\emptyset]\beta$.

Bob's Reasoning, Case 2: Bob only Know Alice's Objective. Suppose now that Bob *does know Alice's objective* and knows that Alice can guarantee the achievement of that objective and will act towards that, but Bob does not know *how exactly* Alice might act. E.g., Bob knows that Alice is coming to campus to pick some book, but does not know where and when.

Then, Bob can only claim that the outcome β will occur for sure if β is true *on every possible course of events ("play") on which α is true.* (E.g., Bob knows that the book is in the library, and β is the event "Alice enters the library building".) This can be expressed as a conditional $\alpha \rightarrow \beta$, in the right context. Depending on how the conditional is interpreted, there are different cases:

- $\alpha \rightarrow \beta$ is a material implication, with unconstrained context.
  This can be expressed in CL simply as $[\emptyset](\alpha \rightarrow \beta)$.
- $\alpha \rightarrow \beta$ is a proper conditional, with a somehow constrained context.
  In general, this cannot be expressed in CL anymore, but it can possibly be expressed in a suitably extended language and in a suitably updated model. Here we will not pursue this line, but will leave it to a follow-up work.

Bob's Reasoning, Case 3: Bob Knows Alice's Objective and Possible Actions. Suppose now Bob not only knows Alice's objective, but also *knows* *all possible actions/strategies* of Alice that can ensure the satisfaction of her objective α, and knows that Alice will commit to one of them, but *does not know to which one.* (E.g., Bob knows that Alice is coming to campus to meet with her supervisor and she can meet with him either in his office, or in the lecture room, or in the café.)

Now, for Bob to claim that the outcome β will occur for sure, he must know that *each* action of Alice that guarantees α will also guarantee β. (E.g., suppose that all possible meeting places are in the main building and β is the event "Alice enters the main building".) This can no longer be expressed in CL and requires introducing a new strategic operator.

Bob's Reasoning, Case 4: Bob Knows Alice's Action. Lastly, suppose that *Bob knows the specific action which she is taking in order to guarantee the achievement of her objective.* Then, Bob can claim that the outcome β will occur for sure, as long as that specific action of Alice guarantees the satisfaction of β. Again, this claim could be interpreted either in the same (original) model, or in a respectively updated one, obtained by preserving only the plays that are enabled by that action if Alice's strategic commitment is assumed. The latter corresponds to *reasoning with strategy contexts* which we will not discuss here, but in a follow-up work.

2.2 Conditional Reasoning of an Agent About Another Agent's Actions

Suppose now that Bob is not just an observer, but also an acting agent, who has the outcome β as his own goal. Suppose also that there may be other agents, besides Alice and Bob, also acting in pursuit of their own goals, and Bob is reasoning about their individual and collective choices of action and the consequences from these choices.

Now, Bob is to decide, based on his reasoning about Alice's (and other agents) possible choices of actions, on his own action in pursuit of β. This calls for an *agent's conditional strategic reasoning* about statements of the type:

"*For some/every action of Alice that guarantees achievement of α, Bob has/does not have an action of his own to guarantee achievement of his objective β*".

We call this *local conditional strategic reasoning*, as it only refers to the immediate actions of the agents, not about their *global strategies.* Respectively, the outcomes from the local action profiles are just successor states, while in the general case they are (finite or possibly infinite) *plays.* The global conditional strategic reasoning will be treated in a follow-up work.

Each of the cases considered in Sect. 2.1 accordingly applies here, too. However, now in the reasoning case 3 the statement

"*Bob knows that whichever way Alice acts towards achieving the objective α, he can act so as to bring about achievement of his objective β.*"

admits two different readings, *de dicto* and *de re,* which we discuss here.

Bob's Reasoning, Case 3: *de dicto* Reading. In the *de dicto* reading, where Bob only knows that Alice has committed to act so as to achieve α, but, as far as he knows, Alice has not yet chosen her action, or her choice will remain unknown to Bob.

In this case Bob must consider all possible courses of events (plays) that can occur as a result of Alice acting towards achieving α and reason about whether he can act *uniformly* for each of them in a way that would bring about β, without knowing which of them will take place. (E.g., in our running story from Sect. 2.1, Bob can choose to wait for Alice at the only entrance of the main building.) Formally speaking, in this case, based on his knowledge Bob considers the set of states in the model which is the union of all sets of outcome states enabled by the specific actions of Alice that would guarantee α, and is looking for an action that will bring about β on each of these outcome states.

Bob's reasoning, case 3: *de re* Reading. This is the reading where *for every action* of Alice that ensures α, Bob is looking for an action of his, *possibly dependent on Alice's action* that would also ensure the occurrence of β (possibly in different ways for the different actions). More formally, each of Alice's actions that would guarantee α generates a set of possible outcome states, and for each of them Bob is looking for an action that will bring about β on that set of outcome states.

For example, suppose Bob knows that Alice has agreed with her friend Charlie on a meeting on campus today and there are two options: to meet in the campus café or to meet in the library; both options are ok for Charlie and Alice is yet to decide on either option. Note, that the sentence "*Alice has decided to meet with Charlie on campus today*" is true in either case. However, the sentence "*Alice has decided to meet with Charlie in the café today, or Alice has decided to meet with Charlie in the library today*" should not be regarded as true (yet). After Alice makes her choice, this sentence becomes true, too. But even then, from Bob's perspective, the same distinction applies depending on whether or not he knows Alice's choice, so he has to take into account both options when deciding for himself on what to do.

3 A Logic for Conditional Strategic Reasoning

3.1 Preliminaries

Multi-agent Game Models. We fix a finite set of **agents** $\mathbb{A}\mathrm{gt} = \{a_1, ..., a_n\}$ and a set of **atomic propositions** Π. Subsets of $\mathbb{A}\mathrm{gt}$ will also be called **coalitions.**

Definition 1 (Multi-agent game model). *A* ***game model***[2] *for* $\mathbb{A}\mathrm{gt}$ *and* Π *is a tuple*

$$\mathcal{M} = (S, \{\Sigma_a\}_{a \in \mathbb{A}\mathrm{gt}}, g, V)$$

[2] These game models are essentially equivalent to concurrent game models used in [4].

where S is a non-empty set of **states;** *each Σ_a is a non-empty set of possible* **actions** *of agent a; $V : \Pi \to \mathcal{P}(S)$ is a* **valuation** *of the atomic propositions from Π in S; and g is a* **game map** *that assigns to each $s \in S$ a strategic game form $g(s) = (\Sigma^s_{a_1}, ...\Sigma^s_{a_n}, o_s)$, where each $\Sigma^s_{a_i} \subseteq \Sigma_{a_i}$ is a non-empty set of actions available to player a_i at s, and*

$$o_s : \Sigma^s_{a_1} \times ... \times \Sigma^s_{a_n} \to S$$

is a **local outcome function** *assigning to any* **action profile** *$\sigma \in \Sigma^s_{a_1} \times ... \times \Sigma^s_{a_n}$ the* **outcome state** *$o_s(\sigma)$ produced by σ when applied at $s \in S$. The set $\Sigma^s_{a_1} \times ... \times \Sigma^s_{a_n}$ of* **action profiles available at** *s will be denoted by Act_s.*

Now, the **global outcome function** *in $\mathcal{M}$ is the partial mapping*

$$O : S \times \Sigma_{a_1} \times ... \times \Sigma_{a_n} \dashrightarrow S$$

defined by $O(s, \sigma) = o_s(\sigma)$, whenever $\sigma \in \mathrm{Act}_s$.

Given a coalition $\mathsf{C} \subseteq \mathbb{A}\mathrm{gt}$, a **joint action** *for C in the model $\mathcal{M}$ is a tuple of individual actions $\sigma_{\mathsf{C}} \in \prod_{a\in\mathsf{C}} \Sigma_a$. For any such joint action σ_{C} that is available at $s \in S$, we define the* **set of outcome states from** *σ_{C}* **at** *s:*

$$\mathsf{Out}[s, \sigma_{\mathsf{C}}] = \{u \in S \mid \exists\sigma \in \mathrm{Act}_s : \sigma|_{\mathsf{C}} = \sigma_{\mathsf{C}} \ \&\ o_s(\sigma) = u\}$$

where $\sigma|_{\mathsf{C}}$ is the restriction of σ to C. Note that the empty tuple $\sigma_\emptyset$ is the only available joint action for the empty coalition $\emptyset$ at any state.

The Basic Logic for Coalitional Strategic Reasoning CL. Coalition Logic (CL) was introduced in [10], cf. also [11]. CL extends the classical propositional logic with *coalitional strategic modal operators* [C], for any coalition of agents C. Formulae of CL:

$$\varphi := p \mid \neg\varphi \mid \varphi_1 \vee \varphi_2 \mid [\mathsf{C}]\,\varphi$$

(We will write [i] instead of [{i}].) The intuitive reading of $[\mathsf{C}]\,\varphi$ is:

"The coalition C *has a joint action that ensures an outcome (state) satisfying* φ, regardless of how all other agents act."

Semantics of CL. The formulae of CL are interpreted in game models (GM). The semantics is defined in terms of **truth of a CL-formula** ψ **at a state** s **of a GM** $\mathcal{M}$, denoted $\mathcal{M}, s \vDash \psi$, by induction on formulae, via the key clause:

$\mathcal{M}, s \models [\mathsf{C}]\,\phi$ iff there exists a joint action σ_{C} available at s, such that $\mathcal{M}, u \models \phi$ for each $u \in \mathsf{Out}[s, \sigma_{\mathsf{C}}]$.

We note that $[\mathsf{C}]\,\phi$ formalises a claim of the ability of the agent/coalition C to choose a suitable (joint) action so as to achieve the goal ϕ *regardless of how all other agents choose to act*, and therefore regardless of whether the agents in C know the goal(s) of the remaining agents. This subsumes Cases 1 and 2 of Bob's reasoning, discussed in Sect. 2.1.

Alternating Bisimulations. The notion of bisimulation that guarantees truth invariance of all CL-formulae was first defined in [10] for the abstract game models defined there, and later (under the name "alternating bisimulations") in [1], to which we refer the reader for the definition and proof of bisimulation invariance of all ATL-formulae (incl. all CL-formulae).

3.2 The Logic of Conditional Strategic Reasoning ConStR

Given coalitions A and B and joint actions σ_A for A and σ_B for B, we say that σ_B **is consistent with** σ_A if σ_B coincides with σ_A on $A \cap B$.

Operators for Conditional Strategic Reasoning of ConStR

We now introduce new operators for conditional strategic reasoning, for any coalitions A and B with intuitive semantics as follows.

(O_c) $\langle\!\langle A \rangle\!\rangle_{\mathsf{c}}(\phi; \langle B \rangle \psi)$: A has a joint action σ_A such that, when applied, it guarantees the truth of ϕ and enables B to apply a joint action σ_B that is consistent with σ_A and guarantees ψ when *additionally* applied by B, in sense that all agents in A act according to σ_A and those in $B \setminus A$ act according to σ_B.

This operator formalises a scenario (not discussed in Sect. 2, but still basic for conditional strategic reasoning, see further) where A knows the objective of B and can choose to cooperate with B by selecting a suitable action.

(O_{dr}) $[A]_{\mathsf{dr}}(\phi; \langle B \rangle \psi)$: for any joint action σ_A of A that guarantees the truth of ϕ, when applied by A there is an action σ_B that is consistent with σ_A and guarantees ψ when additionally applied by B.

This operator formalises a claim of the ability of the agent/coalition B to choose a suitable (joint) action so as to achieve the goal ψ assuming that A acts so as to achieve the goal ϕ, if B is to choose their (joint) action *after* B learns the (joint) action of A. This corresponds to the *de re* reading of Case 3 of Bob's reasoning, discussed in Sect. 2.2. In particular, the case when B is not informed about the goal of A, but has to choose their action after learning the action of A is formalised by $[A]_{\mathsf{dr}}(\top; \langle B \rangle \psi)$.

(O_{dd}) $[A]_{\mathsf{dd}}(\phi; \langle B \rangle \psi)$: $B \setminus A$ has an action $\sigma_{B \setminus A}$ such that if A applies any action that guarantees the truth of ϕ, then $B \setminus A$ can guarantee the truth of ψ by applying additionally the action $\sigma_{B \setminus A}$.

This operator formalises a claim of the ability of the agent/coalition B to choose a suitable (joint) action so as to achieve the goal ψ assuming that A acts so as to achieve the goal ϕ, if B is to choose their (joint) action *before* A chooses their (joint) action, or before B learns the action of A. This corresponds to the *de dicto* reading of Case 3 of Bob's reasoning, discussed in Sect. 2.2. In particular, the case when B is not informed about the goal of A and has to choose a joint action before A has chosen their action is formalised by $[A]_{\mathsf{dd}}(\top; \langle B \rangle \psi)$.

Language of ConStR. We fix a finite nonempty set of agents $\mathbb{A}\mathrm{gt}$ and a countable set of atomic propositions Π.

The formulae of ConStR, where $p \in \Pi$ and $A, B \subseteq \mathbb{A}\mathrm{gt}$ are defined as follows:

$$\phi ::= p \mid \top \mid \neg\phi \mid (\phi \land \phi) \mid \langle\!\langle A \rangle\!\rangle_{\mathsf{c}}(\phi; \langle B \rangle \phi) \mid [A]_{\mathsf{dr}}(\phi; \langle B \rangle \phi) \mid [A]_{\mathsf{dd}}(\phi; \langle B \rangle \phi)$$

Some Definable Operators and Expressions in ConStR. The following can be easily seen from the informal semantics above, and can also be easily verified with the formal semantics introduced further.

- The dual operator $\neg\langle\langle \mathrm{A}\rangle\rangle_{\mathsf{c}}(\phi; \langle \mathrm{B}\rangle\neg\psi)$ says that every joint action of A that, when applied, guarantees the truth of ϕ, would prevent B from acting additionally so as to guarantee ψ. This formalises the conditional reasoning scenario where the goals of A and B are conflicting and where Bob can establish that whichever way A acts towards their goal, that would block B from acting to guarantee achievement of its goal.
- $[\mathrm{A}]_{\mathsf{c}}(\phi|\psi) := [\mathrm{A}]_{\mathsf{dr}}(\phi; \langle\emptyset\rangle\psi)$: for any joint strategy of A, if it guarantees ϕ to be true then it guarantees ψ to be true, too. This operator formalises Case 2 of Bob's reasoning as an observer (rather than an acting agent), discussed in Sect. 2.1.
- $\langle \mathrm{A}\rangle_{\mathsf{c}}(\phi|\psi) := \neg[\mathrm{A}]_{\mathsf{c}}(\phi|\neg\psi)$: there is a joint strategy of A that guarantees ϕ to be true and enables ψ to be true, too. Note that it is equivalent to a special case of the "socially friendly coalitional operator" SF, $[\mathrm{C}](\phi; \psi_1, \dots, \psi_k)$, introduced in [8], viz. $\langle \mathrm{A}\rangle_{\mathsf{c}}(\phi|\psi) \equiv [A](\phi; \psi)$. Moreover, $\langle \mathrm{A}\rangle_{\mathsf{c}}(\phi|\psi)$ is also definable as $\langle\langle \mathrm{A}\rangle\rangle_{\mathsf{c}}(\phi; \langle\overline{\mathrm{A}}\rangle\psi)$, where $\overline{\mathrm{A}} = \mathbb{A}\mathrm{gt} \setminus \mathrm{A}$.
- The coalitional strategic operator [A] from CL is a special case of the above: $[\mathrm{A}]\phi := \langle \mathrm{A}\rangle_{\mathsf{c}}(\phi|\top)$, meaning "A has a joint action to ensure the truth of ϕ"[3].
- $\langle\langle \mathrm{A}\rangle\rangle_{\mathsf{c}}(\phi; \langle \mathrm{B}\rangle\psi)$ is definable in terms of the "group protecting coalitional operator" GIP, introduced in [8]: $\langle\langle \mathrm{A}\rangle\rangle_{\mathsf{c}}(\phi; \langle \mathrm{B}\rangle\psi) \equiv ⦃A \rhd \phi, \mathrm{A} \cup \mathrm{B} \rhd \psi⦄$. Nevertheless, it now has a different motivation and intuitive interpretation.

Semantics of ConStR. Given coalitions $\mathrm{A}, \mathrm{B} \subseteq \mathbb{A}\mathrm{gt}$ and joint actions σ_{A} for A and σ_{B} for B, we define $\sigma_{\mathrm{A}} \uplus \sigma_{\mathrm{B}}$ to be the joint action for $\mathrm{A} \cup \mathrm{B}$ which equals to σ_{A} when restricted to A and equals to $\sigma_{\mathrm{B}}|_{\mathrm{B}\setminus\mathrm{A}}$ when restricted to $\mathrm{B} \setminus \mathrm{A}$. Note $\sigma_{\mathrm{A}} \uplus \sigma_{\emptyset} = \sigma_{\mathrm{A}}$ for any $\mathrm{A} \subseteq \mathbb{A}\mathrm{gt}$.

Now, let $\mathcal{M} = (S, \{\Sigma_a\}_{a\in\mathbb{A}\mathrm{gt}}, g, V)$ be a game model. The semantics of $\mathsf{ConStR_o}$ extends the one of CL to the new operators as follows:

$\mathcal{M}, s \Vdash \langle\langle \mathrm{A}\rangle\rangle_{\mathsf{c}}(\phi; \langle \mathrm{B}\rangle\psi) \;\Leftrightarrow$ A has a joint action σ_{A}, such that $\mathcal{M}, u \Vdash \phi$ for every $u \in \mathsf{Out}[s, \sigma_{\mathrm{A}}]$ and B has a joint action σ_{B} such that $\mathcal{M}, u \Vdash \psi$ for every $u \in \mathsf{Out}[s, \sigma_{\mathrm{A}} \uplus \sigma_{\mathrm{B}}]$.

$\mathcal{M}, s \Vdash [\mathrm{A}]_{\mathsf{dr}}(\phi; \langle \mathrm{B}\rangle\psi) \;\Leftrightarrow$ for every joint action σ_{A} of A such that $\mathcal{M}, u \Vdash \phi$ for every $u \in \mathsf{Out}[s, \sigma_{\mathrm{A}}]$, B has a joint action σ_{B} (generally, dependent on σ_{A}) such that $\mathcal{M}, u \Vdash \psi$ for every $u \in \mathsf{Out}[s, \sigma_{\mathrm{A}} \uplus \sigma_{\mathrm{B}}]$.

$\mathcal{M}, s \Vdash [\mathrm{A}]_{\mathsf{dd}}(\phi; \langle \mathrm{B}\rangle\psi) \;\Leftrightarrow$ B has a joint action σ_{B} such that for every joint action σ_{A} of A, if $\mathcal{M}, u \Vdash \phi$ for each $u \in \mathsf{Out}[s, \sigma_{\mathrm{A}}]$, then $\mathcal{M}, u \Vdash \psi$ for each $u \in \mathsf{Out}[s, \sigma_{\mathrm{A}} \uplus \sigma_{\mathrm{B}}]$.

[3] NB: We have preserved the box-like notation for [A] from CL, even though it is not consistent with ours.

Remark: The semantics of each of the operators above can be re-stated to consider joint actions for $B \setminus A$ rather than the whole B. For instance, for the latter operator, it can be easily verified that $\mathcal{M}, s \Vdash [A]_{dd}(\phi; \langle B \rangle \psi)$ iff $B \setminus A$ has a joint action $\sigma_{B \setminus A}$ such that for every joint action σ_A of A, if $\mathcal{M}, u \Vdash \phi$ for each $u \in \mathsf{Out}[s, \sigma_A]$, then $\mathcal{M}, u \Vdash \psi$ for each $u \in \mathsf{Out}[s, \sigma_A \uplus \sigma_{B \setminus A}]$.

In the Appendix we provide a few simple examples illustrating the semantics of the strategic operators introduced here.

4 Bisimulations and Expressiveness of ConStR

4.1 Bisimulations for CSR

The definition of ConStR-bisimulation involves, besides atomic equivalence, 3 nested Forth and Back conditions, for each of the respective new operators O_c, O_{dr}, and O_{dd}[4]. We only define ConStR-bisimulation within a game model, which generalises to ConStR-bisimulation *between* game models, by treating both as parts of their disjoint union.

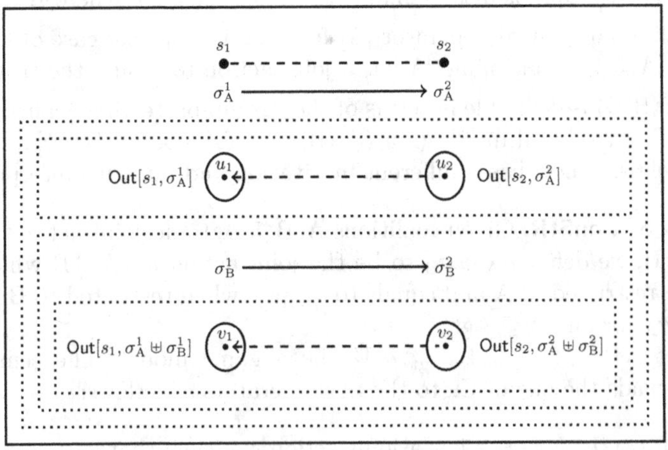

Fig. 1. The A-$\mathbf{Forth}_c$ half of O_c-bisimulation

Definition 2 (ConStR-bisimulation). *Let* $\mathcal{M} = (S, \{\Sigma_a\}_{a \in \mathbb{A}\mathrm{gt}}, g, V)$ *be a game model. A binary relation* $\beta \subseteq S^2$ *is a* ConStR-***bisimulation in*** $\mathcal{M}$ *if it satisfies the following conditions for every pair of states* (s_1, s_2) *such that* $s_1 \beta s_2$ *and for every coalitions* A *and* B*:*

[4] Each of these conditions is a respective variation of the bisimulation conditions for the basic strategic operators in the logics SFCL and GPCL defined in [8].

Atom equivalence: *For every* $p \in \Pi$*:* $s_1 \in V(p)$ *iff* $s_2 \in V(p)$.

O_c**-bisimulation:** *(For illustration, see Fig. 1)*

- **A-Forth$_c$:** *For any joint action* σ^1_{A} *of* A *at* s_1 *there is a joint action* σ^2_{A} *of* A *at* s_2*, such that:*
  - **A-LocalBack$_c$:** *For every* $u_2 \in \mathsf{Out}[s_2, \sigma^2_{\mathrm{A}}]$ *there exists* $u_1 \in \mathsf{Out}[s_1, \sigma^1_{\mathrm{A}}]$ *such that* $u_1 \beta u_2$.
  - **B-Forth$_c$:** *For every joint action* σ^1_{B} *of* B *at* s_1 *there is a joint action* σ^2_{B} *of* B *at* s_2*, such that:*
    - **(A ⊎ B)-LocalBack$_c$:** *For every* $u_2 \in \mathsf{Out}[s_2, \sigma^2_{\mathrm{A}} \uplus \sigma^2_{\mathrm{B}}]$ *there exists* $u_1 \in \mathsf{Out}[s_1, \sigma^1_{\mathrm{A}} \uplus \sigma^1_{\mathrm{B}}]$ *such that* $u_1 \beta u_2$.
- **A-Back$_c$:** *Like* A-***Forth***$_c$*, but with* 1 *and* 2 *swapped.*

O_{dr}**-bisimulation:**

- **A-Forth$_{dr}$:** *For any joint action* σ^1_{A} *of* A *at* s_1 *there is a joint action* σ^2_{A} *of* A *at* s_2*, such that:*
  - **A-LocalBack$_{dr}$:** *For every* $u_2 \in \mathsf{Out}[s_2, \sigma^2_{\mathrm{A}}]$ *there exists* $u_1 \in \mathsf{Out}[s_1, \sigma^1_{\mathrm{A}}]$ *such that* $u_1 \beta u_2$.
  - **B-Back$_{dr}$:** *For every joint action* σ^2_{B} *of* B *at* s_2 *there is a joint action* σ^1_{B} *of* B *at* s_1*, such that:*
    - **(A ⊎ B)-LocalForth$_{dr}$:** *For every* $u_1 \in \mathsf{Out}[s_1, \sigma^1_{\mathrm{A}} \uplus \sigma^1_{\mathrm{B}}]$ *there exists* $u_2 \in \mathsf{Out}[s_2, \sigma^2_{\mathrm{A}} \uplus \sigma^2_{\mathrm{B}}]$ *such that* $u_1 \beta u_2$.
- **A-Back$_{dr}$:** *Like* A-***Forth****, but with* 1 *and* 2 *swapped.*

O_{dd}**-bisimulation:**

- **B-Forth$_{dd}$:** *For any joint action* σ^1_{B} *of* B *at* s_1 *there is a joint action* σ^2_{B} *of* B *at* s_2*, such that:*
  - **A-Back$_{dd}$:** *For every joint action* σ^2_{A} *of* A *at* s_2 *there is a joint action* σ^1_{A} *of* A *at* s_1*, such that:*
    - **(A)-LocalForth$_{dd}$:** *For every* $u_1 \in \mathsf{Out}[s_1, \sigma^1_{\mathrm{A}}]$ *there exists* $u_2 \in \mathsf{Out}[s_2, \sigma^2_{\mathrm{A}}]$ *such that* $u_1 \beta u_2$.
    - **(A ⊎ B)-LocalBack$_{dd}$:** *For every* $u_2 \in \mathsf{Out}[s_2, \sigma^2_{\mathrm{A}} \uplus \sigma^2_{\mathrm{B}}]$ *there exists* $u_1 \in \mathsf{Out}[s_1, \sigma^1_{\mathrm{A}} \uplus \sigma^1_{\mathrm{B}}]$ *such that* $u_1 \beta u_2$.
- **B-Back$_{dd}$:** *Like* B-***Forth****, but with* 1 *and* 2 *swapped.*

States $s_1, s_2 \in \mathcal{M}$ *are* ConStR*-**bisimulation equivalent**, or just* ConStR*-**bisimilar** if there is a bisimulation* β *in* $\mathcal{M}$ *such that* $s_1 \beta s_2$.

Proposition 1 (ConStR-bisimulation invariance). *Let* β *be a* ConStR*-bisimulation in a game model* $\mathcal{M}$*. Then for every* ConStR*-formula* θ *and a pair* $s_1, s_2 \in \mathcal{M}$ *such that* $s_1 \beta s_2$*:* $\mathcal{M}, s_1 \models \theta$ *iff* $\mathcal{M}, s_2 \models \theta$.

Proof. Induction on θ. All boolean cases are straightforward. The cases for the 3 strategic operators are similar, but we will nevertheless check each of them, to ensure that the bisimulation conditions above are correctly defined.

For the strategic operators, we only check here the case of $\theta = \langle\langle \mathrm{A} \rangle\rangle_{\mathsf{c}}(\phi; \langle \mathrm{B} \rangle \psi)$, assuming that the claim holds for ϕ and ψ. The cases of $[\mathrm{A}]_{\mathsf{dr}}(\phi; \langle \mathrm{B} \rangle \psi)$ and $[\mathrm{A}]_{\mathsf{dd}}(\phi; \langle \mathrm{B} \rangle \psi)$ are quite analogous. Their proofs are omitted for lack of space. **(Case O$_c$)** Let $\theta = \langle\langle \mathrm{A} \rangle\rangle_{\mathsf{c}}(\phi; \langle \mathrm{B} \rangle \psi)$, assuming that the claim holds for ϕ and ψ.

Suppose, $\mathcal{M}, s_1 \models \theta$. Then A has a joint action σ^1_{A} at s_1 such that, when applied, it guarantees ϕ and enables B to adopt a joint action σ_{B} that is consistent with σ_{A} and guarantees ψ when additionally applied by B. By **A-Forth**$_{\mathsf{c}}$, there is a joint action σ^2_{A} of A at s_2, such that, by **A-LocalBack**$_{\mathsf{c}}$, for each $u_2 \in \mathsf{Out}[s_2, \sigma^2_{\mathrm{A}}]$ there exists $u_1 \in \mathsf{Out}[s_1, \sigma^1_{\mathrm{A}}]$ such that $u_1 \beta u_2$. By the choice of σ^1_{A}, $\mathcal{M}, u_1 \models \phi$ for each $u_1 \in \mathsf{Out}[s_1, \sigma^1_{\mathrm{A}}]$. It follows, by the inductive hypothesis applied to ϕ, that $\mathcal{M}, u_2 \models \phi$ for each $u_2 \in \mathsf{Out}[s_2, \sigma^2_{\mathrm{A}}]$. Moreover, B has a joint action σ^1_{B} at s_1 such that, when applied by B, in addition to A applying σ^1_{A}, it guarantees ψ, i.e. $\mathcal{M}, u_1 \models \psi$ for each $u_1 \in \mathsf{Out}[s_1, \sigma^1_{\mathrm{A}} \uplus \sigma^1_{\mathrm{B}}]$. By condition **B-Forth**$_{\mathsf{c}}$, there is a joint action σ^2_{B} of B at s_2, such that, by **(A$\uplus$B)-LocalBack**$_{\mathsf{c}}$, for every $u_2 \in \mathsf{Out}[s_2, \sigma^2_{\mathrm{A}} \uplus \sigma^2_{\mathrm{B}}]$ there exists $u_1 \in \mathsf{Out}[s_1, \sigma^1_{\mathrm{A}} \uplus \sigma^1_{\mathrm{B}}]$ such that $u_1 \beta u_2$. Therefore, by the inductive hypothesis applied to ψ, $\mathcal{M}, u_2 \models \psi$ for each $u_2 \in \mathsf{Out}[s_2, \sigma^2_{\mathrm{A}} \uplus \sigma^2_{\mathrm{B}}]$. Thus, $\mathcal{M}, s_2 \models \theta$. The converse is similar, using **A-Back**$_{\mathsf{c}}$.

Proposition 2 (Hennessy-Milner property). *For any finite game model $\mathcal{M}$ there is a* ConStR*-bisimulation β in $\mathcal{M}$, such that for any pair $s_1, s_2 \in \mathcal{M}$, $s_1 \beta s_2$ holds iff s_1 and s_2 are* ConStR*-equivalent (satisfy the same* ConStR*-formulae).*

Proof (Sketch). One direction follows from Proposition 1. For the converse, it suffices to prove that the relation of ConStR-equivalence is itself a ConStR-bisimulation in $\mathcal{M}$. Since $\mathcal{M}$ is finite, there is a mapping χ from $\mathcal{M}$ to the formulae of ConStR that assigns to each state s in $\mathcal{M}$ its characteristic formula $\chi(s)$, such that s_1, s_2 are ConStR-equivalent if and only if s_1 satisfies $\chi(s_2)$ (and vice versa), iff $\chi(s_1) \equiv \chi(s_2)$. Furthermore, $\chi(s_1) \wedge \chi(s_2) \equiv \bot$ whenever s_1 and s_2 are not ConStR-equivalent. Now, for any set of states Z in $\mathcal{M}$ we define $\chi(Z) := \bigvee_{z \in Z} \chi(z)$.

The crucial observation for proving the claim is that every state $s \in \mathcal{M}$ satisfies each of the following formulae, enabling the verification of the respective ConStR-bisimulation conditions:

(1) $\bigwedge_{A,B \subseteq \mathbb{A}\mathrm{gt}} \{\langle\!\langle \mathrm{A} \rangle\!\rangle_{\mathsf{c}}(\chi(Z); \langle \mathrm{B} \rangle \chi(Y)) \mid \exists \sigma \in \mathrm{Act}_s : \mathsf{Out}[s, \sigma|_A] = Z$ and $\mathsf{Out}[s, \sigma|_{(A \cup B)}] = Y\}$

(2) $\bigwedge_{A,B \subseteq \mathbb{A}\mathrm{gt}} \{[\mathrm{A}]_{\mathsf{dr}}(\chi(Z); \langle \mathrm{B} \rangle \chi(Y)) \mid \forall \sigma \in \mathrm{Act}_s : \mathsf{Out}[s, \sigma|_A] \subseteq Z$ implies $\mathsf{Out}[s, \sigma'|_{(A \cup B)}] \subseteq Y$ for some $\sigma' \in \mathrm{Act}_s$ such that $\sigma'|_A = \sigma|_A\}$

(3) $\bigwedge_{A,B \subseteq \mathbb{A}\mathrm{gt}} \{[\mathrm{A}]_{\mathsf{dd}}(\chi(Z); \langle \mathrm{B} \rangle \chi(Y)) \mid \exists \sigma \in \mathrm{Act}_s : \forall \sigma' \in \mathrm{Act}_s$ if $\mathsf{Out}[s, \sigma'|_A] \subseteq Z$ and $\sigma'|_{(\mathrm{B} \setminus \mathrm{A})} = \sigma|_{(\mathrm{B} \setminus \mathrm{A})}$ then $\mathsf{Out}[s, \sigma'|_{(A \cup B)}] \subseteq Y\}$

4.2 Some Remarks on Expressiveness and Definability

Proposition 3. *Let* a, b *be different agents and p, q be different atomic propositions. Then the following hold, where $\equiv$ is the logical equivalence in* ConStR[5].

[5] Even though we state the non-definability claims for CL, they apply likewise even to ATL*, because all formulae of ATL* are invariant under alternating bisimulations.

1. $[\mathsf{a}]_{\mathsf{dr}}(p; \langle \mathsf{b} \rangle q) \not\equiv [\mathsf{a}]_{\mathsf{dd}}(p; \langle \mathsf{b} \rangle q)$.
2. $\langle\langle \mathsf{a} \rangle\rangle_{\mathsf{c}}(p; \langle \mathsf{b} \rangle q)$ *is not definable in* CL.
3. $[\mathsf{a}]_{\mathsf{c}}(p|q)$ *(and, consequently,* $[\mathsf{a}]_{\mathsf{dr}}(p; \langle \emptyset \rangle q)$*) is not definable in* CL.
4. $[\mathsf{b}]_{\mathsf{dd}}(q; \langle \mathsf{a} \rangle p)$ *is not definable in* CL.

Proof. The first 3 claims follow respectively from Examples 3, 4, and 5 in the Appendix. The proof of the last claim is analogous.

The results above generalise to pairwise coalitions in a straightforward way.

5 Concluding Remarks: The Road Ahead

First, we note that, while the new strategic operators introduced here can be expressed in a suitable version of Strategy Logic (cf. [9]), we choose – for both conceptual and computational reasons – to stay within a purely modal framework where actions and strategies are not explicitly referred and quantified over in the language, but are only present in the semantics.

We regard this work as a first step towards developing a rich technical framework for logic-based conditional strategic reasoning of rational agents. The major further steps and directions include:

1. Complete axiomatization and proof of decidability of the logic ConStR (currently under development).
2. Extending the framework to a full-fledged, *long term* conditional strategic reasoning, by extending the language with standard temporal operators, to produce an ATL-like extension of ConStR.
3. The long term conditional strategic reasoning naturally requires considerations about strategic commitments and model updates (cf. [1,2]) and, more generally, requires involving strategy contexts in the semantics ([7]).
4. Adding knowledge, explicitly in the language, and implicitly, in the semantics, by assuming that the agents reason and act under imperfect information.
5. Last, but most important long-term objective of this project is to model and capture by semantically richer logic-based formalism the *mutually conditional strategic reasoning*, where all agents reason about their strategic choices, conditional on the others' strategic choices, conditional on the reasoners' choices, etc., recursively.

Appendix: Some Examples

Example 1. The game model $\mathcal{M}$ below has two players, a and b. Each has two actions at state s_0: a_1, a_2, resp. b_1, b_2.

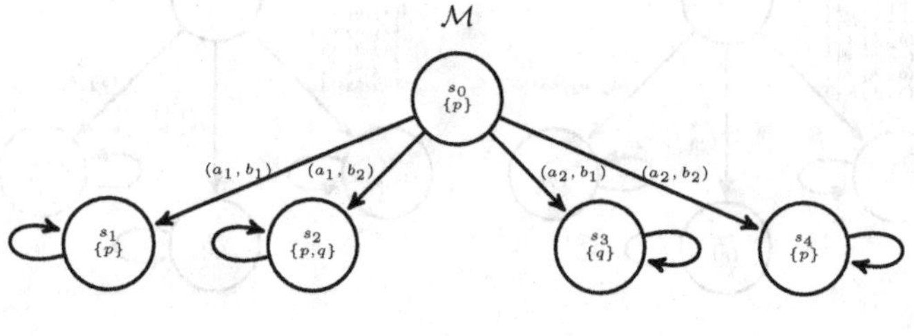

It can be verified that $\mathcal{M}, s_0 \Vdash \langle\!\langle \mathsf{a} \rangle\!\rangle_{\mathsf{c}}(p; \langle \mathsf{b} \rangle q)$, while $\mathcal{M}, s_0 \nVdash [\mathsf{b}]q$. Thus, an agent may have only conditional ability to achieve its goal.

Example 2. The game model $\mathcal{M}$ below has two players, a and b.

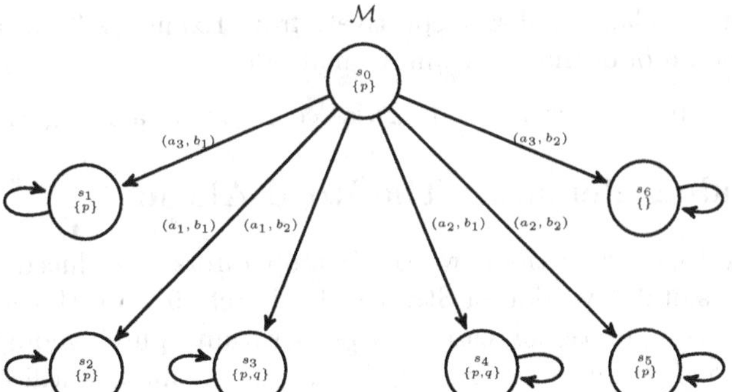

It can be verified that $\mathcal{M}, s_0 \Vdash [\mathsf{a}]_{\mathsf{dr}}(p; \langle \mathsf{b} \rangle q)$. However, $\mathcal{M}, s_0$ does not satisfy the ATL* formula $[\![\mathsf{a}]\!](\mathrm{X}p \to \langle\!\langle \mathsf{b} \rangle\!\rangle \mathrm{X}q)$, hence these are not equivalent.

Also, $\mathcal{M}, s_0 \nVdash [\mathsf{a}]_{\mathsf{dd}}(p; \langle \mathsf{b} \rangle q)$. However, if the outcomes of (a_2, b_1) and (a_2, b_2) are swapped, then $[\mathsf{a}]_{\mathsf{dd}}(p; \langle \mathsf{b} \rangle q)$ becomes true at s_0 in the resulting model.

Example 3. The game model $\mathcal{M}$ below involves two players: a and b. It can be verified that $\mathcal{M}, s_0 \Vdash [\mathsf{a}]_{\mathsf{dr}}(p; \langle \mathsf{b} \rangle q)$ but $\mathcal{M}, s_0 \nVdash [\mathsf{a}]_{\mathsf{dd}}(p; \langle \mathsf{b} \rangle q)$.

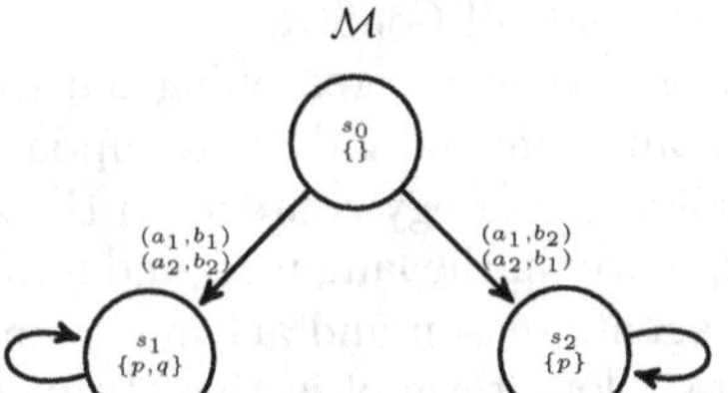

Example 4. The game models $\mathcal{M}_1$ and $\mathcal{M}_2$ below involve three players: a, b, c. It can be verified that:

(1) The relation $\beta = \{(s_i, t_i) \mid i = 0, 1, 2, 3\}$ is an alternating bisimulation between $\mathcal{M}_1$ and $\mathcal{M}_2$ (cf. [1]).
(2) $\mathcal{M}_1, s_0 \Vdash \langle\!\langle \mathsf{a} \rangle\!\rangle_{\mathsf{c}}(p; \langle \mathsf{b} \rangle q)$ but $\mathcal{M}_2, t_0 \nVdash \langle\!\langle \mathsf{a} \rangle\!\rangle_{\mathsf{c}}(p; \langle \mathsf{b} \rangle q)$.

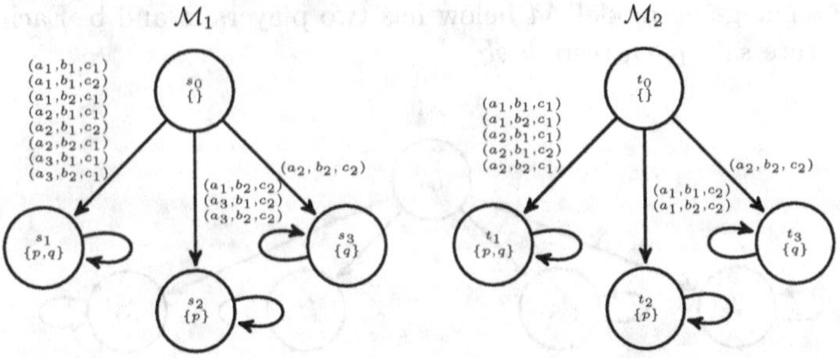

Example 5. The game models $\mathcal{M}_1$ and $\mathcal{M}_2$ below involve two players: a and b. It can be verified that:

(1) The relation $\beta = \{(s_i, t_i) \mid i = 0, 1, 2, 3\}$ is an alternating bisimulation between $\mathcal{M}_1$ and $\mathcal{M}_2$ (cf. [1]).
(2) $\mathcal{M}_1, s_0 \Vdash [\mathsf{a}]_{\mathsf{c}}(p|q)$ but $\mathcal{M}_2, t_0 \nVdash [\mathsf{a}]_{\mathsf{c}}(p|q)$.

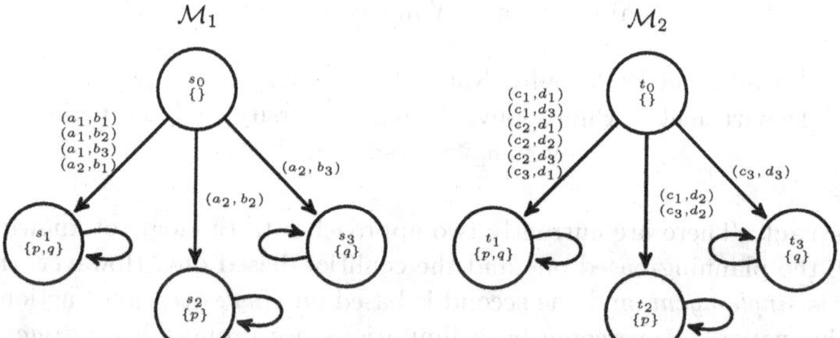

References

1. Ågotnes, T., Goranko, V., Jamroga, W.: Alternating-time temporal logics with irrevocable strategies. In: Samet, D. (ed.) Proceedings of TARK XI, pp. 15–24 (2007)
2. Ågotnes, T., Goranko, V., Jamroga, W.: Strategic commitment and release in logics for multi-agent systems (extended abstract). Technical report IfI-08-01, Clausthal University of Technology (2008)
3. Ågotnes, T., Goranko, V., Jamroga, W., Wooldridge, M.: Knowledge and ability. In: van Ditmarsch, H., Halpern, J., van der Hoek, W., Kooi, B. (eds.) Handbook of Epistemic Logic, pp. 543–589. College Publications, London (2015)
4. Alur, R., Henzinger, T.A., Kupferman, O.: Alternating-time temporal logic. J. ACM **49**(5), 672–713 (2002)
5. van Benthem, J., Ghosh, S., Verbrugge, R. (eds.): Models of Strategic Reasoning - Logics, Games, and Communities. LNCS, vol. 8972. Springer, Heidelberg (2015). https://doi.org/10.1007/978-3-662-48540-8
6. Brafman, R., Tennenholtz, M.: Modeling agents as qualitative decision makers. Artif. Intell. **94**(1), 217–268 (1997)
7. Brihaye, T., Da Costa, A., Laroussinie, F., Markey, N.: ATL with strategy contexts and bounded memory. In: Artemov, S., Nerode, A. (eds.) LFCS 2009. LNCS, vol. 5407, pp. 92–106. Springer, Heidelberg (2008). https://doi.org/10.1007/978-3-540-92687-0_7
8. Goranko, V., Enqvist, S.: Socially friendly and group protecting coalition logics. In: Proceedings of AAMAS 2018, pp. 372–380 (2018)
9. Mogavero, F., Murano, A., Perelli, G., Vardi, M.Y.: Reasoning about strategies: on the satisfiability problem. Logical Methods Comput. Sci. **13**(1), 1–37 (2017)
10. Pauly, M.: Logic for social software. Ph.D. thesis, University of Amsterdam (2001)
11. Pauly, M.: A modal logic for coalitional power in games. J. Logic Comput. **12**(1), 149–166 (2002)

Multi-agent Knowing How via Multi-step Plans: A Dynamic Epistemic Planning Based Approach

Yanjun Li[1] and Yanjing Wang[2](✉)

[1] College of Philosophy, Nankai University, Tianjin, China
[2] Department of Philosophy, Peking University, Beijing, China
y.wang@pku.edu.cn

Abstract. There are currently two approaches to the logic of knowing how: the planning-based one and the coalition-based one. However, the first is *single-agent*, and the second is based on *single-step* joint actions. In this paper, to overcome both limitations, we propose a *multi-agent* framework for the logic of knowing how, based on *multi-step* dynamic epistemic planning studied in the literature. We obtain a sound and complete axiomatization and show that the logic is decidable, although the corresponding multi-agent epistemic planning problem is undecidable.

1 Introduction

Standard epistemic logic has been mainly focusing on the reasoning patterns about *knowing that*. In recent years, the logics of other knowledge expressions such as *knowing whether* [5], *knowing how* [17], *knowing why* [22], and *knowing what* [21] are attracting increasing attention (see [18] for a survey). In particular, the logics of knowing how is well-situated in the common interest of logic, philosophy [14] and artificial intelligence [9].

Currently, there are two main approaches of the logics of *knowing how*. The planning-based approach initiated by Wang [16,18] is inspired by the philosophical discussions on the semantics of knowing how [14] and the ideas of automated planning under uncertainty in AI [6,18]. The semantics of the know-how modality is rendered formally by using the idea of first-order modal logic. In a nutshell, knowing how to achieve φ means that *there exists* a plan σ such that the agent *knows* that σ can definitely guarantee φ eventually ($\exists\sigma\mathcal{K}(\sigma\ \mathtt{guarantees}\ \varphi)$). Depending on the notion of the plan and the corresponding planning problem, various kinds of semantics for the know-how operator were proposed such as [6,8,18]. The other approach is proposed by Naumov and Tao [12], inspired by the tradition of coalition logic and alternating-time temporal logic. According to this approach, a coalition C knows how to achieve φ if and only if there is a joint action a for C such that it is distributed knowledge for C that a can achieve φ no matter what others do. Variants of the basic framework were proposed and discussed, such as knowing another coalition's ability [10], knowing how under the perfect recall assumption [11], knowing how with the degree of uncertainty [13], and so on.

P. Blackburn et al. (Eds.): LORI 2019, LNCS 11813, pp. 126–139, 2019.
https://doi.org/10.1007/978-3-662-60292-8_10

However, the drawback of the planning-based approach so far is that it is based on *single-agent* automated planning. On the other hand, the disadvantage of the coalition-based approach to know-how is that it so far only concerns the simple *single-step* actions.[1] Moreover, both approaches are based on models where the effects of actions are given as abstract transitions but not actions as in practical planning problems. In this paper, we make the first attempt to combine the advantages of the two approaches by a *multi-agent* framework of knowing how based on non-trivial *multi-step* plans of actions. Our framework is based on multi-agent epistemic planning studied in the literature, where arguably the most promising approach uses dynamic epistemic logic (DEL) as the background logic (cf. e.g., [4]). Given a multi-agent pointed epistemic model, a finite set of action models, and a goal formula φ, the epistemic planning problem is to decide whether there is a finite sequence of available action models such that all the resulting states after executing this sequence satisfy the goal formula. It is shown that the single-agent planning problem is decidable but the multi-agent case is in general undecidable (cf. e.g., [2]). Efforts are made to obtain decidability by restricting the action models, e.g., [23].

Coming back to the logic of knowing how, as in [6], in this paper we also have both the knowledge operator $\mathcal{K}_i$ and the know-how operator $\mathcal{K}h_i$ in the language (now indexed with agent names). By having both modalities, we can express interesting things such as: i does not know how to achieve φ but i knows that j knows how ($\neg\mathcal{K}h_i\varphi \wedge \mathcal{K}_i\mathcal{K}h_j\varphi$); i knows how to let j know φ but at the same time prevent k from knowing the truth value of φ ($\mathcal{K}h_i(\mathcal{K}_j\varphi \wedge \neg\mathcal{K}h_k(\mathcal{K}_k\varphi \vee \mathcal{K}_k\neg\varphi))$. We formalize the semantics of the know-how operator based on the above idea of dynamic epistemic planning. In contrast with the transition-system-like models in [6,12], a model in our setting is a standard epistemic model with a set of available action models for each agent. $\mathcal{K}h_i\varphi$ holds at a state s iff there is a finite sequence σ of action models for i such that i knows that φ holds after executing the action sequence on the initial model using the standard product update in DEL. The main technical result is a complete axiomatization of the logic of knowing how under this setting. Through the axioms of our logic we will see clearly the implicit assumptions behind dynamic epistemic planning at a very abstract level, which can be compared to the axioms of knowing how based on other planning notions in AI. Note that, as an initial attempt, we will not consider group notions of know-that and know-how as in [11].

In the rest of the paper, we will first layout the basics about epistemic planning in Sect. 2. Section 3 first introduces the language of knowing how and the semantics based on dynamic epistemic planning, and then gives a proof system. The main technical contributions of the paper are the completeness of the axiomatization and its decidability, which are proved in Sect. 5 based on the alternative 'static' semantics introduced in Sect. 4 inspired by the method developed in [19,20]. Finally, we conclude with further directions in Sect. 6.

[1] It is not the case in the closely related epistemic ATL e.g., [7].

2 Preliminaries

Definition 1 (Epistemic model). *Given a set* **Ag** *of agents, an epistemic model* $\mathcal{N}$ *is a tuple* $\langle W, \{\sim_i \mid i \in \mathbf{Ag}\}, V\rangle$ *where:*

- W *is a non-empty set;*
- $\sim_i \subseteq W \times W$ *is an equivalence relation over* W *for each* $i \in \mathbf{Ag}$*;*
- $V : W \to 2^{\mathbf{P}}$ *is a valuation funtion.*

Given a model $\mathcal{N}$, we may refer to its components as $W^{\mathcal{N}}$, $\sim_i^{\mathcal{N}}$, and $V^{\mathcal{N}}$.

Definition 2 (EL language). *Given a set* **P** *of proposition letters and a set* **Ag** *of agents, the epistemic language* $\mathcal{L}^{\mathsf{EL}}$ *is defined:*

$$\varphi \quad ::= \quad \top \mid p \mid \neg\varphi \mid (\varphi \wedge \varphi) \mid \mathcal{K}_i\varphi$$

where $p \in \mathbf{P}$ *and* $i \in \mathbf{Ag}$.

Definition 3 (EL semantics). *Given a formula* $\varphi \in \mathcal{L}^{\mathsf{EL}}$ *and a pointed epistemic model* $(\mathcal{N}, s)$*, the satisfaction relation on* φ *and pointed model* $(\mathcal{N}, s)$ *is defined:*

$$\begin{array}{c}
\mathcal{N}, s \vDash \top \quad \textit{always} \\
\mathcal{N}, s \vDash p \Leftrightarrow p \in V(s) \\
\mathcal{N}, s \vDash \neg\varphi \Leftrightarrow \mathcal{N}, s \nvDash \varphi \\
\mathcal{N}, s \vDash (\varphi \wedge \psi) \Leftrightarrow \mathcal{N}, s \vDash \varphi \textit{ and } \mathcal{N}, s \vDash \psi \\
\mathcal{N}, s \vDash \mathcal{K}_i\varphi \Leftrightarrow s \sim_i t \textit{ implies } \mathcal{N}, t \vDash \varphi
\end{array}$$

In DEL, there are also action models which capture actions that have factual and epistemic effects. Here we take the definition form [3].

Definition 4 (Action model). *An action model is*

$$\mathcal{E} = \langle E, \{\sim_i \mid i \in \mathbf{Ag}\}, pre, post\rangle$$

where

- E *is a non-empty set of event;*
- $\sim_i \subseteq E \times E$ *is an equivalence relation over* E *for each* $i \in \mathbf{Ag}$*;*
- $pre : E \to \mathcal{L}^{\mathsf{EL}}$ *assigns a precondition to each event;*
- $post : E \to \mathcal{L}^{\mathsf{EL}}$ *assigns a postcondition to each event. For each* $e \in E$, $post(e)$ *is a conjunction of literals over* **P** *(including* $\top$*).*

Given an action model $\mathcal{E}$*, we may refer to its components as* $E^{\mathcal{E}}$, $\sim_i^{\mathcal{E}}$, $pre^{\mathcal{E}}$ *and* $post^{\mathcal{E}}$. *We use* $post^{\mathcal{E}}(e)^+$ *to denote the set of proposition letters that are positive literals of* $post^{\mathcal{E}}(e)$*, and* $post^{\mathcal{E}}(e)^-$ *negatives. For each* $e \in E$*, we require that* $post^{\mathcal{E}}(e)^+ \cap post^{\mathcal{E}}(e)^- = \emptyset$.

For each nonempty finite subset S *of* E*, the pair* $(\mathcal{E}, S)$ *is called an* epistemic action *(or simply an* action*), and the events in* S *are called the* designated events.

Given an epistemic model $\mathcal{N}$ and $X \subseteq W^{\mathcal{N}}$, we use $[X]^i$ to denote the set $\{s \in W^{\mathcal{N}} \mid s \sim_i s' \text{ for some } s' \in X\}$. When X is a singleton $\{s\}$, we write it as $[s]^i$. Similarly, we use the notation $[Y]^i$ where Y is a subset of the domain of an action model $\mathcal{E}$. Given an action $a = (\mathcal{E}, S)$, the associated local action of agent $i \in \mathbf{Ag}$, denoted a^i, is $(\mathcal{E}, [S]^i)$. If $S = [S]^i$, we say a is a local action *of agent i. It is possible that one action is local to multiple agents.*

As an example, consider an action model $\mathcal{E}$ which is depicted as follows:

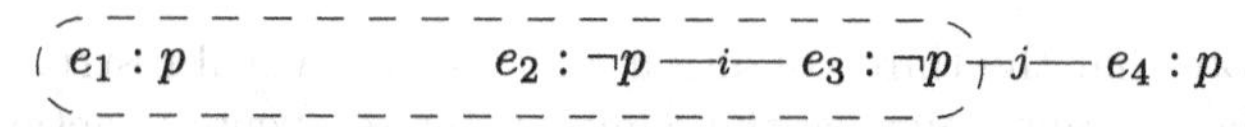

The domain of $\mathcal{E}$ is $E = \{e_1, e_2, e_3, e_4\}$, the precondition of e_1 and e_4 is p, and the precondition of e_2 and e_3 is $\neg p$. The action $\langle \mathcal{E}, \{e_1, e_2\} \rangle$ intuitively means an announcement of the truth value of p, but, if $\neg p$ is the case, for some reason, the agent i cannot make sure the agent j actually hear the announcement or not. Thus, when the action $\langle \mathcal{E}, \{e_1, e_2\} \rangle$ happens, what agent i thought has happened is $\langle \mathcal{E}, \{e_1, e_2, e_3\} \rangle$, i.e. the local action of i but not of j.

Definition 5 (Product update). *Given an epistemic model $\mathcal{N} = \langle W, \{\sim_i \mid i \in \mathbf{Ag}\}, V \rangle$ and an action model $\mathcal{E} = \langle E, \{\sim_i \mid i \in \mathbf{Ag}\}, pre, post \rangle$, the product update is*

$$\mathcal{N} \otimes \mathcal{E} = \langle W', \{\sim'_i \mid i \in \mathbf{Ag}\}, V \rangle$$

where

- $W' = \{(w, e) \in W \times E \mid (\mathcal{N}, w) \vDash pre(e)\}$
- $\sim'_i = \{(w, e), (w', e') \mid w \sim_i w', e \sim_i e'\}$
- $V(w, e) = (V^{\mathcal{N}}(w) \setminus post^{\mathcal{E}}(e)^-) \cup post^{\mathcal{E}}(e)^+$.

In [3], given a (finite) model $\mathcal{N}$ with a non-empty set $X \subseteq W^{\mathcal{N}}$, a linear sequence σ of (finite) actions is a good epistemic plan for an epistemic goal formula φ if σ is *applicable* on X (recursively defined below) and it reaches only φ states eventually by executing the corresponding product updates.

Definition 6 (Applicability [3]). *Given an action $a = (\mathcal{E}, S)$ and an epistemic model $\mathcal{N}$, we say that a is applicable in $X \subseteq W$ if, for each $w \in X$, there is an event $e \in S$ such that $\mathcal{N}, w \vDash pre^{\mathcal{E}}(e)$. An action sequence $(\mathcal{E}_1, S_1) \cdots (\mathcal{E}_{n+1}, S_{n+1})$ for all $n \geq 0$ is applicable on X if $(\mathcal{E}_1, S_1) \cdots (\mathcal{E}_n, S_n)$ is applicable on X and $(\mathcal{E}_{n+1}, S_{n+1})$ is applicable on each state in $(X \times S_1 \cdots \times S_n)|_{W^{\mathcal{N} \otimes \mathcal{E}_1 \cdots \otimes \mathcal{E}_n}}$, i.e., the restriction of $X \times S_1 \cdots \times S_n$ to $W^{\mathcal{N} \otimes \mathcal{E}_1 \cdots \otimes \mathcal{E}_n}$.*

3 Logic of Knowing How

Definition 7 (ELKh language). *Given a set $\mathbf{P}$ of proposition letters and a set $\mathbf{Ag}$ of agents, the language $\mathcal{L}^{ELKh}$ is constructed as follows:*

$$\varphi ::= \top \mid p \mid \neg\varphi \mid (\varphi \wedge \varphi) \mid \mathcal{K}_i\varphi \mid \mathcal{K}h_i\varphi$$

where $p \in \mathbf{P}$ and $i \in \mathbf{Ag}$.

Definition 8 (Dynamic model of ELKh). *A dynamic model is a pair* $\mathcal{M} = \langle \mathcal{N}, \{\mathcal{A}_i \mid i \in \mathbf{Ag}\}\rangle$ *where*

- $\mathcal{N}$ *is an epistemic model;*
- *for each* $i \in \mathbf{Ag}$, $\mathcal{A}_i$ *is a set of local actions of* i;
- *all actions are based on the* same *action model, that is, if there are* $\langle \mathcal{E}_1, S_1\rangle \in \mathcal{A}_i$ *and* $\langle \mathcal{E}_2, S_2\rangle \in \mathcal{A}_j$ *for some* $i, j \in \mathbf{Ag}$ *then* $\mathcal{E}_1 = \mathcal{E}_2$.

We use $\mathcal{A}$ *to denote the set* $\bigcup_{i\in\mathbf{Ag}} \mathcal{A}_i$.

Remark 1. Note that the third condition above is not a real restriction because we can use the disjoint union as the common set of events if actions differ in their own sets of events.

As mentioned before, the idea behind the semantics of $\mathcal{K}h_i\varphi$ here is that there is a plan consisting of available actions in $\mathcal{A}_i$ such that agent i knows that it can guarantee φ, i.e., σ is a good epistemic plan according to [3].

Definition 9 (Dynamic semantics of ELKh). *Given any dynamic model* $\mathcal{M} = \langle \mathcal{N}, \{\mathcal{A}_i \mid i \in \mathbf{Ag}\}\rangle$, *the truth conditions for* $\mathcal{L}^{\mathsf{EL}}$ *formulas are as usual based on* $\mathcal{N}$, *and the dynamic semantics for* $\mathcal{K}h$*-formulas is defined as follows:*

$$\begin{aligned}\mathcal{M}, s \vDash \mathcal{K}h_i\varphi \Leftrightarrow\ & \text{there is an action sequence } (\mathcal{E}, S_1)\cdots(\mathcal{E}, S_n) \in (\mathcal{A}_i)^* :\\ & (1.)\ (\mathcal{E}, S_1)\cdots(\mathcal{E}, S_n) \text{ is applicable on } [s]^i\\ & (2.)\ \mathcal{M}', w \vDash \varphi \text{ for each } w \in D\end{aligned}$$

where $\mathcal{M}' = \langle \mathcal{N} \otimes \mathcal{E}^n, \{\mathcal{A}_i \mid i \in \mathbf{Ag}\}\rangle$ *and* $D = ([s]^i \times S_1 \cdots \times S_n)|_{W^{\mathcal{N}\otimes\mathcal{E}^n}}$ *represent the final model and the final states after executing the action sequence.*

Given a dynamic model $\mathcal{M} = \langle \mathcal{N}, \{\mathcal{A}_i \mid i \in \mathbf{Ag}\}\rangle$, it is not hard to see that, according to [3], $\mathcal{M}, s \vDash \mathcal{K}h_i p$ iff there is an epistemic plan for the goal p w.r.t. $\mathcal{N}$ with $[s]^i \subseteq W^{\mathcal{N}}$, and the action set $\mathcal{A}_i$. For each $\varphi \in \mathcal{L}^{\mathsf{EL}}$, we have that $\mathcal{M}, s \vDash \varphi$ iff $\mathcal{N}, s \vDash \varphi$.

Definition 10. *The proof system* $\mathbb{SLKH}$ *is as follows:*

Axioms

TAUT	*all axioms of propositional logic*
DISTK	$\mathcal{K}_i p \wedge \mathcal{K}_i(p \to q) \to \mathcal{K}_i q$
T	$\mathcal{K}_i p \to p$
4	$\mathcal{K}_i p \to \mathcal{K}_i\mathcal{K}_i p$
5	$\neg\mathcal{K}_i p \to \mathcal{K}_i\neg\mathcal{K}_i p$
AxKtoKh	$\mathcal{K}_i p \to \mathcal{K}h_i p$
AxKhtoKKh	$\mathcal{K}h_i p \to \mathcal{K}_i\mathcal{K}h_i p$
AxKhbot	$\neg\mathcal{K}h_i\bot$
AxKhtoKhK	$\mathcal{K}h_i p \to \mathcal{K}h_i\mathcal{K}_i p$

Rules

MP $\dfrac{\varphi, \varphi \to \psi}{\psi}$ NECK $\dfrac{\varphi}{\mathcal{K}_i\varphi}$

MONOKh $\dfrac{\varphi \to \psi}{\mathcal{K}h_i\varphi \to \mathcal{K}h_i\psi}$ SUB $\dfrac{\varphi(p)}{\varphi[\psi/p]}$

AxKtoKh is valid since we allow empty plan.[2] AxKhtoKKh is valid because we actually start with the set $[s]^i$ in the semantics. The validity of AxKhbot is due

[2] One may find AxKtoKh counter-intuitive under the global view of know-how (cf. [18]): knowing that the safe is open does not imply knowing how to open it in all the possible situations. However, the notion of know-how in this paper is a local one: knowing how to achieve φ under the current circumstance.

to applicability. To see the validity of `AxKhtoKhK` the reader can verify that the set of D in the semantics is i-closed based on the fact that $[s]^i$ is an equivalence class and the product update preserve the i-closeness if S_k are i-closed for all k.

Theorem 1 (Soundness). *$\mathbb{SLKH}$ is sound, namely, if $\vdash \varphi$ then $\vDash \varphi$.*

On the other hand, $\mathcal{K}_i\mathcal{K}h_j\varphi \to \mathcal{K}h_i\varphi$ and $\mathcal{K}h_i\mathcal{K}h_j\varphi \to \mathcal{K}h_i\varphi$ are not valid due to the fact that the available actions to each agent may be different, in contrast with the validity of $\mathcal{K}_i\mathcal{K}_j\varphi \to \mathcal{K}_i\varphi$. Note that compared to [6], we do not have the axiom of $\mathcal{K}h_i\mathcal{K}h_i\varphi \to \mathcal{K}h_i\varphi$ since in contrast to the branching plans in [6], we only allow linear plans here.

4 An Alternative Semantics

Showing the completeness of our system is a highly non-trivial task due to the fact that the semantics is based on non-trivial use of DEL actions, which cannot be expressed directly in our language of know-how, unlike the language of DEL. Moreover, the dynamics in the semantics is hard to handle, thus in this section, inspired by the method in [15,19,20], we first give an equivalent alternative semantics based on *extended models* with both the epistemic relations and the transitions between the states, which intuitively can be viewed as temporal *unravelings* of the epistemic model w.r.t. available actions. Then in Sect. 5 we prove the completeness w.r.t. this alternative semantics and finally obtain the completeness w.r.t. the original semantics.

Definition 11 (Extended model of ELKh). *An extended model is a triple $\mathcal{M}^+ = \langle \mathcal{N}, \{\mathcal{A}_i \mid i \in \mathbf{Ag}\}, \{\xrightarrow{e} \mid e \in S \text{ for some } \langle \mathcal{E}, S\rangle \in \mathcal{A}\}\rangle$ where $\langle \mathcal{N}, \{\mathcal{A}_i \mid i \in \mathbf{Ag}\}\rangle$ is a dynamic model and $\xrightarrow{e}$ is a binary relation on $W^{\mathcal{N}}$. For each $w \in W^{\mathcal{N}}$, $(\mathcal{M}^+, w)$ is a pointed extended model.*

Given an extended model $\mathcal{M}^+$ and an action $a = \langle \mathcal{E}, S\rangle \in \mathcal{A}$, we define the binary relation $\xrightarrow{a}$ as

$$\xrightarrow{a} = \bigcup_{e \in S} \xrightarrow{e}$$

We also write $(w, v) \in \xrightarrow{a}$ as $w \xrightarrow{a} v$. We use $Q(a_1 \cdots a_n)(s)$ to denote the set of states t such that there exist $t_1 \cdots t_n$ satisfying that $s \xrightarrow{a_1} t_1 \cdots \xrightarrow{a_n} t_n$ and $t = t_n$. If X is a set of states, we then use $Q(a_1 \cdots a_n)(X)$ to mean $\bigcup_{s \in X} Q(a_1 \cdots a_n)(s)$, in particular $Q(a_1 \cdots a_n)([s]) = \bigcup_{u \sim s} Q(a_1 \cdots a_n)(u)$. If $Q(a_1 \cdots a_n)(s) \neq \emptyset$, we say that $a_1 \cdots a_n$ is executable on s. We define a notion of *strong executability* over extended models inspired by [18].

Definition 12 (Strong executability). *Let $\mathcal{M}^+$ be an extended model $\langle \mathcal{N}, \{\mathcal{A}_i \mid i \in \mathbf{Ag}\}, \{\xrightarrow{e} \mid e \in S \text{ for some } \langle \mathcal{E}, S\rangle \in \mathcal{A}\}\rangle$, we say an action $a = \langle \mathcal{E}, S\rangle \in \mathcal{A}$ is strongly executable on $X \subseteq W$ if, for each $w \in X$, there is an event $e \in S$ such that $\mathcal{N}, w \vDash pre^{\mathcal{E}}(e)$. An action sequence $a_1 \cdots a_{n+1} \in \mathcal{A}^*$ is strongly executable on X if $a_1 \cdots a_n$ is strongly executable on X and a_{n+1} is strongly executable on each state v such that $w \xrightarrow{a_1 \cdots a_n} v$ for some $w \in X$.*

We can now give the alternative semantics of ELKh over extended models.

Definition 13 (Static semantics of ELKh). *Given an extended model* $\mathcal{M}^+ = \langle \mathcal{N}, \{\mathcal{A}_i \mid i \in \mathbf{Ag}\}, \{\xrightarrow{e} \mid e \in S \text{ for some } \langle \mathcal{E}, S\rangle \in \mathcal{A}\}\rangle$, *the static semantics* ($\Vdash$) *is defined as follows:*

$$\mathcal{M}^+, s \Vdash \mathcal{K}h_i\varphi \Leftrightarrow \text{there is an action sequence } a_1 \cdots a_n \in (\mathcal{A}_i)^* : \\ (1.)\ a_1 \cdots a_n \text{ is strongly executable on } [s]^i \\ (2.)\ \mathcal{M}^+, w \vDash \varphi \text{ for each } w \in Q(a_1 \cdots a_n)([s]^i).$$

Please note that, for each $\varphi \in \mathcal{L}^{\mathsf{EL}}$, we have that $\mathcal{M}^+, s \Vdash \varphi$ iff $\mathcal{N}, s \vDash \varphi$.

Definition 14 (**EA-Bisimulation**). *Let* $\mathcal{M}_1^+$ *and* $\mathcal{M}_2^+$ *be two extended models such that* $\mathcal{A}_i^{\mathcal{M}_1^+} = \mathcal{A}_i^{\mathcal{M}_2^+}$ *for each* $i \in \mathbf{Ag}$. *A binary relation* $Z \subseteq (W^{\mathcal{M}_1^+} \times W^{\mathcal{M}_2^+})$ *is called an* EA-*bisimulation if whenever* $(w, w') \in Z$ *the following hold:*

- *for each* $p \in \mathbf{P}$, $p \in V^{\mathcal{M}_1^+}(w)$ *iff* $p \in V^{\mathcal{M}_1^+}(w')$;
- *for each* $i \in \mathbf{Ag}$,
  - *if* $w \sim_i v$ *in* $\mathcal{M}_1^+$ *for some* v, *then there is* v' *in* $\mathcal{M}_2^+$ *such that* $w' \sim_i v'$ *in* $\mathcal{M}_2^+$ *and* $(v, v') \in Z$;
  - *if* $w' \sim_i v'$ *in* $\mathcal{M}_2^+$ *for some* v', *then there is* v *in* $\mathcal{M}_1^+$ *such that* $w \sim_i v$ *in* $\mathcal{M}_1^+$ *and* $(v, v') \in Z$;
- *for each* $a = \langle \mathcal{E}, S\rangle \in \mathcal{A}$ *and* $e \in S$,
  - *if* $w \xrightarrow{e} v$ *in* $\mathcal{M}_1^+$ *for some* v, *then there is* v' *in* $\mathcal{M}_2^+$ *such that* $w' \xrightarrow{e} v'$ *in* $\mathcal{M}_2^+$ *and* $(v, v') \in Z$;
  - *if* $w' \xrightarrow{e} v'$ *in* $\mathcal{M}_2^+$ *for some* v', *then there is* v *in* $\mathcal{M}_1^+$ *such that* $w \xrightarrow{e} v$ *in* $\mathcal{M}_1^+$ *and* $(v, v') \in Z$.

If it only satisfies the condition for propositional variables and the condition for $\sim_i$, *we call* Z *an E-bisimulation.*

Given two pointed model $\mathcal{M}_1^+, w$ and $\mathcal{M}_2^+, v$, if there is an EA-bisimulaiton Z between $\mathcal{M}_1^+$ and $\mathcal{M}_2^+$ and $(w, v) \in Z$, we will write it as $\mathcal{M}_1^+, w \leftrightarroweq_{EA} \mathcal{M}_2^+, v$.

Proposition 1. *If* $\mathcal{M}_1^+, w \leftrightarroweq_{EA} \mathcal{M}_2^+, w'$, *then we have that* $\mathcal{M}_1^+, w \Vdash \varphi$ *iff* $\mathcal{M}_2^+, w' \Vdash \varphi$ *for each* $\varphi \in$ *ELKh.*

Proof. (Sketch) This can be proved by induction on φ. We will only focus on the case of $\mathcal{K}h_i\varphi$; the other cases are straightforward by IH. For the case of $\mathcal{K}h_i\varphi$, the key of the proof is that, if there is a sequence $w \xrightarrow{e_1} \cdots \xrightarrow{e_n} w_n$ in $\mathcal{M}_1^+$, due to $\mathcal{M}_1^+, w \leftrightarroweq_{EA} \mathcal{M}_2^+, w'$, there exists a sequence $w' \xrightarrow{e_1} \cdots \xrightarrow{e_n} w'_n$ in $\mathcal{M}_2^+$ such that $\mathcal{M}_1^+, w_n \leftrightarroweq_{EA} \mathcal{M}_2^+, w'_n$.

Definition 15. *Given a dynamic model* $\mathcal{M} = \langle \mathcal{N}, \{\mathcal{A}_i \mid i \in \mathbf{Ag}\}\rangle$, *the extended model* $\mathcal{M}^\omega$ *is* $\langle \biguplus_{i \in \mathrm{N}} \mathcal{N} \otimes \mathcal{E}^i, \{\mathcal{A}_i \mid i \in \mathbf{Ag}\}, \{\xrightarrow{e} \mid e \in S \text{ for some } \langle \mathcal{E}, S\rangle \in \mathcal{A}\}\rangle$:

- $\biguplus_{i \in \mathrm{N}} \mathcal{N} \otimes \mathcal{E}^i$ *is the disjoint union of all* $\mathcal{N} \otimes \mathcal{E}^n$, *also written as* $\mathcal{N}^\omega$;
- $\xrightarrow{e} = \{(\xi, \xi') \mid \xi' = (\xi, e), \text{ and } \xi, \xi' \in \mathcal{N}^\omega\}$

As in [19], some conditions of the extended models are given to equate $\vDash$ and $\Vdash$.

Definition 16 (Model Class $\mathbb{C}$). *The model class $\mathbb{C}$ is the set of extended models that satisfy the following conditions:*

Pre *for each $\langle \mathcal{E}, S\rangle \in \mathcal{A}$ and each $e \in S$, w has an e-successor iff $\mathcal{M}^+, w \Vdash pre^{\mathcal{E}}(e)$;*

Pos *for each $\langle \mathcal{E}, S\rangle \in \mathcal{A}$ and each $e \in S$, if $w \xrightarrow{e} v$ then $V(v) = (V(w) \setminus post^{\mathcal{E}}(e)^-) \cup post^{\mathcal{E}}(e)^+$;*

Nm (no miracles) *for each $\langle \mathcal{E}, S\rangle \in \mathcal{A}$ and $e_1, e_2 \in S$, if $w \sim_i w'$, $w \xrightarrow{e_1} v$, $w' \xrightarrow{e_2} v'$, and $e_1 \sim_i^{\mathcal{E}} e_2$, we then have that $v \sim_i v'$;*

Pr (perfect recall) *for each $\langle \mathcal{E}, S\rangle \in \mathcal{A}$ and each $e_1 \in S$, if $w \xrightarrow{e_1} v$ and $v \sim_i v'$, we then have that there exist $w' \in W$ and $e_2 \in S$ such that $w \sim_i w'$, $w' \xrightarrow{e_2} v'$ and $e_1 \sim_i^{\mathcal{E}} e_2$.*

Proposition 2. *Given an extended model $\mathcal{M}^+ = \langle \mathcal{M}, \{\xrightarrow{e} \mid e \in S \text{ for some } \langle \mathcal{E}, S\rangle \in \mathcal{A}\}\rangle$ where $\mathcal{M} = \langle \mathcal{N}, \{\mathcal{A}_i \mid i \in \mathbf{Ag}\}\rangle$, let Z_n be a binary relation on $W^{\mathcal{N}}$ and $W^{\mathcal{N}\otimes\mathcal{E}^n}$ for $i \in \mathbb{N}$, which is defined as $Z_n = \{(t, se_1 \cdots e_n) \in W^{\mathcal{N}} \times W^{\mathcal{N}\otimes\mathcal{E}^n} \mid$ there exist $s_0, \cdots, s_n \in W^{\mathcal{N}}$ such that $s_0 = s$, $s_n = t$ and $s_i \xrightarrow{e_{i+1}} s_{i+1}$ in $\mathcal{M}^+$ for $0 \leq i < n\}$. If $Z_n \neq \emptyset$ then Z_n is an E-bisimulaiton between $\mathcal{N}$ and $\mathcal{N} \otimes \mathcal{E}^n$.*

Lemma 1. *Given a dynamic model $\mathcal{M} = \langle \mathcal{N}, \{\mathcal{A}_i \mid i \in \mathbf{Ag}\}\rangle$ and an extended model $\mathcal{M}^+ = \langle \mathcal{M}, \{\xrightarrow{e} \mid e \in S \text{ for some } \langle \mathcal{E}, S\rangle \in \mathcal{A}\}\rangle$, if $\mathcal{M}^+ \in \mathbb{C}$, we then have that $\mathcal{M}^+, w \leftrightarroweq_{EA} \mathcal{M}^\omega, w$ for all $w \in W^{\mathcal{N}}$.*

Lemma 2. *Given a dynamic model $\mathcal{M} = \langle \mathcal{N}, \{\mathcal{A}_i \mid i \in \mathbf{Ag}\}\rangle$, for each $\varphi \in$ ELKh and each $n \in \mathbb{N}$, we have that $\langle \mathcal{N} \otimes \mathcal{E}^n, \{\mathcal{A}_i \mid i \in \mathbf{Ag}\}\rangle, (we_1 \cdots e_n) \vDash \varphi$ iff $\mathcal{M}^\omega, (we_1 \cdots e_n) \Vdash \varphi$.*

Proofs of Proposition 2 and Lemmas 1 and 2 are omitted due to limited space (cf. [19] for similar results).

Theorem 2. *For each ELKh-formula φ, if $\vDash \varphi$ then $\mathbb{C} \Vdash \varphi$.*

Proof. If $\mathbb{C} \nVdash \varphi$, there exists an extended model $\mathcal{M}^+ \in \mathbb{C}$ and a state s such that $\mathcal{M}^+, s \Vdash \neg\varphi$. By Lemma 1 and Proposition 1, we have that $\mathcal{M}^\omega, s \Vdash \neg\varphi$. By Lemma 2, we then have that $\mathcal{M}, s \vDash \neg\varphi$. Thus we have that $\nvDash \varphi$.

Actually, it is not hard to show that the other way around also holds though we do not need it in the later completeness proof.

5 Completeness

We prove the completeness of $\mathbb{SLKH}$ over extended models in $\mathbb{C}$ w.r.t. $\Vdash$ by constructing a canonical extended model with a canonical action model.

We will construct canonical model on maximal consistent set w.r.t. a closure $cl(\Phi)$ of formulas, which is defined in the following. Let Φ be a subformula-closed set of ELKh-formulas that is finite. Given a set of formulas Δ, let: $\Delta|_{\mathcal{K}_i} = \{\mathcal{K}_i\varphi \mid \mathcal{K}_i\varphi \in \Delta\}$, $\Delta|_{\neg\mathcal{K}_i} = \{\neg\mathcal{K}_i\varphi \mid \neg\mathcal{K}_i\varphi \in \Delta\}$, $\Delta|_{\mathcal{K}h_i} = \{\mathcal{K}h_i\varphi \mid \mathcal{K}h_i\varphi \in \Delta\}$, $\Delta|_{\neg\mathcal{K}h_i} = \{\neg\mathcal{K}h_i\varphi \mid \neg\mathcal{K}h_i\varphi \in \Delta\}$.

Definition 17. *The closure $cl(\Phi)$ is $\Phi \cup \{\mathcal{K}_i\mathcal{K}h_i\varphi, \mathcal{K}_i\varphi \mid \mathcal{K}h_i\varphi \in \Phi\}$.*

From Definition 17, we have that $\mathcal{K}_i\mathcal{K}h_i\varphi, \mathcal{K}_i\varphi \in cl(\Phi)$ for each $\mathcal{K}h_i\varphi \in cl(\Phi)$. Since Φ is finite, so is $cl(\Phi)$. Next we define the maximal consistent sets w.r.t. $cl(\Phi)$ and will use them to build canonical models.

Definition 18 (Atom). *We enumerate the formulas in $cl(\Phi)$ by $\{\psi_0, \cdots, \psi_h\}$ where $h \in \mathbb{N}$. The set of formulas $\Delta = \{Y_i \mid i \leq h\}$ is an atom of $cl(\Phi)$ if*

- *$Y_i = \psi_i$ or $Y_i = \neg\psi_i$ for each $\psi_i \in cl(\Phi)$;*
- *Δ is consistent in $\mathbb{SLKH}$.*

Proposition 3. *Let Δ be an atom of $cl(\Phi)$, and $\mathcal{K}_i\varphi \in cl(\Phi)$. If $\mathcal{K}_i\varphi \notin \Delta$ then there exists Δ' such that $\Delta'|_{\mathcal{K}_i} = \Delta|_{\mathcal{K}_i}$ and $\neg\varphi \in \Delta'$.*

Let $\mathtt{A}(\Phi)$ be the set of all atoms of $cl(\Phi)$. For each $s \in \mathtt{A}(\Phi)$, we need two copies s^- and s^+. and we will construct canonical models based on the set $\{s^+, s^- \mid s \in \mathtt{A}(\Phi)\}$. The only difference between s^+ and s^- is that the know-how formula $\mathcal{K}h_i\varphi \in s$ will hold on s^+ and s^- due to different plans, which will become more clear formally. Also note that besides the propositional variables in Φ, we also use the set of auxiliary propositional variables $\mathbf{P}^{\mathtt{A}(\Phi)} = \{q_{s^+}, q_{s^-} \mid s$ is an atom of $cl(\Phi)\}$ where the propositional letter q_{s^+} will be used to characterize the state s^+, and q_{s^-} characterizes s^-.

Definition 19 (Canonical epistemic model). *Given Φ, the canonical epistemic model $\mathcal{N}^\Phi = \langle W^\Phi, \{\sim_i \mid i \in \mathbf{Ag}\}, V^\Phi\rangle$ is*

- $W^\Phi = \{s^+, s^- \mid s \in \mathtt{A}(\Phi)\}$
- $\sim_i = \{(s^+, t^+), (s^-, t^-) \mid s|_\mathcal{K} = t|_\mathcal{K}\}$
- $V^\Phi(s^\pm) = \{p \in \mathbf{P} \mid p \in s\} \cup \{q_{s^\pm}\}$

We also use the notation $s^\pm$ which means that we do not exactly know what its superscript is. The superscript of $s^\pm$ might be $+$ and also might be $-$. Please note that the equivalence relation $\sim_i$ consists of pairs of states that have the same superscript.

Next, we will define the canonical action model. The intuition behind the idea is that each event e is a *triple*. The first element of e, denoted by $\mathtt{fir}(e)$, and the third element of e, $\mathtt{thi}(e)$, are states of the canonical epistemic model, which means that this event e will link these two states, i.e. $\mathtt{fir}(e) \xrightarrow{e} \mathtt{thi}(e)$, in the final canonical extended model. The second element of e, $\mathtt{sec}(e)$, is a know-how formula with superscript, which represents that this event e will be used to realize the know-how formula.

Definition 20 (Canonical action model). *Given Φ, the canonical action model $\mathcal{E}^\Phi = \langle E^\Phi, \{\sim_i \mid i \in \mathbf{Ag}\}, pre, post\rangle$ is*

- $E^\Phi = \bigcup\{\widehat{\mathcal{K}h_i\varphi^+}, \widehat{\mathcal{K}h_i\varphi^-} \mid \mathcal{K}h_i\varphi \in \Phi, i \in \mathbf{Ag}\}$ *where*

$$\widehat{\mathcal{K}h_i\varphi^+} = \{\langle s^+, \mathcal{K}h_i\varphi^+, t^+\rangle, \langle s^+, \mathcal{K}h_i\varphi^+, t^-\rangle \mid \mathcal{K}h_i\varphi \in s \in \mathtt{A}(\Phi), \mathcal{K}_i\varphi \in t \in \mathtt{A}(\Phi)\}$$
$$\cup\{\langle s^-, \mathcal{K}h_i\varphi^+, t^+\rangle, \langle s^-, \mathcal{K}h_i\varphi^+, t^-\rangle \mid \mathcal{K}h_i\varphi \in s \in \mathtt{A}(\Phi), t \in \mathtt{A}(\Phi)\},$$
$$\widehat{\mathcal{K}h_i\varphi^-} = \{\langle s^+, \mathcal{K}h_i\varphi^-, t^+\rangle, \langle s^+, \mathcal{K}h_i\varphi^-, t^-\rangle \mid \mathcal{K}h_i\varphi \in s \in \mathtt{A}(\Phi), t \in \mathtt{A}(\Phi)\}$$
$$\cup\{\langle s^-, \mathcal{K}h_i\varphi^-, t^+\rangle, \langle s^-, \mathcal{K}h_i\varphi^-, t^-\rangle \mid \mathcal{K}h_i\varphi \in s \in \mathtt{A}(\Phi), \mathcal{K}_i\varphi \in t \in \mathtt{A}(\Phi)\}.$$

- $\sim_i = \{(e, e') \mid \mathtt{fir}(e) \sim_i \mathtt{fir}(e'), \mathtt{sec}(e) = \mathtt{sec}(e') = \widehat{\mathcal{K}h_i\varphi^{\pm}}$ *for some* $\mathcal{K}h_i\varphi \in cl(\Phi), \mathtt{thi}(e) \sim_i \mathtt{thi}(e')\}$
- $pre(e) = q_{\mathtt{fir}(e)}$
- $post(e) = \{\neg p \mid p \notin t\} \cup \{p \mid p \in t\}$ *where* $\mathtt{thi}(e) = t^{\pm}$.

Definition 21 (Canonical dynamic model). *The canonical dynamic model* $\mathcal{M}^{\Phi} = \langle \mathcal{N}^{\Phi}, \{\mathcal{A}_i^{\Phi} \mid i \in \mathbf{Ag}\}\rangle$ *where*

- $\mathcal{N}^{\Phi}$ *is the canonical epistemic model;*
- *for each* $i \in \mathbf{Ag}$, $\mathcal{A}_i^{\Phi} = \{\langle \mathcal{E}^{\Phi}, \widehat{\mathcal{K}h_i\varphi^+}\rangle, \langle \mathcal{E}^{\Phi}, \widehat{\mathcal{K}h_i\varphi^-}\rangle \mid \mathcal{K}h_i\varphi \in \Phi\}$ *where* $\mathcal{E}^{\Phi}$ *is the canonical action model.*

Proposition 4. *For each agent* $i \in \mathbf{Ag}$, *every action* $\langle \mathcal{E}^{\Phi}, S\rangle \in \mathcal{A}_i^{\Phi}$ *is a local action of agent* i.

Definition 22 (Canonical extended model). *The canonical extended model* $\mathcal{M}^{\Phi+} = \langle \mathcal{M}^{\Phi}, \{\xrightarrow{e} \mid e \in S$ *for some* $\langle \mathcal{E}^{\Phi}, S\rangle \in \mathcal{A}^{\Phi}\}\rangle$ *where*

- $\mathcal{M}^{\Phi}$ *is the canonical dynamic model;*
- $\xrightarrow{e} = \{(\mathtt{fir}(e), \mathtt{thi}(e))\}$

Proposition 5. *Given an action* $a \in \mathcal{A}_i^{\Phi}$ *that is executable on* $s^{\pm}$, *if* $\varphi \in t$ *for each* $t^{\pm} \in Q(a)(s^{\pm})$, *we then have that* $\mathcal{K}h_i\varphi \in s$.

Proof. Let a be $\langle \mathcal{E}^{\Phi}, \widehat{\mathcal{K}h_i\psi^{\pm}}\rangle$. There are two cases:

- The superscripts of $\widehat{\mathcal{K}h_i\psi^{\pm}}$ and $s^{\pm}$ are the same with each other. That is, either $a = \langle \mathcal{E}^{\Phi}, \widehat{\mathcal{K}h_i\psi^+}\rangle$ and $s^{\pm} = s^+$, or $a = \langle \mathcal{E}^{\Phi}, \widehat{\mathcal{K}h_i\psi^-}\rangle$ and $s^{\pm} = s^-$.
  Next, we will show that if $\varphi \in t$ for each $t^{\pm} \in Q(\langle \mathcal{E}^{\Phi}, \widehat{\mathcal{K}h_i\psi^+}\rangle)(s^+)$, we then have that $\mathcal{K}h_i\varphi \in s$.
  Firstly, since a is executable on s^+, we then have that there is some $t^{\pm}$ such that $s^+ \xrightarrow{a} t^{\pm}$. This follows that $s^+ \xrightarrow{e} t^{\pm}$ for some $e \in \widehat{\mathcal{K}h_i\psi^+}$. This follows that $e = \langle s^+, \mathcal{K}h_i\psi^+, t^{\pm}\rangle$. By Definition 20, we then have that $\mathcal{K}h_i\psi \in s$ and $\mathcal{K}_i\psi \in t$.
  Secondly, we will show that $\vdash \mathcal{K}_i\psi \to \mathcal{K}_i\varphi$. If not, this follows that $\{\mathcal{K}_i\psi, \neg\mathcal{K}_i\varphi\}$ is consistent. By Lindenbaum's lemma, we then have that there is some $v \in \mathtt{A}(\Phi)$ such that $\{\mathcal{K}_i\psi, \neg\mathcal{K}_i\varphi\} \subseteq v$. This follows that $\mathcal{K}_i\varphi \notin v$. By Proposition 3, we then have that there is some $w \in \mathtt{A}(\Phi)$ such that $\neg\varphi \in w$ and $w|_{\mathcal{K}_i} = v|_{\mathcal{K}_i}$. Since $\mathcal{K}\psi \in w$, this follows that $e' = \langle s^+, \mathcal{K}h_i\psi^+, w^+\rangle \in \widehat{\mathcal{K}h_i\psi^+}$.
  By Definition 22, we have that $s^+ \xrightarrow{e'} w^+$, and then $s^+ \xrightarrow{a} w^+$. This is contradictory with that $\varphi \in t$ for each $t^{\pm} \in Q(a)(s^+)$. therefore, we have shown that $\vdash \mathcal{K}_i\psi \to \mathcal{K}_i\varphi$. By Rule MONOKh, we have that $\vdash \mathcal{K}h_i\mathcal{K}_i\psi \to \mathcal{K}h_i\mathcal{K}_i\varphi$. By Axioms AxKhtoKhK and others, we have that $\vdash \mathcal{K}h_i\psi \to \mathcal{K}h_i\varphi$. Since $\mathcal{K}h_i\psi \in s$, therefore, we have that $\mathcal{K}h_i\varphi \in s$.
  From a similar process, it can be shown that, if $\varphi \in t$ for each state $t^{\pm}$ in $Q(\langle \mathcal{E}^{\Phi}, \widehat{\mathcal{K}h_i\psi^-}\rangle)(s^-)$, we then have that $\mathcal{K}h_i\varphi \in s$.

- The superscripts of $\widehat{\mathcal{K}h_i\psi}^{\pm}$ and $s^{\pm}$ are different from each other. That is, either $a = \langle \mathcal{E}^{\Phi}, \widehat{\mathcal{K}h_i\psi^+} \rangle$ and $s^{\pm} = s^-$, or $a = \langle \mathcal{E}^{\Phi}, \widehat{\mathcal{K}h_i\psi^-} \rangle$ and $s^{\pm} = s^+$.
  Next, we will show that if $\varphi \in t$ for each $t^{\pm} \in Q(\langle \mathcal{E}^{\Phi}, \widehat{\mathcal{K}h_i\psi^+} \rangle)(s^-)$, we then have that $\mathcal{K}h_i\varphi \in s$.
  Firstly, since a is executable on s^-, we can have that $\mathcal{K}h_i\psi \in s$. By Definition 20, we have that $\langle s^-, \mathcal{K}h_i\psi, t^{\pm} \rangle \in \widehat{\mathcal{K}h_i\psi^+}$ for each $t \in \mathtt{A}(\Phi)$. By Definition 22, we have that $s^- \xrightarrow{\langle s^-, \mathcal{K}h_i\psi, t^{\pm} \rangle} t^{\pm}$ for each $t \in \mathtt{A}(\Phi)$. Therefore, $t^{\pm} \in Q(\langle \mathcal{E}^{\Phi}, \widehat{\mathcal{K}h_i\psi^+} \rangle)(s^-)$ for each $t \in \mathtt{A}(\Phi)$. Thus, we have that $\varphi \in t$ for each $t \in \mathtt{A}(\Phi)$. Thus, we have that $\vdash \varphi \leftrightarrow \top$. Therefore, $\vdash \mathcal{K}h_i\varphi \leftrightarrow \mathcal{K}h_i\top \leftrightarrow \top$. Thus, we have that $\mathcal{K}h_i\varphi \in s$.
  From a similar process, it can be shown that, if $\varphi \in t$ for each state $t^{\pm}$ in $Q(\langle \mathcal{E}^{\Phi}, \widehat{\mathcal{K}h_i\psi^-} \rangle)(s^+)$, we then have that $\mathcal{K}h_i\varphi \in s$.

Proposition 6. *Given an action sequence $a_1 \cdots a_n \in \mathcal{A}_i^{\Phi}$ that is strongly executable on $s^{\pm}$, if $n \geq 2$, we then have that $t^+, t^- \in Q(a_1 \cdots a_n)(s^{\pm})$ for each $t \in \mathtt{A}(\Phi)$.*

Due to the space limit, the proof of this proposition is omitted. The idea of the proof is that doing the action $a = \langle \mathcal{E}^{\Phi}, \widehat{\mathcal{K}h_i\varphi^+} \rangle$ on states with superscript $+$, or doing $b = \langle \mathcal{E}^{\Phi}, \widehat{\mathcal{K}h_i\varphi^-} \rangle$ on states with $-$, will reach on $\mathcal{K}_i\varphi$-states both with $+$ and with $-$. However, doing a on states with $-$, or doing b on states with $+$, will reach all states. Thus, if the length of the plan is bigger than or equal to 2, performing the plan will terminate on all states.

Next we will show the truth lemma.

Lemma 3. *For each $\varphi \in cl(\Phi)$, we have that $\mathcal{M}^{\Phi+}, s^{\pm} \Vdash \varphi$ iff $\varphi \in s$.*

Proof. We prove it by induction on φ. We only focus on the case of $\mathcal{K}h_i\varphi \in cl(\Phi)$; the other cases are straightforward, e.g., $\mathcal{K}\varphi$ case can be proved based on Proposition 3. Note that if $\mathcal{K}h_i\varphi \in cl(\Phi)$ then $\mathcal{K}h_i\varphi \in \Phi$, therefore, by Definition 17, $\mathcal{K}_i\varphi \in cl(\Phi)$.

Right to Left: Assuming that $\mathcal{K}h_i\varphi \in s$, we will show $\mathcal{M}^{\Phi+}, s^{\pm} \Vdash \mathcal{K}h_i\varphi$. Firstly, we will show that there is some atom t of $cl(\Phi)$ such that $\mathcal{K}_i\varphi \in t$. Since $\mathcal{K}h_i\varphi \in s$ and s is consistent, this follows that $\mathcal{K}_i\varphi$ is consistent. If not, we then have that $\vdash \mathcal{K}_i\varphi \to \bot$. By Rule `MONOKh` and Axiom `AxKhtoKKh`, we then have that $\vdash \mathcal{K}h_i\varphi \to \mathcal{K}h_i\bot$. By Axiom `AxKhbot`, we then have that $\vdash \neg\mathcal{K}h_i\varphi$. Contradiction! Thus, we have that $\mathcal{K}_i\varphi \in cl(\Phi)$ is consistent. By Lindenbaum's lemma, there is some atom t of $cl(\Phi)$ such that $\mathcal{K}_i\varphi \in t$.

Next, we will show that $\mathcal{M}^{\Phi+}, s^+ \Vdash \mathcal{K}h_i\varphi$. Since t is an atom, this follows that $t^{\pm} \in W^{\Phi}$. We then have that $e = \langle s^+, \mathcal{K}h_i\varphi^+, t^+ \rangle \in \widehat{\mathcal{K}h_i\varphi^+}$. Let a be the action $\langle \mathcal{E}^{\Phi}, \widehat{\mathcal{K}h_i\varphi^+} \rangle$, next we will show that a is strongly executable on $[s^+]^i$ and $\mathcal{M}^{\Phi+}, w^{\pm} \Vdash \varphi$ for each $w^{\pm} \in Q(a)([s^+]^i)$. By Definition 19, we know that the superscript of the state that is in $[s^+]^i$ is $+$. For each $v^+ \in [s^+]^i$, by Definition 19, we then have that $s|_{\mathcal{K}_i} = v|_{\mathcal{K}_i}$. Moreover, since $\vdash \mathcal{K}h_i\varphi \to \mathcal{K}_i\mathcal{K}h_i\varphi$, this follows that $\mathcal{K}_i\mathcal{K}h_i\varphi \in v$ because of $\mathcal{K}h_i\varphi \in s$ and $\mathcal{K}_i\mathcal{K}h_i\varphi \in cl(\Phi)$.

This follows by Axiom $\mathtt{T}$ that $\mathcal{K}h_i\varphi \in v$. By Definition 20, we then have that $e' = \langle v^+, \mathcal{K}h_i\varphi^+, t^+\rangle \in \widehat{\mathcal{K}h_i\varphi^+}$. Further, by Definition 22, we have that $v^+ \xrightarrow{e'} t^+$, and then $v^+ \xrightarrow{a} t^+$. Therefore, a is strongly executable on $[s^+]^i$. For each $w^\pm \in Q(a)([s^+]^i)$, there is some $u^+ \in [s^+]^i$ such that $u^+ \xrightarrow{a} w^\pm$. This follows that there is some $e'' \in \widehat{\mathcal{K}h_i\varphi^+}$ such that $u^+ \xrightarrow{e''} w^\pm$. By Definition 22, we then know that $e'' = \langle u^+, \mathcal{K}h_i\varphi^+, w^\pm\rangle$. By Definition 20, we then have $\mathcal{K}_i\varphi \in w$. By Axiom $\mathtt{T}$, we have that $\varphi \in w$. By IH, we then have that $\mathcal{M}^{\Phi+}, w^\pm \Vdash \varphi$.

We then have shown that $\mathcal{M}^{\Phi+}, s^+ \Vdash \mathcal{K}h_i\varphi$. From a similar process, It can be shown that $\mathcal{M}^{\Phi+}, s^- \Vdash \mathcal{K}h_i\varphi$.

Left to Right: Assuming that $\mathcal{M}^{\Phi+}, s^\pm \Vdash \mathcal{K}h_i\varphi$, we will show $\mathcal{K}h_i\varphi \in s$. It follows by $\mathcal{M}^{\Phi+}, s^\pm \Vdash \mathcal{K}h_i\varphi$ that there is an action sequence $\sigma = a_1 \cdots a_n$ such that σ is strongly executable on $[s^\pm]^i$ and $\mathcal{M}^{\Phi+}, t^\pm \Vdash \varphi$ for each $t^\pm \in Q(\sigma)([s^\pm]^i)$. By IH, we have that $\varphi \in t$ for each $t^\pm \in Q(\sigma)([s^\pm]^i)$. There are three cases: $\sigma = \epsilon$; $n = 1$; and $n \geq 2$.

If $\sigma = \epsilon$, this follows that $\varphi \in t$ for each $t^\pm \in [s^\pm]^i$. This follows that $\mathcal{K}_i\varphi \in s$. If not, it will be contradictory with Proposition 3. It follows by Axiom $\mathtt{AxKtoKh}$ that $\mathcal{K}h_i\varphi \in s$.

If $n = 1$, this follows that a_1 is executable on $s^\pm$. Since $\varphi \in t$ for each $t^\pm \in Q(a_1)([s^\pm]^i)$, by Proposition 5, we then have that $\mathcal{K}h_i\varphi \in s$.

If $n \geq 2$, this follows that $a_1 \cdots a_n$ is executable on $s^\pm$. By Proposition 6, we have that $v^\pm \in Q(a_1 \cdots a_n)(s^\pm)$ for each $v \in \mathtt{A}(\Phi)$. Since $\varphi \in t$ for each $t^\pm \in Q(a_1)(s^\pm)$, this follows that $\varphi \in v$ for all $v \in \mathtt{A}(\Phi)$. This follows that $\vdash \varphi \leftrightarrow \top$. Thus, we have by Rule $\mathtt{MONOKh}$ that $\vdash \mathcal{K}h_i\varphi \leftrightarrow \mathcal{K}h_i\top$. Therefore, we have that $\mathcal{K}h_i\varphi \in s$.

Proposition 7. $\mathcal{M}^{\Phi+} \in \mathbb{C}$.

Theorem 3 (Completeness). $\mathbb{SLKH}$ *is complete w.r.t. dynamic semantics, namely, if* $\vDash \varphi$ *then* $\vdash \varphi$.

Proof. It follows by Lemma 3 and Proposition 7 that if $\mathbb{C} \Vdash \varphi$ then $\vdash \varphi$. Moreover, by Theorem 2, we have that if $\vDash \varphi$ then $\vdash \varphi$.

Theorem 4. $\mathbb{SLKH}$ *is decidable.*

Proof. By Lemma 3, we know that if φ is satisfiable then it is satisfied on the canonical extended model $\mathcal{M}^{\Phi+}$. Moreover, the canonical extended model $\mathcal{M}^{\Phi+}$ is a finite model which is bounded by the length of the formula φ. Please note that $\mathcal{M}^{\Phi+}$ consists of the canonical epistemic model $\mathcal{N}^{\Phi}$ and the canonical action model $\mathcal{E}^{\Phi}$. Let the size of the set $cl(\Phi)$ be n, which is bounded by the length of φ. By Definition 19, we know that the size of $\mathcal{N}^{\Phi}$ is bounded by $O(2^n)$. By Definition 20, we know that the size of $\mathcal{E}^{\Phi}$ is bounded by $O(n2^{2n})$. Thus, the size of $\mathcal{M}^{\Phi+}$ is bounded by $O(n2^{2n})$.

Given φ, we consider the following procedure: (1) we calculate the set Φ, which consists of all subformulas of φ, and the closure $cl(\Phi)$; (2) we take a subset of the power set of $cl(\Phi)$ as all the atoms of $cl(\Phi)$, and we then construct the 'canonical extended model' based on these 'atoms'; (3) we check whether φ is satisfied on this 'canonical extended model', and we put 'yes' out if φ is satisfied; (4) if φ is not satisfied on the 'canonical extended model', we then go to step (3) and continue to try another subset of the power set of $cl(\Phi)$. If we fail to find a model for φ after trying all the subsets of the power set of $cl(\Phi)$, we put 'no' out.

If φ is satisfiable, by Lemma 3, this procedure will put 'yes' out. If this procedure puts 'yes' out, we then have a model that satisfies φ, and thus φ is satisfiable. This procedure will terminate for an input formula φ. The reason is the following: step (1) will terminate due to the fact that $cl(\Phi)$ is bounded by the length of φ; step (2) will terminate because the size of the canonical extended model is bounded; step (3) will terminate if model checking of ELKh is bounded. To show model checking of ELKh is decidable, the key is to show model checking know-how formula $\mathcal{K}h_i\varphi$ over extended model is decidable. It can be shown that if $\mathcal{M}^+, s \Vdash \mathcal{K}h_i\varphi$ then then there is a good plan whose length is bounded by 2^m, where m is the size of $\mathcal{M}^+$. Thus, checking whether a know-how formula $\mathcal{K}h_i\varphi$ is satisfied on an extended model $\mathcal{M}^+$ is decidable.

6 Conclusions

In this paper, we propose a multi-agent framework based on multi-step dynamic epistemic planning. We obtain a sound and complete logic of knowing how. The axioms captures the abstract properties of dynamic epistemic planning. Note that although the multi-agent epistemic planning problem is undecidable, the abstract logic of know-how based on it is decidable.

This is just the beginning of an interesting story, which opens various further directions. First of all, inspired by [11], we can add the group knowledge operators such as distributed knowledge and common knowledge for both know-that and know-how. Secondly, it is also interesting to consider branching epistemic plans as in [1] rather than the linear one. Furthermore, we can discuss the correspondence between the axioms and the assumptions underlying the epistemic planning notion, e.g., when the axiom $\mathcal{K}h\mathcal{K}h\varphi \to \mathcal{K}h\varphi$ becomes valid by restricting the set of action models. Finally, we implicitly assume the set of actions is commonly known to all the agents which can be relaxed in the future.

Acknowledgement. Yanjun Li would like to thank the support from NSSF grant 18CZX062. The authors thank the anonymous reviewers for their detailed comments that improved the presentation of the paper.

References

1. Andersen, M.B., Bolander, T., Jensen, M.H.: Conditional epistemic planning. In: del Cerro, L.F., Herzig, A., Mengin, J. (eds.) JELIA 2012. LNCS (LNAI), vol. 7519, pp. 94–106. Springer, Heidelberg (2012). https://doi.org/10.1007/978-3-642-33353-8_8

2. Aucher, G., Bolander, T.: Undecidability in epistemic planning. In: Proceedings of IJCAI 2013, pp. 27–33 (2013)
3. Bolander, T.: A gentle introduction to epistemic planning: the DEL approach. Electron. Proc. Theor. Comput. Sci. **243**, 1–22 (2017)
4. Bolander, T., Andersen, M.B.: Epistemic planning for single and multi-agent systems. J. Appl. Non-Classical Logics **21**(1), 9–34 (2011)
5. Fan, J., Wang, Y., Van Ditmarsch, H.: Contingency and knowing whether. Rev. Symbolic Logic **8**(1), 75–107 (2015)
6. Fervari, R., Herzig, A., Li, Y., Wang, Y.: Strategically knowing how. In: IJCAI International Joint Conference on Artificial Intelligence, pp. 1031–1038 (2017)
7. Jamroga, W., Ågotnes, T.: Constructive knowledge: what agents can achieve under imperfect information. J. Appl. Non-Classical Logics **17**(4), 423–475 (2007)
8. Li, Y., Wang, Y.: Achieving while maintaining: a logic of knowing how with intermediate constraints. In: Ghosh, S., Prasad, S. (eds.) ICLA 2017. LNCS, vol. 10119, pp. 154–167. Springer, Heidelberg (2017). https://doi.org/10.1007/978-3-662-54069-5_12
9. Moore, R.C.: A formal theory of knowledge and action. Technical report, DTIC Document (1984)
10. Naumov, P., Tao, J.: Second-order know-how strategies. In: 2018 AAMAS (2018)
11. Naumov, P., Tao, J.: Strategic coalitions with perfect recall. In: 2018 AAAI (2018)
12. Naumov, P., Tao, J.: Together we know how to achieve: an epistemic logic of know-how. Artif. Intell. **262**(9), 279–300 (2018)
13. Naumov, P., Tao, J.: Knowing-how under uncertainty. Artif. Intell. **276**, 41–56 (2019)
14. Stanley, J.: Know How. Oxford University Press, Oxford (2011)
15. van Benthem, J., Gerbrandy, J., Hoshi, T., Pacuit, E.: Merging frameworks for interaction. J. Philos. Logic **38**(5), 491–526 (2009)
16. Wang, Y.: A Logic of knowing how. In: Proceedings of LORI-V, pp. 392–405 (2015)
17. Wang, Y.: A logic of goal-directed knowing how. In: Synthese, pp. 1–21 (2016)
18. Wang, Y.: Beyond knowing that: a new generation of epistemic logics. In: van Ditmarsch, H., Sandu, G. (eds.) Jaakko Hintikka on Knowledge and Game-Theoretical Semantics. OCL, vol. 12, pp. 499–533. Springer, Cham (2018). https://doi.org/10.1007/978-3-319-62864-6_21
19. Wang, Y., Aucher, G.: An alternative axiomatization of DEL and its applications. In: IJCAI, pp. 1147–1154 (2013)
20. Wang, Y., Cao, Q.: On axiomatizations of public announcement logic. Synthese **190**(1S), 103–134 (2013)
21. Wang, Y., Fan, J.: Conditionally knowing what. In: Proceedings of AiML, vol. 10, pp. 569–587 (2014)
22. Xu, C., Wang, Y., Studer, T.: A logic of knowing why. Synthese (2019)
23. Yu, Q., Wen, X., Liu, Y.: Multi-agent epistemic explanatory diagnosis via reasoning about actions. In: Proceedings of IJCAI 2013, pp. 1183–1190 (2013)

The Sequent Systems and Algebraic Semantics of Intuitionistic Tense Logics

Kaiyang Lin[1] and Zhe Lin[2,3](✉)

[1] Department of Philosophy, Peking University, Beijing, China
linkaiyang7@gmail.com

[2] Institute of Logic and Intelligence, Southwest University, Chongqing, China
pennyshaq@gmail.com

[3] Institute of Logic and Cognition, Sun Yat-sen University, Guangzhou, China

Abstract. In this paper we consider a weak Ewald's intuitionistic tense logic (wIK.t). We study its sequent system and algebraic semantics. We prove the soundness and the completeness results. We also show that the sequent system for wIK.t introduced in the present paper admits cut elimination. Finally we propose a criterion and prove that all extensions of wIK.t satisfying this criterion have cut free sequent systems.

1 Introduction

The tense operators were first introduced to classical propositional logic in Burgess [2], which yielded the basic classical tense logic. Ewald [5] introduced an intuitionistic tense logic denoted by IK.t which is analogue to the classical tense logic. In this logic Ewald considered four tense operators G (□), H (■), F (◊) and P (♦) on intuitionistic propositional calculus satisfying some mutual conditions. Sequent system and Hilbert style system were given in the same paper. Further Ewald defined a Kripke model for IK.t and proved the completeness and decidability. Although the decidability proof (via finite model property) is shown to be incorrect in [10], the completeness result with respect to kripke model use in the define article remains true. By enriching with different modal axioms, one can obtain intuitionistic counterparts of different classical tense logics. Notice that the S5 extensions of IK.t (IK.t enrich with modal axioms 4, T, B) is coincided with Pior's MIPC.

An algebraic axiomatization for IK.t was given by Figallo and Pelaitay in [6]. A tense algebra for IK.t (IK.t-algebras) $(A, \wedge, \vee, \rightarrow, 0, 1, \Diamond, \Box, \blacklozenge, \blacksquare)$ is a structure such that $(A, \wedge, \vee, \rightarrow, 0, 1)$ is a Heyting algebra and $\Diamond, \Box, \blacklozenge, \blacksquare$ are unary operators on A satisfying the following conditions for all $a, b \in A$

(t1) $\Box 1 = 1$ and $\blacksquare 1 = 1$
(t2) $\Box(a \wedge b) = \Box a \wedge \Box b$ and $\blacksquare(a \wedge b) = \blacksquare a \wedge \blacksquare b$
(t3) $a \leq \blacksquare\Diamond a$ and $a \leq \Box\blacklozenge a$

The work of both authors were supported by Key program of Chongqing's Key Research Institute of Humanities and Social Sciences (No. 14SKB044).

P. Blackburn et al. (Eds.): LORI 2019, LNCS 11813, pp. 140–152, 2019.
https://doi.org/10.1007/978-3-662-60292-8_11

(t4) $\Diamond 0 = 0$ and $\blacklozenge 0 = 0$
(t5) $\Diamond(a \vee b) = \Diamond a \vee \Diamond b$ and $\blacksquare(a \vee b) = \blacksquare a \vee \blacksquare b$
(t6) $\Diamond\blacksquare a \leq a$ and $\blacklozenge\Box a \leq a$
(t7) $\Diamond(a \to b) \leq \Box a \to \Diamond b$ and $\blacklozenge(a \to b) \leq \blacksquare a \to \blacklozenge b$.

$\neg$ is defined as follow: $\neg a = a \to \bot$. It was shown in the same paper that Ewald's intuitionistic tense logic is sound and complete with respect to IK.t-algebras.

Beyond Ewald's setting, Chajda [4] introduced another basic intuitionistic tense algebras by replacing (t7) by a classical view of the modalities: (t8) $\Diamond a = \neg\Box\neg a$ and $\blacklozenge a = \neg\blacksquare\neg a$. This follows the tradition of Bull [1], in which Bull considered an intuitionistic version of modal logic S5 satisfying (t8). Algebraic semantics was given for this logic and the finite model property was proved in the same paper. Obviously in Chajda's tense algebras, (t7) dose not hold, meanwhile in [6], Figallo and Pelaitay argumented that (t8) is not satisfied in IK.t-algebras. Hence both classes of algebras are independent. Thus it is natural to consider classes of algebras included both algebras in [6] and [4]. The definition of these classes of algebras (called weak IK.t-algebras) can be found in Sect. 2 (Definition 2). There is not even any general agreement on what the intuitionistic analogue of the basic tense logic, K.t, is. As we will see, in this paper we propose a new candidate.

In the present paper, we study the logics of weak IK.t-algebras from the proof theoretical point of view. The study of proof theory on different inuitionistic modal and tense logics is vast. Genzten style sequent and natural deduction systems were investigated in the literature, e.g. [5,8]. Labelled, display, deep inferences and nested sequent systems were introduced and studied in [7,9,11]. Particularly to the best of our knowledge, there is not cut free sequent system for IK.t. We introduce a sequent system which is sound and complete with respect to weak IK.t-algebras. We show that this sequent system admits cut elimination. Further we also discuss the sequent systems of extensions of weak IK.t logics and introduce a criterion for these cut free sequent systems. The present work is rooted in the studies of substructural logics. We consider sequent systems with structure operations for tense operators, and various properties of tense connectives are presented in the form of structure rules. This follows the tradition of substructural modal logics. It is worth mentioning that Moorgat's works on multi modal Lambek calculus was probably the beginning of this stream of research. However many other researchers have contributed to this area, and we refer to [3,12,13].

This paper is organized as follows. In next section, we introduce the weak IK.t algebras and the sequent system of the corresponding logic. We also show the soundness and completeness results. In Sect. 3, we provide a cut free proof for the sequent system introduced in Sect. 2. Section 4 discuss a criterion for cut free sequent systems of extensions of weak IK.t logics. As a consequence we show that our results can be extended to weak IK.t logics with any combinations of axioms D, T, B, 4 and 5.

2 Algebraic Semantics and Sequent System

In this section we introduce weak IK.t-algebras and develop a sequent system which is sound and complete with respect to weak IK.t-algebras.

Definition 1. *A Heyting algebra $(A, \wedge, \vee, \to, 0, 1)$ is a structure such that $(A, \wedge \vee, 0, 1)$ is a bounded distributive lattice and $\to$ is a binary operation on A satisfying the following conditions for all $a, b \in A$*

$$\text{(Res)} \quad a \wedge b \leq c \quad \text{iff} \quad b \leq a \to c$$

One usually define $\neg a \equiv a \to 0$. Note that 0 and 1 are the least and greatest elements in A.

Definition 2. *A weak IK.t algebra (wIK.tA) $(A, \wedge, \vee, \to, 0, 1, \Diamond, \Box, \blacklozenge, \blacksquare)$ is a structure such that $(A, \wedge, \vee, \to, 0, 1)$ is a Heyting algebra and $\Diamond, \Box, \blacklozenge, \blacksquare$ are unary operations on A satisfying the following conditions for all $a, b \in A$*

$$(\mathrm{Adj}_{\Diamond\blacksquare}) \quad \Diamond a \leq b \quad \text{iff} \quad a \leq \blacksquare b$$

$$(\mathrm{Adj}_{\blacklozenge\Box}) \quad \blacklozenge a \leq b \quad \text{iff} \quad a \leq \Box b$$

$$(\mathrm{Dual}_{\Diamond\Box}) \quad \Box\neg a \leq \neg\Diamond a$$

$$(\mathrm{Dual}_{\blacklozenge\blacksquare}) \quad \blacksquare\neg a \leq \neg\blacklozenge a$$

It is known that $(\mathrm{Adj}_{\Diamond\blacksquare})$ and $(\mathrm{Adj}_{\blacklozenge\Box})$ follow from the monotonicity for $\Diamond, \blacklozenge, \Box, \blacksquare$, (t3) and (t6). $(\mathrm{Dual}_{\Diamond\Box})$ and $(\mathrm{Dual}_{\blacklozenge\blacksquare})$ are special cases for conditions $\Box(a \to b) \leq \Diamond a \to \Diamond b$ and $\blacksquare(a \to b) \leq \blacklozenge a \to \blacklozenge b$ if $b = \bot$, which are admissible in iK.t algebras. Later we call these conditions WK axioms since they firstly appeared in Wijesekera's intuitionistic modal system. Clearly $(\mathrm{Dual}_{\Diamond\Box})$ and $(\mathrm{Dual}_{\blacklozenge\blacksquare})$ hold in Chajda's intuitionistic tense algebras. Hence IK.t-algebras and Chajda's intuitionistic tense algebras are both weak IK.t-algebras.

Lemma 1. *For any weak IK.t algebra $(A, \wedge, \vee, \to, 0, 1, \Diamond, \Box, \blacklozenge, \blacksquare)$ and $a, b, c \in A$, the following holds*

(1) $1 \leq \Box 1$ and $1 \leq \blacksquare 1$;
(2) $\Diamond 0 \leq 0$ and $\blacklozenge 0 \leq 0$;
(3) $\Box(a \wedge b) = \Box a \wedge \Box b$ and $\blacksquare(a \wedge b) = \blacksquare a \wedge \blacksquare b$;
(4) $\Diamond(a \vee b) = \Diamond a \vee \Diamond b$ and $\blacklozenge(a \vee b) = \blacklozenge a \vee \blacklozenge b$;
(5) $\neg\Diamond a \leq \Box\neg a$ and $\neg\blacklozenge a \leq \blacksquare\neg a$;
(6) $\Box a \leq \neg\Diamond\neg a$ and $\blacksquare a \leq \neg\blacklozenge\neg a$;
(7) $\Diamond\blacksquare a \leq a$ and $\blacklozenge\Box a \leq a$;
(8) $a \leq \blacksquare\Diamond a$ and $a \leq \Box\blacklozenge a$;
(9) $\Diamond a \wedge b \leq 0$ iff $a \wedge \blacklozenge b \leq 0$.

Proof. We only provide the proof of one direction in (9) and skip the details of all other proofs. Assume that $\Diamond a \wedge b \leq 0$. Then by (Res), $\Diamond a \leq \neg b$. By monotonicity for $\blacksquare$, $\blacksquare\Diamond a \leq \blacksquare\neg b$. Since $a \leq \blacksquare\Diamond a$ and $\blacksquare\neg b \leq \neg\blacklozenge b$, $a \leq \neg\blacklozenge b$. Therefore by (Res), $a \wedge \blacklozenge b \leq 0$.

Now we consider a sequent system for the logic of weak IK.t algebras.

Definition 3. *The set of formulas (terms) $\mathcal{F}$ is defined inductively as follows:*

$$\mathcal{F} \ni \alpha ::= p \mid \top \mid \bot \mid \alpha_1 \wedge \alpha_2 \mid \alpha_1 \vee \alpha_2 \mid \alpha_1 \to \alpha_2 \mid \lozenge\alpha \mid \Box\alpha \mid \blacklozenge\alpha \mid \blacksquare\alpha$$

where $p \in \mathsf{Var}$. We use the abbreviation $\neg\alpha := \alpha \to \bot$.

Definition 4. *Let $(,)$ $\circ$ and $\bullet$ be structure operatorions for $\wedge$, $\lozenge$ and $\blacklozenge$ respectively. The set of all formula structures $\mathcal{FS}$ are defined inductively as follows:*

$$\mathcal{FS} \ni \Gamma ::= \alpha \mid \Gamma_1, \Gamma_2 \mid \circ\Gamma \mid \bullet\Gamma$$

A sequent is an expression of the form $\Gamma \Rightarrow \alpha$ where Γ is a formula structure and α is a formula. A context is a formula structure $\Gamma[-]$ with a designated position $[-]$ which can be filled with a formula structure. In particular, a single position $[-]$ is a context. Let $\Gamma[\Delta]$ be formula structure obtained from $\Gamma[-]$ by substituting Δ for $[-]$. By $f(\Gamma)$ we denote the formula obtained from Γ by replacing all structure operatorions by their corresponding formula connectives.

Definition 5. The Gentzen sequent calculus $\mathsf{GwIK.t}$ for the intuitionistic tense logic $\mathsf{wIK.t}$ consists of the following axiom and rules:

(1) Axiom:

$$(\mathrm{Id})\ \alpha \Rightarrow \alpha$$

(2) Logical rules

$$\frac{\Gamma[\top] \Rightarrow \beta}{\Gamma[\Delta] \Rightarrow \beta}(\top) \quad \frac{\Delta \Rightarrow \bot}{\Gamma[\Delta] \Rightarrow \alpha}(\bot)$$

$$\frac{\Gamma[\alpha_1, \alpha_2] \Rightarrow \beta}{\Gamma[\alpha_1 \wedge \alpha_2] \Rightarrow \beta}(\wedge\mathrm{L}) \quad \frac{\Gamma_1 \Rightarrow \alpha_1 \quad \Gamma_2 \Rightarrow \alpha_2}{\Gamma_1, \Gamma_2 \Rightarrow \alpha_1 \wedge \alpha_2}(\wedge\mathrm{R})$$

$$\frac{\Gamma[\alpha_1] \Rightarrow \beta \quad \Gamma[\alpha_2] \Rightarrow \beta}{\Gamma[\alpha_1 \vee \alpha_2] \Rightarrow \beta}(\vee\mathrm{L}) \quad \frac{\Gamma \Rightarrow \alpha_i}{\Gamma \Rightarrow \alpha_1 \vee \alpha_2}(\vee\mathrm{R})(i = 1, 2)$$

$$\frac{\Delta \Rightarrow \alpha_1 \quad \Gamma[\alpha_2] \Rightarrow \beta}{\Gamma[\Delta, \alpha_1 \to \alpha_2] \Rightarrow \beta}(\to\mathrm{L}) \quad \frac{\alpha_1, \Gamma \Rightarrow \alpha_2}{\Gamma \Rightarrow \alpha_1 \to \alpha_2}(\to\mathrm{R})$$

(3) Structural rules:

$$\frac{\Gamma[\Delta_i] \Rightarrow \beta}{\Gamma[\Delta_1, \Delta_2] \Rightarrow \beta}(\mathrm{Wek})(i = 1, 2) \quad \frac{\Gamma[\alpha, \alpha] \Rightarrow \beta}{\Gamma[\alpha] \Rightarrow \beta}(\mathrm{Conf})$$

$$\frac{\Gamma[\circ\Delta_1, \circ\Delta_2] \Rightarrow \beta}{\Gamma[\circ(\Delta_1, \Delta_2)] \Rightarrow \beta}(\mathrm{Con}_\circ) \quad \frac{\Gamma[\bullet\Delta_1, \bullet\Delta_2] \Rightarrow \beta}{\Gamma[\bullet(\Delta_1, \Delta_2)] \Rightarrow \beta}(\mathrm{Con}_\bullet)$$

$$\frac{\Gamma[\Delta_1, \Delta_2] \Rightarrow \beta}{\Gamma[\Delta_2, \Delta_1] \Rightarrow \beta}(\mathrm{Ex}) \quad \frac{\Gamma[\Delta_1, (\Delta_2, \Delta_3)] \Rightarrow \beta}{\Gamma[(\Delta_1, \Delta_2), \Delta_3] \Rightarrow \beta}(\mathrm{As}_1) \quad \frac{\Gamma[(\Delta_1, \Delta_2), \Delta_3] \Rightarrow \beta}{\Gamma[\Delta_1, (\Delta_2, \Delta_3)] \Rightarrow \beta}(\mathrm{As}_2)$$

(4) Cut rule:

$$\frac{\Delta \Rightarrow \alpha \quad \Gamma[\alpha] \Rightarrow \beta}{\Gamma[\Delta] \Rightarrow \beta}(\text{Cut})$$

(5) Modal rules:

$$\frac{\Gamma[\circ\alpha] \Rightarrow \beta}{\Gamma[\Diamond\alpha] \Rightarrow \beta}(\Diamond\text{L}) \quad \frac{\Gamma \Rightarrow \alpha}{\circ\Gamma \Rightarrow \Diamond\alpha}(\Diamond\text{R}) \quad \frac{\Gamma[\bullet\alpha] \Rightarrow \beta}{\Gamma[\blacklozenge\alpha] \Rightarrow \beta}(\blacklozenge\text{L}) \quad \frac{\Gamma \Rightarrow \alpha}{\bullet\Gamma \Rightarrow \blacklozenge\alpha}(\blacklozenge\text{R})$$

$$\frac{\Gamma[\alpha] \Rightarrow \beta}{\Gamma[\circ\blacksquare\alpha] \Rightarrow \beta}(\blacksquare\text{L}), \quad \frac{\circ\Gamma \Rightarrow \alpha}{\Gamma \Rightarrow \blacksquare\alpha}(\blacksquare\text{R}). \quad \frac{\Gamma[\alpha] \Rightarrow \beta}{\Gamma[\bullet\Box\alpha] \Rightarrow \beta}(\Box\text{L}), \quad \frac{\bullet\Gamma \Rightarrow \alpha}{\Gamma \Rightarrow \Box\alpha}(\Box\text{R}).$$

$$\frac{\circ\Delta_1, \Delta_2 \Rightarrow \bot}{\Gamma[\Delta_1, \bullet\Delta_2] \Rightarrow \beta}(\text{Dual}_{\circ\bullet}). \quad \frac{\bullet\Delta_1, \Delta_2 \Rightarrow \bot}{\Gamma[\Delta_1, \circ\Delta_2] \Rightarrow \beta}(\text{Dual}_{\bullet\circ}).$$

A sequent $\Gamma \Rightarrow \beta$ is provable in GwIK.t, notation $\vdash_{\text{GwIK.t}} \Gamma \Rightarrow \beta$, if there is a derivation of $\Gamma \Rightarrow \beta$ in GwIK.t. We write $\vdash_{\text{GwIK.t}} \alpha \Leftrightarrow \beta$ if $\vdash_{\text{GwIK.t}} \alpha \Rightarrow \beta$ and $\vdash_{\text{GwIK.t}} \beta \Rightarrow \alpha$. Hereafter we usually skip the applications of rules (Ex), (As_1) and (As_2) in the derivations.

Lemma 2. *The following holds in* GwIK.t

(1) (res): $\alpha \wedge \beta \Rightarrow \gamma$ *iff* $\alpha \Rightarrow \beta \rightarrow \gamma$;
(2) (adj$_{\Diamond\blacksquare}$): $\Diamond\alpha \Rightarrow \beta$ *iff* $\alpha \Rightarrow \blacksquare\beta$;
(3) (adj$_{\blacklozenge\Box}$): $\blacklozenge\alpha \Rightarrow \beta$ *iff* $\alpha \Rightarrow \Box\beta$;
(4) (dual): $\Box\neg\alpha \Leftrightarrow \neg\Diamond\alpha$ *and* $\blacksquare\neg\alpha \Leftrightarrow \neg\blacklozenge\alpha$.

Proof. We provide the proof for the first sequent in (4). Others can be checked regularly.

$$\dfrac{\dfrac{\dfrac{\dfrac{\dfrac{\alpha \Rightarrow \alpha \qquad \bot \Rightarrow \bot}{\alpha, \neg\alpha \Rightarrow \bot}(\rightarrow \text{L})}{\alpha, \bullet\Box\neg\alpha \Rightarrow \bot}(\Box\ \text{L})}{\circ\alpha, \Box\neg\alpha \Rightarrow \bot}(\text{Dual}_{\bullet\circ})}{\Diamond\alpha, \Box\neg\alpha \Rightarrow \bot}(\Diamond\ \text{L})}{\Box\neg\alpha \Rightarrow \neg\Diamond\alpha}(\rightarrow \text{R}) \qquad \dfrac{\dfrac{\dfrac{\dfrac{\dfrac{\alpha \Rightarrow \alpha}{\circ\alpha \Rightarrow \Diamond\alpha}(\Diamond\ \text{R}) \qquad \bot \Rightarrow \bot}{\circ\alpha, \neg\Diamond\alpha \Rightarrow \bot}(\rightarrow \text{L})}{\alpha, \bullet\neg\Diamond\alpha \Rightarrow \bot}(\text{Dual}_{\circ\bullet})}{\bullet\neg\Diamond\alpha \Rightarrow \neg\alpha}(\rightarrow\text{R})}{\neg\Diamond\alpha \Rightarrow \Box\neg\alpha}(\Box\text{R})$$

Now we show that the sequent system in Definition 4 is sound and complete with respect to weak IK.t algebras. For any algebraic structure $\mathbb{A}$ with domain A, an *assignment* in $\mathbb{A}$ is a function $\sigma : \mathsf{Var} \rightarrow A$. Every assignment σ in $\mathbb{A}$ can be extended homomorphically. Let $\widehat{\sigma}(\alpha)$ be the element in A denoted by α. An *algebraic model* is a pair $(\mathbb{A}, \sigma)$ where $\mathbb{A}$ is an algebraic structure and σ is an assignment in $\mathbb{A}$. A sequent $\Gamma \Rightarrow \beta$ is *true* in an algebraic model $(\mathbb{A}, \sigma)$ notation $\models_{\mathbb{A},\sigma} \Gamma \Rightarrow \beta$, if $\widehat{\sigma}(f(\Gamma)) \leq \widehat{\sigma}(\beta)$. A sequent $\Gamma \Rightarrow \psi$ is *true* in a class of algebraic structures $\mathcal{K}$, notation $\models_{\mathcal{K}} \Gamma \Rightarrow \beta$, if $\models_{\mathbb{A},\sigma} \Gamma \Rightarrow \psi$ for any algebraic model $(\mathbb{A}, \sigma)$ with $\mathbb{A} \in \mathcal{K}$. A sequent rule with premises $\Gamma_1 \Rightarrow \psi_1, \ldots, \Gamma_n \Rightarrow \psi_n$ and conclusion $\Gamma_0 \Rightarrow \psi_0$ *preserves truth* in $\mathcal{K}$, if $\models_{\mathbb{A},\sigma} \Gamma_0 \Rightarrow \psi_0$ whenever $\models_{\mathbb{A},\sigma} \Gamma_i \Rightarrow \psi_i$ for $1 \leq i \leq n$, for any algebraic model $(\mathbb{A}, \sigma)$ with $\mathbb{A} \in \mathcal{K}$.

Definition 6. *A sequent calculus* G *is called* sound *with respect to* $\mathcal{K}$, *if for any sequent* $\Gamma \Rightarrow \psi$, $\vdash_{\mathsf{G}} \Gamma \Rightarrow \psi$ *implies* $\models_{\mathcal{K}} \Gamma \Rightarrow \psi$. *A sequent calculus* G *is called* complete *with respect to* $\mathcal{K}$, *if for any sequent* $\Gamma \Rightarrow \psi$, $\models_{\mathcal{K}} \Gamma \Rightarrow \psi$ *implies* $\vdash_{\mathsf{G}} \Gamma \Rightarrow \psi$.

Definition 7. *Define a formula from a context* $\Gamma[-]$ *and a formula* β *denoted by* $dif(\Gamma\ [-];\beta)$ *recursively as follows*

- $dif(\varepsilon;\beta) = \beta$
- $dif(\Gamma_1, \Gamma_2[-];\beta) = dif(\Gamma_2[-]; f(\Gamma_1) \to \beta)$
- $dif(\Gamma_1[-], \Gamma_2;\beta) = dif(\Gamma_1[-]; f(\Gamma_2) \to \beta)$
- $dif(\circ\Gamma'[-];\beta) = dif(\Gamma'[-]; \blacksquare\beta)$
- $dif(\bullet\Gamma'[-];\beta) = dif(\Gamma'[-]; \Box\beta)$

Lemma 3. $\models_{(\mathfrak{A},\sigma)} \Gamma[\Delta] \Rightarrow \beta$ *iff* $\models_{(\mathfrak{A},\sigma)} \Delta \Rightarrow dif(\Gamma[-],\beta)$.

Proof. We proceed by induction on the complexity of $\Gamma[-]$. If $\Gamma[-] = \varepsilon$, then the claim obviously holds. Let $\Gamma[-] = \Gamma_1, \Gamma_2[-]$. Then

$$\begin{aligned}
\models_{(\mathfrak{A},\sigma)} \Gamma_1, \Gamma_2[\Delta] \Rightarrow \beta \ \text{ iff }\ & \widehat{\sigma}(\Gamma_1, \Gamma_2[\Delta]) \leq \widehat{\sigma}(\beta) \\
\text{iff }\ & \widehat{\sigma}(\Gamma_1) \wedge \widehat{\sigma}(\Gamma_2[\Delta]) \leq \widehat{\sigma}(\beta) \\
\text{iff }\ & \widehat{\sigma}(\Gamma_2[\Delta]) \leq \widehat{\sigma}(\Gamma_1) \to \widehat{\sigma}(\beta) \quad \text{(by (Res))} \\
\text{iff }\ & \models_{(\mathfrak{A},\sigma)} \Gamma_2[\Delta] \Rightarrow f(\Gamma_1) \to \beta \quad \text{(by Def of } \widehat{\sigma}) \\
\text{iff }\ & \models_{(\mathfrak{A},\sigma)} \Delta \Rightarrow dif(\Gamma_2[-]; f(\Gamma_1) \to \beta) \quad \text{(by IH)} \\
\text{iff }\ & \models_{(\mathfrak{A},\sigma)} \Delta \Rightarrow dif(\Gamma[-];\beta) \text{(by Def of } dif).
\end{aligned}$$

Let $\Gamma[-] = \circ\Gamma'[-]$. Then

$$\begin{aligned}
\models_{(\mathfrak{A},\sigma)} \circ\Gamma'[\Delta] \Rightarrow \beta \ \text{ iff }\ & \widehat{\sigma}(\circ\Gamma'[\Delta]) \leq \widehat{\sigma}(\beta) \\
\text{iff }\ & \Diamond\widehat{\sigma}(\Gamma'[\Delta]) \leq \widehat{\sigma}(\beta) \\
\text{iff }\ & \widehat{\sigma}(\Gamma'[\Delta]) \leq \blacksquare\widehat{\sigma}(\beta) \quad \text{(by Adj}_{\Diamond\blacksquare}) \\
\text{iff }\ & \models_{(\mathfrak{A},\sigma)} \Gamma'[\Delta] \Rightarrow \blacksquare\beta \quad \text{(by Def of } \widehat{\sigma}) \\
\text{iff }\ & \models_{(\mathfrak{A},\sigma)} \Delta \Rightarrow dif(\Gamma'[-]; \blacksquare\beta) \quad \text{(by IH)} \\
\text{iff }\ & \models_{(\mathfrak{A},\sigma)} \Delta \Rightarrow dif(\circ\Gamma'[-];\beta) \quad \text{(by Def of } dif).
\end{aligned}$$

The argument is similar if $\Gamma[-] = \bullet\Gamma'[-]$.

Theorem 1. GwIK.t *is sound with respect to weak IK.t algebras.*

Proof. It suffices to show that all axiom and rules are true in weak IK.t algebras. The proof for the case of axiom is obvious. Let's consider rule $(\to \mathrm{L})$. Let $(\mathfrak{A}, \sigma)$ where $\mathfrak{A} \in \mathrm{wIK.tA}$ be a model of GwIK.t. Assume that $\models_{(\mathfrak{A},\sigma)} \Delta \Rightarrow \beta$ and $\models_{(\mathfrak{A},\sigma)} \Gamma[\gamma] \Rightarrow \alpha$. Then by Lemma 3 $\models_{(\mathfrak{A},\sigma)} \gamma \Rightarrow dif(\Gamma[-];\alpha)$.

So $\widehat{\sigma}(\gamma) \leq \widehat{\sigma}(dif(\Gamma[-];\alpha))$ and $\widehat{\sigma}(f(\Delta)) \leq \widehat{\sigma}(\beta)$. Thus $\widehat{\sigma}(\beta) \to \widehat{\sigma}(\gamma) \leq \widehat{\sigma}(\beta) \to \widehat{\sigma}(dif(\Gamma[-];\alpha))$. So by (Res), $\widehat{\sigma}(\beta) \wedge \widehat{\sigma}(\beta) \to \widehat{\sigma}(\gamma) \leq \widehat{\sigma}(dif(\Gamma[-];\alpha))$.

Therefore $\widehat{\sigma}(f(\Delta)) \wedge \widehat{\sigma}(\beta) \rightarrow \widehat{\sigma}(\gamma) \leq \widehat{\sigma}(dif(\Gamma[-];\alpha))$. Hence $\models_{(\mathfrak{A},\sigma)} \Delta, \beta \rightarrow \gamma \Rightarrow dif(\Gamma[-];\alpha)$. Consequently $\models_{(\mathfrak{A},\sigma)} \Gamma[\Delta, \beta \rightarrow \gamma] \Rightarrow \alpha$. Other rules can be checked similarly. Notice that (Dual$_{\circ\bullet}$) and (Dual$_{\circ\bullet}$) preserve true in wIK.tA due to Lemma 1 (9).

Lemma 4. *GwIK.t is complete with respect to weak IK.t algebras.*

Proof. The completeness can be proved by standard construction. Let $|\alpha| = \{\beta| \vdash_{\mathrm{GwIK.t}} \alpha \Leftrightarrow \beta\}$. Let A be the set of all $|\alpha|$. Defined $\wedge, \vee, \rightarrow, \Diamond, \Box, \blacklozenge, \blacksquare$ on A as follows:

$$|\alpha_1| \wedge |\alpha_2| = |\alpha_1 \wedge \alpha_2| \quad |\alpha_1| \vee |\alpha_2| = |\alpha_1 \vee \alpha_2| \quad |\alpha_1| \rightarrow |\alpha_2| = |\alpha_1 \rightarrow \alpha_2|$$

$$\Diamond|\alpha| = |\Diamond\alpha| \quad \blacklozenge|\alpha| = |\blacklozenge\alpha| \quad \Box|\alpha| = |\Box\alpha| \quad \blacksquare|\alpha| = |\blacksquare\alpha|$$

Clearly $\mathfrak{A} = (A, \wedge, \vee, \rightarrow, \Diamond, \Box, \blacklozenge, \blacksquare, |\top|, |\bot|)$ is a Heyting algebra. By Lemma 2 $\mathfrak{A}$ is a wIK.tA. The lattice order is defined as $|\alpha_1| \leq |\alpha_2| = |\alpha_1| \wedge |\alpha_2| = |\alpha_1|$. Thus $|\alpha_1| \leq |\alpha_2|$ iff $\vdash_{\mathrm{GwIK.t}} \alpha_1 \Rightarrow \alpha_2$. Define an assignment $\sigma : \mathrm{Var} \rightarrow A$ such that $\sigma(p) = |p|$. By proceed induction on the complexity of formula, one shows that $\widehat{\sigma}(\alpha) = |\alpha|$ for any formula α. Hence $\nvdash_{\mathrm{GwIK.t}} \Gamma \Rightarrow \alpha$ implies $\not\models_{\mathfrak{A},\sigma} \widehat{\sigma}(f(\Gamma)) \leq \widehat{\sigma}(\beta)$. Consequently $\nvdash_{\mathrm{wGIK.t}} \Gamma \Rightarrow \beta$. This complete the proof.

3 Cut Elimination

In this section we show that GwIK.t admits cut elimination. By the proof length of a derivation, we mean the length of its maximum branch in the derivation tree.

Lemma 5. *If $\vdash_{\mathrm{GwIK.t}} \Gamma[\Delta, \Delta] \Rightarrow \beta$ is derivable without any application of (Cut), then $\vdash_{\mathrm{GwIK.t}} \Gamma[\Delta] \Rightarrow \beta$ is derivable without any application of (Cut).*

Proof. We proceed by induction on the complexity of Δ i.e. the total number of structure operations $(,)$, $\circ$ and $\bullet$. Suppose that there is a cut free derivation of $\Gamma[\Delta, \Delta] \Rightarrow \beta$ in GwIK.t. If $\Delta = \alpha$ for some formula α, then by the assumption and (Conf), one gets $\vdash_{\mathrm{GwIK.t}} \Gamma[\alpha] \Rightarrow \beta$ without any application of (Cut). Otherwise we consider the following cases.

Case 1. Let $\Delta = (\Delta_1, \Delta_2)$ for some Δ_1, Δ_2. Then $\vdash_{\mathrm{GwIK.t}} \Gamma[\Delta_1, \Delta_2, \Delta_1, \Delta_2] \Rightarrow \beta$. Hence one gets $\vdash_{\mathrm{GwIK.t}} \Gamma[(\Delta_1, \Delta_1), (\Delta_2, \Delta_2)] \Rightarrow \beta$. Thus by induction hypothesis, we obtain $\vdash_{\mathrm{GwIK.t}} \Gamma[\Delta_1, \Delta_2] \Rightarrow \beta$, that is $\vdash_{\mathrm{GwIK.t}} \Gamma[\Delta] \Rightarrow \beta$.

Case 2. Let $\Delta = \bullet\Delta'$ for some Δ'. Then $\vdash_{\mathrm{GwIK.t}} \Gamma[\bullet\Delta', \bullet\Delta'] \Rightarrow \beta$. By (Con$_\bullet$) rule we have $\Gamma[\bullet(\Delta', \Delta')] \Rightarrow \beta$ Then by induction hypothesis we have $\Gamma[\bullet\Delta'] \Rightarrow \beta$, that is $\vdash_{\mathrm{GwIK.t}} \Gamma[\Delta] \Rightarrow \beta$. The argument is similar if $\Delta = \circ\Delta'$ for some Δ'.

Theorem 2. *If $\vdash_{\mathrm{GwIK.t}} \Gamma \Rightarrow \beta$, then $\vdash_{\mathrm{GwIK.t}} \Gamma \Rightarrow \beta$ without any application of (Cut).*

Proof. Assume that there is a subderivation of $\Gamma \Rightarrow \beta$ ending with an application of (Cut) as follows:

$$\frac{\vdash \Delta \Rightarrow \alpha \qquad \vdash \Sigma[\alpha] \Rightarrow \beta}{\Sigma[\Delta] \Rightarrow \beta}\,(\text{Cut})$$

It suffices to show that if $\Delta \Rightarrow \alpha$ and $\Sigma[\alpha] \Rightarrow \beta$ are both provable in GwIK.t without any applications of (Cut), then $\Sigma[\Delta] \Rightarrow \beta$ is provable in GwIK.t without any applications of (Cut). We proceed by induction on (I) the complexity of (Cut) formula α. In each case we proceed by induction on (II) the proof length of left premise of (Cut) and (III) the proof length of right premise of (Cut). Assume that $\Delta \Rightarrow \alpha$ is obtained by (R_l) and $\Sigma[\alpha] \Rightarrow \beta$ is obtained by (R_r). We consider the following cases.

(1) α is not introduced by (R_l). Assume that (R_l) contained only one premise. We transform the derivation by first applying (Cut) to a premise of (R_l) and $\Sigma[\alpha] \Rightarrow \beta$. After that we apply ($\text{R}_\text{l}$) to the resulting sequent. The case that (R_l) is a two-premise rule is similar. Take ($\blacklozenge$L) rule as an example to interpret this. The remaining cases can be treated similarly.

(R_l) is ($\blacklozenge$L). Then $\Delta = \Delta'[\blacklozenge\gamma]$. The proof

$$\frac{\dfrac{\Delta'[\bullet\gamma] \Rightarrow \alpha}{\Delta'[\blacklozenge\gamma] \Rightarrow \alpha}\,(\blacklozenge\text{L}) \qquad \Sigma[\alpha] \Rightarrow \beta}{\Sigma[\Delta'[\blacklozenge\gamma]] \Rightarrow \beta}\,(\text{Cut})$$

can be transformed into

$$\dfrac{\dfrac{\Delta'[\bullet\gamma] \Rightarrow \alpha \qquad \Sigma[\alpha] \Rightarrow \beta}{\Sigma[\Delta'[\bullet\gamma]] \Rightarrow \beta}\,(\text{Cut})}{\Sigma[\Delta'[\blacklozenge\gamma]] \Rightarrow \beta}\,(\blacklozenge\text{L})$$

Thus the new applications of (Cut) has lower length of its left premise. By induction hypothesis (II), the claim holds

(2) Let α is introduced by (R_l). Assume that $\Sigma[\alpha] \Rightarrow \beta$ is obtained from $\Sigma[\alpha^n] \Rightarrow \beta$ by (n − 1) times of (Conf) such that $\Sigma[\alpha^n]$ is not obtained from a (Conf) rule with principle formulas in α^n. Suppose that $\Sigma[\alpha^n]$ is obtained by rule (R). Obviously if $n = 1$ then $\text{R}_\text{r} = \text{R}$. Let the derivation ends with

$$\frac{\Delta \Rightarrow \alpha \qquad \dfrac{\dfrac{\cdots}{\Sigma[\alpha^n] \Rightarrow \beta}\,(\text{R})}{\Sigma[\alpha] \Rightarrow \beta}\,(\text{Conf}\times(\text{n}-1))}{\Sigma[\Delta] \Rightarrow \beta}\,(\text{Cut})$$

We consider the following subcases according to (R).

(2.1) α is not introduced by (R) and (R) is a one premise rule. We transform the derivation by first applying n times of (Cut) to a premise of (R) and $\Delta \Rightarrow \alpha$. After that we apply ($R_1$) to the resulting sequent. Then by Lemma 5, one gets the desired end sequent. For instance let (R) is ($Con_\circ$). Suppose that the derivation ends with

$$\dfrac{\Delta \Rightarrow \alpha \qquad \dfrac{\dfrac{\Sigma'[\circ(\Sigma_1, \alpha^{n_1}), \circ(\Sigma_2, \alpha^{n_2})] \Rightarrow \beta}{\Sigma'[\circ(\Sigma_1, \alpha^{n_1}, \Sigma_2, \alpha^{n_2})] \Rightarrow \beta}\,(\mathrm{Con}_\circ)}{\Sigma'[\circ(\Sigma_1, \Sigma_2, \alpha)] \Rightarrow \beta}\,(\mathrm{Conf}\times(\mathrm{n}-1))}{\Sigma'[\circ(\Sigma_1, \Sigma_2, \Delta)] \Rightarrow \beta}\,(\mathrm{Cut})$$

such that $0 \leq n_1 + n_2 \leq n$. Then the derivation can be transformed into

$$\dfrac{\dfrac{\dfrac{\Delta \Rightarrow \alpha \qquad \Sigma'[\circ(\Sigma_1, \alpha^{n_1}), \circ(\Sigma_2, \alpha^{n_2})] \Rightarrow \beta}{\Sigma'[\circ(\Sigma_1, \Delta^{n_1}), \circ(\Sigma_2, \Delta^{n_2})] \Rightarrow \beta}\,((\mathrm{Cut})\times\mathrm{n})}{\Sigma'[\circ(\Sigma_1, \Delta^{n_1}, \Sigma_2, \Delta^{n_2})] \Rightarrow \beta}\,(\mathrm{Con}_\circ)}{\Sigma'[\circ(\Sigma_1, \Sigma_2, \Delta)] \Rightarrow \beta}\,(\mathrm{Lemma\ 5})$$

Clear in the original derivation the proof length of the right premise of (Cut) is $n+1$. Thus the new applications of (Cut) have right premises with proof length lower than $n+1$. Hence By induction hypothesis (III), there is a cut free derivation of $\Sigma'[\circ(\Sigma_1, \Delta^{n_1}, \Sigma_2, \Delta^{n_2})] \Rightarrow \beta$. Then by Lemma 5, one gets $\Sigma'[\circ(\Sigma_1, \Sigma_2, \Delta)] \Rightarrow \beta$ without any applications of (Cut). Thus the claim holds. The remaining cases can be treated similarly.

(2.2) α is not introduced by (R) and (R) is a two premise rule. We consider the following three subcases

Case 1. α only appears in the left premise of (R). We transform the derivation by first applying n times of (Cut) to left premise of (R) and $\Delta \Rightarrow \alpha$. After that we apply ($R_1$) to the resulting sequent and the right premise. Then by Lemma 5, one gets the desired end sequent. For instance let (R) is ($\rightarrow$ L). Suppose that the derivation ends with

$$\dfrac{\Delta \Rightarrow \alpha \qquad \dfrac{\dfrac{\Theta[\alpha^n] \Rightarrow \gamma \qquad \Sigma'[\delta] \Rightarrow \beta}{\Sigma'[\Theta[\alpha^n], \gamma \rightarrow \delta] \Rightarrow \beta}\,(\rightarrow \mathrm{L})}{\Sigma'[\Theta[\alpha], \gamma \rightarrow \delta] \Rightarrow \beta}\,(\mathrm{Conf}\times(\mathrm{n}-1))}{\Sigma'[\Theta[\Delta], \gamma \rightarrow \delta] \Rightarrow \beta}\,(\mathrm{Cut})$$

The derivation can be transform into

$$\dfrac{\dfrac{\dfrac{\Delta \Rightarrow \alpha \qquad \Theta[\alpha^n] \Rightarrow \gamma}{\Theta[\Delta^n] \Rightarrow \gamma}\,((\mathrm{Cut})\times\mathrm{n}) \qquad \Sigma'[\delta] \Rightarrow \beta}{\Sigma'[\Theta[\Delta^n], \gamma \rightarrow \delta] \Rightarrow \beta}\,(\rightarrow \mathrm{L})}{\Sigma'[\Theta[\Delta], \gamma \rightarrow \delta] \Rightarrow \beta}\,(\mathrm{Lemma\ 5})$$

Thus the new applications of (Cut) have lower length of their right premises. By induction hypothesis (III), the claim holds.

Case 2. α only appears in the right premise of (R). We transform the derivation by first applying n times of (Cut) to right premise of (R) and $\Delta \Rightarrow \alpha$. After that we apply (R) to the resulting sequent and the left premise. Then by Lemma 5, one gets the desired end sequent. For instance let (R) is ($\rightarrow$ L). Suppose that the derivation ends with

$$\dfrac{\Delta \Rightarrow \alpha \qquad \dfrac{\dfrac{\Theta \Rightarrow \gamma \qquad \Sigma'[\alpha^n][\delta] \Rightarrow \beta}{\Sigma'[\alpha^n][\Theta, \gamma \rightarrow \delta] \Rightarrow \beta}(\rightarrow \mathrm{L})}{\Sigma'[\alpha][\Theta, \gamma \rightarrow \delta] \Rightarrow \beta}(\mathrm{Conf}\times(\mathrm{n}-1))}{\Sigma'[\Delta][\Theta, \gamma \rightarrow \delta] \Rightarrow \beta}(\mathrm{Cut})$$

The derivation can be transform into

$$\dfrac{\dfrac{\dfrac{\Delta \Rightarrow \alpha \qquad \Sigma'[\alpha^n][\delta] \Rightarrow \beta}{\Theta[\Delta^n] \Rightarrow \gamma}((\mathrm{Cut})\times\mathrm{n}) \qquad \Sigma'[\delta] \Rightarrow \beta}{\Sigma'[\Theta[\Delta^n], \gamma \rightarrow \delta] \Rightarrow \beta}(\rightarrow \mathrm{L})}{\Sigma'[\Theta[\Delta], \gamma \rightarrow \delta] \Rightarrow \beta}(\mathrm{Lemma\ 5})$$

Thus the new applications of (Cut) have lower length of their right premises. By induction hypothesis (III), the claim holds.

Case 3. α appear n_1 times in the left premise of (R) while α appear n_2 times in the right premise of (R). $n = n_1 + n_2$. . We transform the derivation by first applying n_1 and n_2 times of (Cut) to left and right premise of (R) and $\Delta \Rightarrow \alpha$ respectively. After that we apply (R) to the resulting sequents. Then by Lemma 5, one gets the desired end sequent. For instance let (R) is ($\rightarrow$ L). Suppose that the derivation ends with

$$\dfrac{\Delta \Rightarrow \alpha \qquad \dfrac{\dfrac{\Theta, \alpha^{n_1} \Rightarrow \gamma \qquad \Sigma'[\alpha^{n_2}, \delta] \Rightarrow \beta}{\Sigma'[\Theta, \alpha^{(n_1+n_2)}, \gamma \rightarrow \delta] \Rightarrow \beta}(\rightarrow \mathrm{L})}{\Sigma'[\Theta, \alpha, \gamma \rightarrow \delta] \Rightarrow \beta}(\mathrm{Conf}\times(\mathrm{n}-1))}{\Sigma'[\Theta, \Delta, \gamma \rightarrow \delta] \Rightarrow \beta}(\mathrm{Cut})$$

The derivation can be transformed into

$$\dfrac{\dfrac{\dfrac{\Delta \Rightarrow \alpha \qquad \Theta, \alpha^{n_1} \Rightarrow \gamma}{\Theta, \Delta^{n_1} \Rightarrow \gamma}((\mathrm{Cut})\mathrm{s}) \qquad \dfrac{\Delta \Rightarrow \alpha \qquad \Sigma'[\alpha^{n_2}, \delta] \Rightarrow \beta}{\Sigma'[\Delta^{n_2}, \delta] \Rightarrow \beta}((\mathrm{Cut})\mathrm{s})}{\Sigma'[\Theta, \Delta^{n_1+n_2}, \gamma \rightarrow \delta] \Rightarrow \beta}(\rightarrow \mathrm{L})}{\Sigma'[\Theta, \Delta, \gamma \rightarrow \delta] \Rightarrow \beta}(\mathrm{Lemma\ 5})$$

Thus the new applications of (Cut) havd lower length of their right premises. By induction hypothesis (III), the claim holds.

(2.3) α is introduced by (R). We consider the following subcases. Others can be treated similarly.

Case 1. $\alpha = \Diamond\alpha'$. Suppose that the derivation ends with

$$\dfrac{\dfrac{\Delta' \Rightarrow \alpha'}{\circ\Delta' \Rightarrow \Diamond\alpha'}(\Diamond\mathrm{R}) \qquad \dfrac{\dfrac{\Sigma[(\Diamond\alpha')^{n-1}, \circ\alpha'] \Rightarrow \beta}{\Sigma[(\Diamond\alpha')^{n-1}, \Diamond\alpha'] \Rightarrow \beta}(\Diamond\mathrm{L})}{\Sigma[\Diamond\alpha'] \Rightarrow \beta}((\mathrm{Conf})\times(\mathrm{n}-1))}{\Sigma[\circ\Delta] \Rightarrow \beta}(\mathrm{Cut})$$

The derivation can be transformed into

$$\dfrac{\Delta' \Rightarrow \alpha' \qquad \dfrac{\circ\Delta' \Rightarrow \Diamond\alpha' \qquad \Sigma[(\Diamond\alpha')^{n-1}, \circ\alpha'] \Rightarrow \beta}{\Sigma[(\circ\Delta')^{n-1}, \circ\alpha'] \Rightarrow \beta}\,(\mathrm{Cut})\times(n-2)}{\dfrac{\Sigma[(\circ\Delta')^{n-1}, \circ\Delta'] \Rightarrow \beta}{\Sigma[\circ\Delta'] \Rightarrow \beta}\,(\text{Lemma 5})}\,(\mathrm{Cut})$$

Thus the first $n-2$ new applications of (Cut) have lower length of their right premises while the last application of (Cut) has lower complexity of (Cut) formula. By induction hypothesis (I) and (III), the claim holds. Case 2. $\alpha = \Box\alpha'$. Clearly $n = 1$. $\Sigma[\alpha] = \Sigma'[\bullet(\Box\alpha')]$. The proof

$$\dfrac{\dfrac{\bullet\Delta \Rightarrow \alpha'}{\Delta \Rightarrow \Box\alpha'}\,(\Box\mathrm{R}) \qquad \dfrac{\Sigma'[\alpha'] \Rightarrow \beta}{\Sigma'[\bullet\Box\alpha'] \Rightarrow \beta}\,(\Box\mathrm{L})}{\Sigma'[\bullet\Delta] \Rightarrow \beta}\,(\mathrm{Cut})$$

can be transformed into

$$\dfrac{\bullet\Delta \Rightarrow \alpha' \qquad \Sigma'[\alpha'] \Rightarrow \beta}{\Sigma'[\bullet\Delta] \Rightarrow \beta}\,(\mathrm{Cut})$$

Clearly the new application of (Cut) has lower complexity of (Cut) formula. By induction hypothesis (I), the claim holds.

4 Extensions and Structure Criterion

In this section we consider extensions of wIK.t which have cut free sequent systems. Given two formula α, β, we define $\alpha \sim^m \beta$ if β can be obtained from α by changing the positions and numbers of connectives $\Diamond, \blacklozenge$ in α. For instance $\Diamond(p \wedge \blacklozenge q) \sim^m \blacklozenge(\Diamond\Diamond p \wedge q)$. Similarly we define $\Gamma \sim^m \Delta$ if Δ can be obtained from Γ by changing the positions and numbers of structure operations $\circ, \bullet$ in Γ. For instance $\circ(\Delta_1, \bullet\Delta_2) \sim^m \bullet\bullet\Delta_1, \Delta_2$. By wIK.t$\oplus\alpha \Rightarrow \beta$, we mean the extension of wIK.t enriching with axiom $\alpha \Rightarrow \beta$. Two sequents $\alpha \Rightarrow \beta$ and $\alpha' \Rightarrow \beta$ are called equivalence over wIK.t if $\vdash_{\mathrm{wIK.t}\oplus\alpha\Rightarrow\beta} \alpha' \Rightarrow \beta'$ and $\vdash_{\mathrm{wIK.t}\oplus\alpha'\Rightarrow\beta'} \alpha \Rightarrow \beta$.

Definition 8. *A sequent is call primitive sequent if it is of the form $\alpha \Rightarrow \beta$ where both α and β only contains variables, $\top$, $\bot$, $\wedge, \Diamond$ and $\blacklozenge$ and $\alpha \sim^m \beta$. A sequent is called modal structure sequent if it is obtained from a primitive sequent by substituting variables with formulas. We say that a sequent is modal structure display w.r.t. wIK.t if it has a equivalence modal structure sequent over wIK.t.*

A modal structure rule is

$$\dfrac{\Gamma[\Delta] \Rightarrow \alpha}{\Gamma[\Sigma] \Rightarrow \alpha}$$

where $\Delta \sim^m \Sigma$. Clearly every modal structure sequent can be characterized by a structure modal rule. Furthermore one can extend the proof of cut elimination with any new modal structure rule easily. In the proof of cut elimination, suppose that one of the cut premise is obtained from a modal structure rule. One can firstly apply (Cut) rule to the premise of modal structure rule and the other cut premise. Then apply modal structure rule to the conclusion of the new (Cut).

Hence the new application of (Cut) has lower proof length of right or left premise, by induction hypothesis, this (Cut) can be eliminated. This is the general idea of the proof of cut elimination. Here we list some modal axioms which are frequently considered in the literature. Notice that if one add the following WK axiom to wIK.t, then the resulting logic is just IK.t.

Table 1. Modal structure axioms

Name	Axiom
T	$\Box\alpha \Rightarrow \alpha$ and $\blacksquare\alpha \Rightarrow \alpha$
4	$\Box\alpha \Rightarrow \Box\Box\alpha$ and $\blacksquare\alpha \Rightarrow \blacksquare\blacksquare\alpha$
5	$\Diamond\alpha \Rightarrow \Box\Diamond\alpha$ and $\blacklozenge\alpha \Rightarrow \blacksquare\blacklozenge\alpha$
B	$\alpha \Rightarrow \Box\Diamond\alpha$ and $\alpha \Rightarrow \blacksquare\blacklozenge\alpha$
D	$\top \Rightarrow \Diamond\top$ and $\top \Rightarrow \blacklozenge\top$
WK	$\Box(\alpha \to \beta) \Rightarrow \Diamond\alpha \to \Diamond\beta$ and $\blacksquare(\alpha \to \beta) \Rightarrow \blacklozenge\alpha \to \blacklozenge\beta$

Lemma 6. *The modal axioms in Table 1 is equivalence to the following corresponding modal structure axioms over wIK.t:*

T': $\alpha \Rightarrow \Diamond\alpha$ and $\alpha \Rightarrow \blacklozenge\alpha$
$4'$: $\Diamond\Diamond\alpha \Rightarrow \Diamond\alpha$ and $\blacklozenge\blacklozenge\alpha \Rightarrow \blacklozenge\alpha$
$5'$: $\blacklozenge\Diamond\alpha \Rightarrow \Diamond\alpha$ and $\Diamond\blacklozenge\alpha \Rightarrow \blacklozenge\alpha$
B': $\blacklozenge\alpha \Rightarrow \Diamond\alpha$ and $\Diamond\alpha \Rightarrow \blacklozenge\alpha$
D': $\top \Rightarrow \Diamond\top$ and $\top \Rightarrow \blacklozenge\top$
WK': $\alpha \wedge \Diamond\beta \Rightarrow \Diamond(\blacklozenge\alpha \wedge \beta)$ and $\alpha \wedge \blacklozenge\beta \Rightarrow \blacklozenge(\Diamond\alpha \wedge \beta)$

Proof. We provide the proof of the first part for WK′. The rest can be proved easily. First we derive WK′ from WK over wIK.t. Since $\Diamond(\blacklozenge\alpha \wedge \beta) \Rightarrow \Diamond(\blacklozenge\alpha \wedge \beta)$, then by (adj$_{\Diamond\blacksquare}$) one gets $(\blacklozenge\alpha \wedge \beta) \Rightarrow \blacksquare\Diamond(\blacklozenge\alpha \wedge \beta)$. By (res), $\blacklozenge\alpha \Rightarrow \beta \to \blacksquare\Diamond(\blacklozenge\alpha \wedge \beta)$. By (adj$_{\blacklozenge\Box}$), $\alpha \Rightarrow \Box(\beta \to \blacksquare\Diamond(\blacklozenge\alpha \wedge \beta))$. By (WK), $\Box(\beta \to \blacksquare\Diamond(\blacklozenge\alpha \wedge \beta)) \Rightarrow \Diamond\beta \to \Diamond\blacksquare\Diamond(\blacklozenge\alpha \wedge \beta)$. Further $\Diamond\blacksquare\Diamond(\blacklozenge\alpha \wedge \beta) \Rightarrow \Diamond(\blacklozenge\alpha \wedge \beta)$. By (Cut), $\alpha \Rightarrow \Diamond\beta \to \Diamond(\blacklozenge\alpha \wedge \beta)$. Then by (res) $\alpha \wedge \Diamond\beta \Rightarrow \Diamond(\blacklozenge\alpha \wedge \beta)$. Conversely since $\Diamond(\alpha \wedge \alpha \to \beta) \Rightarrow \Diamond\beta$ and $\Diamond(\alpha \wedge \blacklozenge\Box(\alpha \to \beta)) \Rightarrow \Diamond(\alpha \wedge \alpha \to \beta)$, by (Cut) $\Diamond(\alpha \wedge \blacklozenge\Box(\alpha \to \beta)) \Rightarrow \Diamond\beta$. By (WK′), $\Diamond\alpha \wedge \Box(\alpha \to \beta) \Rightarrow \Diamond(\alpha \wedge \blacklozenge\Box(\alpha \to \beta))$. Thus by (Cut), $\Diamond\alpha \wedge \Box(\alpha \to \beta) \Rightarrow \Diamond\beta$. Hence by (res) $\Box(\alpha \to \beta) \Rightarrow \Diamond\alpha \to \Diamond\beta$.

Lemma 7. *T', $4'$, $5'$, B', D' and WK' can be characterized by the following corresponding structure rules respectively:*

$$\frac{\Gamma[\circ\Delta] \Rightarrow \alpha}{\Gamma[\Delta] \Rightarrow \alpha}(\mathrm{T}\circ) \quad \frac{\Gamma[\bullet\Delta] \Rightarrow \alpha}{\Gamma[\Delta] \Rightarrow \alpha}(\mathrm{T}\bullet) \quad \frac{\Gamma[\circ\Delta] \Rightarrow \alpha}{\Gamma[\circ\circ\Delta] \Rightarrow \alpha}(4\circ) \quad \frac{\Gamma[\bullet\Delta] \Rightarrow \alpha}{\Gamma[\bullet\bullet\Delta] \Rightarrow \alpha}(4\bullet)$$

$$\frac{\Gamma[\circ\Delta] \Rightarrow \alpha}{\Gamma[\bullet\circ\Delta] \Rightarrow \alpha}(5\circ) \quad \frac{\Gamma[\bullet\Delta] \Rightarrow \alpha}{\Gamma[\circ\bullet\Delta] \Rightarrow \alpha}(5\bullet) \quad \frac{\Gamma[\bullet\Delta] \Rightarrow \alpha}{\Gamma[\circ\Delta] \Rightarrow \alpha}(\mathrm{B}\circ) \quad \frac{\Gamma[\circ\Delta] \Rightarrow \alpha}{\Gamma[\bullet\Delta] \Rightarrow \alpha}(\mathrm{B}\bullet)$$

$$\frac{\Gamma[\circ\top] \Rightarrow \alpha}{\Gamma[\top] \Rightarrow \alpha}(\mathrm{D}\circ) \quad \frac{\Gamma[\bullet\top] \Rightarrow \alpha}{\Gamma[\top] \Rightarrow \alpha}(\mathrm{D}\bullet)$$

$$\frac{\Gamma[\circ(\bullet\Delta_1, \Delta_2)] \Rightarrow \alpha}{\Gamma[\Delta_1, \circ\Delta_2] \Rightarrow \alpha}(\mathrm{WK}\circ) \quad \frac{\Gamma[\bullet(\circ\Delta_1, \Delta_2)] \Rightarrow \alpha}{\Gamma[\Delta_1, \bullet\Delta_2] \Rightarrow \alpha}(\mathrm{WK}\bullet)$$

By EXT1(GwIK.t), we denote the class of sequent system for extensions of GwIK.t with any combination of structure rules corresponding to axioms in Table 1.

Theorem 3. *For any sequent system* G *where* G ∈ EXT1(GwIK.t), G *admits cut elimination.*

References

1. Bull, R.A.: Some modal calculi based on IC. Stud. Logic Found. Math. **40**, 3–7 (1965)
2. Burgess, J.P.: Basic Tense Logic. Springer, Dordrecht (1984)
3. Buszkowski, W.: Interpolation and FEP for logics of residuated algebras. Logic J. IGPL **19**(3), 437–454 (2018)
4. Chajda, I.: Algebraic axiomatization of tense intuitionistic logic. Cent. Eur. J. Math. **9**(4), 1185–1191 (2011)
5. Ewald, W.: Intuitionistic tense and modal logic. J. Symb. Logic **51**(1), 166–179 (1986)
6. Figallo, A., Pelaitay, G.: An algebraic axiomatization of the Ewald's intuitionistic tense logic. Soft. Comput. **18**(10), 1873–1883 (2014)
7. Fitting, M.: Nested sequents for intuitionistic logics. Notre Dame J. Formal Logic **1**(1), 41–61 (2014)
8. Galmiche, D., Yakoub, S.: Label-free natural deduction systems for intuitionistic and classical modal logics. J. Appl. Non-Class. Logics **20**(4), 373–421 (2010)
9. Gore, R., Postniece, L., Tiu, A.: Cut-elimination and proof search for bi-intuitionistic tense logic. In: Valentin, S., Beklemishev, L., Valentin, G. (eds.) Advances in Modal Logic, pp. 156–177. College Publications, London (2010)
10. Simpson, A.K.: The proof theory and semantics of intuitionistic modal logic (1994)
11. Straßburger, L.: Cut elimination in nested sequents for intuitionistic modal logics. In: Pfenning, F. (ed.) FoSSaCS 2013. LNCS, vol. 7794, pp. 209–224. Springer, Heidelberg (2013). https://doi.org/10.1007/978-3-642-37075-5_14
12. Versmissen, J.: Grammatical composition: modes, models, modalities: logical and linguistic aspects of multimodal categorial grammars (2001)
13. Zhe, L.: Non-associative Lambek calculus with modalities: interpolation, complexity and FEP. Logic J. IGPL **22**(3), 494–512 (2014)

Weakly Aggregative Modal Logic: Characterization and Interpolation

Jixin Liu[1], Yanjing Wang[2](✉), and Yifeng Ding[3]

[1] Department of Philosophy, Sichuan University, Chengdu, China
[2] Department of Philosophy, Peking University, Beijing, China
y.wang@pku.edu.cn
[3] Group in Logic and the Methodology of Science, UC Berkeley, Berkeley, USA

Abstract. Weakly Aggregative Modal Logic (WAML) is a collection of disguised polyadic modal logics with n-ary modalities whose arguments are all the same. WAML has some interesting applications on epistemic logic and logic of games, so we study some basic model theoretical aspects of WAML in this paper. Specifically, we give a van Benthem-Rosen characterization theorem of WAML based on an intuitive notion of bisimulation and show that each basic WAML system $\mathbb{K}_n$ lacks Craig Interpolation.

Keywords: Weakly Aggregative Modal Logic · Bisimulation · van Benthem-Rosen characterzation · Craig Interpolation

1 Introduction

You are invited to a dinner party for married couples after a logic conference in China. The host tells you the following facts:

- at least one person of each couple is a logician and
- at least one person of each couple is Chinese.

Given these two facts, can you infer that at least one person of each couple is a Chinese logician? The answer is clearly negative, since there might be a couple consisting of a foreign logician and a Chinese spouse who is not a logician.

Now, suppose that the host adds another fact:

- at least one person of each couple likes spicy food.

What do you know now? Actually, you can infer that for each couple, one of the two people must be either:

- a Chinese logician, or
- a logician who likes spicy food, or
- a Chinese who likes spicy food.

The main work of the first author was completed during his Ph.D. at Peking University.

P. Blackburn et al. (Eds.): LORI 2019, LNCS 11813, pp. 153–167, 2019.
https://doi.org/10.1007/978-3-662-60292-8_12

This can be verified by the *Pigeonhole Principle*: for each couple, there is a logician, a Chinese, and a fan for spicy food, thus there must be at least one person of the couple who has two of those three properties. This can clearly be generalized to n-tuples of things w.r.t. $n+1$ properties.

Now, going back to logic, if we express "at least one person of each couple has property φ" by $\Box\varphi$ then the above reasoning shows that the following is not valid:

$$\mathtt{C} : \Box p \wedge \Box q \to \Box(p \wedge q).$$

On the other hand, the following should be valid:

$$\mathtt{K}_2 : \Box p \wedge \Box q \wedge \Box r \to \Box((p \wedge q) \vee (p \wedge r) \vee (q \wedge r)).$$

In general, if $\Box\varphi$ expresses "at least one thing of each (relevant) n-tuple of things has property φ" then the following is intuitively valid:

$$\mathtt{K}_n : \Box p_0 \wedge \cdots \wedge \Box p_n \to \Box \bigvee_{(0 \leq i < j \leq n)} (p_i \wedge p_j).$$

Note that $\mathtt{K}_1$ is just $\mathtt{C}$, which is a theorem in the weakest normal modal logic $\mathbb{K}$. $\mathtt{C}$ is sometimes called the *Closure of Conjunction* [11], or *Aggregative Axiom* [20], or *Adjunctive Axiom* [6]. Clearly, when $n \geq 2$, $\mathtt{K}_n$ are weaker versions of $\mathtt{C}$. The resulting logics departing from the basic normal modal logics by using weaker aggregative axioms $\mathtt{K}_n$ instead of $\mathtt{C}$ are called *Weakly Aggregative Modal Logics* (WAML) [32]. There are various readings of $\Box p$ under which it is intuitive to reject $\mathtt{C}$ besides the one we mentioned in our motivating party story. For example, if we read $\Box p$ as "p is obligatory" as in deontic logic, then $\mathtt{C}$ is not that reasonable since one may easily face two conflicting obligations without having any single contradictory obligation [32]. As another example, in epistemic logic of knowing how [16,35], if $\Box p$ expresses "knowing how to achieve p", then it is reasonable to make $\mathtt{C}$ invalid: you may know how to open a door and know how to close the door, but you can never know how to make the door both open and closed.

Coming back to our setting where $\mathtt{K}_n$ are valid, the readings of $\Box\varphi$ in those axioms may sound complicated, but they are actually grounded in a more general picture of *Polyadic Modal Logics* (*PML*) which studies the logics with n-ary modalities. Polyadic modalities arose naturally in the literature of philosophical logic, particularly for the binary ones, such as the *until* modality in temporal logic [21], instantial operators in games-related neighborhood modal logics [34], relativized knowledge operators in epistemic logic [9,36], Routley and Meyer's ternary accessibility relation semantics in relevance logics [29,30], and the conditional operators in the logics of conditionals [8]. Following the notation in [10], we use ∇ for the n-ary generalization of the $\Box$ modality when $n > 1$.[1] The semantics of $\nabla(\varphi_1, \ldots, \varphi_n)$ is based on Kripke models with $n+1$-ary relations R [10,20]:

[1] This is not to be confused with the non-contingency operator, which is also denoted as ∇ in non-contingency or knowing whether logics [14].

$\nabla(\varphi_1, \dots, \varphi_n)$ holds at s iff *for all* $s_1, ..., s_n$ such that $Rss_1 \dots s_n$ *there exists* some $i \in [1, n]$ such that φ_i holds at s_i.

We will call ∇ the normal polyadic modal operator and one should also notice that, by contrast, in those examples with unary operators we just mentioned above, the unary operators are not normal.[2] However, the reading we mentioned for $\Box\varphi$ in our motivating story is simply the semantics for $\nabla(\varphi_1, \dots, \varphi_n)$ where $\varphi_1 = \dots = \varphi_n$: notice how they share the same $\forall\exists$ quantifier alternation pattern. Thus, the formulas $\Box\varphi$ under the new reading can be viewed as special cases of the modal formulas in polyadic modal languages. Due to the fact that the arguments are the same in $\nabla(\varphi, \dots, \varphi)$, we can also call the $\Box$ the *diagonal n-modalities.*[3] In this light, we may call the new semantics for $\Box\varphi$ the *diagonal n-semantics* (given frames with $n + 1$-ary relations).

Diagonal modalities also arise in other settings in disguise. For example, in epistemic logic of *knowing value* [18], the formula $\mathsf{Kv}(\varphi, c)$ says that the agent knows the value of c given φ, which semantically amounts to that for all the pairs of φ worlds that the agent cannot distinguish from the actual worlds, c has the same value. In other words, in *every* pair of the indistinguishable worlds where c has *different* values, *there is* a $\neg\varphi$ world, which can be expressed by $\Box^c\neg\varphi$ with the diagonal 2-modality ($\Box^c$) based on intuitive ternary relations (see details in [18]). As another example in epistemic logic, [13] proposed a local reasoning operator based on models where each agent on each world may have different *frames of mind* (sets of indistinguishable worlds). That one agent believes φ then means that in *one* of his current frame of mind, φ is true *everywhere.* This belief modality can also be viewed as the dual of a diagonal 2-modality (noticing the quantifier alternation $\exists\forall$ in the informal semantics).

Yet another important reason to study diagonal modalities comes from the connection with paraconsistent reasoning established by Schotch and Jennings [32]. In a nutshell, [32] introduces a notion of *n-forcing* where a set of formulas Γ n-forces φ ($\Gamma \vdash_n \varphi$) if for each n-partition of Γ there is a cell Δ such that φ follows from Δ classically w.r.t. some given logic ($\Gamma \vdash \varphi$). This leads to a notion of *n-coherence* relaxing the notion of consistency: $\Gamma \nvdash_n \bot$ (Γ is n-coherent) iff there exists an n-partition of Γ such that all the cells are classically consistent. These notions led the authors of [32] to the discovery of the diagonal semantics for $\Box$ based on frames with $n + 1$-ary relations, by requiring $\Box(u) = \{\varphi \mid u \vDash \Box\varphi\}$ to be an n-theory based on the closure over n-forcing, under some other minor conditions. Since the derivation relation of basic normal modal logic $\mathbb{K}$ can be characterized by a proof system extending the propositional one with the rule $\Gamma \vdash \varphi / \Box(\Gamma) \vdash \Box\varphi$ where $\Box(\Gamma) = \{\Box\varphi \mid \varphi \in \Gamma\}$, it is interesting to ask whether adding $\Gamma \vdash_n \varphi / \Box(\Gamma) \vdash_n \Box\varphi$ characterizes exactly the valid consequences for modal logic under the diagonal semantics based on frames with n-ary relations. Apostoli and Brown answered this question positively in [5] 15 years later, and they characterize $\vdash_n$ by a Gentzen-style sequent calculus based on the compactness of $\vdash_n$ proved by using

[2] One can find a model theoretical survey on PML in [22].

[3] Name mentioned by Yde Venema via personal communications.

a compact result for coloring hypergraphs.[4] Moreover, they show that the WAML proof systems with $\mathtt{K}_n$ are also complete w.r.t. the class of all frames with $n+1$-ary relations respectively. The latter proof is then simplified in [26] without using the graph theoretical compactness result. This completeness result is further generalized to the extensions of WAML with extra one-degree axioms in [4]. The computational complexity issues of such logics are discussed in [1], and this concludes our relatively long introduction to WAML, which might not be that well-known to many modal logicians.

In this paper, we continue the line of work on WAML by looking at the model theoretical aspects. In particular, we mainly focus on the following two questions:

- How to characterize the expressive power of WAML structurally within first-order logic over (finite) pointed models?
- Whether WAML has Craig Interpolation?

For the first question, we propose a notion of bisimulation to characterize WAML within the corresponding first-order logic. The answer for the second question is negative, and we will provide counterexamples in this paper to show WAML (in particular, each $\mathbb{K}_n$ for $n \geq 2$) does not have Craig Interpolation.

In the rest of the paper, we lay out the basics of WAML in Sect. 2, prove the characterization theorem based on a bisimulation notion in Sect. 3, and give counterexamples for the interpolation theorem in Sect. 4 before concluding with future work in Sect. 5.

2 Preliminaries

In this section we review some basic definitions and results in the literature.

2.1 Weakly Aggregative Modal Logic

The language for WAML is the same as the language for basic (monadic) modal logic.

Definition 1. *Given a set of propositional letters Φ and a single unary modality $\Box$, the language of WAML is defined by:*

$$\varphi := p \mid \neg\varphi \mid (\varphi \wedge \varphi) \mid \Box\varphi$$

where $p \in \Phi$. We define $\top$, $\varphi \vee \psi$, $\varphi \to \psi$, and $\Diamond\varphi$ as usual.

However, given n, WAML can be viewed as a fragment of polyadic modal logic with a n-ary modality, since $\Box\varphi$ is essentially $\nabla(\varphi, \ldots, \varphi)$. **Notation:** in the sequel, we use WAML^n, where $n > 1$, to denote the logical framework with the semantics based on n-models defined below:

[4] Other connections between WAML and graph coloring problems can be found in [24] where the four-color problem is coded by the validity of some formulas in the WAML language.

Definition 2 (**n-Semantics**). *An n-frame is a pair $\langle W, R\rangle$ where W is an nonempty set and R is an $n+1$-ary relation over W. An n-model $\mathcal{M}$ is a pair $\langle \mathcal{F}, V\rangle$ where the valuation function V assigns each $w \in W$ a subset of Φ. We say $\mathcal{M}$ is an image-finite model if there are only finitely many n-ary successors of each point. The semantics for $\Box\varphi$ (and $\Diamond\varphi$) is defined by:*

$\mathcal{M}, w \models \Box\varphi$ *iff for all* $v_1, \dots v_n \in W$ *with* $Rwv_1 \dots, v_n, \mathcal{M}, v_i \models \varphi$ *for some* $i \leq n$.
$\mathcal{M}, w \models \Diamond\varphi$ *iff there are* $v_1, \dots v_n \in W$ *st.* $Rwv_1 \dots, v_n$ *and* $\mathcal{M}, v_i \models \varphi$ *for all* $i \leq n$.

According to the above semantics, it is not hard to see that the aggregation axiom $\Box\varphi \wedge \Box\psi \to \Box(\varphi \wedge \psi)$ in basic normal modal logic is not valid on n-frames for any $n > 1$.

[32] proposed the following proof systems $\mathbb{K}_n$ for each n.

Definition 3 (Weakly aggregative modal logic). *The logic $\mathbb{K}_n$ is a modal logic including propositional tautologies, the axiom $\mathtt{K}_n$ and closed under the rules*[5] *$\mathtt{N}$ and $\mathtt{RM}$:*

$$
\begin{array}{ll}
\mathtt{K}_n & \Box p_0 \wedge \dots \wedge \Box p_n \to \Box \bigvee_{(0 \leq i < j \leq n)} (p_i \wedge p_j) \\
\mathtt{RM} & \vdash \varphi \to \psi \implies \vdash \Box\varphi \to \Box\psi \\
\mathtt{N} & \vdash \varphi \implies \vdash \Box\varphi
\end{array}
$$

It is clear that $\mathtt{K}_1$ is just the aggregation axiom $\mathtt{C}$ and thus $\mathbb{K}_1$ is just the normal monadic modal logic $\mathbb{K}$. It can also be shown easily that for each $n > m$, $\mathbb{K}_n$ is strictly weaker than $\mathbb{K}_m$. In fact, many familiar equivalences in normal modal logics, like the equivalence between $\Diamond\top$ and $\Box p \to \Diamond p$, no longer hold in $\mathbb{K}_n$ for $n > 1$. Semantically speaking, while $\Box p \to \Diamond p$'s validity corresponds to seriality on 1-frames (usual Kripke frames), its correspondence on 2-frames is not even elementary ($\Diamond\top$ still corresponds to each point having at least a successor tuple).

After being open for more than a decade, the completeness for $\mathbb{K}_n$ over n-models was finally proved in [5] and [4], by reducing to the n-forcing relation proposed in [32]. In [26], a more direct completeness proof is given using some non-trivial combinatorial analysis to derive a crucial theorem of $\mathbb{K}_n$.

3 Characterization via Bisimulation

In this section, we introduce a notion of bisimulation for WAML and prove the van Benthem-Rosen Characteristic Theorem for WAML.[6]

Definition 4 (**wa^n-bisimulation**). *Let $\mathcal{M} = (W, R, V)$ and $\mathcal{M}' = (W', R', V')$ be two n-models. A non-empty binary relation $Z \subseteq W \times W'$ is called a wa^n-bisimulation between $\mathcal{M}$ and $\mathcal{M}'$ if the following conditions are satisfied:*

inv If wZw', then w and w' satisfy the same propositional letters (in Φ).

[5] This rule can be simplified by the axiom $\Box\top$ since we have $\mathtt{RM}$ here.

[6] We have another proof for the Characterization theorem over arbitrary n-models, using tailored notions of saturation and ultrafilter extension for WAML^n, due to the space limit we only present the proof which also works for finite models.

forth *If* wZw' *and* $Rwv_1, \ldots, v_n$ *then there are* $v'_1, \ldots, v'_n$ *in* W' *s.t.* $R'w'v'_1, \ldots, v'_n$ *and for each* v'_j *there is a* v_i *such that* $v_i Z v'_j$ *where* $1 \leq i, j \leq n$.

back *If* wZw' *and* $R'w'v'_1, \ldots, v'_n$ *then there are* $v_1, \ldots, v_n$ *in* W *s.t.* $Rwv_1, \ldots, v_n$ *and for each* v_i *there is a* v'_j *such that* $v_i Z v'_j$ *where* $1 \leq i, j \leq n$.

When Z *is a bisimulation linking two states* w *in* $\mathcal{M}$ *and* w' *in* $\mathcal{M}'$ *we say that* w *and* w' *are* Φ*-*wa^n*-bisimilar* $(\mathcal{M}, w \leftrightarroweq^n \mathcal{M}', w')$.

Remark 1. Observe the two subtleties in the above definition: i, j in the forth and back conditions are not necessarily the same, thus we may not have an aligned correspondence of each v_i and v'_i; in the second part of the forth condition, we require each v'_j to have a corresponding v_i, not the other way around. Similar in the back condition. This reflects the quantifier alternation in the semantics of $\Box$ in WAML^n.

Example 1. Consider the following two 2-models where $\{\langle w, w_1, w_2\rangle, \langle w, w_2, w_3\rangle\}$ is the ternary relation in the left model, and $\{\langle v, v_1, v_2\rangle\}$ is the ternary relation in the right model.

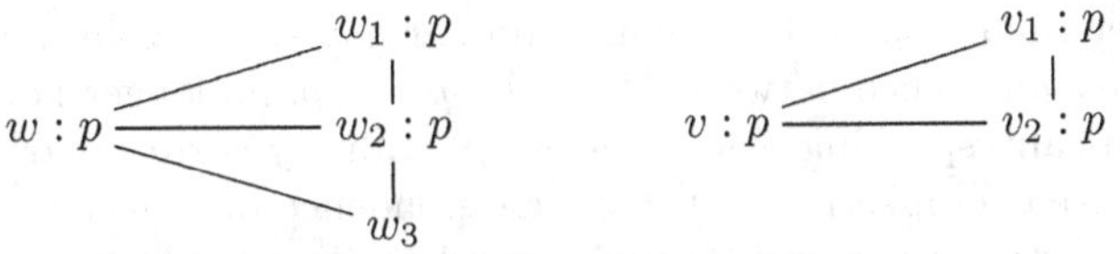

$Z = \{\langle w, v\rangle, \langle w_1, v_1\rangle, \langle w_2, v_2\rangle, \langle w_2, v_1\rangle\}$ is a wa^2-bisimulation. A polyadic modal formula $\neg\nabla\neg(p, \neg p)$, not expressible in WAML^2, can distinguish w and v.

It is easy to verify that $\leftrightarroweq^n$ is indeed an equivalence relation and we show WAML^n is invariant under it.

Proposition 1. *Let* $\mathcal{M} = (W, R, V)$ *and* $\mathcal{M}' = (W', R', V')$ *be two n-models. Then for every* $w \in W$ *and* $w' \in W'$, $w \leftrightarroweq^n w'$ *implies* $w \equiv_{\mathsf{WAML}^n} w'$. *In words,* WAML^n *formulas are invariant under* wa^n*-bisimulation.*

Proof. We consider only the modality case. Suppose that $w \leftrightarroweq^n w'$ and $w \models \Diamond\varphi$. Then there are $v_1, \ldots, v_n$ s.t. $Rwv_1, \ldots, v_n$, and each $v_i \models \varphi$. By the forth condition, there are $v'_1, \ldots, v'_n$ in W' s.t. $Rw'v'_1, \ldots, v'_n$ and for each v'_j there is a v_i such that $v_i Z v'_j$. From the I.H. we have each $v'_i \models \varphi$. As a result, $w' \models \Diamond\varphi$. For the converse direction just use the back condition.

Theorem 1 (Hennessy-Milner Theorem for WAML^n). *Let* $\mathcal{M} = (W, R, V)$ *and* $\mathcal{M}' = (W', R', V')$ *be two image-finite n-models. Then for every* $w \in W$ *and* $w' \in W'$, $w \leftrightarroweq^n w'$ *iff* $w \equiv_{\mathsf{WAML}^n} w'$.

Proof. As in basic modal logic, the crucial part is to show $\equiv_{\mathsf{WAML}^n}$ is indeed a wa^n-bisimulation and we only verify the forth condition. Suppose towards contradiction that $Rwv_1 \dots v_n$ but for each $v'_1 \dots v'_n$ such that $R'w'v'_1 \dots v'_n$ there is a v'_j such that it is not WAML^n-equivalent to any of v_i. In image-finite models we can list such v'_j as $u_1 \dots u_m$. Now for each u_k and v_i we have φ^i_k which holds on v_i but not on u_k. Now we consider the formula $\psi = \Diamond(\bigvee_{1 \le i \le n} \bigwedge_{1 \le k \le m} \varphi^i_k)$. It is not hard to see that ψ holds on w but not w', hence contradiction.

Like in normal modal logic, we can also define a notion of k-bisimulation of WAML^n, by restricting the maximal depth we may go to.

Definition 5 (k-wa^n-bisimulation). *Let $\mathcal{M} = (W, R, V)$ and $\mathcal{M}' = (W', R', V')$ be two n-models. w and w' are 0-wa^n-bisimilar ($w \leftrightarroweq^n_0 w'$) iff $V(v) = V'(v')$. $w \leftrightarroweq^n_{k+1} w'$ iff $w \leftrightarroweq^n_k w'$ and the follow two conditions are satisfied:*

forth If $v \leftrightarroweq^n_{k+1} v'$ and $Rvv_1, \dots, v_n$ then there are $v'_1, \dots, v'_n$ in W' s.t. $R'v'v'_1, \dots, v'_n$ and for each v'_j there is a v_i such that $v_i \leftrightarroweq^n_k v'_j$ where $1 \le i, j \le n$.

back If $v \leftrightarroweq^n_{k+1} v'$ and $R'v'v'_1, \dots, v'_n$ then there are $v_1, \dots, v_n$ in W s.t. $Rvv_1, \dots, v_n$ and for each v_i there is a v'_j such that $v_i \leftrightarroweq^n_k v'_j$ where $1 \le i, j \le n$.

We can translate each WAML^n formula to an equivalent FOL formula with one free variable and one $n+1$-ary relation symbol, thus WAML^n is also compact.

Definition 6 (Standard translation). $ST : \mathit{WAML}^n \to \mathit{FOL}$:

$$
\begin{aligned}
ST_x(p) &= Px\\
ST_x(\neg\varphi) &= \neg ST_x(\varphi)\\
ST_x(\varphi \wedge \psi) &= ST_x(\varphi) \wedge ST_x(\psi)\\
ST_x(\Box\varphi) &= \forall y_1 \forall y_2 \dots \forall y_n (Rxy_1y_2 \dots y_n \to ST_{y_1}(\varphi) \vee \dots \vee ST_{y_n}(\varphi))
\end{aligned}
$$

By following a similar strategy as in [27], we will show a van Benthem-Rosen characterization theorem for WAML^n: a FOL formula is equivalent to the translation of a WAML^n formula (over finite n-models) if and only if it is invariant under wa^n-bisimulations (over finite n-models).

First we need to define a notion of *unraveling* w.r.t. n-ary models similarly to models with binary relations. We use an example of a graph with ternary relations to illustrate the intuitive idea behind the general n-ary unraveling, which is first introduced in [28].

Example 2. Given the 2-model with ternary relations $\langle \{w, v, u, t\}, \{\langle w, u, t\rangle, \langle u, t, u\rangle, \langle t, w, v\rangle\}, V\rangle$. It is quite intuitive to first unravel it into a tree with pairs of states as nodes, illustrated below:

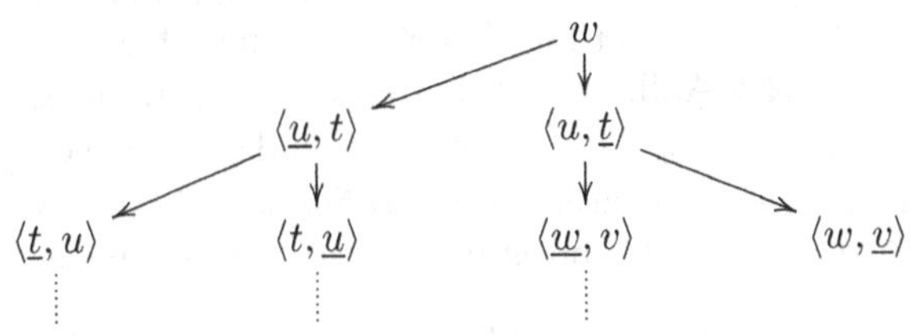

To turn it into a 2-model, we need to define the new ternary relations. For each triple $\langle s_0, s_1, s_2\rangle$ of pairs, $\langle s_0, s_1, s_2\rangle$ is in the new ternary relation iff s_1 and s_2 are successors of s_0 in the above graph and the triple of underlined worlds in s_0, s_1, s_2 respectively is in the original ternary relation, e.g., $\langle u, \underline{t}\rangle, \langle \underline{w}, v\rangle, \langle w, \underline{v}\rangle$ is in the new ternary relation since $\langle t, w, v\rangle$ is in the original ternary relation.

In general, we can use the n-tuples of the states in the original model together with a natural number $k \in [1, n]$ as the basic building blocks for the unraveling of an n-model, e.g., $\langle w, v, u, 2\rangle$ means the second the state is the underlined one. To make the definition uniform, we define the root as the sequence $\langle w, \dots, w, 1\rangle$. Like the unraveling for a binary graph, formally we will use sequences of such building blocks as the nodes in the unraveling of a n-model, e.g., the left-most node $\langle \underline{t}, u\rangle$ in the above example will become $\langle\langle w, w, 1\rangle, \langle u, t, 1\rangle, \langle t, u, 1\rangle\rangle$. This leads to the following definition.

Definition 7. *Given an n-model $\mathcal{M} = \langle W, R, V\rangle$ and $w \in W$, we first define the binary unraveling $\mathcal{M}^b_w$ of $\mathcal{M}$ around w as $\langle W_w, R^b, V'\rangle$ where:*

- *W_w is the set of sequences $\langle\langle \boldsymbol{v}_0, i_0\rangle, \langle \boldsymbol{v}_1, i_1\rangle, \dots, \langle \boldsymbol{v}_m, i_m\rangle\rangle$ where:*
  - *$m \in \mathbb{N}$;*
  - *for each $j \in [0, m]$, $\boldsymbol{v}_j \in W^n$ and $i_j \in [1, n]$ such that $R(\boldsymbol{v}_j[i_j])\boldsymbol{v}_{j+1}$;*
  - *$\boldsymbol{v}_0$ is the constant n-sequence $\langle w, \dots, w\rangle$ and $i_0 = 1$;*
- *$R^b ss'$ iff s' extends s with some $\langle \boldsymbol{v}, i\rangle$*
- *$V'(s) = V(r(s))$, where $r(s) = \boldsymbol{v}_m[i_m]$ if $s = \langle \dots, \langle \boldsymbol{v}_m, i_m\rangle\rangle$.*

The unraveling $\mathcal{M}_w = \langle W_w, R', V'\rangle$ is based on $\mathcal{M}^b_w$ by defining $R's_0s_1\dots s_n$ iff $Rr(s_0)r(s_1)\dots r(s_n)$ and $R^b s_0 s_i$ for all $i \in [1, n]$. Let the bounded unraveling $\mathcal{M}_w|_l$ be the submodel of $\mathcal{M}_w$ up to level l.

Remark 2. The unravelling $\mathcal{M}_w$ itself is not totally "tree-like", since there may be some node w occurs in both an n-tuple successor of x and an n-tuple successor of y for $x \neq y$. But clearly $\mathcal{M}^b_w$ is a tree, and in $\mathcal{M}_w$, if $Rs_0\dots s_n$ then $s_1\dots s_n$ are at the next "level" of s_0. The latter property is crucial in the later proofs, but due to space issues, we have to omit the details here.

r defined above reveals the corresponding state of s in the original model $\mathcal{M}$. It is not hard to show the following.

Proposition 2. *The above r (viewed as a relation) is a wa^n-bisimulation between $\mathcal{M}_w$ and $\mathcal{M}$. Actually r is a p-morphism (over n-models) from $\mathcal{M}_w$ to $\mathcal{M}$.*

Now we have all the ingredients to prove the following characterization theorem. Note that the characterization works with or without the finite model constraints.

Theorem 2. *A first-order formula $\alpha(x)$ is invariant under $\underline{\leftrightarrow}^n$ (over finite models) iff $\alpha(x)$ is equivalent to a WAML^n formula (over finite models).*

Following the general strategy in [27], the only non-trivial part is to show that the FOL formula $\alpha(x)$ that is invariant under wa^n-bisimulation has some locality property w.r.t. its bounded unraveling $\mathcal{M}_w|_l$ for some l. Due to lack of space, we only show the following lemma and give a proof sketch here. For other relatively routine parts of the proof, see [27].

Lemma 1 (locality). *An FOL formula $\alpha(x)$ is invariant under $\underline{\leftrightarrow}$ (over finite models) implies that for some $l \in \mathbb{N}$, for any n-model $\mathcal{M}, w$: $\mathcal{M}, w \Vdash \alpha(x)[w]$ iff $\mathcal{M}_w|_l \Vdash \alpha(x)[(\langle \boldsymbol{w}, 1 \rangle)]$.*

Here we explain the most important ideas behind the proof. First of all, like in [27], we take $l = 2^q - 1$ where q is the quantifier rank of $\alpha(x)$, and build two bigger models $\mathcal{M}^*, w^*$ and $\mathcal{N}^*, v^*$ which are wa^n-bisimilar to $\mathcal{M}, w$ and $\mathcal{M}_w|_l, w$ respectively using our new unraveling notion. Then we show that in the q-round EF game between the bigger n-models Duplicator has a winning strategy. To specify the strategy, which is essentially letting the duplicator to keep some "safe zones" for extensions of partial isomorphisms, we need to define the distance of points in n-models. Let the distance between s and s' (notation $d(s, s')$) be the length of the shortest (undirected) path between s and s' via a new relation binary R^c where $R^c xy$ iff $Rxy_1 \ldots y_n$ and $y = y_i$ for some $i \in [1, n]$. We set $d(s, s') = \omega$ if s and s' are not connected by any such path. It is easy to see that in the unraveling $\mathcal{M}_w$, $d(s, s')$ is exactly the distance in the usual sense between s, s' in the tree $\mathcal{M}^b_w$. Then, the winning strategy looks exactly like the one in [27] for binary models. The key point to show that the same strategy is a winning strategy in the new setting is that when building the correct induction hypothesis, we need to define two "neighborhoods" of a node–a big one and a small one. In particular, first let (a_i, b_i) be the pair selected at i round where each $a_i \in \mathcal{M}^*$ and $b_i \in \mathcal{N}^*$, where by the rule of the game, $a_0 = w^*$ and $b_0 = v^*$. Then define $S(m) = \{a_i \mid i \leq m\}$, $N_i(m)$ to be the neighborhood of a_i within distance of $2^{q-m} - 1$, and $N_i^{'}(m)$ to be the neighborhood of a_i within distance of $2^{q-(m+1)}$. Here the N and $N^{'}$ are the two "neighborhoods". Then finally the induction hypothesis can be correctly formulated as the following.

After m rounds ($0 \leq m \leq q$), the following two hold.

1. The selected points form a partial isomorphism I: $\mathcal{M}^* \to \mathcal{N}^*$.
2. If $m < q$ then there is a sequence $(I_0, \ldots, I_m)$ s.t. for each $i \leq m$,
   (a) $I_i \supseteq I$ is a partial isomorphism with $Dom(I_i) = N_i(m) \cup S(m)$;

(b) $\forall h, j \leq m \forall x \in N_h'(m) \cap N_j'(m)(I_h(x) = I_j(x))$.

In Otto's original proof in [27], the induction hypothesis is not very clear, and we think it is necessary to give such an explicit formulation here.

Remark 3. It is not hard to show that under our distance notion, for each x, y, z in the model, $d(x,z) \geq d(x,y) - d(y,z)$, i.e., $d(x,z) + d(z,y) \geq d(x,y)$ which is a more usual form of the *triangle inequality.* This justifies the new distance notion. To see why a similar strategy like the one in [27] for binary models works, note that our unraveling $\mathcal{M}_w$ is essentially based on a *tree* $\mathcal{M}_w^b$ by definition, and the n-ary relation over such a tree structure has a very special property: if $Rs_0 \ldots s_n$ then $s_1 \ldots s_n$ are immediate successors of s_0 in the binary unraveling as mentioned in Remark 2. This leads to the following crucial property we will use repeatedly: if we already established a partial isomorphism I between S and N (w.r.t. also n-ary relations), and $x \notin S$ is not directly connected to anything in S, and $y \notin N$ is also not directly connected to anything in N then $I \cup \{(x,y)\}$ extending I is again a partial isomorphism.

Finally, the bound $l = 2^q - 1$ in the above proof, which we choose uniformly for every n, is actually not "optimal", since for a larger n, we can have a lower bound. Especially, when $n > q$, even $l = 1$, the Duplicator could have a winning strategy, since any bijection will be a partial isomorphism. So the distance we define here is not a appropriate one for us to find the minimal bound l. Here we conjecture that the bound should be the least integer l s.t. $l \geq (2^q - 1)/n$.

4 Interpolation

By a standard strategy in [19], we know that the basic polyadic modal logics (PML) have the Craig Interpolation theorem. What's more, in [31], the authors proved that the minimal monotonic modal logic M has Uniform Interpolation. Furthermore, we know that the basic modal logic $\mathbb{K}$ also has Uniform Interpolation from [3] and [2]. From the following three aspects we may conjecture that the basic WAML systems $\mathbb{K}_n$ should have interpolation too:

1 WAML can be treated as a fragment of PML.
2 $\mathbb{K}_n$ is regarded as a general version of $\mathbb{K}$, since $\mathbb{K}$ is just $\mathbb{K}_1$.
3 $\mathbb{K}_n$ can be viewed as a special kind of monotonic modal logics.

But in fact no $\mathbb{K}_n$ has the Craig Interpolation Property for $n \geq 2$. The first counterexample for interpolation we found is for $\mathbb{K}_3$, which is relatively easy to understand and can be readily generalized to all $\mathbb{K}_n$ for $n \geq 3$. Later we found a counterexample for $\mathbb{K}_2$, which is slightly more complicated. Here we first give the two counterexamples for $\mathbb{K}_2$ and $\mathbb{K}_3$ and then provide the general construction for $\mathbb{K}_n$ ($n \geq 3$). But before we state the counterexamples, let us first clarify what do we mean by "a counterexample" of the Craig Interpolation Property for $\mathbb{K}_n$.

Lemma 2. *Let n be a non-zero natural number. If there are two pointed n-models $\mathcal{M}, w$ and $\mathcal{N}, v$ and two formulas φ and ψ such that*

1. $\mathcal{M}, w \models \varphi$ *and* $\mathcal{N}, v \models \psi$;
2. $\mathbb{K}_n \vdash \varphi \to \neg\psi$;
3. *letting* Φ' *be the set of all the propositional letters that appear both in* φ *and* ψ, *for any formula* γ *in* WAML *such that only letters in* Φ' *appear,* $\mathcal{M}, w \models \gamma$ *iff* $\mathcal{N}, v \models \gamma$;

then $\mathbb{K}_n$ *lack the Craig Interpolation Property.*

Proof. Assume for contradiction that $\mathbb{K}_n$ has the Craig Interpolation Property. Then since $\mathbb{K}_n \vdash \varphi \to \neg\psi$, there is a interpolant γ such that

- $\mathbb{K}_n \vdash \varphi \to \gamma$ and $\mathbb{K}_n \vdash \gamma \to \neg\psi$;
- only letters in Φ' appear in γ.

Now since $\mathcal{M}, w \models \varphi$ with $\mathcal{M}$ being an n-model and $\mathbb{K}_n \vdash \varphi \to \gamma$, by soundness, $\mathcal{M}, w \models \gamma$. Then $\mathcal{N}, v \models \gamma$ by 3. Then using $\mathbb{K}_n \vdash \gamma \to \neg\psi$ and soundness again, $\mathcal{N}, v \models \neg\psi$, contradicting $\mathcal{N}, v \models \psi$.

Given this proposition, a pair of pointed n-models and a pair of formulas satisfying the antecedent constitute a counterexample of the Craig Interpolation property. Now we proceed to provide them for each $\mathbb{K}_n$ with $n \geq 2$.

Example 3. Consider the following two 2-models where $\{\langle w, w_1, w_1\rangle, \langle w, w_2, w_3\rangle\}$ is the ternary relation in the left model $\mathcal{M}_2$, and $\{\langle v, v_1, v_2\rangle\}$ is the ternary relation in the right model $\mathcal{N}_2$, where the valuations are as in the diagram.

Then set $\varphi_2 = \Box(\neg p \vee \neg q) \wedge \Diamond q$ and $\psi_2 = \Box(p \wedge r) \wedge \Box(p \wedge \neg r)$. It is easy to see that $\mathcal{M}_2, w \models \varphi_2$ and $\mathcal{N}_2, v \models \psi_2$. To see that $\mathbb{K}_2 \vdash \varphi_2 \to \neg\psi_2$, consider the following derivation, where to make long Boolean combinations readable, we write negation of propositional letters as overline, omit $\wedge$ between purely Boolean formulas and replace $\vee$ with $|$.

- $\vdash_2 \Box(\bar{p}|\bar{q}) \wedge \Box rp \wedge \Box \bar{r}p \to \Box(((\bar{p}|\bar{q})rp)|((\bar{p}|\bar{q})\bar{r}p)|rp\bar{r}p)$ — $\mathtt{K}_2$
- $\vdash_2 \Box(\bar{p}|\bar{q}) \wedge \Box rp \wedge \Box \bar{r}p \to \Box p\bar{q}$ — PL,RE
- $\vdash_2 \varphi_2 \wedge \psi_2 \to \Box p\bar{q} \wedge \Diamond q$ — PL
- $\vdash_2 \varphi_2 \wedge \psi_2 \to \Box \bar{q} \wedge \neg\Box\bar{q}$ — PL, RM
- $\vdash_2 \varphi_2 \to \neg\psi_2$ — PL

Here PL means propositional reasoning. Hence we are done with the first two points for this pair of models and formulas to be a counterexamples. For the last point, note that $Z = \{\langle w, v\rangle, \langle w_1, v_1\rangle, \langle w_1, v_2\rangle, \langle w_2, v_1\rangle, \langle w_2, v_2\rangle\}$ is a wa^2-bisimulation when $\Phi = \{p\}$. Hence by Proposition 1, for any formula γ with p the only propositional letter, $\mathcal{M}_2, w \models \gamma$ iff $\mathcal{N}_2, v \models \gamma$. But p is the only common propositional letters in φ_2 and ψ_2. Clearly, now $\mathcal{M}, w$, $\mathcal{N}, v$, φ_2, and ψ_2 form a counterexample to the Craig Interpolation Property for $\mathbb{K}_2$.

Example 4. Consider the following two 3-models where $\{\langle w, w_1, w_2, w_3\rangle\}$ is the relation in $\mathcal{M}_3$ and $\{\langle v, v_1, v_2, v_3\rangle\}$ is the relation in $\mathcal{N}_3$.

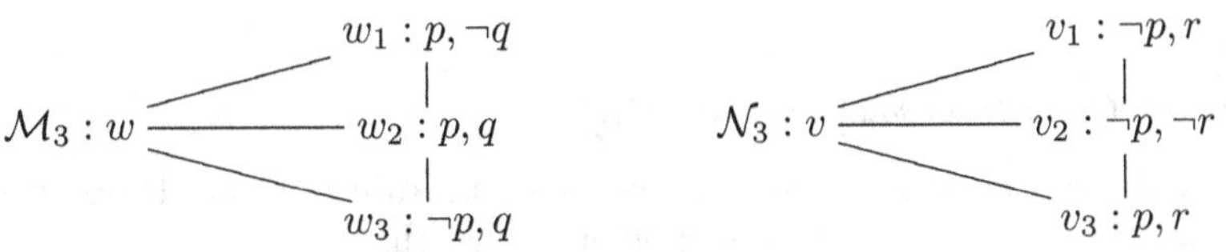

Then set $\varphi_3 = \Box p\bar{q} \wedge \Box pq \wedge \Diamond(p|\bar{p})$, $\psi_3 = \Box \bar{p}r \wedge \Box \bar{p}\bar{r} \wedge \Diamond(p|\bar{p})$. Clearly $\mathcal{M}_3, w \models \varphi_3$ and $\mathcal{N}_3, v \models \psi_3$. Further, $\mathbb{K}_2 \vdash \varphi_3 \rightarrow \neg\psi_3$ since we have the following derivation.

- $\vdash_3 \Box p\bar{q} \wedge \Box pq \wedge \Box \bar{p}r \wedge \Box \bar{p}\bar{r} \rightarrow \Box(p\bar{q}pq|p\bar{q}\bar{p}r|p\bar{q}\bar{p}\bar{r}|pq\bar{p}r|pq\bar{p}\bar{r}|\bar{p}r\bar{p}\bar{r})$ K3
- $\vdash_3 \Box p\bar{q} \wedge \Box pq \wedge \Box \bar{p}r \wedge \Box \bar{p}\bar{r} \rightarrow \Box p\bar{p}$ PL,
- $\vdash_3 \varphi_3 \wedge \psi_3 \rightarrow \Box p\bar{p} \wedge \Diamond(p|\bar{p})$ PL, RM
- $\vdash_3 \varphi_3 \rightarrow \neg\psi_3$ PL

Finally, note that $Z = \{\langle w, v\rangle, \langle w_1, v_3\rangle, \langle w_2, v_3\rangle, \langle w_3, v_1\rangle, \langle w_3, v_2\rangle\}$ is a wa^3-bisimulation if $\Phi = \{p\}$.

The above example can be naturally generalized for each $\mathbb{K}_n$ with $n > 3$. Let m be the least natural number s.t. $2^m \geq n-1$ and pick m many distinct propositional letters $r_1, \ldots, r_m$ from Φ. Then for each i from 1 to $n-1$, we can associate a distinct conjunction of literals ρ_i using r_j's so that $\rho_i \wedge \rho_{i'}$ are incompatible for each $i \neq i'$. Then we can state the general counterexample.

Example 5. Consider the following two n-models where $\{\langle w, w_1, ..., w_n\rangle\}$ is the relation in $\mathcal{M}_n$, and $\{\langle v, v_1, ..., v_n\rangle\}$ is the relation in $\mathcal{N}_n$.

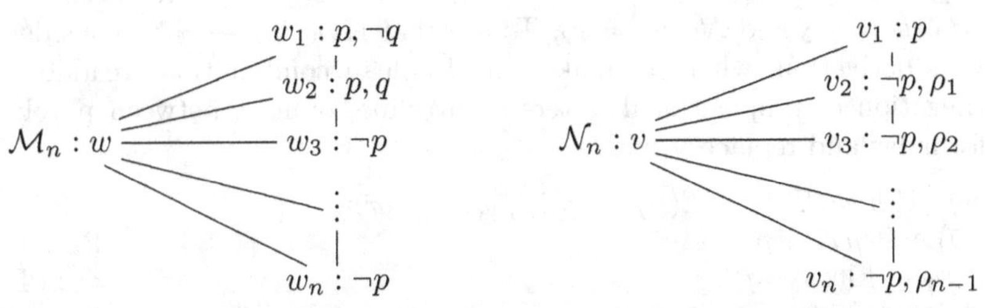

Set $\varphi_n = \Box(p \wedge \neg q) \wedge \Box(p \wedge q) \wedge \Diamond\top$ and $\psi_n = \bigwedge_{i=1}^{n-1} \Box(\neg p \wedge \rho_i) \wedge \Diamond\top$. Clearly $\mathcal{M}_n, w \models \varphi_n$ and $\mathcal{N}_n, v \models \psi_n$. It is also easy to see that by K_n, we can derive $(\Box(p\wedge\neg q)\wedge\Box(p\wedge q)\wedge\bigwedge_{i=1}^{n-1}\Box(\neg p\wedge\rho_i)) \rightarrow \Box\bot$. With this we can then easily derive $\varphi_n \rightarrow \neg\psi_n$ in $\mathbb{K}_n$. Finally, note that p is the only common propositional letter in φ_n and ψ_n and that $Z = \{\langle w, v\rangle, \langle w_1, v_1\rangle, \langle w_2, v_1\rangle\} \cup \{w_3, ..., w_n\} \times \{v_2, ..., v_n\}$ is a wa^n-bisimulation when $\Phi = \{p\}$.

With the examples and Lemma 2, the main theorem of this section follows.

Theorem 3. *For any $n \geq 2$, $\mathbb{K}_n$ does not have the Craig Interpolation Property.*

Remark 4. Note that the Lemma 2 uses only the soundness of the logics. Hence for any extension of $\mathbb{K}_n$ that is sound on $\mathcal{M}_n$ and $\mathcal{N}_n$, it still lacks the Craig Interpolation Property. For example, we may extend $\mathbb{K}_n$ with 4 and our examples still work since 4 is valid on the underlying frames.

5 Conclusion

In this paper, we proved two results about WAML: first, WAML have a van Benthem-Rosen characterization, and second, WAML do not have Craig Interpolation Property (CIP). We conclude with two potentially promising lines of further investigation.

First, the main part of the completeness proof of $\mathbb{K}_n$ over n-models is to solve some combinatorial puzzle [26]. Due to the semantics of WAML there is a natural link between combinatorics and WAML as also shown in the use of graph coloring problem in [5]. As future work, we would like to explore the possibility of using WAML to express interesting combinatorial properties in graph theory, like the one in [25].

Second, even though we proved that WAML do not have Craig Interpolation Property, it doesn't mean that the same must be the case under further constraints (stronger logics). For instance, the counterexample in our paper cannot show that $\mathbb{K}_n \oplus \mathtt{T}$ lacks CIP since the logic is not sound on the frames of the models we provided. What remains to be done then is to chart the map of CIP among the logics extending $\mathbb{K}_n$'s and look for more general methods.

Lastly, if we change all the $\Box$ in $\mathbb{K}_n$ into $\Diamond$, we get the following formula:

$$\mathtt{K}_n^* : \Diamond p_0 \wedge \cdots \wedge \Diamond p_n \to \Diamond \bigvee_{(0 \leq i < j \leq n)} (p_i \wedge p_j).$$

In basic normal modal logics, this formula characterizes frames where each world has at most n accessible worlds and is equivalent (assuming the normality of $\Box$) to what is commonly called the $\mathtt{Alt}_n$ in the literature [33]. It is not too hard to observe that the strategy we gave in Sect. 4 can be used to show that for each $n \geq 3$, normal modal logic $\mathtt{K} \oplus \mathtt{K}_n^*$ lacks CIP. It seems that, more abstractly speaking, the counterexamples exist because the logic can reason about with the help of extra propositional letters, but cannot express directly, whether there are many accessible worlds satisfying a property. Note that counting the number of accessible worlds satisfying a property is intuitively important and has been studied in Description Logics (DL) [7] and Graded Modal Logics (GML) [12,15,17]. There are already some CIP work in those logics, like [23], and we conjecture that CIP may return when we add modalities that talk directly about numbers.

References

1. Allen, M.: Complexity results for logics of local reasoning and inconsistent belief. In: Proceedings of the 10th Conference on Theoretical Aspects of Rationality and Knowledge, pp. 92–108. National University of Singapore (2005)

2. Andréka, H., Németi, I., van Benthem, J.: Modal languages and bounded fragments of predicate logic. J. Philos. Logic **27**(3), 217–274 (1998)
3. Andréka, H., Van Benthem, J., Németi, I.: Back and forth between modal logic and classical logic. Logic J. IGPL **3**(5), 685–720 (1995)
4. Apostoli, P.: On the completeness of first degree weakly aggregative modal logics. J. Philos. Logic **26**(2), 169–180 (1997)
5. Apostoli, P., Brown, B.: A solution to the completeness problem for weakly aggregative modal logic. J. Symbolic Logic **60**(3), 832–842 (1995)
6. Arló Costa, H.: Non-adjunctive inference and classical modalities. J. Philos. Logic **34**(5), 581–605 (2005)
7. Baader, F., Horrocks, I., Sattler, U.: Description logics. Found. Artif. Intell. **3**, 135–179 (2008)
8. Beall, J., et al.: On the ternary relation and conditionality. J. Philos. Logic **41**(3), 595–612 (2012)
9. van Benthem, J., van Eijck, J., Kooi, B.: Logics of communication and change. Inf. Comput. **204**(11), 1620–1662 (2006)
10. Blackburn, P., De Rijke, M., Venema, Y.: Modal Logic, vol. 53. Cambridge University Press, Cambridge (2002)
11. Chellas, B.F.: Modal Logic: An Introduction. Cambridge University Press, Cambridge (1980)
12. De Caro, F.: Graded modalities, ii (canonical models). Studia Logica **47**(1), 1–10 (1988). https://doi.org/10.1007/BF00374047
13. Fagin, R., Halpern, J., Moses, Y., Vardi, M.: Reasoning About Knowledge. MIT Press, Cambridge (1995)
14. Fan, J., Wang, Y., van Ditmarsch, H.: Contingency and knowing whether. Rev. Symbol. Logic **8**, 75–107 (2015)
15. Fattorosi-Barnaba, M., De Caro, F.: Graded modalities. I. Studia Logica **44**(2), 197–221 (1985). https://doi.org/10.1007/BF00379767
16. Fervari, R., Herzig, A., Li, Y., Wang, Y.: Strategically knowing how. In: Proceedings of IJCAI 2017, pp. 1031–1038 (2017)
17. Fine, K., et al.: In so many possible worlds. Notre Dame J. Formal Logic **13**(4), 516–520 (1972)
18. Gu, T., Wang, Y.: "Knowing value" logic as a normal modal logic. In: Proceedings of AiML, vol. 11, pp. 362–381 (2016)
19. Hansen, H., Kupke, C., Pacuit, E.: Neighbourhood structures: bisimilarity and basic model theory. Logical Meth. Comput. Sci. **5**(2) (2009). https://doi.org/10.2168/LMCS-5(2:2)2009
20. Jennings, R.E., Schotch, P.K.: Some remarks on (weakly) weak modal logics. Notre Dame J. Formal Logic **22**(4), 309–314 (1981)
21. Kamp, H.: Tense logic and the theory of linear order. Ph.D. thesis, UCLA (1968)
22. Liu, J.: Model theoretical aspects of polyadic modal logic: an exposition. Stud. Logic **12**(3), 79–101 (2019)
23. Lutz, C., Wolter, F.: Foundations for uniform interpolation and forgetting in expressive description logics. In: Twenty-Second International Joint Conference on Artificial Intelligence (2011)
24. Nicholson, T., Allen, M.: Aggregative combinatorics: an introduction. In: Student Session of 2nd North American Summer School in Language, Logic, and Information, pp. 15–25 (2003)
25. Nicholson, T., Allen, M.: Aggregative combinatorics: an introduction. In: Proceedings of the Student Session, 2nd North American Summer School in Logic, Language, and Information (NASSLLI-03), pp. 15–25 (2003)

26. Nicholson, T., Jennings, R.E., Sarenac, D.: Revisiting completeness for the K_n modal logics: a new proof. Logic J. IGPL **8**(1), 101–105 (2000)
27. Otto, M.: Elementary proof of the van Benthem-Rosen characterisation theorem. Technical report 2342 (2004)
28. de Rijke, M.: Extending modal logic. Ph.D. thesis, ILLC, University of Amsterdam (1993)
29. Routley, R., Meyer, R.K.: The semantics of entailment – II. J. Philos. Logic **1**(1), 53–73 (1972)
30. Routley, R., Meyer, R.K.: The semantics of entailment – III. J. Philos. Logic **1**(2), 192–208 (1972)
31. Santocanale, L., Venema, Y., et al.: Uniform interpolation for monotone modal logic. Adv. Modal Logic **8**, 350–370 (2010)
32. Schotch, P., Jennings, R.: Modal logic and the theory of modal aggregation. Philosophia **9**(2), 265–278 (1980)
33. Segerberg, K.: An Essay in Classical Modal Logic. Filosofiska Föreningen Och Filosofiska Institutionen Vid Uppsala Universitet, Uppsala (1971)
34. Van Benthem, J., Bezhanishvili, N., Enqvist, S., Yu, J.: Instantial neighbourhood logic. Rev. Symbolic Logic **10**(1), 116–144 (2017)
35. Wang, Y.: A logic of goal-directed knowing how. Synthese **195**(10), 4419–4439 (2018)
36. Wang, Y., Fan, J.: Conditionally knowing what. In: Proceedings of AiML, vol. 10, pp. 569–587 (2014). www.aiml.net/volumes/volume10/Wang-Fan.pdf

Dynamic Term-Modal Logic for Epistemic Social Network Dynamics

Andrés Occhipinti Liberman[1(✉)] and Rasmus K. Rendsvig[2]

[1] DTU Compute, Lyngby, Denmark
aocc@dtu.dk

[2] Center for Information and Bubble Studies, University of Copenhagen, Copenhagen, Denmark
rasmus@hum.ku.dk

Abstract. Logics for social networks have been studied in recent literature. This paper presents a framework based on *dynamic term-modal logic* (DTML), a quantified variant of dynamic epistemic logic (DEL). In contrast with DEL where it is commonly known to whom agent names refer, DTML can represent dynamics with uncertainty about agent identity. We exemplify dynamics where such uncertainty and *de re/de dicto* distinctions are key to social network epistemics. Technically, we show that DTML semantics can represent a popular class of hybrid logic epistemic social network models. We also show that DTML can encode previously discussed dynamics for which finding a complete logic was left open. As complete reduction axioms systems exist for DTML, this yields a complete system for the dynamics in question.

Keywords: Social networks · Term-modal logic · Dynamic epistemic logic

1 Introduction

Over recent years, several papers have been dedicated to logical studies of social networks, their epistemics and dynamics [3,11–15,19–23,25,26]. The purpose of this literature typically is to define and investigate some social dynamics with respect to e.g. long-term stabilization or other properties, or to introduce formal logics that capture some social dynamics, or both.

This paper illustrates how *dynamic term-modal logic* (DTML, [2]) may be used for the second purpose. In general, *term-modal logics* are first-order modal logics where the index of modal operators are first-order terms. I.e., the operators double as predicates to the effect that e.g. $\exists x K_x N(x, a)$ is a formula—read, in this paper, as "there there exists an agent that knows of itself that it is a social network neighbor of a". The *dynamic* term-modal logic of [2] extends term-modal logic with suitably generalized action models that can effectuate both factual changes (e.g. to the network structure) as well as epistemic changes. For all the DTML action model encodable dynamics, [2] presents a general sound

P. Blackburn et al. (Eds.): LORI 2019, LNCS 11813, pp. 168–182, 2019.
https://doi.org/10.1007/978-3-662-60292-8_13

and complete reduction axiom-based logic in the style of *dynamic epistemic logic* (DEL, [4,5]). Hence, whenever an epistemic social network dynamics is encodable using DTML, completeness follows. With this in mind, the main goal of this paper is to introduce and illustrate DTML as a formalism for representing epistemic social network dynamics, and to show how it may be used to obtain completeness results.

To this end, the paper progresses as follows. Section 2 sketches some common themes in the logical literature on social networks before introducing DTML and its application to epistemic social networks. Section 2 contains the bulk of the paper, with numerous examples of both static DTML models and action models. The examples are both meant to showcase the scope of DTML and to explain the more non-standard technical details involved in calculating updated models. In Sect. 3, we turn to technical results, where it is shown that DTML may encode popular static hybrid logical models of epistemic networks, as well as the dynamics of [13], for which finding a complete logic was left open. Many details are deferred to an extended version, [1]. Section 4 contains final remarks.

2 Models and Languages for Epistemic Social Networks

To situate DTML in the logical literature on social networks, we cannot but describe the literature in broad terms. We omit both focus, formal details and main results of the individual contributions in favor of a broad perspective. That said, then all relevant literature in one way or other concern *social networks*. In general, a **social network** is a graph (A, N) where A is a set of agents and $N \subseteq A \times A$ is represents a social relation, e.g., being friends on some social media platform. Depending on interpretation, N may be assumed irreflexive and symmetric. Social networks may be augmented with assignments of atomic properties to agents, representing e.g. behaviors, opinions or beliefs. One set of papers investigates such models and their dynamics using fully propositional static languages [14,21,25,26].

A second set of papers combines social networks with a semantically represented epistemic dimension in the style of epistemic logic. In these works, the fundamental structure of interest is (akin to) a tuple

$$(A, W, \{N_w\}_{w \in W}, \sim)$$

with agents A and worlds W, with each world w associated with a network $N_w \subseteq A \times A$, and $\sim : A \to \mathcal{P}(W \times W)$ associating each agent with an indistinguishability (equivalence) relation $\sim_a$. Call such a tuple an **epistemic network structure**.

The existing work on epistemic network structures may be organized in terms of the static languages they work with: *propositional modal logic* [3,15] or *hybrid logic* [10–13,19,20,22,23]. In the former, the social network is described using designated atomic propositions (e.g., N_{ab} for 'b is a neighbor of a'). To produce a model, an epistemic network structure is augmented with a propositional valuation $V : P \to \mathcal{P}(W)$. Semantically, N_{ab} is then true at w iff $(a, b) \in N_w$.

Knowledge is expressed using operators $\{K_a\}_{a \in A}$ as in standard epistemic logic with K_a the Kripke modality for $\sim_a$.

In the hybrid case, the network is instead described using modal operators. The hybrid languages typically include a set of agent nominals Nom (agent names), atoms P and *indexical* modal operators K and N, read "I know that" and "all my neighbors". Some papers additionally include state nominals, hybrid operators ($@_x$, $\downarrow_x$) and/or universal modalities U ("for all agents"). A **hybrid network model** is an epistemic network structures extended with two assignments: a nominal assignment $g : Nom \rightarrow A$ that names agents, and a two-dimensional hybrid valuation $V : P \rightarrow \mathcal{P}(W \times A)$, where $(w, a) \in V(p)$ represents that the indexical proposition p holds of agent a at w. The satisfaction relation is relative to *both* an epistemic alternative w and an agent a, where the noteworthy clause are: $M, w, a \models p$ iff $(w, a) \in V(p)$; $M, w, a \models K\varphi$ iff $M, v, a \models \varphi$ for every $v \sim_a w$; and $M, w, a \models N\varphi$ iff $M, w, b \models \varphi$ for every b such that $N_w(a, b)$. With these semantics, formulas are read indexically. E.g. KNp reads as "I know that all my neighbors are p".

In relation to these two language types, the term-modal approach of this paper lies closer to the former: By including a binary 'neighbor of' relation symbol N in the signature of a term-modal language, the social network component of models is described non-modally. This straightforwardly allows expressing e.g. that that all agents know all their neighbors ($\forall x \forall y(N(x, y) \rightarrow K_x(N(x, y)))$ or that an agent has *de re* vs. *de dicto* knowledge of someone being a neighbor ($\exists x K_{\underline{a}} N(\underline{a}, x)$ vs. $K_{\underline{a}} \exists x N(\underline{a}, x)$). Moreover, hybrid languages can be translated into DTML, in such a way that hybrid formulas such as $@_a p$ ("agent a has property p") become equivalent to $P(\underline{a})$, if $\underline{a}$ is the name of a.

2.1 Term-Modal Logic and Epistemic Network Structures

In general, term-modal languages may be based on any first-order signature, but for the purposes of representing social networks and factual properties of agents, we limit attention to the following:[1]

Definition 1. A **signature** is a tuple $\Sigma = (\mathtt{V}, \mathtt{C}, \mathtt{P}, N, \doteq)$ with $\mathtt{V}$ a countably infinite set of variables, $\mathtt{C}$ and $\mathtt{P}$ countable sets of constants and unary predicates, N a binary relation symbol and $\doteq$ for identity. The **terms** of Σ are $\mathtt{T} := \mathtt{V} \cup \mathtt{C}$. With $t_1, t_2 \in \mathtt{T}, x \in \mathtt{V}$ and $P \in \mathtt{P}$, the **language** $\mathcal{L}(\Sigma)$ is given by

$$\varphi := P(t_1) \mid N(t_1, t_2) \mid (t_1 \doteq t_2) \mid \neg\varphi \mid \varphi \wedge \varphi \mid K_t\varphi \mid \forall x\varphi$$

Standard Boolean connectives, $\top$, $\exists$ and $\hat{K}_t$ are defined per usual. With $\varphi \in \mathcal{L}(\Sigma), t \in \mathtt{T}$, $x \in \mathtt{V}$, the result of replacing all occurrences of x in φ with t is denoted $\varphi(x \mapsto t)$. Formulas from the first three clauses are called **atoms**; if an atom contains no variables, it is **ground**.

[1] The defined are special cases of the setting in [2], which allows general signatures and non-agent terms. [2] also reviews the term-modal literature.

Throughout, $\underline{a}, \underline{b}$, etc. are used for constants and the relation symbol N denotes a social network. The reading of $N(t_1, t_2)$ depends on application. $K_t\varphi$ is a term-indexed epistemic operator which read as "agent t knows that φ". $\mathcal{L}(\Sigma)$ neither enforces nor requires a fixed-size agent set A, in contrast with standard epistemic languages, where the set of operators is given by reference to A. Hence the same language may be used to describe networks of varying size.

To interpret $\mathcal{L}(\Sigma)$, we use *constant-domain* models (the same number of agents in each world) with *non-rigid* constants (names, like predicates and relations, may change extension between worlds; this allows for uncertainty about agent identity). See Figs. 1 and 2 for examples of such models.

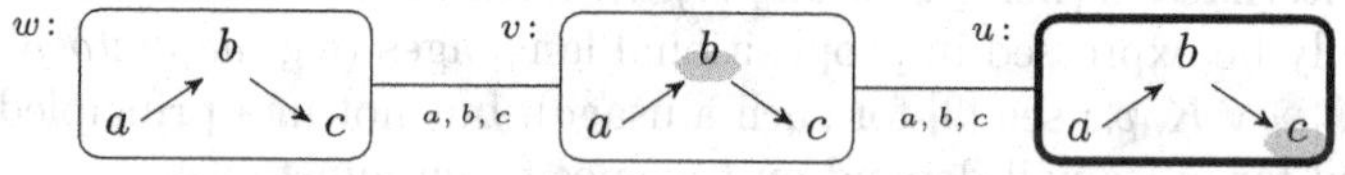

Fig. 1. Example 1, pt. 1 (Server Error). Three agents a,b and c work in a company with a hierarchical command structure, $\longrightarrow$: a is the direct boss of b, who is the direct boss of c. The server has thrown an error after both b and c tampered with it. Either (w) the server failed spontaneously, (v) b made a mistake (marked by gray) or (u) c made a mistake. Lines represent indistinguishability with reflexive and transitive links omitted. There is no uncertainty about the hierarchy, but nobody knows why the server failed. In fact, c made a mistake: the actual world has a thick outline.

Definition 2. An $\mathcal{L}(\Sigma)$-**model** is a tuple $M = (A, W, \sim, I)$ where A is a non-empty **domain of agents**, W is a non-empty set of **worlds**, $\sim : A \rightarrow \mathcal{P}(W \times W)$ assigns to each agent $a \in A$ an **equivalence relation** on W denoted $\sim_a$, and I is an **interpretation** satisfying, for all $w \in W$, 1. for $c \in \mathtt{C}$, $I(c, w) \in A$; 2. for $P \in \mathtt{P}$, $I(P, w) \subseteq A$; 3. $I(N, w) \subseteq A \times A$. A **pointed model** is a pair (M, w) with $w \in W$ called the **actual world**.
A variable **valuation** of Σ over M is a map $g : \mathtt{V} \rightarrow A$. The valuation identical to g except mapping x to a is denoted $g[x \mapsto a]$. The **extension** of the term $t \in \mathtt{T}$ at w in M under g is $[\![t]\!]_w^{I,g} = g(t)$ *for* $t \in \mathtt{V}$ and $[\![t]\!]_w^{I,g} = I(t, w)$ *for* $t \in \mathtt{C}$.
Given the inclusion of N in the signature Σ, each $\mathcal{L}(\Sigma)$-model embeds an epistemic network structure $(A, W, (\sim_a)_{a \in A}, (I(N, w))_{w \in W})$.

Formulas are evaluated over pointed models using a direct combination of first-order and modal semantics:

Definition 3. Let Σ, M and g be given. The **satisfaction** of formulas of $\mathcal{L}(\Sigma)$ is given recursively by

$M, w \vDash_g P(t_1)$ iff $[\![t_1]\!]_w^{I,g} \in I(P, w)$, for $P \in \mathtt{P}$.
$M, w \vDash_g N(t_1, t_2)$ iff $([\![t_1]\!]_w^{I,g}, [\![t_2]\!]_w^{I,g}) \in I(N, w)$.
$M, w \vDash_g (t_1 \doteq t_2)$ iff $[\![t_1]\!]_w^{I,g} = [\![t_2]\!]_w^{I,g}$.
$M, w \vDash_g \neg\varphi$ iff not $M, w \vDash_g \varphi$.
$M, w \vDash_g \varphi \wedge \psi$ iff $M, w \vDash_g \varphi$ and $M, w \vDash_g \psi$.
$M, w \vDash_g \forall x\varphi$ *iff* $M, w \vDash_{g[x \mapsto a]} \varphi$ for all $a \in A$.
$M, w \vDash_g K_t\varphi$ iff $M, w' \vDash_g \varphi$ for all w' such that $w \sim_{[\![t]\!]_w^{I,g}} w'$.

2.2 Knowing Who and Knowledge *De Dicto* and *De Re*

First-order modal languages can represent propositional attitudes *de dicto* (about the statement) and *de re* (about the thing) in principled manners. For example, $K_{\underline{a}}\exists x P(x)$ is a *de dicto* statement: knowledge is expressed about the proposition that a P-thing exists. In contrast, $\exists x K_{\underline{a}} P(x)$ is a *de re* statement: it is expressed that of some thing x, that x is known to be a P-thing. In general, *de re* statements are stronger than *de dicto* statements. The difference has been appreciated in epistemic logic since Hintikka's seminal [17], where he argues that $\exists x K_{\underline{a}}(x \doteq \underline{b})$ expresses that *a knows who b is* (see Fig. 2). Semantically, the formula ensures that the constant $\underline{b}$ refers to the same individual in all $\underline{a}$'s epistemic alternatives (i.e., $\underline{b}$ is *locally rigid*). Both *de dicto* and *de re* statements may partially be expressed in propositional languages (e.g. *de dicto* $K_a(p_b \vee p_c)$ vs. *de re* $K_a p_b \vee K_a p_c$; see [3] for such a usage), but not in a principled manner: the required formulas will depend on the specific circumstances.

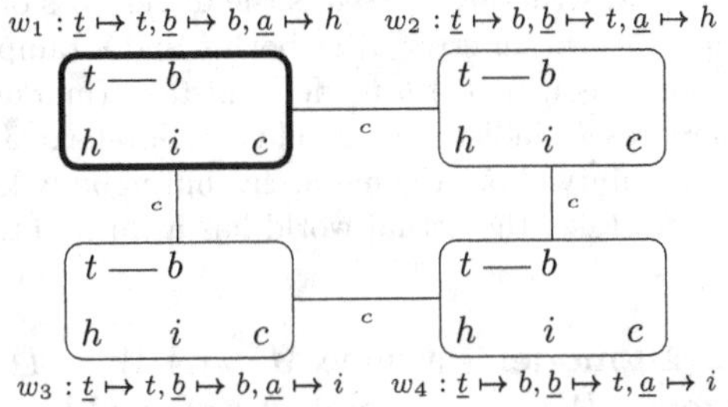

Fig. 2. Example 2, pt. 1 (Knowing Who). Two thieves, t and b, hide in a building with hostages h and i. Outside, a cop, c, waits. To communicate safely, the thieves use *code names* 'Tokyo' and 'Berlin' for each other and 'The Asset' for the specially valuable hostage h. Agents t, b, h and i all know whom the code names denote (the names are rigid for them), but the cop does not. The code names are $\underline{t}$ for t, $\underline{b}$ for b and $\underline{a}$ for h. Known by all, h and i are in fact called $\underline{h}$ and $\underline{i}$. The thief network (—) is assumed symmetric and transitive. The case is modeled using four worlds, identical up to code name denotation, (shown by $\mapsto$). E.g., in the actual world is w_1, $\underline{t}$ names t, but in w_4, it names b. Hence the cop does not know who Tokyo is: $M, w_1 \models_g \neg\exists x K_{\underline{c}}(x \doteq \underline{t})$.

2.3 Dynamics: Action Models and Product Update

To code operations on static models, we use a variant of DEL-style action models, adapted to term-modal logic (see Fig. 3). They include (adapted versions of) *pre-conditions* specifying when an event is executable [4,5], *postconditions* describing the factual effects of events [6,8,16] as well as *edge-conditions* representing how an agent's observation of an action depends on the agent's circumstances [7]—for example their position in a network, cf. Fig. 3. Edge-conditions are non-standard and deserve a remark. With E the set of events, edge-conditions are assigned by a map Q. For each edge $(e, e') \in E \times E$, $Q(e, e')$ is a formula with a single free variable $x^\star$. Given a model M, an agent i cannot distinguish e from e' iff the edge-condition $Q(e, e')$ is true in M when the free variable $x^\star$ is mapped to i. Intuitively, if the situation described by the edge-condition is true for i, the way in which i is observing the action does not allow her to tell whether e or e' is taking place. See Fig. 4 for an example. See [2] for a comparison of this approach to that of [7] and the term-modal action models of [18].

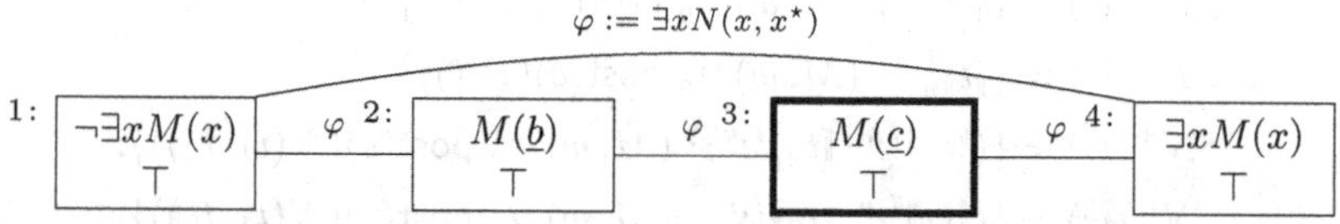

Fig. 3. Example 1, pt. 2 (Edge-Conditions: Announcement to Subgroup). To learn what happened to the server, the top boss a requests its log file. The log holds one of four pieces of information: (1) Nobody made a mistake, (2) b made a mistake (M), (3) c made a mistake or (4) somebody made a mistake. Each box represents one of these events: top lines are preconditions, bottom lines postconditions ($\top$ means no factual change). In fact, the log rats on c. N denotes the hierarchy. The log is sent only to the top boss: the others cannot see its content. This is represented by the edge-condition φ: If you, $x^\star$, have a boss, then you cannot tell (1) from (2) nor (2) from (3) etc. For unillustrated edges, $Q(e, e) = (x^\star \doteq x^\star)$ and $Q(e, e') = \varphi$ when $e \neq e'$.

For simplicity, we here only define action models that take pre-, post, and edge-conditions in the static language $\mathcal{L}(\Sigma)$. However, dynamic conditions are needed for completeness; we refer to [2] for details.

Definition 4. An **action model** *for* $\mathcal{L}(\Sigma)$ *is a tuple* $\Delta = (E, Q, \mathsf{pre}, \mathsf{post})$ *where*

▷ E is a non-empty, finite set of **events**.
▷ $Q : (E \times E) \to \mathcal{L}(\Sigma)$ where each **edge-condition** $Q(e, e')$ has exactly one free variable $x^\star$.
▷ $\mathsf{pre} : E \to \mathcal{L}(\Sigma)$ where each **precondition** $\mathsf{pre}(e)$ has no free variables.
▷ $\mathsf{post} : E \to (\mathtt{GroundAtoms}(\mathcal{L}(\Sigma)) \to \mathcal{L}(\Sigma))$ assigns to each $e \in E$ a **postcondition** for each ground atom (the notation $\mathtt{GroundAtoms}(\mathcal{L}(\Sigma)$ denotes the set of ground atoms of $\mathcal{L}(\Sigma)$).
To preserve the meaning of equality, let $\mathsf{post}(e)(t \doteq t) = \top$ for all $e \in E$.

With no general restrictions on Q, to ensure that all agents' indistinguishability relations continue to be equivalence relations after updating, Q must be chosen with care. Throughout, we assume $Q(e, e) = (x^\star \doteq x^\star)$ for all $e \in E$. To update, *product update* may be altered to fit the edge-condition term-modal setting as below. Figure 4 illustrates the product update of Fig. 1 with 3. The use of postconditions is illustrated in Figs. 7 and 8.

Definition 5. Let $M = (A, W, \sim, I)$ and $\Delta = (E, Q, \mathsf{pre}, \mathsf{post})$ be given. The **product update** of M and Δ is the model $M \otimes \Delta = (A', W', \sim', I')$ where

1. $A' = A$
2. $W' = \{(w, e) \in W \times E : (M, w) \vDash_g \mathsf{pre}(e)\}$ for any g,
3. $(w, e) \sim'_i (w', e')$ iff $w \sim_i w'$ and $M, w \vDash_{g[x^\star \mapsto i]} Q(e, e')$,
4. $I'(c, (w, e)) = I(c, w)$ for all $c \in \mathsf{C}$, and
$I'(X, (w, e)) = (I(X, w) \cup X^+(w)) \setminus X^-(w)$, for $X \in \{P, N\}, P \in \mathsf{P}$, where:

$$P^+(w) := \{[\![t]\!]_w^{I,v} : (M, w) \vDash_g \mathsf{post}(e)(P(t))\};$$
$$P^-(w) := \{[\![t]\!]_w^{I,v} : (M, w) \nvDash_g \mathsf{post}(e)(P(t))\};$$
$$N^+(w) := \{([\![t_1]\!]_w^{I,v}, [\![t_2]\!]_w^{I,v}) : (M, w) \vDash_g \mathsf{post}(e)(N(t_1, t_2))\};$$
$$N^-(w) := \{([\![t_1]\!]_w^{I,v}, [\![t_2]\!]_w^{I,v}) : (M, w) \nvDash_g \mathsf{post}(e)(N(t_1, t_2))\}$$

If $(M, w) \models \mathsf{pre}(e)$, *then* (A, e) is **applicable** to (M, w), and the product update of the two is the pointed model $(M \otimes \Delta, (w, e))$. Else it is undefined.

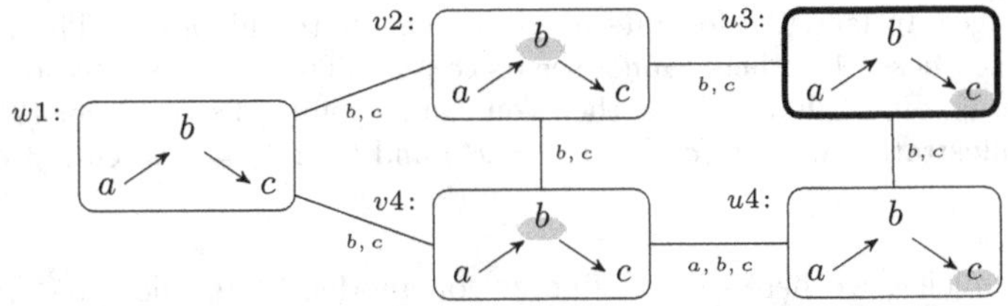

Fig. 4. Example 1, pt. 3 (Product Update: Edge-Conditions). The product update of Figs. 1 and 3. After checking the logs, the boss has learned that c made a mistake, while both b and c are now both uncertain about this, as well as about the boss' information. Worlds are named using by the world-event pair they represent: $w1$ is the child of w and 1, etc. The pair $w2$ is not a world: w did not satisfy the precondition of 1. We have $w1 \sim'_b v2$ as $w \sim_b v$ and $M, w \vDash_{g[x^\star \mapsto a]} Q(1, 2)$—as $M, w \vDash_g \exists x N(x, \underline{b})$. Likewise, $v2 \sim'_b w1$ as $v \sim_b w$ and $M, v \vDash_g \exists x N(x, \underline{b})$. That $w1 \not\sim'_a v2$ follows as $M, w \vDash_g \neg \exists x N(x, \underline{b})$, but $v4 \sim'_a u4$ as $M, v \vDash_g (\underline{a} \doteq \underline{a})$. The same reason, reflexive loops are preserved. The boss now knows that c made a mistake: $K_{\underline{a}} M(\underline{c})$

2.4 Announcements *De Dicto* and *De Re*

With *de dicto* and *de re* statements expressible in DTML, they may be used to define principled announcements, as exemplified in Figs. 5 and 6. The action models are applicable to *any* DTML model for a signature that includes the constant $\underline{a}$ and the predicate M, irrespective of the size of the set of agents. This level of general applicability is not mirrored in standard DEL action models.

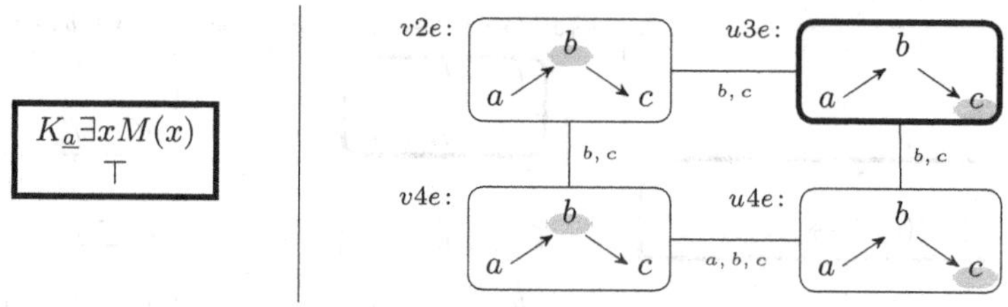

Fig. 5. Example 1, pt. 3 (De Dicto Announcement). The boss breaks the news from the log to b and c piecemeal. ***Left:*** First, a makes a *de dicto* announcement: a knows that somebody made a mistake. ***Right:*** The effect on Fig. 4. Only $w1$ does not survive. In $u3e$, everybody knows *de dicto* that somebody messed up: $\forall x K_x \exists y M(y)$. The boss also knows *de re*, i.e., knows *who*: $u3e \vDash_g \exists x K_a M(x)$, as $u3e \vDash_{g[x \mapsto c]} K_a M(x)$. The employees do not know that a knows *de re*: $u3e \vDash_g \forall x(\exists y N(y,x) \rightarrow \hat{K}_x \neg \exists z K_a M(z))$—since $v4e \vDash_g M(x)$ iff $g(x) = b$, but then $u4e \nvDash_g M(x)$. I.e., there is no *one* object to serve as valuation for x such that $v4e$ and $u4e$ satisfy $M(x)$ simultaneously). The employees are held in suspense!

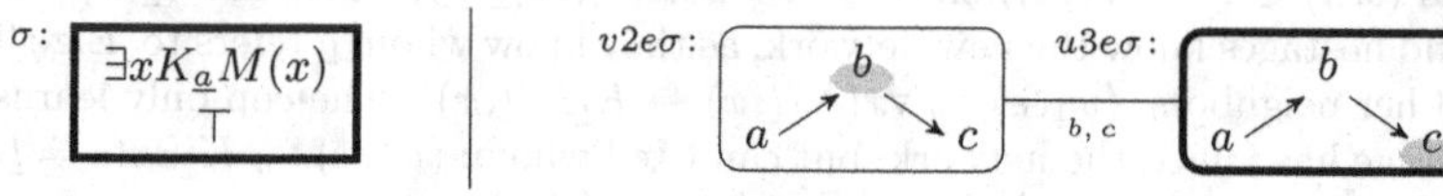

Fig. 6. Example 1, pt. 4 (De Re Announcement). Following a dramatic pause, the boss reveals a stronger piece of information: the boss knows who messed up. This *de re* announcement is on the left, with $Q(e,e) = (x^\star = x^\star)$; its result on Fig. 5 (Right) on the right. In $u3e\sigma$, everybody knows that a has *de re* knowledge: $\forall x K_x \exists y K_a M(y)$, but b and c still only have *de dicto* knowledge: $\forall x((x = b \vee x = c) \rightarrow K_x \exists y M(y) \wedge \neg \exists z K_x M(z))$.

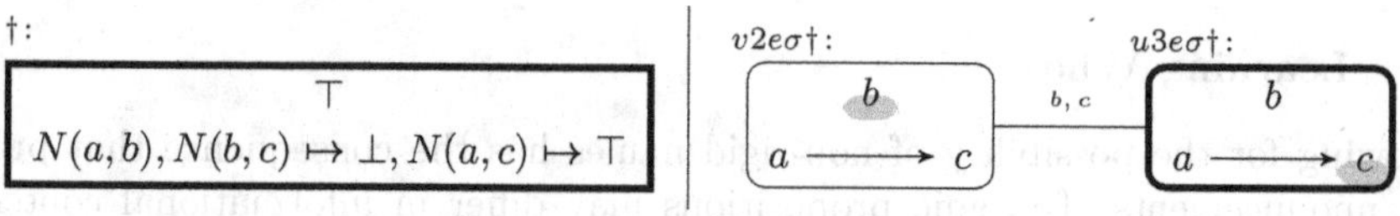

Fig. 7. Example 1, pt. 5 (Getting Fired). The employees are dying to know who messed up the server. But the boss just proclaims: 'b, you are fired! c, you are promoted!' ***Left:*** Action with three instructions for factual change: $\mathsf{post}(\dagger)(N(a,b)) = \bot$, $\mathsf{post}(\dagger)(N(b,c)) = \bot$ and $\mathsf{post}(\dagger)(N(a,c)) = \top$ (illustrated by $\mapsto$). Else $\mathsf{post} = id$. As $u3e\sigma \nvDash \bot$, the first two instructions entail that $(a,b),(b,c) \in N^-(u3e\sigma)$, while the latter implies that $(a,c) \in N^+(u3e\sigma)$. ***Right:*** The network is updated to $I'(N, u3e\sigma\dagger) = (I(N, u3e\sigma) \cup N^+(u3e\sigma)) \backslash N^-(u3e\sigma) = (\{(a,b),(b,c)\} \cup \{(a,c)\}) \backslash \{(a,b),(b,c)\} = \{(a,b)\}$. In $u3e\sigma\dagger$, neither b nor c know who made the mistake. Unrepresented, a thinks that only bad superiors let their employees make mistakes.

2.5 Postconditions and Network Change

Action models with postconditions allows DTML to represent changes to the social network. Such changes may be combined with the general functionality of

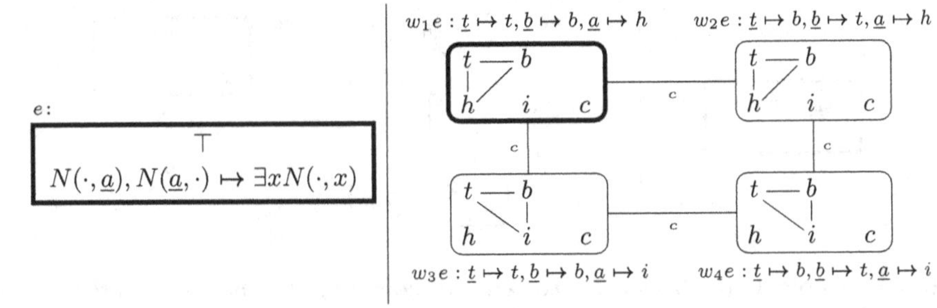

Fig. 8. Example 2, pt. 2 (Becoming Criminal) ***Left:*** The thieves convince The Asset to cooperate with them, in exchange for stolen goods. For simplicity, assume that the action of $\underline{a}$ joining the thief network is noticed by everyone. We model this with the action model, with $\mathsf{post}(e)(N(\cdot,\underline{a})) = \exists x N(\cdot,x)$ and $\mathsf{post}(e)(N(\underline{a},\cdot)) = \exists x N(x,\cdot)$ for $\cdot \in \{\underline{t},\underline{b},\underline{a},\underline{h},\underline{i},\underline{c}\}$. Informally, these say: "If you are a member of the network, then $\underline{a}$ becomes your neighbor". ***Right:*** The effect of event e on Fig. 2: The network has changed in all worlds, but differently. E.g., in w_1, we had $\neg N(\underline{b},\underline{a})$; in (w_1,e), we have $N(\underline{b},\underline{a})$ as $(b,h) \in N^+((w_1,e))$ since $w_1 \vDash_g \mathsf{post}(e)(N(\underline{b},\underline{a}))$—i.e., $\exists x N(\underline{b},x)$. Now all thieves and hostages know the new network, as they know whom $\underline{a}$ refers to. E.g.: Tokyo knows all her neighbors, $(w_1,e) \vDash_g \forall x(N(\underline{t},x) \rightarrow K_{\underline{t}} N(\underline{t},x))$. The cop only learns that *some* hostage has joined the network, but can't tell whom: $(w_1,e) \vDash_g K_{\underline{c}} \exists x(x \not\doteq \underline{t} \wedge x \not\doteq \underline{b} \wedge N(\underline{t},x))$ but $(w_1,e) \nvDash_g \exists x K_{\underline{c}}(x \not\doteq \underline{t} \wedge x \not\doteq \underline{b} \wedge N(\underline{t},x))$.

action models such that some agents may know what changes occur while others remain in the dark. Figure 7 provides a simple example, including the details calculating the updated network. Figure 8 presents an example of how *de re/de dicto* knowledge affects what is learned by a publicly observed network change.

2.6 Learning Who

Allowing for the possibility of non-rigid names has the consequence that public announcements of atomic propositions may differ in informational content depending on the epistemic state of the listener. This can be exploited by the

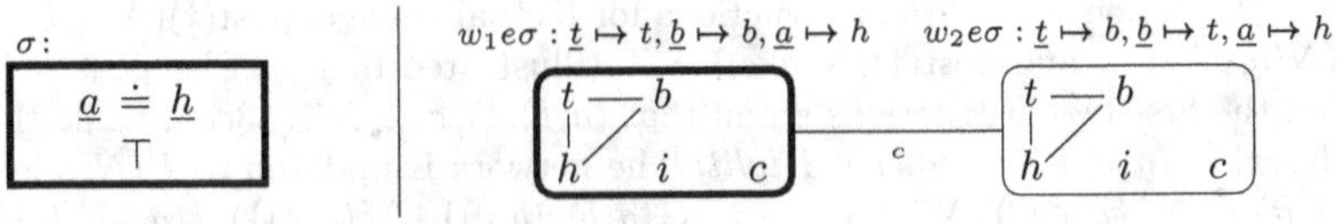

Fig. 9. Example 2, pt. 4 (Revealing the Asset). In the model in Fig. 8 (Right), even a *public announcement* of $N(\underline{t},\underline{a})$ would not inform the cop about who joined the network. To know who joined the network, the cop must learn *who The Asset is*. As the cop knows who $\underline{h}$ is, learning that $\underline{h}$ is The Asset suffices. ***Left:*** The event model σ for the public announcement that $\underline{a} \doteq \underline{h}$, revealing the identity of The Asset. ***Right:*** The product update of Fig. 8 (Right) and event σ. The cop now knows the structure of the network, as a result of the removal of w_3e and w_4e.

thieves of Example 2 to enforce a form of *privacy*—as code names should. The notion of privacy involved is orthogonal to the notion of privacy modeled in DEL using private announcements. Though the message is public in the standard sense of everyone being aware of it and its content, as it involves non-rigid names, its epistemic effects are not the same for all agents. This is in contrast with standard public announcements, which yield the same information to everyone.

3 Embedding Dynamic Social Network Logics in DTML

This section examines relations between the hybrid network models and their languages to DTML. As hybrid languages corresponds to fragments of first-order logic with equality ($\mathsf{FOL}_=$), which term-modal logic extends, it stands to reason that the hybrid languages and models mentioned in Sect. 2 may be embedded in term-modal logic. A precise statement and a proof sketch follows below. Turning to dynamics, things are more complicated. [23] presents a very flexible hybrid framework expressing network dynamics using *General Dynamic Dynamic Logic* (GDDL, [24]). We leave general characterizations of equi-expressive fragments of GDDL and DTML as open question, but remark that all GDDL action-examples of [23] may be emulated using DTML action models, and in many cases via fairly simple ones. More thoroughly, we show that the logic of *Knowledge, Diffusion and Learning* (KDL, [13]) has a complete and decidable system, a question left open in [13]. This is shown by encoding KDL in DTML.

3.1 Embedding Static Languages and Models

The static hybrid languages of [10–13,20,22,23] are all sub-languages of $\mathcal{L}(P, Nom)$, defined and translated into DTML below. [19] also includes state nominals, which our results do not cover. $\mathcal{L}(P, Nom)$ is read indexically, as described in Sect. 2.

Definition 6. *With $p \in P$ and $x \in Nom$, the **language** $\mathcal{L}(P, Nom)$ is given by*

$$\varphi := p \mid \neg\varphi \mid \varphi \wedge \varphi \mid @_x\varphi \mid K\varphi \mid N\varphi \mid U\varphi$$

Denote the fragments without U and $@_x$ by $\mathcal{L}_{-U}(P, Nom)$ and $\mathcal{L}_{-@}(P, Nom)$.

Hybrid logics may be translated into $\mathsf{FOL}_=$; our translation resembles that of [9]. We identify agent nominals with first-order variables, translate the modal operator N to the relation symbol $N(\cdot, \cdot)$, and relativize the interpretation of the indexical K to the nominal/variable x by using the term-indexed operator K_x. Formally, the translation is defined as follows.

Definition 7. *Let $\Sigma_n(P, Nom) = (\mathtt{V}, \mathtt{C}, \mathtt{P}, N, \doteq)$ be the signature with $\mathtt{V} = Nom$, $\mathtt{C} = \{\underline{a}_1, \ldots, \underline{a}_n\}$ and $\mathtt{P} = P$. Translations T_x, T_y both mapping $\mathcal{L}(P, Nom)$ to $\mathcal{L}(\Sigma_n(P, Nom))$ are defined by mutual recursion. It is assumed that two nominals*

x and y are given which do not occur in the formulas to be translated. For $p \in P$ and $i \in Nom$, define T_x by:

$$
\begin{aligned}
T_x(p) &= p(x) & T_x(@_i\varphi) &= T_x(\varphi)(x \mapsto i)\\
T_x(i) &= x \doteq i & T_x(N\varphi) &= \forall y(N(x,y) \rightarrow T_y(\varphi))\\
T_x(\varphi \wedge \psi) &= T_x(\varphi) \wedge T_x(\psi) & T_x(K\varphi) &= K_x T_x(\varphi)\\
T_x(\neg\varphi) &= \neg T_x(\varphi) & T_x(U\varphi) &= \forall x T_x(\varphi)
\end{aligned}
$$

The translation T_y is obtained by exchanging x and y in T_x.

To show the translation truth-preserving, we embed the class of hybrid network models into a class of term-modal models:

Definition 8. *Let $M = (A, W, (N_w)_{w \in W}, \sim, g, V)$ be a hybrid network model for $\mathcal{L}(P, Nom)$. Then the* TML ***image*** *of M is the $\mathcal{L}(\Sigma_n(P, Nom))$* TML *model $\mathsf{T}(M) = (A, W, \sim, I)$ sharing A, W and $\sim$ with M and with I given by*

1. $\forall \underline{c} \in \mathsf{C}, \forall w, v \in W, \forall a, b \in A, (I(\underline{c}, w) = a$ *and* $w \sim_b v \Rightarrow I(\underline{c}, v) = a)$
2. $I(p, w) = \{a \colon (w, a) \in V(p)\}$
3. $I(N, w) = \{(a, b) \in A \times A \colon (a, b) \in N_w\}$

The model $\mathsf{T}(M)$ has the same agents, worlds and epistemic relations as M. The interpretation 1. encodes ***weak rigidity***: if $(w, v) \in \bigcup_{a \in A} \sim_a$, then any constant denotes the same in w and v, emulating the rigid names of hybrid network models; 2. ensures predicates are true of the same agents at the same worlds, and 3. ensures the same agents are networked in the same worlds.

With the translations T_x, T_y and the embedding T, it may be shown that DTML can fully code the static semantics of $\mathcal{L}(P, Nom)$ hybrid network logics:

Proposition 1. *Let $M = (A, W, (N_w)_{w \in W}, \sim, g, V)$ be a hybrid network model. Then for all $\varphi \in \mathcal{L}(P, Nom)$, $M, w, g(\bullet) \models \varphi$ iff $\mathsf{T}(M), w \models_g T_\bullet(\varphi)$, $\bullet = x, y$.*

3.2 KDL Dynamic Transformations and Learning Updates in DTML

We show that KDL [13] dynamics may be embedded in DTML, for finite agent sets (as assumed in [13]). Given Proposition 1, we argue that each KDL model transformer is representable by a DTML action model and that the dynamic KDL language is truth-preservingly translatable into a DTML sublanguage. The logic of the class of KDL models is, up to language translations, the logic of its corresponding class of DTML models. We show that *the logic of this class of* DTML *models can be completely axiomatized, and the resulting system is decidable.* Thus, by embedding KDL in DTML, we find a complete system for the former. Due to space restrictions, technical details and proofs are deferred to the extended version to this paper [1].

In KDL[2], agents are described by *feature propositions* reading "for feature f, I have value z". With F a countable set of features and Z_f a finite set of possible

[2] Notation here is equivalent but different to fit better with the rest of this paper.

values of $\mathsf{f} \in \mathsf{F}$, the set of feature propositions is $\mathsf{FP} = \{(\mathsf{f} \doteqdot \mathsf{z})\colon \mathsf{f} \in \mathsf{F}, \mathsf{z} \in \mathsf{Z}_\mathsf{f}\}$. The static language of [13] is then $\mathcal{L}_{-U}(\mathsf{FP}, Nom)$. The dynamic language $\mathcal{L}_{\mathsf{KDL}}$ extends $\mathcal{L}_{-U}(\mathsf{FP}, Nom)$ with dynamic modalities $[d]$ and $[\ell]$ for *dynamic transformations* d and *learning updates* ℓ:

$$\varphi ::= (\mathsf{f} \doteqdot \mathsf{z}) \mid i \mid \neg\varphi \mid \varphi \wedge \varphi \mid @_i\varphi \mid N\varphi \mid K\varphi \mid [d]\varphi \mid [\ell]\varphi$$

A **dynamic transformation** d changes feature values of agents: each is a pair $d = (\Phi, post)$ where $\Phi \subseteq \mathcal{L}_{\mathsf{KDL}}$ is a non-empty finite set of pairwise inconsistent formulas and $post : \Phi \times \mathsf{F} \to (Z_n \cup \{\star\})$ is a KDL post-condition. Encoded by $post(\varphi, \mathsf{f}) = x$ is the instruction: if $(w, a) \vDash \varphi$, then after d, set f to value x at (w, a), if $x \in Z_n$; if $x = \star$, f is unchanged. A **learning update** cuts accessibility relations: the update with finite $\ell \subseteq \mathcal{L}_{\mathsf{KDL}}$ keeps a $\sim_a$ link between worlds w and v iff, for all $\varphi \in \ell$, $(w, b) \vDash \varphi \Leftrightarrow (v, b) \vDash \varphi$ for all neighbors b of a. See [13] or [1] for formal details. Let D and L be the sets of dynamic transformations and learning updates. The result of applying $\dagger \in \mathsf{D} \cup \mathsf{L}$ to M is denoted $M^\dagger$, and the $[\dagger]$ modality has semantics $M, w, a \models [\dagger]\varphi$ iff $M^\dagger w, a \models \varphi$.

For every $\dagger \in \mathsf{D} \cup \mathsf{L}$, there is a pointed DTML action model $(\Delta^\dagger, e^\dagger)$ with identical effects. As KDL operations may involve formulas with $[\dagger]$-modalities, we must use DTML action models that allow $[\Delta, e]$-modalities in their conditions, and translate $\mathcal{L}_{\mathsf{KDL}}$ into the general DTML language that results, denoted $\mathcal{L}(\Sigma_n(\mathsf{FP}, Nom) + [\Delta])$.[3] This language is interpreted over DTML models with standard action model semantics:

$$(M, w) \vDash_g [\Delta, e]\varphi \text{ iff } M \otimes \Delta, (w, e) \vDash \varphi.$$

The translation is obtained by adding the following clauses to $T_\bullet$, $\bullet = x, y$:

$$T_\bullet([d]\varphi) = [\Delta^d, e^d]T_\bullet(\varphi), \quad T_\bullet([\ell]\varphi) = \bigwedge_{e \in E^\ell} (\mathsf{pre}^\ell(e) \to [\Delta^\ell, e]T_\bullet(\varphi))$$

where $\Delta^\dagger$ is an action model implementation of $\dagger \in \mathsf{D} \cup \mathsf{L}$. Then KDL statics and dynamics can be shown performable in DTML:

Proposition 2. *For any finite agent hybrid network model M with nominal valuation g and $\varphi \in \mathcal{L}_{\mathsf{KDL}}$: $M, w, g(\bullet) \models \varphi$ iff $\mathsf{T}(M), w \models_g T_\bullet(\varphi)$, for $\bullet = x, y$.*

Proof (sketch). By induction on φ, with formulas $[\dagger]\varphi$ the difficult cases, requiring construction of action models. For $\dagger \in \mathsf{D}$, [12] provide reduction axioms showing $\dagger$'s instructions statically encodable in $\mathcal{L}_{\mathsf{KDL}}$. As $\dagger \in \mathsf{D}$ changes atomic truth values under a definable instruction, $\dagger$ may be simulated by an action model with a matching post-condition (a translated instruction). For $\dagger \in \mathsf{L}$, $(\Delta^\dagger, e^\dagger)$ has events e^X, e^Y for any maximal consistent subsets X, Y of $\{\varphi(\underline{c}), \neg\varphi(\underline{c})\colon \varphi \in \ell, \underline{c} \in \mathsf{C}\}$ with edge-condition $Q(e^X, e^Y)$ satisfied for agents for whom all neighbors agree on X and Y. Unsatisfied edge-conditions thereby capture the link cutting mechanism of ℓ. For details, see [1], especially Definitions 6, 9 and Proposition 3 therein.

[3] Defined using double recursion as standard; see [2] for details.

With Proposition 2 embedding KDL in DTML, it remains to show that there is a complete and decidable system for the image of KDL. Up to translation, such a logic is then a logic for the class of KDL models. To state the result, denote the TML image of the class of n-agent KDL models by $\mathsf{T}(\mathsf{KDL}_n)$.

Definition 9. *Let* $\mathsf{F}_n \subseteq \mathcal{L}(\Sigma_n(\mathsf{FP}, Nom) + [\Delta])$ *be the logic extending the term-modal* S5 *logic with static axioms* $\mathsf{Named_n}$ *(there are exactly* n *agents, all named by constants);* $\mathsf{Rig_n}$ *(defining the weak rigidity condition of Definition 8);* Neigh *(N is irreflexive and symmetric); and* KnowNeigh *(all agents know their neighbors); as well as reduction axioms for action models* $(\Delta^\dagger, \mathsf{e}^\dagger)$, $\dagger \in \mathsf{D} \cup \mathsf{L}$.

See [1] for static axioms and [2] for reduction axioms. We then obtain the result:

Theorem 1. *For any* $n \in \mathbb{N}$*, the logic* F_n *is sound, strongly complete and decidable w.r.t.* $\mathsf{T}(\mathsf{KDL}_n)$.

Proof (sketch). In [1], it is shown that F_n statically characterizes $\mathsf{T}(\mathsf{KDL}_n)$. The result then follows from three results from [2]: 1. Any extension of the term-modal logic K with axioms A is strongly complete with respect to the class of frames characterized by A, and 2. If A characterizes a class with finitely many agents, then the logic is also decidable, and 3. Any dynamic DTML formula is provably equivalent to a static DTML formula using reduction axioms.

4 Final Remarks

This paper has showcased DTML as a framework for modeling social networks, their epistemics and dynamics, including examples in which uncertainty about name reference and *de dicto/de re* distinctions are key to modelling information flow and network change correctly. It was shown that DTML may encode the popular hybrid logical models of epistemic networks, and that DTML may be used to obtain completeness for an open-question dynamics through emulation.

We are very interested in learning how DTML relates to GDDL with respect to the encodable dynamics. We have been able to emulate the updates used in the examples of [23], but the general question is open. Further, the statics of frameworks that describe networks using propositional logic [3,15] must be DTML encodable, and we expect the name about their updates, where reduction axioms exist. This raises two questions: if we can show this by a general results instead of piecemeal, and whether principled DTML action models exist for classes of updates. E.g., the *threshold update* of [3] gives an agent's property P if a given fraction of neighbors are P; for a fixed agent set, this is DTML encodable by using the reduction axioms of [3] to provide pre- and postconditions. For a principled update, however, seemingly we need a generalized quantifier (e.g., a Rescher quantifier). If so, the general update form is not DTML encodable. Classification results like these would add valuable insights on network logics.

References

1. Liberman, A.O., Rendsvig, R.K.: Dynamic Term-Modal Logic for Epistemic Social Network Dynamics (Extended). arXiv:1908.09658 (2019)
2. Achen, A., Liberman, A.O., Rendsvig, R.K.: Dynamic term-modal logics for epistemic planning. arXiv:1906.06047 (2019, under review)
3. Baltag, A., Christoff, Z.: Dynamic epistemic logics of diffusion and prediction in social networks. Stud. Logica **107**(3), 489–531 (2018)
4. Baltag, A., Moss, L.S.: Logics for epistemic programs. Synthese **139**(2), 165–224 (2004)
5. Baltag, A., Moss, L.S., Solecki, S.S.: The logic of public announcements, common knowledge, and private suspicions. In: TARK 1998, pp. 43–56 (1998)
6. Benthemvan Benthem, J., Eijckvan Eijck, J., Kooi, B.: Logics of communication and change. Inf. Comput. **204**(11), 1620–1662 (2006)
7. Bolander, T.: Seeing is believing: formalising false-belief tasks in dynamic epistemic logic. In: ECSI, vol. 1283, pp. 87–107 (2014)
8. Bolander, T., Andersen, M.B.: Epistemic planning for single- and multi-agent systems. J. Appl. Non-Class. Log. **21**(1), 9–34 (2011)
9. Brauner, T.: Hybrid Logic and Its Proof-Theory. Springer, Dordrecht (2011)
10. Christoff, Z.: Dynamic logics of networks. Ph.D. thesis, University of Amsterdam (2016)
11. Christoff, Z., Hansen, J.U.: A two-tiered formalization of social influence. In: Grossi, D., Roy, O., Huang, H. (eds.) LORI 2013. LNCS, vol. 8196, pp. 68–81. Springer, Heidelberg (2013). https://doi.org/10.1007/978-3-642-40948-6_6
12. Christoff, Z., Hansen, J.U.: A logic for diffusion in social networks. J. Appl. Log. **13**, 48–77 (2015)
13. Christoff, Z., Hansen, J.U., Proietti, C.: Reflecting on social influence in networks. J. Log. Lang. Inf. **25**, 299–333 (2016)
14. Christoff, Z., Naumov, P.: Diffusion in social networks with recalcitrant agents. J. Log. Comput. **29**(1), 53–70 (2018)
15. Christoff, Z., Rendsvig, R.K.: Dynamic logics for threshold models and their epistemic extension. In: Proceedings of ELISIEM (2014)
16. Ditmarschvan Ditmarsch, H., Kooi, B.: Semantic results for ontic and epistemic change. In: Logic and the Foundations of Game and Decision Theory (LOFT 7). Texts in Logic and Games, vol. 3, pp. 87–117. Amsterdam University Press (2008)
17. Hintikka, J.: Knowledge and Belief: An Introduction to the Logic of the Two Notions. College Publications (1962). 2nd edn. 2005
18. Kooi, B.: Dynamic term-modal logic. In: LORI 2007. Texts in Computer Science 8, pp. 173–185 (2007)
19. Zhen, L., Seligman, J.: A logical model of the dynamics of peer pressure. Electron. Notes Theor. Comput. Sci. **278**, 275–288 (2011)
20. Liu, F., Seligman, J., Girard, P.: Logical dynamics of belief change in the community. Synthese **191**(11), 2403–2431 (2014)
21. Rendsvig, R.K.: Diffusion, influence and best-response dynamics in networks: an action model approach. In: ESSLLI 2014 Student Session, pp. 63–75. arXiv:1708.01477 (2014)
22. Seligman, J., Liu, F., Girard, P.: Logic in the community. In: Banerjee, M., Seth, A. (eds.) ICLA 2011. LNCS (LNAI), vol. 6521, pp. 178–188. Springer, Heidelberg (2011). https://doi.org/10.1007/978-3-642-18026-2_15

23. Seligman, J., Liu, F., Girard, P.: Facebook and the epistemic logic of friendship. In: TARK 2013, pp. 229–238 (2013)
24. Seligman, J., Liu, F., Girard, P.: General dynamic dynamic logic. In: Advances in Modal Logic, vol. 9, pp. 239–260. Springer (2012)
25. Smets, S., Velázquez-Quesada, F.R.: How to make friends: a logical approach to social group creation. In: Baltag, A., Seligman, J., Yamada, T. (eds.) LORI 2017. LNCS, vol. 10455, pp. 377–390. Springer, Heidelberg (2017). https://doi.org/10.1007/978-3-662-55665-8_26
26. Smets, S., Velázquez-Quesada, F.R.: The creation and change of social networks: a logical study based on group size. In: Madeira, A., Benevides, M. (eds.) DALI 2017. LNCS, vol. 10669, pp. 171–184. Springer, Cham (2018). https://doi.org/10.1007/978-3-319-73579-5_11

Analyzing Echo Chambers: A Logic of Strong and Weak Ties

Mina Young Pedersen[1(✉)], Sonja Smets[1,3(✉)], and Thomas Ågotnes[2,3(✉)]

[1] ILLC, University of Amsterdam, Amsterdam, The Netherlands
minaypedersen@gmail.com, S.J.L.Smets@uva.nl
[2] ILI, Southwest University, Chongqing, China
[3] Infomedia, University of Bergen, Bergen, Norway
Thomas.Agotnes@uib.no

Abstract. Echo chamber is a widely used term describing a situation where certain information is reinforced within a closed network. To reason about echo chambers, we introduce a two-sorted hybrid logic of strong and weak ties based on a logic of positive and negative relations known from the literature. We show that some classical property definitions can be formalized and that a known claim from social network analysis is a validity. We also prove that the logic is axiomatizable, sound and strongly complete. We combine our results with research on homophily and social group formation to represent relations between similar agents. Lastly, we add a knowledge modality and dynamic operators to analyze change in these networks.

Keywords: Strong and weak ties · Echo chamber · Friendship logic · Hybrid logic · Logic for social epistemic networks · Homophily

1 Introduction

In its various forms, social media – as a communication tool between peers, a form of direct access to the personal opinions of powerful people or a news outlet – is indisputably a vital source of information in our current world. Most social media platforms give users or platform operators the ability to filter out annoying and/or incompatible voices. The downside is that this can promote the emergence of echo chambers where a shortage of new and opposing information can lead to fragmentation in society [14].

Echo chamber is not a formal terminology with an explicit definition, but is widely and vaguely referred to in several contexts. The phrase is often used as a derogatory term pointing to a situation where certain information within a group is contained and repeated inwards and where challenging opinions are rejected. As an echo reflects sound, an echo chamber reflects similar opinions in a setting closed off from the outside world to a certain degree.

In this paper, we aim to develop a logical framework to analyze echo chambers. To do this, we build upon and combine two known concepts from social

P. Blackburn et al. (Eds.): LORI 2019, LNCS 11813, pp. 183–198, 2019.
https://doi.org/10.1007/978-3-662-60292-8_14

network analysis: *triadic closure* and *homophily*. Triadic closure is meant to formalize the phenomenon where one is likely to know the friends of one's friends. The formalization was made popular by Mark Granovetter in the 1970s as part of his theory of the strength of weak ties [13]. In this theory, relations in a graph-like social network structure are divided into strong and weak ties between nodes that represents agents. Strong ties are meant to represent friends, and weak ties model acquaintances such that one tie cannot be both strong and weak.

The strength of weak ties theory by Granovetter suggests that social networks have a high density of triadic closure. This property is formally represented in this theory as *Strong Triadic Closure*. Strong Triadic Closure is a property of individual agents in the network and holds if and only if its strong ties are subsequently tied together by a weak or a strong tie. As a result of Strong Triadic Closure, a network of strong and weak ties are likely to be made up of clusters of strong ties which are individually tied by weak ties. The argument goes that weak ties have an important role channeling information between clusters of strong ties.

Homophily is known as the tendency of being similar to ones friends [11]. By combining the strength of weak ties theory with the likelihood of homophily, Granovetter's clusters of strong ties are not only comprised of friends, but of agents that are alike. We therefore propose that these clusters are essentially echo chamber-like structures where information to a great degree is preserved within the strong ties.

The angle from which we approach this topic is to represent social networks as two-sorted hybrid frames of a logic we name *tied logic*. Tied logic is inspired by the logic of positive and negative relations [19,20] substituting positive and negative with strong and weak. Tied logic also has hybrid operators in the language, which the logic of positive and negative relations in its original form has not. We analyze echo chambers in this framework by describing social network properties like Strong Triadic Closure as logical formulas and researching their constraints on the network structure.

To further formalize echo chambers, we also explicitly implement tied logic with homophily. Perhaps among many, there are two particularly prominent reasons behind homophily. One is social influence; the habit of becoming like the people we surround ourselves with. Another is social selection; that we form friendships with others that are alike us. We illustrate the latter phenomenon by defining a subclass of tied logic models similar to the threshold models of [17,18]. Where traditional threshold models operate with one threshold, we include two linearly ordered thresholds for weak and strong ties.

Not only are we defining tied logic to analyze echo chambers, but also for the sake of investigating the expressive power of the logic itself. Tied logic is a contribution to the emerging field of social network logic, where friendships between agents in a network are presented as relations on a frame. Like other hybrid logics of social networks (e.g. [8]) we include dynamic operators to investigate change in the networks. We are also intrigued by the logical study of knowledge in social networks, and therefore adopt a knowledge modality inspired by epistemic social network logics like [15,16]. The idea of threshold models in a sociopsychological context has been studied in for instance [4], however here in a different setting.

The structure of the paper will be as follows. We begin in the next section by presenting tied logic (**TL**). We define syntax and semantics, and show that Granovetter's most notable claim is a validity in our logic. We also offer a full axiomatization of **TL** and show that it is sound and strongly complete with respect to the class of what we call tied frames. In the end of this section we combine our results with research on social group formation and introduce homophily into the models. The final part of the paper is devoted to knowledge and dynamics. Here we examine the addition of a knowledge operator and dynamic modalities to tied logic, extending it to *tied epistemic logic* (**TEL**). To reason about the interaction between knowledge and change in echo chambers, we discuss some possible epistemic constraints to assign the agents. Then we present some validities of **TEL** and an example of a tied epistemic threshold model where we discuss what formulas might hold at specific agents depending on what axioms we adopt to restrict the model. We end with conclusion and a brief reflection of future work.

2 Tied Logic

2.1 Language and Semantics

As in other hybrid logics[1], the language of **TL** includes operators $@_i$ and $\downarrow x$. Intuitively, $@_i$ lets us shift the evaluation to the agent where name i is true. $\downarrow x$ names the current agent 'x'. These operators are closely related, but serve different purposes. By including both, we allow formulas where naming agents lets us later return the evaluation to the same agent. The language of **TL** includes the two diamond modalities $\langle S\rangle$ and $\langle W\rangle$. They are read intuitively as $\langle S\rangle\phi$ when the current agent has a strong tie where ϕ holds. A strong tie is replaced by a weak tie for $\langle W\rangle\phi$. We define the syntax, frames and models of **TL** formally as follows.

Definition 1 (Syntax of TL). *Let* At *be a set of propositional atoms and* Nom *be a set of nominals. Further, let* Var *be a set of agent variables. Let* $\mathsf{At}, \mathsf{Nom}$ *and* Var *be countable and pairwise disjoint. We define the well-formed formulas of the language* $\mathcal{L}_{TL}$ *to be generated by the following grammar:*

$$\phi ::= p \mid s \mid \neg\phi \mid (\phi \wedge \phi) \mid \langle S\rangle\phi \mid \langle W\rangle\phi \mid @_s\phi \mid \downarrow x.\phi$$

where $p \in \mathsf{At}$, $s \in \mathsf{Nom} \cup \mathsf{Var}$ *and* $x \in \mathsf{Var}$. *We define propositional connectives like* $\vee, \rightarrow$ *and the formulas* $\top, \bot$ *as usual. Further, we define the duals as standard* $[S] := \neg\langle S\rangle\neg$ *and* $[W] := \neg\langle W\rangle\neg$.

We will denote members of $\mathsf{At} = \{p, q, r, \dots\}$, $\mathsf{Nom} = \{i, j, k, \dots\}$ and $\mathsf{Var} = \{x, y, z, \dots\}$.

[1] For further details on hybrid logics beyond the scope of this paper, we recommend turning to [1].

Definition 2 (Tied Model and Frame). *Let A be a non-empty set of agents and R^S and R^W be two symmetric and non-overlapping binary relations on A where R^S is reflexive and R^W is irreflexive. A tied model is a tuple $\mathbb{M} = \langle A, R^S, R^W, V\rangle$ where $V : A \to \mathcal{P}(\mathsf{At} \cup \mathsf{Nom})$ is a valuation function such that $\forall i \in \mathsf{Nom}$: $\exists a \in A$ such that $i \in V(a)$, and $\forall b, c, \in A$: if $i \in V(b)$ and $i \in V(c)$, then $b = c$.*

We define a tied frame $\mathbb{F} = \langle A, R^S, R^W\rangle$ as a tied model without valuation.

Our two relations R^S and R^W define strong and weak ties, respectively. We assume reflexivity of R^S, irreflexivity of R^W and symmetry of both relations. Moreover, we undertake the property of *non-overlapping*: no two agents can be related by both a strong and a weak tie. For $a, b \in A$ we will denote members of R^S as $(a, b)_S$, members of R^W as $(a, b)_W$. As tied models describe social networks, we will sometimes refer to tied models as networks or social networks. Moreover, we view all formulas as propositions about agents. For instance we read $p \in V(a)$ as proposition p or feature p holds of agent a. Thus we view $V(a)$ as the set of all basic properties that holds of a.

To define truth in a tied model, we need to include an assignment function $g : \mathsf{Var} \to A$ that assigns agents to variables. Further, define the *x-variant* of g to be $g^x_a(x) = a$ and $g^x_a(y) = g(y)$ for all $y \neq x$. Also define $[s]^{\mathbb{M},g}$ for $s \in \mathsf{Nom} \cup \mathsf{Var}$. For $i \in \mathsf{Nom}$, $[i]^{\mathbb{M},g}$ is the state $a \in A$ called 'i', i.e. the unique a such that $i \in V(a)$ For $x \in \mathsf{Var}$, $[x]^{\mathbb{M},g} = g(x)$. We can now present satisfaction in a tied model.

Definition 3 (Semantics of TL). *Let $\mathbb{M} = \langle A, R^S, R^W, V\rangle$ be a tied model, a an agent in A and $g : \mathsf{Var} \to A$ an assignment function. We inductively define the truth conditions as follows:*

$$\mathbb{M}, g, a \Vdash p \text{ iff } p \in V(a) \text{ for } p \in \mathsf{At}$$
$$\mathbb{M}, g, a \Vdash s \text{ iff } a = [s]^{\mathbb{M},g} \text{ for } x \in \mathsf{Nom} \cup \mathsf{Var}$$
$$\mathbb{M}, g, a \Vdash \neg\phi \text{ iff } \mathbb{M}, g, a \nVdash \phi$$
$$\mathbb{M}, g, a \Vdash \phi \wedge \psi \text{ iff } \mathbb{M}, g, a \Vdash \phi \text{ and } \mathbb{M}, g, a \Vdash \psi$$
$$\mathbb{M}, g, a \Vdash \langle S\rangle\phi \text{ iff } \exists b \in A \text{ such that } aR^S b \text{ and } \mathbb{M}, g, b \Vdash \phi$$
$$\mathbb{M}, g, a \Vdash \langle W\rangle\phi \text{ iff } \exists b \in A \text{ such that } aR^W b \text{ and } \mathbb{M}, g, b \Vdash \phi$$
$$\mathbb{M}, g, a \Vdash @_s\phi \text{ iff } \mathbb{M}, g, [s]^{\mathbb{M},g} \Vdash \phi \text{ for } s \in \mathsf{Nom} \cup \mathsf{Var}$$
$$\mathbb{M}, g, a \Vdash \downarrow x.\phi \text{ iff } \mathbb{M}, g^x_a, a \Vdash \phi$$

2.2 Strong Triadic Closure and Local Bridges

As mentioned earlier, the formation of echo chambers is tightly connected to the property of Strong Triadic Closure. A formal definition follows.

Definition 4 (Strong Triadic Closure [11,13]). *Let $\mathbb{M} = \langle A, R^S, R^W, V\rangle$ be a tied model. An agent $a \in A$ has the strong triadic closure property iff $\forall b, c \in A$:*

- *if* aR^Sb *and* aR^Sc*, then* bR^Sc *or* bR^Wc*.*

Strong Triadic Closure is closely related to Euclidicity in the standard Kripke semantics [2]. It is important to note that where Euclidicity is a frame property, Strong Triadic Closure is defined as a property of agents in the network. However, the property of *all* agents being strongly triadic closed is indeed a frame property which we prove in the following lemma.

Lemma 1. *For any tied frame* $\mathbb{F}$, $\mathbb{F} \Vdash \langle S\rangle p \rightarrow [S](\langle S\rangle p \vee \langle W\rangle p)$ *iff all agents in* $\mathbb{F}$ *have the Strong Triadic Closure property.*

Proof. ($\Rightarrow$) Proof by contraposition. Let $\mathbb{F} = \langle A, R^S, R^W\rangle$ be a tied frame such that there exists an agent $a \in A$ that does not have the Strong Triadic Closure property. Then $\exists b, c \in A$ such that aR^Sb and aR^Sc, but $\neg(bR^Sc)$ and $\neg(bR^Wc)$. Consider now a valuation such that $p \in V(b)$, and $\forall x \in A$ such that $x \neq b$: $p \notin V(x)$. Since $\neg(bR^Sc)$ and $\neg(bR^Wc)$, it follows that $(\mathbb{F}, V), c \Vdash \neg\langle S\rangle p \wedge \neg\langle W\rangle p$. Thus, as aR^Sc, we know that $(\mathbb{F}, V), a \Vdash \langle S\rangle(\neg\langle S\rangle p \wedge \neg\langle W\rangle p)$. As aR^Sb, we have that $(\mathbb{F}, V), a \Vdash \langle S\rangle p$. Hence $(\mathbb{F}, V), a \nVdash \langle S\rangle p \rightarrow [S](\langle S\rangle p \vee \langle W\rangle p)$ and we conclude that $\mathbb{F} \nVdash \langle S\rangle p \rightarrow [S](\langle S\rangle p \vee \langle W\rangle p)$.

($\Leftarrow$) Let $\mathbb{F} = \langle A, R^S, R^W\rangle$ be a tied frame where all agents in A have the Strong Triadic Closure property. Fix an arbitrary $a \in A$ and let V be a valuation on $\mathbb{F}$ such that $(\mathbb{F}, V), a \Vdash \langle S\rangle p$. Then $\exists b \in A$ such that aR^Sb and $(\mathbb{F}, V), b \Vdash p$. Let $c \in A$ be an arbitrary agent such that aR^Sc. By the Strong Triadic Closure property of a, it follows that bR^Sc or bR^Wc. Thus $(\mathbb{F}, V), c \Vdash \langle S\rangle p \vee \langle W\rangle p$. Hence, as c was chosen arbitrarily $(\mathbb{F}, V), a \Vdash \langle S\rangle \rightarrow [S](\langle S\rangle p \vee \langle W\rangle p)$. As we fixed $a \in A$ and V arbitrarily too, we have that $\mathbb{F} \Vdash \langle S\rangle \rightarrow [S](\langle S\rangle p \vee \langle W\rangle p)$ which concludes the proof.

Granovetter's theory that networks with a high occurrence of Strong Triadic Closure have a tendency to form weakly tied clusters of strong ties, is demonstrated in a known claim. Before we present this claim, we introduce the concept of a local bridge.

Definition 5 (Local Bridge [11]**).** *Let* $\mathbb{M} = \langle A, R^S, R^W, V\rangle$ *be a tied model. Let* $a, b \in A$*. An edge* $(a, b)_\circ$ *for* $\circ \in \{S, W\}$ *is a local bridge iff* $\forall c \in A$ *such that* $c \neq a$, $c \neq b$: $\neg(aR^Sc)$ *and* $\neg(aR^Wc)$*, or* $\neg(bR^Sc)$ *and* $\neg(bR^Wc)$*.*

A local bridge is tie between two agents such that these two agents have no other friends or acquaintances in common. Agents in a social network that are related by a local bridge are in an important position when it comes to distribution of information. In a clustered network local bridges are essential carriers of outside information. As we argue, clusters create echo chamber-like situations. Local bridges carrying new information are crucial in dissolving dangerous situations as for instance radicalization.

The claim is stated informally as follows.

---

[2] See [7] for details.

Claim ([11,13]). *If an agent in a network satisfies the Strong Triadic Closure property and is connected to other agents by at least two strong ties, then any local bridge it is related to another agent with, must be a weak tie.*[3]

To formalize this claim as a validity in **TL**, we introduce formulas corresponding to the relevant properties. For simplicity, we first define the following abbreviation $\langle S \cup W \rangle \phi := \langle S \rangle \phi \vee \langle W \rangle \phi$. We read $\langle S \cup W \rangle \phi$ as true at an agent a if and only if a is connected to an agent by a strong or a weak tie where ϕ holds. We now present the following formulas and corresponding lemma.

$$\mathbf{STC} :=\downarrow x.[S] \downarrow y.@_x[S](\neg y \rightarrow \langle S \cup W \rangle y)$$

$$\mathbf{S}_2 :=\downarrow x.\langle S \rangle \downarrow y.(\neg x \wedge @_x\langle S \rangle(\neg x \wedge \neg y))$$

$$\mathbf{LB} :=\downarrow x.\langle S \cup W \rangle \downarrow y.(\neg x \wedge \neg\langle S \cup W \rangle(\neg x \wedge \neg y \wedge \langle S \cup W \rangle x))$$

$$\mathbf{LB}_W :=\downarrow x.[S] \downarrow y.(((\langle S \cup W \rangle(\neg x \wedge \neg y \wedge (\langle W \rangle x \vee \langle S \rangle x))$$

Lemma 2. *For any tied model* $\mathbb{M} = \langle A, R^S, R^W, V \rangle$*, agent* $a \in A$ *and assignment function* $g : \mathsf{Var} \rightarrow A$*:*

1. $\mathbb{M}, g, a \Vdash \boldsymbol{STC}$ *iff* a *has the Strong Triadic Closure property;*
2. $\mathbb{M}, g, a \Vdash \boldsymbol{S}_2$ *iff* a *is strongly tied to at least two other unique agents;*
3. $\mathbb{M}, g, a \Vdash \boldsymbol{LB}$ *iff* a *is related to another agent by a local bridge;*
4. $\mathbb{M}, g, a \Vdash \boldsymbol{LB}_W$ *iff any local bridge* a *is related to another agent with is a weak tie.*

We can thus present the following corollary; that Granovetter's claim is a validity of **TL**.

Corollary 1. $(\boldsymbol{STC} \wedge \boldsymbol{S}_2) \rightarrow \boldsymbol{LB}_W$ *is a validity of* **TL**.

Proof. Follows by the original work by Granovetter [13].

2.3 Axiomatization

To account for strong and weak ties in a social network, we assume strong reflexivity and weak irreflexivity, as previously noted. These frame properties are defined by the following two axioms T^S and $IrrT^W$, respectively.

$$i \rightarrow \langle S \rangle i \qquad (T^S)$$

$$i \rightarrow \neg\langle W \rangle i \qquad (IrrT^W)$$

Symmetry of both relations is preserved in the following axiom B^{SW}.

[3] For proof of the claim see Granovetter's original paper [13].

$$i \to ([S]\langle S\rangle i \wedge [W]\langle W\rangle i) \qquad (B^{SW})$$

Non-overlapping can also be defined with the hybrid axiom *NonO*, stated directly below. Note that non-overlapping, like weak irreflexivity is modally undefinable in the standard Kripke semantics considering two-sorted Kripke frames.

$$i \to [W](\langle S\rangle i \to i) \qquad (NonO)$$

The final axiomatization of **TL** is the axiomatization of the standard normal hybrid logic $\mathbf{K}_{\mathcal{H}(@,\downarrow)}$ [1] together with our recently presented axioms; see Table 1.

Table 1. Axiomatization of **TL**, where $\Diamond \in \{\langle S\rangle, \langle W\rangle\}$ and $\Box \in \{[S], [W]\}$.

(CT)	All classical tautologies
$(K_\Box)$	$\vdash \Box(\phi \to \psi) \to \Box\phi \to \Box\psi$
$(K_@)$	$\vdash @_i(\phi \to \psi) \to @_i\phi \to @_i\psi$
$(Selfdual)$	$\vdash @_i\phi \leftrightarrow \neg @_i \neg\phi$
$(Ref_@)$	$\vdash @_i i$
$(Agree)$	$\vdash @_i @_j \phi \leftrightarrow @_j\phi$
$(Intro)$	$\vdash i \to (\phi \leftrightarrow @_i\phi)$
$(Back)$	$\vdash \Diamond @_i\phi \to @_i\phi$
(DA)	$\vdash @_i(\downarrow x.\phi \leftrightarrow \phi[x/i])$
(T^S)	$\vdash i \to \langle S\rangle i$
$(IrrT^W)$	$\vdash i \to \neg\langle W\rangle i$
(B^{SW})	$\vdash i \to ([S]\langle S\rangle i \wedge [W]\langle W\rangle i)$
$(NonO)$	$i \to [W](\langle S\rangle i \to i)$
(MP)	If $\vdash \phi$ and $\vdash \phi \to \psi$ then $\vdash \psi$
(Subst)	If $\vdash \phi$ then $\vdash \phi^\sigma$, for σ a substitution
(Gen$_@$)	If $\vdash \phi$ then $\vdash @_i\phi$
(Gen$_\Box$)	If $\vdash \phi$ then $\vdash \Box\phi$
(Name)	If $\vdash @_i\phi$ and i does not occur in ϕ, then $\vdash \phi$
(BG)	If $\vdash @_i\Diamond j \to @_j\phi, j \neq i$ and j does not occur in ϕ, then $\vdash @_i\Box\phi$

2.4 Soundness and Completeness

We will now prove that **TL** is sound and strongly complete with respect to the class of tied frames.

Theorem 1. ***TL*** *is sound and strongly complete with respect to the class of tied frames.*

Proof (Soundness). Let $\mathcal{F}$ be the class of tied frames. Since $\mathbf{K}_{\mathcal{H}(@,\downarrow)}$ is sound with respect to the class of all hybrid frames, we know that $\mathcal{F} \Vdash \mathbf{K}_{\mathcal{H}(@,\downarrow)}$. Thus it suffices to show the validity of the axioms $T^S, IrrT^W, B^{SW}$ and $NonO$. We leave the details of the proof to the reader.

(Completeness). Note again that **TL** is $\mathbf{K}_{\mathcal{H}(@,\downarrow)} + \{T^S, IrrT^W, B^{SW}, NonO\}$. The Sahlqvist-like theorem proved in [6] states that if Σ is a set of pure $\mathcal{H}(@,\downarrow)$-formulas, then $\mathbf{K}_{\mathcal{H}(@,\downarrow)} + \Sigma$ is strongly complete for the class of frames defined by Σ. It follows directly that if we can show that $\{T^S, IrrT^W, B^{SW}, NonO\}$ is a set of pure $\mathcal{H}(@,\downarrow)$-formulas, then **TL** is strongly complete with respect to the class of tied frames. The result follows straightforwardly from the fact that none of the axioms contain any propositional variables and that they can all be formulated in the language $\mathcal{H}(@,\downarrow)$.

Corollary 2. $\boldsymbol{TL} + \langle S\rangle i \rightarrow [S](\langle S\rangle i \vee \langle W\rangle i)$ *is sound and strongly complete with respect to the class of tied frames where all agents have the Strong Triadic Closure property.*

Proof of Corollary 2 follows from Lemma 1 and Theorem 1. Whereas this is perhaps not a surprising result, completeness of this class of frames is worth taking note of. Recall that according to Granovetter's theory, Strong Triadic Closure is a property we often observe across social networks. We now have the ability to reason and conduct a logical analysis directly within networks with this property; where the following corollary is a favorable example.

Corollary 3. $\boldsymbol{S_2} \rightarrow \boldsymbol{LB_W}$ *is a validity of* $\boldsymbol{TL} + \langle S\rangle i \rightarrow [S](\langle S\rangle i \vee \langle W\rangle i)$.

2.5 Homophily

We now explicitly implement the social selection aspect of homophily into our logical framework. By building upon research on network formation in [17,18], we will define a new class of tied models where relations are restricted according to homophily concepts. In these papers, relations between agents are formed based on the amount of propositional atoms they have in common, with respect to a certain threshold θ. We will follow the same approach, although whereas the original threshold models use a single threshold we need two linearly ordered thresholds for strong and weak ties.

We first introduce the notions *mismatch* and *distance*.

Definition 6 (MSMTCH and DIST). *Let* $\mathbb{M} = \langle A, R^S, R^W, V\rangle$ *be a tied model. We define the set of features distinguishing agents* $a, b \in A$ *in* $\mathbb{M}$ *as:*

$$MSMTCH^{\mathbb{M}}(a,b) := \mathsf{At} \setminus \{p \in \mathsf{At} : p \in V(a) \textit{ iff } p \in V(b)\}.$$

Further, we define the distance between a *and* b *in* $\mathbb{M}$ *to be:*

$$DIST^{\mathbb{M}}(a,b) := |MSMTCH^{\mathbb{M}}(a,b)|.$$

Intuitively, the mismatch of agents a and b in tied model $\mathbb{M}$ is the set of all features or properties that the agents do not share. This also includes properties that none of them have. The distance of a and b with respect to the same tied model is the cardinality of the mismatch, i.e. the number of properties a and b do not share. We read that agents with a small distance are more similar and have a higher degree of homophily than agents with a larger distance. We now define the class of tied threshold models.

Definition 7 (Tied Threshold Model). *Let $\theta_S, \theta_W \in \mathbb{N}$ be two thresholds such that $\theta_S < \theta_W \leq |\mathsf{At}|$. We define a tied threshold model $\mathbb{M}_\theta = \langle A, R^{\theta S}, R^{\theta W}, V\rangle$ where:*

- *A is a non-empty set of agents;*
- *$R^{\theta S}$ is a symmetric and reflexive binary relation on A such that $R^{\theta S} := \{(a,b) \in A \times A : DIST^{\mathbb{M}}(a,b) \leq \theta_S\}$;*
- *$R^{\theta W}$ is a symmetric, irreflexive and $R^{\theta S}$-non-overlapping binary relation on A such that $R^{\theta W} := \{(a,b) \in A \times A : \theta_S < DIST^{\mathbb{M}}(a,b) \leq \theta_W\}$;*
- *$V : A \to \mathcal{P}(\mathsf{At} \cup \mathsf{Nom})$ is a valuation function.*

We read $aR^{\theta S}b$ as agent a and b have enough in common to be connected by a strong tie. Similarly, we read $aR^{\theta W}b$ as a and b have enough in common to be connected by a weak tie, but not enough in common to be connected by a strong tie. To familiarize the reader with the newly defined tied threshold models, we make the following observations.

- $@_i\langle S\rangle j \to (@_i p \leftrightarrow @_j p)$ is valid on tied threshold frames where $\theta_S = 0$;
- $@_i\neg\langle S \cup W\rangle j \to (@_i p \to @_j \neg p)$ is valid on tied threshold frames where $\theta_W = |\mathsf{At}| - 1$.

The first formula expresses that when $\theta_S = 0$, if two agents i and j are strongly tied, then the property p holds at i if and only if p holds at j. When $\theta_W = |At| - 1$, two agents that are neither tied strongly nor weakly do not share any properties. The latter formula asserts that if this is the case for two agents i and j, if the property p holds at i, then it does not hold at j.

3 Adding Knowledge and Dynamics

In this section we add a knowledge modality and dynamic operators to **TL** and in consequence extend it to tied epistemic logic (**TEL**). **TEL** is inspired by other epistemic logics for social networks e.g. [15,16]. Yet, **TEL** differs from these on some notable accounts. Firstly, our valuation function where the range includes nominals depends on epistemic states. This is as we do not want the underlying assumption that every agent in the network knows the name of all other agents. Secondly, the language of **TEL** includes dynamic local adding modalities. These modalities are influenced by local deleting modalities known from sabotage modal logic [2,10], but are modified as related to relation addition, not deletion as is custom. On that note we now introduce syntax and semantics of **TEL**.

3.1 Language and Semantics

Definition 8 (Syntax of TEL). *Let* At *be a set of propositional atoms and* Nom *be a set of nominals. Further, let* Var *be a set of agent variables. Let* At, Nom *and* Var *be countable and pairwise disjoint. We define the well-formed formulas of the language $\mathcal{L}_{TEL}$ to be generated by the following grammar:*

$$\phi ::= p \mid s \mid \neg\phi \mid (\phi \wedge \phi) \mid \langle S\rangle\phi \mid \langle W\rangle\phi \mid [A]\phi \mid K\phi \mid @_s\phi \mid \downarrow x.\phi \mid [\mathbb{M}S]_L\phi \mid [\mathbb{M}W]_L\phi$$

where $p \in$ At, $s \in$ Nom $\cup$ Var and $x \in$ Var. We define propositional connectives like $\vee, \rightarrow$ and the formulas $\top, \bot$ as usual. Further, we again define the duals as standard $[S] := \neg\langle S\rangle\neg$ and $[W] := \neg\langle W\rangle\neg$.

Before we define tied epistemic frames and models, we note that $MSMTCH^{\mathbb{M}}$ and $DIST^{\mathbb{M}}$ are defined as before, but now for each possible world $w \in W$.

Definition 9 (Tied Epistemic Threshold Model and Frame). *Let θ_S, $\theta_W \in \mathbb{N}$ be two thresholds such that $\theta_S < \theta_W \leq |\mathsf{At}|$. A tied epistemic threshold model is a tuple $\mathbb{M} = \langle W, A, \sim, R^{\theta S}, R^{\theta W}, V\rangle$ where:*

- *W is a set of epistemic alternatives,*
- *A is a non-empty set of agents,*
- *$\sim$ is a family of equivalence relations $\sim_a$ on W for every $a \in A$,*
- *$R^{\theta S}$ is a family of symmetric and reflexive relations $R^{\theta S}_w$ on A for each $w \in W$ such that $R^{\theta S}_w := \{(a,b) \in A \times A : \forall u \sim_a w, DIST^{\mathbb{M}}_u(a,b) \leq \theta_S\}$,*
- *$R^{\theta W}$ is a family of $R^{\theta S}$-non-overlapping, irreflexive and symmetric relations $R^{\theta W}_w$ on A for each $w \in W$ such that $R^{\theta W}_w := \{(a,b) \in A \times A : \forall u \sim_a w, \theta_S < DIST^{\mathbb{M}}_u(a,b) \leq \theta_W\}$,*
- *$V : W \times A \rightarrow \mathcal{P}(\mathsf{At} \cup \mathsf{Nom})$ is a valuation function, assigning each agent to a unique name and a set of properties in an epistemic state. I.e, for each $i \in$ Nom and for all $w \in W$ and all $a, b \in A$: if $i \in V(w,a)$ and $i \in V(w,b)$, then $a = b$. Additionally, all names correspond to an agent and an epistemic state. That is $\forall i \in$ Nom: $\exists a \in A$ and $\exists w \in W$ such that $i \in V(w,a)$.*

We define a frame $\mathbb{F} = \langle W, A, \sim, R^{\theta S}, R^{\theta W}\rangle$ in the usual way.

Again, let g : Var $\rightarrow A$ be an assignment function assigning agents to variables. Furthermore, define the *x-variant* of g to be $g^x_a(x) = a$ and $g^x_a(y) = g(y)$ for all $y \neq x$. We now define the semantics of **TEL**.

Definition 10 (Semantics of TEL). *Let $\mathbb{M}$ be a model, a an agent in A, $w \in W$ an epistemic state and g : Var $\rightarrow A$ an assignment function. We inductively define* *the truth conditions as follows:*

$\mathbb{M}, g, w, a \Vdash p$ *iff* $p \in V(w, a)$ *for* $p \in \mathsf{At}$
$\mathbb{M}, g, w, a \Vdash i$ *iff* $i \in V(w, a)$ *for* $i \in \mathsf{Nom}$
$\mathbb{M}, g, w, a \Vdash x$ *iff* $a = g(x)$ *for* $x \in \mathsf{Var}$
$\mathbb{M}, g, w, a \Vdash \neg\phi$ *iff* $\mathbb{M}, g, w, a \not\Vdash \phi$
$\mathbb{M}, g, w, a \Vdash \phi \wedge \psi$ *iff* $\mathbb{M}, g, w, a \Vdash \phi$ *and* $\mathbb{M}, g, w, a \Vdash \psi$
$\mathbb{M}, g, w, a \Vdash \langle S\rangle\phi$ *iff* $\exists b \in A$ *such that* $aR_w^{\theta S}b$ *and* $\mathbb{M}, g, w, b \Vdash \phi$
$\mathbb{M}, g, w, a \Vdash \langle W\rangle\phi$ *iff* $\exists b \in A$ *such that* $aR_w^{\theta W}b$ *and* $\mathbb{M}, g, w, b \Vdash \phi$
$\mathbb{M}, g, w, a \Vdash [A]\phi$ *iff* $\forall b \in A : \mathbb{M}, g, w, b \Vdash \phi$
$\mathbb{M}, g, w, a \Vdash K\phi$ *iff* $\forall v \in W$ *such that* $w \sim_a v : \mathbb{M}, g, v, a \Vdash \phi$
$\mathbb{M}, g, w, a \Vdash @_s\phi$ *iff* $\forall b \in A : \mathbb{M}, g, w, b \Vdash s \rightarrow \phi$ *for* $s \in \mathsf{Nom} \cup \mathsf{Var}$
$\mathbb{M}, g, w, a \Vdash \downarrow x.\phi$ *iff* $\mathbb{M}, g_a^x, w, a \Vdash \phi$
$\mathbb{M}, g, w, a \Vdash [\text{⩕}S]_L\phi$ *iff* $\exists b \in A$ *such that* $a\neg R_w^{\theta S}b$, $a\neg R_w^{\theta W}b$ *and*
$\langle W, A, \sim, R_w^{\theta S} \cup \{(a,b),(b,a)\}, R_w^{\theta W}, V\rangle, g, w, a \Vdash \phi$
$\mathbb{M}, g, w, a \Vdash [\text{⩕}W]_L\phi$ *iff* $\exists b \in A$ *such that* $a\neg R_w^{\theta S}b$, $a\neg R_w^{\theta W}b$ *and*
$\langle W, A, \sim, R_w^{\theta S}, R_w^{\theta W} \cup \{(a,b),(b,a)\}, V\rangle, g, w, a \Vdash \phi$

We observe that there are four new operators in the language. The intuitive reading of them are as follows. We read $[A]\phi$ to hold at the current agent if and only if ϕ is universally true at all agents in the network. $K\phi$ is intuitively read as the current agent knows that ϕ. The two dynamic modalities $[\text{⩕}S]_L$ and $[\text{⩕}W]_L$ are model changing operators, whose semantics, similar as done in Dynamic Epistemic Logic [5,9], is evaluated by taking into account an updated model in which only the relations $R^{\theta S}$ or $R^{\theta W}$ are changed. Intuitively, modality $[\text{⩕}S]_L\phi$ holds at agent a if and only if after adding a strong tie that a previously did not have, ϕ is true at a. $[\text{⩕}W]_L\phi$ is read similarly, although by replacing a strong tie with a weak tie.

3.2 Possible Axioms

We might want to add some axioms to narrow down our class of tied epistemic frames. In this section we consider candidates corresponding to some properties we believe put natural constraints on agents in an epistemic context.

The first property we propose is that an agent knows it when it is strongly tied to another agent.

$$\downarrow x.[S] \downarrow y.@_x K\langle S\rangle y \tag{1}$$

Perhaps a bit less likely is the property that an agent knows it when it is weakly tied to another agent. This property depends on what we assign to the term 'acquaintance' and the meaning we expect of the knowledge modality.

$$\downarrow x.[W] \downarrow y.@_x K\langle W\rangle y \tag{2}$$

Another reasonable attribute to assume is that an agent knows it when they have a property p or a name i, defined by the following two axioms.

$$p \to Kp \tag{3}$$

$$i \to Ki \tag{4}$$

It is also likely to assume that if an agent is strongly tied to another agent whose name is i, they know the other agent's name.

$$\downarrow x.[S] \downarrow y.(i \to @_x K @_y i) \tag{5}$$

A further axiom up for discussion is the one defining the property that an agent knows if its strong tie has the property p. This is a strong assumption that might only be relevant in certain contexts.

$$\downarrow x.[S] \downarrow y.(p \to @_x K @_y p) \tag{6}$$

Similarly, but perhaps a weaker assumption is that agents know the strong ties of their strong ties. This seems likely when defining strong ties as agents' closest friends.

$$\downarrow x.[S] \downarrow y.[S] \downarrow z.@_x K @_y \langle S \rangle z \tag{7}$$

The last axiom we will consider defines the likely property of agents knowing when they are acquainted by a new agent by a weak tie. Note that this would be implied by Axiom (2) and that the strong tie version of this axiom is implied by Axiom (1).

$$\downarrow x.[\not\!W]_L[W] \downarrow y.@_x K \langle W \rangle y \tag{8}$$

3.3 Validities

We look at some validities of **TEL** depending on axioms we choose to embrace to get a better understanding of our logic in relation to echo chambers and related social phenomena. Firstly, if we would adopt Axiom (6) for all $p \in \mathsf{At}$, the following would be a validity.

$$\mathbf{STC} \to K\mathbf{STC}$$

This validity is a result of the homophily-motivated definitions of $R^{\theta S}$ and $R^{\theta W}$ and represents that the current agent knows whether they are strongly triadic closed. Strong Triadic Closure is closely related to echo chamber formation. If the agent knows whether they are strongly triadic closed, then principally they would know whether they could be in an echo chamber-like situation.

$$\downarrow x.((\langle S \rangle \downarrow y.@_x \langle S \rangle \downarrow z.@_x K @_z \neg \langle S \cup W \rangle y) \to K\neg\mathbf{STC})$$

This is a formula valid on tied threshold models where we adopt Axiom (1) and says that "*If I am tied to any two successors y and z by strong ties and I know that y and z do not know each other, then I know I am not strongly* *triadic closed.*" This also shows the relationship between knowledge and echo chambers. If an agent knows that they do not have the Strong Triadic Closure property, they can derive that it is less likely that they are participating in an echo chamber.

$$\downarrow x.(K\neg\langle S \cup W\rangle \downarrow y.(i \wedge [A](\langle S \cup W\rangle y \rightarrow \neg\langle S \cup W\rangle x))) \rightarrow K[\mathbb{W}]_L(\langle W\rangle i \rightarrow \mathbf{LB})$$

This validity states that "*If I know that there is another agent y in the network in which we do not have any friends in common, then I know that if we become acquainted by a weak tie, then I am related to another agent by a weak local bridge.*" Imagine agents find themselves in a place where they suspect an echo chamber has been or were about to be formed. They might have an incentive to get acquainted by a local bridge to receive some new information and hear opposing opinions. Note that the above formula is also valid in the case of strong instead of a weak tie.

3.4 Example

In the concluding section we present an example of a tied epistemic threshold model and discuss what formulas might hold at specific agents depending on what axioms we adopt to restrict the model.

Consider the tied epistemic threshold model $\mathbb{M}_{\theta SW} = \langle W, A, \sim, R^{\theta S}, R^{\theta W}, V\rangle$ in Fig. 1. In particular, we observe that $A = \{a, b, c, d, e, f\}$ and $W = \{w, v\}$. For simplicity the reflexive arrows are omitted for $\sim_x$ for all $x \in A$ as well as for $R^{\theta S}_w$ and $R^{\theta S}_v$.

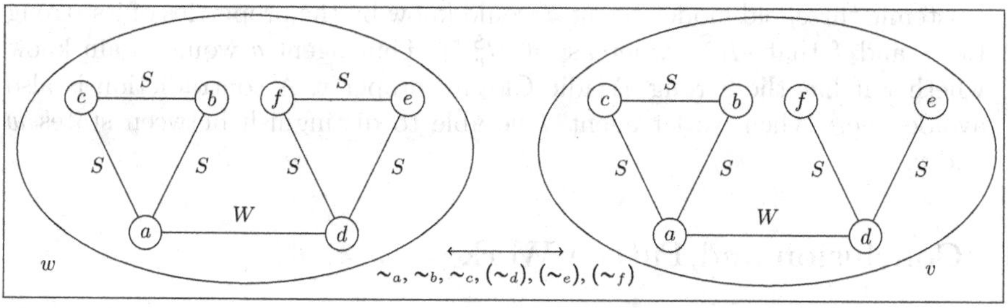

Fig. 1. A tied epistemic threshold model $\mathbb{M}_{\theta SW}$.

We first regard the model $\mathbb{M}_{\theta SW}$ where $(w \sim_x v) \in \sim_x$ for all $x \in A$. Let $a, b, c, d, e, f \in$ **Nom** such that the corresponding 'name' is true for each agent in A in each epistemic state in W. For instance $\mathbb{M}_{\theta SW}, g, w, c \Vdash c$ and $\mathbb{M}_{\theta SW}, g, v, e \Vdash e$ etc. We make, among many, the following observations.

– $\mathbb{M}_{\theta SW}, g, w, a \Vdash \mathbf{KSTC}_L$
  Agent a knows in w that it has the Strong Triadic Closure property.

- $\mathbb{M}_{\theta SW}, g, w, d \Vdash \mathbf{STC}_L \wedge \neg K\mathbf{STC}_L$
  Agent d has the Strong Triadic Closure property in state w, but does not know it.
- $\mathbb{M}_{\theta SW}, g, v, a \Vdash \mathbf{LB}$
  Agent a is related by a local bridge in state v. We see that this also holds in both w and v for agents a and d.
- $\mathbb{M}_{\theta SW}, g, w, f \Vdash \langle S\rangle e \wedge \neg K\langle S\rangle e$
  Agent f is strongly tied to agent e in state w, but does not know it.
- $\mathbb{M}_{\theta SW}, g, v, e \Vdash [\mathbb{A}S]_L(\langle S\rangle f \to @_d K\mathbf{STC}_L) \wedge [\mathbb{A}W]_L(\langle W\rangle f \to @_d K\mathbf{STC}_L)$
  If agent e in state v is strongly or weakly tied to f after adding a strong or weak tie respectively, then agent d will know that it then has the Strong Triadic Closure property.

We now notice the following regarding the axioms in the previous section.

- Axiom (1) does not hold in $\mathbb{M}_{\theta SW}$ in both states at e and f in particular. Adjusting $\sim$ such that $\neg(w \sim_e v)$ and $\neg(w \sim_f v)$ is one way to let Axiom (1) hold at all agents in both epistemic states. The reasoning is as follows. Axiom (1) lets agents e and f know that they are related in state w, and not in state v; $\mathbb{M}_{\theta SW}, g, w, e \Vdash K\langle S\rangle f$ and $\mathbb{M}_{\theta SW}, g, v, e \Vdash K\neg\langle S\rangle f$. Thus if both agents can distinguish between states w and v, the axiom holds.
- Neither Axiom (6) nor Axiom (7) is forced at agent d in either epistemic state. Letting $\neg(w \sim_d v)$ would make either axiom true. Axiom (7) restricts the model such that every agent knows the strong ties of their strong ties. If this is the case then $\mathbb{M}_{\theta SW}, g, w, d \Vdash K\mathbf{STC}$ while $\mathbb{M}_{\theta SW}, g, v, d \Vdash K\neg\mathbf{STC}$. Letting agent d distinguish between w and v would solve this problem. Axiom (6) make agents know if their strong ties has a property p. As $\mathbb{M}_{\theta SW}$ is a tied epistemic threshold model, agent d would know by the properties of its strong ties e and f that $eR^S_w f$ whereas $\neg(eR^S_v f)$. Thus agent d would again know whether it has the Strong Triadic Closure property. A contradiction is also avoided here when we let agent d be able to distinguish between states w and v.

4 Conclusion and Future Work

In this paper, we combined two known concepts from social network theory and proposed a logical analysis of echo chambers. We developed tied logic (**TL**) as well as its epistemic variant tied epistemic logic (**TEL**). We have shown that **TL** is axiomatizable, sound and strongly complete. With homophily and social selection as a motivation, we defined tied threshold models and justified their binary relations by logical theories of network formation. We also discussed possible restrictions to tied epistemic frames and presented essential validities of **TEL** before concluding with an example.

A natural place to continue this work is to explore further technical results of our logics, and in particular assess the potential completeness proof of fragments of **TEL**. It remains to see if the results are directly connected to the open problem of axiomatization of sabotage modal logic [2].

Important to remark is that our presentation of echo chambers as strongly tied clusters is not the only way in which one can formalize the concept. Other formalizations (e.g. [12]) include a filtering mechanism, representing the habit of filtering out extraneous information within echo chambers. Such a mechanism could be engaging to implement into our framework and can possibly be done by taking into account the 'selective learning principle' that is formally explored in [3]. Similarly, we also note that the homophily inspired notions we have included are simply one of many ways of approaching the topic. In future work, one option is to formalize the social influence factor of homophily.

Lastly, this paper motivates a further investigation of change in these social networks. This could be done by exploring other validities, but also by the inclusion of additional dynamic operators. Likely candidates are tie-changing or -deleting modalities. Another possible approach is to extend our multi-agent dynamic framework with features of communication such as public announcements and group knowledge such as common knowledge [5,9].

References

1. Areces, C., ten Cate, B.: Hybrid logics. In: van Benthem, J., Blackburn, P., Wolter, F. (eds.) Handbook of Modal Logic. Elsevier, Amsterdam (2006)
2. Aucher, G., van Benthem, J., Grossi, D.: Modal logics of sabotage revisited. J. Log. Comput. **28**(2), 269–303 (2017)
3. Baltag, A., Boddy, R., Smets, S.: Group knowledge in interrogative epistemology. In: van Ditmarsch, H., Sandu, G. (eds.) Jaakko Hintikka on Knowledge and Game-Theoretical Semantics. Outstanding Contributions to Logic, vol. 12, pp. 131–164. Springer, Cham (2018). https://doi.org/10.1007/978-3-319-62864-6_5
4. Baltag, A., Christoff, Z., Rendsvig, R.K., Smets, S.: Dynamic epistemic logics of diffusion and prediction in social networks. Stud. Logica **107**(3), 1–43 (2018)
5. Baltag, A., Moss, L.S., Solecki, S.: The logic of public announcements, common knowledge, and private suspicions. In: Arló-Costa, H., Hendricks, V., van Benthem, J. (eds.) Readings in Formal Epistemology. Springer Graduate Texts in Philosophy, vol. 1, pp. 773–812. Springer, Cham (2016). https://doi.org/10.1007/978-3-319-20451-2_38
6. Blackburn, P., ten Cate, B.: Pure extensions, proof rules, and hybrid axiomatics. Stud. Logica **84**(2), 277–322 (2006)
7. Blackburn, P., de Rijke, M., Venema, Y.: Modal Logic. Cambridge University Press, Cambridge (2001)
8. Christoff, Z., Hansen, J.U.: A logic for diffusion in social networks. J. Appl. Log. **13**(1), 48–77 (2015)
9. van Ditmarsch, H., van Der Hoek, W., Kooi, B.: Dynamic Epistemic Logic, vol. 337. Springer, Dordrecht (2007). https://doi.org/10.1007/978-1-4020-5839-4
10. Li, D.: Losing Connection: the Modal Logic of Definable Link Deletion, ILLC Technical Notes Series (X-2019-01) (2019)
11. Easley, D., Kleinberg, J.: Networks, Crowds and Markets. Cambridge University Press, Cambridge (2010)

12. Geschke, D., Lorenz, J., Holtz, P.: The triple-filter bubble: using agent-based modelling to test a meta-theoretical framework for the emergence of filter bubbles and echo chambers. Br. J. Soc. Psychol. **58**(1), 129–149 (2019)
13. Granovetter, M.S.: The Strength of Weak Ties. Social Networks, pp. 347–367. Academic Press, New York (1977)
14. Hendricks, V.F., Hansen, P.G.: Infostorms. Springer, Cham (2016). https://doi.org/10.1007/978-3-319-32765-5
15. Seligman, J., Liu, F., Girard, P.: Facebook and the epistemic logic of friendship. In: Proceedings of the TARK conference on Theoretical Aspects of Rationality and Knowledge, pp. 229–238 (2013)
16. Seligman, J., Liu, F., Girard, P.: Logic in the community. In: Banerjee, M., Seth, A. (eds.) ICLA 2011. LNCS (LNAI), vol. 6521, pp. 178–188. Springer, Heidelberg (2011). https://doi.org/10.1007/978-3-642-18026-2_15
17. Smets, S., Velázquez-Quesada, F.R.: The creation and change of social networks: a logical study based on group size. In: Madeira, A., Benevides, M. (eds.) DALI 2017. LNCS, vol. 10669, pp. 171–184. Springer, Cham (2018). https://doi.org/10.1007/978-3-319-73579-5_11
18. Smets, S., Velázquez-Quesada, F.R.: How to make friends: a logical approach to social group creation. In: Baltag, A., Seligman, J., Yamada, T. (eds.) LORI 2017. LNCS, vol. 10455, pp. 377–390. Springer, Heidelberg (2017). https://doi.org/10.1007/978-3-662-55665-8_26
19. Xiong, Z., Ågotnes, T.: On the logic of balance in social networks. J. Log. Lang. Inf. (2019, to appear)
20. Xiong, Z.: On the logic of multicast messaging and balance in social networks. Doctoral dissertation, University of Bergen (2017)

Public Group Announcements and Trust in Doxastic Logic

Elise Perrotin[1](✉), Rustam Galimullin[2], Quentin Canu[3], and Natasha Alechina[2]

[1] IRIT, Toulouse, France
elise.perrotin@irit.fr
[2] University of Nottingham, Nottingham, UK
{rustam.galimullin, natasha.alechina}@nottingham.ac.uk
[3] ENS Paris-Saclay, Cachan, France
quentin.canu@ens-paris-saclay.fr

Abstract. We present a doxastic logic for multi-agent systems with public group announcements. Beliefs are represented using belief bases and a dynamic of trust is introduced in order to handle belief change under contradictory announcements. We provide a complete axiomatization for this logic and illustrate its expressive power with a simple example.

Keywords: Syntactic beliefs · Group announcements · Belief change · Trust

1 Introduction

Receiving contradictory pieces of information from different sources is a common occurrence of daily life. However, deciding what to believe and who to trust as a result of those announcements is a task that is usually not that straightforward. Neither is representing these interactions between announcements, trust and beliefs.

We here propose a simple logic which can express these interactions following group announcements. All of the individual elements we have mentioned have been studied rather extensively, and often with various different approaches. We will now give an overview of this literature and what we are taking from it. As our focus is to allow for interaction of these elements in a single, manageable logic, we will often be choosing the less expressive, but easier to work with, options.

The first thing to consider is representation of beliefs, which may be false, and of their evolution. Two standard approaches exist for this: Dynamic Epistemic Logic (DEL, [10]), in which beliefs are represented using Kripke models with possible worlds, and the AGM approach [3], which uses sets of formulas, or *belief bases*, for that purpose, and from which follows Dynamic Doxastic Logic (DDL),

P. Blackburn et al. (Eds.): LORI 2019, LNCS 11813, pp. 199–213, 2019.
https://doi.org/10.1007/978-3-662-60292-8_15

as introduced in [21]. Public announcements and the resulting reorganization of beliefs, with or without trust, are built into DEL. However, these announcements, while they may be of false beliefs or even lies [9], are given one formula at a time and therefore do not allow for contradiction within that formula. While group announcement logics based on DEL exist [1], they handle announcements of what agents know rather than believe, and therefore these announcements are also necessarily consistent. When dealing with beliefs, plausibility models [4] can be used and the worlds reorganized after announcements, but those reorganizations depend on the order in which announcements are received, and it is not clear how they should function in the case of several simultaneous announcements, especially contradictory.

We will here work with belief bases, as we wish to focus first on what agents should believe after an announcement, rather than delving into all of the expressive ramifications allowed by DEL. The AGM approach is ideal for this, though, as pointed out in [5], it usually does not give much thought to where new information comes from, focusing instead on how to integrate it into existing beliefs. The dynamics expressed by DDL could, however, be interpreted as the results of announcements. AGM and DDL usually deal with only one agent, but multi-agent systems using belief bases, such as in [18], have also been proposed.

Handling contradicting statements in AGM is one of the topics of paraconsistent logic [20], which deals with identifying and isolating contradictions and extracting useful information from a belief base or an announcement. Closely related to this is belief merging [15], the aim of which is to merge several belief bases into one while preserving consistency. While weighted belief merging, as presented in [7], offers a mechanism allowing for different levels of reliability of sources, which can be interpreted as trust, it does not handle the associated evolution of this trust that we wish to represent. We will here avoid many difficulties brought up in paraconsistent logic and belief merging by using simplified belief bases, in which precise sources of contradiction are clearly identifiable.

Representing trust is the subject of yet another rather extensive body of work. The word 'trust' can have many meanings, and many corresponding models (see, e.g., [8,12–14,16,17]). We here focus on trust as belief that what the other says is true, that is, trust in the reliability of another agent. For the sake of simplicity, we restrict ourselves to this definition of trust only, and assume in particular that agents only announce what they actually believe. The closest account of trust is the one given in [17], in which the focus is on whether information given by a source should be believed or not by an agent following the trust of the agent in that source, and in which contradicting announcements lead to loss of trust. Another interesting study of the evolution of trust is given in [14], in which trust is seen to be gained and lost as a result of so-called *trust-positive* and *trust-negative* experiences.

We present here a very basic notion of trust, which is binary: an agent either fully trusts or fully distrusts another agent, with no variation depending on topic (as opposed to [17]) and no gradual trust (as opposed to [14]). We also work with a memory-less trust, once again to present the most basic version of our framework. As such, trust-negative experiences are quite easy to identify (viz. contradictions in announcements), but trust-positive experiences are more tricky to define. For this reason we will here present a framework in which trust can only be lost. While quite basic, we argue that the notion of trust that we use can still be relevant in practice.

In the rest of the paper, we first present our logic, which we call *Syntactic Dynamic Doxastic Logic with Trust* (SDDLT) and the notion of trust that we use, and show that we can express coalition announcements. We then give an axiomatization and show that it is complete. We finish by giving a simple application example in order to justify our choice of trust dynamics and illustrate what can be expressed within our framework, and further discussing some of the choices we have made.

2 Syntax

Let Agt be a finite set of agents, and $\mathcal{P}$ a countable set of propositional variables. Consider $p \in \mathcal{P}$, $a, b \in Agt$ and $G \subseteq Agt$. The language $\mathcal{L}_{\mathsf{SDDLT}}$ of SDDLT is described by the following grammar:

$$\mathcal{L}_B \ni \varepsilon ::= \top \mid p \mid \neg p \mid B_a\varepsilon$$

$$\mathcal{L}_{\mathsf{SDDLT}} \ni \varphi ::= T_{a,b} \mid \varepsilon \mid \neg\varphi \mid \varphi \wedge \varphi \mid [A]\varphi$$

$B_a\varepsilon$ reads "agent a believes ε", $T_{a,b}$ reads "agent a trusts agent b", and $[A]\varphi$ reads "after announcement A (defined below), φ holds".

We also introduce some useful notations for the rest of the paper. The letter l will be used to denote literals (*i.e.* a variable or its negation). Given $a_1, \ldots, a_n \in Agt$ and $p \in \mathcal{P}$, if $\varepsilon_1 = B_{a_1} \ldots B_{a_n} p$ and $\varepsilon_2 = B_{a_1} \ldots B_{a_n} \neg p$, then we denote $\varepsilon_1 = \overline{\varepsilon_2}$ and $\varepsilon_2 = \overline{\varepsilon_1}$. Moreover, we denote the subformula relation by $\preceq$.

Announcements in our setting are *group announcements*, that is, they are public announcements consisting of statements given simultaneously by a group of agents. Given a group of agents $G \subseteq Agt$, an announcement by group G is a collection of pairs (a, A_a) where a is in G and A_a is a subset of $\mathcal{L}_B$ consisting of the formulas announced by agent a. Only one such set of formulas is allowed for each agent of G, and conversely, there is a set of announced formulas for each agent of G, though this set may be empty. We identify an announcement A with the corresponding function, that is, if A is an announcement by a group G and a is in G, we call $A(a)$ the set of formulas such that $(a, A(a))$ is in A.

We add a requirement on announcements: if A is an announcement by a group G, and if a is an agent of G, we require that for any formula ε_a of $A(a)$, no agent b of G simultaneously announces $B_a\overline{\varepsilon_a}$. That is, for all b in G, $B_a\overline{\varepsilon_a} \notin A(b)$. These kinds of announcements could be dealt with so that trust in the announcer is lost, but would lead to a much more complex axiomatization. For the sake of clarity down the line it is simpler to consider that these situations do not happen. This and other choices concerning announcements and trust dynamics will be further discussed in Sect. 7.1.

For any group G, we call Ann_G the set of all possible announcements by agents of G, that is, the subset of $(2^{\mathcal{L}_B})^G$ following the above requirement. We call Ann the set of all possible announcements.

3 Semantics

3.1 Belief States and Trust Models

A belief state is a tuple $s = (\{BB^s_a\}_{a \in Agt}, \{T^s_a\}_{a \in Agt})$, where BB^s_a and T^s_a respectively denote the belief base and trust set of agent a at s. Belief bases are subsets of $\mathcal{L}_B$ and trust sets are subsets of Agt.

We require for all agents to trust at least themselves, and for all belief bases to contain at least $\top$. Belief bases must also be *consistent*, that is, they should not contain both ε and $\overline{\varepsilon}$ for any formula ε of $\mathcal{L}_B$. Furthermore, let $\dashrightarrow^s_a$ be defined in the following manner:

$$\varepsilon_1 \dashrightarrow^s_a \varepsilon_2 \text{ iff } \exists b \in T^s_a, \varepsilon_1 = B_b \varepsilon_2.$$

We denote by $\rightarrow^s_a$ the reflexive and transitive closure of $\dashrightarrow^s_a$. That is, if ε_1 and ε_2 are in $\mathcal{L}_B$, then $\varepsilon_1 \rightarrow^s_a \varepsilon_2$ if and only if there exist agents $a_1, \ldots, a_k$ in T^s_a (for some $k \geq 0$) such that $\varepsilon_1 = B_{a_1} \ldots B_{a_k} \varepsilon_2$. We require BB^s_a to be closed under $\rightarrow^s_a$ for any agent a at any belief state s: if $\varepsilon_1 = B_{a_1} \ldots B_{a_k} \varepsilon_2$, a believes ε_1, and a trusts a_1, then by that trust a should believe $B_{a_2} \ldots B_{a_k} \varepsilon_2$, but then if a trusts a_2, a should also believe $B_{a_3} \ldots B_{a_k} \varepsilon_2$, and so on. More generally, we denote by $Cl^s_a(B)$ the closure of a set B under $\rightarrow^s_a$.

A *trust model* is a pair (S, V) where S is the set of all belief states and $V \subseteq \mathcal{P}$ is a valuation representing the actual state of the world.

We will work with pointed models (M, s) where $s \in S$.

3.2 Contradictions

When announcements are made, integrating the announced formulas to the belief bases of agents as is may render those belief bases inconsistent. Our restricted language, however, allows us to identify contradictions within announcements as well as between announcements and agents' beliefs rather easily.

If a is an agent, we say that two formulas ε_1 and ε_2 are *contradictory according to a at s* if $Cl^s_a(\{\varepsilon_1, \varepsilon_2\})$ is inconsistent. We say that a formula ε *contradicts a's beliefs at s* if there is a formula ε' in the belief base of a at s such that ε and ε' are contradictory according to a at s. Moreover, ε is *supported by a's beliefs at s* if $\overline{\varepsilon}$ contradicts those beliefs, and ε is *neutral w.r.t. a's beliefs at s* if it neither contradicts nor is supported by them.

We introduce the following notation: given $\varepsilon \in \mathcal{L}_B$, $\min^s_a(\varepsilon)$ is the shortest suffix ε' of ε such that $\varepsilon \rightarrow^s_a \varepsilon'$, that is, the shortest formula that a can deduce from ε at s. For example, if a trusts b and not c at s, we have that $\min^s_a(B_b B_c p) = B_c p$.

Lemma 1. *Let ε and ε' be two formulas of $\mathcal{L}_B$, a an agent, and s a belief state.*

1. *ε and ε' are contradictory according to a at s iff $\min^s_a(\varepsilon) = \overline{\min^s_a(\varepsilon')}$.*
2. *ε contradicts a's beliefs at s iff $\overline{\min^s_a(\varepsilon)} \in BB^s_a$, and ε is supported by a's beliefs at s iff $\min^s_a(\varepsilon) \in BB^s_a$.*

Proof. For the first statement, the interesting proof is that of the left-to-right direction. Suppose that $Cl_a^s(\{\varepsilon, \varepsilon'\})$ is inconsistent. This means that there exists a formula ε_0 such that $\varepsilon_0 \in Cl_a^s(\{\varepsilon, \varepsilon'\})$ and $\overline{\varepsilon_0} \in Cl_a^s(\{\varepsilon, \varepsilon'\})$. By definition of Cl_a^s, and because ε_0 and $\overline{\varepsilon_0}$ have different literals, we have either $\varepsilon \rightarrow_a^s \varepsilon_0$ and $\varepsilon' \rightarrow_a^s \overline{\varepsilon_0}$, or $\varepsilon' \rightarrow_a^s \varepsilon_0$ and $\varepsilon \rightarrow_a^s \overline{\varepsilon_0}$. The second case is reducible to the first by replacing ε_0 by $\overline{\varepsilon_0}$, so consider that $\varepsilon \rightarrow_a^s \varepsilon_0$ and $\varepsilon' \rightarrow_a^s \overline{\varepsilon_0}$. By definition of $\min_a^s(\varepsilon')$, the latter implies that $\overline{\varepsilon_0} \rightarrow_a^s \min_a^s(\varepsilon')$, that is, $\varepsilon_0 \rightarrow_a^s \overline{\min_a^s(\varepsilon')}$. By transitivity of $\rightarrow_a^s$, we get that $\varepsilon \rightarrow_a^s \overline{\min_a^s(\varepsilon')}$, and once again by minimality of $\min_a^s(\varepsilon')$, we have that $\min_a^s(\varepsilon) = \overline{\min_a^s(\varepsilon')}$.

For the second statement, suppose that ε contradicts a's beliefs at s. Then there is a formula ε' in BB_a^s such that ε and ε' are contradictory according to a at s. By the first statement, this means that $\min_a^s(\varepsilon)' = \overline{\min_a^s(\varepsilon)}$, and by closure of BB_a^s under $\rightarrow_a^s$, $\min_a^s(\varepsilon')$ is in BB_a^s.

3.3 Update of Trust

Given an announcement A made by a group G at a state s, we wish to define the updated state $s \cdot [A]G$. For this we must define $BB_a^{s \cdot [A]G}$ and $T_a^{s \cdot [A]G}$ for any a. We begin with updates of trust.

When an announcement is made, agents first update their trust in other agents. In our framework, trust can only be lost. An agent a will stop trusting other agents when contradictory information is given in the announcement.

Let s be a state, a an agent, A an announcement by a group G, b and c two agents of G, and let ε_b and ε_c be two formulas such that $\varepsilon_b \in A(b)$ and $\varepsilon_c \in A(c)$. What a is learning is that $B_b\varepsilon_b$ and $B_c\varepsilon_c$, and a problem occurs when these two formulas are contradictory according to a at this state.

In order to choose which of the contradicting agents is no longer to be trusted, a will look at the statements as well as their own beliefs. If $B_b\varepsilon_b$ contradicts a's beliefs at s (which is equivalent to $B_c\varepsilon_c$ being supported by a's beliefs at s), then b is no longer trusted. Otherwise, if none of the statements are supported by a's beliefs, then a has no means of discrimination between the statements and both b and c are no longer trusted.

Finally, the new trust set is defined by:

$$T_a^{s \cdot [A]G} = T_a^s \setminus \{b \in G \mid \exists c \in G, \varepsilon_b \in A(b), \varepsilon_c \in A(c),$$
$$\min_a^s(B_b\varepsilon_b) = \overline{\min_a^s(B_c\varepsilon_c)} \text{ and } \min_a^s(B_b\varepsilon_b) \notin BB_a^w\}$$

3.4 Update of Belief Bases

Once the trust sets are updated, we can update the belief bases. As agents believe in the sincerity of all announcements, all formulas $B_b\varepsilon_b$ where $\varepsilon_b \in A(b)$ will be added to all belief bases. The update of trust ensures that there are no longer any conflicts between these formulas.

In case of conflict not with other statements in the announcement, but with previous beliefs of an agent, priority is given to the new information. We therefore remove formulas in the belief base which still contradict the announcement after updating trust. More formally, we obtain:

$$BB_a^{s\cdot[A]G} = Cl_a^{s\cdot[A]G}((BB_a^s \setminus \{\varepsilon \in BB_a^s \mid \exists b \in G, \exists \varepsilon_b \in A(b), \min_a^{s\cdot[A]G}(\varepsilon) = \min_a^{s\cdot[A]G}(\overline{B_b\varepsilon_b})\}) \cup \{B_b\varepsilon_b \mid b \in G \text{ and } \varepsilon_b \in A(b)\})$$

By removing all conflicts, we have ensured that this new belief base is indeed consistent.

3.5 Examples

To illustrate these dynamics, we study the effects of announcements on an agent a such that $T_a^s = \{a, b, c, d\}$ and $BB_a^s = \{p\}$, where $Agt = \{a, b, c, d, e\}$.

- If b announces $B_d p$ and c announces $\neg p$ ($A_1 = ((b, B_d p), (c, \neg p))$), then b's statement is supported by a's beliefs, and therefore a loses trust in c: $T_a^{s\cdot A_1} = \{a, b, d\}$ and $BB_a^{s\cdot A_1} = \{p, B_b B_d p, B_d p, B_c \neg p\}$.
- If b announces q and c announces $\neg q$ ($A_2 = ((b, q), (c, \neg q))$), a has no way of discriminating between the two announcements and therefore a loses trust in both b and c: $T_a^{s\cdot A_2} = \{a, d\}$ and $BB_a^{s\cdot A_2} = \{p, B_b q, B_c \neg q\}$.
- If c announces $\neg p$ and e announces p ($A_3 = ((c, \neg p), (e, p))$), there is no conflict according to a, because a does not trust e and therefore cannot deduce p from $B_e p$. In this case, because c is trusted by a, c's announcement takes precedence over a's previous beliefs: $T_a^{s\cdot A_3} = \{a, b, c, d\}$ and $BB_a^{s\cdot A_3} = \{\neg p, B_c \neg p, B_e p\}$.

3.6 Semantics

Finally, we can define the semantics of SDDLT. Let (M, s) be a pointed trust model, $p \in \mathcal{P}$ a variable, $a \in Agt$ an agent, and $\varphi, \psi \in \mathcal{L}_{\mathsf{SDDLT}}$. Let A be an announcement by a group $G \subseteq Agt$. We introduce the shorthand $B_G A := \bigwedge_{g \in G} \bigwedge_{\varepsilon_g \in A(g)} B_g \varepsilon_g$. Then,

$$\begin{array}{ll}
(M, s) \models p & \text{iff } p \in V \\
(M, s) \models \neg\varphi & \text{iff } (M, s) \not\models \varphi \\
(M, s) \models \varphi \wedge \psi & \text{iff } (M, s) \models \varphi \text{ and } (M, s) \models \psi \\
(M, s) \models B_a \varepsilon & \text{iff } \varepsilon \in BB_a^s \\
(M, s) \models T_{a,b} & \text{iff } b \in T_a^s \\
(M, s) \models [A]G\varphi & \text{iff } ((M, s) \models B_G A \Rightarrow (M, s \cdot [A]G) \models \varphi)
\end{array}$$

4 Announcements by Groups and Coalitions

Announcements considered in the paper are made by groups of agents. Quantification over such announcements in a setting of epistemic logic has been studied in [1,2]. The resulting formalisms – group announcement logic and coalition announcement logic – expand Public Announcement Logic [19] with operators $\langle G\rangle\varphi$ and $\langle\!\langle G\rangle\!\rangle\varphi$ correspondingly. The former is read as 'there is a joint public announcement by agents from group G such that φ holds in the resulting model,' and the latter means that 'there is a joint public announcement by agents from coalition G such that whatever agents from $A \setminus G$ announce at the same time, φ holds in the resulting model.'

Group and coalition announcements have been so far studied only from the epistemic perspective, i.e. agents in groups and coalitions announce what they *know*. Treatment of these operators in the doxastic setting is an open research problem.

Due to agents' limited reasoning in our framework, it is possible to define group and coalition announcements in SDDLT *provided we restrict the maximal depth of nestings of belief operators in any announcement.* This means that if the depth of nesting is restricted to some number, say 3, then the agents can make announcements of the form $B_aB_bB_ap$, but they are not allowed make announcements of the form $B_aB_bB_aB_bp$. Such a restriction is commonly made when formalising resource-bounded reasoning (see, e.g., [11]). We refer to the restricted logic where the depth of announced formulas cannot be higher than m as SDDLT^m.

We denote by $\mathrm{Lit}(\phi)$ the literals (positive and negative variables) appearing in a formula ϕ, and by $\mathrm{lit}(\varepsilon)$ the single literal appearing in a formula ε. The restriction on the maximal depth of formulas means we can now consider finite numbers of possible announcements. We denote the set of possible announcements by a group G relevant to φ by $PA(G,\varphi) = \{A \in Ann_G \mid \forall b \in G, \forall \varepsilon \in A(b), \mathrm{depth}(\varepsilon) \leqslant m \text{ and } \mathrm{lit}(\varepsilon) \in \mathrm{Lit}(\varphi)\}$, where for any formula ε, $\mathrm{depth}(\varepsilon)$ is the depth of ε defined as the length of the sequence of belief operators in ε.

We now define the group and coalition announcement operators for SDDLT^m:

$$\langle G\rangle\varphi \leftrightarrow \bigvee_{A\in PA(G,\varphi)} \Big(\bigwedge_{\substack{g\in G\\ \varepsilon_g\in A(g)}} B_g\varepsilon_g \wedge [A]\varphi\Big)$$

$$\langle\!\langle G\rangle\!\rangle\varphi \leftrightarrow \bigvee_{A_G\in PA(G,\varphi)} \ \bigwedge_{A_{\overline{G}}\in PA(\overline{G},\varphi)} \Big(\bigwedge_{\substack{g\in G\\ \varepsilon_g\in A_G(g)}} B_g\varepsilon_g \wedge \big(\bigwedge_{\substack{g\in G\\ \varepsilon_g\in A_{\overline{G}}}} B_g\varepsilon_g \rightarrow [A_G]Agt\varphi\big)\Big),$$

where $\overline{G} = Agt \setminus G$ and $A_{Agt} = A_{\overline{G}} \cup A_G = \{(a, A_{\overline{G}}(a) \cup A_G(a)) \mid a \in Agt\}$.

5 Axiomatization

Now that SDDLT is completely defined, the next step is to give a sound and complete axiomatization for it. As group and coalition announcements are definable from the other operators, we do not consider these types of announcements in the axiomatization. Completeness will be proved for the fragment of the logic with no announcements, and we will give axioms reducing SDDLT to that fragment.

5.1 The First Set of Axioms

We give in Table 1 the first nine axioms and the two inference rules of our system, where A represents an announcement by a group G. We will state two more axioms later as they require additional definitions.

Table 1. The first set of axioms of SDDLT

Propositional tautologies	(A0)
$B_a \top$	(A1)
$\varepsilon \to \neg\overline{\varepsilon}$	(A2)
$T_{a,b} \to (B_a B_b \varepsilon \to B_a \varepsilon)$	(A3)
$T_{a,a}$	(A4)
$[A]\neg\varphi \leftrightarrow (B_G A \to \neg[A]\varphi)$	(A5)
$[A](\varphi \wedge \psi) \leftrightarrow [A]\varphi \wedge [A]G\psi$	(A6)
$[A]Gp \leftrightarrow (B_G A_G \to p)$	(A7)
$[A]T_{a,b} \leftrightarrow T_{a,b}$ for $b \notin G$	(A8)
$\vdash \varphi, \varphi \to \psi \Rightarrow \vdash \psi$	(I0)
$\vdash \varphi \Rightarrow \vdash [A]\varphi$	(I1)

The proofs of soundness of these axioms are rather straightforward. In order to have completeness, we need two more reduction axioms. The cases left to deal with are those of $[A]T_{a,b}$ when $b \in G$ and $[A]B_a\varepsilon$.

5.2 Trust and Announcement

We first introduce a few notations. Given $\varepsilon, \varepsilon' \in \mathcal{L}_B$, we call $\max(\varepsilon, \varepsilon')$ the longest common suffix of ε and ε', that is, the longest formula μ such that $\mu \preceq \varepsilon$ and $\mu \preceq \varepsilon'$. This may be the empty formula. If $\varepsilon = B_{a_1} \dots B_{a_n} l$, then for all $1 \leq k \leq n$, we denote $Agt(\varepsilon \setminus B_{a_k} \dots B_{a_n} l) = \{a_1, \dots, a_{k-1}\}$. If $k = 1$, then $Agt(\varepsilon \setminus \varepsilon) = \emptyset$. Moreover, $Agt(\varepsilon \setminus l) = \{a_1, \dots, a_n\}$.

Recall the definition of the updated trust set for agent a after an announcement A by a group G at s:

$$T_a^{s \cdot [A]} = T_a^s \setminus \{b \in G \mid \exists c \in G, \varepsilon_b \in A(b), \varepsilon_c \in A(c),$$
$$\min_a^s(B_b\varepsilon_b) = \overline{\min_a^s(B_c\varepsilon_c)} \text{ and } \min_a^s(B_b\varepsilon_b) \notin BB_a^w\}$$

We need to express $\min_a^s(\varepsilon_b) = \overline{\min_a^s(\varepsilon_c)}$ ("ε_b and ε_c are contradictory from the point of view of a") and $\min_a^s(\varepsilon_b) \notin BB_a^w$ ("a has no previous beliefs backing up b's claim").

We have the following result:

Lemma 2. *Let ε_1 and ε_2 be formulas of $\mathcal{L}_B$, a be an agent, and s a state. Define ε_0 as $\varepsilon_0 = \max(\varepsilon_1, \overline{\varepsilon_2})$. We have that $\min_a^s(\varepsilon_1) = \overline{\min_a^s(\varepsilon_2)}$ iff $\mathrm{lit}(\varepsilon_1) = \mathrm{lit}(\overline{\varepsilon_2})$, $\varepsilon_1 \rightarrow_a^s \varepsilon_0$ and $\varepsilon_2 \rightarrow_a^s \overline{\varepsilon_0}$.*

Using lemma 2, we get the following formula expressing contradiction of ε_1 and ε_2 from the point of view of a:

$$CO(a, \varepsilon_1, \varepsilon_2) = \bot$$

if $lit(\varepsilon_1) \neq \overline{lit(\varepsilon_2)}$, and

$$CO(a, \varepsilon_1, \varepsilon_2) = (\bigwedge_{\alpha \in Agt(\varepsilon_1 \setminus \max(\varepsilon_1, \overline{\varepsilon_2}))} T_{a,\alpha}) \wedge (\bigwedge_{\beta \in Agt(\varepsilon_2 \setminus \max(\varepsilon_2, \overline{\varepsilon_1}))} T_{a,\beta})$$

otherwise.

Now to express that $\min_a^s(\varepsilon_b) \notin BB_a^s$, we use the following result:

Lemma 3. *If ε is a formula of $\mathcal{L}_B$, a is an agent, and s is a state, we have that:*

$$\min_a^s(\varepsilon) \notin BB_a^s \Leftrightarrow (\forall \mu \preceq \varepsilon, \varepsilon \rightarrow_a^s \mu \Rightarrow \mu \notin BB_a^s)$$

Using this we express the condition about previous beliefs:

$$PB(a, \varepsilon) = \bigwedge_{\mu \preceq \varepsilon} ((\bigwedge_{\alpha \in Agt(\varepsilon \setminus \mu)} T_{a,\alpha}) \rightarrow \neg B_a \mu)$$

Finally, we give the reduction axiom ($A9$):

$$[A]T_{a,b} \leftrightarrow T_{a,b} \wedge \neg \bigvee_{\varepsilon_b \in A(b)} \bigvee_{\substack{c \in G \\ \varepsilon_c \in A(c)}} (CO(a, B_b\varepsilon_b, B_c\varepsilon_c) \wedge PB(a, \varepsilon_b)) \qquad \text{(A9)}$$

if $b \in G$.

5.3 Belief and Announcement

The last reduction axiom we need is for $[A]B_a\varepsilon$ where A is an announcement given by a group G, a is an agent, and ε is a formula of $\mathcal{L}_B$. Looking at the semantics, we have:

$$\begin{aligned} BB_a^{s \cdot [A]} = Cl_a^{s \cdot [A]}((BB_a^s \setminus \{\varepsilon \in BB_a^s \mid \exists b \in G, \exists \varepsilon_b \in A(b), \\ \min_a^{s \cdot [A]}(\varepsilon) = \min_a^{s \cdot [A]}(\overline{B_b\varepsilon_b})\}) \\ \cup \{B_b\varepsilon_b \mid b \in G \text{ and } \varepsilon_b \in A(b)\}) \end{aligned}$$

Hence ε is in $BB_a^{s \cdot [A]}$ if:

- There exists a formula ε_b announced by an agent b such that $B_b\varepsilon_b \rightarrow_a^{s \cdot [A]} \varepsilon$ (we say that ε is *successfully announced* to a)

- or ε was believed by a and not contradicted by the announcement from a's point of view after updating trust. Formally, this means that ε is in BB_a^s and for any formula ε_b announced by an agent b, we have that $\min_a^{s\cdot[A]}(\varepsilon) \neq \min_a^{s\cdot[A]}(\overline{B_b\varepsilon_b})$, or equivalently, $\min_a^{s\cdot[A]}(\overline{\varepsilon}) \neq \min_a^{s\cdot[A]}(B_b\varepsilon_b)$. That is, $\min_a^{s\cdot[A]}(\overline{\varepsilon})$ is not successfully announced to a.

We express the fact that ε is successfully announced to a through announcement A by the following formula:

$$Ann(A,\varepsilon,a) = \bigvee_{\substack{b\in G \\ \varepsilon_b\in A(b):\varepsilon\preceq B_b\varepsilon_b}} \bigwedge_{\alpha\in Agt(B_b\varepsilon_b\backslash\varepsilon)} [A]T_{a,\alpha}$$

Using this, we can also express the fact that ε was not contradicted in A from a's point of view. This is equivalent to no formula $\mu \preceq \overline{\varepsilon}$ such that $\overline{\varepsilon} \rightarrow_a^{s\cdot[A]} \mu$ being successfully announced:

$$NC(A,\varepsilon,a) = \bigwedge_{\mu\preceq\overline{\varepsilon}} ((\bigwedge_{\alpha\in Agt(\overline{\varepsilon}\backslash\mu)} [A]T_{a,\alpha}) \rightarrow \neg Ann(A,\mu,a))$$

Finally, the axiom $(A10)$ is defined as:

$$[A]B_a\varepsilon \leftrightarrow Ann(A,\varepsilon,a) \vee (B_a\varepsilon \wedge NC(A,\varepsilon,a)) \qquad \text{(A10)}$$

5.4 Completeness

The reduction axioms follow the semantics quite closely, and therefore we will not dwell on the proofs of soundness of each axiom and instead move on to completeness of our axiom system. As we have reduction axioms allowing us to translate formulas of the full language to that of the static language $\mathcal{L}^*_{\mathsf{SDDLT}}$, it suffices to show completeness of the axiom system constituted by the axioms $(A0)$–$(A4)$ and the inference rule $(I0)$ for the corresponding static logic SDDLT^*, that is, the logic without announcements.

The proof of completeness of SDDLT^* is a standard canonical model proof using maximal consistent sets, as given in [6]. Because of space constraints, we do not detail this proof but only give a few indications of the details specific to this logic. First, the canonical model used is the following:

Definition 1. *Let Γ be a consistent set of formulas of $\mathcal{L}^*_{SDDLT}$ that is maximal for inclusion. The canonical model for Γ is defined as $M^\Gamma = (S, V^\Gamma)$ where $V^\Gamma = \mathcal{P} \cap \Gamma$.*

We consider the state s_Γ such that $BB_a^{s_\Gamma} = \{\varphi \mid B_a\varphi \in \Gamma\}$ and $T_a^{s_\Gamma} = \{b \in Agt \mid T_{a,b} \in \Gamma\}$ for all a in Agt, verifying that it is indeed a belief state. A truth lemma stating that the formulas true at (M^Γ, s_Γ) are exactly the formulas of Γ is then shown by induction. This, conjointly with the Lindenbaum lemma, gives us the completeness of SDDLT^*.

Theorem 1. *For every* $\varphi \in \mathcal{L}^*_{\mathsf{SDDLT}}$*, if* $\models \varphi$*, then* $\vdash_{\mathsf{SDDLT}^*} \varphi$.

Corollary 1 (Completeness of *SDDLT***).** *For every* $\varphi \in \mathcal{L}_{\mathsf{SDDLT}}$*, if* $\models \varphi$*, then* $\vdash_{\mathsf{SDDLT}} \varphi$.

Completeness of SDDLT^m (the logic with the belief depth of announcements bound to a fixed m) is a straightforward corollary, which gives us an axiomatisation of a logic with coalition announcements.

Corollary 2 (Completeness of SDDLT^m**).** *Given* $\varphi \in \mathcal{L}^m_{\mathsf{SDDLT}}$*, we have*

$$\models \varphi \Rightarrow \vdash_{\mathsf{SDDLT}} \varphi$$

6 A Simple Example: Of Bad Influences and the Importance of Speaking Out

We here give a concrete example to argue that our choice of trust dynamics is relevant, even though the agents' reasoning is quite basic. When considering real-life situations, group announcements do not consist of simultaneous announcements, but can be seen as statements proclaimed over a short period of time.

We see many news and articles about the importance of speaking up against bullying or harassment, or about how media is a bad influence to children. Everyone has their own story of something nobody talked to them about when they were a child, which led them to believe ridiculous –in hindsight– ideas they got from the television or magazines. The importance of speaking out against these wrong opinions to prevent the spread of their influence can be illustrated using our framework.

Say we have a group of agents Agt. In this group, there is a bad influence b, a group of gullible agents Gul, and a group of watchers Wat. The gullible agents trust everybody, the watchers only trust other watchers, and the bad influence trusts only themselves. For example, the bad influence could be the television, the gullible demographic the younger audience, and the watchers the parents of this audience. The bad influence could also be someone being uncivil, the gullible group could be foreigners still learning about the local culture, while the watchers would be local bystanders. Another example would be a bully, or a group of bullies acting as one, other students, and teachers at a school.

While all other agents in the group think $\neg p$ (for example, the incivility going on is not normal), the bad influence believes that p. We equate b carrying out the uncivil act to their announcing that p (there is nothing wrong with that action). While there is no risk of the bystanders starting to believe p, the foreigners could be led to believe that this is how things are done in this country. If nobody speaks out against b's actions, this is what the situation will lead to. However, as soon as one person speaks up, they will confirm others' belief that there is a problem, make the gullible agents lose trust in the troublemaker, and ensure that the incivility does not spread.

Formally this situation is described in the following model. Let $Agt = Gul \cup Wat \cup b$ and $M = (S, V)$ for some V (the actual state of the world is not important here). Consider s in S such that:

- $\forall a \in Gul \cup Wat, BB_a^s = \{\neg p\}$
- $BB_b^s = \{p\}$
- $\forall a \in Wat, T_a^s = Wat$
- $\forall a \in Gul, T_a^s = Agt$
- $T_b^s = \{b\}$

Then we have that

$$(M, s) \models [\{(b, \{p\})\}] \bigwedge_{g \in Gul} B_g p$$

and

$$(M, s) \models \bigvee_{a \in Wat} [\{(b, \{p\}), (a, \{\neg p\})\}] \bigwedge_{g \in Gul} (B_g \neg p \wedge \neg T_{g,b})$$

That is, if b announces p and nothing else is said, then gullible agents will start believing that p. However, if any one watcher a states that $\neg p$ as b claims the opposite, then the gullible agents will both retain the belief that $\neg p$ and learn that b is not to be trusted.

7 Discussion

7.1 On the Choice of Dynamics

Throughout this paper we have made choices in order to attempt to find a balance between non-trivial trust dynamics and clarity of the formalization. More complex dynamics could, of course, be envisioned: for example, in our framework, if an agent a receives a false announcement about their own beliefs from a trusted agent b (say, b announces $B_a p$ but a actually believes $\neg p$), they will blindly trust b and change these beliefs, so that a will start to believe p. This kind of blind trust may easily be considered too strong, and we may wish in this situation for a to stop trusting b instead, thus eliminating instances of contradictions with b's announcements. This can be expressed in the update of the trust set, which would become, after announcement A by a group G: if we call $C(a, s, A)$ the set $\{b \in G \mid \exists \varepsilon \notin BB_a^s, B_a \varepsilon \in A(b)\}$, we have

$$T_a^{s \cdot [A]} = T_a^s \setminus (C(a, s, A) \cup \{b \in G \mid \exists c \in G \setminus C(a, s, A), \varepsilon_b \in A(b), \varepsilon_c \in A(c),$$
$$\min_a^s(B_b \varepsilon_b) = \overline{\min_a^s(B_c \varepsilon_c)} \text{ and } \min_a^s(B_b \varepsilon_b) \notin BB_a^w\})$$

We have shown in Sect. 5 how to express as formulas all of the properties necessary to amend the reduction axioms to follow these new dynamics. The constraint on announcements given in Sect. 2 could also be lifted in a similar manner. However, this would lead to longer and less legible reduction axioms. For this reason we have presented dynamics which are more naive, but which suffice to make our point and are adaptable enough to work with more complex dynamics.

7.2 Sequential and Simultaneous Announcements

Say an agent a announces p, and afterwards an agent b announces $\neg p$, and say agent c trust both a and b. This situation can be modeled either as a group announcement $((a, \{p\}), (b, \{\neg p\}))$, which will lead c to lose trust in at least one of the two other agents, or as two announcements $(a, \{p\})$ and $(b, \{\neg p\})$ happening one after the other, leading c to believe first p, then $\neg p$, without losing trust in either a or b. In general, while simultaneous announcements are understandable as is in the context of, say, a search query, it is less clear what they represent when dealing with social interactions. Our understanding of group announcements in this context is that of announcements given over a short amount of time. This notion of short may depend on the agents and the situation, and while there is no clear-cut time stamp we can put on this, we can imagine that if enough time has gone by after a's announcement, c will no longer associate p with a, or they will accept that the situation may have changed since that first announcement, and accept b's announcement without feeling that there is too much of a conflict.

8 Conclusions and Future Work

We have defined a framework for reasoning about the evolution of trust and beliefs as triggered by group announcements, in which contradictions within the group can lead agents to lose trust in the speakers involved, and in which agents' beliefs can help them pick a side in case of conflict. We have given an axiomatization for our logic SDDLT, and shown that it is sound and complete. We have also shown that two operators for coalition announcements are definable in SDDLT.

Our work could be furthered in several directions. First, it would be interesting to expand on announcements. Though we have only considered public announcements here, we believe it would be quite simple to generalize these semantics to private announcements. It would also be interesting to no longer assume sincerity of the agents.

The notion of trust could also be expanded upon, following the existing literature, in particular the trust functions in [14]. For example, we could have several 'degrees' of trust, rather than simply binary trust. It would also be interesting to add a mechanism for gaining trust, the inner workings of which are less clear than those for loss of trust. The latter could also be refined: for instance, in case of conflict where no previous beliefs help the agent choose whom to trust, trust in both agents could be lost "until confirmation" of one of the two theses.

Finally, agents in our setting have very limited reasoning capabilities. Allowing more complex reasoning, and more complex formulas in belief bases, would make identifying contradictions stemming from announcements less straightforward. However, belief merging techniques could perhaps be applied to our framework to allow for these expansions.

Acknowledgements. Support from the ANR project CoPains (Cognitive Planning in Persuasive Multimodal Communication) is gratefully acknowledged.

References

1. Ågotnes, T., Balbiani, P., van Ditmarsch, H., Seban, P.: Group announcement logic. J. Appl. Log. **8**(1), 62–81 (2010)
2. Ågotnes, T., van Ditmarsch, H.: Coalitions and announcements. In: Padgham, L., Parkes, D.C., Müller, J.P., Parsons, S. (eds.) Proceedings of AAMAS 2008, vol. 2, pp. 673–680. IFAAMAS (2008)
3. Alchourrón, C.E., Gärdenfors, P., Makinson, D.: On the logic of theory change: partial meet contraction and revision functions. J. Symbolic Log. **50**(2), 510–530 (1985)
4. Baltag, A., Smets, S.: Dynamic belief revision over multi-agent plausibility models. In: Proceedings of LOFT, vol. 6, pp. 11–24. University of Liverpool (2006)
5. van Benthem, J.: Dynamic logic for belief revision. J. Appl. Non-class. Log. **17**(2), 129–155 (2007)
6. Blackburn, P., de Rijke, M., Venema, Y.: Modal Logic. Cambridge University Press, Cambridge (2001)
7. Delgrande, J.P., Dubois, D., Lang, J.: Iterated revision as prioritized merging. KR **6**, 210–220 (2006)
8. Demolombe, R., Liau, C.J.: A logic of graded trust and belief fusion. In: Proceedings of the 4th Workshop on Deception, Fraud and Trust in Agent Societies, pp. 13–25 (2001)
9. van Ditmarsch, H.: Dynamics of lying. Synthese **191**(5), 745–777 (2014)
10. van Ditmarsch, H., van der Hoek, W., Kooi, B.: Dynamic Epistemic Logic. Synthese Library, vol. 337. Springer, Dordrecht (2008). https://doi.org/10.1007/978-1-4020-5839-4
11. Fisher, M., Ghidini, C.: Exploring the future with resource-bounded agents. J. Logic Lang. Inform. **18**(1), 3–21 (2009)
12. Ghosh, S., Velázquez-Quesada, F.R.: Agreeing to agree: reaching unanimity via preference dynamics based on reliable agents. In: Proceedings of AAMAS 2015, pp. 1491–1499. IFAAMAS (2015)
13. Herzig, A., Lorini, E., Hübner, J.F., Vercouter, L.: A logic of trust and reputation. Log. J. IGPL **18**(1), 214–244 (2010)
14. Jonker, C.M., Treur, J.: Formal analysis of models for the dynamics of trust based on experiences. In: Garijo, F.J., Boman, M. (eds.) MAAMAW 1999. LNCS (LNAI), vol. 1647, pp. 221–231. Springer, Heidelberg (1999). https://doi.org/10.1007/3-540-48437-X_18
15. Konieczny, S., Pérez, R.P.: Logic based merging. J. Philos. Log. **40**(2), 239–270 (2011)
16. Leturc, C., Bonnet, G.: A normal modal logic for trust in the sincerity. In: André, E., Koenig, S., Dastani, M., Sukthankar, G. (eds.) Proceedings of AAMAS 2018, pp. 175–183. IFAAMAS (2018)
17. Liau, C.J.: Belief, information acquisition, and trust in multi-agent systemsa modal logic formulation. Artif. Intell. **149**, 31–60 (2003)
18. Lorini, E.: In praise of belief bases: doing epistemic logic without possible worlds. In: Thirty-Second AAAI Conference on Artificial Intelligence (2018)
19. Plaza, J.: Logics of public communications (reprint of 1989's paper). Synthese **158**(2), 165–179 (2007)

20. Priest, G., Tanaka, K., Weber, Z.: Paraconsistent logic. In: Zalta, E.N. (ed.) The Stanford Encyclopedia of Philosophy. Metaphysics Research Lab, Stanford University, summer 2018 edn. (2018)
21. Segerberg, K.: Belief revision from the point of view of doxastic logic. Log. J. IGPL **3**(4), 535–553 (1995)

Bipolar Argumentation Frameworks, Modal Logic and Semantic Paradoxes

Carlo Proietti[1(✉)], Davide Grossi[2], Sonja Smets[1,3], and Fernando R. Velázquez-Quesada[1]

[1] ILLC, University of Amsterdam, Amsterdam, The Netherlands
carlo.proietti@fil.lu.se
[2] University of Groningen, Groningen, The Netherlands
[3] Logic, Information and Interaction Group, University of Bergen, Bergen, Norway

Abstract. Bipolar Argumentation Frameworks (BAF) are a natural extension of Dung's Argumentation Frameworks (AF) where a relation of *support* between arguments is added to the standard *attack* relation. Despite their interest, BAF present several difficulties and their semantics are quite complex. This paper provides a definition of semantic concepts for BAF in terms of *fixpoints* of the functions of *neutrality* and *defense*, thus preserving most of the fundamental properties of Dung's AF. From this angle it becomes easy to show that *propositional dynamic logic* provides an adequate language to talk about BAF. Finally, we illustrate how this framework allows to encode the structure of the referential discourse involved in semantic paradoxes such as the *Liar*. It turns out that such paradoxes can be seen as BAF without a stable extension.

1 Introduction

Bipolar Argumentation Frameworks (BAF) were introduced by [5] and [6] to enrich Dung's Argumentation Frameworks (AF) [9] with an explicit relation of support.[1] In many respects, the semantics of BAF are more difficult to categorize than those of standard AF. There are two main (related) reasons for this. First of all, at least two different interpretations of support are available:

- *deductive support*: a supports b means "the acceptance of a implies the acceptance of b" [3].

[1] Indeed, the only support available in standard AF is the "defense" relation: argument a supports argument b by attacking one of its attackers. This is too restrictive in most real-life debates, where arguments providing direct support are commonly used.

Carlo Proietti gratefully acknowledges funding received from the European Commission (Marie Skłodowska-Curie Individual Fellowship 2016, 748421) for his research. The authors would like to thank the anonymous reviewers of LORI'19 for many helpful suggestions.

P. Blackburn et al. (Eds.): LORI 2019, LNCS 11813, pp. 214–229, 2019.
https://doi.org/10.1007/978-3-662-60292-8_16

- *necessary support*: a supports b means "a is a necessary condition for the acceptance of b" or, equivalently, "acceptance of b implies the acceptance of a" [18,19].[2]

A further notion is that of *evidential support* [20] that we will not deal with here.[3]

The following scenario provides an example of deductive support (given the background information) from a to b.

Example 1. Suppose that, on the day before the last matchday of Premier League, Liverpool is at the top, one point ahead Manchester. Consider the following arguments:

a. Liverpool wins last match. *c*. Manchester wins Premier League.
b. Liverpool wins Premier League.

This other gives an example of a necessary support from b' to c'.

Example 2. The dark room. Consider one room with no windows that can only be illuminated by an electric light (with no other external sources available). Consider the following arguments:

a'. The switch was turned off last night. *c'*. The room is illuminated.
b'. The switch is on.

The second main problem, as the examples suggest, is that the interaction of support and attack induces several forms of *complex attack*, such as those from a to c (a supports b which attacks c) and from a' to c' (a' attacks b' which supports c'). However, while a complex attack as that of Example 1 is intuitively effective for deductive support, it is not for necessary support. The converse holds for complex attacks as that of Example 2. The presence of complex attacks complicates the criterion of *coherence* for a set of arguments, which for standard AF is encoded by *conflict-freeness*. The literature on BAF provides several characterizations of coherence which, by consequence, multiply the criteria of *admissibility* for sets of arguments. This, in turn, generates a caleidoscope of additional criteria for *acceptable* (*complete, preferred, grounded* and *stable*) extensions.

We define extensions (or *semantics*) for BAF in line with [9] by only using conflict-freeness (and self-defense) w.r.t. to (complex) attacks. Despite this minimal coherence criterion, extensions thus defined turn out to be coherent in the

[2] We limit ourselves to *binary* necessary support. Indeed this notion of support is often introduced as a more general relation between a *set* of arguments and an argument [18,19].

[3] Evidential support can be seen as a special kind of necessary support where an argument cannot be accepted unless it is ultimately supported by "evidence", the latter being a special type argument (also called a *prima facie* argument) that can be neither attacked nor supported by other arguments.

strongest possible sense, and their properties are in line with those of standard AF. To define our extensions, we fix a primitive notion of complex attack for each reading of the support relation,[4] then we use it for defining the *defense function* (*characteristic function* in Dung's original work) and the *neutrality function*. The extensions are then characterised in terms of *(post)fixpoints* of these functions.[5] As a further relevant point, we show that these semantics have a modal representation in the framework of *propositional dynamic logic* [16], which therefore provides an adequate language to talk about BAF.

Directed graphs offer a natural representation of the referential structure of a discourse [8,10–12,23]. In this context, the semantics of Dung's AF provide an interesting tool to understand the nature of paradoxes as "pathological" graphs. This specific link has been established by [10,11], the central result being Fact 9 below, which associates paradoxality with lack of a stable extension. BAF, as a natural expansion of AF, allow to express referential structures in a more compact way, although being equally expressive as standard AF in this respect [1]. Our result in Theorem 3 subsumes Fact 9 as a special case and provides a first bridge from the semantics of BAF to the analysis of paradoxes.

The paper proceeds as follows. Section 2 recalls the basic concepts of AF, introducing BAF with necessary and deductive support and defining their extension concepts. We show that extensions thus defined preserve the fundamental properties of their corresponding AF extensions, and then prove additional results (Theorems 1 and 2). Section 3 introduces a language of *propositional dynamic logic* to talk about BAF (plus a complete axiom system), providing a modal definition of the extension concepts introduced in Sect. 2. Section 4 focuses on the analysis of semantic paradoxes, showing first how to encode the structure of the referential discourse within BAF with necessary support. Based on this we prove our main correspondence result in Theorem 3. Section 5 summarizes the results and mentions open problems for future work.

2 Argumentation Frameworks

A basic AF $\mathcal{A} = (A, \rightarrow)$ is a relational structure, with $A \neq \emptyset$ the set of arguments and $\rightarrow \subseteq A \times A$ a binary relation, where $a \rightarrow b$ is read as "a attacks b". We use the shortenings $X \rightarrow a$ for $\exists x \in X : x \rightarrow a$, $a \rightarrow X$ for $\exists x \in X : a \rightarrow x$, and $X \rightarrow Y$ for $\exists x \in X, \exists y \in Y : x \rightarrow y$. Additionally, for X a set and R a relation, the set $\langle R \rangle X := \{x \mid \exists y \in X \text{ and } xRy\}$ contains the arguments that can R-access some element in X, while $[R]X := \{x \mid \forall y \text{ if } xRy \text{ then } y \in X\}$ contains the arguments that can R-access only elements in X.

The fundamental concept in abstract argumentation is that of an *extension* or *solution*. Intuitively, a set of arguments X is a solution for $\mathcal{A}$ only if it satisfies

[4] A similar strategy was proposed by [7], which already provides some of our results. However, this was done without the use of algebraic and fixpoint notions.

[5] Fixpoint-theoretic notions were of high impact in Dung's original work; since then, they have been scarcely exploited for the study of BAF and for abstract argumentation in general.

certain properties which make it an "acceptable" opinion in the argumentation represented by $\mathcal{A}$. Most solution concepts for AF share two basic properties: *conflict-freeness* and *defense* of their own arguments. There are many equivalent ways to define such properties; here we characterize them in terms of the *neutrality* and *defense* functions (as in [13]).

Definition 3 (Neutrality and defense function). *Let $\mathcal{A} = (A, \rightarrow)$ be an AF. The neutrality function $\mathbf{n}_{\mathcal{A}} : \mathcal{P}(A) \longrightarrow \mathcal{P}(A)$ is:*

$$\mathbf{n}_{\mathcal{A}}(X) = \{x \in A : \ NOT\ X \rightarrow x\}$$

The defense function $\mathbf{d}_{\mathcal{A}} : \mathcal{P}(A) \longrightarrow \mathcal{P}(A)$ is:

$$\mathbf{d}_{\mathcal{A}}(X) = \{x \in A : \forall y \in A : \ IF\ y \rightarrow x\ THEN\ X \rightarrow y\}$$

In other words, $\mathbf{n}_{\mathcal{A}}(X)$ is the set of arguments that are not attacked by X (i.e. to which X is neutral) and $\mathbf{d}_{\mathcal{A}}(X)$ is the set of arguments that are defended by X. The advantage of this characterization is that it provides an insightful and compact definition of solution concepts as *(post)fixpoints* of $\mathbf{n}_{\mathcal{A}}$ and $\mathbf{d}_{\mathcal{A}}$. This will prove useful in the study of BAF.

Definition 4 (Solution concepts). *Given a framework $\mathcal{A}$:*

- *A set X is conflict-free ($\mathtt{Cfr}_{\mathcal{A}}(X)$) iff $X \subseteq \mathbf{n}_{\mathcal{A}}(X)$ (i.e. X is a postfixpoint of $\mathbf{n}_{\mathcal{A}}$).*
- *A set X is self-defended ($\mathtt{Sdf}_{\mathcal{A}}(X)$) iff $X \subseteq \mathbf{d}_{\mathcal{A}}(X)$ (i.e. X is a postfixpoint of $\mathbf{d}_{\mathcal{A}}$).*
- *A set X is an admissible extension ($\mathtt{Adm}_{\mathcal{A}}(X)$) iff X is conflict-free and self-defended.*
- *A set X is a complete extension ($\mathtt{Cmp}_{\mathcal{A}}(X)$) iff $X = \mathbf{d}_{\mathcal{A}}(X)$ and $X \subseteq \mathbf{n}_{\mathcal{A}}(X)$ (i.e. X is admissible and is a fixpoint of $\mathbf{d}_{\mathcal{A}}$).*
- *A set X is a (the) grounded extension ($\mathtt{Grn}_{\mathcal{A}}(X)$) iff X is the smallest fixpoint of $\mathbf{d}_{\mathcal{A}}$.*
- *A set X is a preferred extension ($\mathtt{Prf}_{\mathcal{A}}(X)$) iff X is maximal (for set inclusion) among the admissible (or complete) extensions of $\mathcal{A}$.*
- *A set X is a stable extension ($\mathtt{Stb}_{\mathcal{A}}(X)$) iff $X = \mathbf{n}_{\mathcal{A}}(X)$ (i.e., X is a fixpoint of $\mathbf{n}_{\mathcal{A}}$).*

Fact 5 below recapitulates known facts about solution concepts, with $\mathtt{Adm}_{\mathcal{A}}$ denoting the set of admissible extensions of $\mathcal{A}$ and likewise for other solution concepts.

Fact 5 ([9]). *Let $\mathcal{A}$ be an argumentation framework.*

1. *$\langle \mathtt{Adm}_{\mathcal{A}}, \subseteq \rangle$ is a poset.*
2. *Any upward directed non-empty family in $\mathtt{Adm}_{\mathcal{A}}$ is closed under union.*
3. *$\emptyset \in \mathtt{Sdf}_{\mathcal{A}}$*
4. *$\mathtt{Prf}_{\mathcal{A}} \neq \emptyset$*

5. *The defense function is monotonic and therefore the grounded set always exists.*
6. *A stable extension is not guaranteed to exist.*
7. *If $\rightarrow$ is well-founded*[6] *then $\mathcal{A}$ has exactly one complete extension, which is grounded, preferred and stable.*

The following is worth noticing: 1 and 2 together imply that the set of admissible solutions forms a complete partial order; 3–6 establish the existence, in any argumentation framework, of admissible, complete, grounded and preferred extensions, but that is not the case for stable extensions; 7 entails that all extensions are one and the same when the attack relation is well-founded.

2.1 Bipolar Argumentation Frameworks

A BAF $\mathcal{A} = (A, \rightarrow, \Rightarrow)$ is a birelational directed graph, with A and $\rightarrow$ as before, and $a \Rightarrow b$ indicating "a supports b". BAF like those in Fig. 1 allow to represent Examples 1 and 2. As mentioned in Sect. 1, two complex attacks are represented here: from a to c and from a' to c'. However, their interpretation depends on the specific reading of the support relation. If $\Rightarrow$ is read as deductive support, then the attack from a to c is effective, while the one from a' to c' is not; the opposite holds for necessary support. Hence, the semantics of necessary and deductive support should be treated separately.

Fig. 1. Examples 1 and 2 represented with bipolar argumentation frameworks.

Necessary Support. Two main types of complex attacks are generated by BAF with necessary supports, namely *secondary attacks*, as in Fig. 2(a), and *extended attacks*, in Fig. 2(b) (see [7,18,19]). A secondary attack from a to b holds if there is a path $a \rightarrow b_0 \Rightarrow \cdots \Rightarrow b_n$ with $b = b_n$ for $n \geq 0$; more succinctly, a attacks b iff $a \rightarrow \cdot \Rightarrow^* b$, with $\cdot$ the operation of *composition* and $\Rightarrow^*$ the *reflexive and transitive closure* of $\Rightarrow$ (we shall also write $a \rightarrow\Rightarrow^* b$ for conciseness). An extended attack holds if $a(\Rightarrow^{-1})^* \rightarrow b$, with $\Rightarrow^{-1}$ the converse of $\Rightarrow$. As stressed by [7] (Proposition 6), both types of attacks are special cases of n+-attacks (Fig. 2(c)), which hold whenever $a(\Rightarrow^{-1})^* \rightarrow\Rightarrow^* b$.[7]

[6] We recall that a binary relation is well-founded whenever it does not contain any infinitely descending chain, i.e., in our case, there exists no infinite chain $a_0 \leftarrow a_1 \leftarrow \cdots \leftarrow a_n \leftarrow \ldots$ of attacked arguments.

[7] This is because both relations $\rightarrow\Rightarrow^*$ and $(\Rightarrow^{-1})^* \rightarrow$ are contained in relation $(\Rightarrow^{-1})^* \rightarrow\Rightarrow^*$.

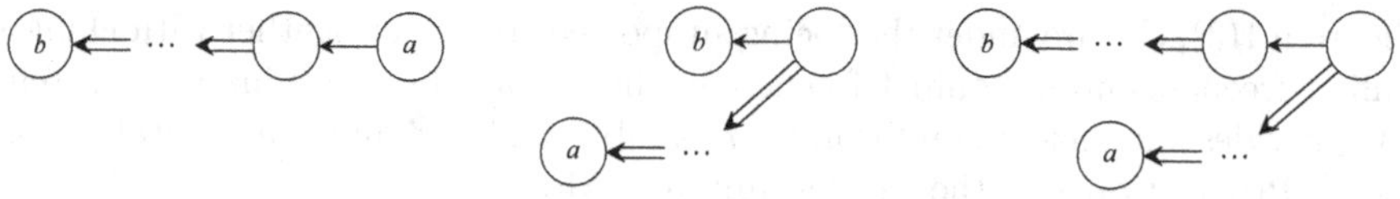

(a) Secondary attack (b) Extended attack (c) n+-attack

Fig. 2. Complex attacks for necessary support

Here we assume "secondary attack" to be our primitive notion of attack for BAF with necessary support. Given a BAF $\mathcal{A} = (A, \rightarrow, \Rightarrow)$, this enables to define the neutrality function $\mathbf{n}^{ns}_{\mathcal{A}} : \mathcal{P}(A) \longrightarrow \mathcal{P}(A)$ as:

$$\mathbf{n}^{ns}_{\mathcal{A}}(X) = \{x \in A : \text{ NOT } X \rightarrow\Rightarrow^* x\}$$

and the defense function $\mathbf{d}^{ns}_{\mathcal{A}} : \mathcal{P}(A) \longrightarrow \mathcal{P}(A)$ as:

$$\mathbf{d}^{ns}_{\mathcal{A}}(X) = \{x \in A : \forall y \in A : \text{ IF } y \rightarrow\Rightarrow^* x \text{ THEN } X \rightarrow\Rightarrow^* y\}$$

This approach has the advantage of anchoring the definitions of the solution concepts to those provided by [9].[8] For example, define

$$\mathtt{Cmp}^{ns}_{\mathcal{A}}(X) \text{ iff } X = \mathbf{d}^{ns}_{\mathcal{A}}(X) \text{ and } X \subseteq \mathbf{n}^{ns}_{\mathcal{A}}(X)$$

It is an immediate consequence of these definitions that all the fundamental results listed in Fact 5(1–7) hold for the new solution concepts. For example, every BAF $\mathcal{A}$ where $\rightarrow\Rightarrow^*$ is well-founded has exactly one complete extension, which is grounded, preferred and stable, by Fact 5(7). The proofs are completely analogous to those provided by [9].

The following theorem establishes key properties of the new solution concepts.

Theorem 1. *Let $\mathcal{A} = (A, \rightarrow, \Rightarrow)$ be a BAF.*

1. *Any $X \subseteq A$ s.t. $\mathtt{Adm}^{ns}_{\mathcal{A}}(X)$ does not contain any n+-attack.*
2. *Any $X \subseteq A$ s.t. $\mathtt{Cmp}^{ns}_{\mathcal{A}}(X)$ is closed for $\Rightarrow^{-1}$.*
3. *If X is closed for $\Rightarrow^{-1}$ and $X \subseteq \mathbf{n}_{\mathcal{A}}(X)$ then $X \subseteq \mathbf{n}^{ns}_{\mathcal{A}}(X)$.*
4. *If $\Rightarrow$ is well-founded then $\mathtt{Stb}^{ns}_{\mathcal{A}}(X)$ iff $X = \mathbf{n}_{\mathcal{A}}(X) \cap [\Rightarrow^{-1}](X)$.*

Proof. See Appendix.

Theorem 1(1) shows that any admissible set is conflict-free w.r.t. any type of complex attack. Therefore, all the defined solution concepts are strongly coherent even though $\mathtt{Cfr}^{ns}_{\mathcal{A}}$ takes only secondary attacks into account. Part (2) demonstrates that all solution concepts stronger than complete (preferred, grounded and stable) are closed under the "being supported" relation. Furthermore, by

[8] In an analogous way we could assume extended or n+-attacks as our primitive notion and define the neutrality and defense function accordingly.

Theorem 1(3), closure under the "being supported" relation together with Dung's conflict-freeness entails conflict-freeness in the extended sense. Finally, Theorem 1(4) provides a sufficient condition for ns-stability of X,[9] and is a generalization of [17], Proposition 1 to the case of infinite BAF.

Deductive Support. BAF with deductive support present two main patterns of complex attacks, namely *supported attacks*, Fig. 3(a), and *mediated attacks*, Fig. 3(b) (see [7]). A supported attack from a to b holds only if there is a path $a \Rightarrow b_0 \Rightarrow \cdots \Rightarrow b_{n-1} \rightarrow b_n$ with $b = b_n$ for $n \geq 0$; more compactly, a attacks b iff $a \Rightarrow^* \cdot \rightarrow b$. A mediated attack instead holds if $a \rightarrow (\Rightarrow^{-1})^* b$. Here again it is not difficult to find a more general pattern of incompatibility, as for n+-attacks, by generalizing the two kinds of attack, i.e. $a \Rightarrow^* \rightarrow (\Rightarrow^{-1})^* b$. We shall call this a d+-attack.

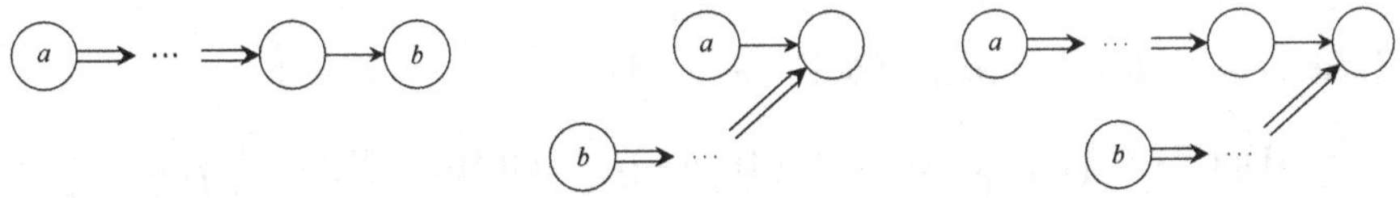

Fig. 3. Complex attacks for deductive support

Deductive support is naturally interpreted as the converse of necessary support, i.e. $\Rightarrow^{-1}$ [7]. According to this reading, mediated attacks under deductive support are nothing more than secondary attacks under necessary support. It therefore makes sense to assume the notion of "mediated attack" as primitive, i.e. a attacks b iff $a \rightarrow (\Rightarrow^{-1})^* b$, defining the neutrality function $\mathbf{n}^{ds}_{\mathcal{A}} : \mathcal{P}(A) \longrightarrow \mathcal{P}(A)$ as:

$$\mathbf{n}^{ds}_{\mathcal{A}}(X) = \{x \in A : \text{ NOT } X \rightarrow (\Rightarrow^{-1})^* x\}$$

and the defense function $\mathbf{d}^{ds}_{\mathcal{A}} : \mathcal{P}(A) \longrightarrow \mathcal{P}(A)$ as:

$$\mathbf{d}^{ds}_{\mathcal{A}}(X) = \{x \in A : \forall y \in A : \text{ IF } y \rightarrow (\Rightarrow^{-1})^* x \text{ THEN } X \rightarrow (\Rightarrow^{-1})^* y\}$$

Here again solution concepts are defined over the new defense and neutrality function and all results resumed in Fact 5 hold. In particular, any BAF $\mathcal{A}$ with deductive support where $\rightarrow (\Rightarrow^{-1})^*$ is well-founded has exactly one complete extension, which is also grounded, preferred and stable.

Two important properties of the solution concepts for BAF with deductive support are the following.

Theorem 2. *Let $\mathcal{A} = (A, \rightarrow, \Rightarrow)$ be a BAF.*

1. *Any $X \subseteq A$ s.t. $\mathtt{Adm}^{ds}_{\mathcal{A}}(X)$ does not contain any d+-attack.*
2. *Any $X \subseteq A$ s.t. $\mathtt{Cmp}^{ds}_{\mathcal{A}}(X)$ is closed for $\Rightarrow$.*

Proof. See Appendix.

[9] We may rephrase this condition as: for any $x \in X$, all of x's attackers are outside X and all x's supporters are inside.

3 Modal Logics for Bipolar Argumentation

Propositional modal logic with a universal modality is expressive enough to talk about standard AF [13,14]. A modal language to express the fundamental concepts of BAF requires instead the more complex resources of *propositional dynamic logic* (PDL) with the global universal modality $[U]$. Our language $\mathcal{L}_U$ It is built over a set of atoms P and a set of four basic actions $\Pi = \{\alpha, \beta, \alpha^{-1}, \beta^{-1}\}$ by the following BNF:

$$\phi ::= p \mid \bot \mid \neg\phi \mid \phi \wedge \phi \mid \langle\pi\rangle\phi \mid \langle U\rangle\phi \qquad \text{for } p \in P$$
$$\pi ::= \alpha \mid \beta \mid \alpha^{-1} \mid \beta^{-1} \mid \pi;\pi \mid \pi^*$$

Define other Boolean connectives (disjunction $\vee$, implication $\supset$ and bi-implication $\equiv$) as usual; take $[\pi]\phi := \neg\langle\pi\rangle\neg\phi$ and $[U]\phi := \neg\langle U\rangle\neg\phi$. The operator $\langle\alpha\rangle$ (resp. $\langle\beta\rangle$) is the "being attacked" (resp. "being supported") modality; e.g., $\langle\alpha\rangle\phi$ indicates that the argument is attacked by some argument labelled ϕ. Action α^{-1} (resp. β^{-1}) is the converse of α (resp. β), so $\langle\alpha^{-1}\rangle$ (resp. $\langle\beta^{-1}\rangle$) express the "attacks" (resp. "supports") modality.

Definition 6 (Bipolar models). *Let $\boldsymbol{P}$ be a set of atoms. A bipolar model is a tuple $\mathcal{M} = \langle\mathcal{A}, \mathcal{V}\rangle$, with $\mathcal{A} = \langle A, \rightarrow, \Rightarrow\rangle$ a BAF and $\mathcal{V} : \boldsymbol{P} \longrightarrow \mathcal{P}(A)$ a valuation function.*

$\mathfrak{M}$ denotes the set of models. The formal semantics of $\mathcal{L}_U$ is expressed via the notion of satisfaction of a formula in a model.

Definition 7 (Satisfaction). *The satisfaction of ϕ by a point a in a bipolar model $\mathcal{M} = \langle\mathcal{A}, \mathcal{V}\rangle$ is defined, for atoms and Boolean operators, in the standard way. For the rest,*

$$\mathcal{M}, a \models \langle\pi\rangle\phi \textit{ IFF } \exists b \in A : aR_\pi b \textit{ AND } \mathcal{M}, b \models \phi, \quad \mathcal{M}, a \models \langle U\rangle\phi \textit{ IFF } \exists b \in A : \mathcal{M}, b \models \phi$$

with R_α and R_β defined as the respective converses *of $\rightarrow$ and $\Rightarrow$, and the remaining R_π defined in the standard way.*[10] *The truth-set of ϕ in $\mathcal{M}$ is $[\![\phi]\!]_\mathcal{M} = \{a \in A \mid \mathcal{M}, a \models \phi\}$;*[11] *the set of valid formulae (those true in every point of every model) is called (logic) K_U.*

As it has been proved, the axiom system of Table 1 is sound and complete for K_U.[12] The first three groups of axioms together with rules (`[π]-Nec`) and (`LI`) provide a standard axiomatization for the PDL modalities [16]. Axioms (`[π]-Conv₁`) through (`[π]-Conv₄`) characterise the fact that $\rightarrow^{-1}$ and $\Rightarrow^{-1}$ are the converse of $\rightarrow$ and $\Rightarrow$ [2,16]. The fifth group consists of `S5` axioms for the universal modality and `Incl`, the latter determining the inclusion of any relation π in the universal accessibility relation.

---

[10] That is, $R_{\alpha^{-1}} = (R_\alpha)^{-1}$, $R_{\beta^{-1}} = (R_\beta)^{-1}$, $R_{\pi;\pi'} = \{(a,b) \in A \times A \mid \exists c \in A : (R_\pi ac \;\&\; R_{\pi'} cb)\}$ and $R_{\pi^*} = \bigcup_{n \geq 0} R_{\pi^n}$ (with $R_{\pi^0} = \{(a,a) \mid a \in A\}$ and $R_{\pi^{n+1}} = R_{\pi^n;\pi}$, for the latter).

[11] Thus, ***(i)*** $\mathcal{M}, a \models \langle\alpha\rangle\phi$ if and only if $\exists b \in A$ with $b \rightarrow a$ and $b \in [\![\phi]\!]_\mathcal{M}$, ***(ii)*** $\mathcal{M}, a \models \langle\beta\rangle\phi$ if and only if $\exists b \in A$ with $b \Rightarrow a$ and $b \in [\![\phi]\!]_\mathcal{M}$, and ***(iii)*** $\mathcal{M}, a \models [U]\phi$ if and only if $[\![\phi]\!]_\mathcal{M} = A$.

[12] See [2,16] for the PDL, converse and $[U]$ fragments. (See [21,22] for PDL+$[U]$).

Table 1. Axiom system for K_U

$\vdash \varphi$ for φ an instance of a propositional tautology		From $\vdash \varphi$ and $\vdash \varphi \supset \psi$ infer $\vdash \psi$	(MP)
$\vdash [\pi](p \supset q) \supset ([\pi]p \supset [\pi]q)$	([π]-Normality)	$\vdash [\pi]p \equiv \neg\langle\pi\rangle\neg p$	([π]-Dual)
$\vdash [\pi;\sigma]p \equiv [\pi][\sigma]p$	([$\pi;\sigma$])	$\vdash [\pi^*]p \equiv p \wedge [\pi][\pi^*]p$	([π^*])
$\vdash p \supset [\alpha]\langle\alpha^{-1}\rangle p$	([π]-Conv$_1$)	$\vdash p \supset [\alpha^{-1}]\langle\alpha\rangle p$	([π]-Conv$_2$)
$\vdash p \supset [\beta]\langle\beta^{-1}\rangle p$	([π]-Conv$_3$)	$\vdash p \supset [\beta^{-1}]\langle\beta\rangle p$	([π]-Conv$_4$)
$\vdash [U](p \supset q) \supset ([U]p \supset [U]q)$	([U]-Normality)	$\vdash [U]p \equiv \neg\langle U\rangle\neg p$	([U]-Dual)
$\vdash p \supset \langle U\rangle p$	([U]-Reflexivity)	$\vdash p \supset [U]\langle U\rangle p$	([U]-Symmetry)
$\vdash \langle U\rangle\langle U\rangle p \supset \langle U\rangle p$	([U]-Transitivity)	$\vdash \langle\pi\rangle p \supset \langle U\rangle p$	(Incl)
From $\vdash \phi \supset [\pi]\phi$ infer $\vdash \phi \supset [\pi^*]\phi$	(LI)	From $\vdash \phi$ infer $\vdash [\pi]\phi$	([π]-Nec)
		From $\vdash \phi$ infer $\vdash [U]\phi$	([U]-Nec)

Interestingly, $\mathcal{L}_U$ can define the class of AF in which a given action π is well-founded. Indeed, $[U]([\pi]p \supset p) \supset p$ holds in a AF if and only if R_π is well-founded [2, chap. 7.1]. Thus, it is possible to isolate the classes of AF and the classes of BAF with necessary (resp. deductive) support where extensions are unique (Fact 5(7)).[13]

Several solution concepts for BAF are expressible within $\mathcal{L}_U$, as those for standard AF are by standard modal logic [13]. In the case of necessary support, the property of not being attacked via a secondary attack (see Fig. 2(a)) by the set $[\![p]\!]$ is expressed by the concatenation $\neg\langle\beta^*;\alpha\rangle p$ which therefore can be taken to be the modal rendering of the neutrality function. Analogously the property of being defended by the set $[\![p]\!]$ is expressed by the concatenation $[\beta^*;\alpha]\langle\beta^*;\alpha\rangle p$, i.e. the defense function. This provides the following list of characterizations.

Proposition 1 (Solution concepts for necessary supports). *For any* $(\mathcal{A}, \mathcal{V}), a$,

$$\mathcal{V}(p) \in \mathtt{Cfr}^{ns}_{\mathcal{A}} \quad \textit{IFF} \quad (\mathcal{A},\mathcal{V}), a \models [U](p \supset \neg\langle\beta^*;\alpha\rangle p)$$
$$\mathcal{V}(p) \in \mathtt{Sdf}^{ns}_{\mathcal{A}} \quad \textit{IFF} \quad (\mathcal{A},\mathcal{V}), a \models [U](p \supset [\beta^*;\alpha]\langle\beta^*;\alpha\rangle p)$$
$$\mathcal{V}(p) \in \mathtt{Adm}^{ns}_{\mathcal{A}} \quad \textit{IFF} \quad (\mathcal{A},\mathcal{V}), a \models [U](p \supset \neg\langle\beta^*;\alpha\rangle p) \wedge [U](p \supset [\beta^*;\alpha]\langle\beta^*;\alpha\rangle p)$$
$$\mathcal{V}(p) \in \mathtt{Cmp}^{ns}_{\mathcal{A}} \quad \textit{IFF} \quad (\mathcal{A},\mathcal{V}), a \models [U](p \supset \neg\langle\beta^*;\alpha\rangle p) \wedge [U](p \equiv [\beta^*;\alpha]\langle\beta^*;\alpha\rangle p)$$
$$\mathcal{V}(p) \in \mathtt{Stb}^{ns}_{\mathcal{A}} \quad \textit{IFF} \quad (\mathcal{A},\mathcal{V}), a \models [U](p \equiv \neg\langle\beta^*;\alpha\rangle p)$$

Furthermore, $[\beta^*;\alpha]\langle\beta^*;\alpha\rangle$ is equivalent to $\neg\langle\beta^*;\alpha\rangle\neg\langle\beta^*;\alpha\rangle p$, so the defense function is the double iteration of the neutrality function (see [13]). Thus, the

[13] Thus, the formulas $[U]([\alpha]p \supset p) \supset p$, $[U]([\beta^*;\alpha]p \supset p) \supset p$ and $[U]([\beta^{-1*};\alpha]p \supset p) \supset p$ characterise, respectively, the well-foundedness of $\rightarrow$, $\rightarrow\Rightarrow^*$ and $\rightarrow (\Rightarrow^{-1})^*$.

fact that $\mathtt{Stb}^{ns}_{\mathcal{A}}(X)$ entails $X = \mathbf{n}_{\mathcal{A}}(X) \cap [\Rightarrow^{-1}](X)$ (Theorem 1(4), right to left) can be restated in modal terms:

Fact 8. *For any bipolar model* $\mathcal{M} = \langle \mathcal{A}, \mathcal{V} \rangle$ *and any* $a \in \mathcal{A}$,

$$\mathtt{Stb}_{\mathcal{A}}([\![p]\!]_{\mathcal{M}}) \text{ entails } \mathcal{M}, a \models [U](p \equiv (\neg\langle\alpha\rangle p \wedge [\beta]p)).$$

For deductive support, the property of not being attacked via a mediated attack (see Fig. 3(b)) by the set $[\![p]\!]$ is expressed by $\neg\langle\beta^{-1*};\alpha\rangle p$. Therefore, by the same mechanism we can provide the following modal definitions for solution concepts of BAF with deductive support.

Proposition 2 (Solution concepts for deductive supports). *For any* $(\mathcal{A}, \mathcal{V}), a$,

$$\mathcal{V}(p) \in \mathtt{Cfr}^{ds}_{\mathcal{A}} \text{ IFF } (\mathcal{A}, \mathcal{V}), a \models [U](p \supset \neg\langle\beta^{-1*};\alpha\rangle p)$$
$$\mathcal{V}(p) \in \mathtt{Sdf}^{ds}_{\mathcal{A}} \text{ IFF } (\mathcal{A}, \mathcal{V}), a \models [U](p \supset [\beta^{-1*};\alpha]\langle\beta^{-1*};\alpha\rangle p)$$
$$\mathcal{V}(p) \in \mathtt{Adm}^{ds}_{\mathcal{A}} \text{ IFF } (\mathcal{A}, \mathcal{V}), a \models [U](p \supset \neg\langle\beta^{-1*};\alpha\rangle p) \wedge [U](p \supset [\beta^{*};\alpha]\langle\beta^{-1*};\alpha\rangle p)$$
$$\mathcal{V}(p) \in \mathtt{Cmp}^{ds}_{\mathcal{A}} \text{ IFF } (\mathcal{A}, \mathcal{V}), a \models [U](p \supset \neg\langle\beta^{-1*};\alpha\rangle p) \wedge [U](p \equiv [\beta^{-1*};\alpha]\langle\beta^{-1*};\alpha\rangle p)$$
$$\mathcal{V}(p) \in \mathtt{Stb}^{ds}_{\mathcal{A}} \text{ IFF } (\mathcal{A}, \mathcal{V}), a \models [U](p \equiv \neg\langle\beta^{-1*};\alpha\rangle p)$$

Here too the concatenation $[\beta^{-1*};\alpha]\langle\beta^{-1*};\alpha\rangle$ is equivalent to $\neg\langle\beta^{-1*};\alpha\rangle\neg\langle\beta^{-1*};\alpha\rangle p$; thus, the defense function is the double iteration of the neutrality function.

4 Bipolarity and Semantic Paradoxes

The *Liar Paradox* consists of any statement of the following kind

$$a := \text{ The statement } a \text{ is false}$$

to which no true or false value can be assigned. Early diagnoses of the problem pointed to the *self-reference* of statement a as the culprit. In many cases, however, self-reference is not direct, as the following paradox shows [24]:

$a :=$ The statement b is true and the statement c is false.
$b :=$ Either the statement a is false or the statement c is true
$c :=$ Both statements a and b are true.

Moreover, Yablo's paradox [25] provides an example with no referential circuits of the above kind. Therefore, although the problem lies clearly in the referential structure of the discourse, it is more complex than what an intuitive understanding of "self-referentiality" and "circularity" may suggest.

An important clue for clarifying this structural problem comes from two relatively new approaches to semantic paradoxes. One of them is the equational approach by [24] and the other is a graph-theoretic one [8, 10–12, 23]. In its bare bones, the equational approach interprets referential discourses of the above kind as systems of boolean equations, or equivalently as sets of biconditionals where referential statements figure as a set of propositional variables A. The Liar is then translated as the biconditional $a \equiv \neg a$, while the second example consists of the three biconditionals $a \equiv b \wedge \neg c$, $b \equiv \neg a \vee c$ and $c \equiv a \wedge b$.

Both examples determine a propositional theory $\mathcal{T}$ that is *paradoxical* insofar as $\mathcal{MOD}(\mathcal{T}) = \emptyset$, where $\mathcal{MOD}(\mathcal{T})$ denotes the set of propositional assignments $v : A \longrightarrow \{1, 0\}$ that satisfy the theory. It has been shown [1,11] that any system of boolean equations $\mathcal{T}$ can be transformed into and equivalent $\mathcal{T}'$ in *digraph normal form*, i.e. a theory consisting of a set $S = \{s_0, \ldots, s_n\}$ of n sentences of the form

$$s_i := x_i \equiv \bigwedge_{x \in X_i} \neg x$$

for $0 \leq i \leq n$, where by convention $\bigwedge \emptyset = 1$. Any such $\mathcal{T}'$ can be represented by a corresponding AF $\mathcal{A}(\mathcal{T}') = (A, \rightarrow)$ defined as follows [10,11]:

$$A = \bigcup_{i \leq n} (\{x_i\} \cup X_i \cup \{\overline{x} \mid x \in X_i \wedge \forall i \leq n : x \neq x_i\})$$

$$\rightarrow = (\bigcup_{i \leq n} \{(x, x_i) \mid x \in X_i\}) \cup \{(x, \overline{x}), (\overline{x}, x) \mid x \in A\}$$

Note that a cannot be true (accepted) if b is true (accepted); hence, an attack $b \rightarrow a$ encodes "$a := b$ is false". Moreover, there are mutual attacks between newly added $\overline{x}$ and those x which would otherwise be unattacked (thus forced to be true). The intuitive meaning of the attack relation is captured by a *complete labelling* [4], defined for any AF $\mathcal{A} = (A, \rightarrow)$ as a (partial) function $l : A \longrightarrow \{1, 0\}$ such that, for every $a \in A$,

1. $l(a) = 1$ iff $\forall b$, $b \rightarrow a$ entails $l(b) = 0$
2. $l(a) = 0$ iff $\exists b$, $b \rightarrow a$ and $l(b) = 1$

For a given $\mathcal{A}(\mathcal{T})$, any such labelling l can be regarded as a propositional assignment to the set $V(\mathcal{T})$ of variables occurring in $\mathcal{T}$. In general, given l, we denote by $l \uparrow_{V(\mathcal{T})}$ the restriction of l to such set and by l^* the valuation of propositional formulas induced by l. By l^1 we denote the set $\{a \in A \mid l(a) = 1\}$. Then the following correspondence holds:

Fact 9 ([10]). *For any theory $\mathcal{T}$ in digraph normal form and any labelling l of $\mathcal{A}(\mathcal{T})$:*

$$l \uparrow_{V(\mathcal{T})} \in \mathcal{MOD}(\mathcal{T}) \text{ iff } l^1 \text{ is a stable extension of } \mathcal{A}(\mathcal{T})$$

An important consequence of this fact is that any paradoxical theory $\mathcal{T}$ corresponds to a graph with no stable extension (the Liar corresponds to a single node with a self-loop) and this provides an interesting structural criterion for understanding paradoxicality.

What is important here is that, *a fortiori*, any propositional theory can also be translated in what one may call a *bipolar digraph normal form* (see [23]), i.e. as a set of sentences of the following form:

$$x_i \equiv \bigwedge_{x \in X_i} \neg x \wedge \bigwedge_{x \in Y_i} y$$

Any such theory gives rise to a corresponding BAF $\mathcal{A}(\mathcal{T}) = (A, \rightarrow, \Rightarrow)$ where A and $\rightarrow$ are as before and

$$\Rightarrow = \bigcup_{i \leq n} \{(y, x_i) \mid y \in Y_i\}$$

Here $b \Rightarrow a$ encodes "$a := b$ is true", since the truth of every conjunct b is a necessary condition for the truth of a. Therefore the bipolar digraph normal form and its corresponding BAF are a natural and more compact way to represent referential discourses with both predicates "true" and "false".

Let us define a labelling l for bipolar graphs as follows:

1. $l(a) = 1$ iff ($\forall b$, $b \rightarrow a$ entails $l(b) = 0$ and $\forall c$, $c \Rightarrow a$ entails $l(c) = 1$)
2. $l(a) = 0$ iff otherwise

Then it is possible to establish the following correspondence

Theorem 3. *Let $\mathcal{T}$ be a theory in bipolar digraph normal form such that $\mathcal{A}(\mathcal{T})$ is well-founded for $\Rightarrow$. Then the following holds for any labelling l:*

$$l \uparrow_{V(\mathcal{T})} \in \mathcal{MOD}(\mathcal{T}) \text{ iff } \mathcal{S}tb^{ns}_{\mathcal{A}(\mathcal{T})}(l^1)$$

Proof. See Appendix.

Stability provides a general clue for understanding several patterns of paradox. For example, consider Yablo's paradox, which consists of a numerable set of biconditionals with infinite conjunctions on the right side, of the form $x_n \equiv \bigwedge_{k>n} \neg x_k$, with $n \in \mathbb{N}$. It is indeed a propositional theory whose corresponding graph, represented in Fig. 4, lacks a stable extension.

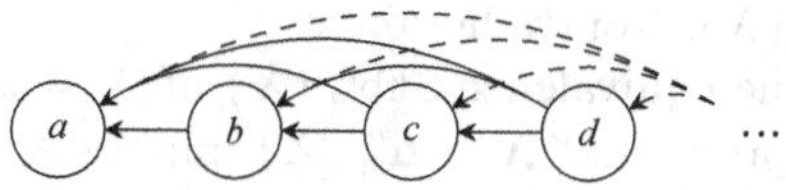

Fig. 4. Yablo's paradox

Interestingly, from the point of view of modal logic a labelling can be seen as a valuation $\mathcal{V}_l : \{\mathbf{0},\mathbf{1}\} \longrightarrow \mathcal{P}(A)$ over the set of propositional letters $\mathbf{0}$ and $\mathbf{1}$, which satisfies the conditions 1 and 2 above. By our remark in Sect. 3, any $\mathcal{A}(\mathcal{T})$ with a well-founded $\Rightarrow$ is a structure such that $\mathcal{A}(\mathcal{T}) \models [U]([\beta]p \supset p) \supset p$. Within this class, the paradoxal structures are those where there is no labelling l such that $(\mathcal{A}(\mathcal{T}), \mathcal{V}_l) \models [U](\mathbf{1} \equiv \neg\langle\beta^{-1*}; \alpha\rangle\mathbf{1})$.

5 Conclusions

This work provides a new approach to the study of BAF where the fundamental solution concepts are introduced by means of the neutrality and the defense function in a systematic way. We also show how PDL provides an adequate modal language to talk about BAF. Finally, BAF with necessary support are employed to encode the referential discourse contained in semantic paradoxes as the Liar. It is shown that the paradoxality of a referential discourse $\mathcal{T}$ corresponds to the absence of a stable solution for the generated BAF $\mathcal{A}(\mathcal{T})$ whenever $\mathcal{A}(\mathcal{T})$ is well-founded for the support relation. A problem however arises when the support relation is not well-founded. This is the case of a propositional theory as the following: $a \equiv \neg a \wedge b$ and $b \equiv a$. Here the labelling $l(a) = 0; l(b) = 0$ provides a model. However, it is easy to ascertain that $l^1 = \emptyset$ is not a ns-stable extension for the corresponding graph. This leaves open the problem of finding an adequate full correspondence. We leave this for future work.

Appendix

Proof of Theorem 1:

1. It suffices to show that no n+-attack is possible. Suppose $\texttt{Adm}^{ns}_{\mathcal{A}}(X)$ and that X contains a and b such that $a(\Rightarrow^{-1})^* \rightarrow\Rightarrow^* b$. Then there is a c such that $a(\Rightarrow^{-1})^* c$ and $c \rightarrow\Rightarrow^* b$ (as in Fig. 2(c)). Therefore c carries a secondary attack towards b. But since X is admissible it defends b against c, i.e. there is $d \in X$ such that $d \rightarrow\Rightarrow^* c$. But then $d \rightarrow\Rightarrow^* a$, i.e. $d \in X$ attacks $a \in X$, against the assumption that $\texttt{Cfr}^{ns}_{\mathcal{A}}(X)$.
2. Suppose $a \in X$ and $b \Rightarrow a$. As $\texttt{Cmp}^{ns}_{\mathcal{A}}(X)$ implies $X = \mathbf{d}^{ns}_{\mathcal{A}}(X)$, it is enough to show that $b \in \mathbf{d}^{ns}_{\mathcal{A}}(X)$. Indeed, if $b \notin \mathbf{d}^{ns}_{\mathcal{A}}(X)$ then $\exists c \rightarrow\Rightarrow^* b$ and not $X \rightarrow\Rightarrow^* c$. But then $c \rightarrow\Rightarrow^* a$ and X does not defend a, from which we get a contradiction by the completeness of X. Therefore $b \in \mathbf{d}^{ns}_{\mathcal{A}}(X)$.
3. Suppose $a \in X$ and $\exists b \in X$ such that $b \rightarrow\Rightarrow^* a$. Then there is a c such that $b \rightarrow c$ and $c \Rightarrow^* a$. Since $a \in X$ we get, by $\Rightarrow^{-1}$-closure, that $c \in X$, which entails that $X \nsubseteq \mathbf{n}_{\mathcal{A}}(X)$. Contradiction.
4. The proof exploits the equivalence $\texttt{Stb}^{ns}_{\mathcal{A}}(X)$ iff $X = \mathbf{n}^{ns}_{\mathcal{A}}(X)$ (Definition 4). It is not difficult to prove that $X = \mathbf{n}^{ns}_{\mathcal{A}}(X)$ implies $X = \mathbf{n}_{\mathcal{A}}(X) \cap [\Rightarrow^{-1}](X)$ even without restriction to well-foundedness of $\Rightarrow$. We skip this part here.

For the other direction we need to prove that $\mathbf{n}^{ns}_{\mathcal{A}}(X) = X$. The only difficult part is $\mathbf{n}^{ns}_{\mathcal{A}}(X) \subseteq X$, the converse inclusion being almost immediate. For this it suffices to show that $a \notin X$ implies $a \notin \mathbf{n}_{\mathcal{A}}(X)$. Suppose $a \notin X$. Therefore, by $X = \mathbf{n}_{\mathcal{A}}(X) \cap [\Rightarrow^{-1}](X)$, either (a) $\exists c_0 \in X$ such that $c_0 \rightarrow a$, in which case $a \notin \mathbf{n}^{ns}_{\mathcal{A}}(X)$ and we are done, or else (b) $\exists b_0 \notin X$ such that $b_0 \Rightarrow a$. The same reasoning applies to b_0: either (a) $\exists c_1 \in X$ such that $c_1 \rightarrow b$, in which case $a \notin \mathbf{n}^{ns}_{\mathcal{A}}(X)$ (since $c_1 \rightarrow\Rightarrow^* a$), or else (b) $\exists b_1 \notin X$ such that $b_1 \Rightarrow b_0$. Alternative (b) can only apply a finite number of times, otherwise it would determine an infinite descending chain of supports, which is excluded by the well-foundedness of $\Rightarrow$. Therefore $a \notin \mathbf{n}^{ns}_{\mathcal{A}}(X)$ and the inclusion is proved.

Proof of Theorem 2:

1. Suppose that $\mathtt{Adm}^{ds}_{\mathcal{A}}(X)$ and X contains both a and b with $a \Rightarrow^{*}\rightarrow (\Rightarrow^{-1})^{*}b$. Then there is a c such that $a \Rightarrow^{*} c$ and $c \rightarrow (\Rightarrow^{-1})^{*}b$. Therefore c carries a mediated attack towards b. Since X is admissible it defends b against c, i.e. there is $d \in X$ such that $d \rightarrow (\Rightarrow^{-1})^{*}c$. But then $d \rightarrow (\Rightarrow^{-1})^{*}a$, against the conflict-freeness of X.
2. Suppose $a \in X$ and $a \Rightarrow b$. As $\mathtt{Cmp}^{ds}_{\mathcal{A}}(X)$ implies $X = \mathbf{d}_{\mathcal{A}}(X)$, it is enough to show that $b \in \mathbf{d}_{\mathcal{A}}(X)$. Suppose $b \notin \mathbf{d}_{\mathcal{A}}(X)$; then $\exists c \rightarrow (\Rightarrow^{-1})^{*}b$ and not $X \rightarrow (\Rightarrow^{-1})^{*}c$. But then clearly $c \rightarrow (\Rightarrow^{-1})^{*}a$ and X does not defend a, a contradiction. Therefore $b \in \mathbf{d}_{\mathcal{A}}(X)$.

Proof of Theorem 3:

1. From right to left. Assume that $\mathtt{Stb}^{ns}_{\mathcal{A}(\mathcal{T})}(l^1)$. Consider any biconditional $\phi := x_i \equiv \bigwedge_{x \in X_i} \neg x \wedge \bigwedge_{x \in Y_i} y$ in the theory. There are two cases: (a) $l(x_i) = 1$, i.e. $x_i \in l^1$. An immediate consequence of this, by Theorem 1(4, left-to-right) is that for all attacker x of x_i: $x \notin l^1$, i.e. $x \in l^0$ by the given definition of labelling, and for all supporter y of x_i: $x \in l^1$ (closure of stable sets under support Theorem 1(2)). This suffices to guarantee that $l^*(\bigwedge_{x \in X_i} \neg x \wedge \bigwedge_{x \in Y_i} y) = 1$ and then $l^*(\phi) = 1$.
   (b) $l(x_i) = 0$. Since l^0 is the complement of l^1, by Theorem 1(4, left-to-right) either some attacker x of x_i: $x \in l^1$, or some supporter y of x_i: $y \in l^0$. By construction of $\mathcal{A}(\mathcal{T})$ all supporters and attackers figure on the right handside of ϕ. As a consequence $l^*(\bigwedge_{x \in X_i} \neg x \wedge \bigwedge_{x \in Y_i} y) = 0$ and then $l^*(\phi) = 1$.
2. From left to right. Assume that l is such that $l^*(\phi) = 1$ for all $\phi \in \mathcal{T}$. In order to show that $\mathtt{Stb}^{ns}_{\mathcal{A}(\mathcal{T})}(l^1)$ we need to prove that $l^1 = \mathbf{n}^{ns}_{\mathcal{A}}(l^1)$. We first prove that
   (a) $l^1 \subseteq \mathbf{n}^{ns}_{\mathcal{A}}(l^1)$. Let $x \in l^1$. We have three cases. (a.1) x is of the form $\overline{y}$. Then by construction x has no supporters and is only attacked by y. Then $y \in l^0$ by condition 1 on labellings. Since, by construction, y is the only (direct or indirect) attacker of x, it follows that $x \in \mathbf{n}^{ns}_{\mathcal{A}}(l^1)$. (a.2) x appears only on the right hand side of a biconditional. Again, by construction, x has no supporters and is only attacked directly by $\overline{x}$, which however is labelled 0. Ergo $x \in \mathbf{n}^{ns}_{\mathcal{A}}(l^1)$. Otherwise suppose that (a.3) $x \in l^1$ appears on the left hand side of some biconditional. If $x \notin \mathbf{n}^{ns}_{\mathcal{A}}(l^1)$ then there is a chain $y_0 \rightarrow y_1 \Rightarrow \cdots \Rightarrow y_n \Rightarrow x$ such that $y_0 \in l^1$, $y_1, \ldots, y_n \in V(\mathcal{T})$, and at least $y_2, \ldots, y_n$ appear on the right hand side of some biconditional. This forces $y_1, \ldots, y_n \in l^1$. But then $l^1 \rightarrow y_1$ against condition 1 on labelling. Therefore $x \in \mathbf{n}^{ns}_{\mathcal{A}}(l^1)$.
   (b) $\mathbf{n}^{ns}_{\mathcal{A}}(l^1) \subseteq l^1$. For this is sufficient to show that for every $x \notin l^1$ there is an $y \in l^1$: $y \rightarrow\Rightarrow^{*} x$. It is straightforward to prove this for the cases where

(b.1) x is of the form $\overline{y}$ or (b.2) x appears only on the right hand side of a biconditional.We consider (b.3) $x = x_i \notin l^1$ appears on the left hand side of some biconditional $\phi := x_i \equiv \bigwedge_{x \in X_i} \neg x \wedge \bigwedge_{x \in Y_i} y$. Since $x_i \in l^0$ (the complement of l^1) and $l^*(\phi) = 1$ by assumption, then either one of the conjuncts $z \in X_i$ is in l^1, in which case $x_i \notin \mathbf{n}^{ns}_{\mathcal{A}}(l^1)$ and we are done, or else one of the conjuncts $y \in Y_i$ is in l^0. If y is as in (b.1) or (b.2) then it is attacked by l^1 and therefore $x_i \notin \mathbf{n}^{ns}_{\mathcal{A}}(l^1)$. Otherwise y is either attacked by l^1 or supported by some y' in l^0. However the chain of supports cannot go on forever because the support relation is well-founded by assumption. Therefore we should finally find some attacker in l^1 and $x_i \notin \mathbf{n}^{ns}_{\mathcal{A}}(l^1)$.

References

1. Bezem, M., Grabmayer, C., Walicki, M.: Expressive power of digraph solvability. Ann. Pure Appl. Logic **162**(3), 200–212 (2012)
2. Blackburn, P., de Rijke, M., Venema, Y.: Modal Logic. Cambridge University Press, Cambridge (2001)
3. Boella, G., Gabbay, D.M., van der Torre, L., Villata, S.: Support in abstract argumentation. In: COMMA 2010, pp. 111–122. IOS Press (2010)
4. Caminada, M., Gabbay, D.: A logical account of abstract argumentation. Studia Logica **93**, 109–145 (2009)
5. Cayrol, C., Lagasquie-Schiex, M.C.: On the acceptability of arguments in bipolar argumentation frameworks. In: Godo, L. (ed.) ECSQARU 2005. LNCS (LNAI), vol. 3571, pp. 378–389. Springer, Heidelberg (2005). https://doi.org/10.1007/11518655_33
6. Cayrol, C., Lagasquie-Schiex, M.C.: Bipolarity in argumentation graphs: towards a better understanding. Int. J. Approx. Reasoning **54**(7), 876–899 (2013)
7. Cayrol, C., Lagasquie-Schiex, M.-C.: An axiomatic approach to support in argumentation. In: Black, E., Modgil, S., Oren, N. (eds.) TAFA 2015. LNCS (LNAI), vol. 9524, pp. 74–91. Springer, Cham (2015). https://doi.org/10.1007/978-3-319-28460-6_5
8. Cook, R.: Patterns of paradox. J. Symbolic Logic **69**(3), 767–774 (2004)
9. Dung, P.M.: On the acceptability of arguments and its fundamental role in nonmonotonic reasoning, logic programming and n-person games. Artif. Intell. **77**(2), 321–357 (1995)
10. Dyrkolbotn, S.: Doing argumentation using theories in graph normal form. In: Rendsvig, R., Katrenko, S. (eds.) ESSLLI 2012 Student Session Proceedings, pp. 13–22 (2012)
11. Dyrkolbotn, S., Walicki, M.: Propositional discourse logic. Ann. Pure Appl. Logic **162**(3), 200–212 (2014)
12. Gaifman, H.: Operational pointer semantics: solution to self-referential puzzles. In: Vardi, M. (ed.) Theoretical Aspects of Reasoning About Knowledge, pp. 43–59 (1988)
13. Grossi, D.: On the logic of argumentation theory. In: van der Hoek, W., Kaminka, G., Lespérance, Y., Sen, S. (eds.) Proceedings of the 9th International Conference on Autonomous Agents and Multiagent Systems (AAMAS 2010), pp. 409–416 (2010)

14. Grossi, D.: Argumentation in the view of modal logic. In: McBurney, P., Rahwan, I., Parsons, S. (eds.) ArgMAS 2010. LNCS (LNAI), vol. 6614, pp. 190–208. Springer, Heidelberg (2011). https://doi.org/10.1007/978-3-642-21940-5_12
15. Grossi, D., Gabbay, D.: When are two arguments the same? Invariance in abstract argumentation. In: Baltag, A., Smets, S. (eds.) Johan van Benthem on Logic and Information Dynamics, pp. 677–701 (2012)
16. Harel, D., Tiuryn, J., Kozen, D.: Dynamic Logic. MIT Press, Cambridge (2000)
17. Nouioua, F., Risch, V.: Bipolar argumentation frameworks with specialized supports. In: 22nd International Conference on Tools with Artificial Intelligence, vol. 1, pp. 215–218 (2010)
18. Nouioua, F., Risch, V.: Argumentation frameworks with necessities. In: Benferhat, S., Grant, J. (eds.) SUM 2011. LNCS (LNAI), vol. 6929, pp. 163–176. Springer, Heidelberg (2011). https://doi.org/10.1007/978-3-642-23963-2_14
19. Nouioua, F.: AFs with necessities: further semantics and labelling characterization. In: Liu, W., Subrahmanian, V.S., Wijsen, J. (eds.) SUM 2013. LNCS (LNAI), vol. 8078, pp. 120–133. Springer, Heidelberg (2013). https://doi.org/10.1007/978-3-642-40381-1_10
20. Oren, N., Norman, T.J.: Semantics for evidence-based argumentation. In: COMMA 2008, pp. 276–284 (2008)
21. Passy, S., Tinchev, T.: PDL with data constants. Inf. Process. Lett. **20**(1), 35–41 (1985)
22. Passy, S., Tinchev, T.: Quantifiers in combinatory PDL: completeness, definability, incompleteness. In: Budach, L. (ed.) FCT 1985. LNCS, vol. 199, pp. 512–519. Springer, Heidelberg (1985). https://doi.org/10.1007/BFb0028835
23. Walicki, M.: Reference, paradoxes and truth. Synthese **171**, 195–226 (2009)
24. Wen, L.: Semantic paradoxes as equations. Math. Intell. **23**(1), 43–48 (2001)
25. Yablo, S.: Paradox without self-reference. Analysis **53**(4), 251–252 (1993)

Improving Judgment Reliability in Social Networks via Jury Theorems

Paolo Galeazzi[1], Rasmus K. Rendsvig[1](✉), and Marija Slavkovik[2]

[1] Center for Information and Bubble Studies, University of Copenhagen, Copenhagen, Denmark
pagale87@gmail.com, rasmus@hum.ku.dk

[2] Department of Information Science and Media Studies, University of Bergen, Bergen, Norway
marija.slavkovik@uib.no

Abstract. Opinion aggregators—such as 'like' or 'retweet' counters—are ubiquitous on social media platforms and often treated as implicit quality evaluations of the entry liked or retweeted, with higher counts indicating higher quality. Many such aggregators are poor quality evaluators as they allow disruptions of the conditions for positive wisdom-of-the-crowds effects. This paper proposes a design of theoretically justified aggregators that improve judgment reliability. Interpreting states of diffusion processes on social networks as implicit voting scenarios, we specify procedures for isolating sets of independent voters in order to use jury theorems to quantify the reliability of network states as quality evaluators. As real-world networks tend to grow very large and independence tests are computationally expensive, a primary goal is to limit the number of such tests. We consider five procedures, each trading a degree of reliability for efficiency, the most efficient requiring a low-degree polynomial number of tests.

1 Introduction

Web2.0 introduced the possibility for internet consumers to also become content creators [1]. This change led to an unprecedented amount of information being shared online, but also to a deterioration in the quality control of the information that is being shared. In some areas, such as medicine, the quality of the information online has been a now long-standing concern (e.g., [2,3]), leading to the development of a code of conduct and certification standards [4]. Recently, we are witnessing the problem of low quality or even damaging content spreading in many areas, with the problem of so-called "fake news" being particularly worrisome [5,6].

With the content quality problem being an all-topic concerning issue, the challenge is to find a way to automatically separate the reliable content from bad quality, unreliable, offensive and even illegal content. As a possible improvement of the current situation, here we propose that a good use of the consumers' behavior can help ascertain the quality of a post.

A ubiquitous feature of social media platforms is the ability of users to directly express their impressions and opinions about media content making

P. Blackburn et al. (Eds.): LORI 2019, LNCS 11813, pp. 230–243, 2019.
https://doi.org/10.1007/978-3-662-60292-8_17

its rounds on the platform. Beyond the option of reporting content for violations of law or community standards, there are three main, widely available channels for such expressions. One is by reaction buttons such as Twitter and Instagram's ♡ buttons, Reddit or YouTube's up- and downvotes, or Facebook's six choices of emoticons. The second is by textual reply, such as Twitter's reply option or Facebook's comments. The third, and the focus of this paper, is by sharing the content with one's social network, as e.g. by Twitter's retweet option.

In this paper, we assume that the aforementioned "judgment" expressions—and especially the decision on whether to further propagate content through sharing—may be seen as quality signifiers, while remaining aware that this is not necessarily always the case in social media, where an upvote may be a herding reaction [7] and sharing (or not) may be influenced more by emotional response than by sound quality judgment [8].

Interpreting decisions to share or not as quality signifiers, we aim to design procedures which aggregate such judgment expressions into a reliable collective quality judgment through majority rule. Yet, a collection of judgment expressions is not sufficient to provide a good and reliable collective judgment. Some users may not be competent at ascertaining content quality, if for no other reason than by not being human [9]. Even when judgment expressions are from competent users, the majority verdict cannot simply be taken as a theoretically sound quality assessment: due to herding effects or the mimicking of salient users, influenced signals may stop carrying information about a competent quality judgment but opaquely repeat already accounted for signals. Such dependencies between judgments invalidate the premises of the Condorcet Jury Theorem [10] and its many generalizations (e.g., [11–14]) and thus disrupt the theoretical foundations of positive wisdom-of-the-crowds effects. When dependent signals are present, majority polling can no longer be trusted to reach the correct evaluation with high probability. To reestablish the positive wisdom-of-the-crowds effects of jury theorems, a main focus of the paper is on the identification of sets of independent users, whose majority vote may be trusted to be correct with high probability.

Our main contribution is the new approach to providing a quality evaluation of a post, given the current state of a diffusion process of the post in a social network. We suggest to obtain quality evaluations by considering a subset (the jury) of agents that have been exposed to the post through its diffusion in the network. The majority vote in the jury together with its correctness probability then constitutes the quality evaluation of the post (Sect. 3). How to select a jury among all agents exposed to the post is the main consideration of the paper. For simple jury theorems to apply (Sect. 3), the jury must be independent. Independence may be established statistically (Sect. 4), given a presumed set of diffusion processes serving as background data (Sect. 2). Yet, independence tests are computationally expensive: as real-world networks tend to grow large, the number of tests required to find a satisfactory jury should ideally be minimized for efficiency. In Sect. 5, we consider five jury selection procedures, each trading a degree of reliability for efficiency, and analyze the number of required tests of each of them, ranging from being exponential in the size of the set of exposed agents to being low-degree polynomial.

In designing the procedures, we need to make several choices regarding our priorities. Ideally, we desire a quality evaluation procedure that (*i*) is theoretically justifiable, (*ii*) is computationally efficient, (*iii*) uses only data endogenous to the social network on which it is implemented, and (*iv*) is conservative in its estimates, so that in choosing between two evils, we rather want a measure overly cautious with respect to estimating an entry as high quality than *vice versa*. The last point entails that we rather ignore the voice of a juror that might be beneficial than include the vote of a juror that may be misleading. Finally, in this paper, we prioritize (*i*) over (*ii*), retaining the use of expensive independence tests. We hope future work on such problems will improve efficiency.

2 Social Networks and Diffusion

A **social network** is specified by a set of links $N \subseteq \mathcal{A}^2$ connecting finitely many **agents** $\mathcal{A}$. We interpret $(a, b) \in N$ to mean that a sees the content shared by b. Let $N(a) = \{b \in \mathcal{A} : (a, b) \in N\}$. We assume no properties of N.

In the following, each network N is associated with a **topic**, identified with a set of **entries** $E = \{e_1, e_2, ...\}$. We think of N as used by the agents to discuss the topic E with the desire to evaluate the quality of each entry. We assume that entries bear an objective quality, or truth value, given by an exogenous **quality valuation** $V : E \to \{\bot, \top\}$, unknown to the agents. We interpret agents as casting votes on the quality of each entry. Given an entry, $\top_a$ is the event that agent a votes for the entry being of high quality, and $\bot_a$ that a votes for it being of low quality. As it is common in jury theorems, under the assumption of independence, each agent $a \in \mathcal{A}$ is assumed to have the same individual **correctness probability** $c \in [0, 1]$ in their assessment, formally expressed in terms of the following conditional probabilities: $\forall a \in \mathcal{A}$,

$$p(\top_a \,|\, \top) = p(\bot_a \,|\, \bot) = c = 1 - p(\top_a \,|\, \bot) = 1 - p(\bot_a \,|\, \top).$$

Remark 1. The homogeneous correctness probability is a fundamental assumption for the jury theorem applied throughout, and deserves comment. First, we use homogeneous correctness probabilities for simplicity: jury theorems heterogeneous correctness probabilities also exist—see e.g. [11]. The procedures introduced below assume knowledge of the correctness probability of the individuals, and that this correctness probability is homogeneous for all users. In this respect, we remark that knowledge of the correctness probability c may be obtained through lab experiments where single individuals are tested in isolation and separately from the network interaction, just as individual utility functions in economics are elicited through lab tests, separately from interactive game-theoretic scenarios (see e.g. [15]). This would provide the aforementioned assumptions with testable behavioral foundations. The details of this process are outside of the scope of this work.

To describe the information flow through the network, define a **state** s of N as a pair (e, L_i) with e an entry and $L_i : \mathcal{A} \to \{\mathtt{S}, \mathtt{R}, \mathtt{U}\}$ a labeling map. Throughout, let $S_i := L_i^{-1}(\mathtt{S})$, $R_i := L_i^{-1}(\mathtt{R})$ and $U_i := L_i^{-1}(\mathtt{U})$. Agents in S_i have chosen to *share* e, those in R_i have been *reached* by e (e.g., by having a neighbor in S_i), and agents in U_i are *unreached* by e. An **initial state** $s_0 = (e, L_0)$ satisfies $S_0 = \emptyset$. A state $s_i = (e, L_i)$ can then **transition** to state $s_j = (e', L_j)$ if and only if

1. $e' = e$
2. $S_i \subseteq S_j$ and $S_j \backslash S_i \subseteq R_i$
3. $R_j = (R_i \backslash S_j) \cup \{a \in U_i : \exists b \in S_j \text{ and } b \in N(a)\}$
4. $U_j = U_i \backslash R_j$

I.e.: 1. the entry stays fixed, 2. agents never un-share and only reached agents can start sharing, 3. sharing agents stop being reached, but agents with a sharing neighbor become reached, and 4. else agents remain unreached.

A **diffusion process** is a sequence of states $d = s_0, s_1, \ldots$ such that (i) s_0 is initial, (ii) all other states are transitions from the previous state, and (iii) if $s_k = s_{k+1}$, then $s_k = s_{k+n}$ for all $n \in \mathbb{N}$. These transition rules ensure that any diffusion process reaches a fixpoint s_k with $s_k = s_{k+n}$ for all $n \in \mathbb{N}$, called the **terminal state** of d.

Proposition 2. *For any network N, for any initial state s_0, any diffusion process $d = s_0, s_1, \ldots$ reaches a fixpoint.*

Proof. For any two states $s_k = (e, L_k), s_{k+1} = (e, L_{k+1})$ for which s_k can transition to s_{k+1}, points 2. and 4. of ensure that $S_k \subseteq S_{k+1}$ and $U_{k+1} \subseteq U_k$. Hence the proposition follows as N is finite.

We interpret a state in a diffusion process as an implicit, possibly partial, cast of votes. For a state $s_n = (e, L_n)$, $n \geq 1$, we take the set of agents that have voted on the quality of e to be the **jury** $J_n = R_{n-1} \cup S_n$ at time n. The jury J_n does not include newly reached agents $R_n \backslash (R_{n-1} \cup S_n)$ as they have not yet had the opportunity to choose whether to share or not. The jury J_n gives rise to a **voting profile** $(v_a)_{a \in J_n}$ with $v_a = \bot_a$ if $a \in R_n$ and $v_a = \top_a$ if $a \in S_n$. Hence, the agents that have been reached in previous states but have not shared vote for the low quality of e, while those that have shared vote for the high quality. Agents in U_n are excluded from the jury: unexposed to e, they have not had the chance to share. We use lower-case j's to refer to jury cardinality, such that $j = |J|, j_n = |J_n|$, etc., for J, J_n given by context. We assume diffusion processes are observed, so we know the progress of each entry.

The model makes at least the following idealized assumptions: All agents (i) pay attention to all entries that reach them, and (ii) to the best of their ability decide to share or not to with the only aim of proliferating high quality content. For (i), moving agents from U to R could be done based on logged screen activity, as social media users may not always pay attention to all content shared by neighbors. The approach suggested here is not apt to extract reliable information if one does not assume (ii).

3 Voting and a Jury Theorem

Jury theorems provide a mathematical argument for larger groups being beneficial in collective decision making. Under some assumptions on correctness probability ($c > 1/2$) and on independence of the voters, the majority vote in larger groups is probabilistically more accurate than in smaller groups (see e.g. [11,16,17]). Accepting social network diffusions as implicit voting scenarios and aiming to extract information on the quality of the entries, jury theorems thus constitute a natural basis for the selection of informative juries.

Given a quality evaluation $V(e)$ and a voting profile $(v_a)_{a \in J}$, for conciseness we write $v_a = 1$ if agent a's vote is correct about e (i.e., if $V(e) = \top$ and $v_a = \top_a$, or $V(e) = \bot$ and $v_a = \bot_a$), and $v_a = 0$ otherwise. For a state s with jury J, the **majority vote** is whichever of $\bot$ and $\top$ that gets more votes (or, in case of a tie, either $\bot$ or $\top$ is chosen by a fair coin toss). The majority vote of voting profile $(v_a)_{a \in J}$ on entry e is **correct** if it coincides with the quality valuation $V(e)$.

At a state s, if all j jurors in J vote independently, the probability that the majority vote is correct is given by $M(j)$ below, where the first term captures the tie-breaking rule, and the second is the probability of correctness of a strict majority.

$$M(j) = \frac{\ell}{2}\binom{j}{\frac{j}{2}}c^{\frac{j}{2}}(1-c)^{\frac{j}{2}} + \sum_{k=m_j}^{j}\binom{j}{k}c^k(1-c)^{j-k}$$

with $\ell = 0$ and $m_j = (j+1)/2$ for j odd, and $\ell = 1$ and $m_j = j/2+1$ for j even.

The simplest jury theorems, to which we stick here, concern the probability that a group makes the correct decision under *majority rule*. One statement of the classic ***Condorcet Jury Theorem*** is: if $c > 1/2$ and all jurors vote independently, then

1. the probability of a correct majority vote goes to one as the jury size goes to infinity: $\lim_{j\to\infty} M(j) = 1$.
2. the probability of a correct majority vote increases under the addition of two jurors: for $j + j'$ with $j' = (0 \mod 2)$, $M(j) < M(j + j')$.

Points 1 and 2 are sometimes referred to as the *asymptotic* and the *non-asymptotic* part of the theorem, respectively, with the former possibly taken to show that huge groups are infallible, while the latter shows that larger groups are better truth-trackers than small ones [18]. Note that the addition of two jurors is essential in the non-asymptotic part: moving from an odd to an even jury by adding a single juror may cause a drop in the probability of correctness due to the tie-breaking rule. As customary, in the following we simplify matters by limiting attention to juries of odd size.

Similar jury theorems also exist for juries which exhibit patterns of dependence and correlation among the voters (see [13,14,19]). Using independent juries for quality assessments may thus cause an information loss: an independent set of jurors can possibly be extended by the addition of dependent jurors while improving the majority vote precision. An alternative to our approach of seeking independent juries is thus to look for juries that exhibit patterns of correlation which do not negatively affect the majority correctness probability. However, the approaches of [13] and [19] requires to calculate the majority correctness probability by an expression exponential in jury size (The results of [14] are not applicable, being for the asymptotic case.) For this reason, we here make use of the simpler jury theorem requiring independent juries and suggest to use the majority vote of an independent jury as quality evaluation, presented together with its correctness probability.

4 Assessing Independence

Given a set of voters, whether their votes are independent and therefore warrant an application of the Condorcet Jury Theorem is an unobservable empirical matter. On this matter, we cannot supply a theoretical guarantee, but only hypothesize from other observables. One could hypothesize from network structure: If no voters in the set are connected, then conclude the set independent. Or one could hypothesize from personal or demographic traits: if all voters in the set are ‘different enough’, then conclude the set independent. These approaches cannot ensure independence: agents may be influenced by another through long chains in the network, and trait differences may lead to negative correlation in voting.

A third is a history-based, statistical approach: if the set of voters have not previously shown stochastically dependent behavior, then conclude the set independent. This approach proceeds via an independence test, e.g. the χ^2 test, to check if the voters’ previously observed votes were stochastically independent.[1] We follow this approach.

The χ^2 test assesses how compatible some observed frequencies are with a theoretical probability distribution. In our case, the theoretical distribution is the distribution of votes obtained under the assumption of independent jurors. Having a network N with correctness probability c, and a set D of diffusion processes with terminal states $T = (t_1, ..., t_m)$ over the topic $E = \{e_1, ..., e_m\}$ valuated by V, we can then use the χ^2 test to compare the theoretical distribution with the observed distribution of votes in T to assess the stochastic independence of any subset of voters $J \subseteq \mathcal{A}$. Again, this does not guarantee independence: not rejecting the null hypothesis that the agents are independent does not prove that they are independent, but at least tells us that they cannot be shown dependent beyond any reasonable doubt (the P-value). The details follow.

Theoretical Distribution. In a terminal state t where $U = \emptyset$, the probability distribution p of vote profiles $v = (v_1, ..., v_n) \in \{0,1\}^{|\mathcal{A}|}$, given independent voters with correctness probability c and quality valuation V, is

$$p(v_1, ..., v_n) = \prod_{i \in \mathcal{A}} x_i \text{ with } x_i = \begin{cases} c & \text{if } v_i = 1 \\ 1 - c & \text{if } v_i = 0. \end{cases}$$

[1] We use the classic χ^2 test just to exemplify our procedure, but other alternatives are also possible, e.g. the G-test.

This distribution does not apply when $U \neq \emptyset$, since agents in U are just unreached by the entry and have not had the chance to vote. For such cases, we have to find the appropriate theoretical distribution by taking the marginal of p on the set of exposed agents $S_t \cup R_t$. For a subset of voters $J \subseteq \mathcal{A}$, let $T_J \subseteq T$ be the set of terminal states such that $J \subseteq S_t \cup R_t$ for all $t \in T_J$. Then the marginal on J of p is

$$p_J(v_J) = \sum_{v_{-J} \in \{0,1\}^{|\mathcal{A} \setminus J|}} p(v_J, v_{-J})$$

with $v_J = (v_i)_{i \in J}$ and $v_{-J} = (v_i)_{i \in \mathcal{A} \setminus J}$. The distribution p_J hence gives the probability that one should expect, under the assumption of independent voters, on the votes by the agents in J who are jointly exposed to the entries in T_J.

The χ^2 Test for Independence. Given the theoretical distribution p, for each subset $J \subseteq \mathcal{A}$, we can then test for independence between its members. Running a χ^2 test amounts to the following: 1. Select a significance (e.g., .1, .05, .01) for rejecting the null hypothesis that the tested variables (votes) are dependent. 2. For each outcome i, $0 \leq i \leq n$, find the number E_i of occurrences estimated by the theoretical distribution given the bounds set by the data. 3. Compare E_i to the number O_i of observed occurrences of i by finding the χ^2 statistic:

$$\chi^2 := \textstyle\sum_{i=0}^{|J|+1} \frac{(O_i - E_i)^2}{E_i}$$

4. Compare χ^2 to the upper-tail critical values of the χ^2 distribution for the selected significance level and the appropriate number of degrees of freedom, and reject the null hypothesis if the P-value is less than the chosen significance level.

[20] shows that there exists an algorithm for testing the independence of j random variables with time complexity

$$O\left(\left(\textstyle\prod_{i=1}^{j} [i]\right)^{1/2} + \textstyle\sum_{i=1}^{j} [i]\right)$$

where $[i]$ denotes the number of possible values that random variable i can take. The exponential complexity arises as a direct consequence of the exponential growth of the number outcomes, as going through the whole outcome space is necessary to assess the independence of a given set of random variables. In our social network scenario, the (votes of the) agents in jury J are the random variables whose mutual independence we are interested in, and the number of possible outcomes of the voting process among the agents in J is the number of possible voting profiles, $2^{|J|}$. Therefore, given the result in [20], the time complexity in our case is $O\left(2^{|J|/2} + 2|J|\right)$.

With independence tests being computationally expensive, we cannot readily design a procedure that both checks independence and is effective: any procedure based on the current state-of-the-art will be exponential in the size of the jury. We then look for procedures that limit the required number of tests.

Remark: Family-Wise Error and Bonferroni Correction. Seeking procedures that limit the required number of tests highlights the expectation that it will often be necessary to run multiple tests before settling on a jury. This implies that we cannot fix a general significance level to be used in all tests, for the risk of committing type I errors. E.g. setting a significance level of 0.1 while running 1000 tests makes the family-wise error rate 1, theoretically guaranteeing one false positive—one jury deemed independent when it is not.

To control the family-wise error rate, we throughout assume the significance level of the individual tests is adjusted using **Bonferroni correction**: with an overall desired significance of α, Bonferroni correction tests the individual hypotheses at α/m, with m the total number of hypotheses to be tested. As each of the procedures discussed below specifies such an m and as the significance level does not affect the complexity of a χ^2-test, Bonferroni correction does not affect the overall test complexity.

Compared to alternative methods, Bonferroni correction fits present purposes well: First, it is conservative in its estimates, fitting well with the conservatism of point *(iv)* from the introduction. Second, it is computationally trivial, compared e.g. to the stronger *Holm-Bonferroni method* which requires finding all m P-values prior to running any tests.

5 Jury Finding Problems

Given a network state s_n with jury J_n, we seek to present the users with a conservative estimate of the quality of the current entry e in s_n. Due to possible dependencies between jurors, the majority decision of J_n need not be as trustworthy as stochastic independence would imply. Therefore, we look for subsets $J^* \subseteq J_n$ that retain independence. However, we also seek to constantly improve correctness of the assessment that we can extract from the agents as the information flows through the network. To this end, we look for a jury J^* in relation to a jury $J' \subseteq J_{n-1}$ assumed found independent in the previous round $n-1$. For the sake of brevity, we refer to J_n, J' and J^* with the above properties implicit throughout this section.

5.1 Optimal Juries

The optimal choice of J^* is any largest set of independent agents that we can find among those in J_n, as this maximizes collective correctness probability. An optimal jury may be found by solving the following problem for increasing values of correctness probability, C:

Problem 3. C-**precise Jury** is the decision problem
Instance: A social network $(\mathcal{A}, N)$, its network state s_n with jury J_n, a jury J' for state s_{n-1} and a number $C \in [c, 1]$.
Question: Is there an independent subset $J^* \subseteq J_n$ such that $M(J^*) \geq C$?

Finding an optimal choice of J^* corresponds to finding a solution to the optimization version of C-precise Jury. Using binary search, we need only solve the decision problem for $O(\log(j_n - j'))$ different values of C, as j' is a lower bound on the jury size of interest and the number of possible values of C is limited to $\{M(k)\colon 1 \leq k \leq j_n\}$.

The C-precise jury problem is NP-hard, as may be shown by reduction to the *Independent Set problem*, cf. e.g. [21]. Given a graph $G = (V, E)$, call a set $A \subseteq V$ `independent`* if no $v, v' \in A$ are connected by an edge in E. Then:

Problem 4. **Independent Set** is the decision problem
Instance: A graph $G = (V, E)$, and an integer $k \leq |V|$.
Question: Does G contain an `independent`* set of size at least k?

Proposition 5. *C-precise Jury is NP-complete.*

Proof. We show hardness by reduction to Independent Set which is NP-hard cf. e.g. [21]. Identify the graph (V, E) with a network $(\mathcal{A}, N)$ given by $\mathcal{A} = V$ and $N = E$. Assume a non-initial diffusion state s_n of $(\mathcal{A}, N)$ such that $J_n = \mathcal{A}$. We make no special use of J', so let $J' = \{a\}$ for some $a \in J$. Checking that an `independent`* set of G of size k exists is a special case of checking whether a jury J^* with $M(j^*) \geq C$ exists. Let $C = M(k)$. It is always possible to find a probability distribution p on voting profiles $(v_i)_{i \in \mathcal{A}}$ such that two agents i,j are not independent under p if and only if they are connected by an edge in N, and where p additionally satisfies that for all $A \subseteq \mathcal{A}$, if all elements of A are pairwise independent, then A' is mutually independent. For such p, `independence`* in $(\mathcal{A}, N)$ implies mutual independence in $(\mathcal{A}, N)$. As A can only be mutually independent if it is pairwise independent, mutual independence also implies `independence`*. Hence an independent jury $J^* \subseteq J_n$ satisfying $M(j^*) \geq C$ is also an `independent`* set of size at least k. *Inclusion:* A simple guess and check algorithm can be constructed: if we guess a subset $J^* \subseteq J_n$, we can check in polynomial time whether $M(J^*) \geq C$.

Finding an optimal jury at state s_n requires finding the maximal C for which the C-precise Jury is solved in the positive. A naive brute force algorithm examining every candidate subset $J \subseteq J_n, j > j'$, in the worst case requires running $2^{j_n - 1}$ independence tests. By the result of [20], finding a optimal jury is in $O\left(2^{j_n - 1} \cdot \left(2^{j_n/2} + 2j_n\right)\right)$. The naive aspect of this complexity may be diminished be the algorithm of [22] which finds a maximum `independent`* set in $O(1.1996^{j_n})$ using polynomial space.

5.2 Error-Diminishing Juries

Due to the exorbitant number of independence tests required, the optimality of the outcome jury has to be foregone in the interest of computational efficiency. A first alternative is to look for improvements in the collective correctness probability, without aiming to identify the best possible jury. In the next three subsections we follow this path.

One option in this direction is to seek a jury that diminishes the error of the current jury by a given percentage:

Definition 6. *A jury $J^* \subseteq J_n$ is* ***error-diminishing*** *by $h\%$ with respect to $J' \subseteq J_{n-1}$ if its probability of an incorrect majority vote is $h\%$ lower than that of J'. I.e., if $1 - M(j^*) \leq \frac{h}{100}(1 - M(j'))$.*

In finding an error-diminishing jury J^*, the required size depends on the size of J', as well as on the correctness probability c and the increment h. The number of necessary tests, however, has not been reduced much by this approach:

Proposition 7. *Whether a jury J^* error-diminishing by $h\%$ exists can be determined by testing at most $\binom{j}{\frac{j+1}{2}}$ subsets for independence. Worst case, none of these tests are redundant.*

Proof. With a fixed error-diminishing degree $h\%$, there is some $j^* \in \mathbb{N}$ given as a function of the size of J' such that only juries of size at least j^* will be precise enough. It then suffices to seek through the $\binom{j}{j^*}$-many size j^* subsets of J for an independent jury: no smaller sets will do, and every larger set will be non-independent if all size j^* are. With $h^* := 1 - \frac{h}{100}(1 - M(j'))$, this is the smallest j^* such that $h^* \leq M(j^*)$. This j^* may be approached from j' using binary search, but checking higher values is more expensive: the inequality needed checked for a value m has a fixed left-hand side, but a right-hand side increasing linearly in m. Finding j^* is thus in $O(j)$. Second, we seek for a suitable $J^* \subseteq J$. Worst case, j^* is $\arg\max \binom{j}{x} = \{\frac{j-1}{2}, \frac{j+1}{2}\}$, providing an upper bound of $\binom{j}{\frac{j+1}{2}}$ tests before concluding. The lower bound is established by the worst case where each of the $\binom{j}{\frac{j+1}{2}}$ subsets may be non-independent due to just one agent. In this case, none of the tests are redundant.

For reference, we remark that $f(x) = \binom{x}{\frac{x+1}{2}}$ is not a slow-growing function. In fact, it grows as fast $a^x, a > 1$.

5.3 Incrementally Improved Juries

Since the required number of tests has not appreciably decreased by looking for error-diminishing juries, we turn to the alternative of improving the collective correctness probability by a fixed percentage. For a 5% increment, for instance, the number of required additional jurors is illustrated in Table 1.

Table 1. The number of agents of correctness probability c (in the rows) needed to reach a certain collective correctness probability (in the columns). Even when the individual correctness c is relatively low, e.g. $c = .6$, only 65 independent jurors are needed to reach a collective correctness probability of 0.95.

	.60	.65	.70	.75	.80	.85	.90	.95
$c = .6$	1	3	7	11	17	27	41	65
$c = .75$	1	1	1	1	3	5	5	9

Definition 8. *A jury $J^* \subseteq J_n$ is* ***incremental*** *by $h\%$ with respect to $J' \subseteq J_{n-1}$ if the probability of a correct majority vote is $h\%$ higher than that of J'. I.e., if $(1 + \frac{h}{100})M(j') \leq M(j^*)$.*

Note that while the error-diminishing requirement above may fail to be satisfied for some jury J' just because there is no sufficiently large independent subset of the currently exposed agents, finding a jury of improved correctness probability by a fixed increment may be impossible also because the current jury's correctness probability cannot be raised by $h\%$ without exceeding 1. One may therefore expect that this could reduce the search for improved juries as compared to the previous case. However, the required number of tests does not change:

Proposition 9. *Whether a jury J^* incremental by $h\%$ exists can be determined by testing at most $\binom{j}{\frac{j+1}{2}}$ subsets for independence. Worst case, none of these tests are redundant.*

Proof. With a fixed desired increment, there is some $j^* \in \mathbb{N}$ given as a function of the size of J' such that only juries of size at least j^* will be precise enough. As in the proof of Proposition 7, finding j^* is in $O(j)$ using binary search, but here we find the smallest j^* satisfying $h^* \leq M(j^*)$ for $h^* := (1 + \frac{h}{100})M(c, j')$ fixed. Again as in the proof of Proposition 7, it suffices to seek through the $\binom{j}{j^*}$-many size j^* subsets the same non-redundancy argument applies.

5.4 Monotonic Juries

Looking thus for even simpler approaches, the next possible simplification is to merely seek any improvement in the correctness probability of the jury. A corresponding formal requirement is then the following:

Definition 10. *A jury $J^* \subseteq J_n$ is* ***monotonic*** *with respect to $J' \subseteq J_{n-1}$ if the probability of a correct majority vote is strictly higher than that of J'. I.e., if $M(j') < M(j^*)$.*

Proposition 11. *Whether a jury J^* monotonic with respect to a fixed jury J' exists can be determined by testing at most $\binom{j}{j'+2}$ subsets for independence. When J' may grow with J, it can be determined by testing at most $\binom{j}{\frac{j+1}{2}}$ subsets for independence. Worst case, none of these tests are redundant.*

Proof. For a fixed size J', searching through juries of size $j^* = j' + 2$ is sufficient: if an independent jury J^* of size j^* is found, it will satisfy $M(c, j') < M(c, j^*)$; if no independent size j^* jury exists, then every larger set will also be non-independent. Testing each of the $\binom{j}{j'+2}$ size $j' + 2$ subsets may also be necessary, as each of the subsets may be non-independent due to just 1 agent. When J' may grow with J, the upper bound is established by the worst case number of tests, $\arg\max \binom{j}{x} = \{\frac{j-1}{2}, \frac{j+1}{2}\}$. The lower bound is established by the argument used for Proposition 9.

5.5 Inflationary Juries

A common cause for the large search space—and hence the many required independence tests—across the hitherto considered procedures is that they make little use of the jury J' assumed found in the previous step: only its *size* matters, used as a lower bound of the size of an improved jury J^*. The simplification we now propose is instead based on the idea of looking only for additions to the current jury J', rather than throwing it away and starting the search anew. In other words, rather than focusing on a general improvement in the correctness probability as we have done in subsections 5.1–5.4, one may instead focus exclusively on extensions of the current jury J'.

On the one hand, this approach may be considered the farthest from Sect. 5.1's search for an optimal jury, in that an unfortunate start might lead to a maximal independent set of agents much smaller than an *de facto* optimal jury. On the other hand, however, its convenience arises precisely from favoring simplicity over optimality. As we have seen in Table 1 above, small independent juries already suffice to achieve a high collective correctness probability, even when the individual correctness c is low. Given nowadays dimensions of social networks, a set of a few tens or hundreds of agents is but a minimal fraction of the total number of users. Hence, even when holding the actual independent jury J' fixed, one may reasonably hope to be able to find another pair of agents independent of J' among the many available users.

The corresponding requirement that the next jury J^* should satisfy is then the following:

Definition 12. *A jury $J^* \subseteq J_n$ is **inflationary** with respect to $J' \subseteq J_{n-1}$ if it extends J' and the probability of a correct majority vote is strictly higher than that of J'. I.e., if $J' \subseteq J^*$ and $M(j') < M(j^*)$.*

By the Condorcet Jury Theorem, to find an inflationary jury, it is sufficient to find a pair of agents $a, a' \in J \backslash J'$ such that $J' \cup \{a, a'\} = J^*$ is a jury of mutually independent agents. We are thus able to greatly reduce the required number of tests, as stated by the following:

Proposition 13. *Whether an inflationary jury J^* exists can be determined by testing at most $\binom{j-j'}{2}$ subsets for independence. Worst case, none of these tests are redundant.*

Proof. Given J', by the Condorcet Jury Theorem, it suffices to find J^* with $j^* = j' + 2$ and $J' \subseteq J^*$. There are $\binom{j-j'}{2}$ candidates of pairs to add to J'. Testing each is sufficient; testing each may also be necessary, as each may be non-independent due to just 1 agent.

As $\binom{n}{2} = \sum_{k=1}^{n-1} k = \frac{(n^2-n)}{2}$, the number of tests required to find an inflationary jury is bounded above by a degree-2 polynomial. The inflationary jury procedure thus considerably reduces the number of required tests.

6 Conclusion

We have considered how states of diffusion processes in social networks may be used as quality evaluations of shared content. We have noted that establishing independence of juries is essential to rely on wisdom-of-the-crowds results from jury theorems and to ensure a theoretically sound evaluation, but that independence testing is computationally expensive. For this reason, we have sought jury selection procedures that reduce the number of necessary independence tests.

Of the five selection procedures introduced here, only the inflationary jury procedure requires a number of tests bounded by a polynomial (of degree 2). While, given the current algorithms, the time complexity of testing for independence remains exponential in the size of the inflationary jury, we have also shown that there is hope for tractably using the inflationary jury procedure in practice, as the number of independent jurors needed to achieve a high collective correctness probability is, even for low individual correctness probability, rather small, cf. Table 1.

Several fundamental questions remain unexplored, and core elements may be chosen differently. One question pertains to the amount of data required to conduct the χ^2 tests. Highly competent voters will often vote alike, wherefore a large set of previous diffusion processes will be required to determine whether their voting pattern significantly differs deform from the theoretical distribution under independence. We do not know how this required data grows with competence, and it may thus introduce computational hindrances. Related is the use of the χ^2 test itself. Possibly, alternative statistical approaches may lead to stronger conclusions about independence. That field should be surveyed, with complexity issues in mind.

The results presented do not tell us much about the practical difficulty of the proposed approach. It could be informative to develop a a randomized algorithm, or applying reduction to SAT to use one the excellent SAT solving algorithms developed in recent years. Currently, we do not know if the problems posed in this paper are highly approximate, or exactly solvable for all practical problems.

Finally, it would be instructive to perform empirical evaluations of several aspects of the proposed approach, to gauge both its efficiency (cf. the above) and its necessity. As real-life social networks tend to grow large, the continuous and global observation assumed here may be unfeasible. Due to the large size of networks, it could also be the case that random sampling of users or other selection methods *de facto* provide a way to obtain a correct aggregated judgment with sufficiently high frequency. Empirical studies could thus be instructive in determining how to best improve judgment reliability in social networks via jury theorems.

Acknowledgments. We thank the reviewers of *LORI-VII* and the participants of the *Social Interactions in Epistemology and in Economics* conference (Cph, 29-31/5/2019) for insightful comments. The Center for Information and Bubble Studies is funded by the Carlsberg Foundation. RKR was partially supported by the DFG-ANR joint project *Collective Attitude Formation* [RO 4548/8-1].

References

1. Brake, D.R.: Are we all online content creators now? Web 2.0 and digital divides*. J. Comput.-Mediated Commun. **19**(3), 591–609 (2014)
2. Gagliardi, A., Jadad, A.R.: Examination of instruments used to rate quality of health information on the internet: chronicle of a voyage with an unclear destination. BMJ **324**(7337), 569–573 (2002)
3. Chumber, S., Huber, J., Ghezzi, P.: A methodology to analyze the quality of health information on the internet: the example of diabetic neuropathy. Diab. Educ. **41**(1), 95–105 (2015)
4. Greenberg, L., D'Andrea, G., Lorence, D.: Setting the public agenda for online health search: a white paper and action agenda. J. Med. Internet Res. **6**(2), e18 (2004)
5. Lazer, D.M.J., et al.: The science of fake news. Science **359**(6380), 1094–1096 (2018)
6. Hendricks, V.F., Vestergaard, M.: Reality Lost: Markets of Attention. Misinformation and Manipulation. Springer, Heidelberg (2019)
7. Muchnik, L., Aral, S., Taylor, S.J.: Social influence bias: a randomized experiment. Science **341**(6146), 647–651 (2013)
8. Berger, J., Milkman, K.L.: What makes online content viral? J. Mark. Res. **49**(2), 192–205 (2012)
9. Ferrara, E., Varol, O., Davis, C., Menczer, F., Flammini, A.: The rise of social bots. Commun. ACM **59**(7), 96–104 (2016)
10. Condorcet, M.M.d.: Essai sur l'Application de l'Analyse à la Probabilité des Décisions Rendues à la Pluralité des Voix. Paris (1785)
11. Grofman, B., Owen, G., Feld, S.L.: Thirteen theorems in search of the truth. Theor. Decis. **15**(3), 261–278 (1983)
12. Ladha, K.K.: Information pooling through majority-rule voting: Condorcet's Jury Theorem with correlated votes. J. Econ. Behav. Organ. **26**(3), 353–372 (1995)
13. Berend, D., Sapir, L.: Monotonicity in Condorcet's Jury Theorem with dependent voters. Soc. Choice Welf. **28**(3), 507–528 (2007)
14. Pivato, M.: Epistemic democracy with correlated voters. J. Math. Econ. **72**, 51–69 (2017)
15. Wakker, P.: Prospect Theory. Cambridge University Press, Cambridge (2010)
16. Ben-Yashar, R., Paroush, J.: A nonasymptotic Condorcet Jury Theorem. Soc. Choice Welf. **17**(2), 189–199 (2000)
17. Berend, D., Sapir, L.: Monotonicity in condorcet jury theorem. Soc. Choice Welf. **24**(1), 83–92 (2005)
18. Dietricht, F., Spiekermann, K.: Jury Theorems. In: Fricker, M., Graham, P.J., Henderson, D., Pedersen, N., Wyatt, J. (eds.) The Routledge Companion to Social Epistemology. Routledge (2019)
19. Kaniovski, S.: Aggregation of correlated votes and Condorcet's Jury Theorem. Theor. Decis. **69**(3), 453–468 (2010)
20. Acharya, J., Daskalakis, C., Kamath, G.: Optimal testing for properties of distributions. In: Cortes, C., Lawrence, N.D., Lee, D.D., Sugiyama, M., Garnett, R. (eds.) Advances in Neural Information Processing Systems 28, pp. 3591–3599. Curran Associates, Inc. (2015)
21. Kleinberg, J., Tardos, E.: Algorithm Design. Pearson (2005)
22. Xiao, M., Nagamochi, H.: Exact algorithms for maximum independent set. Inf. Comput. **255**, 126–146 (2017)

Term-Sequence-Modal Logics

Takahiro Sawasaki[1(✉)], Katsuhiko Sano[2], and Tomoyuki Yamada[2]

[1] Graduate School of Letters, Hokkaido University, Sapporo, Japan
taka.sawasaki562@gmail.com

[2] Faculty of Humanities and Human Sciences, Hokkaido University, Sapporo, Japan
{v-sano,yamada}@let.hokudai.ac.jp

Abstract. Term-modal logics, developed by Fitting et al., enable us to index a modal operator by a term of the first-order logic and even to quantify variables in the index of the modal operator. In this paper, we expand term-modal logics by allowing a modal operator to be indexed by a finite sequence of terms as well as a single term. The expanded logics are generalizations of both term-modal logics and quantified modal logics. We provide sound Hilbert-style axiomatizations (without Barcan-like axioms) for the logics and establish the strong completeness results for some of the logics. We also propose sequent calculi for the logics and show cut elimination theorems and Craig interpolation theorems for some of the calculi.

Keywords: Term-sequence-modal logic · Term-modal logic · Quantified modal logic · Hilbert system · Sequent calculus

1 Introduction

This paper proposes term-sequence-modal logics (TSMLs), i.e., expansions of term-modal logics by allowing the term-modal operator in [21] and [4] to be indexed with a sequence of terms. Our TSMLs overcome a limitation that a term-modality can only be relativized to one term, and also subsume quantified modal logics and term-modal logics.

Term-modal logics, developed by Thalmann [21] and Fitting et al. [4], enable us to index each modal operator by a term and even to quantify variables in the index of the modal operator.[1] They have a number of advantages over propositional

[1] We can find the very idea of term-modal logic even in Hintikka's *Knowledge and Belief*, where for a sentence like "a knows that P" he says "a is a name of a person or [...] a definite description referring to a human being." [6, p. 3] He also considers substitution of such names by equality axioms. [6, ch. 6] For some works related to term-modal logic after [4,21], see [18, ch. 1] and [22].

We would like to thank three reviewers for their constructive comments to our manuscript. The work of all authors was partially supported by the research supported by JSPS Grant-in-Aid for Scientific Research (B) (KAKENHI 17H02258). The work of the second author was partially supported also by JSPS KAKENHI Grant-in-Aid for Scientific Research (C) Grant Number 19K12113 and JSPS Core-to-Core Program (A. Advanced Research Networks).

P. Blackburn et al. (Eds.): LORI 2019, LNCS 11813, pp. 244–258, 2019.
https://doi.org/10.1007/978-3-662-60292-8_18

multi-modal epistemic logics (MELs) when we reason about agents and groups. For instance, term-modal logics can express a particular agent's knowledge about herself by a formula $[a]Pa$ ("a knows that a is P") and even all agents' knowledge about themselves by a formula $\forall x[x]Px$ ("all agents know that they are P"), but either of them is difficult to express in MELs.[2] Since it is possible even to read the term-modal operator $[\cdot]$ in term-modal logics as a deontic operator, term-modal logics provide us a very natural way to formalize sentences in natural languages which indicate deontic modalities relativized to a term. We can also find some developments on term-modal logics in [9,16,17,19].

Nevertheless, term-modal logics still have a limitation on relativization of a modality: a term-modality can only be relativized to *one term*. This limitation should be overcome for philosophical investigations, since some modalities seem to be relativized to two or more terms. One straightforward example is a deontic modality in "**A**da has been obligated by **B**en to **s**hoot **C**harles". Let us interpret the term-modal operator $[\cdot]$ as deontic modality. Then a possible formalization of this sentence in term-modal logics might be like $[a]Sac$, but it fails to capture the relation of Ada's owing the very obligation to Ben. What we need here is at least a modality relativized to *a pair of agents*.

Our TSMLs enable us to use such a deontic operator of the form $[t, t']$, whose first item stands for an agent who owes the obligation and second item agent to whom the first agent's obligation is owed. Then we can naturally formalize the sentence above by a formula $[a, b]Sac$. Similar ideas are in fact found in the literature on modal logic (cf. [10,23]). For example, Yamada [23] has developed a refined multi-agent monadic deontic logic $\mathsf{MDL}^{+}\mathsf{II}$, in which the deontic operator $\mathcal{O}_{(i,j)}$ relativized to a pair of agents i, j is adopted to represent the obligation generated by $j's$ command given to i.

Moreover, our TSMLs can subsume quantified modal logics and term-modal logics in a straightforward way by interpreting a term-sequence-modal operator $[\cdot]$ as $\Box$ when having the empty term-sequence ε and as $[t]$ when having a term-sequence t of length 1, respectively. Note that an assignment of variables we introduce below plays an essential role which a set of "parameters" plays in [4,21]. In addition, cut elimination theorems for the sequent calculi (without labels) for some TSMLs are proved in a proof-theoretic way. Our proofs are worthwhile since [4] gives only *semantic* proofs of cut elimination theorems for the sequent calculi provided in [4] and [16] gives proof-theoretic proofs of cut elimination theorems for the *labeled* sequent calculi provided in [16].

The paper proceeds as follows. We first introduce the language of TSMLs in Sect. 2 and then provide Kripke semantics and Hilbert systems for TSMLs in Sects. 3 and 4. In Sect. 5 we show the strong completeness results for some important class of the Hilbert systems for TSMLs, which includes an S5 version of term-modal logic (Theorems 1 and 2). In Sect. 6 we provide the sequent calculi which are equipollent to all the Hilbert systems in Sect. 5. We give therein proof-

---

[2] Strictly speaking, a's knowledge that a is P is not the self knowledge that she would express by saying "I am P" as a might not know that she is a. For early discussions on this and related issues, see [6, ch. 6] and [3].

theoretic proofs for cut elimination theorems and Craig interpolation theorems for some of the calculi (Theorems 3 and 4).

2 Syntax of Term-Sequence-Modal Logics

The *language* $\mathcal{L}$ of *term-sequence-modal logics* (TSMLs) consists of a countably infinite set $\mathsf{Var} = \{x, y, \dots\}$ of *variables*, finite or countably infinite sets $\mathsf{Con} = \{c, d, \dots\}$ of *constants* and $\mathsf{Fn} = \{f, g, \cdots\}$ of *function symbols* each of which has a fixed finite arity more than zero, a countably infinite set $\mathsf{Pred} = \{P, Q, \dots\}$ of *predicate symbols* each of which has a fixed finite arity, and *logical symbols*, $\neg, \supset, \forall$ and $[\cdot]$. We do not include equality symbol. The set Term of *terms* is defined as usual and the set Form of formulas of $\mathcal{L}$ is defined recursively by

$$\mathsf{Form} \ni A ::= Pt_1 \dots t_n \mid \bot \mid \neg A \mid (A \supset A) \mid \forall x A \mid [t_1, \dots, t_n]A,$$

where P is a predicate symbol with arity n and $t_1, \dots, t_n$ are terms. Boolean connectives $\wedge, \vee$ are defined as usual and $\langle t_1, \dots, t_n\rangle A$ is defined by $\neg[t_1, \dots, t_n]\neg A$. We often write $\vec{s_n}$ or $\vec{s}$ instead of $s_1, \dots, s_n$ for short. We define $\Box A := [\varepsilon]A$ and $\Diamond A$ as its dual, where ε is an empty sequence.

Definition 1. We define $\mathcal{L}_n$ as the sublanguage of $\mathcal{L}$ such that the set of all the term-sequences in term-sequence-modal operators is Term^n, i.e., the set of all term-sequences of length n.

So, $\mathcal{L}_0$ can be regarded as the language of the ordinary quantified modal logic (cf. [7]) and $\mathcal{L}_1$ is the language of the term-modal logic developed by [4,21].

We define the sets $\mathsf{FV}(t), \mathsf{FV}(A), \mathsf{FV}(\Gamma)$ of free variables in a term t, a formula A and a set Γ of formulas as usual, except that $\mathsf{FV}([t_1, \dots, t_n]A) = \mathsf{FV}(t_1) \cup \dots \cup \mathsf{FV}(t_n) \cup \mathsf{FV}(A)$. We also define substitutions $s(t/x)$ and $A(t/x)$ of a term t for a variable x in a term s and a formula A, respectively, as usual, except that $([s_1, \dots, s_n]A)(t/x) \equiv [s_1(t/x), \dots, s_n(t/x)]A(t/x)$, where a notation $A \equiv B$ means that A is identical with B as a string of symbols.

3 Kripke Semantics for Term-Sequence-Modal Logics

In this section we introduce Kripke semantics for our syntax and establish the definability results for TSMLs. With respect to interpretations of constants and function symbols, we follow Thalmann [21] and Fitting et al. [4] just for making easy the comparison between our TSMLs and original term-modal logics. An assignment we introduce below plays a role which a set of "parameters" plays in [4,21].

Let $\mathcal{D}$ be a nonempty set which is called the *domain* and whose elements are assumed to be *agents* in this paper. A *frame over* $\mathcal{D}$ is a tuple $\langle W, D, R\rangle$, where W is a nonempty set whose elements are called *states*; D is a W-indexed family $(D_w)_{w \in W}$ of a non-empty subset of $\mathcal{D}$;[3] R is a subset of $W \times \mathcal{D}^{<\omega} \times W$

[3] A frame over $\mathcal{D}$ for $\mathcal{L}$ must satisfy that $\bigcap_{w \in W} D_w \neq \varnothing$ if $\mathsf{Con} \neq \varnothing$ in $\mathcal{L}$.

where $\mathcal{D}^{<\omega}$ is the set of all finite sequences of $\mathcal{D}$ and R satisfies the *monotonicity condition* or the *inclusion requirement*: if $(w, \vec{d}, v) \in R$ then $D_w \subseteq D_v$ for all $w, v \in W$ and all $\vec{d} \in \mathcal{D}^{<\omega}$. We often write $(w, \vec{d}, v) \in R$ as $wR_{\vec{d}}v$. A *model over* $\mathcal{D}$ is a tuple (F, V, I), where F is a frame, V is a *valuation* that maps each state w and each predicate symbol P of arity n to a subset $V_w(P) \subseteq D_w^n$, and I is an *interpretation* that maps each constant c to an element $I(c) \in \bigcap_{w \in W} D_w$ and each function symbol f of arity n to an n-place function $I(f) : \mathcal{D}^n \to \mathcal{D}$ such that $I(f)(d_1, \ldots, d_n) \in D_w$ for all $(d_1, \ldots, d_n) \in D_w^n$. An *assignment* α is a function from Var to $\mathcal{D}$ and its domain can be extended to Term by defining $\alpha(c) := I(c)$ and $\alpha(f(t_1, \ldots, t_n)) := I(f)(\alpha(t_1), \ldots, \alpha(t_n))$. We denote $\alpha(x|d)$ as the same assignment as α except for assigning d to x.

Given a model $M = (W, D, R, V, I)$ over $\mathcal{D}$, $w \in W$, a formula A, and an assignment α such that $\alpha(x) \in D_w$ for all $x \in \mathsf{FV}(A)$, the satisfaction relation $M, w, \alpha \models A$ is defined as follows.

1. $M, w, \alpha \models Pt_1 \ldots t_n$ iff $(\alpha(t_1), \ldots, \alpha(t_n)) \in V_w(P)$
2. $M, w, \alpha \not\models \bot$
3. $M, w, \alpha \models \neg B$ iff $M, w, \alpha \not\models B$
4. $M, w, \alpha \models B \supset C$ iff $M, w, \alpha \models B$ implies $M, w, \alpha \models C$
5. $M, w, \alpha \models \forall x B$ iff $M, w, \alpha(x|d) \models B$ for any $d \in D_w$
6. $M, w, \alpha \models [t_1, \ldots, t_n]B$ iff $M, v, \alpha \models B$ for any $v \in W$ such that $wR_{(\alpha(t_1), \ldots, \alpha(t_n))}v$.

Below we often write $\alpha(t_1, \ldots, t_n)$ instead of $(\alpha(t_1), \ldots, \alpha(t_n))$ for short.

Definition 2 (Validity). Let M be a model over $\mathcal{D}$, w a state in W, and A be a formula. We say that A *is valid at* w *in* M (written: $M, w \models A$) if $M, w, \alpha \models A$ for all assignments α such that $\alpha(x) \in D_w$ for all $x \in \mathsf{FV}(A)$. We also say that A *is valid in* M (written: $M \models A$) if $M, w \models A$ for all $w \in W$, that A *is valid in a frame* F (written: $F \models A$) if $M \models A$ for all models M based on F, and that A *is valid in a class* $\mathbb{F}$ *of frames* if $F \models A$ for all $F \in \mathbb{F}$. Finally, we write $M, w, \alpha \models \Gamma$ if $M, w, \alpha \models B$ for all $B \in \Gamma$ and say that A *is a consequence from a set* Γ of formulas in $\mathbb{F}$ (written: $\Gamma \models_{\mathbb{F}} A$) if $M, w, \alpha \models \Gamma$ implies $M, w, \alpha \models A$ for all models M based on any $F \in \mathbb{F}$, all states w, all assignments α such that $\alpha(x) \in D_w$ for all $x \in \mathsf{FV}(\Gamma, A)$.

Definition 3 (Frame Properties). Let $n \in \mathbb{N}$ and $F = \langle W, D, R \rangle$ be a frame over $\mathcal{D}$.

1. F is *n-serial* if for all $w \in W$, $\vec{d} \in D_w^n$, there is a $v \in W$, $wR_{\vec{d}}v$.
2. F is *n-reflexive* if for all $w \in W$ and $\vec{d} \in D_w^n$, $wR_{\vec{d}}w$.
3. F is *n-symmetric* if for all $w, v \in W$ and $\vec{d} \in D_w^n$, if $wR_{\vec{d}}v$ then $vR_{\vec{d}}w$.
4. F is *n-transitive* if for all $w, v, u \in W$ and $\vec{d} \in D_w^n$, if $wR_{\vec{d}}v$ and $vR_{\vec{d}}u$ then $wR_{\vec{d}}u$.

If $n = 0$ and so $D^0 = \{ \varepsilon \}$, each frame is serial, reflexive, symmetric and transitive in the ordinary sense by identifying R_ε with a binary relation R.

Definition 4. A set Γ of formulas *defines* a class $\mathbb{F}$ of frames when the equivalence $F \models \Gamma$ iff $F \in \mathbb{F}$ holds for each frame F. If $\Gamma = \{ A \}$, we say that A defines $\mathbb{F}$. We denote by $\mathbb{F}_\Gamma$ the class of all frames defined by Γ.

Proposition 5. *If Γ_1 and Γ_2 define $\mathbb{F}_1$ and $\mathbb{F}_2$ respectively, then $\Gamma_1 \cup \Gamma_2$ defines $\mathbb{F}_1 \cap \mathbb{F}_2$.*

Proposition 6. *Each formula in Table 1 defines the class of frames satisfying the corresponding property.*

Example 1. When $n = 0$ and we restrict our attention to $\mathcal{L}_0$, we can obtain from Proposition 6 the frame definability results for the ordinary quantified modal logic. For example, D_0 means $\Box P \supset \Diamond P$ and it defines seriality of R in the ordinary sense. When $n = 1$ and we focus on the sublanguage $\mathcal{L}_1$ (the syntax for term-modal logic), Propositions 6 and 5 tell us that $\{ [x]P \supset P, [x]P \supset [x][x]P, P \supset [x]\langle x \rangle P \}$ defines the class of Kripke frames $F = (W, R, D)$ for term-modal logics (cf. [4,21]) where F is 1-reflexive, 1-transitive and 1-symmetric.

Table 1. Frame definability of all properties in Definition 3 ($n \in \mathbb{N}$)

D_n $[\vec{x_n}]P \supset \langle \vec{x_n} \rangle P$	T_n $[\vec{x_n}]P \supset P$	B_n $P \supset [\vec{x_n}]\langle \vec{x_n} \rangle P$	4_n $[\vec{x_n}]P \supset [\vec{x_n}][\vec{x_n}]P$
n-seriality	n-reflexivity	n-symmetry	n-transitivity

4 Hilbert Systems for Term-Sequence-Modal Logics

The Hilbert system $\mathsf{H}(\mathbf{tK})$ for the minimal term-sequence-modal logic $\mathbf{tK}$ is given in Table 2. Axiom schemata A1 to A5 and inference rules MP and Gen are just from first-order logic. Only the axiom tK and the inference rule Nec are additional parts for TSMLs. The Hilbert systems with additional axiom schemata are defined as follows.

Definition 7. Given a set $\Sigma \subseteq \bigcup_{n \in \mathbb{N}} \{ \mathrm{D}_n, \mathrm{T}_n, \mathrm{B}_n, 4_n \}$, Hilbert system $\mathsf{H}(\mathbf{tK}\Sigma)$ for $\mathbf{tK}\Sigma$ consists of all axioms and inference rules of $\mathsf{H}(\mathbf{tK})$ as well as all substitution instances of a formula in Σ with respect to variables $\vec{x_n}$ and 0-arity predicate symbol P in the formula. The Hilbert system $\mathsf{H}(\mathbf{tK}\Sigma \restriction \mathcal{L}_n)$ is the restriction of $\mathsf{H}(\mathbf{tK}\Sigma)$ to the sublanguage $\mathcal{L}_n$ of $\mathcal{L}$.

When Σ is a familiar combination of axioms, say $\Sigma = \{ \mathrm{T}_n, 4_n \}$ or $\{ \mathrm{T}_n, 4_n, \mathrm{B}_n \}$, we simply write, say, $\mathsf{H}(\mathbf{tS4}_n)$ or $\mathsf{H}(\mathbf{tS5}_n)$, as naturally expected. The notion of proof in a Hilbert system is defined as usual.

Table 2. Hilbert system $\mathsf{H}(\mathbf{tK})$ for the minimal term-sequence-modal logic

Axiom Schemata	
A1 $A \supset (B \supset A)$	A4 $\bot \supset A$
A2 $(A \supset (B \supset C)) \supset ((A \supset B) \supset (A \supset C))$	A5 $\forall x A \supset A(t/x)$
A3 $(\neg B \supset \neg A) \supset (A \supset B)$	tK $[\vec{t}](A \supset B) \supset ([\vec{t}]A \supset [\vec{t}]B)$

Inference Rules		
$\dfrac{A \supset B \qquad A}{B}$ MP	$\dfrac{A \supset B(y/x)}{A \supset \forall x B}$ Gen†	$\dfrac{A}{[\vec{t}]A}$ Nec
†: $y \notin \mathsf{FV}(A, \forall x B)$.		

Example 2. A combination Σ of axiom schemata on term-sequence-modal operators $[\vec{t}]$ of different length is possible. For example, let us consider $\Sigma = \{ \mathrm{D}_0, \mathrm{T}_1 \}$ and suppose that $[t]$ is read as an epistemic modality and $\Box$ as a deontic modality. Then the resulting Hilbert system $\mathsf{H}(\mathbf{tKD_0T_1})$ has axiom schemata of $\Box A \supset \Diamond A$ and $[t]A \supset A$. Moreover, we can establish that $[c]\Box P \supset \Diamond P$ ("if c knows that it is obligatory that P, then it is permissible that P") is provable in $\mathsf{H}(\mathbf{tKD_0T_1})$ but it is neither provable in $\mathsf{H}(\mathbf{tKD_0})$ nor $\mathsf{H}(\mathbf{tKT_1})$. When we consider the sublanguage $\mathcal{L}_0$, $\mathsf{H}(\mathbf{tKD_0} \restriction \mathcal{L}_0)$ is the Hilbert system of the smallest normal quantified modal logic (without Barcan axioms: $\forall x \Box A \supset \Box \forall x A$) which contains the axiom schema D (the Hilbert system LPC + KD in [7, p. 282]). When we focus on the sublanguage $\mathcal{L}_1$ and $\Sigma \subseteq \{ \mathrm{T}_1, \mathrm{D}_1, 4_1 \}$, then we can provide the previously unknown Hilbert systems $\mathsf{H}(\mathbf{tK}\Sigma \restriction \mathcal{L}_1)$ for all the corresponding term-modal logics studied in [4]. When $\Sigma = \{ \mathrm{T}_1, 4_1, \mathrm{B}_1 \}$, the system $\mathsf{H}(\mathbf{tKT_1 4_1 B_1} \restriction \mathcal{L}_1)$ (or $\mathsf{H}(\mathbf{tS5_1} \restriction \mathcal{L}_1)$ simply) is the Hilbert system of the term-modal logic expansion of **S5**, which was not studied at all in [4].

Proposition 8. *Let* $\Sigma \subseteq \bigcup_{n \in \mathbb{N}} \{ \mathrm{D}_n, \mathrm{T}_n, \mathrm{B}_n, 4_n \}$ *such that* $\mathrm{B}_k \in \Sigma$. *Then,*

$$\forall x [t_1, \ldots, t_k] A \supset [t_1, \ldots, t_k] \forall x A$$

is provable in $\mathsf{H}(\mathbf{tK}\Sigma)$ *provided* $x \notin \mathsf{FV}(t_1) \cup \cdots \cup \mathsf{FV}(t_k)$.

The condition $x \notin \mathsf{FV}(t_1) \cup \cdots \cup \mathsf{FV}(t_k)$ is crucial in Proposition 8 as Kooi pointed out in [9] for term-modal logic. By Proposition 8, whenever $\mathrm{B}_k \in \Sigma$, every model for $\mathbf{tK}\Sigma$ has a "local" constant domain in the sense that $wR_{\vec{d}_k}v$ implies $D_w = D_v$.

The soundness results for TSMLs are easy to establish by Proposition 6.

Proposition 9 (Soundness). *Let* $\Sigma \subseteq \bigcup_{n \in \mathbb{N}} \{ \mathrm{D}_n, \mathrm{T}_n, \mathrm{B}_n, 4_n \}$. *Then if a formula* A *is provable in* $\mathsf{H}(\mathbf{tK}\Sigma)$ *then* A *is valid in the class* $\mathbb{F}_\Sigma$ *of frames defined by* Σ.

5 Strong Completeness of Term-Sequence-Modal Logics

Given $\Sigma \subseteq \bigcup_{n\in\mathbb{N}}\{\mathrm{D}_n, \mathrm{T}_n, \mathrm{B}_n, 4_n\}$, we say that $\mathsf{H}(\mathbf{tK}\Sigma)$ is strongly complete with respect to a class $\mathbb{F}$ of frames just in case if $\Gamma \models_{\mathbb{F}} A$ then $\Gamma \vdash_{\mathsf{H}(\mathbf{tK}\Sigma)} A$ for any set $\Gamma \cup \{A\}$ of formulas. In this section we show the strong completeness results of Hilbert systems for TSMLs without any B_n axiom schema and the Hilbert system for a TSML which can be seen as an S5 version of *term-modal logic*. More precisely: for any $\Sigma \subseteq \bigcup_{n\in\mathbb{N}}\{\mathrm{D}_n, \mathrm{T}_n, 4_n\}$, we show the strong completeness results of $\mathsf{H}(\mathbf{tK}\Sigma)$ and $\mathsf{H}(\mathbf{tS5}_1 \restriction \mathcal{L}_1)$. The strong completeness result of $\mathsf{H}(\mathbf{tS5}_1 \restriction \mathcal{L}_1)$ is important, because the strong completeness result of an S5 version of term-modal logic corresponding to our $\mathsf{H}(\mathbf{tS5}_1 \restriction \mathcal{L}_1)$ is not provided in [4,21].[4]

We stipulate necessary notations and terminology. Let $\Sigma \subseteq \bigcup_{n\in\mathbb{N}}\{\mathrm{D}_n, \mathrm{T}_n, \mathrm{B}_n, 4_n\}$. A set Γ of formulas is $\mathbf{tK}\Sigma$-*inconsistent* if $\Gamma \vdash_{\mathsf{H}(\mathbf{tK}\Sigma)} \bot$, and Γ is $\mathbf{tK}\Sigma$-*consistent* if Γ is not $\mathbf{tK}\Sigma$-inconsistent. The set Γ is a $\mathbf{tK}\Sigma$-*maximally consistent set* ($\mathbf{tK}\Sigma$-MCS for short) if Γ is $\mathbf{tK}\Sigma$-consistent and $A \in \Gamma$ or $\neg A \in \Gamma$ for any formula $A \in \mathsf{Form}(\Gamma)$, where $\mathsf{Form}(\Gamma) = \{A \mid \mathsf{FV}(A) \subseteq \mathsf{FV}(\Gamma)\}$. The set Γ has $\forall$-*property* if for any formula of the form $\forall x A$ in $\mathsf{Form}(\Gamma)$, there is a variable $y \in \mathsf{FV}(\Gamma)$ such that $A(y/x) \supset \forall x A \in \Gamma$. We define $\mathcal{L}^+ := \mathcal{L} \cup \mathsf{Var}^+$ and $\mathcal{L}_1^+ := \mathcal{L}_1 \cup \mathsf{Var}^+$, where Var^+ is a countably infinite set of variables disjoint from those in $\mathcal{L}$. We also say that Γ is *modest* if $\mathsf{Var}^+ \backslash \mathsf{FV}(\Gamma)$ is infinite and that Γ is *full* if $\mathsf{FV}(\Gamma) = \mathsf{Var}^+$. The former notion plays a role to prove the strong completeness results of $\mathsf{H}(\mathbf{tK}\Sigma)$ where $\Sigma \subseteq \bigcup_{n\in\mathbb{N}}\{\mathrm{D}_n, \mathrm{T}_n, 4_n\}$ and the latter notion plays a role to prove the strong completeness result of $\mathsf{H}(\mathbf{tS5}_1 \restriction \mathcal{L}_1)$.

5.1 Strong Completeness of Hilbert Systems for TSMLs Without Any $\mathbf{B}_n$ axiom schema

Let $\Sigma \subseteq \bigcup_{n\in\mathbb{N}}\{\mathrm{D}_n, \mathrm{T}_n, 4_n\}$ and $\Lambda := \mathbf{tK}\Sigma$ throughout this subsection.

Lemma 1 (Lindenbaum Lemma). *Let Γ be a modest Λ-consistent set in $\mathcal{L}^+$. There is a modest Λ-MCS Γ^+ with $\forall$-property in $\mathcal{L}^+$ such that $\Gamma \subseteq \Gamma^+$.*

Lemma 2. *Let Γ be a modest Λ-MCS with $\forall$-property in $\mathcal{L}^+$. Then, for any formula of the form $[\vec{t}]A$ in $\mathsf{Form}(\Gamma)$, if $[\vec{t}]A \notin \Gamma$ then there is a modest Λ-MCS Δ with $\forall$-property in $\mathcal{L}^+$ such that $\{B \mid [\vec{t}]B \in \Gamma\} \cup \{\neg A\} \subseteq \Delta$.*

Definition 10 (Canonical Λ-Model). Let $\mathcal{D}$ be the set of terms in $\mathcal{L}^+$. The *canonical Λ-model $M^\Lambda = (W^\Lambda, D^\Lambda, R^\Lambda, V^\Lambda, I^\Lambda)$ over $\mathcal{D}$* is defined by

- $W^\Lambda := \{\Gamma \mid \Gamma$ is a modest Λ-MCS with $\forall$-property in $\mathcal{L}^+\}$;

[4] The strong completeness results of $\mathsf{H}(\mathbf{tK}\Sigma)$ for *all* $\Sigma \subseteq \bigcup_{n\in\mathbb{N}}\{\mathrm{D}_n, \mathrm{T}_n, 4_n, \mathrm{B}_n\}$ such that $\mathrm{B}_k \in \Sigma$ for some $k \in \mathbb{N}$ are not presented in this paper. For example, the Hilbert system $\mathsf{H}(\mathbf{tKB}_1)$ is not yet proved to be strongly complete, as the ordinal canonical model construction is not so straightforward for $\mathbf{tKB}_1$. The step-by-step method introduced in [2, p. 223] might be applicable for the strong completeness results of the Hilbert systems for such logics, but we have not done yet.

- $D^{\Lambda}_{\Gamma} := \mathsf{Term}(\Gamma)$;
- $\Gamma R^{\Lambda}_{\vec{t_n}} \Delta$ iff $\vec{t_n} \in (D^{\Lambda}_{\Gamma})^n$ and $\left([\vec{t_n}]A \in \Gamma \text{ implies } A \in \Delta\right)$ for all formulas A in $\mathcal{L}^+$;
- $(t_1, \ldots, t_n) \in V^{\Lambda}_{\Gamma}(P)$ iff $Pt_1 \ldots t_n \in \Gamma$;
- $I^{\Lambda}(c) = c$;
- $I^{\Lambda}(f)(t_1, \ldots, t_n) = f(t_1, \ldots, t_n)$;

where $\mathsf{Term}(\Gamma)$ is the set of terms whose variables are restricted to $\mathsf{FV}(\Gamma)$.

Proposition 11. *The canonical Λ-model M^{Λ} is a model.*

Lemma 3 (Truth Lemma). *Let M^{Λ} be the canonical Λ-model and ι be the canonical assignment defined by $\iota(x) = x$. For any formula A in $\mathcal{L}^+$ and any $\Gamma \in W^{\Lambda}$ such that $A \in \mathsf{Form}(\Gamma)$, $A \in \Gamma$ iff $M^{\Lambda}, \Gamma, \iota \models A$.*

Theorem 1 (Strong Completeness of $\mathsf{H}(\mathbf{tK}\Sigma)$). *Let $\Sigma \subseteq \bigcup_{n \in \mathbb{N}} \{\mathrm{D}_n, \mathrm{T}_n, 4_n\}$. The Hilbert system $\mathsf{H}(\mathbf{tK}\Sigma)$ is strongly complete with respect to the class $\mathbb{F}_{\Sigma}$ of frames defined by Σ.*

Proof (Sketch). Recall $\Lambda = \mathbf{tK}\Sigma$ and suppose $\Gamma \not\vdash_{\mathsf{H}(\Lambda)} A$. By Lemma 1 we construct a modest Λ-MCS Γ^+ with $\forall$-property in $\mathcal{L}^+$ such that $\Gamma \cup \{\neg A\} \subseteq \Gamma^+$. Construct the canonical Λ-model M^{Λ} and the canonical assignment ι. By Lemma 3 we then get $M^{\Lambda}, \Gamma^+, \iota \models \Gamma \cup \{\neg A\}$. This establishes $\Gamma \not\models_{\mathbb{F}_{\Sigma}} A$, since the frame of M^{Λ} has the frame properties defined by Σ. □

Corollary 12. *Given $\Sigma \subseteq \{\mathrm{D}_1, \mathrm{T}_1, 4_1\}$, $\mathsf{H}(\mathbf{tK}\Sigma \restriction \mathcal{L}_1)$ are sound and strongly complete with respect to the class of frames defined by Σ.*

5.2 Strong Completeness of $\mathsf{H}(\mathbf{tS5}_1 \restriction \mathcal{L}_1)$

We let $\Lambda := \mathbf{tS5}_1 \restriction \mathcal{L}_1$ throughout this subsection.

Lemma 4. *Let Γ be a full Λ-MCS with $\forall$-property in $\mathcal{L}^+_1$. Then, for any formula of the form $[t]A$ in $\mathsf{Form}(\Gamma)$, if $[t]A \notin \Gamma$ then there is a full Λ-MCS Δ with $\forall$-property in $\mathcal{L}^+_1$ such that $\{B \mid [t]B \in \Gamma\} \cup \{\neg A\} \subseteq \Delta$.*

Proof (Sketch). We basically follows the strategy developed in [7]. Fix any formula of the form $[t]A$ in $\mathsf{Form}(\Gamma)$ such that $[t]A \notin \Gamma$ and let $[t]^{-1}\Gamma = \{B \mid [t]B \in \Gamma\} \cup \{\neg A\}$. Enumerate Var^+ and all formulas of the form $\forall xC$ in $\mathsf{Form}(\Gamma)$. We first define a chain $(\Delta_n)_{n \in \mathbb{N}}$ of sets of formulas by $\Delta_0 := [t]^{-1}\Gamma$ and $\Delta_{n+1} := \Delta_n \cup \{(C(y_m/x) \supset \forall xC_n)\}$, where m is the first number in the enumeration on Var^+ such that Δ_{n+1} is Λ-consistent. Then, the following argument guarantees that there is always a variable y such that $\Delta_n \cup \{(C(y_m/x) \supset \forall xC_n)\}$ is Λ-consistent. Suppose for contradiction that such a y does not exist. Following the strategy developed in [7], we can claim

(1) $\Gamma \vdash [t](D \supset \neg(C(y/x) \supset \forall x C_n))$ for each variable y
(2) $\Gamma \vdash [t](D \supset \neg(C(y/x) \supset \forall x C_n)) \supset \forall z[t](D \supset \neg(C(z/x) \supset \forall x C_n))$ for some variable y,

where $D \equiv \neg A \wedge D_1 \wedge \cdots \wedge D_m$ for some $D_1, \dots, D_m \in \Delta_n$ and a variable z does not occur in t, D nor $\forall x C_n$. Then, we can show $[t]\neg D \in \Gamma$ by (1), (2) and Proposition 8; hence $\Delta_n \cup \{ (C(y_m/x) \supset \forall x C_n) \}$ is Λ-inconsistent and so a contradiction occurs. Thus the construction above is possible. From $\bigcup_{n \in \mathbb{N}} \Delta_n$ we can construct a full Λ-MCS Δ with $\forall$-property in $\mathcal{L}_1^+$ such that $[t]^{-1}\Gamma \subseteq \bigcup_{n \in \mathbb{N}} \Delta_n \subseteq \Delta$. □

Definition 13 (Canonical Λ-model). Let $\mathcal{D}$ be the set Term^+ of terms in $\mathcal{L}_1^+$. The *canonical Λ-model* $M = (W, D, R, V, I)$ *over* $\mathcal{D}$ *is defined by*

- $W := \{ \Gamma \mid \Gamma$ is a full Λ-MCS with $\forall$-property in $\mathcal{L}_1^+ \}$;
- $D_\Gamma := \mathsf{Term}^+$;
- $\Gamma R_t \Delta$ iff $\left([t]A \in \Gamma \text{ implies } A \in \Delta\right)$ for all formulas A in $\mathcal{L}_1^+$;
- $(t_1, \dots, t_n) \in V_\Gamma(P)$ iff $Pt_1 \dots t_n \in \Gamma$;
- $I(c) = c$;
- $I(f)(t_1, \dots, t_n) = f(t_1, \dots, t_n)$.

Proposition 14. *The canonical Λ-model M is a model.*

Lemma 5 (Truth Lemma). *Let M be the canonical Λ-model and ι be the canonical assignment defined by $\iota(x) = x$. For any formula A in $\mathcal{L}_1^+$ and any $\Gamma \in W$, $A \in \Gamma$ iff $M, \Gamma, \iota \models A$.*

Theorem 2 (Strong Completeness of $\mathsf{H}(\mathbf{tS5}_1 \upharpoonright \mathcal{L}_1)$). *The Hilbert system $\mathsf{H}(\mathbf{tS5}_1 \upharpoonright \mathcal{L}_1)$ is strongly complete with respect to the class of frames defined by* $\{ \mathrm{T}_1, \mathrm{B}_1, 4_1 \}$.

Proof (Sketch). Recall $\Lambda = \mathbf{tS5}_1 \upharpoonright \mathcal{L}_1$ and define $\Sigma := \{ \mathrm{T}_1, \mathrm{B}_1, 4_1 \}$. Suppose $\Gamma \not\vdash_{\mathsf{H}(\Lambda)} A$. As in the proof of Theorem 1, construct a full Λ-MCS Γ^+ with $\forall$-property in $\mathcal{L}^+$ such that $\Gamma \cup \{ \neg A \} \subseteq \Gamma^+$ and the canonical Λ-model M and the canonical assignment ι. By Lemma 5 we then get $M, \Gamma^+, \iota \models \Gamma \cup \{ \neg A \}$. This establishes $\Gamma \not\models_{\mathbb{F}_\Sigma} A$, since the frame of M^Λ has the frame properties defined by Σ. □

6 Sequent Calculi for Term-Sequence-Modal Logics

In this section, we provide sequent calculi for all Hilbert systems in Sects. 5.1 and 5.2. Let $\Sigma \subseteq \bigcup_{n \in \mathbb{N}} \{ \mathrm{D}_n, \mathrm{T}_n, 4_n \}$ and Γ, Δ be finite multisets of formulas. We also define $[\vec{t}]\Gamma := \{ [\vec{t}]A \mid A \in \Gamma \}$, $\Sigma_n := \{ X_n \mid X_n \in \Sigma \}$ and call an expression $\Gamma \Rightarrow \Delta$ a *sequent.* A sequent calculus G for first-order logic consists of initial sequents, structural rules and logical rules displayed in Table 3. The sequent calculus $\mathsf{G}(\mathbf{tK}\Sigma)$ and $\mathsf{G}(\mathbf{tS5}_1 \upharpoonright \mathcal{L}_1)$ are defined as follows.

Definition 15. Given $\Sigma \subseteq \bigcup_{n\in\mathbb{N}}\{\mathrm{D}_n, \mathrm{T}_n, 4_n\}$, the sequent calculus $\mathsf{G}(\mathbf{tK}\Sigma)$ is obtained from G by adding all additional rules for Σ_n for each $n \in \mathbb{N}$ displayed in Table 3. The sequent calculus $\mathsf{G}(\mathbf{tS5}_1 \restriction \mathcal{L}_1)$ for the sublanguage $\mathcal{L}_1$ of $\mathcal{L}$ is obtained from the restriction of G to $\mathcal{L}_1$ by adding all additional rules for $\{\mathrm{T}_1, \mathrm{B}_1, 4_1\}$ displayed in Table 3. We call $\mathsf{G}^-(\mathbf{tK}\Sigma)$ to be the calculus obtained by removing *Cut* from $\mathsf{G}(\mathbf{tK}\Sigma)$.

The notion on a derivation in a sequent calculus is defined as usual.

Example 3.

1. Recall a formula $[c]\Box P \supset \Diamond P$ in Example 2. Let us consider $\Sigma = \{\mathrm{D}_0, \mathrm{T}_1\}$. Then, a derivation of the formula in $\mathsf{G}(\mathbf{tKD}_0\mathbf{T}_1)$ is given as follows:

$$\dfrac{\dfrac{\dfrac{\dfrac{\dfrac{P \Rightarrow P}{P, \neg P \Rightarrow}\ {\scriptstyle \neg\Rightarrow}}{\Box P, \Box\neg P \Rightarrow}\ {\scriptstyle \Box \mathrm{D}_0}}{\Box P \Rightarrow \neg\Box\neg P}\ {\scriptstyle \Rightarrow\neg}}{[c]\Box P \Rightarrow \neg\Box\neg P}\ {\scriptstyle \Box \mathrm{T}_1}}{\Rightarrow [c]\Box P \supset \neg\Box\neg P}\ {\scriptstyle \Rightarrow\supset}$$

2. We give another example. Let us consider $\Sigma = \{\mathrm{T}_1, \mathrm{B}_1, 4_1\}$. Then, $P \supset [t]\langle t\rangle P$ is derivable in $\mathsf{G}(\mathbf{tS5}_1 \restriction \mathcal{L}_1)$ as follows:

$$\dfrac{\dfrac{\dfrac{\dfrac{\dfrac{P \Rightarrow P}{\neg P, P \Rightarrow}\ {\scriptstyle \neg\Rightarrow}}{[t]\neg P, P \Rightarrow}\ {\scriptstyle \Box \mathrm{T}_1}}{P \Rightarrow \neg[t]\neg P}\ {\scriptstyle \Rightarrow\neg} \qquad \dfrac{\dfrac{\dfrac{[t]\neg P \Rightarrow [t]\neg P}{\Rightarrow \neg[t]\neg P, [t]\neg P}\ {\scriptstyle \Rightarrow\neg}}{\Rightarrow [t]\neg[t]\neg P, [t]\neg P}\ {\scriptstyle \Box \mathrm{tS5}_1}}{\neg[t]\neg P \Rightarrow [t]\neg[t]\neg P}\ {\scriptstyle \neg\Rightarrow}}{P \Rightarrow [t]\neg[t]\neg P}\ {\scriptstyle Cut}}{\Rightarrow P \supset [t]\neg[t]\neg P}\ {\scriptstyle \Rightarrow\supset}$$

Note that an application of *Cut* is necessary in any derivation of $P \supset [t]\langle t\rangle P$ in $\mathsf{G}(\mathbf{tS5}_1 \restriction \mathcal{L}_1)$, as pointed out in [13, p. 124] and [14, p. 222].[5]

Proposition 16 (Equipollence). *Let* $\Sigma \subseteq \bigcup_{n\in\mathbb{N}}\{\mathrm{D}_n, \mathrm{T}_n, 4_n\}$. *It holds that* $\vdash_{\mathsf{H}(\mathbf{tK}\Sigma)} A$ *iff* $\vdash_{\mathsf{G}(\mathbf{tK}\Sigma)} \Rightarrow A$. *It also holds that* $\vdash_{\mathsf{H}(\mathbf{tS5}_1\restriction\mathcal{L}_1)} A$ *iff* $\vdash_{\mathsf{G}(\mathbf{tS5}_1\restriction\mathcal{L}_1)} \Rightarrow A$.

Theorem 3 (Cut Elimination). *Let* $\Sigma \subseteq \bigcup_{n\in\mathbb{N}}\{\mathrm{D}_n, \mathrm{T}_n, 4_n\}$. *If* $\vdash_{\mathsf{G}(\mathbf{tK}\Sigma)} \Gamma \Rightarrow \Delta$ *then* $\vdash_{\mathsf{G}^-(\mathbf{tK}\Sigma)} \Gamma \Rightarrow \Delta$.

Proof (Sketch). We can prove cut elimination theorem for $\mathsf{G}(\mathbf{tK}\Sigma)$ by the method of the extended rule Cut^* of *Cut* introduced in [8, 15]:

[5] Thalmann and Fitting's method for proving the strong completeness results of term-modal logics only works for cut-free systems, which seems to be the reason why they do not provide the strong completeness result of an S5 version of term-modal logic in [4, 21].

Table 3. Sequent calculi G and additional rules

Initial Sequents of G	
$A \Rightarrow A$	$\bot \Rightarrow$

Structural Rules of G	
$\dfrac{\Gamma \Rightarrow \Delta}{\Gamma \Rightarrow \Delta, A} \Rightarrow w$	$\dfrac{\Gamma \Rightarrow \Delta}{A, \Gamma \Rightarrow \Delta}\ w \Rightarrow$
$\dfrac{\Gamma \Rightarrow \Delta, A, A}{\Gamma \Rightarrow \Delta, A} \Rightarrow c$	$\dfrac{A, A, \Gamma \Rightarrow \Delta}{A, \Gamma \Rightarrow \Delta}\ c \Rightarrow$
$\dfrac{\Gamma \Rightarrow \Delta, A \qquad A, \Theta \Rightarrow \Sigma}{\Gamma, \Theta \Rightarrow \Delta, \Sigma}\ Cut$	

Logical Rules of G	
$\dfrac{A, \Gamma \Rightarrow \Delta}{\Gamma \Rightarrow \Delta, \neg A} \Rightarrow \neg$	$\dfrac{\Gamma \Rightarrow \Delta, A}{\neg A, \Gamma \Rightarrow \Delta}\ \neg \Rightarrow$
$\dfrac{A, \Gamma \Rightarrow \Delta, B}{\Gamma \Rightarrow \Delta, A \supset B} \Rightarrow \supset$	$\dfrac{\Gamma \Rightarrow \Delta, A \qquad B, \Theta \Rightarrow \Sigma}{A \supset B, \Gamma, \Theta \Rightarrow \Delta, \Sigma}\ \supset \Rightarrow$
$\dfrac{\Gamma \Rightarrow \Delta, A(y/x)}{\Gamma \Rightarrow \Delta, \forall x A} \Rightarrow \forall^{\dagger}$	$\dfrac{A(t/x), \Gamma \Rightarrow \Delta}{\forall x A, \Gamma \Rightarrow \Delta}\ \forall \Rightarrow$
†: y does not occur in $\Gamma, \Delta, \forall x A$.	

Σ_n	Additional rules
$\varnothing$	$\dfrac{\Gamma \Rightarrow A}{[\vec{t_n}]\Gamma \Rightarrow [\vec{t_n}]A}\ \Box\mathrm{tK}_n$
$\{\mathrm{D}_n\}$	$\dfrac{\Gamma \Rightarrow A}{[\vec{t_n}]\Gamma \Rightarrow [\vec{t_n}]A}\ \Box\mathrm{tK}_n \quad \dfrac{\Gamma \Rightarrow}{[\vec{t_n}]\Gamma \Rightarrow}\ \Box\mathrm{D}_n$
$\{\mathrm{T}_n\}$	$\dfrac{\Gamma \Rightarrow A}{[\vec{t_n}]\Gamma \Rightarrow [\vec{t_n}]A}\ \Box\mathrm{tK}_n \quad \dfrac{A, \Gamma \Rightarrow \Delta}{[\vec{t_n}]A, \Gamma \Rightarrow \Delta}\ \Box\mathrm{T}_n$
$\{4_n\}$	$\dfrac{\Gamma, [\vec{t_n}]\Gamma \Rightarrow A}{[\vec{t_n}]\Gamma \Rightarrow [\vec{t_n}]A}\ \Box 4_n$
$\{\mathrm{D}_n, 4_n\}$	$\dfrac{\Gamma, [\vec{t_n}]\Gamma \Rightarrow}{[\vec{t_n}]\Gamma \Rightarrow}\ \Box\mathrm{D}_n \quad \dfrac{\Gamma, [\vec{t_n}]\Gamma \Rightarrow A}{[\vec{t_n}]\Gamma \Rightarrow [\vec{t_n}]A}\ \Box 4_n$
$\{\mathrm{T}_n, 4_n\}$	$\dfrac{A, \Gamma \Rightarrow \Delta}{[\vec{t_n}]A, \Gamma \Rightarrow \Delta}\ \Box\mathrm{T}_n \quad \dfrac{[\vec{t_n}]\Gamma \Rightarrow A}{[\vec{t_n}]\Gamma \Rightarrow [\vec{t_n}]A}\ \Box 4_n$

$\{\mathrm{T}_1, \mathrm{B}_1, 4_1\}$	$\dfrac{A, \Gamma \Rightarrow \Delta}{[t]A, \Gamma \Rightarrow \Delta}\ \Box\mathrm{T}_1 \quad \dfrac{[t]\Gamma \Rightarrow [t]\Delta, A}{[t]\Gamma \Rightarrow [t]\Delta, [t]A}\ \Box\mathrm{tS5}_1$

$$\frac{\Gamma \Rightarrow \Delta, A^m \qquad A^n, \Theta \Rightarrow \Sigma}{\Gamma, \Theta \Rightarrow \Delta, \Sigma}\ Cut^*,$$

where m, n can be zero; each A is called *cut formula*; $\Gamma \Rightarrow \Delta, A^m$ and $A^n, \Theta \Rightarrow \Sigma$ are called the *left premise* and *right premise* of Cut^*, respectively. The outline of proof is as follows: let $\mathsf{G}^*(\mathbf{tK}\Sigma)$ be $\mathsf{G}(\mathbf{tK}\Sigma)$ in which Cut is replaced by Cut^*. As Cut is an instance of Cut^*, it suffices to show that if $\vdash_{\mathsf{G}^*(\mathbf{tK}\Sigma)} \Gamma \Rightarrow \Delta$ then $\vdash_{\mathsf{G}^-(\mathbf{tK}\Sigma)} \Gamma \Rightarrow \Delta$. We say that a derivation $\mathfrak{D}$ in $\mathsf{G}^*(\mathbf{tK}\Sigma)$ is of the *Cut^*-bottom form* if the last applied rule in $\mathfrak{D}$ is Cut^* and there are no other applications of Cut^* in $\mathfrak{D}$. We also let the *weight* of a derivation of Cut^*-bottom form be the number of sequents occurring in $\mathfrak{D}$ except for its root. Then, given a derivation of Cut^*-bottom form of a sequent $\Gamma' \Rightarrow \Delta'$ in $\mathsf{G}^*(\mathbf{tK}\Sigma)$, by double induction on complexity of the cut formula and weight of the derivation we can construct a derivation of $\Gamma' \Rightarrow \Delta'$ with no applications of Cut^* in $\mathsf{G}^-(\mathbf{tK}\Sigma)$. This proves that if $\vdash_{\mathsf{G}^*(\mathbf{tK}\Sigma)} \Gamma \Rightarrow \Delta$ then $\vdash_{\mathsf{G}^-(\mathbf{tK}\Sigma)} \Gamma \Rightarrow \Delta$. □

From Theorem 3, we can prove the consistency of $\mathsf{G}(\mathbf{tK}\Sigma)$ in a purely proof-theoretic way.

Corollary 17. *Let $\Sigma \subseteq \bigcup_{n\in\mathbb{N}}\{\mathrm{D}_n, \mathrm{T}_n, 4_n\}$. Then $\nvdash_{\mathsf{G}(\mathbf{tK}\Sigma)} \Rightarrow \bot$.*

Proof. Suppose for contradiction that $\vdash_{\mathsf{G}(\mathbf{tK}\Sigma)} \Rightarrow \bot$. It follow from $\vdash_{\mathsf{G}(\mathbf{tK}\Sigma)} \bot \Rightarrow$ that $\vdash_{\mathsf{G}(\mathbf{tK}\Sigma)} \Rightarrow$ by Cut. We deduce from Theorem 3 that $\vdash_{\mathsf{G}^-(\mathbf{tK}\Sigma)} \Rightarrow$ in the cut-free system, which cannot be the case from all the rules of the cut-free system $\mathsf{G}^-(\mathbf{tK}\Sigma)$. □

We can also prove Craig interpolation theorem for $\mathsf{G}(\mathbf{tK}\Sigma)$ whose language $\mathcal{L}$ is restricted so that $\mathsf{Con} = \mathsf{Fn} = \varnothing$. We prove this by the Maehara method introduced in [11, 12]. The outline of proof is as follows. Given a sequent $\Gamma \Rightarrow \Delta$, we call an expression $((\Gamma_1; \Gamma_2), (\Delta_1; \Delta_2))$ a *partition of* $\Gamma \Rightarrow \Delta$, where $\Gamma \equiv \Gamma_1, \Gamma_2$ and $\Delta \equiv \Delta_1, \Delta_2$. By $\mathsf{Pred}(\Gamma)$ we denote the set of predicate symbols in a set Γ of formulas. We can show the following:

Lemma 6. *Let $\Sigma \subseteq \bigcup_{n\in\mathbb{N}}\{\mathrm{D}_n, \mathrm{T}_n, 4_n\}$ and restrict $\mathcal{L}$ so that $\mathsf{Con} = \mathsf{Fn} = \varnothing$. If $\vdash_{\mathsf{G}^-(\mathbf{tK}\Sigma)} \Gamma \Rightarrow \Delta$, then for any partition $((\Gamma_1; \Gamma_2), (\Delta_1; \Delta_2))$ of $\Gamma \Rightarrow \Delta$, there is a formula F such that*

1. $\vdash_{\mathsf{G}^-(\mathbf{tK}\Sigma)} \Gamma_1 \Rightarrow \Delta_1, F$ *and* $\vdash_{\mathsf{G}^-(\mathbf{tK}\Sigma)} F, \Gamma_2 \Rightarrow \Delta_2$;
2. $\mathsf{Pred}(F) \subseteq \mathsf{Pred}(\Gamma_1, \Delta_1) \cap \mathsf{Pred}(\Gamma_2, \Delta_2)$;
3. $\mathsf{FV}(F) \subseteq \mathsf{FV}(\Gamma_1, \Delta_1) \cap \mathsf{FV}(\Gamma_2, \Delta_2)$.

We say that F is an interpolant of $((\Gamma_1; \Gamma_2), (\Delta_1; \Delta_2))$.

Proof (Sketch). Suppose $\vdash_{\mathsf{G}^-(\mathbf{tK}\Sigma)} \Gamma \Rightarrow \Delta$. By induction on height of the derivation $\Gamma \Rightarrow \Delta$, we show that there is an interpolant for any partition $\pi = ((\Gamma_1; \Gamma_2), (\Delta_1; \Delta_2))$ of $\Gamma \Rightarrow \Delta$. Since if either $\Gamma_1 \cup \Delta_1 = \varnothing$ or $\Gamma_2 \cup \Delta_2 = \varnothing$ then $\bot \supset \bot$ or $\bot$ is an interpolant of π respectively, we may assume that $\Gamma_1 \cup \Delta_1 \neq \varnothing$ and $\Gamma_2 \cup \Delta_2 \neq \varnothing$ for any partition π in what follows. Among the inductive cases, the only cases on term-sequence-modal operators are crucial. For example, consider the case that $\Gamma \Rightarrow \Delta$ is obtained by $\Box$tK. In this case, we have the derivation of $[\vec{t}]\Gamma' \Rightarrow [\vec{t}]A$ whose premise is $\Gamma' \Rightarrow A$, where $\Gamma \equiv [\vec{t}]\Gamma'$ and $\Delta \equiv [\vec{t}]A$. Fix any partition π of $[\vec{t}]\Gamma' \Rightarrow [\vec{t}]A$. Then, either $\pi = (([\vec{t}]\Gamma'_1; [\vec{t}]\Gamma'_2), ([\vec{t}]A; \varnothing))$ or $\pi = (([\vec{t}]\Gamma'_1; [\vec{t}]\Gamma'_2), (\varnothing; [\vec{t}]A))$. Since similar arguments are available in either case, we consider only the former case. By inductive hypothesis we get an interpolant C of $((\Gamma'_1; \Gamma'_2), (A; \varnothing))$, so we can show $\vdash [\vec{t}]\Gamma'_1 \Rightarrow [\vec{t}]A, \neg[\vec{t}]\neg C$ and $\vdash \neg[\vec{t}]\neg C, [\vec{t}]\Gamma'_2 \Rightarrow$. Then, since our assumption guarantees $\Gamma'_2 \neq \varnothing$ (by $\Delta'_2 = \varnothing$ and $\Gamma'_2 \cup \Delta'_2 \neq \varnothing$ in our partition π), we have $\mathsf{FV}(\neg[\vec{t}]\neg C) \subseteq \mathsf{FV}([\vec{t}]\Gamma'_1, [\vec{t}]A) \cap \mathsf{FV}([\vec{t}]\Gamma'_2)$. So similarly for the condition on predicate symbols. Thus $\neg[\vec{t}]\neg C$ is an interpolant of π. □

By cut elimination theorem and Lemma 6, Craig interpolation theorem follows.

Theorem 4 (Craig Interpolation). *Let* $\Sigma \subseteq \bigcup_{n\in\mathbb{N}}\{\mathrm{D}_n, \mathrm{T}_n, 4_n\}$ *and restrict* $\mathcal{L}$ *so that* $\mathsf{Con} = \mathsf{Fn} = \varnothing$. *If* $\vdash_{\mathsf{G}(\mathbf{tK}\Sigma)} A \Rightarrow B$, *then there is a formula* C *such that*

1. $\vdash_{\mathsf{G}(\mathbf{tK}\Sigma)} A \Rightarrow C$ *and* $\vdash_{\mathsf{G}(\mathbf{tK}\Sigma)} C \Rightarrow B$;
2. $\mathsf{Pred}(C) \subseteq \mathsf{Pred}(A) \cap \mathsf{Pred}(B)$;
3. $\mathsf{FV}(C) \subseteq \mathsf{FV}(A) \cap \mathsf{FV}(B)$.

7 Conclusion

We list some directions for further research. One of our next tasks may be to pursue a possibility of dropping the monotonicity assumption inherited from [4]. Common sense modal predicate logic, presented by van Benthem in [1, pp. 120–121] and further developed by Seligman in [20], seems to be of much interest in this regard. It allows the domain of each world to be different from those of others in any way as long as it remains non-empty, and only requires individual quantifiers to range over the local domain of objects existing in the current world.[6] On this treatment, some of the instances of Barcan schema are not, but all of the instances of its converse are, valid. Thus, it seems of much interest to incorporate this treatment of individual quantifiers into TSMLs.

In addition, it would be worth developing TSMLs with equality, since there are cases in which we would like to distinguish x and y in a formula $[x, y]Pxy$. Fitting et al. [4] does not consider any term-modal logics with equality. In [19] Rendsvig develops an epistemic term-modal logic with equality, but he adopts the constant domain semantics (whose domain always has at least n agents) and interprets constant symbols as rigid designators unlike Fitting et al. [4] and our settings. Neither of [4, 19], however considers the sequent calculi for term modal logics with the equality symbol. Thus, the task to develop TSMLs with equality from proof-theoretic perspective remains interesting.

[6] This treatment is also found in the definition of the semantics of modal predicate logic given by Gamut in [5, pp. 59–60].

Another immediate task may be the dynamification of TSMLs. While in [9] Kooi develops a dynamic epistemic predicate logic by introducing the notion of the product update into term-modal logic, we are interested in developing a dynamic deontic TSML that characterizes the effects of various speech acts. As the language of the dynamic deontic logic of acts of commanding and promising developed in [24] has a set of deontic operators of the form $\mathcal{O}_{(i,j.k)}$, where i, j and k stand for the agent who owes the obligation, the agent to whom i's obligation is owed, and the agent who creates the obligation respectively, it is natural to recast and extend this logic in a deontic TSML having a term-sequence-modal operator $\mathcal{O}_{(x,y,z)} := [x, y, z]$. Then, for example, a formula $\mathcal{O}_{(x,y,x)}Pxy$ is read as "it is obligatory for x with respect to y by the name x to see to it that Pxy". Thus, by defining in the deontic TSML a model updating operation for act of, say, promising $[\mathsf{Prom}(x, y, A)]$ which is read as "x promises y to see to it that A", it becomes possible for us to state the truth conditions for formulas such as $\forall x \forall y[\mathsf{Prom}(x, y, Pxy)]\mathcal{O}_{(x,y,x)}Pxy$ ("for any agents x, y, whenever x promises y to see to it that Pxy, it is obligatory for x with respect to y by the name of x to see to it that Pxy"). In addition to this, we may extend this dynamified deontic TSML further by adding epistemic modalities to it. This will enable us to recast and extend the dynamified epistemic deontic logic developed in [25], which deals with the effects of acts of requesting and asserting along with acts of commanding and promising.

References

1. van Benthem, J.: Modal Logic for Open Minds. CSLI Publications, Stanford (2010)
2. Blackburn, P., de Rijke, M., Venema, Y.: Modal Logic. Cambridge University Press, Cambridge (2001). Fourth Printing with Corrections 2010
3. Castaneda, H.N.: Review: Jaakko Hintikka, Knowledge and belief. An introduction to the logic of the two notions. J. Symbolic Log. **29**(3), 132–134 (1964)
4. Fitting, M., Thalmann, L., Voronkov, A.: Term-modal logics. Stud. Logica. **69**, 133–169 (2001)
5. Gamut, L.T.F.: Logic, Language, and Meaning. Intensional Logic and Logical Grammar, vol. 2. The University of Chicago Press, Chicago (1991)
6. Hintikka, J.: Knowledge and Belief. Cornell University Press, Ithaca (1962)
7. Hughes, G.E., Cresswell, M.J.: A New Introduction to Modal Logic. Routledge, New York (2001)
8. Kashima, R.: Mathematical Logic. Asakura Publishing Co. Ltd., Osaka (2009). (in Japanese)
9. Kooi, B.: Dynamic term-modal logic. In: van Benthem, J., Ju, S., Veltman, F. (eds.) A Meeting of the Minds. Proceedings of the Workshop on Logic, Rationality and Interaction, Beijing, 2007, Texts in Computing Computer Science 8, pp. 173–185. College Publications (2008)
10. Kooi, B., Tamminga, A.: Moral conflicts between groups of agents. J. Philos. Log. **37**, 1–21 (2008)
11. Maehara, S.: On interpolation theorem of Craig. Sugaku **12**(4), 235–237 (1961). (in Japanese)
12. Maehara, S., Takeuti, G.: A formal system of first-order predicate calculus with infinitely long expressions. J. Math. Soc. Jpn. **13**, 357–370 (1961)

13. Ohnishi, M., Matsumoto, K.: Gentzen method in modal calculi. Osaka Math. J. **9**(2), 113–130 (1957)
14. Ono, H.: Proof-theoretic methods in nonclassical logic - an introduction. In: Theories of Types and Proofs, pp. 207–254 (1998)
15. Ono, H., Komori, Y.: Logics without the contraction rule. J. Symbolic Log. **50**(1), 169–201 (1985)
16. Orlandelli, E., Corsi, G.: Decidable term-modal logics. In: Belardinelli, F., Argente, E. (eds.) EUMAS/AT 2017. LNCS (LNAI), vol. 10767, pp. 147–162. Springer, Cham (2018). https://doi.org/10.1007/978-3-030-01713-2_11
17. Padmanabha, A., Ramanujam, R., Wang, Y.: Bundled fragments of first-order modal logic: (un)decidability. In: Ganguly, S., Pandya, P. (eds.) 38th IARCS Annual Conference on Foundations of Software Technology and Theoretical Computer Science, FSTTCS 2018, Schloss Dagstuhl, Ahmedabad, India, pp. 43:1–43:20, December 2018
18. Priest, G.: Towards Non-Being: The Logic and Metaphysics of Intentionality. Oxford University Press, Oxford (2005)
19. Rendsvig, R.K.: Epistemic term-modal logic. In: Slavkovik, M. (ed.) Proceedings of the 15th Student Session of the European Summer School in Logic, Language and Information, pp. 37–46 (2010)
20. Seligman, J.: Common sense modal predicate logic. In: Non-classical Modal and Predicate Logics: The 9th International Workshop on Logic and Cognition, Guangzhou, China, 4 December 2017, Presentation 2017
21. Thalmann, L.: Term-modal logic and quantifier-free dynamic assignment logic. Ph.D. thesis, Uppsala University (2000)
22. Wang, Y., Seligman, J.: When names are not commonly known: epistemic logic with assignments. In: Bezhanisshvili, G., D'Agostino, G., Metcalfe, G., Studer, T. (eds.) Advances in Modal Logic, vol. 12, pp. 611–628. College Publications, London (2018)
23. Yamada, T.: Logical dynamics of commands and obligations. In: Washio, T., Satoh, K., Takeda, H., Inokuchi, A. (eds.) JSAI 2006. LNCS (LNAI), vol. 4384, pp. 133–146. Springer, Heidelberg (2007). https://doi.org/10.1007/978-3-540-69902-6_13
24. Yamada, T.: Acts of promising in dynamified deontic logic. In: Satoh, K., Inokuchi, A., Nagao, K., Kawamura, T. (eds.) JSAI 2007. LNCS (LNAI), vol. 4914, pp. 95–108. Springer, Heidelberg (2008). https://doi.org/10.1007/978-3-540-78197-4_11
25. Yamada, T.: Assertions and commitments. Philos. Forum **47**(3–4), 475–493 (2016)

A Logical and Empirical Study of Right-Nested Counterfactuals

Katrin Schulz[1], Sonja Smets[1,2], Fernando R. Velázquez-Quesada[1], and Kaibo Xie[1](✉)

[1] Institute for Logic, Language and Computation, Universiteit van Amsterdam, Amsterdam, The Netherlands
{K.Schulz,S.J.L.Smets,F.R.VelazquezQuesada,K.Xie}@uva.nl

[2] Department of Information Science and Media Studies, University of Bergen, Bergen, Norway

Abstract. The paper focuses on a recent challenge brought forward against the interventionist approach to the meaning of counterfactual conditionals. According to this objection, interventionism cannot in general account for the interpretation of right-nested counterfactuals, the problem being its strict interventionism. We will report on the results of an empirical study supporting the objection, and we will extend the well-known logic of actual causality with a new operator expressing an alternative notion of intervention that does not suffer from the problem (and thus can account for some critical examples). The core idea of the alternative approach is a new notion of intervention, which operates on the evaluation of the variables in a causal model, and not on their functional dependencies. Our result provides new insights into the logical analysis of causal reasoning.

1 Introduction

The meaning of counterfactual conditionals, sentences of the form *"If A were/had been the case, then B would be/have been the case"*, bears an intrinsic relation to a number of central scientific problems, like the nature of reasoning, the possibility of knowledge, and the status of laws of nature. Therefore, this topic has fascinated many thinkers from various disciplines: philosophy, logic, psychology and others. But despite a lot of effort, no consensus has been reached yet about how the meaning of these sentences needs to be approached.

Following the similarity approach of Stalnaker and Lewis [1,2], which still is the dominant approach in the philosophical literature, counterfactuals are evaluated as follows. Given the antecedent A and the context of evaluation, we select certain (hypothetical) situations in which the antecedent is true, then checking whether they make the consequent B true as well. The question is how to define the relevant selection function correctly. According to Lewis and Stalnaker, the selection is based on similarity: we select those hypothetical situations that are most similar to the actual world. But this proposal is known to be problematic: among other things, it appears to be too flexible.

P. Blackburn et al. (Eds.): LORI 2019, LNCS 11813, pp. 259–272, 2019.
https://doi.org/10.1007/978-3-662-60292-8_19

In recent years the interventionist approach to counterfactuals became very popular ([3–7] and others). This approach describes the truth conditions of counterfactuals with respect to a representation of the relevant causal dependencies, building on Causal Models as introduced in [8,9]. The approach got its name from the way it describes the selection function. The antecedent is made true by intervention on the given causal dependencies: it is cut off its causal parents and stipulated to be true by law.[1]

Recently, this approach has been criticized by Fisher [12]. He claims that interventionism makes incorrect predictions for right-nested counterfactuals. According to Fisher, the problem is a particular property of the interventionist approach, *strict interventionism*, which he argues needs to be dropped in a proper account. We will argue, using the results of an empirical study, that Fisher is right in his critique. But this does not mean that the interventionist approach needs to be given up. We will propose a variation of the approach that drops strict interventionism and can account for Fisher's core-observations. We will also make precise how this new proposal relates to the classical interventionist approach as spelled out in [6]. We will do so by providing an axiomatization of the new operator for counterfactual reasoning that we introduce. As it will turn out, this new operator can be already defined in terms of the classical intervention operator. Furthermore: to a large extend, they both make the same counterfactuals true. So, our proposal, though formalizing a slightly different take on what intervention means, is in terms of logical properties a very conservative change of the original interventionist approach.

2 The Interventionist Approach to Counterfactuals

Our presentation of the interventionist approach to counterfactuals is based on the one proposed by Briggs in [13]; still, we will only introduce the parts that are relevant for the discussion at hand. The two central ingredients of the approach are ***(i)*** the causal model, which contains information about the relevant causal dependencies, and ***(ii)*** the operation of intervention involved in the definition of the selection function, which maps a given causal model onto a class of models that make the antecedent of a given counterfactual true.

Causal models represent the causal dependencies between a given finite set of variables. For each variable V we fix its range $\mathcal{R}(V)$, the set of possible values the variable can take. The variables are sorted into the set $\mathcal{U}$ of *exogenous* variables (those whose value is independent from the value of other variables in the system), and the set $\mathcal{V}$ of *endogenous* variables (those whose value causally depends on the value of other variables in the system). Given a set of variables $\mathcal{U} \cup \mathcal{V}$, a *causal model* over $\mathcal{U} \cup \mathcal{V}$ is a tuple $\langle \mathcal{S}, \mathcal{A} \rangle$. The first component, $\mathcal{S}$, fixes the causal dependencies between the variables by assigning to each $V \in \mathcal{V}$ a function F_V that maps the values of a set of variables $PA_V \subseteq \mathcal{U} \cup \mathcal{V}$ (the *parents* of V) to a value of the variable V. The second component, $\mathcal{A}$, is a valuation

[1] It turns out that for recursive causal models the interventionist selection function can be understood as just one particular way to make similarity precise [10,11].

function, assigning a value $\mathcal{A}(V) \in \mathcal{R}(V)$ to every $V \in \mathcal{U} \cup \mathcal{V}$ in a way that *complies with the causal dependencies in* $\mathcal{S}$: for all variables V, if $V \in \mathcal{V}$, then $\mathcal{A}(V) = F_V(\mathcal{A}(PA_V))$. Thus, if the values of the exogenous variables (those in $\mathcal{U}$) are given, the values of the variables in $\mathcal{V}$ can be calculated from these values and $\mathcal{S}$.[2] Finally, for talking about causal models, we use a simple propositional language extended with an operator for counterfactual conditionals.[3]

Definition 2.1 (Language $\mathcal{L}_{\Box\!\!\to}$). Formulas ϕ of the language $\mathcal{L}_{\Box\!\!\to}$ over $\mathcal{U} \cup \mathcal{V}$ are given by

$$\phi ::= V = v \mid \neg\phi \mid \phi \wedge \phi \mid (\overrightarrow{V} = \overrightarrow{v}) \Box\!\!\to \phi \quad \text{for } V \in \mathcal{V},\ v \in \mathcal{R}(V),\ \overrightarrow{V} = (V_1, \dots, V_n) \in \mathcal{V}^n,\ n \in \mathbb{N},\ V_i \neq V_j \text{ for } i \neq j,\ \overrightarrow{v} = (v_1, \dots, v_n) \text{ with } v_i \in \mathcal{R}(V_i)$$

Sentences of the form $(\overrightarrow{V} = \overrightarrow{v}) \Box\!\!\to \phi$ should be read as *"if the variables in $\overrightarrow{V}$ were to be set to $\overrightarrow{v}$, then ϕ would hold"*.

The second important ingredient of the interventionist approach is the notion of intervention involved in the interpretation rule for counterfactual sentences $(\overrightarrow{V} = \overrightarrow{v}) \Box\!\!\to \alpha$. Given a causal model $\mathcal{M} = \langle \mathcal{S}, \mathcal{A} \rangle$ and an antecedent $\overrightarrow{V} = \overrightarrow{v}$, we need to define a model that makes the antecedent true;[4] in order to evaluate the consequent there. In the interventionist approach, this model is built by cutting the variables $\overrightarrow{V}$ off their causal parents $PA_{\overrightarrow{V}}$, forcing their value to be the one given by the antecedent $\overrightarrow{V} = \overrightarrow{v}$, as Definition 2.2 below details.[5]

Definition 2.2 (Intervention). Let $\langle \mathcal{S}, \mathcal{A} \rangle$ be a causal model. The semantic interpretation of the Boolean operators in $\mathcal{L}_{\Box\!\!\to}$-formula is as usual; for the rest,

$$\begin{array}{lll} \langle \mathcal{S}, \mathcal{A} \rangle \models V = v & \text{iff}_{def} & \mathcal{A}(V) = v \\ \langle \mathcal{S}, \mathcal{A} \rangle \models (\overrightarrow{V} = \overrightarrow{v}) \Box\!\!\to \phi & \text{iff}_{def} & \langle \mathcal{S}_{\overrightarrow{V} = \overrightarrow{v}}, \mathcal{A}^{\mathcal{S}_{\overrightarrow{V} = \overrightarrow{v}}} \rangle \models \phi \end{array}$$

with $\langle \mathcal{S}_{\overrightarrow{V} = \overrightarrow{v}}, \mathcal{A}^{\mathcal{S}_{\overrightarrow{V} = \overrightarrow{v}}} \rangle$ the causal model where

---

2 This is true for *recursive* causal models, the only ones that this paper will discuss. For their definition: from $\mathcal{S}$, define a relation $\rightarrowtail$ on the set of variables $\mathcal{U} \cup \mathcal{V}$ by writing $X \rightarrowtail Y$ if and only if X is among the parents of Y (the structure $\langle \mathcal{U} \cup \mathcal{V}, \rightarrowtail \rangle$ is called $\mathcal{S}$'s induced *causal graph*). Let $\rightarrowtail^+$ be the transitive closure of $\rightarrowtail$ (so $X \rightarrowtail^+ Y$ indicates that Y is *causally dependent* on X). A causal model is said to be *recursive* when $\rightarrowtail^+$ is a strict partial order (there are no circular dependencies between the variables).

3 This language $\mathcal{L}_{\Box\!\!\to}$ extends the basic causal language (e.g., [6]) by allowing right-nested counterfactuals. Still, it is only a fragment of the language used in [13], as it does not allow Boolean combinations of atoms in the antecedent (which are not relevant to the discussion here).

4 In [13]'s general setting, the selection function returns a set of models. However, for the possible antecedents of counterfactuals considered in our fragment of her language, the selected model is uniquely defined.

5 Our language is a fragment of that in [13]. Thus, here we only recall the tools from [13] that are needed for our formulas' semantic interpretation.

(i) $\mathcal{S}_{\vec{V}=\vec{v}}$ is as $\mathcal{S}$ except that, for each variable $V_i \in \vec{V}$, the function F_{V_i} is replaced by a constant function F'_{V_i} assigning the value v_i (i.e., $F'_{V_i} := v_i$).
(ii) $\mathcal{A}^{\mathcal{S}_{\vec{V}=\vec{v}}}$ is the assignment to causal variables that is identical to $\mathcal{A}$ with respect to exogenous variables, and it complies with the causal dependencies in $\mathcal{S}_{\vec{V}=\vec{v}}$ for the endogenous ones.

Thus, the proposal is that the selection function f discussed in the introduction should be defined as

$$f(\langle \mathcal{S}, \mathcal{A} \rangle, \vec{V} = \vec{v}) := \langle \mathcal{S}_{\vec{V}=\vec{v}}, \mathcal{A}^{\mathcal{S}_{\vec{V}=\vec{v}}} \rangle.$$

It is worthwhile to emphasise that, in the model $\langle \mathcal{S}_{\vec{V}=\vec{v}}, \mathcal{A}^{\mathcal{S}_{\vec{V}=\vec{v}}} \rangle$, the valuation $\mathcal{A}^{\mathcal{S}_{\vec{V}=\vec{v}}}$ *complies* with the model's causal dependencies, $\mathcal{S}_{\vec{V}=\vec{v}}$: for every $V \in \mathcal{V}$ we have $\mathcal{A}^{\mathcal{S}_{\vec{V}=\vec{v}}}(V) = F'_V(\mathcal{A}^{\mathcal{S}_{\vec{V}=\vec{v}}}(PA'_V))$. So, intervention happens at the level of $\mathcal{S}_{\vec{V}=\vec{v}}$, and this change affects the valuation $\mathcal{A}^{\mathcal{S}_{\vec{V}=\vec{v}}}$. In Sect. 5 we will introduce a notion of intervention that changes $\mathcal{A}$ directly and leaves $\mathcal{S}$ unaffected.

3 Fisher's Criticism

Fisher [12] criticizes the approach described above. More concretely, he claims that it makes incorrect predictions for right-nested counterfactuals. Concretely, he discusses the examples (1) and (2) below.[6]

- **Match.** I hold up a match and strike it, but it does not light. I say

(1) If the match had lit, then (even) if it had not been struck, it would have lit.

- **Headlamp.** I hold up a headlamp in good working condition. I say

(2) If the headlamp were emitting light, then if it had had no batteries, the headlamp would be emitting light.

Both examples involve a model of the form shown in Fig. 1, where A_1 stands for the variable the first antecedent talks about and A_2 for the variable of the second antecedent.[7]

---

[6] Fisher also considers another example, involving the counterfactual "If the match were struck and it lit, then if it hadn't been struck, it would have lit". This is not a good example to make his point, as it contains a conjunction of cause (striking the match) and effect (the match lights) in the antecedent. For the counterexample to work, Fisher needs this conjunction to be interpreted as two independent interventions. However, it could be that *"and"* is interpreted causally in this case: *"If the match were struck and because of that it lit, ..."*. But then the fact that the match lights would be introduced as a causal consequent of the striking of the match and not as an independent intervention.

[7] We ignore other possible variables, as they will not affect the relevant predictions made.

Fig. 1. A causal model for **Match** and **Headlamp**, before and after interpreting the counterfactuals.

Following the interventionist approach the evaluation of the first antecedent produces a causal model where A_1 is forced to a particular value, and where the causal connection between A_2 and A_1 has been erased. Evaluating the second antecedent forces A_2 to a particular value too, but this will no longer affect A_1. Hence, the counterfactuals (1) and (2) are predicted to be true, but intuitively, according to Fisher, they should be false. Fisher traces the problem back to the property of *strict interventionism* (SI).

(SI) "When a variable V is intervened on so that it is made to take a value v, V remains set to v unless it is intervened upon again per an iterated application of the interventionist recipe." ([12]:4939).

Interventionist approaches have this property because their selection function maps a given causal model M and an antecedent A to a new causal model in which a causal variable V occurring in the antecedent A has lost all connections to its causal parents. Any later intervention that might affect V's (former) causal parents will no longer affect V itself. So, as long as ψ does not assign a new value to V, the counterfactual $(V = v) \mathbin{\Box\!\!\rightarrow} (\psi \mathbin{\Box\!\!\rightarrow} V = v)$ will always come out as true.

To solve this problem Fisher proposes that we have to give up strict interventionism. More concretely, he proposes the following adequacy condition for approaches to the meaning of counterfactuals: "A causal model semantics for counterfactuals should admit cases in which the variables implicated in the antecedent of a counterfactual remain causally sensitive to their parents throughout the evaluation procedure." ([12]:4942). However, he does not propose an alternative approach that has this property.[8] In the rest of the paper we want to do two things. First of all, we need to confirm Fishers judgments concerning the target examples (1) and (2) with an actual survey. After that, we will develop an alternative interventionist approach to the meaning of counterfactual conditionals that is not strictly interventionist.

4 An Empirical Study on Fisher's Counterexamples

A possible objection against Fisher's observations and the conclusions he derives from them is that he confuses judging a sentence false with rejecting it as not well-formed. Maybe we are inclined to say "No" to the counterfactuals in (1) and (2), because they are very strange counterfactual sentences. To exclude this we

[8] Fisher discusses in [12] an alternative definition of intervention, dubbed "side-constrained intervention", but admits that this variation is not really targeting the root of the problem.

conducted a small empirical study in which we did not only ask the participants to judge the counterfactuals (1) and (2), but also their counterparts (3-a) and (3-b). If participants judge the sentences (1) and (2) false because they consider the sentences defective, they should judge (3-a) and (3-b) to be false as well.

(3) a. If the match had lit, then if it had not been struck, it would not have lit.
b. If the headlamp were emitting light, then if it had had no batteries, the headlamp would not have been emitting light.

4.1 Method and Participants

We used the scenarios **Match** and **Headlamp** in Sect. 3 and a third scenario containing a counterfactual $\varphi \mathrel{\Box\!\!\rightarrow} (\psi \mathrel{\Box\!\!\rightarrow} \xi)$ with ξ talking about a causal effect of φ. For each scenario we asked the participants to judge 3 counterfactuals: the target right-nested counterfactual, the counterfactual with the opposite final consequent and a filler item to check whether the participants where paying attention and understood the presented scenario correctly. This resulted in 9 questions that the participants had to answer. The order of question was randomized. The participants had to judge the truth value of the counterfactual using a slider bar with 5 values from 0 to 4. They were told that 0 means the sentence is false, 4 it is true and 2 that the truth value is unclear. The values 1 and 2 allowed them to indicate that they find a sentence weakly false or true.

The study was implemented in Qualtrics, a web-based survey tool. Participants were recruited via Prolific.ac, an online platform aimed at connecting researchers and participants willing to fill in surveys and questionnaires in exchange for compensation for their time [14]. We recruited native English speakers (British and American English). Fifty-two participants completed the task. Eight participants were excluded. Two participants did not answer the filler question for the match scenario correctly, seven participants did not answer the filler question for the headlamp scenario correctly, one also failed the match scenario. Thus, forty-four responses were included in the analyses reported below. Thirteen participants failed the control question for the third scenario we used. Because of the high number we concluded that there was a problem with the material used and excluded this scenario from the evaluations.

4.2 Results and Discussion

The table in Fig. 2 states the results of the study. We counted both values 3 and 4 on the scale as judging the sentence true and 0 an 1 as judging the sentence false. The graph in Fig. 2 plots the percentages of the different answers first for both scenario's separately and then combined. The results show that first of all a majority of the participants agree with the intuitions reported by Fisher [12]. Furthermore, the results for the opposite counterfactuals (3-a) and (3-b) support the conclusion that the judgements are for the most part judgements about truth values and not well-formedness of the counterfactuals under consideration.

Sentence	True	False	Unclear
(1)	4%	80%	16%
(3-a)	64%	13%	23%
(2)	7%	84%	9%
(3-b)	77%	9%	14%
(1)+(2)	6%	82%	12%
(3-a)+(3-b)	71%	11%	18%

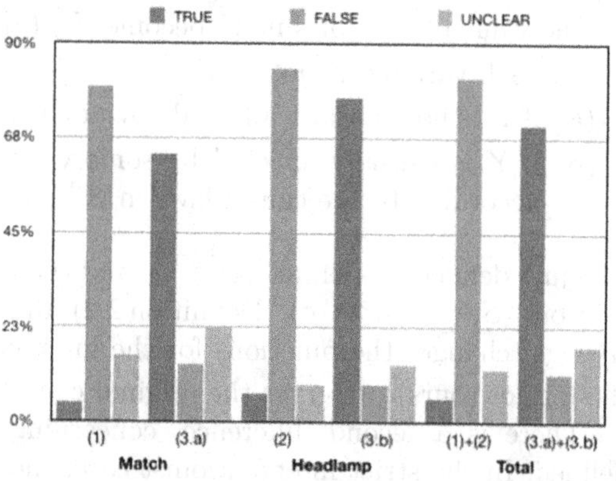

Fig. 2. Results of the 1st study.

Hence, we conclude with Fisher that these nested counterfactuals present a problem for the interventionist approach to their meaning. Fisher discussed the possibility to defend the approach by arguing that the conditionals under discussion are interpreted according to a different (epistemic) reading of counterfactuals and eventually dismisses it. We agree with Fisher and only want to add that such a move does not make sense as long as there is no explanation for why the interventionist reading isn't available for the counterfactuals in question.

But does that mean that we need to give up the interventionist approach to counterfactuals? We don't think so. We can give up the property of strict interventionism responsible for the problematic predictions, but still keep the general idea and all the strong predictions of the interventionist approach. The big conceptual step that needs to be taken is to apply intervention to the valuation $\mathcal{A}$ instead of the representation of the causal dependencies $\mathcal{S}$. In the next section we develop this idea in detail.

5 The Non-strict-intervention

The goal is, then, to find a notion of intervention that coincides with [8]'s account for non-nested cases (so it 'inherits' the good behaviour of the strict interventionism approach in those situations), but also satisfies Fisher's adequacy condition (thus agreeing with the results from our study). The definition below meets all these requirements. Its crucial idea is, again, that counterfactual assumptions might modify the value of causal variables, but preserve causal relationships.

Definition 5.1. Let $M = \langle \mathcal{S}, \mathcal{A} \rangle$ be a recursive causal model, and $\overrightarrow{V} = \overrightarrow{v}$ an intervention; let $\overrightarrow{V_d}$ be the variables in $\overrightarrow{V}$ whose current value (as given by $\mathcal{A}$) is different from their intended new value (as indicated by $\overrightarrow{V} = \overrightarrow{v}$). The selection function f is defined as $f(\langle \mathcal{S}, \mathcal{A} \rangle, \overrightarrow{V} = \overrightarrow{v}) := \langle \mathcal{S}, \mathcal{A}^{\overrightarrow{V}=\overrightarrow{v}} \rangle$, with the new assignment $\mathcal{A}^{\overrightarrow{V}=\overrightarrow{v}}$ calculated in the following way.

1. The value of variables in $\vec{V}$ becomes $\vec{v}$ (as indicated by the intervention).
2. For each variable Y not in $\vec{V}$,
   (a) if Y is not causally affected[9] by any variable in $\vec{V_d}$, keep its value as in $\mathcal{A}$.
   (b) if Y is causally affected by some variables in $\vec{V_d}$, its value is calculated according to the causal laws in $\mathcal{S}$.[10]

The just defined model, $\langle \mathcal{S}, \mathcal{A}^{\vec{V}=\vec{v}} \rangle$, and the one that results from a strict intervention, $\langle \mathcal{S}_{\vec{V}=\vec{v}}, \mathcal{A}^{\mathcal{S}_{\vec{V}=\vec{v}}} \rangle$ (Definition 2.2), differ in their causal laws. The latter (Briggs) changes the functions for the intervened variables (producing $\mathcal{S}_{\vec{V}=\vec{v}}$); the former (ours) preserves the original causal information (the 'old' $\mathcal{S}$).[11]

There is a second difference, concerning the way the new assignment is defined. In the strict interventionist case, the values of *all* non-intervened variables are recalculated according to the (recall: new) causal rules. In our case, the only non-intervened variables for which the recalculation takes place (recall: with respect to the original causal laws) are those that are causally affected by variables whose *value* is directly affected by the intervention.[12]

Note that $\langle \mathcal{S}, \mathcal{A}^{\vec{V}=\vec{v}} \rangle$ can be equivalently defined as follows:

Proposition 5.1. *Let $\langle \mathcal{S}, \mathcal{A} \rangle$ be a causal model and $\vec{V} = \vec{v}$ an intervention. Let*

- *$\vec{V_d}$ be as before: the causal variables in $\vec{V}$ whose value (as given by $\mathcal{A}$) differs from their intended new value (as indicated by $\vec{V} = \vec{v}$);*
- *$\vec{Z}$ be the endogenous variables not causally affected by variables in $\vec{V_d}$, with $\vec{z}$ their values according to $\mathcal{A}$.*

Then, the assignment $\mathcal{A}^{\vec{V}=\vec{v}}$ (Definition 5.1) can be equivalently defined as the (unique) assignment that is identical with $\mathcal{A}$ with respect to exogenous variables, and complies with the causal dependencies in $\mathcal{S}_{(\vec{V}=\vec{v}, \vec{Z}=\vec{z})}$ (see Definition 2.2)[13].

---

[9] "Y is causally affected by Z" intuitively means changing the value of Z may change the value of Y under some setting of variables. Formally, it means there exists some variables $\vec{V}$, $\vec{v} \in \mathcal{R}(\vec{v})$, and some distinct value $y, y' \in \mathcal{R}(Y)$, such that the value of Z forced by setting $\vec{V}, Y$ to $\vec{v}, y$ is different from its value forced by setting $\vec{V}, \vec{Y}$ to $\vec{v}, y'$.

[10] Recall: the model is recursive. Hence, $\mathcal{S}$'s induced causal graph induces, in turn, a chain of sets of variables $S_0 \subsetneq \cdots \subsetneq S_n$ such that $S_0 = \mathcal{U} \cup \vec{V}$, $S_n = \mathcal{U} \cup \mathcal{V}$ and, for any S_i and S_{i+1}, the value of variables in $S_{i+1} \setminus S_i$ can be calculated from the causal dependencies and the value of variables in S_i.

[11] Note: $\mathcal{A}^{\vec{V}=\vec{v}}$ may not comply with the causal dependencies in $\mathcal{S}$.

[12] When the original assignment $\mathcal{A}$ complies with the causal dependencies in $\mathcal{S}$, both strategies produce the same result. This is the only case relevant for Briggs' purposes.

[13] Proofs were omitted due to space limitations, but are available online https://www.dropbox.com/s/0i0xy416rs5dmor/Lori_Proofs.pdf?dl=0.

We will redefine the logic of counterfactuals, using this new notion of intervention for the semantic interpretation of $\Box\!\!\rightarrow$. However, the strict intervention operator will still be useful, in particular, for axiomatizing the non-strict intervention. Thus, it will appear in the language as well, albeit under a different symbol ([ ]).

Definition 5.2. Formulas ϕ of the language $\mathcal{L}_{\Box\!\rightarrow,[\,]}$ over $\mathcal{U} \cup \mathcal{V}$ are given by

$$\phi ::= V = v \mid \neg\phi \mid \phi \wedge \phi \mid (\overrightarrow{V} = \overrightarrow{v}) \Box\!\!\rightarrow \phi \mid [\overrightarrow{V} = \overrightarrow{v}]\phi$$

for $V \in \mathcal{V}$, $v \in \mathcal{R}(V)$, $\overrightarrow{V} = (V_1, \ldots, V_n) \in \mathcal{V}^n$, $n \in \mathbb{N}$, $V_i \neq V_j$ for $i \neq j$, $\overrightarrow{v} = (v_1, \ldots, v_n)$ with $v_i \in \mathcal{R}(V_i)$.

For $\mathcal{L}_{\Box\!\rightarrow,[\,]}$'s semantics, atoms and Boolean operators are evaluated as before. The cases for the intervention operators [ ] and $\Box\!\!\rightarrow$ are as follows.

Definition 5.3. (Intervention). Let $\langle \mathcal{S}, \mathcal{A} \rangle$ be a causal model. Then,

$\langle \mathcal{S}, \mathcal{A} \rangle \models [\overrightarrow{V} = \overrightarrow{v}]\phi$ iff$_{def}$ $\langle \mathcal{S}_{\overrightarrow{V}=\overrightarrow{v}}, \mathcal{A}^{\mathcal{S}_{\overrightarrow{V}=\overrightarrow{v}}} \rangle \models \phi$ (see Definition 2.2)

$\langle \mathcal{S}, \mathcal{A} \rangle \models (\overrightarrow{V} = \overrightarrow{v}) \Box\!\!\rightarrow \phi$ iff$_{def}$ $\langle \mathcal{S}, \mathcal{A}^{\overrightarrow{X}=\overrightarrow{x}} \rangle \models \phi$ (see Definition 5.1)

If $\langle \mathcal{S}, \mathcal{A} \rangle$ is a model without causal violations (i.e., $\mathcal{A}$ complies with $\mathcal{S}$), then the assignment created by our intervention ($\mathcal{A}^{\overrightarrow{V}=\overrightarrow{v}}$) coincides with the one created by a strict intervention ($\mathcal{A}^{\mathcal{S}_{\overrightarrow{V}=\overrightarrow{v}}}$). Thus, our proposal does extend the original causal modelling semantics [8], providing a non-strict-interventionist approach for nested counterfactuals.

5.1 Fisher's Counter-Examples Revisited

The semantics for counterfactuals proposed here can deal with the examples **Match** and **Headlamp** discussed in Sects. 3 and 4. For reasons of space we will only discuss **Match** (**Headlamp** works analogously).

- **Match.** I hold up a match and strike it, but it does not light. I say

(4) If the match had lit, then (even) if it had not been struck, it would have lit.

First, we need to define the causal model $M_1 = \langle \mathcal{S}, \mathcal{A} \rangle$ with respect to which the counterfactual (4) is interpreted. We define $\mathcal{V} = \{S, L\}$ and $\mathcal{U} = \{U\}$, with S indicating whether the match has been struck (1: yes, 0: no), L indicating whether the match has lit (1: yes, 0: no). The exogenous variable U represents external factors causally responsible for S.[14] Furthermore, we define $\mathcal{S} = (S := U, L := S)$ and $\mathcal{A} = (U = 1, S = 1, L = 1)$ (model M_1 in Fig. 3). We need to account for the observation that the counterfactual $(L = 1) \Box\!\!\rightarrow ((S = 0) \Box\!\!\rightarrow L = 1)$ is intuitively false with respect to this model,

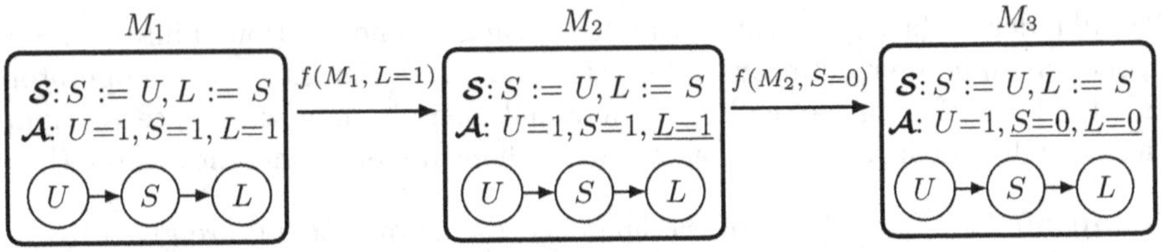

Fig. 3. The evaluation of the **Match** example with the selection function f

a prediction that a strict interventionist approach, as discussed in Sect. 3, is unable to make.

The sentence contains nested counterfactuals, so we need to intervene twice: first, with $L = 1$ (the antecedent of the main counterfactual), and then, with $S = 0$ (the antecedent of the embedded counterfactual). On the resulting model, we should check whether $L = 1$ (the consequent of the embedded counterfactual) is true. The first intervention, $L = 1$, produces model M_2 in Fig. 3 (Definition 5.1), affecting the original assignment but preserving the original causal dependencies. For evaluating the embedded counterfactual $(S = 0) \,\square\!\!\rightarrow L = 1$, we apply the second intervention, $S = 0$, to M_2. This results in the model M_3 in Fig. 3, with $S = 0$ as the intervention requires, and $L = 0$, as L's value is still causally sensitive to S. In this final model, the innermost consequent $L = 1$ fails; thus,

$$M_1 \not\models (L = 1) \,\square\!\!\rightarrow ((S = 0) \,\square\!\!\rightarrow L = 1).$$

We correctly predict that the counterfactual (4) is false in the given context.

5.2 The Axiomatization for the Logic

The modified notion of intervention can be axiomatized (Table 1) with the help of the axioms for the strict intervention operator [ ] (see [15]) plus additional axioms for $\square\!\!\rightarrow$. Axioms **A1** through **A9** characterise the behaviour of the strict intervention operator [ ].[15]. From axioms **A10**–**A11**, every variable has exactly one value,[16] and axiom **A12** states that our modified version of intervention is still deterministic. Axioms **A13** and **A14** are the crucial ones, as they describe the relationship between the two forms of intervention. Axiom **A13** relies on Proposition 5.1 to describe the assignment after a non-strict intervention $\square\!\!\rightarrow$ in terms of the assignment after a (different) strict intervention [ ]. It states that,

[14] If the model allows interventions on *exogenous* variables, the example can be modelled with only two variables: the exogenous one S and the endogenous one L. We use the additional U, as in the literature it is common to allow interventions only on *endogenous* variables.

[15] More precisely, **A1**–**A8** are the axioms for non-nested intervention from [15], and **A9** deals with nested strict-intervention [13, 16].

[16] In [15] there are no causal violations; thus, $V = v$ is equivalent to $[\,](V = v)$, and axioms **A1** and **A2** suffice. This is not the case in our setting, as causal violations might occur; hence the need of **A10**–**A11**.

Table 1. Axiom system for $\mathcal{L}_{\Box\!\!\rightarrow,[\,]}$ w.r.t. causal models.

A0 Propositional tautologies — **MP** from ϕ and $\phi \rightarrow \psi$ infer ψ

A1 $[\vec{V}=\vec{v}](Y=y) \rightarrow \neg[\vec{V}=\vec{v}](Y=y')$ — for $y, y' \in \mathcal{R}(Y)$ with $y \neq y'$

A2 $\bigvee_{y\in\mathcal{R}(Y)}[\vec{V}=\vec{v}](Y=y)$

A3 $\big([\vec{V}=\vec{v}](Y=y) \wedge [\vec{V}=\vec{v}](Z=z)\big) \rightarrow [\vec{V}=\vec{v}, Y=y](Z=z)$

A4 $[\vec{V}=\vec{v}, Y=y](Y=y)$

A5 $\big([\vec{V}=\vec{v}, Y=y](Z=z) \wedge [\vec{V}=\vec{v}, Z=z](Y=y)\big) \rightarrow [\vec{V}=\vec{v}](Z=z)$ — for $Y \neq Z$

A6 $(V_0 \rightsquigarrow V_1 \wedge \cdots \wedge V_{k-1} \rightsquigarrow V_k) \rightarrow \neg(V_k \rightsquigarrow V_0)$ [a]

A7 $[\vec{V}=\vec{v}](\phi \wedge \psi) \leftrightarrow ([\vec{V}=\vec{v}]\phi \wedge [\vec{V}=\vec{v}]\psi)$ **A8** $[\vec{V}=\vec{v}]\neg\phi \leftrightarrow \neg[\vec{V}=\vec{v}]\phi$

A9 $([\vec{V}=\vec{v}][\vec{Y}=\vec{y}]\psi) \leftrightarrow [\vec{V'}=\vec{v'}, \vec{Y}=\vec{y}]\psi$ — for $\vec{V'} = \vec{V} \setminus \vec{Y}$

A10 $(V=v) \rightarrow \neg(V=v')$ — for $v, v' \in \mathcal{R}(V)$ and $v \neq v'$

A11 $\bigvee_{v\in\mathcal{R}(V)}(V=v)$

A12 $(\vec{V}=\vec{v}) \mathrel{\Box\!\!\rightarrow} \bigvee_{y\in\mathcal{R}(Y)}(Y=y)$

A13 $\bigwedge \left\{ \begin{array}{l} \bigwedge_{V_i\in\vec{V_d}}(V_i \neq v_i), \\ \bigwedge_{V_i\in\vec{V}\setminus\vec{V_d}}(V_i = v_i), \\ \bigwedge_{Z\in\vec{Z}} \neg\bigvee_{V\in\vec{V_d}}(V \rightsquigarrow Z), \\ \bigwedge_{Y\in\mathcal{V}\setminus\vec{Z}} \bigvee_{V\in\vec{V_d}}(V \rightsquigarrow Y) \end{array} \right\} \rightarrow \big(((\vec{V}=\vec{v}) \mathrel{\Box\!\!\rightarrow} X=x) \leftrightarrow [\vec{V}=\vec{v}, \vec{Z}=\vec{z}]X=x\big)$

A14 $((\vec{V}=\vec{v}) \mathrel{\Box\!\!\rightarrow} [\vec{X}=\vec{x}]\phi) \leftrightarrow ([\vec{X}=\vec{x}]\phi)$

A15 $\big((\vec{V}=\vec{v}) \mathrel{\Box\!\!\rightarrow} (\phi \wedge \psi)\big) \leftrightarrow ((\vec{V}=\vec{v}) \mathrel{\Box\!\!\rightarrow} \phi \wedge (\vec{V}=\vec{v}) \mathrel{\Box\!\!\rightarrow} \psi)$

A16 $((\vec{V}=\vec{v}) \mathrel{\Box\!\!\rightarrow} \neg\phi) \leftrightarrow \neg((\vec{V}=\vec{v}) \mathrel{\Box\!\!\rightarrow} \phi)$

[a] With $Y \rightsquigarrow Z$, indicating that *"Y has causal effect on Z"*, given (see, e.g., [15]) by

$$\bigvee_{\vec{v}\in\mathcal{R}(\vec{V}),\{y,y'\}\subseteq\mathcal{R}(Y),y\neq y',\{z,z'\}\in\mathcal{R}(Z),z\neq z'} [\vec{V}=\vec{v}, Y=y](Z=z) \wedge [\vec{V}=\vec{v}, Y=y'](Z=z').$$

if $\vec{V_d}$ contains exactly the variables in $\vec{V}$ whose value would change (conjuncts 1 and 2 in the antecedent), and $\vec{Z}$ contains exactly the variables that are not causally affected by those in $\vec{V_d}$ (conjuncts 3 and 4 in the antecedent), then a non-strict-intervention with $\vec{V} = \vec{v}$ coincides with a strict intervention with $\vec{V} = \vec{v}$, $\vec{Z} = \vec{z}$. Axiom **A14** then uses strict intervention to state that causal relationships are invariant under non-strict interventions. Finally, axioms **A15**–**A16** are the rules for Boolean operators.

Theorem 5.1. *This axiom system is sound and strongly complete with respect to recursive causal models (see Footnote 13).*

6 Discussion and Conclusions

In this paper we proposed a new approach to the semantics of counterfactual conditionals. Our proposal builds on the well-known interventionist approach, but uses a different approach to intervention. There are two separate steps that we took in defining our proposal. First, we made a substantial conceptual shift in what we understand to be the object of intervention. We propose that intervention does not take place at the level of structural dependencies, but at the level of the (incidental) valuations of the variables. Conceptually, this means that we see intervention not as a hypothetical modification of the underlying laws of nature, but as the hypothetical assumption of exceptions to the laws (see [4,17] for a similar move). As a consequence, no information on causal dependencies in the actual world is lost. The second part of the proposal lies in how exactly we define the valuation resulting from intervention. We propose that the value of all variables not causally affected by those variables that we intervene on remain unchanged and that then the value of the remaining variables is calculated from this information (as the model is assumed to be recursive) and the unchanged causal dependencies (see Definition 5.1). This approach allows us to satisfy our objectives: (i) the predictions made for the truth conditions of counterfactuals that are not right-nested are the same as made in [13] and (ii) the approach correctly deals with the counterexamples brought forward in [12].

But does that mean that this way all problems with the interventionist approach to counterfactuals are solved? Certainly not. First of all, notice that we target here only the issue of right-nested counterfactuals. But even if we only focus on right-nested counterfactuals, there are still open questions. This approach was specifically designed to deal with the examples and intuitions reported on in [12] and confirmed in Sect. 4. Fisher suggest that the observations he makes generalize to arbitrary right-nested counterfactuals where variables in the first antecedent causally depend on variables in the second antecedent. But whether this is true has to be investigated first. We performed a second study to test whether Fishers expectations are confirmed when using slightly larger models containing a third variable C (see the two scenarios in Fig. 4). While we could confirm, using the same method as before, that still the majority of the participants consider counterfactual of the form (i) $B \mathrel{\Box\!\!\rightarrow} (\neg A \mathrel{\Box\!\!\rightarrow} B)$ false (left diagram in Fig. 4), this effect becomes weaker when the consequent is substituted with the third variable C (the counterfactual becomes (iii) $B \mathrel{\Box\!\!\rightarrow} (\neg A \mathrel{\Box\!\!\rightarrow} C)$) and basically disappears in combination with scenario 2 (right diagram in Fig. 4). In a third study focusing in particular on this scenario and counterfactuals of the form (iii) we could not find any difference between the number of participants that consider this sentence true and those that considered its counterpart (iv) true.

Based on the work of Fisher [12] and the empirical results presented here it seems clear that the first part of our proposal is on the right track: sometimes we need to be able to recall causal dependencies after an intervention has violated them. This means that the structural information about these dependencies should not be the locus of the intervention. So, what we certainly want to defend here is the proposed step from intervention on the causal dependencies to intervention on the valuation of the variables. Whether the exact form we then gave to intervention on the valuation is correct needs to be studied in future work. In some cases, like the examples discussed in [12], it seems to be exactly what is needed, in other cases it is still unclear what we should predict.

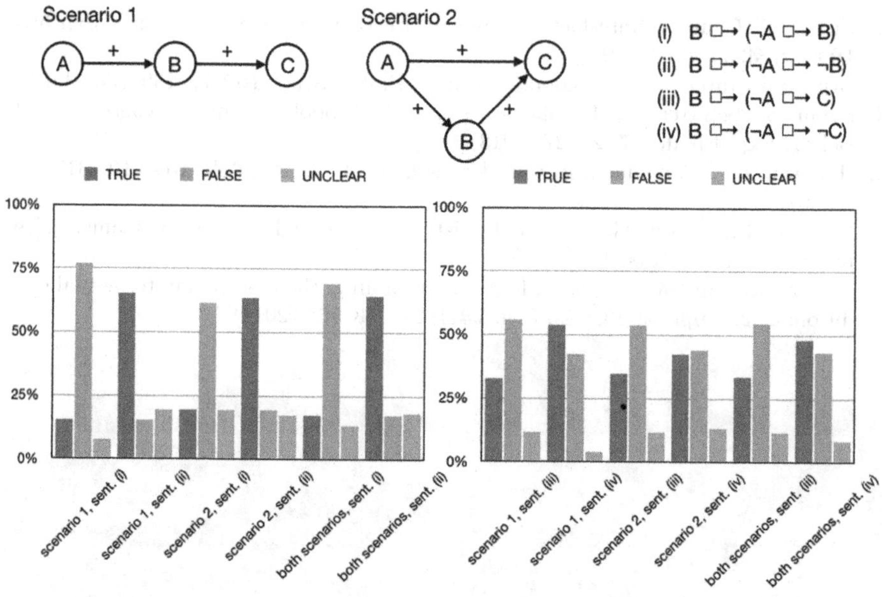

Fig. 4. Overview of the results of the second study; the sentences (i)–(iv) are those that we asked participants to judge in the two scenarios.

References

1. Lewis, D.: Counterfactuals and comparative possibility. In: Harper, W.L., Stalnaker, R., Pearce, G. (eds.) IFS. WONS, vol. 15, pp. 57–85. Springer, Dordrecht (1973). https://doi.org/10.1007/978-94-009-9117-0_3
2. Stalnaker, R.C.: A theory of conditionals. In: Harper, W.L., Stalnaker, R., Pearce, G. (eds.) IFS. WONS, vol. 15, pp. 41–55. Springer, Dordrecht (1968). https://doi.org/10.1007/978-94-009-9117-0_2
3. Pearl, J.: Structural counterfactuals: a brief introduction. Cogn. Sci. **37**, 977–985 (2013)
4. Schulz, K.: "If you wiggle A, then B will change". Causality and counterfactual conditionals. Synthese **179**(2), 239–251 (2011)
5. Kaufmann, S.: Causal premise semantics. Cogn. Sci. **37**, 1136–1170 (2013)
6. Halpern, J.Y.: Actual Causality. MIT Press, Cambridge (2016)
7. Ciardelli, I., Zhang, L., Champollion, L.: Two switches in the theory of counterfactuals. A study of truth conditionality and minimal change. Linguist. Philos. **41**(6), 577–621 (2018)
8. Pearl, J.: Causality. Models, Reasoning, and Inference. Cambridge University Press, Cambridge (2000)
9. Spirtes, P., et al.: Causation, Prediction, and Search. MIT Press, Cambridge (2000)
10. Halpern, J.Y.: From causal models to counterfactual structures. Rev. Symbolic Log. **6**(2), 305–322 (2013)
11. Marti, J., Pinosio, R.: Similarity orders from causal equations. In: Fermé, E., Leite, J. (eds.) JELIA 2014. LNCS (LNAI), vol. 8761, pp. 500–513. Springer, Cham (2014). https://doi.org/10.1007/978-3-319-11558-0_35

12. Fisher, T.: Causal counterfactuals are not interventionist counterfactuals. Synthese **194**(12), 4935–4957 (2017)
13. Briggs, R.: Interventionist counterfactuals. Philos. Stud. **160**(1), 139–166 (2012)
14. Palan, S., Schitter, C.: Prolific.ac — a subject pool for online experiments. J. Behav. Exp. Finan. **17**, 22–27 (2018)
15. Halpern, J.Y.: Axiomatizing causal reasoning. J. Artif. Intell. Res. **12**, 317–337 (2000)
16. Barbero, F., Sandu, G.: Interventionist counterfactuals on causal teams. arXiv preprint arXiv:1901.00593 (2019)
17. Schulz, K.: Minimal models vs. logic programming: the case of counterfactual conditionals. J. Appl. Non-Class. Log. **24**(1–2), 153–168 (2014)

First Degree Entailment with Group Attitudes and Information Updates

Igor Sedlár[1(✉)], Vít Punčochář[1,2], and Andrew Tedder[1]

[1] Institute of Computer Science, The Czech Academy of Sciences, Prague, Czech Republic
{sedlar,puncochar,tedder}@cs.cas.cz

[2] Institute of Philosophy, The Czech Academy of Sciences, Prague, Czech Republic

Abstract. We extend the epistemic logic with De Morgan negation by Fagin et al. (Artif. Intell. 79, 203–240, 1995) by adding operators for universal and common knowledge in a group of agents, and with a formalization of information update using a generalized version of the left division connective of the non-associative Lambek calculus. We provide sound and complete axiomatizations of the basic logic with the group operators and the basic logic with group operators and updates. Both logics are shown to be decidable.

Keywords: Common knowledge · Epistemic logic · First Degree Entailment · Information update · Lambek Calculus

1 Introduction

Belnap's epistemic interpretation of First Degree Entailment [3,4] shows that FDE is useful for reasoning about incomplete and potentially inconsistent information. FDE is not, however, an epistemic logic in the standard sense since its language does not contain operators expressing epistemic attitudes of agents. Such an extension of FDE was provided by Levesque [22] and brought closer to classical epistemic logic by Fagin et al. [17]. These frameworks were originally put forward as an attempt to avoid the logical omniscience problem of classical epistemic logic, and so, to keep unnecessary complications out of the picture, they do not contain any additional operators utilized in the successful applications of classical epistemic logic, such as group epistemic operators [16] or operators expressing various kinds of information update [6,13].

In this paper we extend the framework of Fagin et al. [17] with operators expressing universal and common knowledge in a group of agents (Sect. 2) and with a conditional operator, coming from the Non-associative Lambek Calculus, expressing information update (Sect. 3). These two basic logics are axiomatized and shown to be decidable; extensions are briefly mentioned, but are mostly left for future work (which is discussed in Sect. 4.)

P. Blackburn et al. (Eds.): LORI 2019, LNCS 11813, pp. 273–285, 2019.
https://doi.org/10.1007/978-3-662-60292-8_20

Related Work. Non-classical modal logic with epistemic and information-dynamic operators is underdeveloped. Girard and Tanaka [18] study a paraconsistent logic containing explicit revision operators. (This paper follows up on [28] and [23], but these do not discuss Hintikka-style epistemic logics with revision operators; rather, they consider paraconsistent versions of AGM-style belief revision.) Rivieccio [30] studies an FDE-based version of Public Announcement Logic. Both of these papers contain only single-agent epistemic operators. An FDE-based group epistemic logic with universal and common knowledge is a fragment of paraconsistent Propositional Dynamic Logic studied in [31,32]. Bílková et al. [10] outline an extension of their substructural epistemic framework with common knowledge, but completeness is left for future research. The relation between substructural logic and classical information dynamics is studied in [5,7] and [1], for example; [15,27] discuss an information-dynamic interpretation of the Routley–Meyer semantics for some substructural logics. Restall [27] considers a ternary relation between sets of situations, but the framework considered in Sect. 3 is original to this paper.

2 FDE with Group Epistemic Operators

In this section, we add to the framework of FDE with material implication, based on [17], modal operators representing universal knowledge in groups of agents ("everyone knows that ...") and common knowledge. Firstly, we provide the basic definitions (Subsect. 2.1), then we discuss the informal interpretation of the framework (Subsect. 2.2) and our technical results, namely, a weakly complete axiomatization and a decidability result for the basic logic of the framework (Subsect. 2.3). The proof is given in the technical appendix.

2.1 Group Language and Group Frames

Fix a finite non-empty set Ag ("agents") and a countable set $Prop$ of propositional variables. The language $\mathcal{L}_{Gr}$ of FDE with material implication and group modalities contains

- unary connective $\sim$ (De Morgan negation)
- binary connectives $\wedge, \vee$ and $\supset$ (lattice conjunction and disjunction, material implication)
- unary operators K_G, K^*_G for each non-empty $G \subseteq Ag$ (group epistemic modalities)

Fix any $p \in Prop$ and define $\top := p \supset p$, $\bot := {\sim}\top$ and $\neg\varphi := \varphi \supset \bot$. Formulas ${\sim}\varphi$ are read "φ is false" and $\neg\varphi$ as "φ is not true"; in our setting, these will not be equivalent. Sets $G \subseteq Ag$ represent groups of agents; $K_G\varphi$ is read "Every agent in G knows that φ" and $K^*_G\varphi$ as "It is common knowledge in G that φ". We define $K_a\varphi := K_{\{a\}}\varphi$ and read this as "Agent a knows that φ".

Group frames are $\langle S, \{R_a\}_{a\in Ag}, \star\rangle$ where each R_a is a reflexive binary relation on S and $\star$ is a unary function of period two (that is, $\star(\star(x)) = x$ for all $x \in S$). We usually write $x^\star$ instead of $\star(x)$. Moreover, we define

$$R_G := \bigcup_{a\in G} R_a \qquad\qquad R^*_G := (R_G)^*$$

Group models add to group frames a valuation function $v : Prop \to \mathscr{P}(S)$. For each model with v, we define the satisfaction relation $\models_v$ as usual when it comes to propositional variables and Boolean connectives; moreover, we require that

$$\begin{aligned} &x \models_v {\sim}\varphi \text{ iff } x^\star \not\models_v \varphi\\ &x \models_v K_G\varphi \text{ iff } \forall y(R_G xy \implies y \models_v \varphi)\\ &x \models_v K^*_G\varphi \text{ iff } \forall y(R^*_G xy \implies y \models_v \varphi)\end{aligned}$$

We sometimes use the notation $v(\varphi) = \{x \mid x \models_v \varphi\}$. Formula φ is valid in a model with S and v iff $v(\varphi) = S$; it is valid in a frame iff it is valid in all models based on the frame and it is valid in a class of frames iff it is valid in all frames in the class. This notion of validity will be used throughout the paper. For any language $\mathcal{L}$, the $\mathcal{L}$-theory of a class of frames is the set of all $\mathcal{L}$-formulas valid in the class of frames.

It is easily seen that $v(\top) = S$ and so $v(\neg\varphi) = S \setminus v(\varphi)$. Hence, even though Boolean negation is not a primitive connective of our language, it can be expressed using material implication and De Morgan negation.

2.2 Informal Interpretation

In group frames, elements of S are called *situations* and can be seen as situations in the sense of Barwise and Perry [2], either concrete ones (parts of the world) of abstract ones (representations of parts of the world, either accurate or inaccurate). Mares [24] discusses situations in the presence of De Morgan negation and we follow his interpretation, according to which situations may be incomplete (some φ is neither true nor false, i.e. neither φ nor ${\sim}\varphi$ is satisfied in the situation) and inconsistent (some φ is both true and false); we note that Barwise and Perry also allow "incoherent" situations [2, 96]. Levesque [22] uses the concept of a situation in a similar way; we note that this interpretation of the elements of S is consistent with Belnap's interpretation in terms of "simple databases" [3,4]. Existence of incomplete and inconsistent situations follows from our truth condition for ${\sim}\varphi$ in terms of "the Routley star" $\star$, which is thought of as an operation assigning to each situation its *dual*; intuitively, the dual situation of x makes true everything that is not made false in x and vice versa. In general, we read $x \models_v \varphi$ as "φ is true in situation x (on v)", or "The information that φ is supported by x (on v)".

The informal interpretation of "epistemic accessibility relations" R_a differs only slightly from the standard reading of Kripke models for classical epistemic logic. Our basic idea is that, for each situation x and each agent a, there is a *part of x that is available to a* in the sense that a knows that it is a part of x. For instance, of the situation comprising the building in which my department is situated, only the part comprising my office is available to me at the moment, but upon receiving information from a colleague about something happening on a different floor, a bigger part of the situation becomes available to me. The fact that $R_a xy$ is taken to mean, informally, that the part of x available to a is included in y. Hence, our truth condition for $K_a\varphi$ means that $K_a\varphi$ is supported (true) in x iff each situation that contains the part of x available to a supports φ—we may say that $K_a\varphi$ is supported in x iff the information available to a in x supports φ.

A note of caution is in order here, however. The elements of our models correspond to *prime* situations in the sense that x supports a disjunction iff it supports one of its disjuncts. "Parts" of situations, as we use the term, may not be prime in this sense. For instance, each prime situation containing the fact that Ann has one sibling contains the fact that Ann has one brother or the fact that Ann has one sister, but only the information that Ann has one sibling may be available to me, without me knowing if the sibling is male or female. A disjunction may be supported by a part of a situation without either disjunct being supported by *that* part. "Parts" of situations in this sense are not necessarily elements of the model, but they may be represented by *sets* of elements of the model; intuitively, the set representing a particular "partial" situation comprises all prime situations in the model that contain all the information in the partial situation. For instance, the partial situation supporting only the information that Ann has one sibling can be represented by the set comprising two prime situations differing in the gender of the sibling. See [3,4] for details. Hence, we may speak of $R_a(x) := \{y \mid R_a xy\}$ as representing the part of x available to a—it follows from reflexivity of R_a that each φ supported by all situations in $R_a(x)$ is supported by x.

Let us turn now to the relations used in the satisfaction clauses for group operators. The fact that $R_G xy$ means that y contains the part of x available to some $a \in G$. Hence, $K_G\varphi$ is supported in x iff *all* agents in G have information that supports φ. The fact that $R^*_G xy$ means that (x, y) is in the reflexive transitive closure of R_G. (In fact, speaking of transitive closure is sufficient as all the relations are reflexive; we speak of reflexive transitive closure out of custom). In other words, there is a finite path $z_0 = x, z_1, \ldots, z_{n-1}, z_n = y$ such that, for all $k \in \{0, \ldots, n-1\}$, $(z_k, z_{k+1}) \in R_a$ for some $a \in G$. Note that $(x, z) \in R_a$ and $(z, y) \in R_b$ means that z contains the a-part of x and y contains the b-part of y. This means that $K_a K_b\varphi$ is supported in x iff the a-part of x "says" that the b-part of x supports φ. In other words, a knows that b knows that φ. Hence, $K^*_G\varphi$ is supported in x iff, in a standard manner, each agent in G knows that all the agents know that ... all the agents know that φ.

Belnap [3,4] motivated FDE as a logic useful for reasoning about simple databases containing potentially inconsistent information; this reasoning involved only information formulated using $\sim, \land$ and $\lor$. The epistemic extension of FDE by Fagin et al. [17] can be seen as a logic for reasoning about potentially inconsistent databases where the relevant information may involve K_a, that is, where *information about information* available to individual agents is involved. Here inconsistency may be encountered at least on two levels. Firstly, a database may contain inconsistent information about the information of agent a, that is, it may contain $K_a\varphi$ and $\sim K_a\varphi$ for some φ. In contrast to epistemic logic based on classical logic, the framework of [17] allows to reason with such databases without "explosion", i.e. without inferring any ψ whatsoever. Secondly, a database may contain information that the information of agent a is inconsistent, that is, it may contain $K_a\varphi$ and $K_a{\sim}\varphi$ for some φ. In contrast to classical epistemic logic, the framework of [17] does not force the conclusion that, in this case, $K_a\psi$ holds for any ψ whatsoever. The upshot of our group FDE is that these features are lifted to group epistemic notions—we have here a logic useful for reasoning about potentially inconsistent information, including information about information available to groups of agents that may turn out to be inconsistent on the two levels mentioned above in connection to individual knowledge operators.

2.3 Completeness and Decidability

The axiom system *GrFDE* contains the following axiom schemata and rules ($X \in \{K, K^*\}$):

(A0) Any fixed axiomatization of the $\{\wedge, \vee, \supset\}$-fragment of classical propositional logic
(A1) $\varphi \supset {\sim\sim}\varphi$
(A2) ${\sim\sim}\varphi \supset \varphi$
(A3) $({\sim}\varphi \wedge {\sim}\psi) \supset {\sim}(\varphi \vee \psi)$
(A4) ${\sim}(\varphi \wedge \psi) \supset ({\sim}\varphi \vee {\sim}\psi)$
(A5) $X_G\varphi \wedge X_G\psi \supset X_G(\varphi \wedge \psi)$
(A6) $X_G\varphi \supset \varphi$
(A7) $K_G\varphi \supset\subset \bigwedge_{a\in G} K_a\varphi$
(A8) $K^*_G\varphi \supset K_G(\varphi \wedge K^*_G\varphi)$
(R0) Modus Ponens
(R1) $\dfrac{\varphi \supset \psi}{{\sim}\psi \supset {\sim}\varphi}$
(R2) $\dfrac{\varphi}{X_G\varphi}$
(R3) $\dfrac{\varphi \supset K_G(\psi \wedge \varphi)}{\varphi \supset K^*_G\psi}$

Theorem 1. *GrFDE is a sound and weakly complete axiomatization of the $\mathcal{L}_{Gr}$-theory of all group frames. The theory is decidable.*

Since Boolean negation is expressible in our language, Theorem 1 can be established using the standard technique ([16, Ch. 3.1]). In the technical appendix, we give an alternative "modular" proof, based on [26], that does not invoke Boolean negation and, as such, can be used in a setting where Boolean negation is not expressible (e.g. when specific weaker negations are used instead of De Morgan negation; see Sect. 4).

3 Almost Arbitrary Information Updates

In this section, we extend our framework with a formalization of information update. Instead of focusing on one specific notion of update, such as public announcements, belief revision or the various notions of belief upgrade, we provide a somewhat more general account. Taking inspiration from van Benthem [8], we add to our semantics an *abstract representation of updates* and we study the general framework arising from this addition. (See also [19] for a nicely generalizable framework, based on abstract update relations, for the fragment of Public Announcement Logic closed under substitution; both frameworks bear some similarity to the general semantics for conditional logics [12].) An interesting endeavour is to relate the abstract semantics to known notions of update via special cases of the general framework, but we leave such investigations for future work.

Similarly to the framework of [8], information updates are represented as binary relations between elements of the model indexed by subsets of the model. Instead of pointed models in van Benthem's "update universe", elements of our models are prime situations. This feature of the model derives from the goal of formulating a general representation of information update on an inconsistency-tolerant background. The indexing set of situations, "the proposition triggering the update" [8, 32], corresponds to the information content of the update. We do not assume the content of an update to correspond to a prime situation; typically the "incoming" information corresponds to a part of a prime situation. (Recall that parts of prime situations are represented in our framework by sets of prime situations.)

Hence, an *update relation* on a set of situations S is a function from the power set of S (all possible "triggering propositions") to binary relations on S ("situation transitions"). Equivalently, we may represent an update relation by $R \subseteq (S \times \mathscr{P}(S) \times S)$ ($RxYz$ iff (x, y) is in the transition determined by the triggering proposition Y). In what follows, *group update frames* are $\langle S, \{R_a\}_{a \in Ag}, R, \star \rangle$ where R is such an update relation.

In modal logics of information update we typically have formulas specifying the results of information update depending on the nature of the "triggering proposition"; in general, the interesting feature is whether updates of a certain kind are guaranteed to lead to outputs satisfying specific formulas. Here we will distinguish updates with based on information supported by the "triggering proposition".

The language $\mathcal{L}_{GrUp}$ extends $\mathcal{L}_{Gr}$ with a binary connective $\backslash$; formulas $\varphi \backslash \psi$ are read "After updating with any information supporting φ, ψ will hold". *Group update models* add a valuation function v to group update frames and the satisfaction relation $\vDash_v$ is defined as usual; for $X \subseteq S$, $X \vDash_v \varphi$ means that $x \vDash_v \varphi$ for all $x \in X$. The new clause in the definition of $\vDash_v$ is the following:

$$x \vDash_v \varphi \backslash \psi \text{ iff } (\forall Y)(\forall z)\big((RxYz \;\;\&\;\; Y \vDash_v \varphi) \Rightarrow z \vDash_v \psi\big)$$

Validity is defined as before. Note that $\backslash$ is a generalized version of the left division operator of the Non-Associative Lambek Calculus [14,20,29]. There the truth condition uses individual situations y, not sets of situations.

We read $RxYz$ as "Updating x with the partial situation Y may result in z". Hence, $\varphi \backslash \psi$ is true in x iff ψ holds in every possible result of updating x by a partial situation that supports φ.

The proof system *GrUpFDE* extends *GrFDE* with

(A9) $(\chi\backslash\varphi \wedge \chi\backslash\psi) \supset \chi\backslash(\varphi \wedge \psi)$

(R4) $\dfrac{\varphi_1 \supset \psi_1 \quad \varphi_2 \supset \psi_2}{\psi_1\backslash\varphi_2 \supset \varphi_1\backslash\psi_2}$

(R5) $\dfrac{\varphi}{\psi\backslash\varphi}$

Theorem 2. *GrUpFDE is a sound and weakly complete axiomatization of the $\mathcal{L}_{GrUp}$-theory of all group update frames. The theory is decidable.*

Using Boolean negation, we may define a "diamond version" of the update operator $\backslash$ as $\varphi \circ \psi := \neg(\varphi\backslash\neg\psi)$. It is clear that

$$x \models_v \varphi \circ \psi \text{ iff } (\exists Y)(\exists z)\big(RxYz \;\&\; Y \models_v \varphi \;\&\; z \models_v \psi\big)$$

Note that the connective $\circ$ is not what is usually called *fusion* in the literature on substructural logic; the update operator $\backslash$ is not a residual of $\circ$. An axiomatization of the theory of all group update frames in languages where $\circ$ is present as a primitive operator and Boolean negation is not expressible is an open problem. (This is the case even for the language $\{\wedge, \vee, \backslash, \circ\}$ and the $\langle S, R\rangle$-reducts of group update frames.)

4 Conclusion

In this paper we outlined two FDE-based epistemic logics, the basic logic with universal and common knowledge, and its extension with a generalized left division operator of the Non-associative Lambek Calculus, formalizing an abstract notion of information update. We established axiomatization and decidability results for these logics.

Among topics that we leave out of the present paper is a study of axiomatic extensions of *GrFDE* and *GrUpFDE*. It is especially natural to consider extensions of *GrFDE* by various introspection axioms, such as positive introspection $K_a\varphi \supset K_aK_a\varphi$, Boolean negative introspection $\neg K_a\varphi \supset K_a\neg K_a\varphi$ and De Morgan negative introspection ${\sim}K_a\varphi \supset K_a{\sim}K_a\varphi$. Regarding extensions of *GrUpFDE*, it is interesting to take a look at how our framework accommodates some typical properties of special cases of information update (e.g. monotonicity $\varphi\backslash\chi \supset \varphi\backslash(\psi\backslash\chi)$ or "success" $\varphi\backslash\psi \supset \varphi\backslash(\varphi \wedge \psi)$; the latter seems to require an extension of our frames with a partial order on the set of situations in the style of the Routley–Meyer semantics for substructural logics [29].)

Another topic for future research are specific language extensions of our logics. A particular instance is related to the *iterated update* operator $\backslash^*$, where $\varphi\backslash^*\psi$ is read as "ψ holds after any finite number of updates by φ". A natural semantics for this operator is obtained by defining

$$R^1xYz := RxYz \qquad R^{n+1}xYz := \exists Uv(RxUv \;\&\; R^nvYz)$$

and

$$R^* := \{\langle x, Y, z\rangle \mid (\exists n \in \mathbb{N})(R^n xYz)\}$$

and requiring that

$$x \models_v \varphi \backslash^* \psi \text{ iff } \forall Yz((R^* xYz \;\&\; Y \models_v \varphi) \implies z \models_v \psi)$$

We conjecture that a complete axiomatization of the theory of all group update frames with R^* is obtained by adding to $GrUpFDE$ the following:

(A10) $(\chi\backslash^*\varphi \wedge \chi\backslash^*\psi) \supset \chi\backslash^*(\varphi \wedge \psi)$

(A11) $\varphi\backslash^*\psi \supset (\varphi\backslash\psi \wedge \varphi\backslash(\varphi\backslash^*\psi))$

(A12) $(\varphi\backslash(\varphi\backslash^*\psi)) \supset (\varphi\backslash^*\psi)$

(R6) $\dfrac{\varphi_1 \supset \psi_1 \quad \varphi_2 \supset \psi_2}{\psi_1\backslash^*\varphi_2 \supset \varphi_1\backslash^*\psi_2}$

(R7) $\dfrac{\varphi \supset \psi\backslash\varphi}{\varphi \supset \psi\backslash^*\varphi}$

(On some assumptions concerning the update relation R, $\varphi\backslash^*\psi$ can be expressed in a language containing fusion and the Kleene star operator; see [11]. Our setting intends to be more general. Also, in the presence of Boolean negation, some of these assumptions concerning R lead to undecidability; see [21]. It was shown in [25] that the classical Public Announcement Logic with an operator for iterated announcements is undecidable. Hence, the question is, which notions of update admit a decidable logic with iterated updates? Our general setting is especially suitable for such investigations, but they need to be left for future research.)

Another interesting topic are generalizations of the framework using weaker notions of negation than De Morgan negation used here. In general, negation can be seen as a negative modal operator with the satisfaction clause

$$x \models_v {\sim}\varphi \text{ iff } \forall y(R_{\sim} xy \implies y \not\models_v \varphi)$$

using an arbitrary binary relation $R_{\sim}$. If this relation is not serial, then Boolean negation cannot be expressed and some of the standard techniques used in completeness proofs for logics with common knowledge (and other fixpoint) operators cannot be used.

Acknowledgements. This work was supported by the Czech Science Foundation grant GJ18-19162Y for the project *Non-classical logical models of information dynamics*. The authors are grateful to three anonymous referees for their feedback.

A Proofs

Let L be any set of formulas containing all substitution instances of propositional tautologies in $\{\wedge, \vee, \supset\}$ that is closed under Modus Ponens and Uniform substitution. We say that a set of formulas Δ is *L-derivable* from a set of formulas Γ, notation $\Gamma \vdash_L \Delta$, iff there is $\gamma = \bigwedge \Gamma' \subseteq \Gamma$ and $\delta = \bigvee \Delta' \subseteq \Delta$ such that $\gamma \supset \delta$ is in L. We note that $\bigwedge \emptyset := \top$, so if Δ contains an element of L, then $\Gamma \vdash_L \Delta$ for all Γ. We say that $\langle \Gamma, \Delta\rangle$ is an *independent L-pair* iff $\Gamma \not\vdash_L \Delta$.

A *prime L-theory* is any set of formulas Γ that (i) contains L, (ii) is closed under $\varphi \supset \psi \in L$ (that is, if $\varphi \supset \psi \in L$ and $\varphi \in \Gamma$, then $\psi \in \Gamma$) and (iii) contains $\varphi \vee \psi$ only if it contains φ or ψ. A prime L-theory is called *non-trivial* iff it is not the set of all formulas.

Theorem 3 (Pair Extension). *If $\langle \Gamma, \Delta \rangle$ is an independent L-pair, then there is a non-trivial prime L-theory Σ extending Γ that is also disjoint from Δ.*

Proof. Essentially [29, 92–95]. We note that $\bigvee \emptyset := \bot$, so Σ cannot contain any χ such that $\chi \supset \bot$ is in L; hence Σ has to be non-trivial.

We note that in order for the Pair Extension Theorem to hold it is crucial to define L-derivability in a "finitary" way; see [9].

Theorem 1. *GrFDE is a sound and weakly complete axiomatization of the $\mathcal{L}_{Gr}$-theory of all group frames. The theory is decidable.*

Proof. Soundness is left to the reader as an exercise. Completeness is established using a variant of the standard finite canonical model construction (see e.g. [16, Ch. 3.1]). The argument used here is based on [26].

Assume that φ_0 is not provable in $GrFDE$. Let the closure of φ_0, $Cl(\varphi_0)$, be the smallest set of formulas that is closed under subformulas such that (1) it contains φ_0; (2) it contains $\top$; (3) if $K^*_G\psi \in Cl(\varphi_0)$, then $K_G(\psi \wedge K^*_G\psi) \in Cl(\varphi_0)$; and (4) if $K_G\psi \in Cl(\varphi_0)$, then $K_a\psi \in Cl(\varphi_0)$ for all $a \in G$. Formula ψ is a *negated formula* iff ψ is of the form ${\sim}\chi$ for some formula χ. We define $\tilde{\psi} := {\sim}\psi$ in case ψ is not a negated formula and $\widetilde{{\sim}\chi} := \chi$. Let $Cl^{\sim}(\varphi_0) = Cl(\varphi_0) \cup \{\tilde{\psi} \mid \psi \in Cl(\varphi_0)\}$. It can be shown easily that $Cl^{\sim}(\varphi_0)$ is finite. We denote $Cl(\varphi_0)$ as Φ and $Cl^{\sim}(\varphi_0)$ as Φ' in the rest of the proof.

We define a finite canonical model as follows. The set of situations S is the set of all independent $GrFDE$-pairs $x = \langle x_{in}, x_{out} \rangle$ such that $x_{in} \cup x_{out} = \Phi'$. It can be shown that each independent $GrFDE$-pair $\langle \Gamma, \Delta \rangle$ can be extended to an independent $GrFDE$-pair $\langle \Gamma', \Delta' \rangle$ such that $\Phi' \subseteq (\Gamma' \cup \Delta')$. Note that, for all x, x_{in} contains always at least $\top$. Otherwise $\top \in x_{out}$ and $x_{in} = \emptyset$, but then $\bigwedge x_{in} \supset \bigvee x_{out}$ is provable and so x is not an independent pair.

The rest of the model is defined as follows. The Routley star is defined by $x^\star := \langle x^\star_{in} = \{\psi \in \Phi' \mid \tilde{\psi} \in x_{out}\}, (\Phi' \setminus x^\star_{in}) \rangle$. It is easily seen that $x^\star$ is an independent $GrFDE$-pair and thus an element of S in the finite canonical model. Let us show that the canonical Routley star is of period two. It is clear that $\tilde{\tilde{\psi}} = \psi$ for all ψ. Therefore, $x^{\star\star}_{in} = \{\psi \in \Phi' \mid \tilde{\psi} \in x^\star_{out}\} = \{\psi \in \Phi' \mid \tilde{\tilde{\psi}} \notin x^\star_{out}\}$ $= x_{in}$.

Next, we define R_axy iff $\{\psi \mid K_a\psi \in x_{in}\} \subseteq y_{in}$. R_a is reflexive thanks to (A6). The group relations R_G and R^*_G are defined as in ordinary group models. The canonical valuation is $v : p \mapsto \{x \mid p \in x_{in}\}$ for $p \in \Phi$ and $v : p \mapsto \emptyset$ otherwise.

It remains to be shown that, for all $\psi \in \Phi$, $\psi \in x_{in}$ iff $x \vDash_v \psi$ (the Truth Lemma). For propositional variables, this holds by definition. It is easily seen that $\varphi \wedge \psi \in x_{in}$ iff both $\varphi, \psi \in x_{in}$ and $\varphi \vee \psi \in x_{in}$ iff at least one of φ, ψ is in x_{in}, from which the claims for conjunctions and disjunctions follow. The claim for $\supset$ is similarly easy (it follows from from the fact that FDE proves all positive classical tautologies—including $\top \supset (\varphi \vee (\varphi \supset \psi))$—and the fact that $\top \in x_{in}$ for all $x \in S$).

The claims for the modal operators are established as follows. If $K_a\varphi \in \Phi$, then $K_a\varphi \in x_{in}$ implies $x \vDash_v K_a\varphi$ by definition of R_a. To establish the converse implication, it is sufficient to observe that, if $K_a\varphi \notin x_{in}$, then $\langle\{\psi \mid K_a\psi \in x_{in}\}, \{\varphi\}\rangle$ is an independent pair. Hence, it can be extended to a pair $\langle\Gamma, \Delta\rangle$ such that $\Phi' \subseteq (\Gamma \cup \Delta)$. Take $y_{in} = \Gamma \cap \Phi'$ and $y_{out} = \Delta \cap \Phi'$. It is clear that $y = \langle y_{in}, y_{out}\rangle$ is an element of the canonical model such that $R_a xy$ and that $y \nvDash_v \varphi$ (by the induction hypothesis).

The case of $K_G\varphi \in \Phi$ where G is not a singleton follows from (A7), the definition of R_G and the induction hypothesis (for $K_a\varphi$, $a \in G$; note that we may use the hypothesis as $K_G\varphi \in \Phi$ implies $K_a\varphi \in \Phi$ for all $a \in G$).

Finally, take $K^*_G\varphi \in \Phi$. If $K^*_G\varphi \in x_{in}$, then $x \vDash_v K^*_G\varphi$ by the fact that $K_G(\varphi \wedge K^*_G\varphi) \in \Phi$ and the induction hypothesis for K_G. The converse entailment is established as follows. For each non-empty $Z \subseteq S$ and $y \in S$ of the finite canonical model, define

$$\varphi_y := \bigwedge y_{in} \qquad \varphi_Z := \bigvee_{y \in Z} \varphi_y$$

We sometimes write y instead of φ_y and Z instead of φ_Z. Take $Z := \{y \mid R^*_G xy\}$ and assume that $\varphi \in y_{in}$ for all $y \in Z$. We have to prove that $K^*_G\varphi \in x_{in}$.

Lemma 1. *GrFDE proves* $Z \supset K_G(\varphi \wedge Z)$.

Before proving the lemma, we show how it is applied. Using the Induction Rule (R3) and the fact that $x \in Z$, we obtain $\vdash x \supset K^*_G\varphi$. Hence, $K^*_G\varphi$ must be in x_{in}, otherwise x would not be an independent pair.

Proof of Lemma 1. We write $X_G Y$ if, for all $x \in X$, if $R_G xy$, then $y \in Y$. We prove the following claim.

Lemma 2. *If* $X_G Y$, *then GrFDE proves* $X \supset K_G Y$.

Proof of Lemma 2. We prove that if $X_G Y$, then $GrFDE$ proves $X \supset K_a Y$ for all $a \in G$; the desired result then follows by applying axiom (A7). The proof is by reductio ad absurdum. Assume that $GrFDE$ does not prove $X \supset K_a Y$. Then there is $w \in X$ such that $GrFDE$ does not prove $w \supset K_a Y$. This means that $\langle\{\psi \mid K_a\psi \in w_{in}\}, \{\varphi_z \mid z \in Y\}\rangle$ is an independent pair extendible to $\langle\Gamma, \Delta\rangle$ such that $\Gamma \cup \Delta$ contains Φ'. Take $y = \langle\Gamma \cap \Phi', \Delta \cap \Phi'\rangle$. It is clear that $R_a wy$ and so, by our assumption, $y \in Y$. However, this means that $GrFDE$ proves $\varphi_y \supset \varphi_Y$ and so y cannot be an independent pair. This is a contradiction. Hence, Lemma 2 is established.

We continue the proof of Lemma 1. Note that $Z_G Z$, so $GrFDE$ proves $Z \supset K_G Z$ by Lemma 2. Moreover, our assumption that $\varphi \in y_{in}$ for all $y \in Z$ implies that $GrFDE$ proves $Z \supset \varphi$. Using monotonicity and regularity of K_G, we infer that $GrFDE$ proves $K_G Z \supset K_G(\varphi \wedge Z)$. Hence, $GrFDE$ proves $Z \supset K_G(\varphi \wedge Z)$ as desired. This concludes the proof of Lemma 1 and completeness is established.

Our proof shows that the $\mathcal{L}_{Gr}$-theory of group frames is recursively axiomatizable and the axiomatization is complete with respect to a recursively enumerable set of models (models based in finite group frames). Hence, the theory is decidable.

Theorem 2. *$GrUpFDE$ is a sound and weakly complete axiomatization of the $\mathcal{L}_{GrUp}$-theory of all group update frames. The theory is decidable.*

Proof. Assume that φ_0 is not provable in $GrUpFDE$. Define the finite canonical model based on the closure of φ_0, Φ, and the $\sim$-closure of Φ, which we denote Φ', as in the proof of Theorem 1. Moreover, let $RxYz$ iff there are prime $GrUpFDE$-theories Γ, Σ and Δ_i for $i \in I$ such that

(a) for all $\varphi\backslash\psi$, if $\varphi\backslash\psi \in \Gamma$ and $\varphi \in \bigcap_{i\in I} \Delta_i$, then $\psi \in \Sigma$; and
(b) $x_{in} \subseteq \Gamma$, $(\Sigma \cap \Phi') \subseteq z_{in}$ and, for all Δ_i there is $y_j \in Y$ such that $(y_j)_{in} \subseteq \Delta_i$.

(A similar definition appears in [10].) We have to show only that the Truth Lemma holds for $\varphi\backslash\psi \in \Phi$. If $\varphi\backslash\psi \in x_{in}$, $RxYz$ and $\varphi \in \bigcup\{y_{in} \mid y \in Y\}$, then $\psi \in z_{in}$ by the definition of the canonical R. Conversely, we reason similarly as in [29, 256]. First, assume that $\varphi\backslash\psi \in x_{out}$. Extend x to a prime theory Γ. Second, extend the independent pair $\langle\{\chi \mid \varphi\backslash\chi \in \Gamma\}, \{\psi\}\rangle$ to a prime theory Σ. (The proof that it is an independent pair uses (A9) and (R4); the case $\{\chi \mid \varphi\backslash\chi \in \Gamma\} = \emptyset$ uses (R5).) Third, take the set $\Lambda = \{\alpha \mid \exists\beta(\beta \notin \Sigma \ \& \ \alpha\backslash\beta \in \Gamma\}$. For each $\alpha_i \in \Lambda$, $\varphi \supset \alpha_i$ is not provable. (If some $\varphi \supset \alpha_i$ were provable, then $\beta_i \in \Sigma$ by (R4).) Hence, extend each pair $\langle\{\varphi\}, \{\alpha_i\}\rangle$ to a prime theory Δ_i. It follows from the construction of Σ and Δ_i that (a) holds for $\Gamma, \{\Delta_i\}_{i\in I}, \Sigma$. (If $\Lambda = \emptyset$, then $\{\Delta_i\}_{i\in I} = \emptyset$ and so each $\alpha \in \bigcap_{i\in I} \Delta_i$; but in this case also $\alpha\backslash\beta \in \Gamma$ implies $\beta \in \Sigma$.) Moreover, $\varphi \in \bigcap_{i\in I} \Delta_i$ by the construction of Δ_i. Finally, take $(y_i)_{in} = \Delta_i \cap \Phi'$ and $(y_i)_{out}$ its complement relative to Φ' and similarly for z and Σ. It is clear that $\varphi \in (y_i)_{in}$ for all $i \in I$ and $\psi \notin z_{in}$. This concludes the proof of the Truth Lemma.

Our proof establishes that the $\mathcal{L}_{GrUp}$-theory of group update frames is recursively axiomatizable and the axiomatization is complete with respect to a recursively enumerable set of models (models based on finite group update frames). Hence, the theory is decidable.

References

1. Aucher, G.: Dynamic epistemic logic in update logic. J. Logic Comput. **26**(6), 1913–1960 (2016). https://doi.org/10.1093/logcom/exw002
2. Barwise, J., Perry, J.: Situations and Attitudes. MIT Press, Cambridge (1983)
3. Belnap, N.: A useful four-valued logic. In: Dunn, J.M., Epstein, G. (eds.) Modern Uses of Multiple-Valued Logic. Episteme (A Series in the Foundational, Methodological, Philosophical, Psychological, Sociological, and Political Aspects of the Sciences, Pure and Applied), vol. 2, pp. 5–37. Springer, Dordrecht (1977). https://doi.org/10.1007/978-94-010-1161-7_2
4. Belnap, N.: How a computer should think. In: Ryle, G. (ed.) Contemporary Aspects of Philosophy. Oriel Press Ltd., Stocksfield (1977)

5. van Benthem, J.: Language in Action: Categories, Lambdas and Dynamic Logic. Elsevier Science Publishers, Amsterdam (1991)
6. van Benthem, J.: Logical Dynamics of Information and Interaction. Cambridge University Press, New York (2011)
7. van Benthem, J.: Logical dynamics meets logical pluralism? The Australas. J. Logic **6** (2008). https://doi.org/10.26686/ajl.v6i0.1801, https://ojs.victoria.ac.nz/ajl/article/view/1801
8. Benthem, J.: Two logical faces of belief revision. In: Trypuz, R. (ed.) Krister Segerberg on Logic of Actions. OCL, vol. 1, pp. 281–300. Springer, Dordrecht (2014). https://doi.org/10.1007/978-94-007-7046-1_13
9. Bílková, M., Cintula, P., Lávička, T.: Lindenbaum and pair extension lemma in infinitary logics. In: Moss, L.S., de Queiroz, R., Martinez, M. (eds.) WoLLIC 2018. LNCS, vol. 10944, pp. 130–144. Springer, Heidelberg (2018). https://doi.org/10.1007/978-3-662-57669-4_7
10. Bílková, M., Majer, O., Peliš, M.: Epistemic logics for sceptical agents. J. Logic Comput. **26**(6), 1815–1841 (2016)
11. Bimbó, K., Dunn, J.M.: Relational semantics for Kleene logic and action logic. Notre Dame J. Formal Logic **46**(4), 461–490 (2005). https://doi.org/10.1305/ndjfl/1134397663
12. Chellas, B.F.: Basic conditional logic. J. Philos. Logic **4**(2), 133–153 (1975). https://doi.org/10.1007/BF00693270
13. van Ditmarsch, H., van der Hoek, W., Kooi, B.: Dynamic Epistemic Logic. Springer, Netherlands (2008). https://doi.org/10.1007/978-1-4020-5839-4
14. Došen, K.: A brief survey of frames for the Lambek Calculus. Math. Logic Q. **38**(1), 179–187 (1992). https://doi.org/10.1002/malq.19920380113
15. Dunn, J.M.: Ternary relational semantic and beyond: programs as data and programs as instructions. Logical Stud. **7** (2001)
16. Fagin, R., Halpern, J.Y., Moses, Y., Vardi, M.Y.: Reasoning About Knowledge. MIT Press, Cambridge (1995)
17. Fagin, R., Halpern, J.Y., Vardi, M.: A nonstandard approach to the logical omniscience problem. Artif. Intell. **79**, 203–240 (1995). https://doi.org/10.1016/0004-3702(94)00060-3
18. Girard, P., Tanaka, K.: Paraconsistent dynamics. Synthese **193**(1), 1–14 (2016). https://doi.org/10.1007/s11229-015-0740-2
19. Holliday, W.H., Hoshi, T., Icard III, T.F.: A uniform logic of information dynamics. In: Bolander, T., Braüner, T., Ghilardi, S., Moss, L. (eds.) Advances in Modal Logic 2012, pp. 348–367. College Publications (2012)
20. Kurtonina, N.: Frames and Labels. A Modal Analysis of Categorial Inference. Ph.D Thesis, Utrecht University (1994)
21. Kurucz, Á., Németi, I., Sain, I., Simon, A.: Decidable and undecidable logics with a binary modality. J. Logic Lang. Inf. **4**(3), 191–206 (1995). https://doi.org/10.1007/BF01049412
22. Levesque, H.: A logic of implicit and explicit belief. In: Proceedings of AAAI, 1984, pp. 198–202 (1984)
23. Mares, E.D.: A paraconsistent theory of belief revision. Erkenntnis **56**(2), 229–246 (2002). https://doi.org/10.1023/A:1015690931863
24. Mares, E.D.: Relevant Logic: A Philosophical Interpretation. Cambridge University Press, Cambridge (2004)
25. Miller, J.S., Moss, L.S.: The undecidability of iterated modal relativization. Stud. Logica **79**(3), 373–407 (2005). https://doi.org/10.1007/s11225-005-3612-9

26. Nishimura, H.: Semantical analysis of constructive PDL. Publ. Res. Inst. Math. Sci. **18**(2), 847–858 (1982). https://doi.org/10.2977/prims/1195183579
27. Restall, G.: Information flow and relevant logic. In: Seligman, J., Westershåhl, D. (eds.) Logic, Language and Computation: The 1994 Moraga Proceedings, pp. 463–477. CSLI Press, Stanford (1995)
28. Restall, G., Slaney, J.: Realistic belief revision. In: Proceedings of the Second World Conference on Foundations of Artificial Intelligence, pp. 367–378 (1995)
29. Restall, G.: An Introduction to Substrucutral Logics. Routledge, London (2000)
30. Rivieccio, U.: Bilattice public announcement logic. In: Goré, R., Kooi, B., Kurucz, A. (eds.) Advances in Modal Logic 2014, pp. 459–477. College Publications, London (2014)
31. Sedlár, I.: Propositional dynamic logic with Belnapian truth values. In: Advances in Modal Logic, vol. 11. College Publications, London (2016)
32. Sedlár, I., Punčochář, V.: From positive PDL to its non-classical extensions. Logic J. IGPL (2019).https://doi.org/10.1093/jigpal/jzz017

Knowledge in Topological Argumentation Models

Chenwei Shi(✉)

Department of Philosophy, Tsinghua University, Beijing, China
cshi@mail.tsinghua.edu.cn

Abstract. We formalise justification in the topological argumentation model and define knowledge and belief based on justification. In addition to revealing that the notions of knowledge and belief based on justification satisfy Stalnaker's BK system except for the closure principle, the main contribution of this paper is a counter-intuitive result about the notion of knowledge based on justification, which is related to the no false lemmas.

Keywords: Stalnaker's BK system · Topological evidence model · Abstract argumentation theory · No false lemmas

1 Introduction

In epistemology, justification and knowledge are so closely related that almost no theory of knowledge can avoid justification. However, there is no consensus on what kind of justification is adequate for knowledge. After Gettier's paper [7], so many attempts have been made to explicate what the adequacy of justification for knowledge requires. Among these attempts, logicians also make their contribution. To the best of the author's knowledge, there are mainly two approaches to the logic of justification, justification logic [1] and evidence logic [3]. Based on justification logic, there are works directly dedicated to the issue on the adequacy of justification for knowledge, for example, [6]. There are also works generalising evidence logic in which justification is semantically modelled, for example, [2].

In [2], a notion of knowledge requiring a correct justification is proposed and it is shown how this notion of knowledge can be connected to other theories of knowledge, for example, the no false lemmas theory of knowledge [4], the defeasibility theory of knowledge [12,13] and Stalnaker's theory of knowledge and belief [17].

This paper is a follow-up of a series of papers [14–16] on combining the topological evidence model in [2] and abstract argumentation theory in [5]. The integrated setting, called "topological argumentation model", naturally extends

The research in this paper is supported by the Major Program of the National Social Science Foundation of China (NO. 17ZDA026).

P. Blackburn et al. (Eds.): LORI 2019, LNCS 11813, pp. 286–296, 2019.
https://doi.org/10.1007/978-3-662-60292-8_21

the topological evidence model and leads to a more fine-grained formalisation of justification and belief. A remaining question is whether the notion of knowledge defined in [2] has its more fine-grained counterpart in the topological argumentation model. In this paper, we show that such a counterpart exits in the topological argumentation model.

Roughly speaking, in the topological argumentation model, P is known if and only if there is a correct justification for P. This is exactly the way knowledge is defined in the topological evidence model. However, in the topological argumentation model, we have a more fine-grained notion of justification, which we will call "ground". Thus the notion of knowledge in the topological argumentation model has a more fine-grained structure to be analysed. We will do such an analysis, which is related to the no false lemmas theory of knowledge. It leads us to a counter-intuitive result about knowledge. The key to explaining away the counter-intuitiveness lies in the distinction between argument, justification and ground, which can be made precise in the topological argumentation model.

We will also explore the relationship between knowledge and belief defined in the topological argumentation model. With the help of the logic of argument, belief and knowledge (LABK), which was firstly studied in [16], we will show that almost all the epistemic-doxastic axioms proposed by [17] are satisfied except the principle of closure under implication for knowledge. For example, believing P is the epistemic possibility of knowing P and believing P implies believing that you know P.

The paper is structured as follows. We first introduce the topological argumentation model and how ground is formalised in it in Sect. 2. Then in Sect. 3 we introduce the logic of argument, belief and knowledge and logically analyse the relationship between knowledge and belief. We show how the more fine-grained representation of ground enables us to conduct a further semantical analysis of knowledge in Sect. 4. The analysis resulted in a counter-intuitive theorem about knowledge. After a brief discussion and summary, we point out some possible future work in Sect. 5.

2 Grounds in Topological Argumentation Model

In this section, we first introduce the topological argumentation model and how justification can be formalised in it, which were first proposed in [15].

The topological argumentation model is an extension of the topological evidence model introduced in [2]. Let At be a set of atomic propositions.

Definition 1 (Topological argumentation model [15]**).** *A topological argumentation (TA) model* $(W, \mathcal{E}_0, \tau_{\mathcal{E}_0}, \hookleftarrow, V)$ *is a tuple where* W *is a set of possible states,* $\mathcal{E}_0 \subseteq \wp(W)$ *is a set of non-empty subsets of* W *which includes* W *itself,* $\tau_{\mathcal{E}_0}$[1] *is the topology generated by* $\mathcal{E}_0$,[2] *and* $V : \mathsf{At} \mapsto \wp(W)$ *is a evaluation function mapping each atomic proposition to a subset of* W.

[1] When no confusion arises, $\tau_{\mathcal{E}_0}$ will be denoted simply by τ.

[2] A *topology* over a non-empty domain X is a family $\tau \subseteq 2^X$ containing both X and $\varnothing$, and closed under both *finite* intersections and *arbitrary* unions. The elements of a topology are called *open sets*. The *topology generated by a given* $\mathcal{Y} \subseteq 2^X$ is the smallest topology $\tau_{\mathcal{Y}}$ over X such that $\mathcal{Y} \subseteq \tau_{\mathcal{Y}}$.

The attack relation *on* τ, $\leftarrow\!\!\prec\, \subseteq (\tau \times \tau)$ *with* $T_1 \leftarrow\!\!\prec T_2$ *read as* "T_2 attacks T_1", *is required to satisfy the following conditions*:

1. *for every* $T_1, T_2 \in \tau$: $T_1 \cap T_2 = \varnothing$ *if and only if* $T_1 \leftarrow\!\!\prec T_2$ *or* $T_2 \leftarrow\!\!\prec T_1$;
2. *for every* $T, T_1, T_1' \in \tau$: *if* $T_1 \leftarrow\!\!\prec T$ *and* $T_1' \subseteq T_1$, *then* $T_1' \leftarrow\!\!\prec T$;
3. *for every* $T \in \tau \setminus \{\varnothing\}$: $\varnothing \leftarrow\!\!\prec T$ *and* $T \not\leftarrow\!\!\prec \varnothing$.

Note that $(W, \mathcal{E}_0, \tau, V)$ constitutes a topological evidence model, in which $\mathcal{E}_0$ is interpreted as the agent's set of basic evidence and $\tau_{\mathcal{E}_0}$ is the topology where *non-empty* open sets are taken as arguments in the sense of being generated by taking finite intersection and arbitrary union of pieces of basic evidence. The relation $\leftarrow\!\!\prec$ is thus the attack relation between arguments.

As arguments are not ensured to be coherent (it is possible that for some $T, T' \in \tau$, $T \cap T' = \varnothing$), having one argument $T \in \tau$ supporting a proposition $P \subseteq W$ $(T \subseteq P)$ is not reasonable enough for the agent to believe P, especially when the argument is attacked but not *defended* at all. Then which arguments are qualified to justify the agent's belief? In order to answer this question, we appeal to the abstract argumentation theory [5].

Definition 2 (Defence function). *Let* $M = (W, \mathcal{E}_0, \tau, \leftarrow\!\!\prec, V)$ *be a TA model. A subset* $\sigma \subseteq \tau$ *is said to* defend $T \in \tau$ *if and only if any open* $T' \in \tau$ *attacking* T *(i.e., for all* $T' \in \tau$ *such that* $T \leftarrow\!\!\prec T'$*) is itself attacked by some open in* σ *(i.e., there is* $T'' \in \sigma$ *such that* $T' \leftarrow\!\!\prec T''$*). The* defence function *of* M*, denoted by* d_τ*, receives a set of opens* $\sigma \subseteq \tau$ *and returns the set of opens that* σ *defends:*

$$d_\tau(\sigma) := \{T \in \tau \mid T \textit{ is defended by } \sigma\}$$

The defence function d_τ is monotonic [5, Lemma 19]; hence, its least fixed point LFP_τ (the smallest subset $\mathsf{FP} \subseteq \tau$ satisfying $\mathsf{FP} = d_\tau(\mathsf{FP})$, i.e., exclusively self-defensive) always exists [11,18]. We call elements in the least fixed point of d_τ "grounds".

The least fixed point can be built up from the bottom. Let

$$\sigma_0 = \varnothing, \quad \sigma_{\alpha+1} = d(\sigma_\alpha), \quad \sigma_\beta = \bigcup_{\alpha<\beta} \sigma_\alpha \text{whenever } \beta \text{ is a limit ordinal.}$$

There must be an ordinal α such that $\mathsf{LFP}_\tau = \sigma_\alpha$. This process of building up the least fixed point captures the process of selecting the arguments which are qualified enough as grounds for the agent's beliefs. First, $\sigma_1 = d(\sigma_0)$ picks out those arguments which are not attacked at all. Then $d(\sigma_1)$ adds the arguments defended by σ_1. The whole process adds up at each step the arguments defended by those arguments added up in the previous steps. The process stops growing after hitting the ceiling LFP_τ.

Here are some facts about the least fixed point, which will be useful for the later discussion.

Proposition 1. *Given a TA model,*

1. *For any* $F, F' \in \tau$ *such that* $F \subseteq F'$, *if* $F \in \mathsf{LFP}$ *then* $F' \in \mathsf{LFP}$;
2. *For any* $F, F' \in \mathsf{LFP}$, $F \cap F' \neq \varnothing$.

In order to study several notions based on grounds, in the next section, we will present a logical system, which was first proposed in Chapter 2 of [14] and [16]. In [14] and [16], the main focus of the logic is on belief. In this paper we will go further and see how this logic helps us with issues on knowledge.

3 Logic of Argument, Belief and Knowledge

3.1 Syntax and Semantics

We first present the syntax and semantics of the logic (abbreviated as LABK).

Definition 3 (Language $\mathcal{L}_{\Box,\mathrm{U},\mathrm{K}}$). *The language* $\mathcal{L}_{\Box,\mathrm{U},\mathrm{K}}$ *is generated by:*

$$\varphi ::= p \mid \neg\varphi \mid \varphi \wedge \varphi \mid \mathrm{U}\,\varphi \mid \Box\varphi \mid \mathrm{K}\,\varphi$$

with $p \in \mathsf{At}$.

U is a universal modality, which is included for technical reasons and can be interpreted as infallible knowledge. Formulas of the form $\Box\varphi$ read "there is a *factive argument* supporting φ". K is the modality for (fallible) knowledge. Other Boolean operaleftrighttors ($\vee$, $\rightarrow$, $\leftrightarrow$) as well as the modal duals of U, $\Box$ and K are defined as usual (for the latter: $\widehat{\mathrm{U}}\,\varphi := \neg\,\mathrm{U}\,\neg\varphi$, $\Diamond\,\varphi := \neg\,\Box\,\neg\varphi$, and $\widehat{\mathrm{K}}\,\varphi := \neg\,\mathrm{K}\,\neg\varphi$).

Formulas in $\mathcal{L}_{\Box,\mathrm{U},\mathrm{K}}$ are semantically evaluated in *pointed TA* models, pairs (M, w) with $M = (W, \mathcal{E}_0, \tau, \leftarrow\!\!\!\!\prec, V)$ a *TA* model and $w \in W$ a world in it. The semantic interpretation of atoms and Boolean operators is as usual. For the modal operators,

Table 1. Axiom system $\mathsf{L}_{\Box,\mathrm{U},\mathrm{K}}$, for $\mathcal{L}_{\Box,\mathrm{U},\mathrm{K}}$ w.r.t. *TA* models.

Propositional Tautologies and Modus Ponens	
The S5 axioms and rules for U	The S4 axioms and rules for $\Box$
$\vdash \mathrm{K}\top$	$\vdash \mathrm{K}\varphi \rightarrow \mathrm{K}\mathrm{K}\varphi$
$\vdash \mathrm{K}\varphi \rightarrow \varphi$	From $\vdash \varphi \rightarrow \psi$ infer $\vdash \mathrm{K}\varphi \rightarrow \mathrm{K}\psi$
$\vdash \widehat{\mathrm{U}}\mathrm{K}\varphi \rightarrow \neg\widehat{\mathrm{U}}\mathrm{K}\neg\varphi$	$\vdash (\mathrm{K}\varphi \wedge \mathrm{U}\psi) \rightarrow \mathrm{K}(\varphi \wedge \mathrm{U}\psi)$
$\vdash \mathrm{K}\varphi \rightarrow \Box\varphi$	
$\vdash \mathrm{K}\varphi \rightarrow \mathrm{U}(\Box\varphi \rightarrow \mathrm{K}\varphi)$	$\vdash \mathrm{U}\Diamond\Box\varphi \rightarrow \widehat{\mathrm{U}}\mathrm{K}\varphi$
$\vdash (\widehat{\mathrm{U}}\mathrm{K}\varphi \wedge \neg\widehat{\mathrm{U}}\mathrm{K}\psi \wedge \mathrm{U}((\varphi\wedge\psi) \rightarrow \Box(\varphi\wedge\psi))) \rightarrow \widehat{\mathrm{U}}\Box(\varphi \wedge \neg\psi)$	

$$M,w \models \mathrm{U}\,\varphi \quad \mathit{iff}_{def} \quad W \subseteq [\![\varphi]\!]$$
$$M,w \models \Box\varphi \quad \mathit{iff}_{def} \quad \text{there exists } T \in \tau_{\mathcal{E}_0} \setminus \{\varnothing\} \text{ such that } w \in T \text{ and } T \subseteq [\![\varphi]\!]$$
$$M,w \models \mathrm{K}\,\varphi \quad \mathit{iff}_{def} \quad \text{there exists } F \in \mathsf{LFP}_\tau \text{ such that } w \in F \text{ and } F \subseteq [\![\varphi]\!]$$

with $[\![\varphi]\!] = \{w \in W \mid M,w \models \varphi\}$. A formula is valid (notation: $\models \varphi$) when $M,w \models \varphi$ holds for every world w of every *TA* model M.

The sound and complete axiom system $\mathcal{L}_{\Box,\mathrm{U},\mathrm{K}}$ is presented in Table 1 [14,16].

3.2 Knowledge and Belief Based on Ground

In the remaining part of this section, we will show that the logical relationship between belief and knowledge as advocated in [17] makes sense in the logic presented above. Our contribution is fitting the notion of ground seamlessly into the whole picture.

We start with zooming in on the semantical meaning of knowledge in LABK. It says that the agent knows φ if and only if there is a true ground for φ. Note that this definition of knowledge does not involve any notion of belief directly. Instead, it is based on truth and ground.

If we follow what is suggested in [17] and define belief as $\mathrm{B}\,\varphi := \widehat{\mathrm{K}}\,\mathrm{K}\,\varphi$, then we can prove the following proposition on the semantial meaning of belief in an *TA* model.

Proposition 2. *Given a pointed TA model (M,w), $M,w \models \mathrm{B}\,\varphi$ if and only if there exists $F \in \mathsf{LFP}_\tau$ such that $F \subseteq [\![\varphi]\!]$.*

Proof. First observe that (M,w), $M,w \models \widehat{\mathrm{U}}\,\mathrm{K}\,\varphi$ if and only if there exists $F \in \mathsf{LFP}_\tau$ such that $F \subseteq [\![\varphi]\!]$.

Then we prove that $\vdash \widehat{\mathrm{K}}\,\mathrm{K}\,\varphi \leftrightarrow \widehat{\mathrm{U}}\,\mathrm{K}\,\varphi$ in $\mathsf{L}_{\Box,\mathrm{U},\mathrm{K}}$. The direction from left to right follows from $\vdash \widehat{\mathrm{K}}\,\varphi \rightarrow \widehat{\mathrm{U}}\,\varphi$. The direction from right to left goes as follows (we sketch the deduction):

(1) $\vdash \widehat{\mathrm{U}}\,\mathrm{K}\,\varphi \rightarrow \widehat{\mathrm{U}}\,\mathrm{K}\,\mathrm{K}\,\varphi$	axiom 4 for K; necessitation rule for $\widehat{\mathrm{U}}$
(2) $\vdash \widehat{\mathrm{U}}\,\mathrm{K}\,\mathrm{K}\,\varphi \rightarrow \neg\widehat{\mathrm{U}}\,\mathrm{K}\,\neg\mathrm{K}\,\varphi$	axiom D for $\widehat{\mathrm{U}}\,\mathrm{K}$
(3) $\vdash \neg\widehat{\mathrm{U}}\,\mathrm{K}\,\neg\mathrm{K}\,\varphi \rightarrow \mathrm{U}\,\widehat{\mathrm{K}}\,\mathrm{K}\,\varphi$	the definitions of $\widehat{\mathrm{U}}$ and $\widehat{\mathrm{K}}$
(4) $\vdash \widehat{\mathrm{U}}\,\mathrm{K}\,\varphi \rightarrow \mathrm{U}\,\widehat{\mathrm{K}}\,\mathrm{K}\,\varphi$	(1)(2)(3) and modus ponens
(5) $\vdash \widehat{\mathrm{U}}\,\mathrm{K}\,\varphi \rightarrow \widehat{\mathrm{K}}\,\mathrm{K}\,\varphi$	(4), axiom T for U and modus ponens

By the soundness of the axiom system (Table 1), we can reach the conclusion that $M,w \models \mathrm{B}\,\varphi$ if and only if there exists $F \in \mathsf{LFP}_\tau$ such that $F \subseteq [\![\varphi]\!]$.

Belief is thus also based on ground. The only difference between belief and knowledge in LABK is the truth of their grounds.

The fact that we can define belief in terms of knowledge in LABK does not imply the conceptual priority of knowledge over belief as it is defended in [19]. In fact, we can also define knowledge using belief and argument in LABK.

Table 2. Axiom system BK

$\mathrm{K}(\varphi \to \psi) \to (\mathrm{K}\,\varphi \to \mathrm{K}\,\psi)$	$\mathrm{K}\,\varphi \to \varphi$
$\mathrm{K}\,\varphi \to \mathrm{K}\,\mathrm{K}\,\varphi$	$\mathrm{B}\,\varphi \to \neg\mathrm{B}\neg\varphi$
$\mathrm{B}\,\varphi \to \mathrm{K}\,\mathrm{B}\,\varphi$	$\neg\mathrm{B}\,\varphi \to \mathrm{K}\neg\mathrm{B}\,\varphi$
$\mathrm{K}\,\varphi \to \mathrm{B}\,\varphi$	$\mathrm{B}\,\varphi \to \mathrm{B}\,\mathrm{K}\,\varphi$
Modus Ponens	from φ infer $\mathrm{K}\,\varphi$

Proposition 3.

$$\models \mathrm{K}\,\varphi \leftrightarrow (\mathrm{B}\,\varphi \wedge \Box\varphi)$$

Proof. We prove that $\vdash \mathrm{K}\,\varphi \leftrightarrow (\mathrm{B}\,\varphi \wedge \Box\varphi)$ and then use the soundness of the axiom system.

The direction from left to right is easy. So we just sketch the deduction of the other direction which mainly use the axiom $\vdash \mathrm{K}\,\varphi \to \mathrm{U}(\Box\varphi \to \mathrm{K}\,\varphi)$ and axioms 5 and T for U.

$$\begin{aligned}
&(1) \vdash \hat{\mathrm{U}}\,\mathrm{K}\,\varphi \to \hat{\mathrm{U}}\,\mathrm{U}(\Box\varphi \to \mathrm{K}\,\varphi)\\
&(2) \vdash \hat{\mathrm{U}}\,\mathrm{U}(\Box\varphi \to \mathrm{K}\,\varphi) \to \mathrm{U}(\Box\varphi \to \mathrm{K}\,\varphi)\\
&(3) \vdash \mathrm{U}(\Box\varphi \to \mathrm{K}\,\varphi) \to (\Box\varphi \to \mathrm{K}\,\varphi)\\
&(4) \vdash \hat{\mathrm{U}}\,\mathrm{K}\,\varphi \to (\Box\varphi \to \mathrm{K}\,\varphi)\\
&(5) \vdash (\hat{\mathrm{U}}\,\mathrm{K}\,\varphi \wedge \Box\varphi) \to \mathrm{K}\,\varphi
\end{aligned}$$

Defining knowledge as belief with true ground distinguishes knowledge from justified true belief $(\mathrm{B}\,\varphi \wedge \varphi)$ in LABK.

The following validity shows that the notion of belief defined in LABK is full belief as [17] calls it.

Proposition 4.

$$\models \mathrm{B}\,\varphi \to \mathrm{B}\,\mathrm{K}\,\varphi$$

Proof. The deduction of $\vdash \mathrm{B}\,\varphi \to \mathrm{B}\,\mathrm{K}\,\varphi$ involves the use of axiom 4 for K.

Actually, it can be verified that all axioms and rules in the system BK about knowledge and belief in [17] are valid in LABK except the closure under implication for both knowledge and belief, namely $\mathrm{K}(\varphi \to \psi) \to (\mathrm{K}\,\varphi \to \mathrm{K}\,\psi)$ and $\mathrm{B}(\varphi \to \psi) \to (\mathrm{B}\,\varphi \to \mathrm{B}\,\psi)$.

To see why the closure principle fails, the following counterexample is helpful.

Example 1. Let $\mathsf{At} = \{p, t, b\}$ be a set of atomic propositions. We build the following *TA* model.

$$(\{1,2,3\}, \mathcal{E}_0 = \{\{1\},\{2\},\{3\},\{1,2\},\{2,3\}\}, \tau = 2^{\{1,2,3\}}, \leftarrow, V) \qquad (1)$$

with $V = \{(p,\{1,2\}), (t,\{2,3\}), (b,\{3,4\})\}$ and $\leftarrow$ given by the union of ***(i)*** singletons attacking one another, ***(ii)*** every open attacking the empty set, and

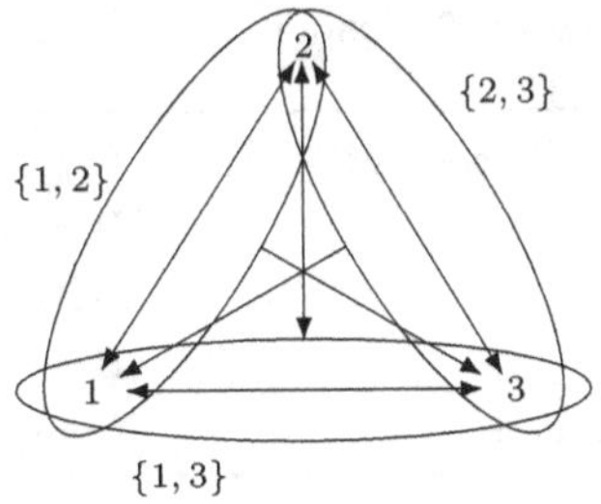

Fig. 1. Grounded beliefs are not closed under conjunction.

(iii) $\{\{3\} \leftarrow\!\!\prec \{1,2\}, \{1\} \leftarrow\!\!\prec \{2,3\}, \{2\} \leftarrow\!\!\prec \{1,3\}, \{1,3\} \leftarrow\!\!\prec \{2\}\}$, as shown in Fig. 1.[3] Following its definition, $\mathsf{LFP}_\tau = \{\{1,2\}, \{2,3\}, \{1,2,3\}\}$ (a set that is not closed under intersection).

Assume that 2 is the actual world. Then $2 \models \mathrm{K}p$ and $2 \models \mathrm{K}t$, but $2 \not\models \mathrm{K}(p \wedge t)$. The same also applies to B.

The example in fact shows that the failure of closure under implication is due to the failure of closure under conjunction introduction.

For knowledge, the failure of closure under conjunction introduction is not a problem at all. It is even advantageous in the sense of playing the key role in solving the scepticism puzzle. We refer readers who are interested to [10]. For belief, a discussion about the implication of the failure of closure under conjunction introduction can be found in Section 5 of [15].

More information on LABK and the *TA* models can be found in [14–16]. In the next section, we will do a deeper semantical analysis of knowledge in the *TA* models.

4 A Fine-Grained Analysis of No False Lemmas

In [7], two counterexamples are presented against the definition of knowledge as justified true belief (JTB theory of knowledge). These two counterexamples share the same structure: the agent has belief of a proposition, for example, $P \vee Q$, where P is false but Q is true; but the agent has justification for P and infers $P \vee Q$ from P; so the agent has justified true belief but intuitively not knowledge.

The problem in the counterexamples is obviously the false justification which supports a true proposition. This situation can be represented in a *TA* model by taking a ground F in LFP_τ where $w \notin F$ and expanding F to a certain extent so that w is included in the expansion.

There are a lot of attempts to remedy the JTB theory of knowledge, for example, [4,8,13]. The proposal in [4] – if the false ground causes the problem, then require the truth of the ground – is similar to our definition of knowledge

[3] Attack edges involving the empty set are not drawn.

in LABK as belief plus true ground. However, [4] goes further, requiring that the true ground also has a true ground and the true ground for the true ground also has a true ground and so on. Moreover, the relationship between each level of ground can be non-deductive. Rather than specifying what the non-deductive relationship is, [4] gives an example.

We are not going to analyse what the non-deductive relationship reflected in [4]'s example can be. Our focus is on the further requirement of a chain of true grounds, which is usually called "no false lemmas". The requirement of a chain of true grounds seems to indicate that the notion of knowledge in LABK which is equivalent to belief plus a true ground is not strong enough.

Recall the building process of the least fixed point LFP_τ, which is the set of grounds. It seems possible that a true ground F which is included in a certain step of the process, for example, $d_{n+1}(\sigma_n)$, would not have been in it, if we took out some false grounds in σ_n. In other words, the membership of F in LFP_τ seems to be dependent on the membership of some false grounds. The requirement of no false lemmas excludes such cases. So seemingly we can improve our definition of knowledge as follows.

Let $t_\tau : \wp(W) \mapsto \wp(W)$ be a function such that $t_\tau(T) = \{T \in d_\tau(T) \mid w \in T\}$. Then let $\sigma_0^w = \varnothing$, $\sigma_{\alpha+1}^w = t_\tau(\sigma_\alpha^w)$, and $\sigma_\beta^w = \bigcup_{\alpha<\beta} \sigma_\alpha^w$ whenever β is a limit ordinal. Since the function t_τ is monotonic, so its least fixed point must exist and there must be an ordinal α such that σ_α^w is its least fixed point. We use TG^w to denote this least fixed point for t_τ. During the process of building the least fixed point TG^w, we essentially excludes in every step those false grounds. Note that $\mathsf{TG}^w \subseteq \{T \in \mathsf{LFP}_\tau \mid w \in T\}$.

Definition 4 (Well-grounded knowledge). *Given a pointed TA model (M, w), $M, w \models \mathcal{K}^g \varphi$ if and only if there exists $F \in \mathsf{TG}^w$ such that $F \subseteq [\![\varphi]\!]$.*

Because well-grounded knowledge requires a chain of true grounds, intuitively, we get a stronger notion of knowledge, namely $\models \mathcal{K}^g \varphi \to \mathrm{K}\,\varphi$. However, it turns out that in *TA* models, we also have $\models \mathrm{K}\,\varphi \to \mathcal{K}^g \varphi$.

Theorem 1.

$$\models \mathcal{K}^g \varphi \leftrightarrow \mathrm{K}\,\varphi$$

Proof. We prove the direction $\models \mathrm{K}\,\varphi \to \mathcal{K}^g \varphi$, for which we just need to prove that $\{T \in \mathsf{LFP}_\tau \mid w \in T\} \subseteq \mathsf{TG}^w$.

We first prove that for any ordinal β, $\{T \in d(\sigma_\beta) \mid w \in T\} \subseteq t_\tau(\sigma_\beta^w)$.

Obviously when $\alpha = 0$, $t_\tau(\sigma_0^w) = \{T \in d(\sigma_0^w) \mid w \in T\} = \{T \in d(\sigma_0) \mid w \in T\}$ because $\sigma_0^w = \sigma_0 = \varnothing$.

First assume that β is not a limit ordinal, that is $\beta = \alpha + 1$. We prove that $\{T \in d(\sigma_{\alpha+1}) \mid w \in T\} \subseteq t_\tau(\sigma_{\alpha+1}^w)$, assuming that $\{T \in d(\sigma_\alpha) \mid w \in T\} \subseteq t_\tau(\sigma_\alpha^w)$.

Take an argument $F \in d_\tau(\sigma_{\alpha+1})$ such that $w \in F$. If there is $T \in \tau$ such that $F \leftarrow\!\!\prec T$, then there must be $F' \in \sigma_{\alpha+1} = d_\tau(\sigma_\alpha)$ such that $T \leftarrow\!\!\prec F'$ and $F' \not\leftarrow\!\!\prec T$. (Otherwise, we can prove by induction on the ordinal α of the building process, that $F \notin d_\tau(\sigma_{\alpha+1})$.)

Now take the union of F and F', i.e. $F'' = F \cup F'$. By condition 2 of $\leftarrow\!\!\!\prec$ and $F'' \not\leftarrow\!\!\!\prec T$, which follows from the fact that $F' \not\leftarrow\!\!\!\prec T$ and the first fact in Proposition 1, it follows that $T \leftarrow\!\!\!\prec F''$.

Because $F' \subseteq F''$ and $F' \in d_\tau(\sigma_\alpha)$, it follows that $F'' \in d_\tau(\sigma_\alpha)$. By the hypothesis $\{T \in d(\sigma_\alpha) \mid w \in T\} \subseteq t_\tau(\sigma_\alpha^w)$, it follows that $F'' \in t_\tau(\sigma_\alpha^w)$. Because $\tau(\sigma_\alpha^w) \subseteq \tau(\sigma_{\alpha+1}^w)$, we get that $F'' \in \tau(\sigma_{\alpha+1}^w)$.

Second, assume that β is a limit ordinal. It is not hard to see that $\{T \in d(\sigma_\beta \mid w \in T\} \subseteq t_\tau(\sigma_\beta^w)$, assuming that $\{T \in d(\sigma_\alpha) \mid w \in T\} \subseteq t_\tau(\sigma_\alpha^w)$ for any $\alpha < \beta$.

Therefore, we have proved that $\{T \in d(\sigma_\beta) \mid w \in T\} \subseteq t_\tau(\sigma_\beta^w)$ for any ordinal β.

Because there must be an ordinal α such that $\mathsf{LFP}_\tau = \sigma_\alpha$. Then $\{T \in \mathsf{LFP}_\tau \mid w \in T\} \subseteq t_\tau(\sigma_\alpha^w)$ for some ordinal α. Since $\sigma_\alpha^w \subseteq \mathsf{TG}^w$ where TG^w is the least fixed point of t_τ, we have proved that $\{T \in \mathsf{LFP}_\tau \mid w \in T\} \subseteq \mathsf{TG}^w$.

Now assume that $\mathrm{K}\varphi$ holds in a given possible world w of a given *TA* model M. Then there is $F \in \{T \in \mathsf{LFP}_\tau \mid w \in T\}$ such that $F \subseteq [\![\varphi]\!]$. It follows by $\{T \in \mathsf{LFP}_\tau \mid w \in T\} \subseteq \mathsf{TG}^w$ that $F \in \mathsf{TG}^w$. So we have $M, w \models \mathcal{K}^g \varphi$

This implies that $M, w \models K\varphi$.

While this result seems counter-intuitive (and paradoxical in the sense of following from a series of seemingly intuitive and reasonable formalisations, for example, the topological structure of arguments, the three conditions on the attack relation and so on), it can be explained away. One possible reason for us to feel counter-intuitive could be that we confuse arguments with grounds and applied our intuition about arguments wrongly to grounds. The membership of factive arguments in a certain set of arguments can be dependent on false arguments, but the set cannot be the least fixed point and thus the arguments are not grounds for the agent, as the following example illustrates.

Example 2. Let $M = (W = \{1,2,3,4\}, \mathcal{E}_0 = \{W, \{1,2\}, \{2,3\}, \{4\}\}, \tau_{\mathcal{E}_0}, \leftarrow\!\!\!\prec, V)$ be a *TA* model. The topology τ consists of W, $\varnothing$ and elements in Fig. 2. The attack relation is illustrated partially (leaving out attack towards the empty open) in Fig. 2.

Assume that $1 \in W$ is the actual world. Starting with a false argument $\{2,3\}$ and applying the defence function d to it, we can build up a set of arguments $d(\{2,3\}) = \{\{2,3\}, \{1,2\}, \{2\}, \{1,2,3\}, W\}$. We can see that the true arguments $\{1,2\}$ is only defended by $\{2,3\}$. So within the set $d(\{2,3\})$, true arguments can

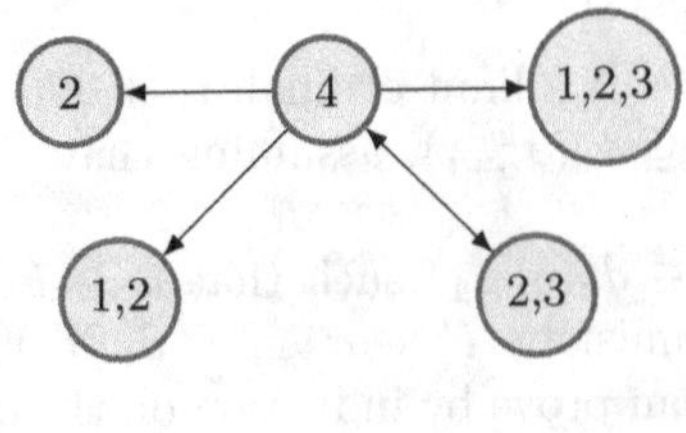

Fig. 2. The attack relation in Example 2

be dependent on false arguments. But $d(\{2,3\})$ is not the least fixed point of the model, which is $\{W\}$. Arguments $\{1,2\}$ and $\{2,3\}$ are not grounds.

5 Discussion and Future Work

In this paper, we proposed a notion of knowledge based on true grounds in topological argumentation models. We have showed that its relationship with belief in LABK satisfy almost all of Stalnaker's epistemic-doxastic axioms in [17] except the closure principle for knowledge. Furthermore, we have revealed that each true ground is well-grounded, i.e. defended by a chain of true grounds.

We mentioned that the notion of knowledge K in LABK is different from U in the sense of being fallible. That is, when given more information, the agent might lose her knowledge, as the famous barn-facades example [9] shows. In the example, Henry happens to see a real barn among a lot of barn-facades. Although from Henry's perspective he cannot distinguish the real barn from barn-facades, he has a true ground by his perception and thus has knowledge according to our definition. However, if Henry is told that all other "barns" he saw are barn-facades, a new argument may be available to him, supporting that the object he saw is not real barn either. This new argument may attack his original argument so that the original argument is not qualified as a ground any more. So his knowledge is lost.

We notice that in the example, although the new information given to Henry is true, it is misleading in the sense of creating new false argument (All the other objects are barn-facades; if all the other objects are barn-facades, then the one seen is also a barn-facade; so the one seen is a barn-facade).

In [2], it is shown that their notion of (fallible) knowledge defined in the topological evidence model is infallible if all information received by the agent is non-misleading. The notion of "non-misleading" can be precisely modelled in the topological evidence model. By using conditional belief, it can also model the property of being infallible.

Is our notion of knowledge also infallible in the same sense? We can define "non-misleading" argument in the *TA* model. But it is more involved to define conditional belief in the *TA* model, because we need to configure not only new arguments conditional on new information but also a new attack relation conditional on new information. How can the notion of conditional belief be defined in the *TA* model? We leave it as our future work.

Acknowledgement. The author is grateful to Fernando R. Velázquez-Quesada for his valuable feedback on the first draft of this paper and to the three anonymous referees of this paper for their comments and remarks.

References

1. Artemov, S.N.: The logic of justification. Rev. Symbolic Logic **1**(4), 477–513 (2008)

2. Baltag, A., Bezhanishvili, N., Özgün, A., Smets, S.: Justified belief and the topology of evidence. In: Väänänen, J., Hirvonen, Å., de Queiroz, R. (eds.) WoLLIC 2016. LNCS, vol. 9803, pp. 83–103. Springer, Heidelberg (2016). https://doi.org/10.1007/978-3-662-52921-8_6
3. van Benthem, J., Fernández-Duque, D., Pacuit, E.: Evidence logic: a new look at neighborhood structures. In: Bolander, T., Braüner, T., Ghilardi, S., Moss, L. (eds.) Proceedings of Advances in Modal Logic Volume 9, vol. 9, pp. 97–118. King's College Press (2012)
4. Clark, M.: Knowledge and grounds: a comment on Mr. Gettier's paper. Analysis **24**, 46–48 (1963)
5. Dung, P.M.: On the acceptability of arguments and its fundamental role in nonmonotonic reasoning, logic programming and n-person games. Artif. Intell. **77**, 321–357 (1995)
6. Egré, P., Marty, P., Renne, B.: Knowledge, Justification, and Reason-based Belief (2014)
7. Gettier, E.: Is justified true belief knowledge? Analysis **23**(6), 121–123 (1963)
8. Goodman, A.: A causal theory of knowing. J. Philos. **64**(12), 357–372 (1967)
9. Goodman, A.: Discrimination and perceptual knowledge. J. Philos. **73**(20), 771–791 (1967)
10. Holliday, W.H.: Knowing what follows: epistemic closure and epistemic logic. Ph.D. thesis, Stanford University (2012)
11. Knaster, B.: Un théorème sur les fonctions d'ensembles. Annales de la Société Polonaise de Mathématiques **6**, 133–134 (1928)
12. Lehrer, K.: Theory of Knowledge, 1st edn. Routledge, London (1990)
13. Lehrer, K., Paxson, T.: Knowledge: undefeated justified true belief. J. Philos. **66**, 225–237 (1969)
14. Shi, C.: Reason to Believe. Ph.D. thesis, Institute for Logic, Language and Computation, Amsterdam, The Netherlands, June 2018. https://1drv.ms/b/s!AjGu4GHCs1Jcp98Gj925XsdiK-pypg, iLLC Dissertation series DS-2018-10
15. Shi, C., Smets, S., Velázquez-Quesada, F.R.: Argument-based belief in topological structures. In: Lang, J. (ed.) Proceedings Sixteenth Conference on Theoretical Aspects of Rationality and Knowledge, TARK 2017, University of Liverpool, Liverpool, UK, 24–26 June 2017. Electronic Proceedings in Theoretical Computer Science, vol. 251, pp. 489–503. Open Publishing Association (2017). http://www.illc.uva.nl/Research/Publications/Reports/PP-2017-18.text.pdf
16. Shi, C., Smets, S., Velázquez-Quesada, F.R.: Beliefs based on evidence and argumentation. In: Moss, L.S., de Queiroz, R., Martinez, M. (eds.) WoLLIC 2018. LNCS, vol. 10944, pp. 289–306. Springer, Heidelberg (2018). https://doi.org/10.1007/978-3-662-57669-4_17
17. Stalnaker, R.: On logics of knowledge and belief. Philos. Stud. **128**(1), 169–199 (2006)
18. Tarski, A.: A lattice-theoretical theorem and its applications. Pac. J. Math. **5**(2), 285–309 (1955)
19. Williamson, T.: Knowledge and Its Limits. Oxford University Press, Oxford (2000)

Towards a Logical Formalisation of Theory of Mind: A Study on False Belief Tasks

Anthia Solaki(✉) and Fernando R. Velázquez-Quesada(✉)

Institute for Logic, Language and Computation,
Universiteit van Amsterdam, Amsterdam, The Netherlands
{a.solaki2,F.R.VelazquezQuesada}@uva.nl

Abstract. Theory of Mind, the cognitive capacity to attribute internal mental states to oneself and others, is a crucial component of social skills. Its formal study has become important, witness recent research on reasoning and information update by intelligent agents, and some proposals for its formal modelling have put forward settings based on Epistemic Logic (*EL*). Still, due to intrinsic idealisations, it is questionable whether *EL* can be used to model the high-order cognition of 'real' agents. This manuscript proposes a mental attribution modelling logical framework that is more in-line with findings in cognitive science. We introduce the setting and some of its technical features, and argue why it does justice to empirical observations, using it for modelling well-known False-Belief Tasks.

Keywords: Theory of Mind · Mental state attribution · False belief tasks · Temporal model · Dynamic epistemic logic

1 Introduction

An important feature of how people function in social scenarios is that of *Theory of Mind* (ToM), the cognitive capacity to attribute internal mental states, such as knowledge and beliefs, to oneself and others [1].[1] Theory of Mind is a crucial component of social skills: someone who understands that others might have mental states different from hers, and can reason about those states, is much better suited to understand their behaviour, and thus act and react appropriately.

Theory of Mind is slowly developed in the course of our lives [3,4] (and at different speed for different types of persons [5,6]), starting with the ability to make *first-order* attributions (e.g., someone knowing/believing that *"Mary believes that the ball is in the bag"*) and progressing through attributions of

[1] There has been a debate on how this understanding of others' mental states is achieved (see, e.g., [2]). Some argue that it is by acquiring a *theory* of commonsense psychology (*theory theory*); some others argue that it comes from a direct *simulation* of others' mental states (*simulation theory*). We will use the term ToM without endorsing any of these views, as such discussion falls outside the scope of this proposal.

P. Blackburn et al. (Eds.): LORI 2019, LNCS 11813, pp. 297–312, 2019.
https://doi.org/10.1007/978-3-662-60292-8_22

second-order mental states (e.g., someone knowing/believing that *"Mary believes that John believes that the ball is in the closet"*). When testing one's ToM, an extensively used experiment is the *Sally-Anne* False-Belief Task.

Example 1. (THE *Sally-Anne* (*SA*) TASK). The following is adapted from [3].

> *Sally and Anne are in a room in which there are a basket and a box. Sally is holding a marble. Then, after putting the marble into the basket, Sally leaves the room. While Sally is away, Anne transfers the marble to the box. Then Sally comes back.*

To pass the test, the subject should answer correctly the question *"where does Sally believe the marble to be?"*. This requires for the subject to distinguish between her own true belief (*"the marble is in the box"*) and Sally's *false* belief (*"the marble is in the basket"*). Experiments have shown that, while children older than 4 years old tend to answer correctly, younger children (or children on the autism spectrum) tend to fail the test, reporting their own belief [3]. (But see [7].) ◀

In the enterprise of studying and understanding ToM, there has been a growing interest on the use of formal frameworks. A seemingly natural choice is Epistemic Logic (*EL*) [8,9], as it provides tools for representing not only the knowledge/beliefs agents have about ontic facts, but also the knowledge/beliefs they have about their own and others' knowledge/beliefs. However, using *EL* has some drawbacks. First, within *EL*'s standard relational 'Kripke' semantics, knowledge/beliefs are closed under logical consequence (the *logical omniscience* problem; [10]). Moreover, the extra relational requirements for 'faithful' representations of knowledge and beliefs turn them into S5 and KD45 modal logics, respectively, thus yielding fully (positive and negative) introspective agents.

There is an even more fundamental reason why *EL* might not be well-suited for representing realistic high-order attributions. Semantically, both knowledge and beliefs correspond to a universal quantification (ϕ is known/believed iff it is the case in *all* the alternatives the agent considers possible); still, for real agents, these notions involve more elaborate considerations (e.g., observation, communication, reasoning). This 'simple' universal quantification works because *EL* uses a loaded model, which contains not only the (maximally consistent) alternatives the agent considers possible, but also every other alternative *every other agent* considers possible.[2] In a few words, the semantic interpretation of (high-order) knowledge/beliefs formulas is simple because the model is complex. Real agents might not be able to have such a loaded structure 'in their mind', and thus it is questionable whether the use of traditional *EL* can provide a proper picture of the way real agents deal with mental attribution scenarios.

---

[2] Frameworks for representing acts of private communication [11] make this clear. Their additional structures, *action models*, have one 'event' for each different perspective the agents might have about the communication, and the model after the communication contains roughly one copy of the original model for each one of these perspectives.

In light of these issues, one could even wonder whether it makes sense to use logical tools for dealing with results of empirical research. Indeed, it has been argued that psychological experiments and logic are essentially different[3], understanding the former as the study of empirical findings on the behaviour of real 'fallible' agents, and the latter as a normative discipline studying what 'rational' agents *should* do. However, other authors (e.g., [14,15]) have justified why bridging these two views is a worthwhile endeavour that also has promising applications (especially on reasoning and information update by intelligent agents). Indeed, empirical research benefits from using formal tools to explain their discoveries and understand their consequences, and logical frameworks become richer and more 'useful' when they capture human limitations and prescribe behaviour attainable by real agents.

This work seeks a ToM's logical setting that is more in-line with the findings in cognitive science, with non-trivial and competent agents whose underlying reasoning is reflected in the syntax and semantics.[4] To that end, we aim at the converse direction to that of *EL*. Our structures are simple, encoding only basic facts, and thus resembling the 'frugal' way real agents keep information stored. However, interpretations of mental state attributions show that agents engage in the, oftentimes strenuous, process of recalling these facts and deriving further information on their basis.

Outline. The text is organised as follows. Section 2 introduces the *temporal visibility* framework, presenting its model and formal language, and also discussing some of its technical aspects. Then, Sect. 3 relates the features of the setting with findings in the cognitive science literature, using it to model well-known mental attribution tasks in detail, and comparing it with other related formal settings. Section 4 closes, recapitulating the highlights, discussing ways in which the framework can be extended, and suggesting lines for further research.

2 Visibility in a Temporal Setting

In most mental attribution tasks, beliefs[5] are, at their lower (ontic) order, about the location of certain objects (e.g., the marble's location in the Sally-Anne Task). We do take objects as the main entities about which agents have mental attitudes; still, for simplicity, we will work with these objects' *colours*. Let $A \neq \varnothing$ be the set of agents $(a, b, \ldots)$, and $O \neq \varnothing$ be the set of objects $(o, p, q, \ldots)$. For each $o \in O$, the set R_o contains the colours the object might have; define $R_O := \bigcup_{o \in O} R_o$. The model is a temporal structure, with each stage (*state*) fully described by both the colour of each object and the objects and agents each agent sees.

---

[3] *Anti-Psychologism* (e.g., [12]) has long been against attempts to reconcile the two [13].

[4] In particular, one goal is to find a system that provides a plausible answer on why people find mental attribution tasks increasingly difficult as their order increases.

[5] Following the common parlance in the literature describing the tasks we later model, the term *belief* will be used for referring to an agent's mental state.

Definition 2.1 (Temporal visibility model). A *temporal visibility (TV) model* is a tuple $\langle n, S, \tau, \kappa, \nu\rangle$ with ***(i)*** $n \in \mathbb{N}$ the index of the 'most recent' (current) stage; ***(ii)*** S a finite set of states with $|S| = n$; ***(iii)*** $\tau : S \to \{1..n\}$ the temporal index (bijective) function, indicating the temporal index $\tau(s) \in \{1..n\}$ of each state $s \in S$; ***(iv)*** $\kappa : S \to (O \to R_O)$ the *colouring* function, with $\kappa(s, o)$ (abbreviated as $\kappa_s(o)$) the colour object o has at state s;[6] ***(v)*** $\nu : S \to (A \to \wp(A \cup O))$ the *visibility* function, with $\nu(s, a)$ (abbreviated as $\nu_s(a)$) the *entities* (agents and objects) agent a sees at state s.[7] Given a *TV* model, let $s_{last} \in S$ be its (unique) state satisfying $\tau(s_{last}) = n$. ◀

Example 2. Take the Sally-Anne Task, with Sally (*Sa*), Anne (*An*) and the marble (*mar*). Consider a two-state model M with ***(i)*** s_1 the initial state, where both agents see all agents and objects ($\nu_{s_1}(Sa) = \nu_{s_1}(An) = \{Sa, An, mar\}$) and the object is black ($\kappa_{s_1}(mar) = black$, read as 'the marble is in Sally's hands'), and ***(ii)*** s_2 the 'next' state, where both agents still see everything, but now the object is white ($\kappa_{s_2}(mar) = white$, read as 'the marble is in the basket'). The model is depicted as

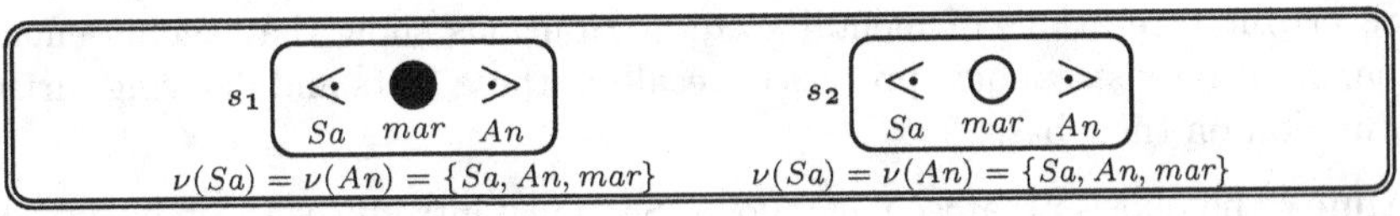

◀

Representing Actions. A *TV* model contains not only a state representing the current situation (the state $\tau^{-1}(n)$) but also states indicating how the situation was in the past (up to the initial $\tau^{-1}(1)$). One can provide operations that *extend* the current model with a state depicting the outcome of a certain activity (the way the situation *will* be). In the Sally-Anne Task, some acts modify the colour of objects (Sally puts the marble into the basket) and some others modify the agents' visibility (Sally leaves the room). Here are operations for them.

Definition 2.2 (Colour change). Let $M = \langle n, S, \tau, \kappa, \nu\rangle$ be a *TV* model, with $s_{new} \notin S$; take a set of objects $\{p_1, \ldots, p_k\} \subseteq O$, with $c_i \in R_{p_i}$ a proper colour for each p_i. The colour assignment $[p_1{:=}c_1, \ldots, p_k{:=}c_k]$ produces the *TV* model

$$M_{[p_1:=c_1,\ldots,p_k:=c_k]} = \langle n+1, S \cup \{s_{new}\}, \tau', \kappa', \nu'\rangle$$

in which ***(i)*** τ' preserves the temporal position of states in S, making s_{new} the most recent (so $\tau'(s) := \tau(s)$ for $s \in S$, and $\tau'(s_{new}) := n+1$); ***(ii)*** κ' is exactly as κ for states in S, with the new s_{new} taking the colouring of s_{last} for objects not mentioned by the assignment, and following the assignment for the colour of the objects it mentions (so, for any $o \in O$, define $\kappa'_s(o) := \kappa_s(o)$ for $s \in S$, with $\kappa'_{s_{new}}(o) := \kappa_{s_{last}}(o)$ when $o \notin \{p_1, \ldots, p_k\}$, and $\kappa'_{s_{new}}(p_j) := c_j$ when $o = p_j$);

[6] Each object has a proper colour: $\kappa_s(o) \in R_o$ holds for all $s \in S$ and $o \in O$.
[7] Every agent can see herself in every state: $a \in \nu_s(a)$ holds for all $s \in S$ and all $a \in A$.

(iii) ν' preserves the visibility assignment for states in S, with visibility in s_{new} exactly as in s_{last} (so, for any $a \in A$, define $\nu'_s(a) := \nu_s(a)$ for $s \in S$, and $\nu'_{s_{new}}(a) := \nu_{s_{last}}(a)$). ◀

Definition 2.3 (Visibility change). Let $M = \langle n, S, \tau, \kappa, \nu \rangle$ be a TV model, with $s_{new} \notin S$; take a set of agents $\{b_1, \ldots, b_k\} \subseteq A$, and let $X_i \subseteq A \cup O$ be a set of agents and objects for every b_i, satisfying $b_i \in X_i$. The visibility assignment $[b_1 \leftarrow X_1, \ldots, b_k \leftarrow X_k]$ produces the TV model

$$M_{[b_1 \leftarrow X_1, \ldots, b_k \leftarrow X_k]} = \langle n+1, S \cup \{s_{new}\}, \tau', \kappa', \nu' \rangle$$

in which ***(i)*** τ' preserves the temporal position of states in S, making s_{new} the most recent (so $\tau'(s) := \tau(s)$ for $s \in S$, and $\tau'(s_{new}) := n+1$); ***(ii)*** κ' preserves the colouring assignment for states in S, with the colouring in s_{new} exactly as in s_{last} (so, for any $o \in O$, define $\kappa'_s(o) := \kappa_s(o)$ for $s \in S$, and $\kappa'_{s_{new}}(o) := \kappa_{s_{last}}(o)$); ***(iii)*** ν' is exactly as ν for states in S, with the new s_{new} taking the visibility of s_{last} for agents not mentioned by the assignment, and following the assignment for those agents it mentions (so, for any $a \in A$, define $\nu'_s(a) := \nu_s(a)$ for $s \in S$, with $\nu'_{s_{new}}(a) := \nu_{s_{last}}(a)$ when $a \notin \{b_1, \ldots, b_k\}$, and $\nu'_{s_{new}}(b_j) := X_j$ when $a = b_j$). ◀

The operations describe a change in the current situation; in this sense, they are analogous to model operations in *Dynamic Epistemic Logic* (*DEL*; [16,17]). Still, there is an important difference. Typically, *DEL* models describe only the current situation, so model operations return a structure representing also a single situation (the 'next' one). In contrast, while a TV model describes how the situation is at the current stage (the state $\tau^{-1}(n)$), it might also describe how the situation was in the past (the other states). Thus, while the operations add a state describing the situation the action produces, they also retain the states of the original model, hence keeping track of the past. In this sense, the TV setting can be understood as a 'dynamic temporal': an underlying temporal structure that can be *extended* by dynamic 'model change' operations. Other proposals using similar ideas include [18] (cf. [19,20]), which redefines the operation representing acts of (public and) private communication [11] to preserve previous stages, and [21], whose models 'remember' the initial epistemic situation.

A Formal Language. The language $\mathcal{L}$, for describing TV models, contains basic formulas expressing the (high-order) beliefs agents have about the colour of an object, and it is closed under both the standard Boolean operators as well as modalities for describing what will be the case after an action takes place.

Definition 2.4 (Language $\mathcal{L}$). Given A, O and $\{R_o\}_{o \in O}$, formulas ϕ of the language $\mathcal{L}$ are given by

$$\phi ::= B_{a_1} \cdots B_{a_k}(o \lhd c) \mid \neg\phi \mid \phi \wedge \phi \mid [\alpha]\,\phi \qquad \text{for } k \geqslant 1, \{a_1, \ldots, a_k\} \subseteq A, o \in O, c \in R_o$$

$$\alpha ::= p_1{:=}c_1, \ldots, p_i{:=}c_i \mid b_1 \leftarrow X_1, \ldots, b_j \leftarrow X_j \qquad \text{for } i \geqslant 1, \{p_1, \ldots, p_i\} \subseteq O, c_i \in R_{p_i},$$
$$j \geqslant 1, \{b_1, \ldots, b_j\} \subseteq A, X_i \subseteq A \cup P \text{ with } b_i \in X_i$$

Formulas of the form $B_{a_1} \cdots B_{a_k}(o \lhd c)$, called *mental attribution formulas*, are read as "*agent a_1 believes that ... that agent a_k believes that o has colour c*". Other Boolean connectives ($\vee, \rightarrow, \leftrightarrow$) are defined in the standard way. ◀

Formulas in $\mathcal{L}$ are evaluated in a TV model with respect its last state s_{last}, the fullest representation of the scenario available up that point. Nevertheless, as the definition shows, the truth-value of formulas is influenced by earlier states.

Definition 2.5 (Semantic interpretation). Let $M = \langle n, S, \tau, \kappa, \nu \rangle$ be a temporal visibility model. The following definitions will be useful.

- Take $\chi := B_{a_1} \cdots B_{a_k}(o \lhd c)$. Its *visibility condition* on $s \in S$, denoted by $\mathsf{vis}_\chi(s)$, and listing the requirements for χ to be evaluated at s (agent a_1 can see agent a_2, ..., agent a_{k-1} can see agent a_k, agent a_k can see object o), is given by

$$\mathsf{vis}_\chi(s) \quad \textit{iff}_{\,def} \quad a_2 \in \nu_s(a_1) \;\&\; \dots \;\&\; a_k \in \nu_s(a_{k-1}) \;\&\; o \in \nu_s(a_k).$$

- Take $s \in S$ and $t \leqslant \tau(s)$. The *t-predecessor* of s, denoted by $[s]_{-t}$, is the (unique) state appearing exactly t stages before s,[8] and it is formally defined as

$$[s]_{-t} := \tau^{-1}(\tau(s) - t)$$

For evaluating $\chi := B_{a_1} \cdots B_{a_k}(o \lhd c)$, the process starts from s_{last}, going 'back in time' one step at the time, looking for a state satisfying χ's visibility condition. If such s' is reached, χ's truth-value depends only on whether o has colour c at s'; otherwise, χ is false. Formally, and by using "⅋" for a natural-language disjunction (just as "&" stands for a natural-language conjunction), the satisfaction relation $\Vdash$ between a TV model and a mental attribution formula is given by

$$M \Vdash B_{a_1} \cdots B_{a_k}(o \lhd c) \quad \textit{iff}_{\,def} \quad \underset{i=0}{\overset{\tau(s_{last})-1}{⅋}} \left(\begin{array}{c} \overbrace{\underset{j=1}{\overset{i}{\&}} \text{not } \mathsf{vis}_{B_{a_1} \cdots B_{a_k}(o \lhd c)}([s_{last}]_{-(j-1)})}^{\mathsf{no-latter-vis}} \\ \& \\ \underbrace{\mathsf{vis}_{B_{a_1} \cdots B_{a_k}(o \lhd c)}([s_{last}]_{-i})}_{\mathsf{vis}} \;\&\; \underbrace{\kappa_{[s_{last}]_{-i}}(o) = c}_{\mathsf{col}} \end{array} \right)$$

Thus, $B_{a_1} \cdots B_{a_k}(o \lhd c)$ holds at M when there is a state (the quantification indicated by the main disjunction) in which the visibility condition is satisfied (the vis part), the object has the indicated colour (the col part), and there is no 'more recent' state satisfying the visibility condition (the no−latter−vis part).

Boolean operators are interpreted as usual. For 'action' modalities,

$$M \Vdash [\alpha]\,\phi \qquad \textit{iff}_{\,def} \qquad M_{[\alpha]} \Vdash \phi$$ ◀

[8] In particular, $[s]_{-0} = s$. Note also how $[s]_{-t}$ is undefined for $t > \tau(s)$.

Before an example of the framework at work, there are four points worthwhile to emphasise. ***(i)*** The semantic interpretation of $\chi := B_{a_1} \cdots B_{a_k}(o \lhd c)$ captures the discussed intuitive idea. On the one hand, if the visibility condition fails at every state, the formula is false (every disjunct fails in its vis part). On the other hand, if some states satisfy the visibility condition, let s' be the time-wise latest (i.e., $s' := \tau^{-1}(\max\{\tau(s) \mid \mathsf{vis}_\chi(s)\})$); then, $M \Vdash \chi$ iff $\kappa_{s'}(o) = c$. ***(ii)*** For the sake of simplicity, we assume that, when an agent a sees an agent b, and b sees an object o, then a in fact sees b *seeing* o, as it should be intuitively the case in order for a formula like $B_a\, B_b(o \lhd c)$ to be evaluated.[9] ***(iii)*** The term 'belief' here does not have the strong *EL* reading; it is rather understood as *"truth according to the agent's current information about what has happened so far"* (a form of *default reasoning* [24,25]: the agent assumes that things remain the way she saw them last). ***(iv)*** Attributions to oneself boil down to the col part of the interpretation, given the properties of ν, thus giving any agent full positive introspection.

Example 3. Recall the Sally-Anne Task, with its first two stages represented by the model M in Example 2. The story continues with Sally leaving the room, after which she can see neither Anne nor the marble anymore, and Anne can only see the marble. This is represented by an operation extending the model with a new state (s_3) in which both *Sa*'s and *An*'s visibility have changed, yielding the model $M_{[Sa \leftarrow \{Sa\}, An \leftarrow \{An, mar\}]} = M'$ below.

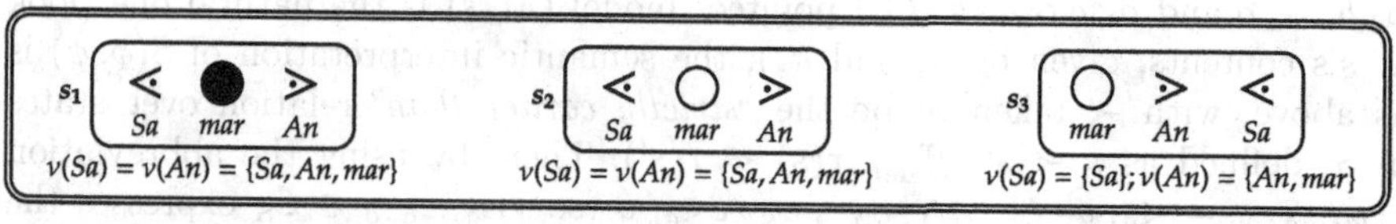

- Does Anne believe that the marble is white? Intuitively, the answer should be *"yes"*, and the system agrees: $M' \Vdash B_{An}(mar \lhd white)$ holds, as at s_{last} Anne sees the marble ($mar \in \nu_{s_3}(An)$), and the marble is indeed white ($\kappa_{s_3}(mar) = white$).
- Does Sally believe that the marble is white? The answer is *"yes"*, but for a different reason: $M' \Vdash B_{Sa}(mar \lhd white)$ holds because ***(i)*** although *Sa* cannot see *mar* now (at s_3), ***(ii)*** the last time she saw it (s_2), *mar* was white.
- Does Anne believe that Sally believes that the marble is white? The relevant state is the last time Anne saw Sally looking at the marble, i.e., s_2. Since *mar* is white at s_2, indeed $M' \Vdash B_{An}\, B_{Sa}(mar \lhd white)$.
- Finally, does Sally believe that Anne believes that the marble is white? As before, we can verify that $M' \Vdash B_{Sa}\, B_{An}(mar \lhd white)$. ◀

***TV* Models from a Modal Perspective.** Readers familiar with modal logic [26] will have noticed that a *TV* model is just a domain with a predecessor

[9] Notice that visibility of each agent is not 'common knowledge': knowledge relies on visibility, and an agent can see without being seen (Subsect. 3.1). Additionally, our simplifying assumption might be a problem for attributions under (semi-)private actions. Work of [22,23] can be especially relevant in that respect.

relation (more precisely, a finite linear temporal structure); thus, it can also be described by more standard modal languages. This will be made precise now, in order to make explicit what the semantic evaluation of mental attribution formulas boils down to. For simplicity, the focus will be $\mathcal{L}'$: the fragment of $\mathcal{L}$ that does not include the dynamic modalities $[p_1{:=}c_1, \ldots, p_i{:=}c_i]$ and $[b_1{\leftarrow}X_1, \ldots, b_j{\leftarrow}X_j]$.

A modal language for describing a *TV* model requires special atoms for agents' visibility and objects' colour. For the modalities, evaluating mental attribution formulas might require visiting previous states, so temporal operators are needed. A suitable one for expressing what mental attribution formulas encode is the *since* operator $\mathrm{S}(\phi, \psi)$ [27] (more precisely, its *strict* version, found also in, e.g., [28]), read as *"since ϕ was true, ψ has been the case"*.[10] Given a linear structure $\mathsf{M} = \langle W, \prec, V \rangle$ and $w \in W$, the formula is interpreted as follows.[11]

$(\mathsf{M}, w) \Vdash \mathrm{S}(\phi, \psi)$ *iff*$_{def}$ there is $u \in W$ with ***(i)*** $u \prec w$, ***(ii)*** $(\mathsf{M}, u) \Vdash \phi$, and ***(iii)*** $(\mathsf{M}, v) \Vdash \psi$ for every $v \in W$ such that $u \prec v \prec w$.

Thus, let $\mathcal{L}_\mathrm{S}$ be the modal language whose formulas are given by

$$\phi ::= \lhd_a b \mid \lhd_a o \mid o \lhd c \mid \neg\phi \mid \phi \wedge \phi \mid \mathrm{S}(\phi, \phi)$$

for $a, b \in A$, $o \in O$ and $c \in R_o$. The semantic interpretation of the atoms $\lhd_a b$, $\lhd_a o$ and $o \lhd c$ over a *TV* 'pointed' model (M, s) is the natural one (look at s's contents, given by ν_s and κ_s); the semantic interpretation of $\mathrm{S}(\phi, \psi)$ is as above, with $\prec$ taken to be the *"strictly earlier than"* relation over states in S, defined as $s \prec s'$ *iff*$_{def}$ $\tau(s) < \tau(s')$. Then, by using the abbreviation $\mathrm{vis}_{a_1 \cdots a_n o} := \lhd_{a_1} a_2 \wedge \cdots \wedge \lhd_{a_{k-1}} a_k \wedge \lhd_{a_k} o$ (so $\mathrm{vis}_{a_1 \cdots a_n o} \in \mathcal{L}_\mathrm{S}$ expresses the visibility condition of the formula $B_{a_1} \cdots B_{a_k}(o \lhd c)$), the translation $tr : \mathcal{L}' \to \mathcal{L}_\mathrm{S}$ is defined as

$$tr(B_{a_1} \cdots B_{a_k}(o \lhd c)) := (\mathrm{vis}_{a_1 \cdots a_n o} \wedge o \lhd c) \vee (\neg\, \mathrm{vis}_{a_1 \cdots a_n o} \wedge \mathrm{S}(\mathrm{vis}_{a_1 \cdots a_n o} \wedge o \lhd c, \neg\, \mathrm{vis}_{a_1 \cdots a_n o})),$$

$$tr(\neg\phi) := \neg tr(\phi), \qquad tr(\phi \wedge \psi) := tr(\phi) \wedge tr(\psi).$$

Then, $M \Vdash \phi$ iff $(M, s_{last}) \Vdash tr(\phi)$ holds for any *TV* model M and any $\phi \in \mathcal{L}'$. The crucial case, for mental attribution formulas, is apparent:

---

[10] Note: a single 'predecessor' modality is insufficient, as the number of back steps the recursive exploration requires is *a priori* unknown. A modality for its reflexive and transitive closure is still not enough: it takes care of the recursive search for a state satisfying the visibility condition, but on its own cannot indicate that every state up to that point should *not* satisfy it. More on the adequacy of *since* can be found in [27].

[11] Within propositional dynamic logic [29], and in the presence of the converse $\succ$, the *since* modality can be defined as $\mathrm{S}(\phi, \psi) := \langle(\succ; (?\phi \cup ?(\neg\phi \wedge \psi)))^+\rangle\, \phi$, with "?" indicating relational test, ";" indicating sequential composition, "$\cup$" indicating non-deterministic choice, and "+" indicating one or more iterations.

$tr(B_{a_1} \cdots B_{a_k}(o \lhd c))$ holds at s_{last} in M if and only if either the visibility condition holds and the object has the indicated colour ($\text{vis}_{a_1 \cdots a_n o} \wedge o \lhd c$), or else the visibility condition fails ($\neg\, \text{vis}_{a_1 \cdots a_n o}$) and there is a state in the past where both visibility and colour were satisfied, and since then visibility has failed ($\text{S}(\text{vis}_{a_1 \cdots a_n o} \wedge o \lhd c, \neg\, \text{vis}_{a_1 \cdots a_n o})$). This is exactly what the semantic interpretation of $B_{a_1} \cdots B_{a_k}(o \lhd c)$ in M requires.

Bisimulation. The translation *tr* provides an insight on the semantic clause for mental attribution formulas. Equally illuminating is a bisimulation for $\mathcal{L}'$.

Definition 2.6 (***TV*-BISIMULATION**)**.** Two *TV* models $M = \langle n, S, \tau, \kappa, \nu \rangle$ and $M' = \langle m, S', \tau', \kappa', \nu' \rangle$ (with s_{last} and s'_{last} their respective 'last' states) are said to be *TV-bisimilar* (notation: $M \leftrightarrows M'$) if and only if, for any mental attribution formula $\chi := B_{a_1} \cdots B_{a_k}(o \lhd c)$, **(I) Forth:** if there is $t \in S$ such that ***(i)*** $\text{vis}_\chi(t)$ holds, ***(ii)*** $\text{vis}_\chi(r)$ fails for every $r \in S$ with $\tau(t) < \tau(r) \leqslant \tau(s_{last})$, and ***(iii)*** $\kappa_t(o) = c$, then there is $t' \in S'$ such that ***(i)*** $\text{vis}_\chi(t')$ holds, ***(ii)*** $\text{vis}_\chi(r')$ fails for every $r' \in S$ with $\tau'(t') < \tau'(r') \leqslant \tau'(s'_{last})$, and ***(iii)*** $\kappa_{t'}(o) = c$. **(II) Back:** vice versa.

◀

It can be proved that, whenever M and M' are *TV*-bisimilar, both models satisfy the same $\mathcal{L}'$-formulas.[12] The colour of an object is relevant only if some agent can see it (so, no 'atom' clause is needed). Note also how two *TV* models satisfying the same $\mathcal{L}'$-formulas might differ in their cardinality, and also make the same formula true in different ways (e.g., $\neg\, B_a(o \lhd c)$ holds in M because, at s_{last}, agent a sees o having a colour other than c, but it holds in M' because, as far as M' is concerned, agent a has never seen o). Finally, notice how, although *TV*-bisimulation implies $\mathcal{L}'$-equivalence, it does not imply $\mathcal{L}$-equivalence. Take $A = \{a\}$ and $O = \{o\}$, with s_1 a state in which a sees o being white, and s_2 one in which a does not see o. Take M to be the model with only s_1, and M' to be the model with both s_1 and s_2. The models are *TV*-bisimilar, hence $\mathcal{L}'$-equivalent. Yet, they can be distinguished by the formula $[o{:=}black]\, B_a(o \lhd black)$ (true in M, false in M'): the different reasons why $\mathcal{L}'$-formulas are made true in bisimilar models become salient when actions enter the picture. For a bisimulation for $\mathcal{L}_\text{S}$, it is enough to consider the mutual satisfaction of atoms in bisimilar points, and suitable *Since* conditions, as the ones discussed in [30, p. 413].

3 On Modelling Mental Attribution Scenarios

The *TV* framework aims to model belief attributions in a more cognitively plausible way (compared with *EL*), revealing features thought of as crucial ingredients of social cognition. Let's justify these claims.

---

12 Since $\mathcal{L}'$-formulas are evaluated with respect to a *TV* model's last state, it is enough for a bisimulation to establish a connection between those states, as the definition does.

Informational Economy. On the one hand, a state in a *TV* model contains a bare informational 'minimum': only basic facts regarding objects and agents' visibility. The operations on the model also induce 'minimal' changes, in accordance to the criterion of informational economy in belief revision [31]. On the other hand, the non-standard semantic clause for belief is complex, as the state representing the current situation might not have all information necessary to evaluate a complex belief attribution, and thus the information at other (previous) stages might be needed. A 'backtracking' process might be difficult and time-consuming, depending on how many different states an agent needs to 'remember', and our clause is sensitive to this observation, unlike the usual modal interpretations. The level of complexity that one finds on the *TV* framework for both representing a situation (low) and evaluating mental attributions (high) can be contrasted with what *EL* does, as discussed in Sect. 1.

Perspective Shifting. Another important feature, identified in analyses of ToM and formalisations of False-Belief Tasks (*FBT*s), is *perspective shifting* [32]. Successful performance in the tasks (i.e., making correct attributions) requires a perspective shift: stepping into the shoes of another agent.[13] Asking for the visibility condition ensures precisely that agents change perspectives, even if that means having to recall earlier stages. Making multiple shifts, e.g. in complex high-order attributions, may be difficult compared to plainly attributing one's own belief to others, capturing why agents might fail in the tasks.

Principle of Inertia. A further crucial notion is the *principle of inertia* [6,33,34]: an agent's beliefs are preserved unless there is reason to the contrary. In our case, reason to the contrary amounts to the satisfaction of visibility; if this is not satisfied in the state of evaluation, then, essentially, the agent maintains beliefs formed in earlier stages, where necessary information was available.

Dual Process Theories of Reasoning. Besides ToM, the *TV* setting is in agreement with the literature supporting the *dual process theories of reasoning* [35–37]. According to them, there are two systems underlying human reasoning. System 1 (the *fast* mode) is quick, unconscious and automatic, often governed by habit, biases and heuristics developed in the course of evolution. System 2 (the *slow* mode) is gradual, deliberate and rule-based, and requires cognitive effort. System 1 is at play most of the time, constructing our idea of the world with elementary cues and avoiding cognitive overload. When rule-based calculations become necessary, e.g. in face of a demanding task, System 2 takes over, building on inputs of System 1 to slowly produce an output in a step-wise fashion.

We argue that agents' higher order reasoning roughly follows this pattern. System 1 keeps track only of a bare-minimum of information (basic facts), without overloading memory with information that can be later inferred. Whenever a task requires more than what is stored (as higher-order attributions), System 2 takes over, using the inputs of System 1. This is precisely the pattern of our

[13] In fact, unsuccessful performance, e.g. of autistic children, is often connected with a failure in perspective shifting, resulting in the subject reporting her own beliefs [6,33].

semantics, with our models and updates encoding only basic facts. Whenever a demanding task appears, such as the evaluation of a mental attribution, our agents follow the cognitively hard calculations of our semantic clause.[14] On the basis of elementary facts regarding whom/what they observed, they test certain conditions and trace back earlier states. It is only after this slow and effortful process that they can determine whether a higher-order attribution holds.

3.1 Detailed Examples

False-Belief Tasks use stories to test the ability to attribute mental states to others. In what follows, we provide formal representations of some of these storylines, to the level of abstraction allowed by our framework's constructions.

Example 4 (FIRST-ORDER *FBT*: THE *Sally-Anne* (*SA*) TASK). The full storyline (Example 1) can be represented within the *TV* framework, modulo minor changes, as already hinted at. *(1)* Sally and Anne are in a room, with Sally holding the marble (the model with only state s_1 in Example 2). *(2)* Sally puts the marble into the basket (the full model in Example 2). *(3)* Sally leaves the room (the model in Example 3). *(4)* Anne transfers the marble to the box (the model in Fig. 1). The task's last step, Sally coming back to the room, prepares the audience for the crucial question: *"where does Sally believe the marble is?"*. The action changes Sally's visibility (she can see Anne now), but it does not change the crucial fact that she cannot see the marble. Thus, it is not relevant for our purposes.

So, which are Anne's and Sally's final high-order beliefs? According to the framework, with M the model in Fig. 1 (top): $M \Vdash B_{Sa}(mar \lhd white) \wedge B_{An}(mar \lhd green)$, and $M \Vdash B_{Sa} B_{An}(mar \lhd white) \wedge B_{An} B_{Sa}(mar \lhd white)$. ◀

Example 5 (SECOND-ORDER *FBT*: THE *chocolate* (*C*) TASK). Adapted from [39], the task is as follows. *(1)* Mary and John are in a room, with a chocolate bar in the room's table. *(2)* John puts the chocolate into the drawer, then *(3)* leaving the room. *(4)* Mary transfers the chocolate to the box. *(5)* John peeks into the room, without Mary noticing, and sees the chocolate in the box.

The *TV* modelling works stepwise, with the initial situation represented by s_1 (*black* indicates the chocolate is on the table), and each subsequent action adding a state. By putting the chocolate into the drawer (*white*), John produces s_2, and by leaving the room he produces s_3. Mary creates s_4 when she moves the chocolate to the box (*green*), and finally s_5 emerges when John peeks into the room. In the final model, displayed in Fig. 1 (bottom), we have the following: *(i)* $M \Vdash B_{Ma}(cho \lhd green) \wedge B_{Jo}(cho \lhd green)$,

[14] Although it is always possible to evaluate attributions of any length (like in possible-worlds semantics), our semantic clause offers a mechanism to account for human reasoning limitations, indicated by empirical research, e.g. on working memory [38]. It allows us to trace how many states need to be held in working memory, and therefore explain why attribution-making might fail from some point on.

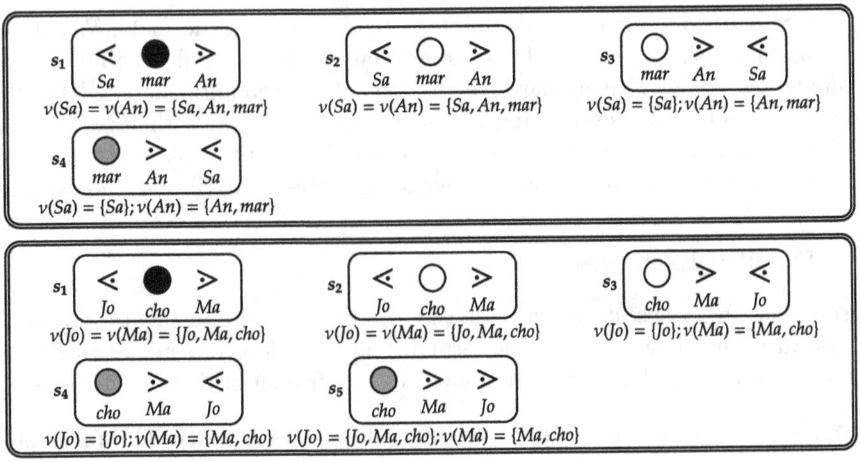

Fig. 1. *TV* representations of Sally-Anne Task (top) and Chocolate Task (bottom). (Color figure online)

(ii) $M \Vdash B_{Ma}B_{Jo}(cho\lhd white) \land B_{Jo}B_{Ma}(cho\lhd green)$, and ***(iii)*** $M \Vdash B_{Ma}B_{Jo}B_{Ma}(cho\lhd white) \land B_{Jo}B_{Ma}B_{Jo}(cho\lhd white)$. ◀

Other *FBT*s (the *Ice Cream Task* [40], the *Puppy Task* [41] and the *Bake-sale* task [42]) can be also represented in the *TV* framework, their crucial ToM features still preserved. Still, some sources of change in zero- or higher- order information in such dynamic scenarios might not be captured by our operations. While conceptually similar examples can fit into our setting, up to some level of abstraction, different operations might be required for other scenarios (Sect. 4).

3.2 Comparison with Other Proposals for Mental Attributions

Through a relational 'preference' framework for modelling different degrees of belief, [43] studies three kinds of agents (including agents on the autism spectrum), each endowed with specific "properties" as higher-order reasoners. Our attempt does not focus on agents with specific strategies when evaluating belief attributions, working instead on *any* agent's reasoning behind such process.

In [6], the authors provide a non-monotonic, closed-world reasoning formalization of first-order *FBT*s, implemented within logical programming. They use *event calculus*, with belief treated as a predicate, and rely on the principle of inertia. While we design a different formalism, we still account for these features without restricting ourselves to specific types of agents or orders of beliefs.

Another interesting logical formalization of *FBT*s is given in [32–34]. These papers use a proof-theoretic Hybrid Logic system for identifying perspective shifts, while using inertia. The straightforward difference is that our approach is rather semantic, with models keeping track of the actions involved, and in which the evaluation of mental attributions reflects their cognitive difficulty.

The framework of [44] uses *EL*-beliefs plus special atoms indicating the location of objects and the agents' visibility, then representing changes in the situation as action-model-based acts of (private) communication that rely on agents' visibility.[15] The differences between our proposal and [44] have been discussed: the contrast between complex models that simplify answering mental attribution questions (*EL*) and simple states that require a complex process for deciding high-order belief issues (here). The representation of actions also differs: while [44] uses (a variation of) the heavy action models machinery (for private communication), the actions of visibility and colour change presented here simply modify atomic information. Finally, [44] also proposes two criteria of success in formalizing *FBT*s: ***(i)*** *robustness* (being able to deal with as many *FBT*s as possible, with no strict limit on the order of belief attribution), and ***(ii)*** *faithfulness* (each action of the story should correspond to an action in the formalism in a natural way). The *TV* framework fulfils these requirements: it is robust enough to deal with different *FBT*s (see Subsect. 3.1 and the discussion therein), and the actions in the stories have a straightforward representation.

4 Summary and Ongoing/Future Work

This paper has introduced a temporal framework suitable for capturing 'real' agents' mental state attributions. Its most important feature is the contrast between a 'simple' semantic model (encoding only objects' colours and agents' visibility) and a 'complex' clause for interpreting mental state attributions (essentially a temporal *"since"* operator). We have argued for its adequacy towards representing important features of social cognition, as informational economy, perspective shifting, inertia, and connections with dual process theories, with these points exemplified through the modelling of common *FBT*s.

This project presents several lines for further research. On the technical side, there are still aspects of the logical setting to be investigated (e.g., axiomatisation). Equally interesting is the exploration of extensions for modelling more empirical findings. The main points made above on the adequacy of the framework make for a suitable basis for such extensions. Here are two possibilities.

A Perspective Function. The setting can be fine-tuned to capture special types of high-order reasoning (see case-studies of [16]). For example, autistic children tend to fail the *FBT*s because they attribute their own beliefs to others [5]. This and other similar situations can be accommodated through the introduction of a *perspective* function $\pi : A \to (A \to A)$ (with $\pi_a(b) = c$ understood as *"agent a considers agent b to have the perspective of agent c"*), which then can be used to define an appropriate variation of the visibility condition. In this way, an autistic agent a would be one for which $\pi_a(x) = a$ for any $x \in A$, essentially relying only on her own information, and thus attributing her own belief to others.

[15] For example, the act through which, in the absence of Sally, Anne moves the marble from the basket to the box, is understood as a private announcement through which only Anne is informed about the marble's new location.

Different States for Different Agents at the Same Stage. Another extension is towards capturing scenarios involving communicative actions, including lying and spread of misinformation (e.g., the Puppy Task, the Bake Sale Task) and other manifestations of social cognition (e.g., negotiations, games). With them, it makes sense to include different states for different agents at the same stage, each one of them representing the (potentially different) information different agents might have about the situation at the same stage.

References

1. Premack, D., Woodruff, G.: Does the chimpanzee have a theory of mind? Behav. Brain Sci. **1**(4), 515–526 (1978)
2. Carruthers, P., Smith, P.K.: Theories of Theories of Mind. Cambridge University Press, Cambridge (1996)
3. Wimmer, H., Perner, J.: Beliefs about beliefs: representation and constraining function of wrong beliefs in young children's understanding of deception. Cognition **13**(1), 103–128 (1983)
4. Wellman, H.M.: From desires to beliefs: acquisition of a theory of mind. In: Natural Theories of Mind: Evolution, Development and Simulation of Everyday Mindreading, pp. 19–38. Basil Blackwell, Cambridge (1991)
5. Baron-Cohen, S., Leslie, A.M., Frith, U.: Does the autistic child have a "theory of mind"? Cognition **21**(1), 37–46 (1985)
6. Stenning, K., van Lambalgen, M.: Human Reasoning and Cognitive Science. MIT Press, Cambridge (2008)
7. Setoh, P., Scott, R.M., Baillargeon, R.: Two-and-a-half-year-olds succeed at a traditional false-belief task with reduced processing demands. Proc. Natl. Acad. Sci. **113**(47), 13360–13365 (2016)
8. Hintikka, J.: Knowledge and Belief. Cornell University Press, Ithaca (1962)
9. Fagin, R., Halpern, J.Y., Moses, Y., Vardi, M.Y.: Reasoning About Knowledge. The MIT Press, Cambridge (1995)
10. Stalnaker, R.: The problem of logical omniscience, I. Synthese **89**(3), 425–440 (1991)
11. Baltag, A., Moss, L.S., Solecki, S.: The logic of public announcements and common knowledge and private suspicions. In: Gilboa, I. (ed.) Proceedings of the 7th Conference on Theoretical Aspects of Rationality and Knowledge (TARK 1998), Evanston, IL, USA, 22–24 July 1998, pp. 43–56. Morgan Kaufmann (1998)
12. Frege, G.: Grundlagen der Arithmetik. Wilhelm Koebner, Breslau (1884)
13. Pelletier, F.J., Elio, R., Hanson, P.: Is logic all in our heads? From naturalism to psychologism. Stud. Logica **88**(1), 3–66 (2008)
14. Verbrugge, R.: Logic and social cognition. J. Philos. Logic **38**(6), 649–680 (2009)
15. van Benthem, J.: Logic and reasoning: do the facts matter? Stud. Logica **88**(1), 67–84 (2008)
16. van Ditmarsch, H., van der Hoek, W., Kooi, B.: Dynamic Epistemic Logic. Synthese Library Series, vol. 337. Springer, Dordrecht (2008). https://doi.org/10.1007/978-1-4020-5839-4
17. van Benthem, J.: Logical Dynamics of Information and Interaction. Cambridge University Press, New York (2011)

18. Yap, A.: Dynamic epistemic logic and temporal modality. In: Girard, P., Roy, O., Marion, M. (eds.) Dynamic Formal Epistemology. Synthese Library (Studies in Epistemology, Logic, Methodology, and Philosophy of Science), vol. 351, pp. 33–50. Springer, Dordrecht (2011). https://doi.org/10.1007/978-94-007-0074-1_3
19. Sack, J.: Temporal languages for epistemic programs. J. Logic Lang. Inf. **17**(2), 183–216 (2008)
20. Renne, B., Sack, J., Yap, A.: Logics of temporal-epistemic actions. Synthese **193**(3), 813–849 (2016)
21. Baltag, A., Özgün, A., Vargas Sandoval, A.L.: APAL with memory is better. In: Moss, L.S., de Queiroz, R., Martinez, M. (eds.) WoLLIC 2018. LNCS, vol. 10944, pp. 106–129. Springer, Heidelberg (2018). https://doi.org/10.1007/978-3-662-57669-4_6
22. Gasquet, O., Goranko, V., Schwarzentruber, F.: Big brother logic: visual-epistemic reasoning in stationary multi-agent systems. Auton. Agents Multi Agent Syst. **30**(5), 793–825 (2016)
23. Charrier, T., Herzig, A., Lorini, E., Maffre, F., Schwarzentruber, F.: Building epistemic logic from observations and public announcements. In: Proceedings of the Fifteenth International Conference on Principles of Knowledge Representation and Reasoning, KR 2016, pp. 268–277. AAAI Press (2016)
24. Reiter, R.: A logic for default reasoning. Artif. Intell. **13**(1–2), 81–132 (1980)
25. Ben-David, S., Ben-Eliyahu-Zohary, R.: A modal logic for subjective default reasoning. Artif. Intell. **116**(1–2), 217–236 (2000)
26. Blackburn, P., de Rijke, M., Venema, Y.: Modal logic. Cambridge University Press, Cambridge (2001)
27. Kamp, H.: Tense logic and the theory of linear order. Ph.D. thesis, University of California (1968)
28. Burgess, J.P.: Axioms for tense logic. I. "since" and "until". Notre Dame J. Formal Logic **23**(4), 367–374 (1982)
29. Harel, D., Kozen, D., Tiuryn, J.: Dynamic Logic. MIT Press, Cambridge (2000)
30. Kurtonina, N., De Rijke, M.: Bisimulations for temporal logic. J. Logic Lang. Inf. **6**(4), 403–425 (1997)
31. Gärdenfors, P.: Knowledge in Flux. Modelling the Dynamics of Epistemic States. MIT Press, Cambridge (1988)
32. Braüner, T.: Hybrid-logical reasoning in the smarties and Sally-Anne tasks. J. Logic Lang. Inf. **23**(4), 415–439 (2014)
33. Braüner, T.: Hybrid-logical reasoning in the smarties and Sally-Anne tasks: what goes wrong when incorrect responses are given? In: Proceedings of the 37th Annual Meeting of the Cognitive Science Society, Pasadena, California, USA, pp. 273–278. Cognitive Science Society (2015)
34. Braüner, T., Blackburn, P., Polyanskaya, I.: Second-order false-belief tasks: analysis and formalization. In: Väänänen, J., Hirvonen, Å., de Queiroz, R. (eds.) WoLLIC 2016. LNCS, vol. 9803, pp. 125–144. Springer, Heidelberg (2016). https://doi.org/10.1007/978-3-662-52921-8_9
35. Kahneman, D.: Thinking, Fast and Slow. Farrar, Straus and Giroux, New York (2011)
36. Evans, J.: Dual process theories. In: Ball, L., Thompson, V. (eds.) The Routledge International Handbook of Thinking and Reasoning, pp. 151–164. Routledge, London (2018)
37. Stanovich, K.E., West, R.F.: Individual differences in reasoning: implications for the rationality debate? Behav. Brain Sci. **23**(5), 645–665 (2000)

38. Cowan, N.: The magical number 4 in short-term memory: a reconsideration of mental storage capacity. Behav. Brain Sci. **24**, 87–114 (2001)
39. Flobbe, L., Verbrugge, R., Hendriks, P., Krämer, I.: Children's application of theory of mind in reasoning and language. J. Logic Lang. Inf. **17**(4), 417–442 (2008)
40. Perner, J., Wimmer, H.: "John Thinks that Mary Thinks that..." attribution of second-order beliefs by 5- to 10-year-old children. J. Exp. Child Psychol. **39**(3), 437–471 (1985)
41. Sullivan, K., Zaitchik, D., Tager-Flusberg, H.: Preschoolers can attribute second-order beliefs. Dev. Psychol. **30**, 395–402 (1994)
42. Hollebrandse, B., van Hout, A., Hendriks, P.: Children's first and second-order false-belief reasoning in a verbal and a low-verbal task. Synthese **191**(3), 321–333 (2014)
43. Ditmarsch, H.V., Labuschagne, W.: My beliefs about your beliefs: a case study in theory of mind and epistemic logic. Synthese **155**(2), 191–209 (2007)
44. Bolander, T.: Seeing is believing: formalising false-belief tasks in dynamic epistemic logic. In: van Ditmarsch, H., Sandu, G. (eds.) Jaakko Hintikka on Knowledge and Game-Theoretical Semantics. OCL, vol. 12, pp. 207–236. Springer, Cham (2018). https://doi.org/10.1007/978-3-319-62864-6_8

A Two-Layer Partition Awareness Structure

Pengfei Song and Wei Xiong(✉)

Department of Philosophy, Institute of Logic and Cognition,
Sun Yat-sen University, Guangzhou, China
songpf@mail2.sysu.edu.cn, hssxwei@mail.sysu.edu.cn

Abstract. With the idea of representing knowledge and awareness by binary relations, we establish a two-layer partition epistemic structure, a syntax-free model, to formalize a non-trivial notion of awareness. We provide two axiomatic systems with awareness modality on our structure and prove soundness and completeness for them. Moreover, we show that our structure and the syntactic awareness structure are equivalent under certain conditions, which helps us understand the connections between knowledge and awareness.

Keywords: Awareness · Epistemic logic · Knowledge · Partition

1 Introduction

The standard models of epistemic logic suffer from the well-known *logical omniscience* problem that agents know all tautologies and all the logical consequences of their knowledge [13]. Indeed, logic omniscience is unrealistic to the extent that agents are assumed to be resource-bounded and of limited reasoning abilities. Consequently, standard models with logical omniscience problem are not appropriate to formalize knowledge in such cases. To avoid this problem, a lot of effort has been devoted in the last few decades. One of the valuable approaches is to introduce a formal notion of awareness by which an agent's knowledge representation and reasoning ability are restricted within the scope of his awareness. This convention dates back to Levesque [14], where knowledge is first classified into *explicit knowledge* and *implicit knowledge*.

There are mainly two methods, the *syntactic* approach and the *set-theoretical* approach, for formalizing awareness in the literature. The former adds the *awareness correspondence function* to the standard models, so as to associate with each agent in each state a set of formulas representing what the agent is aware of [6]. Thus, an agent explicitly knowing a formula is supposed to satisfy two conditions: (a) the formula is true in every accessible state, regarded as implicit knowledge; (b) the formula is included in the awareness set of the current state. As such it proposes a model that is not syntax-free. Halpern [7] gives an axiomatization without the implicit knowledge modal operator by interpreting the explicit knowledge directly into semantics. On the other hand, the main idea

P. Blackburn et al. (Eds.): LORI 2019, LNCS 11813, pp. 313–325, 2019.
https://doi.org/10.1007/978-3-662-60292-8_23

of the set-theoretical approach is that awareness can be defined by knowledge. More specifically, we say that an agent is aware of something if and only if he knows it or he knows that he doesn't know it [15]. Yet it has been shown that a set-theoretical awareness operator satisfying some basic axioms must be trivial in a standard state-space model [2]. To remove this defect, Modica and Rustichini [16] propose a partitional information structure that associates with each state a sub-language generated by a subset of the primitive propositions. Subsequently, Heifetz *et al.* [11,12] extend it to a multi-agent setting. It has however been shown that the model given by Modica and Rustichini can be viewed as a special case of Fagin and Halpern's model [7], and an analogous equivalence result has also been proved in the multi-agent setting [8]. Recently, researchers try to combine two approaches for enlarging the expressivity of epistemic logic involving awareness [3–5,9,10].

One common ground of the approaches is that they all treat awareness as a distinctive notion to the traditional relational semantics. In this paper, we formalize knowledge and awareness merely by means of states with full vocabulary and binary relations, and construct a two-layer partition structure consisting of knowledge partitions and awareness partitions. As a consequence, our structure reveals some connections between knowledge and awareness and is set-theoretical as Aumann's seminal theory [1].

The remainder of this paper is organized as follows. In Sect. 2 we present the basic idea of the paper, and provide some basic definitions and a semantics for the language formalizing awareness. Section 3 proves an equivalence result to establish a connection between our framework and the syntactic approach. Section 4 offers two axiomatic systems for knowledge, and provides a sound and complete axiomatization for each one. Section 5 concludes the paper.

2 Basic Idea and Semantics

Before presenting our basic idea, let us first consider the following example. Suppose that my 5-year-old nephew and I find a mushroom in a forest. Both of us do not know whether it is poisonous or not; accordingly we lack knowledge about its toxicity. Nevertheless, I know that the mushroom could be poisonous. By contrast, my nephew is so young that he might not have a conception about poisonousness in his mind at present. In other words, he is unaware of that a mushroom could be poisonous or nontoxic. The two different epistemic structures for capturing our knowledge and awareness can be essentially depicted as Fig. 1.

The left of Fig. 1 shows my nephew's epistemic status. The two states are both included in a single knowledge partition and awareness partition. The structure underlines that my nephew does not know whether the mushroom is poisonous or not and is not aware of such states. Similarly, the knowledge partition in the right of Fig. 1, my epistemic structure, includes the two states, showing that we share our knowledge. It is important to note, however, that there are two awareness partitions in my epistemic structure. As such, it indicates that, in contrast with my nephew, I can be aware of the two states. Although I cannot

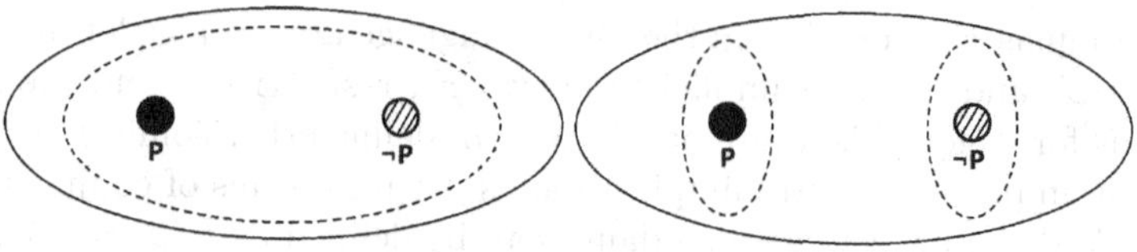

Fig. 1. The solid dot denotes the state at which p is true, where p stands for "the mushroom is poisonous". And the barred dot indicates the state at which $\neg p$ is true, where $\neg p$ stands for "it is not the case that the mushroom is poisonous". The solid ovals indicate our knowledge partitions, while the dotted ovals inside represent our awareness partitions.

discriminate between the two states, I know that p and $\neg p$ fall into different awareness partitions.

The above example inspires us to construct an epistemic structure involving two partitions.

Throughout the contribution, given are a nonempty set $\mathsf{Ag} = \{1, ..., n\}$ of agents and a nonempty set Φ of primitive propositions.

Definition 1. *A two-layer partition awareness frame is a tuple*

$$\mathcal{F} = (\mathsf{S}, (R_i)_{i \in \mathsf{Ag}}, (T_i)_{i \in \mathsf{Ag}}), \textit{where}$$

- S *is a nonempty set of states;*
- *For each agent* $i \in \mathsf{Ag}, \mathsf{R}_i \subseteq \mathsf{S} \times \mathsf{S}$ *is an equivalence relation representing the accessibility relation of agent* i. *We say "agent* i *considers the state* t *possible at state* s *according to his knowledge", if* $(s, t) \in \mathsf{R}_i$;
- *For each agent* $i \in \mathsf{Ag}, \mathsf{T}_i \subseteq \mathsf{S} \times \mathsf{S}$ *is an equivalence relation representing the accessibility relation of agent* i *according to his awareness, and* $\mathsf{T}_i \subseteq \mathsf{R}_i$.

In our awareness frame, each agent i's knowledge and awareness can be represented by the equivalence relations R_i and T_i respectively. Further, we can obtain partitions of the set S under these equivalence relations, called knowledge partitions and awareness partitions, respectively. The requirement $\mathsf{T}_i \subseteq \mathsf{R}_i$ ensures the partitions formed by T_i are finer than or equal to those formed by R_i, and any equivalence class generated by T_i is included in an equivalence class generated by R_i.

With such a frame, we can define a model, called a two-layer partition awareness model, as a pair $\mathcal{M} = (\mathcal{F}, \mathsf{V})$, where $\mathcal{F}$ is a two-layer partition awareness frame and V is the valuation function: $\mathsf{S} \times \Phi \rightarrow \{\mathsf{true}, \mathsf{false}\}$ assigning to each state and primitive proposition a truth value.

Definition 2. *Formula* $\varphi \in \mathcal{L}^{KLA}\Phi$ *is defined as follows, where* $p \in \Phi$ *and* $i \in \mathsf{Ag}$,

$$\varphi ::= \top \mid p \mid \neg\varphi \mid \varphi \wedge \psi \mid K_i\varphi \mid L_i\varphi \mid A_i\varphi$$

The formulas $K_i\varphi$ and $L_i\varphi$ in the above language are read as "agent i explicitly knows φ" and "agent i implicitly knows φ", respectively, and the formula $A_i\varphi$ stands for "agent i is aware of φ". We can define other connectives such as disjunction, implication, and biimplication as usual in terms of conjunction and negation. And the unawareness modality can be defined as $U_i\varphi := \neg A_i\varphi$.

We write $\mathsf{Pr}(\varphi)$ to indicate the set of primitive propositions appearing in φ. It then can be defined inductively as follows.

- $\mathsf{Pr}(\top) := \emptyset$;
- $\mathsf{Pr}(p) := p$, for all $p \in \Phi$;
- $\mathsf{Pr}(\neg\varphi) := \mathsf{Pr}(\varphi)$;
- $\mathsf{Pr}(\varphi \wedge \psi) := \mathsf{Pr}(\varphi) \cup \mathsf{Pr}(\psi)$;
- $\mathsf{Pr}(K_i\varphi) := \mathsf{Pr}(\varphi)$;
- $\mathsf{Pr}(L_i\varphi) := \mathsf{Pr}(\varphi)$;
- $\mathsf{Pr}(A_i\varphi) := \mathsf{Pr}(\varphi)$.

Definition 3. *Given a two-layer partition awareness model $\mathcal{M} = (\mathcal{F}, \mathsf{V})$, a formula ϕ being true in $\mathcal{M}$ at state s is defined inductively as follows (the subscript "TP" means two-layer partition models).*

- $(\mathcal{M}, s) \models_{\text{TP}} \top$ *for all* $s \in \mathsf{S}$;
- $(\mathcal{M}, s) \models_{\text{TP}} p$ *iff* $\mathsf{V}(s, p) = \mathsf{true}$ *for* $p \in \Phi$;
- $(\mathcal{M}, s) \models_{\text{TP}} \varphi \wedge \psi$ *iff* $(\mathcal{M}, s) \models_{\text{TP}} \varphi$ *and* $(\mathcal{M}, s) \models_{\text{TP}} \psi$;
- $(\mathcal{M}, s) \models_{\text{TP}} \neg\varphi$ *iff* $(\mathcal{M}, s) \not\models_{\text{TP}} \varphi$;
- $(\mathcal{M}, s) \models_{\text{TP}} A_i\varphi$ *iff for all* $p \in \mathsf{Pr}(\varphi)$, *and all* t *and* u *with* $(s, t) \in \mathsf{R}_i$ *and* $(t, u) \in \mathsf{T}_i$, *we have* $(\mathcal{M}, t) \models_{\text{TP}} p$ *iff* $(\mathcal{M}, u) \models_{\text{TP}} p$;
- $(\mathcal{M}, s) \models_{\text{TP}} L_i\varphi$ *iff* $(\mathcal{M}, t) \models_{\text{TP}} \varphi$ *for all* t *with* $(s, t) \in \mathsf{R}_i$;
- $(\mathcal{M}, s) \models_{\text{TP}} K_i\varphi$ *iff* $(\mathcal{M}, s) \models_{\text{TP}} A_i\varphi$ *and* $(\mathcal{M}, t) \models_{\text{TP}} \varphi$, *for all* t *with* $(s, t) \in \mathsf{R}_i$.

$A_i\varphi$ being true is interpreted in our model as that all primitive propositions occurring in φ having the uniform value in any awareness partition of agent i overlapping the possibility set of agent i. As is shown in Fig. 1, the finer awareness partitions are, the more conceptions an agent is possibly aware of.

3 A Comparison

This section provides a comparison between our model and the one proposed by Fagin and Halpern [6].

Definition 4. *Let $\mathcal{M} = (\mathsf{S}, \mathsf{R}_i, \mathsf{T}_i, \mathsf{V})$ be a two-layer partition awareness model. Given a state s, a formula φ is awareness independent to ψ at the state s in the model $\mathcal{M}$ for agent i (φ is i-s awareness independent to ψ for short) if there is a primitive proposition $p \in \mathsf{Pr}(\varphi)$ and a model $\mathcal{M}' = (\mathsf{S}, \mathsf{R}_i, \mathsf{T}'_i, \mathsf{V})$ such that $(\mathcal{M}', s) \models_{\text{TP}} \neg A_i p \wedge A_i\psi$.*

Given the possibility set of an agent at state s, a distinctive feature of our structure is that awareness of a formula may result in awareness of another formula, even if there is no common primitive proposition constituting the two formulas. *Awareness independence* in Definition 4 means that awareness of a formula do not cause awareness of another formula necessarily. The following three lemmas are apparent results from Definition 4.

Lemma 1. *Let* $\mathcal{M} = (\mathsf{S}, \mathsf{R}_i, \mathsf{T}_i, \mathsf{V})$ *be a two-layer partition awareness model. Given a state* s, φ *is not* i-s *awareness independent to* ψ *if for any* T'_i *in the model* $\mathcal{M}' = (\mathsf{S}, \mathsf{R}_i, \mathsf{T}'_i, \mathsf{V})$ *and any* $p \in \mathsf{Pr}(\varphi)$, $(\mathcal{M}', s) \models_{\mathtt{TP}} A_i\psi \rightarrow A_i p$.

Proof. It is immediate from Definition 4.

Lemma 2. *Let* $\mathcal{M} = (\mathsf{S}, \mathsf{R}_i, \mathsf{T}_i, \mathsf{V})$ *be a two-layer partition awareness model. If* $\mathsf{Pr}(\varphi) \subseteq \mathsf{Pr}(\psi)$, *then* φ *is not* i-s *awareness independent to* ψ.

Proof. It is straightforward by Definition 4.

Lemma 3. *Let* $\mathcal{M} = (\mathsf{S}, \mathsf{R}_i, \mathsf{T}_i, \mathsf{V})$ *be a two-layer partition awareness model. Given agent* i *at state* s, *if a primitive proposition* p *has uniform value in every* i*-accessible state, then the formulas constituted only by* p *are not* i-s *awareness independent to any propositions.*

Proof. Since p has the same value in every i-accessible state, it is impossible that p and $\neg p$ are both included in any awareness partition. Consequently, $A_i\varphi$ is always true if φ is constituted only by p. According to Definition 4, φ is not i-s awareness independent to any proposition.

Lemma 3 establishes that agent i has to be aware of p if he implicitly knows p.

Definition 5. *A model* $\mathcal{M} = (\mathsf{S}, \mathsf{R}_i, \mathsf{T}_i, \mathsf{V})$ *is* i-s *awareness free if for any formulas* φ *and* ψ *with* $\mathsf{Pr}(\varphi) \nsubseteq \mathsf{Pr}(\psi)$, φ *is* i-s *awareness independent to* ψ.

Definition 5 provides a description of the situation when awareness of any formula does not necessarily result in awareness of another formula. The following two lemmas are properties related to Definition 5.

Lemma 4. *If a model* $\mathcal{M} = (\mathsf{S}, \mathsf{R}_i, \mathsf{T}_i, \mathsf{V})$ *is* i-s *awareness free, then there is no primitive proposition that has the same value in every* i*-accessible state.*

Proof. It is an immediate consequence of Lemma 3.

Lemma 5. *A model* $\mathcal{M} = (\mathsf{S}, \mathsf{R}_i, \mathsf{T}_i, \mathsf{V})$ *is* i-s *awareness free iff for any primitive proposition* p, p *is* i-s *awareness independent to* φ $(p \notin \mathsf{Pr}(\varphi))$.

Proof. The direction from left to right is immediate. To prove the other direction, assume that for any primitive proposition p, p is i-s awareness independent to φ, $p \notin \mathsf{Pr}(\varphi)$. Suppose again that ψ and η satisfy $\mathsf{Pr}(\psi) \nsubseteq \mathsf{Pr}(\eta)$ and ψ is not i-s awareness independent to η. Since $\mathsf{Pr}(\psi) \nsubseteq \mathsf{Pr}(\eta)$, there is a primitive proposition q satisfying $q \in \mathsf{Pr}(\eta)$ and $q \notin \mathsf{Pr}(\psi)$. By the assumption and Definition 4, there is a model $\mathcal{M}' = (\mathsf{S}, \mathsf{R}_i, \mathsf{T}'_i, \mathsf{V})$ in which we have $(\mathcal{M}', s) \models_{\mathtt{TP}} \neg A_i q \wedge A_i\eta$. By $q \in \mathsf{Pr}(\psi)$ and $(\mathcal{M}', s) \models_{\mathtt{TP}} \neg A_i q$, we get $(\mathcal{M}', s) \models_{\mathtt{TP}} \neg A_i\psi \wedge A_i\eta$. It contradicts the assumption that ψ is not i-s awareness independent to η.

Lemma 5 states that when judging whether a model is i-s awareness free, we only need to check the i-s awareness independence of primitive propositions.

Lemma 6. *Given a two-layer partition awareness model* $\mathcal{M} = (\mathsf{S}, \mathsf{R}_i, \mathsf{T}_i, \mathsf{V})$ *and agent* i *at state* s, *given a primitive proposition* p *and a formula* φ ($p \notin \mathsf{Pr}(\varphi)$), *the following hold. (a) if* $p \leftrightarrow \varphi$ *is true in each* i*-accessible state, then* p *is not* i-s *awareness independent to* φ; *(b) if* p *is not* i-s *awareness independent to* φ, *then there is a formula* ψ *such that* $\mathsf{Pr}(\psi) \subseteq \mathsf{Pr}(\varphi)$ *and* $p \leftrightarrow \psi$ *is true in each* i*-accessible state.*

Proof. Part (a). Consider the case that p has a uniform value in each i-accessible state. If the uniform value is true, then $(\mathcal{M}, t) \models_{\mathrm{TP}} p \leftrightarrow \top$ for all t satisfying $(s, t) \in \mathsf{R}_i$. For any model $\mathcal{M}' = (\mathsf{S}, \mathsf{R}_i, \mathsf{T}'_i, \mathsf{V})$, we have $(\mathcal{M}', s) \models_{\mathrm{TP}} A_i\top$ and $(\mathcal{M}', s) \models_{\mathrm{TP}} A_i p$, so $(\mathcal{M}', s) \models_{\mathrm{TP}} A_i\top \rightarrow A_i p$. We can similarly prove it if the uniform value is false. Regard the other case that p does not have a uniform value in all i-accessible state. For any model $\mathcal{M}' = (\mathsf{S}, \mathsf{R}_i, \mathsf{T}'_i, \mathsf{V})$, if $(\mathcal{M}', s) \models_{\mathrm{TP}} A_i\varphi$, then in any awareness partition inside the current knowledge partition in $\mathcal{M}'$ and any $q \in \mathsf{Pr}(\varphi)$, q has uniform value. Thus φ has a uniform value in each awareness partition. Since $p \leftrightarrow \varphi$ is true in each i-accessible state, p also has a uniform value in each awareness partition. So we have $(\mathcal{M}', s) \models_{\mathrm{TP}} A_i\varphi \rightarrow A_i p$.

Part (b). Case 1. If p has a uniform value in each i-accessible state, then it is easy to establish a tautology or a contradiction, being equivalent to p in every i-accessible state, constituted only by propositions from $\mathsf{Pr}(\varphi)$.

Case 2. Consider that p does not have a uniform value in all i-accessible state. Assume that $\mathsf{Pr}(\varphi) = \{q_1, \ldots, q_n\}$, and there is no ψ such that $\mathsf{Pr}(\psi) \subseteq \mathsf{Pr}(\varphi)$ and $p \leftrightarrow \psi$ is true in each i-accessible state. We establish the awareness partitions of $\mathcal{M}' = (\mathsf{S}, \mathsf{R}_i, \mathsf{T}'_i, \mathsf{V})$ with equivalence relations T'_i in the following steps. Firstly, we construct at most two partitions to *just* classify the truth value of q_1. Then by the assumption p is not equivalent to q_1 or $\neg q_1$. As a consequence, the states s and s' appear in the same partition, where $(\mathcal{M}', s) \models_{\mathrm{TP}} p$ and $(\mathcal{M}', s') \models_{\mathrm{TP}} \neg p$. Secondly, we establish at most four partitions based on the previous step to *just* classify the truth value of q_2. Since p is not equivalent to formulas constructed by q_1 and q_2, the states s and s' must appear in the same partition, where $(\mathcal{M}', s) \models_{\mathrm{TP}} p$ and $(\mathcal{M}', s') \models_{\mathrm{TP}} \neg p$. Repeat the procedure until we make at most 2^n partitions to *just* distinguish q_k from $\neg q_k$ ($k \in \{1, \ldots, n\}$). According to the assumption, the states of p being ture at $\mathcal{M}'$ and states of $\neg p$ being ture at the model have to appear in the same partition, which contradicts that p is not i-s awareness independent to φ.

The principal significance of Lemma 6 is that it builds a bridge connecting awareness independence and knowledge of equivalence formulas. In view of this, it provides a technical foundation for the rest of this paper.

Lemma 7. *A model* $\mathcal{M} = (\mathsf{S}, \mathsf{R}_i, \mathsf{T}_i, \mathsf{V})$ *is* i-s *awareness free iff there is no* $p \leftrightarrow \varphi$ *being true in each* i*-accessible state, where* $p \notin \mathsf{Pr}(\varphi)$.

Proof. The direction from right to left is a consequence of the part (b) of Lemma 6. We now prove the other direction. By Lemma 5 and the part (a) of Lemma 6, for any proposition p and φ satisfying $p \notin \mathsf{Pr}(\varphi)$, it is not the case that $p \leftrightarrow \varphi$ is true in each i-accessible state.

In terms of knowledge of equivalence formulas, Lemma 7 gives the equivalent condition of the model being i-s awareness free.

Next we shall explore the relationship between our model and the model proposed by Fagin and Halpern. To this end, let us recall some basic concepts in their model.

Definition 6. *A Fagin and Halpern's awareness model (FHAM for short) is a tuple* $\mathcal{M} = (\mathsf{S}, (\mathsf{R}_i)_{i\in\mathsf{Ag}}, (A_i)_{i\in\mathsf{Ag}}, \mathsf{V})$, *where*

- S *is a nonempty set of states;*
- *For each* $i \in \mathsf{Ag}, \mathsf{R}_i \subseteq \mathsf{S} \times \mathsf{S}$ *is a binary relation on* S *representing the accessibility relation of agent* i*;*
- *For each* $i \in \mathsf{Ag}$, A_i *is a function associating a set of formulas with each state* $s \in \mathsf{S}$*;*
- V *is the valuation function:* $\mathsf{S} \times \Phi \rightarrow \{\mathsf{true}, \mathsf{false}\}$ *which assigns to each state and primitive proposition a truth value.*

As presented in previous sections, the binary relations R_i and T_i are equivalence relations. In view of this, we shall study the relationship between our model and FHAM in the case that the accessibility relations are equivalence relations. Before doing this, we define a property in the FHAM.

Definition 7. *Awareness captured in an FHAM is called restricted awareness if it satisfies the following conditions.*

(i) Awareness is generated by primitive propositions iff for all $i \in \mathsf{Ag}$ *and* $s \in \mathsf{S}$, $\varphi \in \mathsf{A}_i(s)$ *iff* $\mathsf{Pr}(\varphi) \subseteq \mathsf{A}_i(s)$*;*
(ii) Agents know what they are aware of iff for all $i \in \mathsf{Ag}$ *and* $s, s' \in \mathsf{S}$, $(s, s') \in \mathsf{R}_i$ *implies* $\mathsf{A}_i(s') = \mathsf{A}_i(s)$*;*
(iii) Awareness relies on knowledge of equivalences iff for all $i \in \mathsf{Ag}$ *and* $s \in \mathsf{S}$, $p \leftrightarrow \varphi$ *is true in all* s' *with* $(s, s') \in \mathsf{R}_i$ *implies* $A_i\varphi \rightarrow A_ip$ *is true in* s.

The first two restrictions on the awareness functions are jointly called *propositionally determined awareness*, under which the FH awareness model is equivalent to the set-theoretical model proposed by Heifetz *et al.* [11,12]. We call an FHAM satisfying the above three restrictions as a *restricted* FHAM (RFHA for abbreviation), and establish an equivalence result as follows.

Theorem 1. *The following statements hold.*

(a) Given a two-layer partition awareness model $\mathcal{M} = (\mathsf{S}, \mathsf{R}_i, \mathsf{T}_i, \mathsf{V})$, *there is a RFHA,* $\mathcal{M}' = (\mathsf{S}, \mathsf{R}_i, \mathsf{A}_i, \mathsf{V})$, *such that for all formulas* $\varphi \in \mathcal{L}^{KLA}\Phi$, *we have*

$$(\mathcal{M}, s) \models_{\mathtt{TP}} \varphi \textit{ iff } (\mathcal{M}', s) \models_{\mathtt{RFHA}} \varphi.$$

(b) *Given a RFHA,* $\mathcal{M} = (\mathsf{S}, \mathsf{R}_i, \mathsf{A}_i, \mathsf{V})$*, there is a two-layer partition awareness model* $\mathcal{M}' = (\mathsf{S}, \mathsf{R}_i, \mathsf{T}_i, \mathsf{V})$ *such that for all formulas* $\varphi \in \mathcal{L}^{KLA}\Phi$*, we have*

$$(\mathcal{M}, s) \models_{\mathtt{RFHA}} \varphi \textit{ iff } (\mathcal{M}', s) \models_{\mathtt{TP}} \varphi.$$

Proof. We first prove part (a). Given a two-layer partition awareness structure $\mathcal{M} = (\mathsf{S}, \mathsf{R}_i, \mathsf{T}_i, \mathsf{V})$, for each state $s \in \mathsf{S}$, we define $\mathsf{A}_i(s) = \mathcal{L}^{KLA}(\{p \,|\, \text{for all } t \text{ and } u \text{ with } (s,t) \in \mathsf{R}_i \text{ and } (t,u) \in \mathsf{T}_i, (\mathcal{M}, t) \models_{\mathtt{TP}} p \text{ iff } (\mathcal{M}, u) \models_{\mathtt{TP}} p\})$. Let $\mathcal{M}' = (\mathsf{S}, \mathsf{R}_i, \mathsf{A}_i, \mathsf{V})$. It is clear that $\mathcal{M}'$ is a propositionally determined awareness partitional structure. By Lemma 6, $\mathcal{M}'$ satisfies the third condition in Definition 7. Thus, $\mathcal{M}'$ is an RFHA.

We prove the part by induction on the structure of formulas. If φ is a primitive proposition p, then it is immediate from the definitions. Suppose now that it holds for φ_1 and φ_2.

Case 1. φ is a form $\varphi_1 \wedge \varphi_2$ or $\neg\varphi_1$. It follows immediately from the induction hypothesis and the truth conditions.

Case 2. φ is a form $A_i\varphi_1$. By the truth conditions, we have $(\mathcal{M}, s) \models_{\mathtt{TP}} A_i\varphi_1$ iff for all $p \in \mathsf{Pr}(\varphi_1)$, all t and u with $(s,t) \in \mathsf{R}_i$ and $(t,u) \in \mathsf{T}_i$, $(\mathcal{M}, t) \models_{\mathtt{TP}} p$ iff $(\mathcal{M}, u) \models_{\mathtt{TP}} p$. By the definition of $\mathcal{M}'$, $(\mathcal{M}', s) \models_{\mathtt{RFHA}} A_i\varphi_1$ iff for all $p \in \mathsf{Pr}(\varphi_1)$, $p \in \mathsf{A}_i(s)$. Since awareness in $\mathcal{M}'$ is propositionally determined, we get $(\mathcal{M}, s) \models_{\mathtt{TP}} A_i\varphi_1$ iff $(\mathcal{M}', s) \models_{\mathtt{RFHA}} A_i\varphi_1$.

Case 3. φ is a form $L_i\varphi_1$. Part (a) is straightforward by the definition of $\mathcal{M}'$.

Case 4. φ is a form $K_i\varphi_1$. Since $K_i\varphi_1$ is equivalent to $L_i\varphi_1 \wedge A_i\varphi_1$, we have $(\mathcal{M}, s) \models_{\mathtt{TP}} K_i\varphi_1$ iff $(\mathcal{M}', s) \models_{\mathtt{RFHA}} K_i\varphi_1$ by Case 2 and Case 3.

We shall now prove part (b). Given an RFHA, $\mathcal{M} = (\mathsf{S}, \mathsf{R}_i, \mathsf{A}_i, \mathsf{V})$, we prove that there is a model $\mathcal{M}' = (\mathsf{S}, \mathsf{R}_i, \mathsf{T}_i, \mathsf{V})$ such that $(\mathcal{M}', s) \models_{\mathtt{TP}} A_i\varphi$ iff $(\mathcal{M}, s) \models_{\mathtt{RFHA}} A_i\varphi$. If not, then there is a primitive proposition $p \notin \mathsf{A}_i(s)$ that is not i-s awareness independent to φ, where φ is given by $\mathsf{Pr}(\varphi) = \mathsf{Pr}(\mathsf{A}_i(s))$. By part (b) of Lemma 6, there is a formula ψ formed by a subset of $\mathsf{Pr}(\varphi)$ such that $(\mathcal{M}', t) \models_{\mathtt{TP}} p \leftrightarrow \psi$ for all t satisfying $(s,t) \in \mathsf{R}_i$. However, by Definition 7, for any primitive proposition $p \notin \mathsf{A}_i(s)$, if for all t satisfying $(s,t) \in \mathsf{R}_i$, $(\mathcal{M}, t) \models_{\mathtt{RFHA}} (p \leftrightarrow \psi)$, then we have $\psi \not\subseteq \mathsf{A}_i(s)$. Contradiction.

Similar with the proof of part (a), we can establish part (b) in other cases.

The above theorem shows that the equivalence result holds when the FHAM is an RFHA. Nevertheless, our model differs mainly from the FHAM in the following respects. Our model is a syntax-free model, in contrast to the FHAM. In addition, the third condition in Definition 7 is essentially a provable property in our model, whereas it is an external condition for the awareness function in the FHAM. As such, our model provides a natural connection between knowledge and awareness, and thus it provides some insight on how an agent's knowledge can influence his awareness.

4 Axiomatization

In this section we present complete axiomatizations $\mathbf{L^E}$ and $\mathbf{L^L}$ for the logic of explicit knowledge ($\mathrm{L^E}$) and the logic of implicit knowledge ($\mathrm{L^L}$), respectively.

They are interpreted on the two-layer partition awareness models. As $K_i\varphi \leftrightarrow L_i\varphi \wedge A_i\varphi$, we omit the explicit knowledge modality in $\mathbf{L^L}$.

Axiom System $\mathbf{L^E}$

Prop. All substitution instances of tautologies of propositional logic, including the formula $\top$.
AT. $A_i\top$
KA. $K_i\varphi \rightarrow A_i\varphi$ (Knowledge implies Awareness)
AS. $A_i\varphi \leftrightarrow A_i\neg\varphi$ (Awareness Symmetry)
AC. $A_i(\varphi \wedge \psi) \leftrightarrow A_i\varphi \wedge A_i\psi$ (Awareness Conjunction)
AKR. $A_i\varphi \leftrightarrow A_iK_j\varphi$ (Awareness Knowledge Reflection)
AR. $A_i\varphi \leftrightarrow A_iA_j\varphi$ (Awareness Reflection)
AI_E. $A_i\varphi \rightarrow K_iA_i\varphi$ (Awareness Explicit Introspection)
K_E. $(K_i\varphi \wedge K_i(\varphi \rightarrow \psi)) \rightarrow K_i\psi$ (Explicit Distribution Axiom)
T_E. $K_i\varphi \rightarrow \varphi$ (Explicit Knowledge Truth Axiom)
4_E. $K_i\varphi \rightarrow K_iK_i\varphi$ (Positive Introspection Axiom)
5_A. $\neg K_i\varphi \wedge A_i\varphi \rightarrow K_i\neg K_i\varphi$ (Weak Negative Introspection Axiom)
MP. From φ and $\varphi \rightarrow \psi$ infer ψ (Modus Ponens)
Gen_A. From φ infer $A_i\varphi \rightarrow K_i\varphi$ (Modified Knowledge Generalization)
Irr. If no primitive propositions in φ appear in ψ, then from $\neg A_i\varphi \rightarrow \psi$ infer ψ.

Axiom System $\mathbf{L^L}$

Prop. All substitution instances of tautologies of propositional logic, including the formula $\top$.
AT. $A_i\top$
AS. $A_i\varphi \leftrightarrow A_i\neg\varphi$ (Awareness Symmetry)
AC. $A_i(\varphi \wedge \psi) \leftrightarrow A_i\varphi \wedge A_i\psi$ (Awareness Conjunction)
ALR. $A_i\varphi \leftrightarrow A_iL_j\varphi$ (Awareness Implicit Knowledge Reflection)
AR. $A_i\varphi \leftrightarrow A_iA_j\varphi$ (Awareness Reflection)
AI_L. $A_i\varphi \rightarrow L_iA_i\varphi$ (Awareness Implicit Introspection)
UI_L. $\neg A_i\varphi \rightarrow L_i\neg A_i\varphi$ (Unawareness Introspection)
ARKE. $L_i(\varphi \leftrightarrow p) \rightarrow (A_i\varphi \rightarrow A_ip)$ (Awareness Relies on Knowledge of Equivalence)
K_L. $(L_i\varphi \wedge L_i(\varphi \rightarrow \psi)) \rightarrow L_i\psi$ (Implicit Distribution Axiom)
T_L. $L_i\varphi \rightarrow \varphi$ (Implicit Knowledge Truth Axiom)
4_L. $L_i\varphi \rightarrow L_iL_i\varphi$ (Implicit Positive Introspection Axiom)
5. $\neg L_i\varphi \rightarrow L_i\neg L_i\varphi$ (Implicit Negative Introspection Axiom)
MP. From φ and $\varphi \rightarrow \psi$ infer ψ (Modus Ponens)
Gen. From φ infer $L_i\varphi$ (Implicit Knowledge Generalization)

Definition 8. *The canonical model for* $\mathrm{L^E}$ *is a tuple* $\mathcal{M}^c = (\mathsf{S}^c, (\mathsf{R}^c_i), (\mathsf{T}^c_i), \mathsf{V}^c)$, *where*

- S^c *consists of all maximal consistent sets of formulas in* $\mathrm{L^E}$*;*
- $(s,t) \in \mathsf{R}^c_i$ *iff* $\{\varphi \mid K_i\varphi \in s\} \cup \{A_ip \mid A_ip \in s, p \in \Phi\} \cup \{\neg A_ip \mid \neg A_ip \in s, p \in \Phi\} \subseteq t$*;*

- $(s,t) \in \mathsf{T}_i^c$ *iff if* $A_i p \in s (p \in \Phi)$, *then* $p \in t$ *iff* $p \in s$, *and* $A_i p \in t$; *if* $\neg A_i p \in s$, *then* $\neg A_i p \in t$;
- $\mathsf{V}^c(p) = \{s \in \mathsf{S}^c \mid p \in s\}$ *for each* $p \in \Phi$.

Lemma 8. *In the canonical model for* L^{E}, *for every* $i \in \mathsf{Ag}$, R_i^c *and* T_i^c *are reflexive, transitive and Euclidean, and* $\mathsf{T}_i^c \subseteq \mathsf{R}_i^c$.

Proof. By the definition of T_i^c, it is easy to see that $(s,s) \in \mathsf{T}_i^c$. So, T_i^c is reflexive. Suppose $(s,t) \in \mathsf{T}_i^c$, $(t,u) \in \mathsf{T}_i^c$, and $A_i p \in s$ $(p \in \Phi)$, then we have $A_i p \in t$, and $p \in t$ iff $p \in s$, and then $A_i p \in u$, and $p \in u$ iff $p \in t$. If $\neg A_i p \in s$, then $\neg A_i p \in t$, and then $\neg A_i p \in u$. So $(s,u) \in \mathsf{T}_i^c$. Hence T_i^c is transitive. Similarly, we can obtain that T_i^c is Euclidean.

Suppose $(s,s) \notin \mathsf{R}_i^c$. By the definition of R_i^c, there exists a proposition φ satisfying $K_i\varphi \in s$ and $\varphi \notin s$, which contradicts the axiom T. Suppose $(s,t) \in \mathsf{R}_i^c$, $(t,u) \in \mathsf{R}_i^c$ and $(s,u) \notin \mathsf{R}_i^c$. By the definition of R_i^c, we have $A_i p \in s$ iff $A_i p \in u$ $(p \in \Phi)$, and there exists a proposition φ such that $K_i\varphi \in s$ and $\varphi \notin u$. Following the axiom 4_E and the properties of maximally consistent set, we obtain $K_i K_i \varphi \in s$. Again, since $(s,t) \in \mathsf{R}_i^c$ and $(t,u) \in \mathsf{R}_i^c$, we have $K_i\varphi \in t$ and $\varphi \in u$, which contradicts $\varphi \notin u$.

Suppose $(s,t) \in \mathsf{R}_i^c$, $(s,u) \in \mathsf{R}_i^c$ and $(t,u) \notin \mathsf{R}_i^c$. By the definition of R_i^c, we have $A_i p \in t$ iff $A_i p \in u$ $(p \in \Phi)$, and there exists a proposition φ satisfying $K_i\varphi \in t$ and $\varphi \notin u$. Therefore, we have $K_i\varphi \notin s$, $A_i\varphi \in s$, $A_i\varphi \in t$ and $A_i\varphi \in u$. By the properties of maximally consistent set, we obtain that $\neg K_i\varphi \in s$. Again, it follows from the axiom 5_A that $K_i \neg K_i \varphi \in s$. Since $(s,t) \in \mathsf{R}_i^c$, we have $\neg K_i \varphi \in t$, which contradicts $K_i\varphi \in t$.

If $(s,t) \in \mathsf{T}_i^c$, and $(s,t) \notin \mathsf{R}_i^c$, then we have $A_i p \in s$ iff $A_i p \in t$ $(p \in \Phi)$, and there exists a proposition φ such that $K_i\varphi \in s$ and $\varphi \notin t$. By the axiom KA and T, it follows that $A_i\varphi \in s$ and $\varphi \in s$. Since $(s,t) \in \mathsf{T}_i^c$, we have that $p \in s$ iff $p \in t$ for all $p \in \mathsf{Pr}(\varphi)$. So $\varphi \in t$, which contradicts $\varphi \notin t$.

Lemma 9. *For all formulas* $\varphi \in \mathcal{L}^{KA}\Phi$ *and all maximal consistent sets* $s \in \mathsf{S}^c$, $\varphi \in s$ *iff* $(\mathcal{M}^c, s) \models_{\mathrm{TP}} \varphi$.

Proof. If φ is a primitive proposition, it is immediate from Definitions 4 and 8. Suppose that it holds for φ_1 and φ_2.

Case 1. φ is a form $\varphi_1 \wedge \varphi_2$ or $\neg\varphi_1$. It follows easily from the properties of maximal consistent sets.

Case 2. φ is a form $A_i\varphi_1$. Suppose that $A_i\varphi_1 \in s$ and $p \in \mathsf{Pr}(\varphi_1)$. Then we have $A_i p \in s$, since $\mathbf{L^E} \vdash A_i\varphi_1 \to A_i p$. If $(s,t) \in \mathsf{T}_i^c$, then by the definition of the canonical model we have $p \in t$ iff $p \in s$. Since $p \in s$ iff $(\mathcal{M}^c, s) \models_{\mathrm{TP}} p$, $p \in t$ iff $(\mathcal{M}^c, t) \models_{\mathrm{TP}} p$, we have $(\mathcal{M}^c, s) \models_{\mathrm{TP}} p$ iff $(\mathcal{M}^c, t) \models_{\mathrm{TP}} p$. Further, by the definition of R_i^c and $A_i\varphi_1 \in s$, we have $A_i\varphi_1 \in t$ for all t with $(s,t) \in \mathsf{R}_i^c$. Hence, $(\mathcal{M}^c, u) \models_{\mathrm{TP}} p$ iff $(\mathcal{M}^c, t) \models_{\mathrm{TP}} p$ for all u with $(t,u) \in \mathsf{T}_i^c$. By the truth condition of $A_i\varphi_1$, we have $(\mathcal{M}^c, s) \models_{\mathrm{TP}} A_i\varphi_1$.

Now prove the other direction for this case. Suppose that $(\mathcal{M}^c, s) \models_{\mathrm{TP}} A_i\varphi_1$ and $A_i\varphi_1 \notin s$. By the properties of maximally consistent set, we have $\neg A_i\varphi_1 \in s$.

It is clear that

$$\mathbf{L^E} \vdash \bigwedge_{p \in \mathsf{Pr}(\varphi_1)} A_i p \to A_i \varphi_1.$$

Hence,

$$\mathbf{L^E} \vdash \neg A_i \varphi_1 \to \neg \bigwedge_{p \in \mathsf{Pr}(\varphi_1)} A_i p.$$

Moreover, $\neg \bigwedge_{p \in \mathsf{Pr}(\varphi_1)} A_i p$ iff there is $p' \in \mathsf{Pr}(\varphi_1)$ and $\neg A_i p'$. Hence, we have $\neg A_i p' \in s$. As such, there is t with $(s,t) \in \mathsf{T}_i^{\mathsf{c}}$ according to the definition of $\mathsf{T}_i^{\mathsf{c}}$, which implies $p' \in s$ and $p' \notin t$, or $p' \notin s$ and $p' \in t$. On the other hand, as $(s,s) \in \mathsf{R}_i^{\mathsf{c}}$, by the truth condition of $A_i \varphi_1$, for all $p \in \mathsf{Pr}(\varphi_1)$ and all t with $(s,t) \in \mathsf{T}_i^{\mathsf{c}}$, we have $(\mathcal{M}^{\mathsf{c}}, t) \models_{\mathbf{TP}} p$ iff $(\mathcal{M}^{\mathsf{c}}, s) \models_{\mathbf{TP}} p$. Moreover, by Case 1, $p \in s$ iff $(\mathcal{M}^{\mathsf{c}}, s) \models_{\mathbf{TP}} p$, and $p \in t$ iff $(\mathcal{M}^{\mathsf{c}}, t) \models_{\mathbf{TP}} p$, which imply $p \in t$ iff $p \in s$. Contraction! So we have $A_i \varphi_1 \in s$.

Case 3. φ is a form $K_i \varphi_1$. Then similar with the proof in [7] we can prove the result.

Definition 9. *The canonical model for* $\mathrm{L^L}$ *is a tuple* $\mathcal{M}^{\mathrm{cl}} = (\mathsf{S}^{\mathrm{cl}}, (\mathsf{R}_i^{\mathrm{cl}}), (\mathsf{T}_i^{\mathrm{cl}}), \mathsf{V}^{\mathrm{cl}})$*, where*

- S^{cl} *consists of all maximal consistent sets of formulas in* $\mathrm{L^L}$*;*
- $(s,t) \in \mathsf{R}_i^{\mathrm{cl}}$ *iff* $\{\varphi \mid L_i \varphi \in s\} \cup \{A_i p \mid A_i p \in s, p \in \Phi\} \cup \{\neg A_i p \mid \neg A_i p \in s, p \in \Phi\} \subseteq t$*;*
- $(s,t) \in \mathsf{T}_i^{\mathrm{cl}}$ *iff if* $A_i p \in s, p \in \Phi$*, then* $p \in t$ *iff* $p \in s$*, and* $A_i p \in t$*; if* $\neg A_i p \in s$*, then* $\neg A_i p \in t$*; and* $(s,t) \in \mathsf{R}_i^{\mathrm{cl}}$*;*
- $\mathsf{V}^{\mathrm{cl}}(p) = \{s \in \mathsf{S}^{\mathrm{c}} \mid p \in s\}$ *for each* $p \in \Phi$.

Lemma 10. *Given a canonical model for* $\mathrm{L^L}$*, for every* $i \in \mathsf{Ag}$*,* $\mathsf{R}_i^{\mathrm{cl}}$ *and* $\mathsf{T}_i^{\mathrm{cl}}$ *are reflexive, transitive and Euclidean, and* $\mathsf{T}_i^{\mathrm{cl}} \subseteq \mathsf{R}_i^{\mathrm{cl}}$.

Proof. The proof of $\mathsf{T}_i^{\mathrm{cl}}$ being reflexive, transitive and Euclidean stays the same with that in the canonical model for $\mathrm{L^E}$. And the proof of $\mathsf{R}_i^{\mathrm{cl}}$ being reflexive, transitive and Euclidean is the same as the proof in the standard S5 system. Finally, $\mathsf{T}_i^{\mathrm{cl}} \subseteq \mathsf{R}_i^{\mathrm{cl}}$ has been defined in the canonical model.

Lemma 11. *For all formulas* $\varphi \in \mathcal{L}^{LA}\Phi$ *and all maximal consistent sets* $s \in \mathsf{S}^{\mathrm{cl}}$*,* $\varphi \in s$ *iff* $(\mathcal{M}^{\mathrm{cl}}, s) \models_{\mathbf{TP}} \varphi$.

Proof. It is similar with the proof for the analogous lemma for $\mathcal{M}^{\mathsf{c}}$.

Theorem 2. *Axiom system* $\mathrm{L^E}$ *is sound and complete for* $\mathcal{L}^{KA}\Phi$ *with respect to two-layer partition awareness models.*

Proof. Soundness of $\mathrm{L^E}$ can be proved by showing that the axioms are valid in all two-layer partition awareness models and the inference rules preserve validity. And we can prove completeness by Lemma 9.

Theorem 3. *Axiom system* L^{L} *is sound and complete for* $\mathcal{L}^{LA}\Phi$ *with respect to two-layer partition awareness models.*

Proof. Soundness of L^{L} can be proved by showing that the axioms are valid in all two-layer partition awareness models and the inference rules preserve validity. Also we can prove completeness by Lemma 11.

5 Conclusion

An appropriate model for formalizing awareness should have some provable properties that explore the connections between knowledge and awareness. To establish such model, we have proposed a two-layer partition structure being a syntax-free model in this paper. We have also provided two sound and complete axiomatic systems for reasoning about awareness.

Acknowledgements. Pengfei Song's research was supported by the Major project of Key Research Institutes of Humanities and Social Sciences of Ministry of Education of China (No. 17JJD720008), and Wei Xiong's research was supported by the Key Project of National Social Science Foundation of China (No. 16AZX017).

References

1. Aumann, R.: Agreeing to disagree. Ann. Stat. **4**, 1236–1239 (1976)
2. Dekel, E., Lipman, B., Rustichini, A.: Standard state-space models preclude unawareness. Econometrica **66**, 159–173 (1998)
3. van Ditmarsch, H., French, T.: Becoming aware of propositional variables. In: Banerjee, M., Seth, A. (eds.) ICLA 2011. LNCS (LNAI), vol. 6521, pp. 204–218. Springer, Heidelberg (2011). https://doi.org/10.1007/978-3-642-18026-2_17
4. van Ditmarsch, H., French, T.: Semantics for knowledge and change of awareness. J. Logic Lang. Inf. **23**(2), 169–195 (2014)
5. van Ditmarsch, H., French, T., Velázquez-Quesada, F., Wang, Y.: Implicit, explicit and speculative knowledge. Artif. Intell. **256**, 35–67 (2018)
6. Fagin, R., Halpern, J.: Belief, awareness, and limited reasoning. Artif. Intell. **34**(1), 39–76 (1988)
7. Halpern, J.: Alternative semantics for unawareness. Games Econ. Behav. **37**(2), 321–339 (2001)
8. Halpern, J., Rêgo, L.: Interactive unawareness revisited. Games Econ. Behav. **62**(1), 232–262 (2008)
9. Halpern, J., Rêgo, L.: Resoning about knowledge of unawareness. Games Econ. Behav. **67**(2), 503–525 (2009)
10. Halpern, J., Rêgo, L.: Resoning about knowledge of unawareness revisited. Math. Soc. Sci. **65**(2), 73–84 (2013)
11. Heifetz, A., Meier, M., Schipper, B.: Interactive unawareness. J. Econ. Theory **130**, 78–94 (2006)
12. Heifetz, A., Meier, M., Schipper, B.: A canonical model for interactive unawareness. Games Econ. Behav. **62**, 304–324 (2008)
13. Hintikka, J.: Knowledge and Belief. Cornell University Press, Ithaca (1962)

14. Levesque, H.: A logic of implicit and explicit belief. In: Proceedings of AAAI 1984, pp. 198–220. AAAI Press (1984)
15. Modica, S., Rustichini, A.: Awareness and partitional information structures. Theory Decis. **37**, 107–124 (1994)
16. Modica, S., Rustichini, A.: Unawareness and partitional information structures. Games Econ. Behav. **27**, 265–298 (1999)

First-Order Intuitionistic Epistemic Logic

Youan Su[1(✉)] and Katsuhiko Sano[2]

[1] Graduate School of Letters, Hokkaido University, Sapporo, Japan
ariyasu613@gmail.com
[2] Faculty of Humanities and Human Sciences, Hokkaido University, Sapporo, Japan
v-sano@let.hokudai.ac.jp

Abstract. Intuitionistic epistemic logic (**IEL**), introduced by Artemov and Protopopescu (2016), accepts the co-reflection axiom: "$A \supset KA$" in terms of Brouwer-Heyting-Kolmogorov interpretation. There are two variants for **IEL**, one of which has the axiom "$KA \supset \neg\neg A$", while the other does not have it. The aim of this paper is to study the first-order expansions of these two **IELs**. Hilbert systems and sequent calculi of the first-order expansion of these two intuitionistic epistemic logic are provided to be proved sound and complete for the intended semantics. We also prove the cut-elimination theorems for both systems. Furthermore, the Craig interpolation theorems of both systems are established by Maehara's method as consequences of cut-elimination theorems.

Keywords: Intuitionistic logic · Epistemic logic · Sequent calculus · Craig interpolation · Cut elimination · BHK interpretation

1 Introduction

In his famous *Knowledge and Belief*, Hintikka [4] introduced the epistemic logic based on the classical logic, which is widely studied nowadays. In the following decades, several studies have been made on the topic of intuitionistic epistemic logic. Williamson [15] studied a system in which the intuitionistic conception of truth of a formula A is equivalent to the possibility of knowledge of A, that is, A is equivalent to $\Diamond KA$. Proietti [11] distinguished the implicit and explicit knowledge in Kripke semantics for his intuitionistic epistemic logic. Ma et al. [9] studied an intuitionistic public announcement logic, i.e., an expansion of intuitionistic modal logic with public announcement operators. Jäger and Marti studied the common and distributed knowledge operators over intuitionistic logics [5,6].

As is well-known, the Brouwer-Heyting-Kolmogorov interpretation is the conceptual ground for the intuitionistic logic. The interpretation states that how a complex proof can be constructively built from simpler proofs. Then, it becomes reasonable to ask what is a proper BHK reading of the knowledge operator in the intuitionistic epistemic logic. Recently, Artemov and Protopopescu [1] gave a system named as intuitionistic epistemic logic (**IEL**), where they proposed the following BHK interpretation for KA: a proof of a formula KA (it is known that

P. Blackburn et al. (Eds.): LORI 2019, LNCS 11813, pp. 326–339, 2019.
https://doi.org/10.1007/978-3-662-60292-8_24

A) is the conclusive verification of the existence of a proof of A. Let us recall that a proof of $A \supset B$ is a construction such that given a proof of A the construction gives us a proof of B. Then a proof of A itself can be regarded as the verification of the existence of the proof of A. In this sense, we always have such a proof of $A \supset KA$, i.e., $A \supset KA$ is valid. But T-axiom: $KA \supset A$ (usually called *factivity* or *reflection*) is not valid, since the verification does not always give a proof.

In addition to the intuitionistic propositional tautologies and the K-axiom: $K(A \supset B) \supset (KA \supset KB)$, the system of **IEL**$^-$ have the coreflection axiom: $A \supset KA$. The system of **IEL** is obtained from **IEL**$^-$ by adding the intuitionistic reflection: $KA \supset \neg\neg A$, which is shown to be equivalent to $\neg K\bot$ in [1].

The study of intuitionistic epistemic logic by [1] also trivialized the knowability paradox, also known as Fitch's paradox [2]. In this paradox, we can classically derive the omniscience principle $A \supset KA$ from the knowability principle $A \supset \Diamond KA$ (cf. [13]). While $A \supset KA$ is classically read as every truth is known, however, as we have shown, $A \supset KA$ can be intuitionistically accepted according to the BHK interpretation in [1]. As a result, if we can accept the intuitionistic epistemic logic proposed by Artemov and Protopopescu, the paradox is not a problem at all.

Proof-theoretical studies of **IEL** have been investigated. In Krupski and Yatmanov [8], the sequent calculi of **IEL**$^-$ and **IEL** have been given, though an inference rule corresponding to $KA \supset \neg\neg A$ in their system for **IEL** does not satisfy a desired syntactic property, i.e., the subformula property. In Protopopescu [12], a Gödel-McKinsey-Tarski translation from the intuitionistic epistemic propositional logic to the bimodal expansion of the classical modal logic **S4** has been studied.

In this paper, we study the first-order expansions **QIEL** and **QIEL**$^-$ of intuitionistic epistemic logic of **IEL** and **IEL**$^-$. Artemov and Protopopescu mentioned that the notion of the intuitionistic knowledge captures both mathematical knowledge and empirical knowledge. When we consider the mathematical knowledge, quantifiers become inevitable. Moreover when we are concerned with the empirical knowledge, we recall that Hintikka had given arguments for first-order epistemic logic [4]. He mentioned that if we want to deal with the locutions like "knows who," "knows when," "knows where," we can translate these expressions into a language with quantifiers. For example, about "who" we can have variables ranging over the human being, about "where" over the location in space. In this sense, our first-order expansions can provide a fundamental basis when we concern the intuitionistic mathematical and empirical knowledge.

We also propose the sequent calculi for **QIEL** and **QIEL**$^-$. The sequent calculi for **IEL**$^-$ and **IEL** has been given by Krupski and Yatmanov [8]. As we have seen above, the subformula property does not always hold in their systems. Unlike their system for **IEL**, we emphasize that our sequent system for **QIEL** satisfies the subformula property.

The outline of this paper is as follows. In Sect. 2, we provide the semantics and Hilbert systems for **QIEL** and **QIEL**$^-$. In Sect. 3, we give sequent calculi of **QIEL** and **QIEL**$^-$, which are shown to be equivalent to the corresponding Hilbert systems. Moreover, we establish the cut-elimination theorems, from which we demonstrate the disjunction property, the existence property and Craig interpolation theorem of each of **QIEL** and $\mathbf{QIEL}^-$. In Sect. 4, we prove the completeness theorems for both sequent calculi and Hilbert systems of **QIEL** and $\mathbf{QIEL}^-$.

2 Kripke Semantics and Hilbert Systems of First-Order Intuitionistic Epistemic Logics

We fix a countably infinite set of variables as $\mathcal{V} = \{v_n | n \in \mathbb{N}\}$, a countably infinite set of constant symbols as $\mathcal{C} = \{c_n | n \in \mathbb{N}\}$. A *term*, denoted by t etc., is a variable or a constant symbol. Then we fix a countably infinite set $\Sigma = \{P_m\}_{m \in \mathbb{N}}$ of predicate symbols, each of which has a fixed finite arity. The set of formulas of the language $\mathcal{L}(\mathcal{C} \cup \mathcal{V})$ is defined inductively as:

$$A := P(\overrightarrow{u}) \,|\, \bot \,|\, A \wedge A \,|\, A \vee A \,|\, A \supset A \,|\, \forall x A \,|\, \exists x A \,|\, KA,$$

where $P \in \Sigma$, $P(\overrightarrow{u})$ denotes $P(u_1, u_2, \ldots, u_n)$ and u_i is a term for $1 \leq i \leq n$. We define $\neg A \equiv A \supset \bot$. We define *bound variables* and *free variables* as usual. If a formula does not contain any free variables, the formula is called *closed formula.* In the rest of this paper, we assume that free variables and bound variables are always distinguished. We use $A(t/x)$ to mean the result of substitution (avoiding variable clash) of all occurrences of free variable x in A by term t, and let $\Gamma(t/x) := \{A(t/x) | A \in \Gamma\}$.

Definition 1. *A model is a tuple $M = (W, \leq, R, D, I)$ where*

- *$(W, \leq)$ is a preorder,*
- *R is a binary relation on W such that:*
  1. *$R \subseteq \leq$, i.e., for any $w, v \in W$, if wRv then $w \leq v$, and*
  2. *$\leq; R \subseteq R$, i.e., for any $w, v, u \in W$, if $w \leq v$ and vRu then wRu,*
- *D is a function which assigns a nonempty domain $D(w)$ to $w \in W$ such that, for any $w, v \in W$, if $w \leq v$ then $D(w) \subseteq D(v)$,*
- *I is an interpretation such that $I(c) \in D(w)$ for all $w \in W$ for any constant symbol c and $I(P, w) \subseteq D(w)^n$ for every $w \in W$ and every n-arity predicate P such that if $u \leq v$ then $I(P, u) \subseteq I(P, v)$ for all $u, v \in W$.*

The class of all models is denoted as $\mathbb{M}_{\text{all}}$. Furthermore, $\mathbb{M}_{\text{ser}}$ is the class of all models in which R is serial, that is for any $w \in W$ there is a $v \in W$ such that wRv.

Let $\underline{d}$ be the syntactic name of $d \in D(w)$. Given a $w \in W$, let $\mathcal{L}(\mathcal{C} \cup D(w), \mathcal{V})$ be the language $\mathcal{L}(\mathcal{C}, \mathcal{V})$ expanded with the constant symbols $\{\underline{d} | d \in D(w)\}$ for $D(w)$. Given a model $M = (W, \leq, R, D, I)$, a state $w \in W$ and a closed formula

A in $\mathcal{L}(\mathcal{C} \cup D(w), \mathcal{V})$, we inductively define the satisfaction relation $M, w \models A$ as follows:

$$
\begin{array}{lll}
M, w \models P(\underline{d_1}, ..., \underline{d_m}) & \iff & (d_1, ..., d_m) \in I(P, w) \\
M, w \not\models \bot & & \\
M, w \models A \wedge B & \iff & M, w \models A \text{ and } M, w \models B \\
M, w \models A \vee B & \iff & M.w \models A \text{ or } M, w \models B \\
M, w \models A \supset B & \iff & \text{for all } v \in W, w \leq v \text{ and } M, v \models A \text{ imply } M, v \models B \\
M, w \models KA & \iff & \text{for all } v \in W, wRv \text{ implies } M, v \models A \\
M, w \models \forall x A & \iff & \text{for all } v \in W, w \leq v \text{ implies: } M, v \models A(\underline{d}/x) \text{ for all } d \in D(v) \\
M, w \models \exists x A & \iff & \text{there are some } d \in D(w) \text{ such that } M, w \models A(\underline{d}/x).
\end{array}
$$

Proposition 2 (Persistency). *Let M be a model. For any $w, v \in W$, for any closed formulas A in $\mathcal{L}(\mathcal{C} \cup D(w), \mathcal{V})$ if $w \leq v$ and $M, w \models A$ then $M, v \models A$.*

Proof. By induction on the complexity A. Suppose $M, w \models KA$ and $w \leq v$, we show $M, v \models KA$. Fix an arbitrary $s \in W$, suppose vRs, then from $\leq ; R \subseteq R$, we have wRs. By $M, w \models KA$, we have $M, s \models A$. □

Definition 3. *For a formula A and a class $\mathbb{M}$ of models, we say A is valid in $\mathbb{M}$ (denoted by $\mathbb{M} \models A$), if for any $M \in \mathbb{M}$, for any $w \in W$ of M, $M, w \models A(\underline{d_1}/a_1) \cdots (\underline{d_n}/a_n)$, where $(d_1, \ldots, d_n) \in D(w)^n$, where $\{a_1, \ldots, a_n\}$ be the set of all free variables in the formula A.*

Let us introduce Hilbert systems $\mathcal{H}(\mathbf{QIEL}^-)$ and $\mathcal{H}(\mathbf{QIEL})$ as in the following table and define the notion of derivation in these systems as usual.

Theorem 1 (Soundness).

1. *For any formula A, if $\mathcal{H}(\mathbf{QIEL}^-) \vdash A$ then $\mathbb{M}_{\text{all}} \models A$;*
2. *For any formula A, if $\mathcal{H}(\mathbf{QIEL}) \vdash A$ then $\mathbb{M}_{\text{ser}} \models A$.*

Let us recall from Troelstra and van Dalen [14, p.10] the BHK-interpretations for quantifiers as follows:

- a proof of $\forall x A(x)$ is a construction which transforms a proof of $d \in D$ (D the intended range of the variable x) into a proof of $A(d)$.
- a proof of $\exists x A(x)$ is given by providing $d \in D$, and a proof of $A(d)$.

Artemov and Protopopescu [1] explained a counterexample against the validity of T-axiom ($KA \supset A$):

> "Somebody stole your wallet in the subway. You have all evidence for this: the wallet is gone, your backpack has a cut in the corresponding pocket, but you have no idea who did it. You definitely know that 'there is a person who stole my wallet' (in logical form, $\exists x S(x)$, where $S(x)$ stands for 'x stole my wallet') so you have a justification p of $K(\exists x S(x))$. If $K(\exists x S(x)) \to \exists x S(x)$ held intuitionistically, you would have a constructive proof q of $\exists x S(x)$. However a constructive proof of the existential sentence $\exists x S(x)$ requires a witness a for x and a proof b that $S(a)$ holds. You are nowhere near meeting this requirement. $K(\exists x S(x)) \to \exists x S(x)$ does not hold intuitionistically." [1, p.273] (Here, the right arrow denotes the implication.)

Hilbert system $\mathcal{H}(\mathbf{QIEL}^-)$	
(∧-Ax)	$A_1 \wedge A_2 \supset A_i (i = 1\, or\, 2)$
	$A \supset (B \supset (A \wedge B))$
(∨-Ax)	$A_i \supset A_1 \vee A_2 (i = 1\, or\, 2)$
	$(A \supset C) \supset ((B \supset C) \supset (A \vee B \supset C))$
(⊃-Ax)	$A \supset (B \supset A)$
	$(A \supset (B \supset C)) \supset ((A \supset B) \supset (A \supset C))$
(⊥-Ax)	$\bot \supset A$
(∀-Ax)	$\forall x A \supset A(t/x)$
(∃-Ax)	$A(t/x) \supset \exists x A$
(K)	$K(A \supset B) \supset (KA \supset KB)$
(CR)	$A \supset KA$
(∀-Rule)	From $A \supset B(u/x)$, infer $A \supset \forall x B$†
(∃-Rule)	From $B(u/x) \supset A$, infer $\exists x B \supset A$†
(MP)	From A and $A \supset B$, infer B
	† u is fresh in the conclusion.
Hilbert system $\mathcal{H}(\mathbf{QIEL})$	
All the axioms and rules of $\mathcal{H}(\mathbf{QIEL}^-)$ plus:	
(IR)	$KA \supset \neg\neg A$

A proof of $K(\exists x Ax) \supset \exists x Ax$ states that when a conclusive verification of the existence of proof of the provision of $d \in D$ and a proof of $A(d)$, then we can provide a $d \in D$ and give a proof of $A(d)$. As a result, this formula is not valid. Our Kripke semantics captures the BHK-interpretation of the quantifier in the first-order extension of **IEL**.

Proposition 4. *The formula* $K(\exists x q(x)) \supset \exists x q(x)$ *is not derivable in* $\mathcal{H}(\mathbf{QIEL})$.

Proof. Define $M = (W, \leq, R, D, I)$ where $W = \{1, 2, 3\}$, $\leq = \{(1,3), (1,2), (2,3)\} \cup \{ (x,x) \mid x \in W \}$, $R = \{(1,1), (1,3), (2,3), (3,3)\}$, $D(1) = \{s\}$, $D(2) = \{s, d\}$ and $D(3) = \{s, d\}$. $I(p,1) = \{s\}$. $I(p,2) = \{s\}$. $I(p,3) = \{s, d\}$. $I(q,3) = \{s\}$. In the following picture, solid lines stand for R and dotted lines stand for $\leq$.

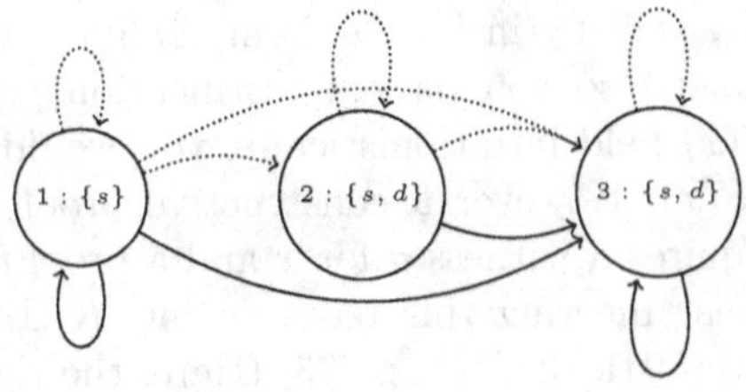

It is easy to see that M *is* a model and it also belongs to $\mathbb{M}_{ser}$. Since $M, 2 \not\models K(\exists x q(x)) \supset \exists x q(x)$, we have $\mathbb{M}_{ser} \not\models K(\exists x q(x)) \supset \exists x q(x)$. By Theorem 1, we have $\mathcal{H}(\mathbf{QIEL}) \nvdash K(\exists x q(x)) \supset \exists x q(x)$. □

We show that Barcan formula is not derivable as follows.

Proposition 5. *The formula* $\forall x K p(x) \supset K \forall x p(x)$ *is not derivable in* $\mathcal{H}(\mathbf{QIEL})$.

Proof. By the same counterexample as in the proof of Proposition 4, we have $M, 1 \not\models \forall x K p(x) \supset K \forall x p(x)$. □

Table 1. Sequent Calculi $\mathcal{G}(\mathbf{QIEL}^-)$, $\mathcal{G}(\mathbf{QIEL})$.

Sequent Calculus $\mathcal{G}(\mathbf{QIEL}^-)$: Δ contains at most one formula below.

Initial Sequents $A \Rightarrow A \quad \bot \Rightarrow$

Structural Rules

$$\frac{\Gamma \Rightarrow \Delta}{A, \Gamma \Rightarrow \Delta}\,(LW) \qquad \frac{\Gamma \Rightarrow}{\Gamma \Rightarrow C}\,(RW) \qquad \frac{A, A, \Gamma \Rightarrow \Delta}{A, \Gamma \Rightarrow \Delta}\,(LC)$$

$$\frac{\Gamma \Rightarrow A \quad A, \Gamma' \Rightarrow \Delta}{\Gamma, \Gamma' \Rightarrow \Delta}\,(Cut)$$

Logical Rules

$$\frac{\Gamma \Rightarrow A_1 \quad \Gamma \Rightarrow A_2}{\Gamma \Rightarrow A_1 \wedge A_2}\,(R\wedge) \qquad \frac{A_i, \Gamma \Rightarrow \Delta}{A_1 \wedge A_2, \Gamma \Rightarrow \Delta}\,(L\wedge)$$

$$\frac{\Gamma \Rightarrow A_i}{\Gamma \Rightarrow A_1 \vee A_2}\,(R\vee) \qquad \frac{A_1, \Gamma \Rightarrow \Delta \quad A_2, \Gamma \Rightarrow \Delta}{A_1 \vee A_2, \Gamma \Rightarrow \Delta}\,(L\vee)$$

$$\frac{A, \Gamma \Rightarrow B}{\Gamma \Rightarrow A \supset B}\,(R\supset) \qquad \frac{\Gamma \Rightarrow A \quad B, \Gamma' \Rightarrow \Delta}{A \supset B, \Gamma, \Gamma' \Rightarrow \Delta}\,(L\supset)$$

$$\frac{\Gamma \Rightarrow A(u/x)}{\Gamma \Rightarrow \forall x A}\,(R\forall)\dagger \qquad \frac{A(t/x), \Gamma \Rightarrow \Delta}{\forall x A, \Gamma \Rightarrow \Delta}\,(L\forall)$$

$$\frac{\Gamma \Rightarrow A(t/x)}{\Gamma \Rightarrow \exists x A}\,(R\exists) \qquad \frac{A(u/x), \Gamma \Rightarrow \Delta}{\exists x A, \Gamma \Rightarrow \Delta}\,(L\exists)\dagger$$

Modal Rule

$$\frac{\Gamma_1, \Gamma_2 \Rightarrow A}{\Gamma_1, K\Gamma_2 \Rightarrow KA}\,(K_{\mathbf{IEL}^-})$$

†: u does not occur in the lower sequent.

Sequent Calculus $\mathcal{G}(\mathbf{QIEL})$: Δ contains at most one formula below.

Replace $(K_{\mathbf{IEL}^-})$ of $\mathcal{G}(\mathbf{QIEL}^-)$ with the following rule:

Modal Rules

$$\frac{\Gamma_1, \Gamma_2 \Rightarrow \Delta}{\Gamma_1, K\Gamma_2 \Rightarrow K\Delta}\,(K_{\mathbf{IEL}})$$

3 Sequent Calculi for First-Order Intuitionistic Epistemic Logics

3.1 The Systems of $\mathcal{G}(\mathbf{QIEL}^-)$ and $\mathcal{G}(\mathbf{QIEL})$

A *sequent*, denoted by $\Gamma \Rightarrow \Delta$, Γ and Δ are a pair of finite multisets such that Δ contains at most one formula. The multiset Γ is the *antecedent* of $\Gamma \Rightarrow \Delta$, while Δ is the *succedent* of the sequent $\Gamma \Rightarrow \Delta$. We define $K\Gamma = \{KA | A \in \Gamma\}$, if Γ is empty then $K\Gamma$ is also empty.

In Krupski and Yatmanov [8], the sequent calculus of **IEL** is obtained from Gentzen's sequent calculus **LJ** (with structural rules of weakening and contraction) for the intuitionistic logic plus the following two inference rules on the knowledge operator:

$$\frac{\Gamma_1, \Gamma_2 \Rightarrow A}{\Gamma_1, K\Gamma_2 \Rightarrow KA} (KI) \qquad \frac{\Gamma \Rightarrow K\bot}{\Gamma \Rightarrow A.} (U)$$

It is remarked, however, that this system does not enjoy the subformula property. That is, in the rule of (U), we have a formula $K\bot$ which might not be a subformula of a formula in the lower sequent of the rule (U). Compared to their work, we let the succedent of $(K_{\mathbf{IEL}})$ be a possibly empty multiset of formulas, then our modal rule in the sequent calculi of **QIEL** have subformula property.

Definition 6. *Let* $\Lambda \in \{\mathbf{QIEL}, \mathbf{QIEL}^-\}$. *Let* $\mathcal{G}(\Lambda)$ *be one of systems of Table 1,* $\mathcal{G}^-(\Lambda)$ *be the systems without the* (Cut) *rule. A derivation* $\mathcal{D}$ *in* $\mathcal{G}(\Lambda)$ *(or* $\mathcal{G}^-(\Lambda)$*) is a finite tree generated by rules of* $\mathcal{G}(\Lambda)$ *(or* $\mathcal{G}^-(\Lambda)$*, respectively) from initial sequents.*

Definition 7 (Context and Principal Formula). *The* Γ *and* Δ *in an inference rule of* $\mathcal{G}(\Lambda)$ *except* $(K_{\mathbf{IEL}})$ *and* $(K_{\mathbf{IEL}-})$*, are called the* context. *In the conclusion of each rule of* $\mathcal{G}(\Lambda)$ *except* $(K_{\mathbf{IEL}})$ *and*$(K_{\mathbf{IEL}-})$*, the formula(s) not in the context is called the* principal formula(s). *For the rule of* $(K_{\mathbf{IEL}})$ *and* $(K_{\mathbf{IEL}-})$*, all formulas in the conclusion except in* Γ_1 *of these rules are called the* principal formulas.

Definition 8. *Given a sequent* $\Gamma \Rightarrow \Delta$, Γ_* *denotes the conjunction of all formulas in* Γ *(*$\Gamma_* \equiv \top$ *if* Γ *is empty) and* Δ^* *denotes the unique formula in* Δ *if* Δ *is non-empty; it denotes* $\bot$ *otherwise. We say that a sequent* $\Gamma \Rightarrow \Delta$ *is valid in a class* $\mathbb{M}$ *of models (denoted by* $\mathbb{M} \models \Gamma \Rightarrow \Delta$*), if* $\mathbb{M} \models \Gamma_* \supset \Delta^*$.

By induction on derivation, we obtain the following.

Theorem 2. *Let* $\Gamma \Rightarrow \Delta$ *be any sequent.*

1. *If* $\mathcal{G}(\mathbf{QIEL}^-) \vdash \Gamma \Rightarrow \Delta$ *then* $\mathbb{M}_{\text{all}} \models \Gamma \Rightarrow \Delta$.
2. *If* $\mathcal{G}(\mathbf{QIEL}) \vdash \Gamma \Rightarrow \Delta$ *then* $\mathbb{M}_{\text{ser}} \models \Gamma \Rightarrow \Delta$.

Proposition 9. *For any sequent* $\Gamma \Rightarrow \Delta$,

1. *if* $\mathcal{G}(\mathbf{QIEL}^-) \vdash \Gamma \Rightarrow \Delta$ *then* $\mathcal{H}(\mathbf{QIEL}^-) \vdash \Gamma_* \supset \Delta^*$.
2. *if* $\mathcal{G}(\mathbf{QIEL}) \vdash \Gamma \Rightarrow \Delta$ *then* $\mathcal{H}(\mathbf{QIEL}) \vdash \Gamma_* \supset \Delta^*$.

Theorem 3 (Equipollence). *Let* $\Lambda \in \{\mathbf{QIEL}^-, \mathbf{QIEL}\}$. *For any formula* A, $\vdash_{\mathcal{H}(\Lambda)} A$ *iff* $\vdash_{\mathcal{G}(\Lambda)} \Rightarrow A$.

Proof. The direction from the right to the left for both cases can be proved by applying Proposition 9, in which we let the antecedent Γ be empty. The direction from the left to the right for both cases can be proved by induction on the derivation of A. We only show the case that the axiom (IR) is derivable in $\mathcal{G}(\mathbf{QIEL})$.

$$\dfrac{\dfrac{\dfrac{\dfrac{\dfrac{A \Rightarrow A \qquad \bot \Rightarrow}{A, \neg A \Rightarrow}\,(L\supset)}{KA, \neg A \Rightarrow}\,(K_{\mathbf{IEL}})}{KA, \neg A \Rightarrow \bot}\,(RW)}{KA \Rightarrow \neg\neg A}\,(R\supset)}{\Rightarrow KA \supset \neg\neg A}\,(R\supset)$$

□

3.2 The Cut Elimination Theorems

We show that the cut elimination theorems hold for $\mathcal{G}(\mathbf{QIEL}^-)$ and $\mathcal{G}(\mathbf{QIEL})$, and we employ our proof-theoretic arguments from [7, 10].

Proposition 10. *Let $\Lambda \in \{\mathbf{QIEL}^-, \mathbf{QIEL}\}$. If $\Gamma \Rightarrow \Delta$ is derivable in $\mathcal{G}(\Lambda)$ by $\mathcal{D}$, then $\Gamma(t/x) \Rightarrow \Delta(t/x)$ is also derivable in $\mathcal{G}(\Lambda)$ with the same height as the derivation $\mathcal{D}$.*

Theorem 4 (Cut Elimination Theorem). *Let $\Lambda \in \{\mathbf{QIEL}^-, \mathbf{QIEL}\}$. If $\mathcal{G}(\Lambda) \vdash \Gamma \Rightarrow \Delta$, then $\mathcal{G}^-(\Lambda) \vdash \Gamma \Rightarrow \Delta$.*

Proof. Let $\Lambda = \mathbf{QIEL}$. For the sake of the contraction rules, we show the elimination of the extended form of (Cut) as $(Ecut)$ where $(Ecut)$ has the following form:

$$\dfrac{\dfrac{\vdots\, \mathcal{D}_1}{\Gamma \Rightarrow A}\,rule(\mathcal{D}_1) \qquad \dfrac{\vdots\, \mathcal{D}_2}{A^n, \Gamma' \Rightarrow \Delta}\,rule(\mathcal{D}_2)}{\Gamma, \Gamma' \Rightarrow \Delta,}\,(Ecut)$$

where A^n $(n \geqslant 0)$ means n-times repetition of the formula A and the formula A is called an *Cut formula* simply. It is noted that the ordinary (Cut) becomes a particular instance of $(Ecut)$. We show that, if an $(Ecut)$ only appear in the end of a derivation $\mathcal{D}$, then there is a derivation in which no $(Ecut)$ appears and ended with the same conclusion as $\mathcal{D}$. This can be proved by double induction on the *complexity* (the number of logical connectives of the cut formulas) and the *weight*, i.e., the number of all the sequents in the derivation. We only show the case where both rules above the last application of $(Ecut)$ are rules of $K_{\mathbf{IEL}}$, i.e.,

$$\dfrac{\dfrac{\overset{\mathcal{D}_1}{\Gamma_1, \Gamma_2 \Rightarrow A}}{\Gamma_1, K\Gamma_2 \Rightarrow KA}\,(K_{\mathbf{IEL}}) \qquad \dfrac{\overset{\mathcal{D}_2}{(A)^m, (KA)^n, \Gamma_3, \Gamma_4 \Rightarrow \Delta}}{(KA)^m, (KA)^n, \Gamma_3, K\Gamma_4 \Rightarrow K\Delta}\,(K_{\mathbf{IEL}})}{\Gamma_1, K\Gamma_2, \Gamma_3, K\Gamma_4 \Rightarrow K\Delta}\,(Ecut)$$

Then it suffices for us to transform this derivation into the following.

$$\dfrac{\dfrac{\begin{matrix}\mathcal{D}_1 \\ \Gamma_1, \Gamma_2 \Rightarrow A\end{matrix} \quad \dfrac{\begin{matrix}\mathcal{D}_1 \\ \Gamma_1, K\Gamma_2 \Rightarrow KA\end{matrix} \quad \begin{matrix}\mathcal{D}_2 \\ (A)^m, (KA)^n, \Gamma_3, \Gamma_4 \Rightarrow \Delta\end{matrix}}{(A)^m, \Gamma_1, K\Gamma_2, \Gamma_3, \Gamma_4 \Rightarrow \Delta}\,(Ecut)}{\dfrac{\Gamma_1, \Gamma_2, \Gamma_1, K\Gamma_2, \Gamma_3, \Gamma_4 \Rightarrow \Delta}{\Gamma_1, K\Gamma_2, \Gamma_1, K\Gamma_2, \Gamma_3, K\Gamma_4 \Rightarrow K\Delta}\,(K_{\mathrm{IEL}})}\,(Ecut)}{\Gamma_1, K\Gamma_2, \Gamma_3, K\Gamma_4 \Rightarrow K\Delta.}\,(LC^*)$$

where (LC^*) means finitely many applications of the rule (LC). □

Then we obtain the following proof-theoretic results.

Corollary 1 (Disjunction Property). *Let $\Lambda \in \{\mathbf{QIEL}^-, \mathbf{QIEL}\}$. For any formulas A and B, if $\Rightarrow A \vee B$ is derivable in $\mathcal{G}(\Lambda)$ then either $\Rightarrow A$ or $\Rightarrow B$ is derivable in $\mathcal{G}(\Lambda)$.*

Corollary 2 (Existence Property). *Let $\Lambda \in \{\mathbf{QIEL}^-, \mathbf{QIEL}\}$. For any formula of the form $\exists x A$, if $\Rightarrow \exists x A$ is derivable in $\mathcal{G}(\Lambda)$ then there exists a term t such that $\Rightarrow A(t/x)$ is derivable in $\mathcal{G}(\Lambda)$.*

Next we prove Craig interpolation theorems by Maehara's mathod. In what follows, $\mathsf{Fv}(\Gamma)$ ($\mathsf{Pr}(\Gamma)$ or $\mathsf{C}(\Gamma)$) denotes the set of all free variables (predicate symbols or constant symbols, respectively) in a formula in Γ. We say that $\langle \Gamma_1; \Gamma_2 \rangle$ is a *partition* of Γ if Γ is Γ_1, Γ_2.

Lemma 1. *Let $\Lambda \in \{\mathbf{QIEL}^-, \mathbf{QIEL}\}$. If $\Gamma \Rightarrow \Delta$ is derivable in $\mathcal{G}(\Lambda)$, then for any partition $\langle \Gamma_1; \Gamma_2 \rangle$ of Γ there exists a formula C (interpolant formula) such that both $\Gamma_1 \Rightarrow C$ and $C, \Gamma_2 \Rightarrow \Delta$ are also derivable in $\mathcal{G}(\Lambda)$, $\mathsf{Fv}(C) \subseteq \mathsf{Fv}(\Gamma_1) \cap \mathsf{Fv}(\Gamma_2, \Delta)$, $\mathsf{Pr}(C) \subseteq \mathsf{Pr}(\Gamma_1) \cap \mathsf{Pr}(\Gamma_2, \Delta)$ and $\mathsf{C}(C) \subseteq \mathsf{C}(\Gamma_1) \cap \mathsf{C}(\Gamma_2, \Delta)$.*

Proof. Let $\Lambda \in \{\mathbf{QIEL}^-, \mathbf{QIEL}\}$. If $\vdash_{\mathcal{G}(\Lambda)} \Gamma \Rightarrow \Delta$, then $\vdash_{\mathcal{G}^-(\Lambda)} \Gamma \Rightarrow \Delta$ by Theorem 4. We prove by induction on a cut-free derivation of $\Gamma \Rightarrow \Delta$. We show the case where the last rule in the derivation is (K_{IEL}):

$$\dfrac{\Gamma, \Gamma' \Rightarrow \Delta}{\Gamma, K\Gamma' \Rightarrow K\Delta}\,(K_{\mathrm{IEL}})$$

Let us consider the partition $\langle \Gamma_1, K\Gamma_1'; \Gamma_2, K\Gamma_2' \rangle$ of $\Gamma, K\Gamma'$. From the induction hypothesis, there exists a formula C such that $\Gamma_1, \Gamma_1' \Rightarrow C$ and $C, \Gamma_2, \Gamma_2' \Rightarrow \Delta$ are derivable and C satisfies the required conditions. By the following derivations:

$$\dfrac{\Gamma_1, \Gamma_1' \Rightarrow C}{\Gamma_1, K\Gamma_1' \Rightarrow KC}\,(K_{\mathrm{IEL}}) \qquad \dfrac{C, \Gamma_2, \Gamma_2' \Rightarrow \Delta}{KC, \Gamma_2, K\Gamma_2' \Rightarrow K\Delta}\,(K_{\mathrm{IEL}})$$

we can have an interpolant formula as KC where the conditions of free variables, predicate symbols and constant symbols hold trivially. □

Theorem 5 (Craig Interpolation Theorem). *Let $\Lambda \in \{\mathbf{QIEL}^-, \mathbf{QIEL}\}$. If $\Rightarrow A \supset B$ is derivable in $\mathcal{G}(\Lambda)$, then there exists a formula C such that $\Rightarrow A \supset C$ and $\Rightarrow C \supset B$ are derivable in $\mathcal{G}(\Lambda)$ and that $\mathsf{Fr}(C) \subseteq \mathsf{Fr}(A) \cap \mathsf{Fr}(B)$, $\mathsf{Pr}(C) \subseteq \mathsf{Pr}(A) \cap \mathsf{Pr}(B)$ and $\mathsf{C}(C) \subseteq \mathsf{C}(A) \cap \mathsf{C}(B)$.*

4 Completeness Theorems of First-Order Intuitionistic Epistemic Logics

In this section, even if Θ is a (possibly infinite) set of formulas, we regard $\Theta \Rightarrow \Delta$ as a sequent, in order to prove the completeness of *cut-free* sequent systems $\mathcal{G}^-(\Lambda)$ for $\Lambda \in \{\mathbf{QIEL}^-, \mathbf{QIEL}\}$ with the help of a method from Hermant [3]. Derivability of a possibly infinite sequent is reduced to that of an ordinary sequent as follows.

Definition 11. *Let $\Lambda \in \{\mathbf{QIEL}^-, \mathbf{QIEL}\}$, let $\Theta \cup \Delta$ be a set of formulas, and Δ contains at most one formula. We say $\Theta \Rightarrow \Delta$ is derivable in $\mathcal{G}(\Lambda)$ if the there is a finite subset $\Omega \subseteq \Theta$ such that $\Omega \Rightarrow \Delta$ is derivable in $\mathcal{G}(\Lambda)$.*

Definition 12 (Δ-saturated). *Let $\Lambda \in \{\mathbf{QIEL}^-, \mathbf{QIEL}\}$, let $\Theta \cup \Delta$ be a set of formulas in $\mathcal{L}(\mathcal{C}, \mathcal{V})$, and Δ contains at most one formula. We say that*

- *Θ is Δ-consistent in $\mathcal{G}^-(\Lambda)$ if $\mathcal{G}^-(\Lambda) \nvdash \Theta \Rightarrow \Delta$;*
- *Θ is Δ -complete in $\mathcal{G}^-(\Lambda)$ if for any formulas B in $\mathcal{L}(\mathcal{C}, \mathcal{V})$, either $\mathcal{G}^-(\Lambda) \vdash \Theta \cup \{B\} \Rightarrow \Delta$ or $B \in \Theta$;*
- *Θ is said to admit Δ-Henkin witness in $\mathcal{G}^-(\Lambda)$ for any formula of the form $\exists x B$ in $\mathcal{L}(\mathcal{C}, \mathcal{V})$ if whenever we have $\mathcal{G}^-(\Lambda) \nvdash \Theta \cup \{\exists x B\} \Rightarrow \Delta$, there exists a variable y such that $B(y/x) \in \Theta$.*

If Θ is Δ-consistent, Δ-complete and admitting Δ-Henkin witness in $\mathcal{G}^-(\Lambda)$, we say Θ is Δ-saturated in $\mathcal{G}^-(\Lambda)$.

By the same argument as in Hermant [3, p.225], we obtain the following.

Lemma 2. *Let $\Lambda \in \{\mathbf{QIEL}^-, \mathbf{QIEL}\}$, let $\Gamma \cup \Delta$ be a set of formula in $\mathcal{L}(\mathcal{C}, \mathcal{V})$ and Δ contains at most one formula. If $\mathcal{G}^-(\Lambda) \nvdash \Gamma \Rightarrow \Delta$, then there exists a set Θ of formulas in an expanded language $\mathcal{L}(\mathcal{C}, \mathcal{V} \cup \mathcal{V}')$ with a countable set $\mathcal{V}'$ of fresh variables such that $\Gamma \subseteq \Theta$ and Θ is Δ-saturated in $\mathcal{G}^-(\Lambda)$.*

Definition 13. *Given any set Θ of formulas, $K^-(\Theta) := \{A \mid KA \in \Theta\}$.*

Lemma 3. *Let $\Lambda \in \{\mathbf{QIEL}^-, \mathbf{QIEL}\}$, let $\Theta \cup \Delta$ be a set of formula in $\mathcal{L}(\mathcal{C}, \mathcal{V})$ and Δ contains at most one formula, and Θ be Δ-saturated in $\mathcal{G}^-(\Lambda)$. Then the following hold:*

1. *if $B \wedge C \in \Theta$ then $B \in \Theta$ and $C \in \Theta$.*
2. *if $B \vee C \in \Theta$ then $B \in \Theta$ or $C \in \Theta$.*
3. *if $B \supset C \in \Theta$ then either $C \in \Theta$ either $\mathcal{G}^-(\Lambda) \nvdash \Theta \Rightarrow B$.*
4. *if $\forall x B \in \Theta$ then $B(t/x) \in \Theta$ for any term t in $\mathcal{C} \cup \mathcal{V}$.*
5. *if $\exists x B \in \Theta$ then $B(t/x) \in \Theta$ for some term t in $\mathcal{C} \cup \mathcal{V}$.*
6. *if $\mathcal{G}^-(\Lambda) \nvdash \Theta \Rightarrow B \wedge C$ then $\mathcal{G}^-(\Lambda) \nvdash \Theta \Rightarrow B$ or $\mathcal{G}^-(\Lambda) \nvdash \Theta \Rightarrow C$.*
7. *if $\mathcal{G}^-(\Lambda) \nvdash \Theta \Rightarrow B \vee C$ then $\mathcal{G}^-(\Lambda) \nvdash \Theta \Rightarrow B$ and $\mathcal{G}^-(\Lambda) \nvdash \Theta \Rightarrow C$.*
8. *if $\mathcal{G}^-(\Lambda) \nvdash \Theta \Rightarrow B \supset C$ then $\mathcal{G}^-(\Lambda) \nvdash \Theta, B \Rightarrow C$.*
9. *if $\mathcal{G}^-(\Lambda) \nvdash \Theta \Rightarrow \exists x B$ then for any v in $\mathcal{C} \cup \mathcal{V}$, $\mathcal{G}^-(\Lambda) \nvdash \Theta \Rightarrow B(v/x)$.*

10. *if* $\mathcal{G}^-(\Lambda) \nvdash \Theta \Rightarrow KB$ *then* $\mathcal{G}^-(\Lambda) \nvdash \Theta \cup K^-(\Theta) \Rightarrow B$ *where recall that* $K^-(\Theta) = \{A \,|\, KA \in \Theta\}$.

Proof. We prove item 10 alone. Suppose $\mathcal{G}^-(\Lambda) \nvdash \Theta \Rightarrow KB$. Our goal is to show that $\mathcal{G}^-(\Lambda) \nvdash \Theta \cup K^-(\Theta) \Rightarrow B$. Suppose for contradiction that $\mathcal{G}^-(\Lambda) \vdash \Theta \cup K^-(\Theta) \Rightarrow B$. Then we have the follwoing derivation:

$$\dfrac{\dfrac{\Theta, K^-(\Theta) \Rightarrow B}{\Theta, KK^-(\Theta) \Rightarrow KB}\,(K_{IEL^-})}{\Theta \Rightarrow KB}\,(LC)^*$$

where we have a contradiction with $\mathcal{G}^-(\Lambda) \nvdash \Theta \Rightarrow KB$. □

Definition 14. *Given a language* $\mathcal{L}(\mathcal{C}, \mathcal{V})$ *and a mutually disjoint denumerable family* $\{\mathcal{V}_n\}_{n \in \mathbb{N}}$ *of a denumerable sets of fresh variables such that* $\mathcal{V}_0 = \mathcal{V}$. *We define* $(\mathcal{L}_n)_{n \in \mathbb{N}}$ *inductively as follows: let* $\mathcal{L}_0 = \mathcal{L}(\mathcal{C}, \mathcal{V})$, *and* $\mathcal{L}_n = \mathcal{L}(\mathcal{C}, \bigcup_{0 \leq i \leq n} \mathcal{V}_i)$.

Definition 15. *Let* $\Lambda \in \{\mathbf{QIEL}^-, \mathbf{QIEL}\}$. *The canonical model* $M^\Lambda = (W, \leq, R, D, I)$ *is defined as follows:*

- $W := \{\Theta : \Theta$ *is* Δ*-saturated in* $\mathcal{G}^-(\Lambda)$ *for some set* Δ *of formulas such that* $\Theta \cup \Delta$ *is in some* $\mathcal{L}_i$ *and* Δ *contains at most one formula*$\}$;
- $\Theta \leq \Omega$ *iff* $\Theta \subseteq \Omega$.
- $\Theta R\, \Omega$ *iff* $K^-(\Theta) \cup \Theta \subseteq \Omega$ *where recall that* $K^-(\Theta) = \{B | KB \in \Theta\}$.
- $D(\Theta)$ *is the set of all terms in* Θ.
- $I(c)$ *is the constant* c *in* Θ.
- $I(P, \Theta)$ *is defined as:* $(u_1, ..., u_m) \in I(P, \Theta)$ *iff* $P(u_1, ..., u_m) \in \Theta$.

Lemma 4. *Let* $\Lambda \in \{\mathbf{QIEL}^-, \mathbf{QIEL}\}$. *The canonical model* $M^\Lambda = (W, \leq, R, D, I)$ *is a model and* $M^{\mathbf{QIEL}} = (W, \leq, R, D, I)$ *satisfies the seriality for* R.

Proof. We show the latter statement alone, i.e., for any $w \in W$, there is a $v \in W$ such that wRv. Fix any w as Θ such that Θ is Δ-saturated for some Δ containing at most one formula, then we have $\mathcal{G}^-(\mathbf{QIEL}) \nvdash \Theta \Rightarrow \Delta$ in $\mathcal{L}_i$. Then $\mathcal{G}^-(\mathbf{QIEL}) \nvdash \Theta \cup K^-(\Theta) \Rightarrow$. Then we can construct an $\emptyset$-saturated Ω in $\mathcal{G}^-(\mathbf{QIEL})$ and $\mathcal{L}_{i+1}$ which obtained from $\mathcal{L}_i$ by adding an countably infinite set $\mathcal{V}_{i+1}$ of fresh variables. Let us take this Ω for v. Therefore $\Theta \cup K^-(\Theta) \subseteq \Omega$ hence $\Theta R\, \Omega$. □

Lemma 5. *Let* $\Lambda \in \{\mathbf{QIEL}^-, \mathbf{QIEL}\}$ *and let* Δ *be a set of formula, containing at most one formula. Suppose that* Θ *be* Δ*-saturated in* $\mathcal{L}_i = \mathcal{L} \cup \mathcal{V}_i$. *For every formula* A *in* $\mathcal{L}_i$,

1. *if* $A \in \Theta$ *then* $M^\Lambda, \Theta \models A$;
2. *if* $\mathcal{G}^-(\Lambda) \nvdash \Theta \Rightarrow A$ *then* $M^\Lambda, \Theta \not\models A$.

Proof. We proceed by induction on the complexity of A. Here we deal with the cases where A is of the form KC and $\forall x B$ alone.

- Suppose that A is of the form KC. For item 1, suppose $KC \in \Theta$. We show that $M^{\Lambda}, \Theta \models KC$. Fix any Ω such that $\Theta R \Omega$. Now our goal is to show $M^{\Lambda}, \Omega \models C$. From $\Theta R \Omega$ we have $K^{-}(\Theta) \subseteq \Omega$. Since $C \in K^{-}(\Theta)$, we have $C \in \Omega$. So we conclude that $M^{\Lambda}, \Omega \models C$ by induction hypothesis.
  For item 2, let us suppose $\mathcal{G}^{-}(\Lambda) \nvdash \Theta \Rightarrow KC$. We show $M^{\Lambda}, \Theta \not\models KC$. By item 10 of Lemma 3, we have $\mathcal{G}^{-}(\Lambda) \nvdash \Theta \cup K^{-}(\Theta) \Rightarrow C$. Since $\Theta \cup K^{-}(\Theta) \cup \{C\}$ is in $\mathcal{L}_i$, Lemma 2 enables us to find a $\{C\}$-saturated set Ω in $\mathcal{L}_{i+1}$ such that $\Theta \cup K^{-}(\Theta) \subseteq \Omega$ and $\mathcal{G}^{-}(\Lambda) \nvdash \Omega \Rightarrow C$. From $\Theta \cup K^{-}(\Theta) \subseteq \Omega$ we have $\Theta R \Omega$. It follows from induction hypothesis that $M^{\Lambda}, \Omega \not\models C$. Therefore $M^{\Lambda}, \Theta \not\models KC$. holds.
- Suppose that A is of the form $\forall x B$. For item 1, suppose $\forall x B \in \Theta$ and fix any Ω such that $\Theta \subseteq \Omega$. It is clear that $\forall x B \in \Omega$ holds. It follows from Proposition 3 that $B(t/x) \in \Omega$ for any term t. By induction hypothesis, $\mathcal{G}^{-}(\Lambda), \Theta \models B(t/x)$ holds. Thus we have $M^{\Lambda}, \Theta \models \forall x B$, as desired.
  For item 2, we assume $\mathcal{G}^{-}(\Lambda) \nvdash \Theta \Rightarrow \forall x B$. Let $\mathcal{L}_i$ be the underlying language of $\Theta \cup \{\forall x B\}$. Take an infinite set $\mathcal{V}_{i+1}$ of fresh variables to consider a fresh $v \in \mathcal{L}_{i+1}$. Then it is easy to prove that $\mathcal{G}^{-}(\Lambda) \nvdash \Theta \Rightarrow B(v/x)$ since v is fresh. Then take another infinite set $\mathcal{V}_{i+2}$ of fresh variables to apply Lemma 2 to construct a $\{B(v/x)\}$-saturated Ω in $\mathcal{L}_{i+2}$ such that $\Theta \subseteq \Omega$. It follows from induction hypothesis that $M^{\Lambda}, \Omega \not\models B(v/x)$ hence $M^{\Lambda}, \Theta \not\models \forall x B$ holds.

□

Theorem 6 (Completeness of Cut-free Systems). *Let $\Gamma \Rightarrow \Delta$ be a sequent.*

1. If $\mathbb{M}_{\mathrm{all}} \models \Gamma \Rightarrow \Delta$ then $\mathcal{G}^{-}(\mathbf{QIEL}^{-}) \vdash \Gamma \Rightarrow \Delta$.
2. If $\mathbb{M}_{\mathrm{ser}} \models \Gamma \Rightarrow \Delta$ then $\mathcal{G}^{-}(\mathbf{QIEL}) \vdash \Gamma \Rightarrow \Delta$.

Proof. We only show the second item and so let $\Lambda = \mathcal{G}^{-}(\mathbf{QIEL})$. Suppose $\mathcal{G}^{-}(\mathbf{QIEL}) \nvdash \Gamma \Rightarrow \Delta$. By Lemma 2, we can have a Δ-saturated Θ in a new language such that $\Gamma \subseteq \Theta$ and $\mathcal{G}^{-}(\mathbf{QIEL}) \nvdash \Theta \Rightarrow \Delta$. Therefore, Θ is a state in $M^{\Lambda} \in \mathbb{M}_{ser}$. It follows from Lemma 5 that $M^{\Lambda}, \Theta \models \Gamma \Rightarrow \Delta$. □

Corollary 3. *Let $\Lambda \in \{\mathbf{QIEL}, \mathbf{QIEL}^{-}\}$. The following are all equivalent.*

1. $\mathbb{M}_{\Lambda} \models A$, *2.* $\mathcal{G}^{-}(\Lambda) \vdash \Rightarrow A$, *3.* $\mathcal{G}(\Lambda) \vdash \Rightarrow A$, *4.* $\mathcal{H}(\Lambda) \vdash A$,

where $\mathbb{M}_{\mathbf{IEL}^{-}} := \mathbb{M}_{\mathrm{all}}$ *and* $\mathbb{M}_{\mathbf{IEL}} := \mathbb{M}_{\mathrm{ser}}$.

Proof. The direction from item 1 to item 2 can be obtained from Theorem 6. The direction from item 2 to item 3 is trivial. The direction from item 3 to item 4 can be obtained from the equivalence of the Hilbert systems and the sequent calculi (Theorem 3). Finally, the direction from item 4 to item 1 can be obtained from the soundness of the Hilbert systems for the semantics (Theorem 1). □

In particular, we can also prove the cut elimination theorems semantically by Theorem 2 (the soundness result of the sequent calculi, i.e., the direction from item 3 to item 1) and Theorem 6.

5 Conclusion and Further Direction

There are the following two directions for further research on the first-order intuitionistic epistemic logics. Firstly, the double negation translation (such as Kuroda translation, Kolmogorov translation, and Gödel and Gentzen translation) provides a method to embedding the first-order classical logic into the first-order intuitionistic logic. Then we can ask: what first-order classical epistemic logic can be embedded into $\mathbf{QIEL}^-$ and **QIEL**? Secondly, Hintikka [4] has shown that the equality symbol is useful in the first-order epistemic logic. Furthermore, it is very natural to have the equality symbol in the study of knowledge of mathematics. Therefore, adding the equality symbol into $\mathbf{QIEL}^-$ and **QIEL** can be an important direction.

Acknowledgement. We would like to thank the reviewers for their helpful comments. The work of all authors was partially supported by JSPS KAKENHI Grant-in-Aid for Scientific Research (C) Grant Number 19K12113. The work of the second author was partially supported also by JSPS KAKENHI Grant-in-Aid for Scientific Research (B) Grant Number 17H02258 and JSPS Core-to-Core Program (A. Advanced Research Networks).

References

1. Artemov, S., Protopopescu, T.: Intuitionistic epistemic logic. Rev. Symbolic Logic **9**, 266–298 (2016). https://doi.org/10.1017/S1755020315000374
2. Fitch, F.: A logical analysis of some value concepts. J. Symbolic Logic **28**(2), 135–142 (1963). https://doi.org/10.2307/2271594
3. Hermant, O.: Semantic cut elimination in the intuitionistic sequent calculus. In: Urzyczyn, P. (ed.) TLCA 2005. LNCS, vol. 3461, pp. 221–233. Springer, Heidelberg (2005). https://doi.org/10.1007/11417170_17
4. Hintikka, J.: Knowledge and Belief. College Publications, London (2005)
5. Jäger, G., Marti, M.: A canonical model construction for intuitionistic distributed knowledge. In: Advances in Modal Logic, pp. 420–434 (2016)
6. Jäger, G., Marti, M.: Intuitionistic common knowledge or belief. J. Appl. logic **18**, 150–163 (2016)
7. Kashima, R.: Mathematical logic. Asakura Publishing Co. Ltd., Tokyo (2009). (in Japanese)
8. Krupski, V.N., Yatmanov, A.: Sequent calculus for intuitionistic epistemic logic IEL. In: Artemov, S., Nerode, A. (eds.) LFCS 2016. LNCS, vol. 9537, pp. 187–201. Springer, Cham (2016). https://doi.org/10.1007/978-3-319-27683-0_14
9. Ma, M., Palmigiano, A., Sadrzadeh, M.: Algebraic semantics and model completeness for intuitionistic public announcement logic. Ann. Pure Appl. Logic **165**(4), 963–995 (2014)
10. Ono, H., Komori, Y.: Logics without the contraction rule. J. Symbolic Logic **50**(1), 169–201 (1985)
11. Proietti, C.: Intuitionistic epistemic logic, Kripke models and Fitch's paradox. J. Philos. Logic **41**, 877–900 (2012). https://doi.org/10.1007/s10992-011-9207-1

12. Protopopescu, T.: Intuitionistic epistemology and modal logics of verification. In: van der Hoek, W., Holliday, W.H., Wang, W. (eds.) LORI 2015. LNCS, vol. 9394, pp. 295–307. Springer, Heidelberg (2015). https://doi.org/10.1007/978-3-662-48561-3_24
13. Su, Y., Sano, K.: Double negation translation in logics for knowability paradox. In: Proceedings of the 4th Asian Workshop on Philosophical Logic, pp. 3–15. Tsinghua University, Beijing, 20 October 2018
14. Troelstra, A., van Dalen, D.: Constructivism in Mathematics, Studies in Logic and the Foundations of Mathematics, vol. 1. Elsevier Science, Amsterdam (1988)
15. Williamson, T.: On intuitionistic modal epistemic logic. J. Philos. Logic **21**, 63–89 (1992)

A Neutral Temporal Deontic STIT Logic

Kees van Berkel(✉) and Tim Lyon

Institut für Logic and Computation, Technische Universität Wien, Vienna, Austria
{kees,lyon}@logic.at

Abstract. In this work we answer a long standing request for temporal embeddings of deontic STIT logics by introducing the multi-agent STIT logic TDS. The logic is based upon atemporal utilitarian STIT logic. Yet, the logic presented here will be neutral: instead of committing ourselves to utilitarian theories, we prove the logic TDS sound and complete with respect to relational frames not employing any utilitarian function. We demonstrate how these neutral frames can be transformed into utilitarian temporal frames, while preserving validity. Last, we discuss problems that arise from employing binary utility functions in a temporal setting.

Keywords: Deontic logic · Logics of agency · Modal logic · Multi-agent STIT logic · Temporal logic · Utilitarianism

1 Introduction

With the increasing integration of automated machines in our everyday lives, the development of formal decision-making tools, which take into account moral and legal considerations, is of critical importance [2,9,10]. Unfortunately, one of the fundamental hazards of incorporating ethics into decision-making processes, is the apparent incomparability of quantitative and qualitative information—that is, moral problems most often resist quantification [16].

In contrast, utility functions are useful quantitative tools for the formal analysis of decision-making. Initially formulated in [5], the influential theory of *utilitarianism* has promoted utility calculation as a ground for *ethical deliberation*: in short, those actions generating highest utility, are the morally right actions. For this reason, utilitarianism has proven itself to be a fruitful approach in the field of formal deontic reasoning and multi-agent systems (e.g. [1,12,15]).

In particular, in the field of STIT logic—agency logics developed primarily for the formal analysis of multi-agent choice-making—the utilitarian approach has received increased attention (e.g. [1,15]). Unfortunately, each available utility function comes with its own (dis)advantages, giving rise to several puzzles (some of them addressed in [12,13]). To avoid such problems, we provide an alternative approach: instead of settling these philosophical issues, we develop a neutral formalism that can be appropriated to different utilitarian value assignments.

Work funded by the projects WWTF MA16-028, FWF I2982 and FWF W1255-N23.

P. Blackburn et al. (Eds.): LORI 2019, LNCS 11813, pp. 340–354, 2019.
https://doi.org/10.1007/978-3-662-60292-8_25

The paper's contributions can be summed up as follows: First, we provide a temporal deontic STIT logic called TDS (Sect. 2). With this logic, we answer a long standing request for temporal embeddings of deontic STIT [4,12,15]. Second, although TDS is based upon the atemporal utilitarian STIT logic from [15], the semantics of TDS will be neutral: instead of committing to utilitarianism, we prove soundness and completeness of TDS with respect to relational frames not employing any utilitarian function (Sect. 3). This approach also extends the results in [3,11,14] by showing that TDS can be characterized without using the traditional branching-time (BT+AC) structures (cf. [4]). Third, we show how neutral TDS frames can be transformed into utilitarian frames, while preserving validity (Sect. 4). Last, we discuss the philosophical ramifications of employing available utility functions in the extended, temporal setting. In particular, we will argue that binary utility assignments can turn out to be problematic.

2 A Neutral Temporal Deontic STIT Logic

In this section, we introduce the language, semantics, and axiomatization of the temporal deontic STIT logic TDS. In particular, we provide neutral relational frames characterizing the logic, which omit mention of specific utility functions. The logic will bring together atemporal deontic STIT logic, presented in [15], and the temporal STIT logic from [14].

Definition 1 (The Language $\mathcal{L}_{\mathsf{TDS}}$). *Let $Ag = \{1, 2, \ldots, n\}$ be a finite set of agent labels and let $Var = \{p_1, p_2, p_3 \ldots\}$ be a countable set of propositional variables. The language $\mathcal{L}_{\mathsf{TDS}}$ is given by the following BNF grammar:*

$$\phi ::= p \mid \neg\phi \mid \phi \wedge \phi \mid \Box\phi \mid [i]\phi \mid [Ag]\phi \mid \mathsf{G}\phi \mid \mathsf{H}\phi \mid \otimes_i \phi$$

where $i \in Ag$ and $p \in Var$.

The logical connectives disjunction $\vee$, implication $\rightarrow$, and bi-conditional $\leftrightarrow$ are defined in the usual way. Let $\bot$ be defined as $p\wedge\neg p$ and define $\top$ to be $p\vee\neg p$. The language consists of single agent STIT operators $[i]$, which are choice-operators describing that 'agent i sees to it that', and the grand coalition operator $[Ag]$, expressing 'the grand coalition of agents sees to it that'. Furthermore, it contains a settledness operator $\Box$, which holds true of a formula that is settled true at a moment, and thus, holds true regardless of the choices made by any of the agents at that moment. The operators G and H have, respectively, the usual temporal interpretation 'always going to be' and 'always has been'. Last, the operator $\otimes_i$ expresses 'agent i ought to see to it that'. We define $\Diamond$, $\langle i\rangle$, $\langle Ag\rangle$ and $\ominus_i$ as the *duals* of $\Box, [i], [Ag]$ and $\otimes_i$, respectively (i.e. $\Diamond$ iff $\neg\Box\neg\phi$, etc.). Furthermore, let $\mathsf{F}\phi$ iff $\neg\mathsf{G}\neg\phi$ and $\mathsf{P}\phi$ iff $\neg\mathsf{H}\neg\phi$, expressing 'ϕ holds somewhere in the future' and 'ϕ holds somewhere in the past', respectively. Finally, deliberative STIT and deliberative ought are obtained accordingly: $[i]^d\phi$ iff $[i]\phi \wedge \Diamond\neg\phi$ and $\otimes_i^d\phi$ iff $\otimes_i\phi \wedge \Diamond\neg\phi$. For a discussion of these operators we refer to [12,14].

In line with [3,6,11,14], we provide relational frames for TDS instead of introducing the traditionally employed, BT+AC frames (cf. [4]). Explanations of the individual frame properties of Definition 2 can be found below.

Definition 2 (Relational TDS Frames and Models). *A* TDS*-frame is defined as a tuple* $F = (W, \mathcal{R}_{\Box}, \{\mathcal{R}_{[i]} \mid i \in Ag\}, \mathcal{R}_{[Ag]}, \mathcal{R}_{\mathsf{G}}, \mathcal{R}_{\mathsf{H}}, \{\mathcal{R}_{\otimes_i} \mid i \in Ag\})$. *Let* $\mathcal{R}_{[\alpha]}(w) := \{v \in W | (w,v) \in R_{[\alpha]}\}$ *for* $[\alpha] \in \mathsf{Boxes}$ *where* $\mathsf{Boxes} := \{\Box, \mathsf{G}, \mathsf{H}, [Ag]\} \cup \{[i] \mid i \in Ag\} \cup \{\otimes_i \mid i \in Ag\}$. *Let* W *be a non-empty set of worlds* $w, v, u \ldots$ *and:*

- *For all* $i \in Ag$, $\mathcal{R}_{\Box}, \mathcal{R}_{[i]}, \mathcal{R}_{[Ag]} \subseteq W \times W$ *are equivalence relations such that:*
  - **(C1)** $\mathcal{R}_{[i]} \subseteq \mathcal{R}_{\Box}$.
  - **(C2)** *For all* $u_1, ..., u_n \in W$, *if* $\mathcal{R}_{\Box} u_i u_j$ *for all* $1 \leq i,j \leq n$, *then* $\bigcap_i \mathcal{R}_{[i]}(u_i) \neq \emptyset$.
  - **(C3)** *For all* $w \in W$, $\mathcal{R}_{[Ag]}(w) \subseteq \bigcap_{i \in Ag} \mathcal{R}_{[i]}(w)$.
- $\mathcal{R}_{\mathsf{G}} \subseteq W \times W$ *is a transitive and serial binary relation and* $\mathcal{R}_{\mathsf{H}}$ *is the converse of* $\mathcal{R}_{\mathsf{G}}$, *such that:*
  - **(T4)** *For all* $w, u, v \in W$, *if* $\mathcal{R}_{\mathsf{G}} wu$ *and* $\mathcal{R}_{\mathsf{G}} wv$, *then* $\mathcal{R}_{\mathsf{G}} uv$, $u = v$, *or* $\mathcal{R}_{\mathsf{G}} vu$.
  - **(T5)** *For all* $w, u, v \in W$, *if* $\mathcal{R}_{\mathsf{H}} wu$ *and* $\mathcal{R}_{\mathsf{H}} wv$, *then* $\mathcal{R}_{\mathsf{H}} uv$, $u = v$, *or* $\mathcal{R}_{\mathsf{H}} vu$.
  - **(T6)** $\mathcal{R}_{\mathsf{G}} \circ \mathcal{R}_{\Box} \subseteq \mathcal{R}_{[Ag]} \circ \mathcal{R}_{\mathsf{G}}$ *(relation composition* $\circ$ *is defined as usual).*
  - **(T7)** *For all* $w, u \in W$, *if* $u \in \mathcal{R}_{\Box}(w)$, *then* $u \notin \mathcal{R}_{\mathsf{G}}(w)$.
- *For all* $i \in Ag$, $\mathcal{R}_{\otimes_i} \subseteq W \times W$ *are binary relations such that:*
  - **(D8)** $\mathcal{R}_{\otimes_i} \subseteq \mathcal{R}_{\Box}$.
  - **(D9)** *For all* $w \in W$ *there exists a* $v \in W$ *such that* $\mathcal{R}_{\Box} wv$ *and for all* $u \in W$, *if* $\mathcal{R}_{[i]} vu$ *then* $\mathcal{R}_{\otimes_i} wu$.
  - **(D10)** *For all* $w, v, u, z \in W$, *if* $\mathcal{R}_{\Box} wv, \mathcal{R}_{\Box} wu$ *and* $\mathcal{R}_{\otimes_i} uz$, *then* $\mathcal{R}_{\otimes_i} vz$.
  - **(D11)** *For all* $w, v \in W$, *if* $\mathcal{R}_{\otimes_i} wv$ *then there exists* $u \in W$ *s.t.* $\mathcal{R}_{\Box} wu$, $\mathcal{R}_{[i]} uv$, *and for all* $z \in W$, *if* $\mathcal{R}_{[i]} uz$ *then* $\mathcal{R}_{\otimes_i} wz$.

A TDS*-model is a tuple* $M = (F, V)$ *where* F *is a* TDS*-frame and* V *is a valuation mapping propositional variables to subsets of* W, *that is,* $V \colon Var \rightarrow \mathcal{P}(W)$.

We label the properties of Definition 2 referring to choice **(Ci)**, those relating to temporal aspects **(Ti)**, and those capturing deontic properties **(Di)**. Observe that, since $\mathcal{R}_{\Box}$ is an equivalence relation, we obtain equivalence classes $\mathcal{R}_{\Box}(w) = \{v \mid (w,v) \in \mathcal{R}_{\Box}\}$. Furthermore, by condition **(C1)** we know that $\mathcal{R}_{[i]}$ is an equivalence relation partitioning the equivalence classes of $\mathcal{R}_{\Box}$. We call $\mathcal{R}_{\Box}(w)$ a *moment* and for each v in a moment $\mathcal{R}_{\Box}(w)$, we refer to $\mathcal{R}_{[i]}(v)$ as a *choice-cell* for agent i at moment $\mathcal{R}_{\Box}(w)$. In the following, we shall frequently refer to moments and choices in the above sense. Condition **(C2)** captures the pivotal *independence of agents* principle for STIT logics, ensuring that at every moment, any combination of different agents' choices is consistent: i.e., simultaneous choices are independent (see [4, 7C.4]). **(C3)** ensures that all agents acting together is a necessary condition for the grand coalition of agents acting.[1]

[1] In future work, we aim to study condition (C3) strengthened to equality, as in [14]. In such a setting, completeness is obtained by proving that each TDS-frame can be transformed into a frame (satisfying the same formulae) with strengthened (C3); hence, showing that the logic does not distinguish between the two frame classes.

The conditions on $\mathcal{R}_{\mathsf{G}}$ and $\mathcal{R}_{\mathsf{H}}$ establish that the frames we consider are irreflexive, temporal orderings of *moments*. Properties **(T4)** and **(T5)** guarantee that *histories*—i.e., maximally ordered paths of worlds passing through moments—are linear. Condition **(T6)** ensures the STIT principle of *no choice between undivided histories*: if two time-lines remain undivided at the next moment, no agent has a choice that realizes one time-line and excludes the other (see [4, 7C.3]). Consequently, this principle also ensures that the ordering of moments is linearly closed with respect to the past and allows for branching with respect to the future: in other words, TDS-frames are *treelike*.[2] Last, **(T7)** ensures the temporal irreflexivity of moments; i.e., the future excludes the present. For an elaborate discussion of the temporal frame conditions we refer to [14].

Last, the criteria **(D8)**-**(D11)** guarantee an essentially agentive characterization of the obligation operator $\otimes_i$ (cf. the impartial 'ought to be' operator in [12]). Condition **(D8)** ensures that ideal worlds are confined to moments: i.e., the ideal worlds accessible at a moment neither lie in the future nor in the past. **(D9)** ensures that, for each agent there is at every moment a choice available that is an ideal choice (cf. the corresponding 'ought implies can' axiom *A*14). Furthermore, **(D10)** expresses that, for each agent, if a world is ideal from the perspective of a particular world at a moment, that world is ideal from the perspective of any world at that moment: i.e., ideal worlds are settled upon moments. Condition **(D11)** captures the idea that every ideal world extends to a complete ideal choice: i.e., no choice contains both ideal and non-ideal worlds. Last, note that conditions **(C2)** and **(D9)** together ensure that every combination of distinct agents' ideal choices is consistent, i.e., non-empty.

Definition 3 (Semantics for $\mathcal{L}_{\mathsf{TDS}}$). *Let M be a* TDS*-model and let $w \in W$ of M. The* satisfaction *of a formula $\phi \in \mathcal{L}_{\mathsf{TDS}}$ in M at w is defined accordingly:*

1. $M, w \models p$ *iff* $w \in V(p)$
2. $M, w \models \neg\phi$ *iff* $M, w \not\models \phi$
3. $M, w \models \phi \land \psi$ *iff* $M, w \models \phi$ *and* $M, w \models \psi$
4. $M, w \models \Box\phi$ *iff* $\forall u \in \mathcal{R}_{\Box}(w)$, $M, u \models \phi$
5. $M, w \models [i]\phi$ *iff* $\forall u \in \mathcal{R}_{[i]}(w)$, $M, u \models \phi$
6. $M, w \models \otimes_i\phi$ *iff* $\forall u \in \mathcal{R}_{\otimes_i}(w)$, $M, u \models \phi$
7. $M, w \models [Ag]\phi$ *iff* $\forall u \in \mathcal{R}_{[Ag]}(w)$, $M, u \models \phi$
8. $M, w \models \mathsf{G}\phi$ *iff* $\forall u \in \mathcal{R}_{\mathsf{G}}(w)$, $M, u \models \phi$
9. $M, w \models \mathsf{H}\phi$ *iff* $\forall u \in \mathcal{R}_{\mathsf{H}}(w)$, $M, u \models \phi$

Global truth, validity, and semantic entailment are defined as usual (see [7]).

The axiomatization of TDS is a composition of [15], together with [14]. (Note that in the language $\mathcal{L}_{\mathsf{TDS}}$ each agent label represents a distinct agent.)

Definition 4 (Axiomatization of TDS). *For each $i \in Ag$ we have,*

A0 *All propositional tautologies.*
A1 $\Box(\phi \to \psi) \to (\Box\phi \to \Box\psi)$,
A2 $\Box\phi \to \phi$
A3 $\Diamond\phi \to \Box\Diamond\phi$
A4 $[i](\phi \to \psi) \to ([i]\phi \to [i]\psi)$
A5 $[i]\phi \to \phi$
A6 $\langle i\rangle\phi \to [i]\langle i\rangle\phi$
A7 $[Ag](\phi \to \psi) \to ([Ag]\phi \to [Ag]\psi)$
A8 $[Ag]\phi \to \phi$
A9 $\langle Ag\rangle\phi \to [Ag]\langle Ag\rangle\phi$

[2] The main reason why the grand coalition operator $[Ag]$ is added to our language, is because it will allow us to axiomatize the *no choice between undivided histories* principle (see A25 of Definition 4). For a discussion of $[Ag]$ we refer to [14].

A10 $\bigwedge_{0 \leq i \leq n} \Diamond [i]\phi_k \rightarrow \Diamond \bigwedge_{0 \leq i \leq n} [i]\phi_k$	*A20* $\mathsf{H}(\phi \rightarrow \psi) \rightarrow (\mathsf{H}\phi \rightarrow \mathsf{H}\psi)$
A11 $\bigwedge_{1 \leq i \leq n} [i]\phi_i \rightarrow [Ag] \bigwedge_{1 \leq i \leq n} \phi_i$	*A21* $\phi \rightarrow \mathsf{GP}\phi$
A12 $\otimes_i(\phi \rightarrow \psi) \rightarrow (\otimes_i \phi \rightarrow \otimes_i \psi)$	*A22* $\phi \rightarrow \mathsf{HF}\phi$
A13 $\Box\phi \rightarrow ([i]\phi \wedge \otimes_i \phi)$	*A23* $\mathsf{FP}\phi \rightarrow \mathsf{P}\phi \vee \phi \vee \mathsf{F}\phi$
A14 $\otimes_i \phi \rightarrow \Diamond [i]\phi$	*A24* $\mathsf{PF}\phi \rightarrow \mathsf{P}\phi \vee \phi \vee \mathsf{F}\phi$
A15 $\Diamond \otimes_i \phi \rightarrow \Box \otimes_i \phi$	*A25* $\mathsf{F}\Diamond\phi \rightarrow \langle Ag \rangle \mathsf{F}\phi$
A16 $\Box([i]\phi \rightarrow [i]\psi) \rightarrow (\otimes_i \phi \rightarrow \otimes_i \psi)$	*R0* $\vdash_{\mathsf{TDS}} (\psi \rightarrow \phi)$ *and* $\vdash_{\mathsf{TDS}} \psi$ *implies* $\vdash_{\mathsf{TDS}} \phi$
A17 $\mathsf{G}(\phi \rightarrow \psi) \rightarrow (\mathsf{G}\phi \rightarrow \mathsf{G}\psi)$	*R1* $\vdash_{\mathsf{TDS}} \phi$ *implies* $\vdash_{\mathsf{TDS}} [\alpha]\phi$, $[\alpha] \in \{\Box, \mathsf{G}, \mathsf{H}\}$
A18 $\mathsf{G}\phi \rightarrow \mathsf{GG}\phi$	*R2* $\vdash_{\mathsf{TDS}} (\Box\neg p \wedge \Box(\mathsf{G}p \wedge \mathsf{H}p)) \rightarrow \phi$ *implies* $\vdash_{\mathsf{TDS}} \phi$, *given* $p \notin \phi$
A19 $\mathsf{G}\phi \rightarrow \mathsf{F}\phi$	

A derivation of ϕ *in* TDS *from a set* Γ, *written* $\Gamma \vdash_{\mathsf{TDS}} \phi$, *is defined in the usual way (See [7, Def. 4.4]). When* $\Gamma = \emptyset$, *we say* ϕ *is a* theorem, *and write* $\vdash_{\mathsf{TDS}} \phi$.

The axioms, *A*1–*A*3, *A*4–*A*6 and *A*7–9 express the S5 behavior of $\Box$, $[i]$ (for each $i \in Ag$) and $[Ag]$, respectively. *A*10 is the *independence of agents* axiom. *A*11 captures that 'all agents acting together implies the grand coalition of agents acting'. *A*13 is a bridge axiom linking $\otimes_i$ to $\Box$ and $[i]$ to $\Box$ (cf. (C1) and (D8) of Definition 2). *A*14 corresponds to the 'ought implies can' principle (cf. (D9) of Definition 2). *A*15 ensures that, when possible, obligatory choices are settled upon moments (cf. (D10) of Definition 2). *A*16 can be understood as a conditional monotonicity principle for ideal choices (cf. (D11) of Definition 2). Axioms *A*12 and *A*13, together with the necessitation rule *R*1, ensure that $\otimes_i$ is a normal modal operator.

With respect to the temporal axioms, *A*17–*A*19 capture the KD4 behavior of G, whereas, axioms *A*21 and *A*22 ensure that H is the converse of G. *A*23 and *A*24 capture *connectedness* of histories through moments and *A*25 characterizes *no choice between undivided histories*. Last, *R*2 is a variation of Gabbay's irreflexivity rule (the proofs of Theorem 1 and 2 give an indication of the rule's functions).

3 Soundness and Completeness of TDS

In this section, we prove that TDS is sound and complete relative to the class of TDS-frames. In the next section, we show how such frames are transformable into frames employing utility assignments. This allows one to model and reason about utilitarian scenarios in a more fine-grained manner, while obtaining completeness of the logic without commitment to particular utility functions.

Unless stated otherwise, all proofs in this section can be found in the online appended version (available at http://arxiv.org/abs/1907.03265).

Theorem 1. (SOUNDNESS OF TDS) $\forall \phi \in \mathcal{L}_{\mathsf{TDS}}$, $\vdash_{\mathsf{TDS}} \phi$ *implies* $\models \phi$.

We prove completeness by constructing maximal consistent sets belonging to a special class and build a canonical TDS model adopting methods from [8,14].

Definition 5. *A set of formulae* $\Gamma \subseteq \mathcal{L}_{\mathsf{TDS}}$ *is a* maximally consistent set (MCS) *iff (i)* $\Gamma \not\vdash_{\mathsf{TDS}} \bot$, *and (ii) for any set* $\Gamma' \subseteq \mathcal{L}_{\mathsf{TDS}}$, *if* $\Gamma \subset \Gamma'$, *then* $\Gamma' \vdash_{\mathsf{TDS}} \bot$.

Definition 6 (CANONICAL MODEL FOR TDS). *Let* $[\alpha] \in \mathsf{Boxes}$ *and let* $\langle\alpha\rangle$ *be the operator dual to* $[\alpha]$. *We define the* canonical model *to be the tuple* $M^{dt} := (W^{dt}, \mathcal{R}^{dt}_{\Box}, \{\mathcal{R}^{dt}_{[i]} \mid i \in Ag\}, \mathcal{R}^{dt}_{[Ag]}, \mathcal{R}^{dt}_{\mathsf{G}}, \mathcal{R}^{dt}_{\mathsf{H}}, \{\mathcal{R}^{dt}_{\otimes_i} \mid i \in Ag\}, V^{dt})$ *such that:*

- $W^{dt} := \{\Gamma \subset \mathcal{L}_{\mathsf{TDS}} \mid \Gamma \text{ is an } MCS\}$;
- *for all* $\Gamma, \Delta \in W^{dt}$, $(\Gamma, \Delta) \in \mathcal{R}^{dt}_{[\alpha]}$ *iff for all* $\phi \in \mathcal{L}_{\mathsf{TDS}}$, *if* $[\alpha]\phi \in \Gamma$, *then* $\phi \in \Delta$ *(for each* $[\alpha] \in \mathsf{Boxes}$*);*
- V^{dt} *is a valuation function s.t.* $\forall p \in Atom$, $V^{dt}(p) := \{\Delta \in W^{dt} \mid p \in \Delta\}$.

Definition 7 (DIAMOND SATURATED SET [14]). *Let* X *be a set of MCSs and let* $\langle\alpha\rangle$ *be dual to* $[\alpha] \in \mathsf{Boxes}$. *We say that* X *is a* diamond saturated set *iff for all* $\Gamma \in X$, *for each* $\langle\alpha\rangle\phi \in \Gamma$ *there exists a* $\Delta \in X$ *such that* $\mathcal{R}_{[\alpha]}\Gamma\Delta$ *and* $\phi \in \Delta$.

In order to ensure that our canonical model will be irreflexive, we introduce a mechanism that allows us to encode MCSs with information that impedes reflexive points in the model. We call these encoded sets IRR-theories and restrict our canonical model to consist of these sets only. Last, we use the notation $M|_X$ to indicate a model M whose domain is restricted to the set X (see [8, Ch.6]).

Lemma 1. *Let* X *be a diamond saturated set with* $\Gamma \in X$, $\phi \in \mathcal{L}_{\mathsf{TDS}}$, *and let* $M^{dt}|_X$ *be the canonical model restricted to* X. *Then,* $M^{dt}|_X, \Gamma \models \phi$ *iff* $\phi \in \Gamma$.

Proof. Proven in the usual manner by induction on ϕ (see [7, Lem. 4.70]).

Following [14], we let IRR-theories be those sets of TDS formulae that (i) are maximally consistent, (ii) contain a label $name(p) := \Box\neg p \wedge \Box(\mathsf{G}p \wedge \mathsf{H}p)$, uniquely labeling a *moment* and (iii) for any world that is reachable through any 'zig-zagging' sequence of diamond operators, that is, every zig-zagging formula ϕ of the form,

$$\langle\alpha_1\rangle(\phi_1 \wedge \langle\alpha_2\rangle(\phi_2 \wedge \ldots \wedge \langle\alpha_n\rangle\phi_n)) \ldots)$$

where $\langle\alpha_i\rangle$ is dual to $[\alpha_i] \in \mathsf{Boxes}$ with $1 \leq i \leq n$, there exists a corresponding zig-zagging formula $\phi(q)$ (where q is a propositional variable) of the form,

$$\langle\alpha_1\rangle(\phi_1 \wedge \langle\alpha_2\rangle(\phi_2 \wedge \ldots \wedge \langle\alpha_n\rangle(\phi_n \wedge \Box\neg q \wedge \Box(\mathsf{G}q \wedge \mathsf{H}q))) \ldots)$$

labeling reachable worlds. Let us make the above formally precise:

Definition 8 (IRR-THEORY) [14]. *Let* Zig *be the set of all zig-zagging formulae in* $\mathcal{L}_{\mathsf{TDS}}$ *and let* name(p)$:= \Box\neg p \wedge \Box(\mathsf{G}p \wedge \mathsf{H}p)$ *where* p *is a propositional variable. A set of formulae* Γ *is called an* IRR-theory *iff the following hold:*

- Γ *is a MCS and* $name(p) \in \Gamma$, *for some propositional variable* p;
- *if* $\phi \in \Gamma \cap \mathsf{Zig}$, *then* $\phi(q) \in \Gamma$, *for some propositional variable* q.

Henceforth, we refer to IRR *as the set of all IRR-theories in* $\mathcal{L}_{\mathsf{TDS}}$.

We now present lemmata relevant to the use of IRR-theories in canonical models.

Lemma 2. *Let* $\phi \in \mathcal{L}_{\mathsf{TDS}}$ *be a consistent formula. Then, there exists an IRR-theory* Γ *such that* $\phi \in \Gamma$.

Lemma 3 (EXISTENCE LEMMA). *Let Γ be an IRR-theory and let $\langle \alpha \rangle$ be dual to $[\alpha] \in \mathsf{Boxes}$. For each $\langle \alpha \rangle \phi \in \Gamma$ there exists an IRR-theory Δ such that $\mathcal{R}_{[\alpha]}\Gamma\Delta$.*

Subsequently, it must be shown that the canonical model *restricted* to the set IRR of IRR-theories (i.e., $M^{dt}|_{\mathsf{IRR}}$) is in fact a TDS model (henceforth, we use W^{dt} and IRR interchangeably). First, we provide lemmata ensuring that the model satisfies the desired temporal and deontic properties of Definition 2. The first two follow from [14] and the latter four results are proven in the online appended version (available at http://arxiv.org/abs/1907.03265).

Lemma 4 ([14])(PROPERTY (C2)). *Let $\Gamma_1, \ldots, \Gamma_n \in \mathsf{IRR}$ such that $\mathcal{R}^{dt}_{\Box}\Gamma_i\Gamma_j$ for all $1 \leq i, j \leq n$. Then, there exists a $\Delta \in \mathsf{IRR}$ such that $\mathcal{R}^{dt}_1\Gamma_1\Delta, \ldots, \mathcal{R}^{dt}_n\Gamma_n\Delta$.*

Lemma 5 ([14])(PROPERTY (T6)). *Let $\Gamma, \Sigma, \Pi \in \mathsf{IRR}$ such that $\mathcal{R}^{dt}_{\mathsf{G}}\Gamma\Sigma$ and $\mathcal{R}^{dt}_{\Box}\Sigma\Pi$. Then, there exists a $\Delta \in \mathsf{IRR}$ such that $\mathcal{R}^{dt}_{[Ag]}\Gamma\Delta$ and $\mathcal{R}^{dt}_{\mathsf{G}}\Delta\Pi$.*

Lemma 6 (PROPERTY (D9)). *Let $\Gamma \in \mathsf{IRR}$. Then, there exists a $\Delta \in \mathsf{IRR}$ such that $\mathcal{R}^{dt}_{\Box}\Gamma\Delta$ and for every $\Sigma \in \mathsf{IRR}$, if $\mathcal{R}^{dt}_{[i]}\Delta\Sigma$, then $\mathcal{R}^{dt}_{\otimes_i}\Gamma\Sigma$.*

Lemma 7 (PROPERTY (D11)). *Let $\Gamma, \Delta \in \mathsf{IRR}$ such that $\mathcal{R}^{dt}_{\otimes_i}\Gamma\Delta$. Then, there exists a $\Sigma \in \mathsf{IRR}$ such that $\mathcal{R}^{dt}_{\Box}\Gamma\Sigma$, $\mathcal{R}^{dt}_{[i]}\Sigma\Delta$, and for all $\Pi \in \mathsf{IRR}$, if $\mathcal{R}^{dt}_{[i]}\Sigma\Pi$, then $\mathcal{R}^{dt}_{\otimes_i}\Gamma\Pi$.*

Lemma 8. *The canonical model $M^{dt}|_{\mathsf{IRR}}$ belongs to the class of* TDS *models.*

Theorem 2 (COMPLETENESS). *If $\phi \in \mathcal{L}_{\mathsf{TDS}}$ is a consistent formula, then ϕ is satisfiable on a* TDS*-model.*

4 Transformations to Utilitarian Models

In this section, we investigate a truth preserving transformation from TDS models to *utilitarian* STIT models, embedded in a temporal language. In particular, we are concerned with the semantic characterization of the *dominant ought* [12, Ch.4]. We start with defining the semantic machinery needed to treat these oughts. In particular, we will introduce a utility function *util* that maps natural numbers (i.e. utilities) to worlds in our domain. In contrast to [12,15], we do not restrict the assignment of utilities to complete histories where all worlds on a maximal linear path have identical utility. The reason will be addressed at the end of the section, where we discuss a problem related to utility assignments over histories, arising in temporal extensions of STIT.

The pivotal notion involved in the dominant ought is that of a *state*: Agent i cannot influence the choices of all other agents and, for this reason, one can regard the joint interaction of all agents excluding i, as a state (of nature) for i. To be more precise, we define a *state $\mathcal{R}^s_{[i]}(v)$ for i at v* accordingly,

$$\mathcal{R}^s_{[i]}(v) = \bigcap_{k \in Ag \setminus \{i\}} \mathcal{R}_k(v)$$

Consequently, all possible combinations of choices available to the agents $Ag\backslash\{i\}$, are the different states available at that moment to agent i.

Subsequently, we define a *preference order* $\leq$ over choices (and subsets thereof). Let $\mathcal{R}_{[i]}(v), \mathcal{R}_{[i]}(z) \subseteq \mathcal{R}_{\Box}(w)$, then weak preference is defined accordingly,

$$\mathcal{R}_{[i]}(v) \leq \mathcal{R}_{[i]}(z) \iff \forall v^* \in \mathcal{R}_{[i]}(v), \forall z^* \in \mathcal{R}_{[i]}(z), util(v^*) \leq util(z^*)$$

That is, for an agent a choice is weakly preferred over another, when all values of the possible outcomes of the former are at least as high as those of the latter (where $util(v)$ is the number assigned to v, etc.). Strict preference is defined as,

$$\mathcal{R}_{[i]}(v) < \mathcal{R}_{[i]}(z) \iff \mathcal{R}_{[i]}(v) \leq \mathcal{R}_{[i]}(z) \wedge \mathcal{R}_{[i]}(z) \not\leq \mathcal{R}_{[i]}(v)$$

Next, a *dominance order* $\preceq$ over choices $\mathcal{R}_{[i]}(v), \mathcal{R}_{[i]}(z) \subseteq \mathcal{R}_{\Box}(w)$ is defined as,

$$\mathcal{R}_{[i]}(v) \preceq \mathcal{R}_{[i]}(z) \iff \forall \mathcal{R}^s_{[i]}(x) \subseteq \mathcal{R}_{\Box}(w), \mathcal{R}_{[i]}(v) \cap \mathcal{R}^s_{[i]}(x) \leq \mathcal{R}_{[i]}(z) \cap \mathcal{R}^s_{[i]}(x)$$

We say an agent's choice weakly dominates another, if the values of the outcomes of the former are weakly preferred to those of the latter choice, *given any possible state available to that agent.* For a discussion of dominance orderings see [12, Ch. 4]. Again, in the usual way we obtain *strict dominance*,

$$\mathcal{R}_{[i]}(v) \prec \mathcal{R}_{[i]}(z) \iff \mathcal{R}_{[i]}(v) \preceq \mathcal{R}_{[i]}(z) \wedge \mathcal{R}_{[i]}(z) \not\preceq \mathcal{R}_{[i]}(v)$$

On the basis of the above, we now formally introduce temporal *utilitarian* STIT frames and models, defined over *relational* Kripke frames.

Definition 9 (Relational TUS Frames and Models). *Let* $\mathcal{R}_{[\alpha]}(w) := \{v \in W | (w, v) \in R_\alpha\}$ *for* $[\alpha] \in \{\Box, [Ag], \mathsf{G}, \mathsf{H}\} \cup \{[i] | i \in Ag\}$. *A relational* Temporal Utilitarian STIT frame (TUS-frame) *is defined as a tuple* $F = (W, \mathcal{R}_{\Box}, \{\mathcal{R}_{[i]} | i \in Ag\}, \mathcal{R}_{[Ag]}, \mathcal{R}_{\mathsf{G}}, \mathcal{R}_{\mathsf{H}}, util)$ *where* W *is a non-empty set of worlds* $w, v, u \ldots$ *and:*

- *For all* $i \in Ag$, $\mathcal{R}_{\Box}$, $\mathcal{R}_{[i]}$, $\mathcal{R}_{[Ag]} \subseteq W \times W$ *are equivalence relations for which conditions* **(C1)-(C3)** *of Definition 2 hold.*
- $\mathcal{R}_{\mathsf{G}} \subseteq W \times W$ *is a transitive and serial binary relation, whereas* $\mathcal{R}_{\mathsf{H}}$ *is the converse of* $\mathcal{R}_{\mathsf{G}}$, *and the conditions* **(T4)-(T7)** *of Definition 2 hold.*
- $util : W \mapsto \mathbb{N}$ *is a utility function assigning each world in* W *to a natural.*

A TUS*-model is a tuple* $M = (F, V)$ *where* F *is a* TUS*-frame and* V *is a valuation function assigning propositional variables to subsets of* W*: i.e.,* $V\colon Var \mapsto \mathcal{P}(W)$.

Notice that the above TUS frames only differ from TDS frames through replacing the relations $\mathcal{R}_{\otimes_i}$ and corresponding conditions **(D8)-(D11)** (for each $i \in Ag$) with the utility function *util*. We observe that the assignment of utilities to worlds is agent-independent. Nevertheless, since the choices of an agent depend on which worlds are inside the choice-cells available to the agent, the resulting obligations are in fact agent-dependent. Let us define the new semantics:

Definition 10 (Semantics of TUS models). *Let M be a* TUS*-model, $w \in W$ of M and $\|\phi\|_M = \{w \mid M, w \models \phi\}$. We define* satisfaction *of a formula $\phi \in \mathcal{L}_{\mathsf{TDS}}$ as follows:*

- *Clause (1)-(10) are the same as those from Definition 3, with the exception of clause (7), which we replace by the following clause (7*):*

$$M, w \models \otimes_i\phi \text{ iff } \forall \mathcal{R}_{[i]}(v) \subseteq \mathcal{R}_{\Box}(w) \text{ if } \mathcal{R}_{[i]}(v) \not\subseteq \|\phi\| \text{ then } \exists \mathcal{R}_{[i]}(z) \subseteq \mathcal{R}_{\Box}(w) \text{ s.t. }$$
$$(i)\ \mathcal{R}_{[i]}(v) \prec \mathcal{R}_{[i]}(z),\ (ii)\ \mathcal{R}_{[i]}(z) \subseteq \|\phi\| \text{ and}$$
$$(iii)\ \forall \mathcal{R}_{[i]}(x) \subseteq \mathcal{R}_{\Box}(w), \mathcal{R}_{[i]}(z) \preceq \mathcal{R}_{[i]}(x) \text{ implies } \mathcal{R}_{[i]}(x) \subseteq \|\phi\|$$

Clause (7*) is interpreted accordingly: Agent i ought to see to it that ϕ iff for every choice $\mathcal{R}_{[i]}(v)$ available to i that does not guarantee ϕ there (i) exists a strictly dominating choice $\mathcal{R}_{[i]}(z)$ that (ii) does guarantee ϕ and (iii) every weakly dominating choice $\mathcal{R}_{[i]}(x)$ over $\mathcal{R}_{[i]}(z)$ also guarantees ϕ. In other words, all choices not guaranteeing ϕ are strictly dominated only by choices guaranteeing ϕ. (We note that clause (7*) is obtained through an adaption of the definition provided in [12] to relational frames.) We show that the logic TDS is also sound and complete with respect to the class of TUS-frames.

Theorem 3 (SOUNDNESS). *$\forall\phi \in \mathcal{L}_{\mathsf{TDS}}$, if $\vdash_{\mathsf{TDS}} \phi$, then $C_f^u \models \phi$.*

Proof. We prove by induction on the given derivation of ϕ in TDS. The argument for axioms A0-A6 and A12 is the same as in Theorem 1. The validity of the axioms A7-A11 can be easily checked by applying semantic clause (7*) of Definition 9.

We now prove that the class C_f^u of TUS-frames characterizes the same set of formulae as the class C_f^d of TDS frames. We prove both directions separately:

Theorem 4. *$\forall\phi \in \mathcal{L}_{\mathsf{TDS}}$ we have $C_f^u \models \phi$ implies $C_f^d \models \phi$.*

Proof. We prove by contraposition assuming $C_f^d \not\models \phi$. Hence, there is a TDS-model, $\mathcal{M}^d = (\mathcal{W}, \mathcal{R}_{\Box}, \{\mathcal{R}_i | i \in Ag\}, \mathcal{R}_{\mathsf{H}}, \mathcal{R}_{\mathsf{G}}, \mathcal{R}_{Ag}, \{\mathcal{R}_{\otimes_i} | i \in Ag\}, \mathcal{V})$ such that $\mathcal{M}^d, w \models \neg\phi$ for some $w \in \mathcal{W}$. We use $\mathcal{M}^d$ to construct a model M in C_f^u, such that:

$$\mathsf{M} = (\mathsf{W}, \mathsf{R}_{\Box}, \{\mathsf{R}_i | i \in Ag\}, \mathsf{R}_{\mathsf{G}}, \mathsf{R}_{\mathsf{H}}, \mathsf{R}_{Ag}, \mathsf{util}, \mathsf{V})$$

We show that $\mathsf{M}, w' \models \neg\phi$ for some $w' \in \mathsf{W}$. To define M let $\mathsf{W} := \mathcal{W}$, $\mathsf{R}_{\Box} := \mathcal{R}_{\Box}$, $\mathsf{R}_i := \mathcal{R}_i$, $\mathsf{R}_{\mathsf{H}} := \mathcal{R}_{\mathsf{H}}$, $\mathsf{R}_{\mathsf{G}} := \mathcal{R}_{\mathsf{G}}$, $\mathsf{R}_{Ag} := \mathcal{R}_{Ag}$, $\mathsf{V}(p) := \mathcal{V}(p)$ and let util be a function assigning each $w \in \mathsf{W}$ to a natural number, satisfying the following criteria:

1. $\forall i \in Ag, \forall w, v, z \in \mathcal{W}$, if $v, z \in \mathcal{R}_{\Box}(w)$, $v \in \mathcal{R}_i^s(w) \setminus \mathcal{R}_{\otimes_i}(w)$, and $z \in \mathcal{R}_i^s(w) \cap \mathcal{R}_{\otimes_i}(w)$, then $\mathsf{util}(v) \leq \mathsf{util}(z)$;
2. $\forall w, v, z \in \mathcal{W}$, if $v \in \mathcal{R}_{\Box}(w) \setminus \mathcal{R}_{\otimes_{Ag}}(w)$ and $z \in \mathcal{R}_{\otimes_{Ag}}(w)$, then $\mathsf{util}(v) < \mathsf{util}(z)$;
3. $\forall w, u, z \in W$, if $v, z \in \mathcal{R}_i^s(w) \cap \mathcal{R}_{\otimes_i}(w)$, then $\mathsf{util}(v) = \mathsf{util}(z)$;

Let $\mathcal{R}_{\otimes_{Ag}} := \bigcap_{i \in Ag} \mathcal{R}_{\otimes_i}$, we call $\mathcal{R}_{[i]}(v) \subseteq \mathcal{R}_{\otimes_i}(w)$ an *optimal choice* for agent i. (It can be easily checked that the function util can be constructed.)

We state the following useful lemma (the proof of which is found in the online appended version: http://arxiv.org/abs/1907.03265):

Lemma 9. *The following holds for any* TDS *frame:*
(1) $\forall v \in \mathcal{R}_{\Box}(w), \mathcal{R}_{\Box}(w) = \mathcal{R}_{\Box}(v)$; (2) $\forall v \in \mathcal{R}_i(w), \mathcal{R}_i(w) = \mathcal{R}_i(v)$;
(3) $\forall v \in \mathcal{R}_i^s(w), \mathcal{R}_i^s(w) = \mathcal{R}_i^s(v)$; (4) $\forall v \in \mathcal{R}_{\Box}(w)$ *we get* $\mathcal{R}_{\otimes_i}(v) = \mathcal{R}_{\otimes_i}(w)$;
(5) $\forall \mathcal{R}_{[i]}(z) \subseteq \mathcal{R}_{\Box}(w)$, *either* $\mathcal{R}_{[i]}(z) \subseteq \mathcal{R}_{\otimes_i}(w)$ *or* $\mathcal{R}_{[i]}(z) \cap \mathcal{R}_{\otimes_i}(w) = \emptyset$.

We observe that conditions **(C1)**–**(C3)** and **(T4)**–**(T7)** will be satisfied in M since all of the relations of $\mathcal{M}^d$, with the exception of $\mathcal{R}_{\otimes_i}$, are identical to those in M. Moreover, util complies with Definition 9 and so M is in fact a TUS model. The desired claim will follow if we additionally show that $\forall \psi \in \mathcal{L}_{\mathsf{TDS}}$ and $\forall w \in \mathcal{W}$:

$$\mathcal{M}^d, w \models \psi \iff \mathsf{M}, w \models \psi$$

We prove the claim by induction on the complexity of ψ.

Base Case. Let ψ be a propositional variable p. By the definition of V in M it follows directly that $\mathcal{M}^d, w \models p$ iff $w \in \mathcal{V}$ iff $w \in \mathsf{V}$ iff $\mathsf{M}, w \models p$.

Inductive Step. The cases for the propositional connectives and the modalities $[\alpha] \in \{\Box, \mathsf{H}, \mathsf{G}, [Ag]\} \cup \{[i] | i \in Ag\}$ are straightforward. We consider the non-trivial case when ψ is of the form $\otimes_i \phi$. Let us first prove the left to right direction.

($\Longrightarrow$) Assume $\mathcal{M}^d, w \models \otimes_i \phi$. We show that $\mathsf{M}, w \models \otimes_i \phi$. By the semantics for $\otimes_i$ (Definition 9) it suffices to prove that: $\forall \mathsf{R}_i(v) \subseteq \mathsf{R}_{\Box}(w)$ if $\mathsf{R}_i(v) \not\subseteq \|\phi\|_{\mathsf{M}}$, then $\exists \mathsf{R}_i(u) \subseteq \mathsf{R}_{\Box}(w)$ such that the following three clauses hold: (i) $\mathsf{R}_i(v) \prec \mathsf{R}_i(u)$; (ii) $\mathsf{R}_i(u) \subseteq \|\phi\|_{\mathsf{M}}$; and (iii) $\forall \mathsf{R}_i(x) \subseteq \mathsf{R}_{\Box}(w)$, $\mathsf{R}_i(u) \preceq \mathsf{R}_i(x)$ implies $\mathsf{R}_i(x) \subseteq \|\phi\|_{\mathsf{M}}$.

Let $\mathsf{R}_i(v) \subseteq \mathsf{R}_{\Box}(w)$ be arbitrary and assume that $\mathsf{R}_i(v) \not\subseteq \|\phi\|_{\mathsf{M}}$. We prove that there is a $\mathsf{R}_i(u) \subseteq \mathsf{R}_{\Box}(w)$ for which conditions (i)-(iii) hold. First, we prove the existence of such a $\mathsf{R}_i(u) \subseteq \mathsf{R}_{\Box}(w)$: By **(C1)** and **(D9)** of Definition 2, we know,

$$\exists u \in \mathcal{W} \text{ such that } \mathcal{R}_i(u) \subseteq \mathcal{R}_{\Box}(w) \text{ and } \mathcal{R}_i(u) \subseteq \mathcal{R}_{\otimes_i}(w). \qquad (1)$$

We also know by **(D9)** that $\forall j \in Ag \backslash \{i\}, \exists u_j \in \mathcal{R}_{\Box}(w)$ such that $\mathcal{R}_j(u_j) \subseteq \mathcal{R}_{\otimes_j}(w)$. By (IOA) we know that $\bigcap_{j \in Ag \backslash \{i\}} \mathcal{R}_j(u_j) \cap \mathcal{R}_i(u) \neq \emptyset$, i.e., there exists a $u^* \in \bigcap_{j \in Ag \backslash \{i\}} \mathcal{R}_j(u_j) \cap \mathcal{R}_i(u)$. Consequently, we obtain the following statement,

$$u^* \in \bigcap_{j \in Ag \backslash \{i\}} \mathcal{R}_{\otimes_j}(w) \cap \mathcal{R}_{\otimes_i}(w) = \mathcal{R}_{\otimes_{Ag}}(w). \qquad (2)$$

Last, by construction of M we know $\mathcal{R}_i(u) = \mathsf{R}_i(u)$. We show that (i)-(iii) hold:

(i) We show $\mathsf{R}_i(v) \prec \mathsf{R}_i(u)$, that is, (a) $\mathsf{R}_i(v) \preceq \mathsf{R}_i(u)$ and (b) $\mathsf{R}_i(u) \not\preceq \mathsf{R}_i(v)$:

(a) Recall, $\mathsf{R}_i(v) \not\subseteq \|\phi\|_{\mathsf{M}}$, we know $\exists v^* \in \mathsf{R}_i(v)$ s.t. $\mathsf{M}, v^* \not\models \phi$. By definition of M, $v^* \in \mathcal{R}_i(v)$ and by (IH) we get $\mathcal{M}^d, v^* \not\models \phi$. Consequently, by the assumption that $\mathcal{M}^d, w \models \otimes_i \phi$, and the fact that $\mathcal{M}^d, v^* \not\models \phi$, it follows that $v^* \notin \mathcal{R}_{\otimes_i}(w)$. Hence, we know that $\mathcal{R}_i(v) \not\subseteq \mathcal{R}_{\otimes_i}(w)$, which implies $\mathcal{R}_{\otimes_i}(w) \cap \mathcal{R}_i(v) = \emptyset$ by Lemma 9-(5). Therefore, by this fact along with statement (1) above, we know that,

For all $x, u^{\blacktriangledown}, v^{\blacktriangledown} \in \mathcal{W}$, if $v^{\blacktriangledown} \in \mathcal{R}_i^s(x) \cap \mathcal{R}_i(v)$ and $u^{\blacktriangledown} \in \mathcal{R}_i^s(x) \cap \mathcal{R}_i(u)$, then $v^{\blacktriangledown} \in \mathcal{R}_i^s(x) \backslash \mathcal{R}_{\otimes_i}(w)$ and $u^{\blacktriangledown} \in \mathcal{R}_i^s(x) \cap \mathcal{R}_{\otimes_i}(w)$.

Let $x, u^{\blacktriangledown}, v^{\blacktriangledown} \in Wd$ be arbitrary and assume that $v^{\blacktriangledown} \in \mathcal{R}_i^s(x) \cap \mathcal{R}_i(v)$ and $u^{\blacktriangledown} \in \mathcal{R}_i^s(x) \cap \mathcal{R}_i(u)$. By the statement above, it follows that $v^{\blacktriangledown} \in \mathcal{R}_i^s(x) \setminus \mathcal{R}_{\otimes_i}(w)$ and $u^{\blacktriangledown} \in \mathcal{R}_i^s(x) \cap \mathcal{R}_{\otimes_i}(w)$, which in conjunction with criterion 1 on the function util implies that $\mathsf{util}(v^{\blacktriangledown}) \leq \mathsf{util}(u^{\blacktriangledown})$. Therefore, the following holds,

For all $x, u^{\blacktriangledown}, v^{\blacktriangledown} \in \mathcal{W}$, if $v^{\blacktriangledown} \in \mathcal{R}_i^s(x) \cap \mathcal{R}_i(v)$ and $u^{\blacktriangledown} \in \mathcal{R}_i^s(x) \cap \mathcal{R}_i(u^{\blacktriangledown})$, then $\mathsf{util}(v^{\blacktriangledown}) \leq \mathsf{util}(u)$.

It follows that $\forall \mathcal{R}_i^s(x) \subseteq \mathcal{R}_\Box(w)$, $\mathcal{R}_i^s(x) \cap \mathcal{R}_i(v) \leq \mathcal{R}_i^s(x) \cap \mathcal{R}_i(u)$. Hence, by the definition of $\preceq$ and the definition of M, we obtain $\mathsf{R}_i(v) \preceq \mathsf{R}_i(u)$.

(b) We need to show $\mathsf{R}_i(u) \npreceq \mathsf{R}_i(v)$. By definition of $\preceq$, it suffices to show that $\exists x, \exists u^{\blacktriangledown}, \exists v^{\blacktriangledown} \in \mathsf{W}$ s.t. $\mathsf{R}_i(x) \subseteq \mathsf{R}_\Box(w)$, $u^{\blacktriangledown} \in \mathsf{R}_i(u) \cap \mathsf{R}_i^s(x)$, $v^{\blacktriangledown} \in \mathsf{R}_i(v) \cap \mathsf{R}_i^s(x)$ and $\mathsf{util}(v^{\blacktriangledown}) < \mathsf{util}(u^{\blacktriangledown})$. Consider $\bigcap_{j \in Ag \setminus i} \mathcal{R}_j(u_j) \cap \mathcal{R}_i(u) \neq \emptyset$ from statement (2). Let $\mathsf{R}_i^s(x) := \bigcap_{j \in Ag \setminus i} \mathsf{R}_j(u_j)$. Clearly, $\mathsf{R}_i^s(x) \subseteq \mathsf{R}_\Box(w)$. By (IOA) we know that $\mathcal{R}_i^s(x) \cap \mathcal{R}_i(v) \neq \emptyset$ (where $\mathcal{R}_i^s(x) = \bigcap_{j \in Ag \setminus i} \mathcal{R}_j(u_j)$), and so, $\mathsf{R}_i^s(x) \cap \mathsf{R}_i(v) \neq \emptyset$ by the definition of M. Therefore, $\exists v^{\blacktriangledown} \in \mathsf{R}_i^s(x) \cap \mathsf{R}_i(v)$. Since $u^* \in \bigcap_{j \in Ag \setminus i} \mathcal{R}_j(u_j) \cap \mathcal{R}_i(u)$ (see paragraph above statement (2)), we know that $u^* \in \bigcap_{j \in Ag \setminus i} \mathsf{R}_j(u_j) \cap \mathsf{R}_i(u)$, implying that $u^* \in \mathsf{R}_i^s(x) \cap \mathsf{R}_i(u)$. Since also $\mathcal{R}_i(v) \cap \mathcal{R}_{\otimes_{Ag}}(w) = \emptyset$, as derived in part (i), we obtain $v^{\blacktriangledown} \in \mathcal{R}_\Box(w) \setminus \mathcal{R}_{\otimes_{Ag}}(w)$. By criterion 2 of util, and the facts $v^{\blacktriangledown} \in \mathcal{R}_\Box(w) \setminus \mathcal{R}_{\otimes_{Ag}}(w)$ and $u^* \in \mathcal{R}_{\otimes_{Ag}}(w)$, by statement (2), we have that $\mathsf{util}(v^{\blacktriangledown}) < \mathsf{util}(u^*)$. Therefore, $\mathsf{R}_i(u) \npreceq \mathsf{R}_i(v)$.

(ii) By assumption $\mathcal{R}_{\otimes_i}(w) \subseteq \|\phi\|_{\mathcal{M}^d}$ and statement (1) we get $\mathcal{R}_i(u) \subseteq \mathcal{R}_{\otimes_i}(w)$. By IH we have $\|\phi\|_{\mathcal{M}^d} = \|\phi\|_{\mathsf{M}}$ and since $\mathcal{R}_i(u) = \mathsf{R}_i(u)$ we know $\mathsf{R}_i(u) \subseteq \|\phi\|_{\mathsf{M}}$.

(iii) We prove the case by contraposition and show that $\forall \mathsf{R}_i(x) \subseteq \mathsf{R}_\Box(w)$, if $\mathsf{R}_i(x) \nsubseteq \|\phi\|$, then $\mathsf{R}_i(u) \npreceq \mathsf{R}_i(x)$. Let $\mathsf{R}_i(x)$ by an arbitrary choice-cell in $\mathsf{R}_\Box(w)$ and assume that $\mathsf{R}_i(x) \nsubseteq \|\phi\|_{\mathsf{M}}$. We aim to prove that $\mathsf{R}_i(u) \npreceq \mathsf{R}_i(x)$. By definition of $\preceq$ it suffices to show that $\exists \mathsf{R}_i^s(y) \subseteq \mathsf{R}_\Box(w)$ such that $\exists u^{\blacktriangledown} \in \mathsf{R}_i(u) \cap \mathsf{R}_i^s(y)$, $\exists x^{\blacktriangledown} \in \mathsf{R}_i(x) \cap \mathsf{R}_i^s(y)$, and $\mathsf{util}(x^{\blacktriangledown}) < \mathsf{util}(u^{\blacktriangledown})$.

By the assumption that $\mathsf{R}_i(x) \nsubseteq \|\phi\|_{\mathsf{M}}$, we know $\exists x^{\blacktriangledown} \in \mathsf{R}_i(x)$ such that $\mathsf{M}, x^{\blacktriangledown} \not\models \phi$. Clearly, $x^{\blacktriangledown} \in \mathcal{R}_i(x)$, and by (IH) we know that $\mathcal{M}^d, x^{\blacktriangledown} \not\models \phi$. Since $\mathcal{M}^d, w \models \otimes_i \phi$, we obtain $(w, x^{\blacktriangledown}) \notin \mathcal{R}_{\otimes_i}$, and by Lemma 9-(5) we obtain $\mathcal{R}_i(x) \nsubseteq \mathcal{R}_{\otimes_i}(w)$.

By statement (2) we had $u^* \in \mathcal{R}_{\otimes_{Ag}}(w)$ and $u^* \in \mathcal{R}_{\otimes_i}(w)$. Also, we know $u^* \in \mathcal{R}_i(u)$ by paragraph preceding statement (2). Since, $u^* \in \bigcap_{j \in Ag \setminus \{i\}} \mathcal{R}_j(u_j) \cap \mathcal{R}_i(u)$, we also have $u^* \in \bigcap_{j \in Ag \setminus \{i\}} \mathcal{R}_j(u_j)$. Let $\mathcal{R}_i^s(u^*) := \bigcap_{j \in Ag \setminus \{i\}} \mathcal{R}_j(u_j)$. By (IOA) we obtain $\mathcal{R}_i(x) \cap \mathcal{R}_i^s(u^*) \neq \emptyset$, implying that there exists some $x^{\blacktriangledown} \in \mathcal{R}_i(x) \cap \mathcal{R}_i^s(u^*)$. It follows from **(D9)** and the fact $\mathcal{R}_i(x) \nsubseteq \mathcal{R}_{\otimes_i}(w)$ that $x^{\blacktriangledown} \notin \mathcal{R}_{\otimes_{Ag}}(w)$, which with the fact $u^* \in \mathcal{R}_{\otimes_{Ag}}(w)$, implies by definition of util (criterion 2) that $\mathsf{util}(x^{\blacktriangledown}) < \mathsf{util}(u^*)$. By the definition of M, we have $x^{\blacktriangledown} \in \mathsf{R}_i(x) \cap \mathsf{R}_i^s(u^*)$, $u^* \in \mathsf{R}_i(u) \cap \mathsf{R}_i^s(u^*)$ and $\mathsf{util}(x^{\blacktriangledown}) < \mathsf{util}(u^*)$, which implies the desired claim.

($\Longleftarrow$) We now prove the right to left direction: Assume $\mathsf{M}, w \models \otimes_i \phi$. We reason towards a contradiction by assuming $\mathcal{M}^d, w \not\models \otimes_i \phi$. Hence, there exists a world $v \in \mathcal{R}_{\otimes_i}(w)$ such that $\mathcal{M}^d, v \not\models \phi$. By **(D11)** we obtain $\mathcal{R}_{[i]}(v) \subseteq \mathcal{R}_{\otimes_i}(w)$ and hence $\mathcal{R}_{[i]}(v) \nsubseteq \|\phi\|_{\mathcal{M}^d}$. By (IH) and the definition of M, we obtain $\mathsf{R}_i(v) \nsubseteq \|\phi\|_{\mathsf{M}}$. This fact, in conjunction with the assumption $\mathsf{M}, w \models \otimes_i \phi$, implies that there exists some $\mathsf{R}_i(z) \subseteq \mathsf{R}_\Box(w)$ such that the following holds: (i) $\mathsf{R}_i(v) \prec \mathsf{R}_i(z)$; (ii) $\mathsf{R}_i(z) \subseteq \|\phi\|_\mathsf{M}$; and (iii) $\forall \mathsf{R}_i(x) \subseteq \mathsf{R}_\Box(w)$, $\mathsf{R}_i(z) \preceq \mathsf{R}_i(x)$ implies $\mathsf{R}_i(x) \subseteq \|\phi\|_\mathsf{M}$.

By Lemma 9-(5) and the fact that $\mathsf{R}_i(z) = \mathcal{R}_i(z)$, we know that either **(a)** $\mathcal{R}_i(z) \subseteq \mathcal{R}_{\otimes_i}(w)$ holds or **(b)** $\mathcal{R}_i(z) \cap \mathcal{R}_{\otimes_i}(w) = \emptyset$ holds.

Assume **(a)**. We know $\mathsf{R}_i(v) \prec \mathsf{R}_i(z)$ and therefore, $\mathsf{R}_i(z) \not\preceq \mathsf{R}_i(v)$. Hence, $\exists \mathsf{R}^s_i(x) \subseteq \mathsf{R}_\Box(w), \exists z^* \in \mathsf{R}_i(z) \cap \mathsf{R}^s_i(x), \exists v^* \in \mathsf{R}_i(v) \cap \mathsf{R}^s_i(x)$ such that $\mathsf{util}(v^*) < \mathsf{util}(z^*)$. We also know $\mathcal{R}_i(v) \subseteq \mathcal{R}_{\otimes_i}(w)$ and $\mathcal{R}_i(z) \subseteq \mathcal{R}_{\otimes_i}(w)$ and thus we obtain $z^*, v^* \in \mathcal{R}_{\otimes_i} \cap \mathcal{R}^s_i(x)$. Consequently, by the definition of util (criterion 3), we get $\mathsf{util}(v^*) = \mathsf{util}(z^*)$. Contradiction.

Assume **(b)**. We know $\mathsf{R}_i(v) \prec \mathsf{R}_i(z)$ and therefore, $\mathsf{R}_i(z) \not\preceq \mathsf{R}_i(v)$. Hence, $\exists \mathsf{R}^s_i(x) \subseteq \mathsf{R}_\Box(w), \exists z^* \in \mathsf{R}_i(z) \cap \mathsf{R}^s_i(x), \exists v^* \in \mathsf{R}_i(v) \cap \mathsf{R}^s_i(x)$ such that $\mathsf{util}(z^*) \not\leq \mathsf{util}(v^*)$. Then, by definition of util (criterion 1), either (I) $z^* \notin \mathcal{R}^s_i(x) \backslash \mathcal{R}_{\otimes_i}(w)$ or (II) $v^* \notin \mathcal{R}^s_i(x) \cap \mathcal{R}_{\otimes_i}(w)$. Suppose (I), since $z^* \in \mathsf{R}^s_i(x)$ we infer $z^* \in \mathcal{R}^s_i(x)$ and thus conclude $z^* \in \mathcal{R}_{\otimes_i}(w)$. However, by earlier assumption $\mathcal{R}_i(z) \cap \mathcal{R}_{\otimes_i}(w) = \emptyset$ we obtain $z^* \notin \mathcal{R}_{\otimes_i}(w)$. Contradiction. Suppose (II), then since $v^* \in \mathcal{R}^s_i(x)$ we infer $v^* \notin \mathcal{R}_{\otimes_i}(w)$. However, $\mathcal{R}_{[i]}(v) \subseteq \mathcal{R}_{\otimes_i}(w)$. Contradiction.

Corollary 1 (COMPLETENESS). *$\forall \phi \in \mathcal{L}_{\mathsf{TDS}}$, if $C^u_f \models \phi$, then $\vdash_{\mathsf{TDS}} \phi$.*

Proof. Follows from Theorem 4 above, together with Theorem 2.

Theorem 5. *$\forall \phi \in \mathcal{L}_{\mathsf{TDS}}$, we get $C^d_f \models \phi$ implies $C^u_f \models \phi$.*

Proof. Follows from Theorem 2 together with Theorem 3.

The Problem with Two-Valued Utility Functions. A well studied candidate function for assigning utilities to *histories*, is the *two-valued* approach where the range of utilities is $\{0, 1\}$ (e.g. [12,15]). As a concluding remark of the present section, we briefly discuss the philosophical ramifications of using binary utility functions in a temporal setting.

Observe that, at a moment where all worlds have a utility of 1 (or all 0), every obligation becomes vacuously satisfied by definition—in such a scenario we would have $\otimes_i \phi$ iff $\Box \phi$—and every choice for each agent will ensure all optimal outcomes (see clause (7*) of Definition 10).[3] If in such a scenario, following [12,15], utilities are assigned to complete histories and thus remain constant through time, all obligations will also be vacuously satisfied at every future moment from thereon (namely, as one moves into the future, the set of histories passing through a moment can only decrease or stay the same). That at such moments all obligations are vacuously satisfied means that no obligation can be violated. Unfortunately, this also implies that at such moments *contrary-to-duty* (CTD) reasoning—i.e., reasoning about obligations that come into being when a previous obligation has been violated—becomes impossible because CTD obligations require the possibility to violate one's obligations in the first place (e.g. see [17]).

[3] This also holds when all intersections of choices of agents contain both a 1 and a 0.

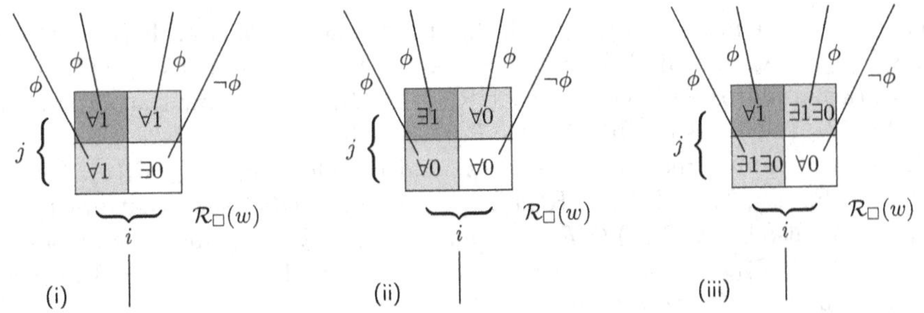

Fig. 1. The only three scenarios where $\otimes_i \phi \wedge \otimes_j \phi \wedge \neg \Box \phi$ holds true at $\mathcal{R}_\Box(w)$ (for $Ag = \{i, j\}$ with 2 choices). Choices of i are vertically presented, those of j horizontally. The symbol $\forall n$ means every history is assigned value n, and $\exists n$ means that some history is assigned n, for $n \in \{0, 1\}$. Optimal choices are shaded and darker shaded when overlapping. At all $\forall k$ outcomes (with $k \in \{0, 1\}$), CTD reasoning becomes impossible.

In order to reason with CTD obligations in *temporal* utilitarian STIT logics, we need to ensure that obligations can be violated, that is, we must consider deliberative obligations: $\otimes_i^d \phi := \otimes_i \phi \wedge \neg \Box \phi$. This means that, for an obligation $\otimes_i^d \phi$ to hold, there exists a choice that does not guarantee ϕ and, by definition, the latter choice must be strictly dominated by (only) ϕ choices. In the binary setting this means that for all optimal choices, there is at least one outcome with a strictly higher utility (which must be 1). Unfortunately, this has a drawback since at such moments *at least* one of the following holds: (1) Worlds in the intersection of all agents acting in accordance with their duty *all have value 1.* (2) Worlds in the intersection of all agents violating their duty *all have value* 0.

Relative to the aforementioned, Fig. 1 illustrates the (only) three scenarios possible in a two-agents, two-choices setting: Sub-figure (i) implies the impossibility of future CTD reasoning in all cases in which at least one agent satisfies its obligation. Sub-figure (ii) implies that there is no future CTD possible in every case witnessing at least one agent violating its obligation. Last, sub-figure (iii) indicates that future CTD obligations can only occur if one of the agents satisfies her obligation if and only if the other violates his. (With the impossibility of future CTD reasoning we mean that from that moment onward, all obligations will be vacuously satisfied.) All three cases are undesirable since they do not allow for future recuperation in those situations in which they clearly should.

The above exhibits that, although $\otimes_i$ does not depend on any temporal aspect (e.g. [15]), we can identify utility functions that are less suitable for temporal extensions of STIT. Binary functions relative to moments only, do not cause these problems, although they have their own issues [12]. In the case where the function ranges over the set of reals, it is possible to assign utilities in such a way that there is always CTD reasoning possible. In future work, we aim to specify such utility functions, making particular use of temporal aspects of TDS-frames.

5 Conclusion and Future Work

In this paper, we extended deontic STIT logic [15] to the temporal setting, incorporating the logic from [14]. In doing so, we answered a long standing open question for temporal embeddings of deontic STIT (e.g. see [4,12,15]). We showed that the resulting logic TDS is sound and complete with respect to its class of frames. We dubbed these frames *neutral* since they allowed us to obtain adequacy of the calculus, while allowing us to refrain from committing to specific utility functions. Subsequently, we showed how these neutral frames can be transformed into particular utilitarian models, while preserving truth. We also briefly argued that in a temporal setting, binary value assignments to histories can generate undesirable behavior with respect to contrary-to-duty obligations.

For future work, we leave open the problem of whether temporal STIT (from [14]) and its deontic extension TDS are decidable. Furthermore, we aim to investigate alternative utility assignments that explicitly exploit the temporal aspects of TDS; e.g., it might be interesting to consider a dynamic approach taking into account that natural agents have limited foresight relative to (future) utilities.

References

1. Abarca, A.I.R., Broersen, J.: A logic of objective and subjective oughts. In: Calimeri, F., Leone, N., Manna, M. (eds.) JELIA 2019. LNCS (LNAI), vol. 11468, pp. 629–641. Springer, Cham (2019). https://doi.org/10.1007/978-3-030-19570-0_41
2. Arkoudas, K., Bringsjord S., Bello, P.: Toward ethical robots via mechanized deontic logic. In: AAAI Fall Symposium on Machine Ethics, pp. 17–23 (2005)
3. Balbiani, P., Herzig, A., Troquard, N.: Alternative axiomatics and complexity of deliberative STIT theories. J. Philos. Logic **37**(4), 387–406 (2008)
4. Belnap, N., Perloff, M., Xu, M.: Facing the Future: Agents and Choices in our Indeterminist World. Oxford University Press on Demand, Oxford (2001)
5. Bentham, J.: An Introduction to the Principles of Morals and Legislation (1789)
6. van Berkel, K., Lyon, T.: Cut-free calculi and relational semantics for temporal STIT logics. In: Calimeri, F., Leone, N., Manna, M. (eds.) JELIA 2019. LNCS (LNAI), vol. 11468, pp. 803–819. Springer, Cham (2019). https://doi.org/10.1007/978-3-030-19570-0_52
7. Blackburn, P., de Rijke, M., Venema, Y.: Modal Logic. Cambridge University Press, Cambridge (2001)
8. Gabbay, D.M., Hodkinson, I., Reynolds, M.: Temporal Logic: Mathematical Foundations and Computational Aspects. Oxford University Press, Oxford (1994)
9. Gerdes, J.C., Thornton, S.M.: Implementable ethics for autonomous vehicles. In: Maurer, M., Gerdes, J.C., Lenz, B., Winner, H. (eds.) Autonomes Fahren, pp. 87–102. Springer, Heidelberg (2015). https://doi.org/10.1007/978-3-662-45854-9_5
10. Goodall, N.J.: Machine ethics and automated vehicles. In: Meyer, G., Beiker, S. (eds.) Road Vehicle Automation. LNM, pp. 93–102. Springer, Cham (2014). https://doi.org/10.1007/978-3-319-05990-7_9
11. Herzig, A., Schwarzentruber, F.: Properties of logics of individual and group agency. In: Advances in Modal Logic, vol. 7, pp. 133–149. College Publications, London (2008)
12. Horty, J.: Agency and Deontic Logic. Oxford University Press, Oxford (2001)

13. Horty, J., Pacuit, E.: Action types in STIT semantics. Rev. Symbolic Logic **10**(4), 617–637 (2017)
14. Lorini, E.: Temporal STIT logic and its application to normative reasoning. J. Appl. Non-Classical Logics **23**(4), 372–399 (2013)
15. Murakami, Y.: Utilitarian deontic logic. In: Advances in Modal Logic, vol. 5, pp. 211–230. King's College Publications, London (2005)
16. Nayebpour, M., Koehn, D.: The ethics of quality: problems and preconditions. J. Bus. Ethics **44**(1), 37–48 (2003)
17. Prakken, H., Sergot, M.: Contrary-to-duty obligations. Studia Logica **57**(1), 91–115 (1996)

Evaluating Networks of Arguments: A Case Study in Mīmāṃsā Dialectics

Kees van Berkel[1(✉)], Agata Ciabattoni[1], Elisa Freschi[2], and Sanjay Modgil[3(✉)]

[1] TU Wien, Vienna, Austria
{kees,agata}@logic.at
[2] ÖAW and UniWien, Vienna, Austria
elisa.freschi@gmail.com
[3] King's College London, London, UK
sanjay.modgil@kcl.ac.uk

Abstract. We formalize networks of authored arguments. These networks are then mapped to $ASPIC^+$ theories that subsequently instantiate Extended Argumentation Frameworks. Evaluation of arguments in the latter determines the status of the arguments in the source networks. The methodology is illustrated through a collaboration between scholars of South Asian philosophy, logicians and formal argumentation theorists, analyzing excerpts of Sanskrit texts concerning a controversial normative debate within the philosophical school of Mīmāṃsā.

Keywords: Instantiated arguments · Extended argumentation frameworks · $ASPIC^+$ · Argument networks · Dialectics · Mīmāṃsā philosophy

1 Introduction

Dung's seminal theory of argumentation [7] provides foundations for dynamic and distributed nonmonotonic reasoning [3]. Given a set of logical formulae, one defines arguments (sets of logical formulae and their inferred conclusions) and a binary (attack) relation amongst them, encoding that one argument is a counter-argument to another. The status of arguments in the resulting argumentation framework (AF) is then evaluated, and the claims of the winning arguments identify the non-monotonic inferences from the 'instantiating' set of logical formulae. However, whereas the above procedure is often static, argumentation in practice is typically dynamic and dialectic, where arguments are authored incrementally rather than being defined by a given, fixed set of formulae. Moreover, in practice not only attacks, but also collective attacks [15], support relations [1,16], attacks on attacks [10] etc. are specified as holding between arguments, thus defining networks of authored arguments (see [11] for more details).

Work partially funded by the projects WWTF MA16-028 and FWF W1255-N23.

P. Blackburn et al. (Eds.): LORI 2019, LNCS 11813, pp. 355–369, 2019.
https://doi.org/10.1007/978-3-662-60292-8_26

In this paper, we formalise a methodology that was informally proposed in [11], and that accounts for the above described dynamic authoring of argument networks. Here, structured arguments—i.e., arguments whose internal logical structure is specified—are first related by attacks and supports, and can express preferences over arguments. These arguments are then mapped to their constituent formulae and rules so as to define an $ASPIC^+$ theory [14] (a promising formal approach to structured-argumentation [9]). The $ASPIC^+$ theory, subsequently, instantiates an Extended Argumentation Framework (EAF) [10] (an extension of Dung's AF that accommodates arguments expressing preferences over other arguments through attacking attack relations). Finally, the evaluation of these arguments in the resulting EAF then determines the status of arguments in the source network, and consequently the inferences defined by the constituent formulae and rules in the original authored arguments. As argued in [11], this methodology is proposed as a more rigorous approach to evaluating arguments—and the defined inferences—in authored networks, as compared with directly evaluating the arguments in the source network (as typically done by scholars defining networks relating arguments by relations other than attacks).

The usefulness of the proposed formalization, corresponding to the aforementioned methodology, will be exemplified through a case study which resulted from a collaboration between scholars in South Asian philosophy, argumentation theorists and logicians. It involves a formal analysis of some excerpts of (Sanskrit) commentaries by philosophers of the school of Mīmāṃsā, and their application to South Asian jurisprudence. This school originated in ancient India more than two millennia ago and was devoted to the analysis of normative statements in the Vedas, the sacred texts of the so-called Hinduism. The dialectic nature of Mīmāṃsā argumentation, its structured analyses and its use of abstract logical principles, makes it particularly suitable for exhibiting the formal extensions introduced in this paper. In particular, we analyzed a portion of the debate on the immolation of widows on their husbands' funeral pyre, i.e., the so-called *satī* ritual. This debate has had deep socio-political implications in South Asia since the 9th c. until today (e.g., see [4,18]) and has been broadly dealt with by South Asian jurists and philosophers, primarily of the Mīmāṃsā school.

Plan of the Paper: We assume familiarity with abstract argumentation and briefly recap EAFs and $ASPIC^+$ instantiations of EAFs in Sect. 2 (for a review see [2,17]). In Sect. 3, we define $ASPIC^+$ argument networks which represent dynamically authored arguments and their relations, as specified by domain experts using some putative authoring—i.e. argument diagramming—tool. We also define the mapping of these networks to $ASPIC^+$ theories. We use the Mīmāṃsā debate on *satī* as a case study to exemplify the formalised methodology in Sect. 4.

2 Background: EAFs and $ASPIC^+$

Extended Argumentation Frameworks. Along with the usual binary attack relation ($\mathcal{C}$) over arguments, *Extended Argumentation Frameworks* (EAFs) [10] extend Argumentation Frameworks (*AF*s) [7] to also include a *pref-attack* relation $\mathcal{D}$: i.e. an argument expresses that Y is *preferred* to X attacks the binary attack from X to Y, so that the latter attack does not succeed as a defeat.

Definition 1. *An* **EAF** *is a tuple* $(\mathcal{A}, \mathcal{C}, \mathcal{D})$, $\mathcal{A}$ *is a set of arguments,* $\mathcal{C} \subseteq \mathcal{A}\times\mathcal{A}$, $\mathcal{D} \subseteq \mathcal{A}\times\mathcal{C}$, *and if* $(Z,(X,Y))$, $(Z',(Y,X)) \in \mathcal{D}$ *then* (Z,Z'), $(Z',Z) \in \mathcal{C}$.

Notice that the constraint on the relation $\mathcal{D}$ ensures that if arguments Z and Z' respectively pref-attack (X,Y) and (Y,X), then Z and Z' express contradictory preferences—i.e., Y is preferred to X, respectively X is preferred to Y—and so themselves symmetrically (i.e. mutually) attack each other. Henceforth, we focus on *bounded hierarchical EAF*s, stratified so that attacks at a level i are only pref-attacked by arguments at the next level (preserving rationality [13]).

Definition 2. $\Delta = (\mathcal{A},\mathcal{C},\mathcal{D})$ *is a* bounded hierarchical EAF **(bh-EAF)** *iff there exists a partition* $\Delta_H = (((\mathcal{A}_1,\mathcal{C}_1),\mathcal{D}_1),\ldots,((\mathcal{A}_n,\mathcal{C}_n),\mathcal{D}_n))$ *s.t.* $D_n = \emptyset$, *and:*

- $\mathcal{A} = \bigcup_{i=1}^n \mathcal{A}_i$, $\mathcal{C} = \bigcup_{i=1}^n \mathcal{C}_i$, $\mathcal{D} = \bigcup_{i=1}^n \mathcal{D}_i$, *and for* $1 \leq i \leq n$, $(\mathcal{A}_i,\mathcal{C}_i)$ *is an* AF
- $(C,\ (A,B)) \in \mathcal{D}_i$ *implies* $(A,B) \in \mathcal{C}_i$, $C \in \mathcal{A}_{i+1}$

The notion of a successful attack (i.e. *defeat*) is then parameterised with respect to the preferences specified by some given set S of arguments: i.e., Y *defeats*$_S$ X (denoted $Y \rightarrow^S X$) *iff* $(Y,X) \in \mathcal{C}$ and $\neg\exists Z \in S$ s.t. $(Z,\ (Y,X)) \in \mathcal{D}$.

Then, a set S is *EAF conflict free* when it does not admit arguments that symmetrically attack, but S can contain some Y and X such that Y *asymmetrically* attacks X, given a $Z \in S$ that pref-attacks the attack from Y to X.

Furthermore, since attacks can themselves be attacked, these attacks need to be reinstated (defended) by attacking arguments that pref-attack. That is, the acceptability of an argument X w.r.t. a set S requires that there is a *reinstatement set* for any reinstating defeat:

Definition 3. *Let* $S \subseteq \mathcal{A}$ *in* $(\mathcal{A},\mathcal{C},\mathcal{D})$. *Let* $R_S = \{X_1 \rightarrow^S Y_1,\ldots,X_n \rightarrow^S Y_n\}$ *s.t. for* $1 \leq i \leq n$, $X_i \in S$. *We call* R_S *a reinstatement set for* $A \rightarrow^S B$, *iff* $A \rightarrow^S B \in R_S$, *and* $\forall X \rightarrow^S Y \in R_S$, $\forall Y'$ *s.t.* $(Y',\ (X,Y)) \in \mathcal{D}$, $\exists X' \rightarrow^S Y' \in R_S$.

Furthermore, X *is acceptable w.r.t.* $S \subseteq \mathcal{A}$ *iff for all* $Y \in \mathcal{A}$ *s.t.* $Y \rightarrow^S X$, *there is a* $Z \in S$ *s.t.* $Z \rightarrow^S Y$, *and there is a* reinstatement set *for* $Z \rightarrow^S Y$.

For *bh-EAF*s, the semantic extensions are defined as for *AF*s. That is, let S be a conflict free set: S is an *admissible* extension iff all arguments in S are acceptable w.r.t. S; S is *complete* iff it is admissible and all arguments acceptable w.r.t. S are in S; S is *preferred* iff it is a set inclusion maximal complete extension; S is the (unique) *grounded* extension iff it is the set inclusion minimal complete extension; S is *stable* iff $\forall Y \notin S$, $\exists X \in S$ s.t. $X \rightarrow^S Y$. Lastly, for $e \in$ {complete, preferred, grounded, stable}, $X \in \mathcal{A}$ is *credulously* (*sceptically*) justified under the e semantics, if X belongs to at least one (all) e extension(s).

ASPIC$^+$ **Instantiations of EAFs.** *ASPIC*$^+$ [14] is a general framework in which one is free to choose a logical language $\mathcal{L}$. One is also free to specify defeasible and strict inference rules, as well as 'axiom' and 'ordinary' premises for construction of arguments. Furthermore, it facilitates preference relations over arguments, used to determine when attacks succeed as defeats. Defeasible rules are typically domain specific, while strict rules may either encode domain specific infallible inferences or inference rules of some deductive logic. In this system, only the fallible ordinary premises and fallible consequents of defeasible rules can be attacked. Axiom premises are infallible and conclusions of strict rules cannot be attacked. A partial function assigns names (wff in $\mathcal{L}$) to defeasible rules, so that applications of defeasible rules can be invalidated by arguments that claim the negation of the rule name. Finally, *ASPIC*$^+$ allows one to specify a contrary function specifying when formulae in $\mathcal{L}$ are said to be in conflict. In this paper, we assume that such conflicts are symmetric.

ASPIC$^+$ poses constraints on the above choices to ensure that the outcomes of evaluating the Dung frameworks instantiated by *ASPIC*$^+$ arguments and defeats, are rational [5]. In this work, the following review of *ASPIC*$^+$ [14] suffices:

Definition 4. *An* argumentation theory *is a tuple* $AT = (\mathcal{L}, ^-, \mathcal{R}, n, \mathcal{K})$ *where* $\mathcal{L}$ *is a logical language, and:*

- $\mathcal{R} = \mathcal{R}_s \cup \mathcal{R}_d$ *is a set of strict* ($\mathcal{R}_s$) *and defeasible* ($\mathcal{R}_d$) *inference rules, respectively of the form* $\varphi_1, \ldots, \varphi_n \rightarrow \varphi$ *and* $\varphi_1, \ldots, \varphi_n \Rightarrow \varphi$ *(where* φ_i *and* φ *are metavariables ranging over wff in* $\mathcal{L}$*);*
- $n : R_d \mapsto \mathcal{L}$ *is a partial naming function;*
- $\mathcal{K} = \mathcal{K}_n \cup \mathcal{K}_p$ *where* $\mathcal{K} \subseteq \mathcal{L}$, $\mathcal{K}_n$ *is a set of* axiom *premises,* $\mathcal{K}_p$ *is a set of* ordinary *premises, and* $\mathcal{K}_n \cap \mathcal{K}_p = \emptyset$.
- *for all wff* ϕ *in* $\mathcal{L}$, *if* $\varphi \in \overline{\psi}$ *then* $\psi \in \overline{\varphi}$. *(In this case, we say that* ψ *and* φ *are* contradictories, *which is denoted by* $\varphi = -\psi$.*)*

Henceforth, for convenience we write '$\delta : \varphi_1, \ldots, \varphi_n \Rightarrow \varphi$' instead of explicitly declaring that n assigns the wff δ to the defeasible rule $\varphi_1, \ldots, \varphi_n \Rightarrow \varphi$.

Definition 5. *An ASPIC*$^+$ argument A *on the basis of an AT* $(\mathcal{L}, ^-, \mathcal{R}, n, \mathcal{K})$ *is:*

1. φ *if* $\varphi \in \mathcal{K}$ *with:* $\texttt{Prem}(A) = \{\varphi\}$; $\texttt{Conc}(A) = \varphi$; $\texttt{Sub}(A) = \{\varphi\}$; $\texttt{Rules}(A) = \emptyset$; $\texttt{DefRules}(A) = \emptyset$; $\texttt{TopRule}(A) =$ *undefined.*
2. $A_1, \ldots A_n \rightarrow/\Rightarrow \psi$ *if* $A_1, \ldots, A_n$ *are arguments such that there exists a strict/defeasible rule* $\texttt{Conc}(A_1), \ldots, \texttt{Conc}(A_n) \rightarrow/\Rightarrow \psi$ *in* $\mathcal{R}_s/\mathcal{R}_d$, *with:*
$\texttt{Prem}(A) = \texttt{Prem}(A_1) \cup \ldots \cup \texttt{Prem}(A_n)$; $\texttt{Conc}(A) = \psi$;
$\texttt{Sub}(A) = \texttt{Sub}(A_1) \cup \ldots \cup \texttt{Sub}(A_n) \cup \{A\}$;
$\texttt{Rules}(A) = \bigcup_{i=1}^{n} \texttt{Rules}(A_i) \cup \{\texttt{Conc}(A_1), \ldots, \texttt{Conc}(A_n) \rightarrow / \Rightarrow \psi\}$;
$\texttt{TopRule}(A) = \texttt{Conc}(A_1), \ldots \texttt{Conc}(A_n) \rightarrow/\Rightarrow \psi$;
$\texttt{DefRules}(A) = \bigcup_{i=1}^{n} \texttt{DefRules}(A_i) \cup \{A_1, \ldots A_n \Rightarrow \psi\}$ *if* $\texttt{TopRule}(A) = A_1, \ldots A_n \Rightarrow \psi$ *and* $\bigcup_{i=1}^{n} \texttt{DefRules}(A_i)$ *otherwise.*

The notation in Definition 5 is generalised to sets of arguments in the usual way: e.g., letting $E = \{A_1, \ldots, A_n\}$, then $\texttt{DefRules}(E) = \bigcup_{i=1}^{n} \texttt{DefRules}(A_i)$.

Definition 6. *Let A, B and B' be ASPIC$^+$ arguments.*

- *A undercuts argument B (on B') iff $\texttt{Conc}(A) \in \overline{n(r)}$ for some $B' \in \texttt{Sub}(B)$ such that $\texttt{TopRule}(B') = r$.*
- *A rebuts argument B on (B') iff $\texttt{Conc}(A) = -\varphi$ for some $B' \in \texttt{Sub}(B)$ of the form $B_1'', \ldots, B_n'' \Rightarrow \varphi$.*
- *A undermines B (on $B' = \varphi$) iff $\texttt{Conc}(A) = -\varphi$ for some $\varphi \in \texttt{Prem}(B) \setminus \mathcal{K}_n$.*

When *ASPIC*$^+$ arguments instantiate an *AF*, a preference relation over the arguments is used to decide whether rebut or undermine attacks succeed as defeats; i.e., an attack from A to B succeeds only if $A \not\prec B'$. Undercuts succeed as defeats independently of preferences. Following [13], we will instantiate *bh-EAF*s in such a way that *ASPIC*$^+$ arguments may themselves conclude preferences over arguments (rather than assuming a given strict ordering $\prec$ over arguments). Pref-attacks originating from these arguments may then target binary attacks, denying the success of the latter as defeats. As in [12], we assume a function $\mathcal{P}$ that maps the conclusion of an individual argument to strict preferences over other arguments; e.g., given A and B with respective defeasible rules $\{r_1\}$ and $\{r_2, r_3\}$, if argument C concludes $(r_1 < r_2) \wedge (r_1 < r_3)$, then $\mathcal{P}(\texttt{Conc}(C)) = A \prec B$ (under the *Elitist* set ordering of [14]):

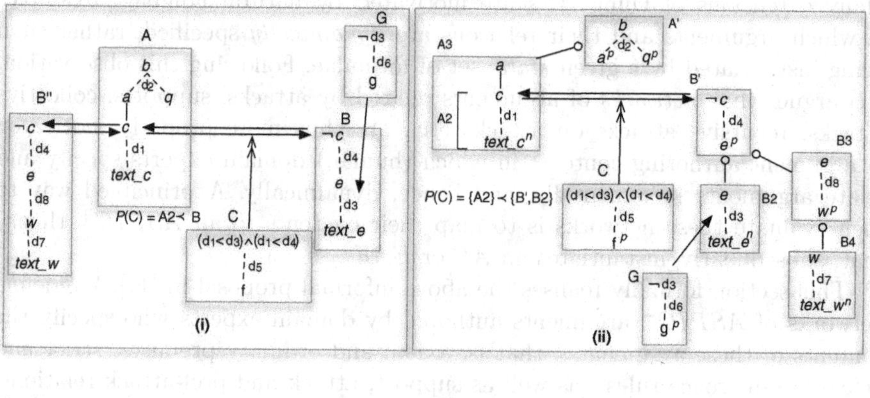

Fig. 1. Figure (i) depicts the *bh−EAF* instantiated by the *ASPIC*$^+$ theory of Example 1. Figure (ii) represents an *ASPIC*$^+$ *EANS* mapped to the *ASPIC*$^+$ theory of Example 1. Solid and dashed lines denote the application of strict and defeasible rules (resp.). We illustrate $(X, Y) \in \mathcal{C}$ with $X \blacktriangleright Y$, and $(Z, (X, Y)) \in \mathcal{D}$ with $Z \twoheadrightarrow (X \blacktriangleright Y)$.

Definition 7. *Let $\mathcal{A}$ be a set of ASPIC$^+$ arguments, $A, B \in \mathcal{A}$ and $\mathcal{C} \subseteq \mathcal{A} \times \mathcal{A}$, s.t. $(A, B) \in \mathcal{C}$ iff A rebuts, undermines or undercuts B. Let $\mathcal{P} : \mathcal{L} \mapsto \prec$, where $\prec \subseteq \mathcal{A} \times \mathcal{A}$ is a strict partial ordering over $\mathcal{A}$. Then an ASPIC$^+$ instantiated EAF is a tuple $(\mathcal{A}, \mathcal{C}, \mathcal{D})$ defined as in Definition 1, where $(C, (A, B)) \in \mathcal{D}$ iff A rebuts or undermines B on B', and $A \prec B' \in \mathcal{P}(\texttt{Conc}(C))$.*

Example 1. To illustrate the above, suppose the *argumentation theory* $AT = (\mathcal{L}, ^{-}, \mathcal{R}, n, \mathcal{K})$, with propositional language $\mathcal{L}$, a strict priority relation $<$ and:

- $\mathcal{K}_n = \{text\_c, text\_e, text\_w\}$; $\mathcal{K}_p = \{f, g, q\}$; $\mathcal{R}_s = \{c \rightarrow a\}$;
  $\mathcal{R}_d = \{d_1\colon text\_c \Rightarrow c;\ d_2\colon a, q \Rightarrow b;\ d_3\colon text\_e \Rightarrow e;\ d_4\colon e \Rightarrow \neg c;\ d_5\colon f \Rightarrow (d_1 < d_3) \wedge (d_1 < d_4);\ d_6\colon g \Rightarrow \neg d_3;\ d_7\colon text\_w \Rightarrow w;\ d_8\colon w \Rightarrow e\}$.
- $\varphi = -\psi$ just in case $\varphi = \neg\psi$ or $\psi = \neg\varphi$

We obtain the instantiated *bh*-EAF (see Fig. 1-i), consisting of the following:

$$\mathcal{A} = \left\{ \begin{array}{l} A1 = [text\_c], A2 = [A1 \Rightarrow c], A3 = [A2 \rightarrow a], A4 = [q], A = [A3, A4 \Rightarrow b] \\ B1 = [text\_e], B2 = [B1 \Rightarrow e], B = [B2 \Rightarrow \neg c], \\ C1 = [f], C = [C1 \Rightarrow (d_1 < d_3) \wedge (d_1 < d_4)], G1 = [g], G = [g \Rightarrow \neg d_3], \\ B5 = [text\_w], B4 = [B5 \Rightarrow w], B3 = [B4 \Rightarrow e], B'' = [B3 \Rightarrow \neg c] \end{array} \right\}$$

$\mathcal{C} = \{(B, A)^*, (B, A2), (A2, B), (G, B), (G, B2), (B'', A)^\dagger, (B'', A2), (A2, B'')\}$
$\mathcal{D} = \{(C, (A2, B))\}$ (NB. $(*)$ B attacks A on $A2$ and $(\dagger)$ B'' attacks A on $A2$.)

The single *grounded* extension of this EAF is the set $E = \{G1, G, C1, C, B1, A1, A4, B5, B4, B3\}$. The two *preferred*/*stable* extensions are $E \cup \{B''\}$ and $E \cup \{A, A2, A3\}$.

3 Towards Formalizing Networks of Authored Arguments

Many extensions of Dung *AF*s are motivated by natural language examples in which arguments and their relations are *dynamically* specified, rather than being instantiated by a given *static* set of formulae. Following this observation, [11] argues that *networks* of arguments related by attacks, supports, collective attacks, recursive attacks on attacks etc., are thus more properly motivated in argument authoring contexts in which (human) domain experts specify and relate arguments incrementally, and hence, dynamically. A principled way to then evaluate these networks is to map their contents to an $ASPIC^+$ theory that, subsequently, instantiates an AF or EAF.

This section formally realises the above informal proposal in [11]. We define networks of $ASPIC^+$ arguments authored by domain experts who specify the contents of these arguments—that is, axiom and ordinary premises, strict and defeasible inference rules—as well as support, attack and pref-attack relations. An argument Y is used to support X only if Y supplies the rationale (argument) for an ordinary (i.e. fallible) premise in X; axiom premises, which typically encode empirically validated information and so cannot be challenged, need not be supported. Hence, when authoring arguments one must distinguish between ordinary and axiom premises (with the respective superscripts p and n). We first define networks related by attacks and supports (Definition 8), and then hierarchies of such networks that include pref-attacks (Definition 10):

Definition 8. *An* $ASPIC^+$ *ANS (Argument Network with Support) is a tuple* $\langle \mathcal{A}, \mathcal{C}, \mathcal{S} \rangle$*, where:*

- $\mathcal{A}$ *are* $ASPIC^+$ *arguments such that for all* $X \in \mathcal{A}$, *and for all* $\alpha \in \mathsf{prem}(X)$, α *is labelled by* p *or* n*;*
- $(X, Y) \in \mathcal{C}$ *implies* (X, Y) *is an* $ASPIC^+$ *attack as defined in Definition 6, and;*
- $(X, Y) \in \mathcal{S}$ *implies* $\exists \alpha^p \in \mathtt{Prem}(Y)$ *such that* $\mathtt{Conc}(X) = \alpha$*, in which case we say that* X *supports* Y *on* α*. We may write* $X \multimap Y$ *to denote* $(X, Y) \in \mathcal{S}$*.*

Since we assume authoring of ANSs by humans, we account for the possibility that not all valid attacks may be explicity authored. Thus, Definition 8 accommodates that although for some $X, Y \in \mathcal{A}$, X attacks Y according to Definition 6, this attack might not be diagrammed as such, and so $(X, Y) \notin \mathcal{C}$.

Next we define a chain of supporting arguments, and the 'collapsing' of a chain into a single argument, with each supported premise replaced by its supporting argument.

Definition 9. *Let* $\Delta = \langle \mathcal{A}, \mathcal{C}, \mathcal{S} \rangle$ *be an* $ASPIC^+$ ANS*. Then* $\mathcal{S}_{chain}$ *is a set of ordered sets defined as follows:*

$\mathcal{S}_{chain}(\Delta) = \{\{A_1, \ldots, A_n\} | \bigcup_{i=1}^{n} A_i \subseteq \mathcal{A}, \neg\exists X, \neg\exists Y \in \mathcal{A}$ *s.t.* $(A_1, X), (Y, A_n) \in \mathcal{S}$, *and for* $i = 1 \ldots n-1, (A_{i+1}, A_i) \in \mathcal{S}\}$

The function coll *takes as input a chain of supporting arguments* Γ*, and returns a single argument if* $|\Gamma| = 2$*, else it returns a chain of supporting arguments* Γ' *in the case that* $|\Gamma| > 2$*:*

- $coll(\{A_1, A_2\}) = A$*, where* A_2 *supports* A_1 *on* α*, and* A *is the argument* A_1 *with* A_2 *replacing premise* α *in* A_1*;*
- $coll(\{A_1, \ldots, A_n\}) = coll(\{A1, \ldots, coll(A_{n-1}, A_n)\})$ *if* $n > 2$*.*

We now define bounded hierarchies of networks of attacking and supporting arguments, in which *pref*-attacks are directed at attacks in the next level down the hierarchy. Since arguments may be 'backward extended' by supporting arguments, so as to define chains, we propose a definition of pref-attacks originating from arguments whose conclusion is specified as mapping (via a function $\mathcal{P}_{set}$) to a preference over *chains of supporting* arguments; i.e., a preference ordering over sets rather than single arguments (cf. Definition 7).

Definition 10. *An* $ASPIC^+$ $EANS$ *(Extended Argument Network with Support) is a tuple* $\Delta = \langle \mathcal{A}, \mathcal{C}, \mathcal{S}, \mathcal{D} \rangle$ *iff there exists a partition* $\Delta_H = \langle ((\mathcal{A}_1, \mathcal{C}_1, \mathcal{S}_1), \mathcal{D}_1), \ldots, ((\mathcal{A}_n, \mathcal{C}_n, \mathcal{S}_n), \mathcal{D}_n) \rangle$ *such that* $D_n = \emptyset$*, and:*

- $\mathcal{A} = \bigcup_{i=1}^{n} \mathcal{A}_i$, $\mathcal{C} = \bigcup_{i=1}^{n} \mathcal{C}_i$, $\mathcal{S} = \bigcup_{i=1}^{n} \mathcal{S}_i$, $\mathcal{D} = \bigcup_{i=1}^{n} \mathcal{D}_i$*, and for* $i = 1 \ldots n$, $\langle \mathcal{A}_i, \mathcal{C}_i, \mathcal{S}_i \rangle$ *is an* $ASPIC^+$ ANS*.*
- $(C, (A, B)) \in \mathcal{D}_i$ *iff* $C \in \mathcal{A}_{i+1}$, $(A, B) \in \mathcal{C}_i$*, where* A *undermine or rebut attacks* B *on* B'*, and* $\exists \{B', \ldots, B_m\}, \exists \{A, \ldots, A_n\} \in \mathcal{S}_{chain}((\mathcal{A}_i, \mathcal{C}_i, \mathcal{S}_i))$ *s.t.* $(\{A, \ldots, A_n\} \prec \{B', \ldots, B_m\}) \in \mathcal{P}_{set}(\mathsf{conc}(C))$*, where* $\mathcal{P}_{set} : \mathcal{L} \mapsto \prec_s$*, and* $\prec_s \subseteq 2^{\mathcal{A}} \times 2^{\mathcal{A}}$ *is a strict partial ordering over sets of arguments.*

Finally, we define a mapping from an $ASPIC^+$ $EANS$ to an $ASPIC^+$ theory and the corresponding instantiation of a *bh-EAF*, which allows us to calculate the theory's extensions. Notice that, if an argument X (not of the form $[\alpha]$) is available to support a premise α^p, then α is not included as a premise in the $ASPIC^+$ theory (given that a rationale has been provided for why α holds). Also observe that, in line with our remark on attacks following Definition 8, X may not have been explicitly moved to support α.

Definition 11. *Let* $\Delta = \langle \mathcal{A}, \mathcal{C}, \mathcal{S}, \mathcal{D} \rangle$ *be an* ASPIC$^+$ $EANS$, $\mathcal{P}_{set}$ *a user specified function s.t.* $\mathcal{P}_{set} : \mathcal{L} \mapsto \prec_s$, *with strict partial ordering* $\prec_s \subseteq 2^{\mathcal{A}} \times 2^{\mathcal{A}}$, *and* $\mathcal{L}$ *and* $^-$ *a given language and contrary function, respectively. Then* $AT_\Delta = \langle \mathcal{L}, ^-, \mathcal{R}, n, \mathcal{K} = \mathcal{K}_n \cup \mathcal{K}_p \rangle$ *is defined as follows:*

1. $\mathcal{R} = \mathtt{Rules}(\mathcal{A})$;
2. $\mathcal{K}_n = \{\alpha |\ \alpha^n \in \mathsf{prem}(\mathcal{A})\}$;
3. $\mathcal{K}_p = \{\alpha |\ \alpha^p \in \mathsf{prem}(\mathcal{A}), \neg\exists X \in \mathcal{A}$ *s.t.* $\mathsf{conc}(X) = \alpha$ *and* X *is not of the form* $[\alpha]\}$;
4. $\forall r \in \mathtt{DefRules}(\mathcal{A})$, $n(r) = \alpha$, *where* α *does not appear in* $\mathcal{K}_n \cup \mathcal{K}_p$, *and* α *does not appear in the antecedent or consequent of a rule in* $\mathcal{R}$.

Let $\mathcal{A}'$ *be a set of* ASPIC$^+$ *arguments defined by* AT_Δ. *Then:*

- $\forall X, A, B \in \mathcal{A}'$, $\mathcal{P}(\mathsf{conc}(X)) = A \prec B$ *iff* $\exists Y \in \mathcal{A}$ *s.t.* $\mathsf{conc}(X) = \mathsf{conc}(Y)$, *and* $\mathcal{P}_{set}(\mathsf{conc}(Y)) = \Gamma_A \prec \Gamma_B$, $A = coll(\Gamma_A)$ *and* $B = coll(\Gamma_B)$.

Let $\mathcal{C}'$ *be the attack relation defined over* $\mathcal{A}'$, *such that* $\forall (A, B) \in \mathcal{C}$, $(A, B) \in \mathcal{C}$ *iff* A *rebuts, undermines or undercuts* B. *Then* $(\mathcal{A}, \mathcal{C}, \mathcal{D})$ *is defined as in Definition 1, where* $(C, (A, B)) \in \mathcal{D}$ *iff* A *rebut or undermines* B *on* B' *and* $A \prec B' \in \mathcal{P}(\mathtt{Conc}(C))$.

Example 2. Consider the network of arguments in Fig. 1-ii (mapped to the $ASPIC^+$ theory of Example 1) as authored by one or more users in the consecutive order $A', A3, B', B2, C, G, B3, B4$. Note that $B2$ supplies the rationale (i.e. argument) for the premise e^p in B', and so supports B'. Hence e is not included as an ordinary premise in the $ASPIC^+$ theory of Example 1. In Fig. 1-i the $bh-EAF$ instantiated by the theory is presented and we obtain the credulously justified arguments $A, A2, A3$ and B'' (under preferred and stable semantics), and so the conclusions c, a, b and $\neg c$ are credulously supported.

4 Case Study: The *satī* Ritual

We now apply the methodology formalized in Sect. 3, and analyze (part of) the controversy surrounding widows immolating themselves on their husbands' funeral pyre (the *satī* ritual). Despite the numerous arguments available, for space reasons, we limit our analysis to a single Mīmāṃsā author, namely, Medhātithi (9th–10th c. Kashmir). The analysis captures the arguments (in the form of an $ASPIC^+$ $EANS$) as they are successively elucidated (and augmented with contextual information in the form of basic reasoning principles). As will be seen, Medhātithi argues that *satī* should not be performed.

Basic Mīmāṃsā Principles. Over the last two millennia, philosophers of the Mīmāṃsā school have thoroughly analyzed prescriptive statements in the Vedas. They distinguish between three classes of normative statements (see, e.g., [8]): obligations, recommendations, and prohibitions. Prohibitions lead to no result if respected but to a sanction if not observed; recommendations, which are driven by a desire, lead to a result if fulfilled and to no sanction otherwise; obligations lead to a result if fulfilled and to a sanction if disregarded. Hence, for instance, if something is obligatory, it is not recommended. For our formalization of the *satī* debate, we can rely on some basic reasoning principles, which are either explicitly formulated or implicitly endorsed by all Mīmāṃsā authors, and are strict or defeasible. The list of principles is presented in Definition 12 below.

Last, when dealing with Mīmāṃsā we distinguish between two levels of normative statements: the ones that can be directly found in the Vedas or in *smṛti* texts based on the Vedas, and those obtained from applying metarules identified by Mīmāṃsā authors. We will refer to the former as *prima facie* norms and to the latter as *derived* normative statements.

Definition 12. *The following list of principles are Mīmāṃsā metarules:*

1. STRICT CONTEXTUAL PRINCIPLES:

D1 *Prima facie prohibitions and prima facie obligations are mutually exclusive.*††

D2 *Prima facie recommendations and prima facie obligations are mutually exclusive.*†

D3 *Prohibitions and obligations are mutually exclusive.*††

2. DEFAULT (DEFEASIBLE) CONTEXTUAL PRINCIPLES:

D4 *An obligation/prohibition/recommendation on the prima facie level, is also an obligation/prohibition/recommendation on the derived level.*

D5 *If an obligatory/prohibited action necessarily presupposes some (other) action, then that action is also obligatory/prohibited.*†

D6 *An argument supported by a rationale (i.e. a justification) is the* preferred *argument in case of a conflict between equipollent claims.*

D7 *If the Vedas/smṛtis prescribe an obligation/prohibition/recommendation, then we take the obligation/prohibition/recommendation to hold prima facie.*††

D8 *a) If two actions cause identical effects and have equal normative status, they are analogous. b) Conclusions drawn for one case apply to analogue cases.* ††

D9 *If the Vedas/smṛtis explicitly mention a reward for a prescribed action, then a) the action brings about that result and b) it is prima facie recommended.*†

D10 *If an action causes some effect, which subsequently implies another effect, then the action causes the second effect as well.*††

The symbol † indicates that the principle is explicitly stated by Mīmāṃsā authors, those with †† are not stated as rules, yet explicitly applied in Mīmāṃsā reasoning. The remaining rules are implicit assumptions that other metarules presuppose.

(A) [OPENING] The performance of *satī* causes a widow to take her life. The latter is, as an act of violence, prohibited for women as it is for men.

(B) [OBJECTION] Performance of *satī* is obligatory, because this prescription is derived from an explicit occurrence in a *smṛti* text.

(C) [FIRST REPLY] The referred *smṛti* text prescribing *satī* mentions a result, namely heaven, and therefore *satī* is recommended, not obligatory.

(D) [SECOND REPLY] The ritual of *satī* is similar to the *śyena* sacrifice; that is, (i) both are performed due to the desire for their respective results and (ii) the performance of each transgresses a prohibition, namely, that of committing violence. By analogy, since *śyena* is prohibited due to a prohibition being violated, *satī* is prohibited too.

(E) [ADDITIONAL ARGUMENT] The claim that the mentioned *smṛti* prescribes the performance of *satī*, expressed in (B), is based on a misinterpretation of the text. Hence, it does not follow from the *smṛti* that *satī* is obligatory.

Fig. 2. Summary of Medhātithi's Argument Against *satī*

The Argument Against *satī*. A synopsis of Medhātithi's argument against *satī*, as found in the Sanskrit source is presented in Fig. 2. The arguments presented here are translated and interpreted by Sanskritists.[1] We will elaborate on the separate steps of the argument, identifying the involved rules and premises, as well as the individual arguments and their relations. We process the above as it consecutively appears in the source, thus capturing the pivotal dialectic aspect of Mīmāṃsā argumentation in an $ASPIC^+$ network.

The formal language used in our case study consists of unary predicates $O(X)$ to express 'X is obligatory', and similarly predicates F and R expressing prohibitions and recommendations, respectively. We reserve $*$ as a superscript for *prima facie* norms (e.g., $O^*(sati)$), whereas the absence of $*$ indicates a *derived* norm. Furthermore, we interpret $cs(X,Y)$ as 'X causes Y'; $eff(X,Y)$ as 'X has Y as an effect'; and we read $txtO^*(X)$ as 'the authoritative texts state that X is obligatory'. Also, $sim(X,Y)$ expresses that 'X and Y are similar', and $mis(X)$ express that 'X has been misinterpreted'. We chronologically label arguments with $A, B, C, ...$ etc. The usage of the other terms will be clear from the context: e.g., we use *sati* for *satī* and *hvn* for 'heaven'. Note also that predicate names for defeasible inference rules, will take as arguments the variables and constants that appear in the rule named. Last, the *contrary function* is defined so that $\phi = -\psi$ iff $\phi = \neg\psi$ or $\psi = \neg\phi$, and $F(X) = -O(X), F^*(X) = -O^*(X), R^*(X) = -O^*(X)$ (the latter three correspond to D1–D3 of Definition 12). Recall that $-$ determines *contraries* (Definition 4). Hence, for example $F(X) = -O(X)$ denotes that obligations and prohibitions are mutually exclusive: i.e., 'if X is obligatory, then X is not forbidden

[1] Different interpretations of these arguments might be implemented in $ASPIC^+$, and compared and evaluated on their logical consequences.

and vice versa' (observe that $^{-}$ is not to be confused with logical negation $\neg$). Let us proceed to the first sub-argument, put forward by Medhātithi.

Argument **(A)** (shown in Fig. 3) claims that *satī* is prohibited because *satī* is a form of taking one's life, which equates with self-violence: $cs(sati, s\_vio)$. Furthermore, self-violence is an instance of violence in general: $eff(s\_vio, vio)$. Any performance of violence, however, is *prima facie* prohibited: $F^*(vio)$. Hence, it is concluded, *satī* must be prohibited too: $F(sati)$. In the above, concrete instances of the following generic rules were applied:

$$\mathcal{R}_d(A) = \left\{ \begin{array}{c} d_{10}(Act, Eff_1, Eff_2)\colon cs(Act, Eff_1), eff(Eff_1, Eff_2) \Rightarrow cs(Act, Eff_2); \\ d_5(Act, Eff)\colon cs(Act, Eff), F^*(Eff) \Rightarrow F(Act); \end{array} \right\}$$

In the corresponding formal argument A, the variables Act, Eff_1 and Eff_2 are respectively substituted by $sati$, s_vio and vio. Note that the labelling of the rules in $\mathcal{R}_d(A)$ corresponds to the list of Mīmāṃsā principles presented in Definition 12.

Subsequently, in argument **(B)**, an opponent objects to **(A)** by asserting that *satī* is instead obligatory—$O(sati)$—since the obligation is *prima facie*: $O^*(sati)$. Claim $O^*(sati)$ is itself supported by argument **(B')** referencing the passage with the prescription 'the widow should die after her husband': $txtO^*(sati)$. The above reasoning uses instantiated applications of the following rules:

$$\mathcal{R}_d(B) = \{d_{4\text{-}O}(X)\colon O^*(X) \Rightarrow O(X); \quad d_7(X)\colon txtO^*(X) \Rightarrow O^*(X).\}$$

In reply to **(B)**, argument **(C)** asserts that *satī* is instead a *prima facie* recommendation: $R^*(sati)$. The claim is substantiated by the observation that, (i) the *smṛti* passage mentioning *satī* explicitly relates the performance of *satī* to a specific reward, namely the reward of heaven: $txt\_Act\_Rew^*(sati, hvn)$. (ii) Explicit mention of a reward identifies a norm as a *prima facie* recommendation:

$$\mathcal{R}_d(C) = \left\{ \begin{array}{l} d_9^a(Act, Rew)\colon txt\_Act\_Rew^*(Act, Rew) \Rightarrow act(Act) \rightsquigarrow rew(Rew)^*; \\ d_9^b(Act, Rew)\colon act(Act) \rightsquigarrow rew(Rew)^* \Rightarrow R^*(Act). \end{array} \right\}$$

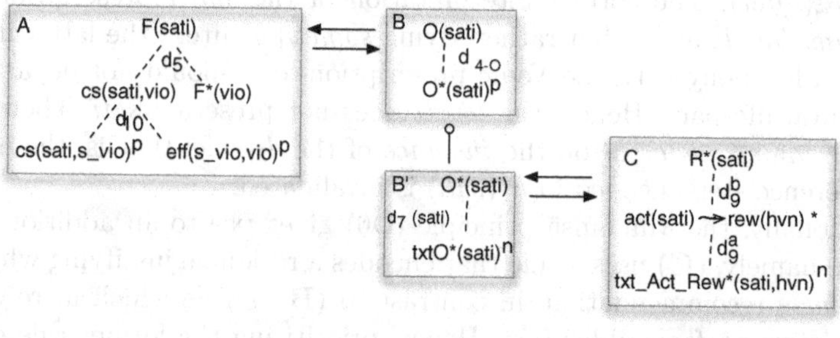

Fig. 3. Arguments A, B, B', C. Rules d_i are shown without instantiated variables.

The contrary function $^{-}$ implies a symmetric attack between arguments C and B'. Figure 3 shows argument A, B, B' and C, where B supports B' and

A and B symmetrically attack each other on the basis of the defined contrary function.

Argument **(D)** is a reply to **(B)**, claiming that *satī* is in fact prohibited: $F(sati)$. This claim follows from the assertions that (i) *satī* is similar to the *śyena* sacrifice—i.e., $sim(sye, sati)$—and (ii) the performance of *śyena* is prohibited: $F(sye)$. Note, the *śyena* sacrifice is a controversial Vedic ritual which results in the death of one's enemy; e.g. see [6].) By *analogy*, since a performance of *satī* violates the prohibition of violence too, we conclude that *satī* must also be prohibited. Clearly, B and D symmetrically attack each other.

A successive argument **(D')** then supports the premise $sim(sye, sati)$ of **(D)**: Both *śyena* and *satī* are recommendations due to fact that they depend on a desired result: $R(sye)$ and $R(sati)$. In particular, $R(sati)$ is justified given the earlier argument **(C)**, whose claim $R^*(sati)$ is again included as a premise (i.e., lemma) in **(D')**, and is used to infer $R(sati)$ via the principle (D4) (i.e., rule d_{4-R}). Subsequently, the performance of *śyena* implies violence—i.e., $cs(sye, vio)$—as does the performance of *satī*. Note that *satī* causing violence was justified earlier in **(A)**; hence, in **(D')** this fact is included as a premise rather than repeated as an argument (cf. the use of lemmas). Therefore, *śyena* and *satī* are similar.

In support of the premise $F(sye)$ of **(D)** the argument **(D")** is added, explaining that the *śyena* sacrifice is prohibited because performing *śyena* implies violence – $cs(sye, vio)$ – and violence is *prima facie* forbidden: $F^*(vio)$. The rules applied in the corresponding formal arguments D, D' and D'' are as follows:

$$\mathcal{R}_d(D) = \left\{ \begin{array}{r} d_5(Act, Eff)\colon cs(Act, Eff), F^*(Eff) \Rightarrow F(Act); \\ d_8^b(Act_1, Act_2)\colon sim(Act_1, Act_2), F(Act_1) \Rightarrow F(Act_2); \\ d_8^a(Act_1, Act_2, Eff)\colon cs(Act_1, Eff), cs(Act_2, Eff), R(Act_1), R(Act_2) \\ \Rightarrow sim(Act_1, Act_2); \\ d_{4-R}(X)\colon R^*(X) \Rightarrow R(X). \end{array} \right\}$$

Lastly, Medhātithi argues in **(E)** that the interpretation of the *smṛti* prescribing *satī*, as purported in **(B')**, is based on a misinterpretation of the word 'after': $mis(after)$. The correct interpretation of the *smṛti* prescription is not 'dying *immediately* after', but rather 'dying *sometime* after'; the latter interpretation is in harmony with the Vedic prescription 'one should not depart before one's natural lifespan'. Hence, the *smṛti* does not prescribe *satī*. Therefore, E attacks B' via an *undercut* on the *instance* of the d_7-rule; that is, the interpretative inference step encoded in $d_7(sat\bar{\imath})$ is invalidated.

Additionally, the Mīmāṃsā principle (D6) gives rise to an additional argument **(F)**: namely, **(C)** uses a rule that encodes a rationale justifying why *satī* is a *prima facie* recommendation, in contrast to **(B')**'s rule which merely claims that *satī* is *prima facie* obligatory. Hence, prioritising the former rule over the latter licenses $\mathcal{P}_{set}(\mathsf{conc}(\mathbf{F})) = \{B'\} \prec \{C\}$. The rules applied in the formal correspondents E and F are, respectively:

$$\mathcal{R}_d(E) = \{ d_E(after, sati)\colon mis(after) \Rightarrow \neg d_7(sati). \}$$
$$\mathcal{R}_d(F) = \{ d_F(sati, hvn)\colon True \Rightarrow d_7(sati) < d_9^b. \}$$

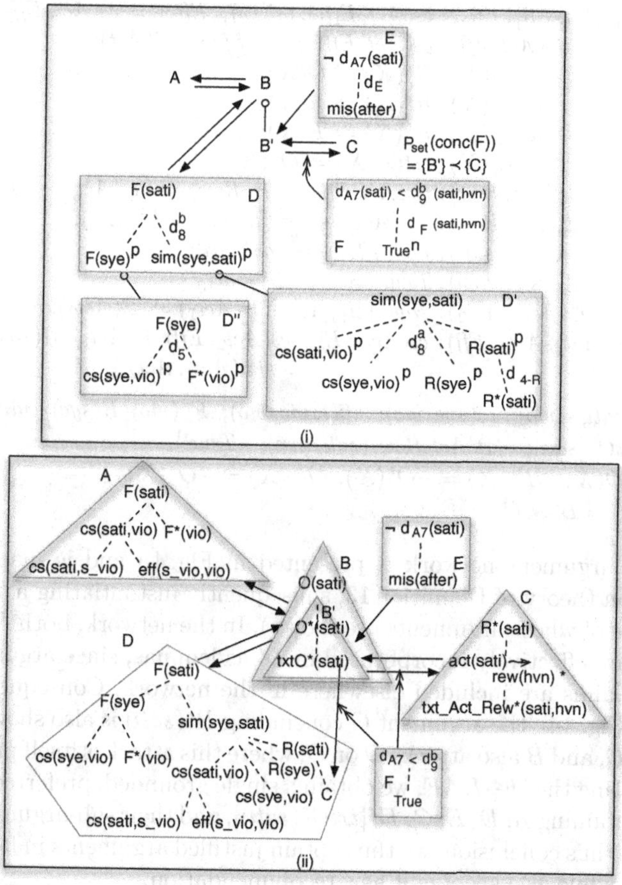

Fig. 4. (i) Authored $ASPIC^+$ network generated from Medhātithi's analysis of *satī* (ii) Some of the arguments and attacks in EAF constructed from mapping of (i) to an $ASPIC^+$ theory.

The resulting formal theory of Medhātithi's argument is defined accordingly:

Definition 13. *The following presents the $ASPIC^+$ Argumentation Theory of Medhātithi's argument against satī:*
1. $\mathcal{R}_s = \emptyset$

$$2.\ \mathcal{R}_d = \left\{ \begin{array}{r} d_{10}(Act, Eff_1, Eff_2): cs(Act, Eff_1), eff(Eff_1, Eff_2) \Rightarrow cs(Act, Eff_2); \\ d_5(Act, Eff): cs(Act, Eff), F^*(Eff) \Rightarrow F(Act); \\ d_{4-O}(X): O^*(X) \Rightarrow O(X); \\ d_{4-R}(X): R^*(X) \Rightarrow R(X); \\ d_{4-F}(X): F^*(X) \Rightarrow F(X); \\ d_7(X): txtO^*(X) \Rightarrow O^*(X); \\ d_9^a(Act, Rew): txt\_Act\_Rew^*(Act, Rew) \Rightarrow act(Act) \rightsquigarrow rew(Rew)^*; \\ d_9^b(Act, Rew): act(Act) \rightsquigarrow rew(Rew)^* \Rightarrow R^*(Act); \\ d_E(after, sati): mis(after) \Rightarrow \neg d_7(sati); \\ d_F(sati, hvn): True \Rightarrow d_7(sati) < d_{10}^b; \\ d_8^b(Act_1, Act_2): sim(Act_1, Act_2), F(Act_1) \Rightarrow F(Act_2); \\ d_8^a(Act_1, Act_2, Eff): cs(Act_1, Eff), cs(Act_2, Eff), R(Act_1), R(Act_2) \\ \Rightarrow sim(Act_1, Act_2). \end{array} \right\}$$

3. $\mathcal{K}_p = \{\, cs(sati, s\_vio);\ cs(sye, vio);\ eff(s\_vio, vio);\ F^*(vio); R(sye);\ mis(after) \,\}$
4. $\mathcal{K}_n = \{\ txtO^*(sati);\ \ txt\_Act\_Rew^*(sati, hvn);\ \ True \}$
5. $F(X) = -O(X),\ \ F^*(X) = -O^*(X),\ \ R^*(X) = -O^*(X)$
6. $\mathcal{P}(conc(F)) = B' \prec C$

The final argument network is presented in Fig. 4-i and is mapped to the argumentation theory of Definition 13, subsequently instantiating a $bh{-}EAF$ in Fig. 4-ii (some of whose arguments are shown). In the network, both $R^*(sat\bar{\imath})$ and $cs(sye, vio)$, are effectively incorporated in D' as lemmas, since arguments justifying these claims are included elsewhere in the network. Consequently, in the $bh{-}EAF$ in Fig. 4-ii, the argument C concluding $R^*(sat\bar{\imath})$ is also shown as a sub-argument of D, and B also attacks D on C, where this attack is itself pref-attacked by F. Evaluating the $bh{-}EAF$, we obtain a single grounded, preferred and stable extension containing A, D, E, C, F, $[txtO^*(sati)]$, and their sub-arguments. In line with Medhātithi's conclusion, we thus obtain justified arguments in favour of prohibiting *satī*, while also keeping it as a recommendation.

Concluding Remark. The above case study highlights the advantages of providing formal argumentative support for scholars: helping to reveal and clarify the structure of the dialectical commentaries being studied, as well as disclosing implicitly used assumptions and rendering these explicit for further analysis (including assumptions as to why some arguments are preferred to others). It also testifies to the utility, and hence promising future developments, of computational tools enabling the authoring of networks, their mapping to $ASPIC^+$ theories, and evaluation of instantiated EAFs. While in this work the authoring, mapping and evaluation was done by hand, our future aim is to provide automated support for each step.

References

1. Amgoud, L., Cayrol, C., Lagasquie-Schiex, M., Livet, P.: On bipolarity in argumentation frameworks. Int. J. Intell. Syst. **23**(10), 1062–1093 (2008)
2. Baroni, P., Caminada, M., Giacomin, M.: An introduction to argumentation semantics. Knowl. Eng. Rev. **26**(4), 365–410 (2011)

3. Bondarenko, A., Dung, P.M., Kowalski, R.A., Toni, F.: An abstract, argumentation-theoretic approach to default reasoning. Artif. Intell. **93**, 63–101 (1997)
4. Brick, D.: The dharmaśāstric debate on widow-burning. J. Am. Orient. Soc. **130**(2), 203–223 (2010)
5. Caminada, M., Amgoud, L.: On the evaluation of argumentation formalisms. Artif. Intell. **171**(5–6), 286–310 (2007)
6. Ciabattoni, A., Freschi, E., Genco, F.A., Lellmann, B.: Mīmāṃsā deontic logic: proof theory and applications. In: De Nivelle, H. (ed.) TABLEAUX 2015. LNCS (LNAI), vol. 9323, pp. 323–338. Springer, Cham (2015). https://doi.org/10.1007/978-3-319-24312-2_22
7. Dung, P.M.: On the acceptability of arguments and its fundamental role in non-monotonic reasoning, logic programming and n-person games. Artif. Intell. **77**(2), 321–358 (1995)
8. Freschi, E., Ollett, A., Pascucci, M.: Duty and Sacrifice. A Logical Analysis of the Mīmāṃsā Theory of Vedic Injunctions. History and Philosophy of Logic (2019, forthcoming)
9. Hunter, A., et al.: Tutorials on structured argumentation. Argument Comput. **5**(1), 1–4 (2014)
10. Modgil, S.: Reasoning about preferences in argumentation frameworks. Artif. Intell. **173**(9–10), 901–934 (2009)
11. Modgil, S.: Revisiting abstract argumentation frameworks. In: Black, E., Modgil, S., Oren, N. (eds.) TAFA 2013. LNCS (LNAI), vol. 8306, pp. 1–15. Springer, Heidelberg (2014). https://doi.org/10.1007/978-3-642-54373-9_1
12. Modgil, S.: Towards a general framework for dialogues that accommodate reasoning about preferences. In: Black, E., Modgil, S., Oren, N. (eds.) TAFA 2017. LNCS (LNAI), vol. 10757, pp. 175–191. Springer, Cham (2018). https://doi.org/10.1007/978-3-319-75553-3_13
13. Modgil, S., Prakken, H.: Reasoning about preferences in structured extended argumentation frameworks. In: Proceedings of the COMMA 2010, pp. 347–358 (2010)
14. Modgil, S., Prakken, H.: A general account of argumentation with preferences. Artif. Intell. **195**, 361–397 (2013)
15. Nielsen, S.H., Parsons, S.: A generalization of dung's abstract framework for argumentation: arguing with sets of attacking arguments. In: Maudet, N., Parsons, S., Rahwan, I. (eds.) ArgMAS 2006. LNCS (LNAI), vol. 4766, pp. 54–73. Springer, Heidelberg (2007). https://doi.org/10.1007/978-3-540-75526-5_4
16. Oren, N., Norman, T.J.: Semantics for evidence-based argumentation. In: Computational Models of Argument (COMMA 2008), pp. 276–284. IOS Press (2008)
17. Prakken, H.: Historical overview of formal argumentation. In: Baroni, P., Gabbay, D., Giacomin, M., van der Torre, L. (eds.) Handbook of Formal Argumentation, pp. 75–144. College Publications, London (2018)
18. Sakuntala, N.: Sati, Widow Burning in India. Viking, Doubleday, New Delhi (1992)

Who Should Be My Friends?
Social Balance from the Perspective of Game Theory

Wiebe van der Hoek[1], Louwe B. Kuijer[1], and Yì N. Wáng[2](✉)

[1] Department of Computer Science, University of Liverpool, Liverpool, UK
{Wiebe.van-der-Hoek,Louwe.Kuijer}@liverpool.ac.uk
[2] Department of Philosophy, Zhejiang University, Hangzhou, People's Republic of China
ynw@xixilogic.org

Abstract. We define *balance games*, which describe the formation of friendships and enmity in social networks. We show that if the agents give high priority to future profits over short term gains, all Pareto optimal strategies will eventually result in a balanced network. If, on the other hand, agents prioritize short term gains over the long term, every Nash equilibrium eventually results in a network that is stable but that might not be balanced.

Keywords: Structural balance theory · Game theory · Nash equilibrium · Pareto optimality

1 Introduction

A *social network* consists of a number of agents and positive or negative relations between them. The agents could be countries, individuals or groups. A positive relation represents a friendship or alliance, while a negative relation represents an enmity or rivalry. Structural balance theory describes such networks, and was introduced by Heider [15,16] and later generalized by Cartwright and Harary [3, 11,12]. It argues that certain patterns are likely to occur while other patterns are unlikely; the likely patterns are referred to as *balanced* while the unlikely ones are *unbalanced*. There is also empirical support for the assertion that networks tend towards balance, see for example [25,27], though a fully balanced network is not always (nor easily) reached [18].

Usually, balance theory describes a network as a whole; it is claimed (quite convincingly) that networks usually become more balanced over time, but relatively little attention is paid to the actions and motivations of individual agents on the way towards balance. Here, we take a different, game-theoretical approach: we explicitly treat the tendency towards balance as evidence for a preference by agents for balanced states over unbalanced ones. This allows us to take a detailed look at how this tendency follows the result of rational choices by the individual agents.

P. Blackburn et al. (Eds.): LORI 2019, LNCS 11813, pp. 370–384, 2019.
https://doi.org/10.1007/978-3-662-60292-8_27

We introduce a class of *balance games*, which are multi-stage games where in each stage one agent updates their relationship with someone else, and all agents prefer being involved in balanced relations over unbalanced ones. We show that if the agents are sufficiently patient (i.e., if the discount factor δ is high enough), any Pareto optimal strategy profile will, with probability 1, eventually result in a balanced network. If the agents are less patient, the end result may not be a balanced network. In fact, we show that for sufficiently impatient agents (i.e., if the discount factor δ is low enough), any subgame perfect Nash equilibrium strategy profile will, with probability 1, result in a network that need not be balanced but that is *stable*. Stability was defined by Van der Hoek et al. [17] and is related to but strictly weaker than balance.

The structure of the paper is as follows. We first give definitions for balance, stability and the balance game in Sect. 2, where we also present a few useful lemmas, give an example, and discuss related work. Then, in Sect. 3 we consider the case of patient agents, and show that for them every Pareto optimal strategy profile results in balance. In Sects. 4 we study the cases of impatient agents. In Sect. 5 we discuss some generalizations as well as some limitations of our results. We conclude in Sect. 6.

2 Definitions and Preliminaries

In this section we first provide definitions of social balance theory, including structural balance and stability. Most of these are from the literature (mainly [3] and [17]). We give examples and introduce some results which will be used in later proofs. We then move on to define a class of balance games and some relevant notions. We use an example to explain the idea of balance games. We then discuss related approaches.

2.1 Structural Balance and Stability

A *(social) network* is an irreflexive, complete, signed and undirected graph, i.e., a pair (A, E) such that A is a finite set of agents (represented by vertices of a graph), and $E : \{\{i, j\} \subseteq A \mid i \neq j\} \rightarrow \{+, -\}$ is an edge function that assigns to each unordered pair of different agents a positive $(+)$ or a negative $(-)$ edge. For simplicity, for pairs of agents we write ij, ik, etc, and for triads we write ijk, ijl, etc. We only consider graphs with at least three agents.

Balance. Given a network $N = (A, E)$, a triad ijk of N is called *balanced*, if the labels of its edges are of one of the types $+++$ or $+--$ up to isomorphism. So in a balanced triad there is an even number of negative edges. The *unbalanced* triads therefore have either of the other two types: $++-$ or $---$. A network is *balanced*, if all of its triads are balanced, and *unbalanced* otherwise.

In a triad of the type $---$, all three agents are enemies of one another. In that situation, it is likely that two of them will set aside their differences and unite against their common foe. Doing so would turn the triad into $+--$, which is balanced. In a triad $++-$, there is one agent i that is friends with both j and k, while j and k are enemies. It is then likely that one of two things will happen: either the mutual friendship with i will form a basis for reconciliation between j and k, resulting in the balanced triad $+++$, or the tension between j and k will force i to end its friendship with one of them, resulting in the balanced triad $+--$. So both types of unbalanced triad have a tendency to evolve into a balanced triad.

Stability. In addition to balance, we will also use the weaker notion of stability, which is defined in terms of mutual and anti-mutual ties. For a pair ij of a network $N = (A, E)$, a *mutual tie* of ij is an agent k of N such that k is a mutual friend or mutual enemy of i and j, i.e., either $E(ik) = E(jk) = +$ or $E(ik) = E(jk) = -$.

An *anti-mutual tie* of ij is an agent k of N such that k is either a friend of i and an enemy of j, or an enemy of i and a friend of j, i.e., if one of the following is true:

- $E(ik) = +$ and $E(jk) = -$
- $E(ik) = -$ and $E(jk) = +$.

We say an pair ij is *stable*, if it is one of the following cases:

- $E(ij) = +$ and ij has at least as many mutual ties as anti-mutual ties;
- $E(ij) = -$ and ij has at least as many anti-mutual ties as mutual ties.

Finally, a network is *stable*, if all of its pairs are stable.

A mutual tie is a reason to stay or become friends, while an anti-mutual tie is a reason to stay or become enemies. A network is therefore stable if every pair of friends has at least as many reasons to remain friends as to become enemies, and every pair of enemies has at least as many reasons to remain hostile as to become friends.

Balance vs. Stability. If ijk is a balanced triad and $E(ij) = +$, then k is a mutual tie for ij. Specifically, if ijk is of type $+++$ then k is a mutual friend, and if ijk is of type $+--$ then k is a mutual foe. Likewise, if ijk is balanced and $E(ij) = -$, then k is an anti-mutual tie for ij. A balanced network is therefore a stable network with the additional property that for all pairs ij, if $E(ij) = +$ then ij has only mutual ties and if $E(ij) = -$ then ij has only anti-mutual ties.

Not all stable networks are balanced, however. Two typical examples of stable networks that are not balanced are illustrated in Fig. 1.

In Fig. 1a, one can verify that every pair has an equal number of mutual and anti-mutual ties. For instance, pair $\{1, 3\}$ has two mutual ties (i.e., agents 4 and 5) and two anti-mutual ties (i.e., agents 2 and 6). It is therefore stable, and so is the entire network. Yet the network is not balanced, for, e.g., the triad $\{1, 2, 3\}$ is not balanced. Similarly, the network of Fig. 1b is also stable but not balanced.

The benefit of the latter network is that it can be generalized to a class of stable and unbalanced networks illustrated in Fig. 1c. For each natural number

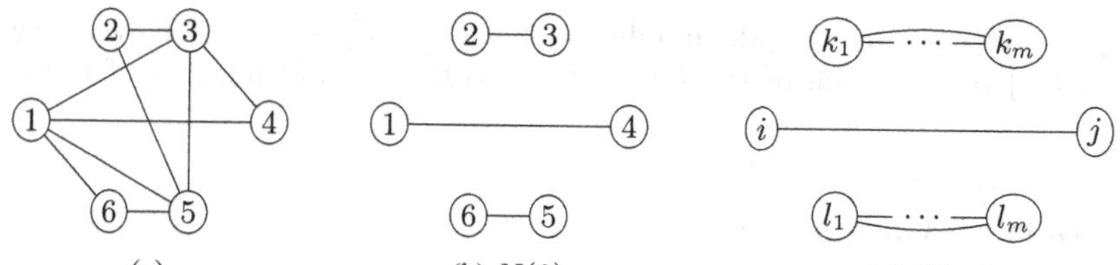

(a) (b) $N(2)$ (c) $N(m)$

Fig. 1. Stable networks that are unbalanced, where a solid line stands for a positive edge and the lack of a line for a negative edge.

$m \geq 2$, the network $N(m)$ can be divided into three cliques: the $\{k_1, \ldots, k_m\}$-party (k-party for short), the $\{l_1, \ldots, l_m\}$-party (l-party for short) which are of equal size, and a small, third party $\{i, j\}$. Agents are friendly towards members of their own clique and hostile towards members of other cliques. The network shown in Fig. 1b is $N(2)$.

One can verify that for any pair $\{k_x, k_y\}$, $\{l_x, l_y\}$ or $\{i, j\}$ in the same party, there are $2m$ mutual ties (i.e., all others are their mutual ties), and is therefore stable. Any pair $\{k_x, l_x\}$ across the two major parties are stable, as there are 2 mutual ties (i.e., i and j) and $(2m-2)$ anti-mutual ties. Any pair $\{i, k_x\}$, $\{i, l_x\}$, $\{j, k_x\}$ or $\{j, l_x\}$ across the third party and a major party has an equal number (i.e., m) of mutual and anti-mutual ties, and is thus stable as well. For every $m \geq 2$, the network $N(m)$ is therefore stable. It is not balanced, however, because it contains triads of the type $---$.

Let us consider a few technical lemmas that will be useful later on. The first lemma is well known in balance theory, and follows immediately from the fact that a triad is balanced if and only if it contains an even number of negative edges.

Lemma 1. *If a triad ijk is balanced, then flipping (the sign of) any single edge of the triad will make it unbalanced. Likewise, if ijk is unbalanced then flipping any single edge of the triad will make it balanced.*

A pair ij is stable if and only if it is part of at least as many balanced triads as unbalanced triads. The following lemma therefore follows from Lemma 1.

Lemma 2. *If a pair ij is stable, then flipping $E(ij)$ does not increase the number of balanced triads containing i, nor does it decrease the number of unbalanced triads containing i.*

If a pair ij is unstable, then flipping $E(ij)$ will strictly increase the number of balanced triads in the network.

Finally, we need a lemma that is new in this paper.

Lemma 3. *For any network, if there is an unbalanced triad, then all agents occur in an unbalanced triad.*

Proof. If ijk contains an odd number of negative edges, then for every agent $l \notin \{i, j, k\}$ at least one of lij, ljk or lik also has an odd number of negative edges.

2.2 Balance Games

We study structural balance from the viewpoint of game theory, by introducing a *balance game* which is a type of multi-stage game of infinitely many stages. All the agents in a network are players of a balance game. Each agent is better off if it is involved in more balanced triads.

Valuation. Given a network N, the valuation for an agent i in that network is the number of balanced triads i is part of minus the number of unbalanced triads it is part of. This valuation is denoted $val_i(N)$.

Actions. At every stage of the game, a single agent (chosen uniformly at random) will be given an opportunity to change one of its relations. This agent can choose to change its relation to one other agent, or it can choose to *pass* and leave all relations unchanged. Note that an agent can only change those relations that it is involved in. Agent i can decide to become enemies with j, but i cannot choose to create an enmity between j and k—although i might be able to create a situation where j and k have an incentive to become enemies.

In a balanced network all triads are balanced, so balance is a *global* optimum of val_i for every i. In a stable network no single change to any relation ij would result in an increase in the number of balanced triads for either i or j (see Lemma 2). So stability is a *local* optimum of val_i for every i.

Cost of Change. If an agent decides to change a relation, it will incur a cost of change. This cost represents the effort and social cost associated with changing one's relation to another agent. For example, deciding to end an enmity might require an apology and a good bottle of wine, whereas ending a friendship may reduce one's social capital.

The exact value that this cost of change should have can be debated. We believe that it should lie in the open interval $(0, 2)$. In order to keep all calculations as simple as possible we prefer to have an integer cost of change, so we set it to be 1. See Sect. 5 for a discussion of why we believe that the cost of change should be between 0 and 2, and an overview of how any cost of change in the interval $[0, \infty)$ would influence our results.

Discount Factor. At every stage of the game, the agents immediately receive utility equal to their valuation of the current network. This rewards them for having more balanced relations and punishes them for unbalanced ones. Additionally, they receive utility from future game stages. A reward today is worth more than the same reward tomorrow, however, so the agents multiply their future utility by a discount factor $\delta \in (0, 1)$. The value of δ indicates the kind of agents that are being modeled; patient agents place (relatively) high value on the future and therefore have a high value for δ, impatient agents prioritize short term gain and therefore have a low value for δ. The utility for agent i in a network N therefore equals $val_i(N)$ plus δ times the expected utility in the successor network (minus the cost of change, if applicable).

We consider only memoryless pure strategies, so a strategy for an agent i can be represented by a function that maps every network to either a single change in a relation for i or to no change. Below we introduce the formal definitions. We assume a fixed set of agents $A = \{1, \ldots, n\}$ with $n \geq 3$, and use $\mathcal{N}$ to denote the set of all social networks over A.

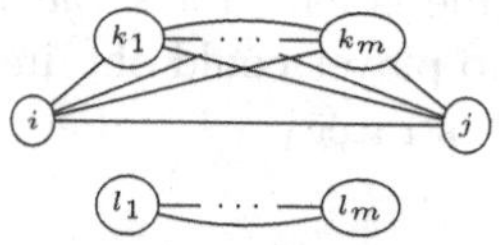

(a) A balanced outcome of $N(m)$ where i and j take the same side.

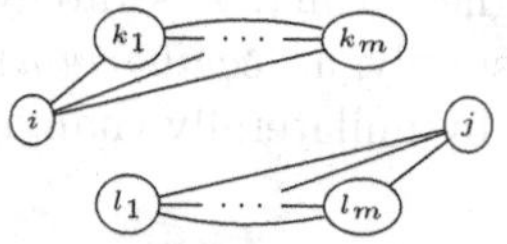

(b) A balanced outcome of $N(m)$ where i and j take different sides.

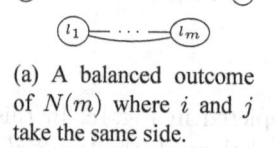

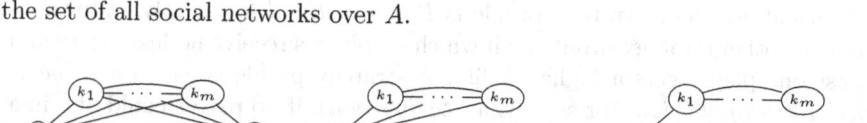

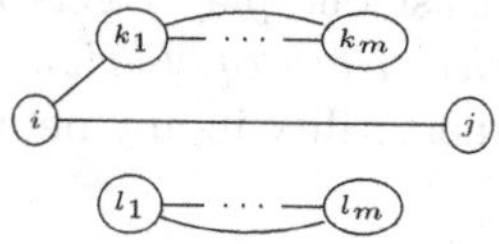

(c) A successor of $N(m)$ where i and k_1 become friends.

Fig. 2. Possible evolutions of the network $N(m)$ from Fig. 1c.

Definition 1. *The* balance game *over a network $N = (A, E)$ is a pair (N, s) given by*

- (Players) *A is the set of players.*
- (Strategies) *$s = (s_1, \ldots, s_n)$ is a strategy profile, such that for every player i, $s_i : \mathcal{N} \rightarrow \{(+, i, j), (-, i, j) \mid j \in A \backslash \{i\}\}$ is a strategy for i.*
- (Outcomes) *The outcome of (N, s) is one of $\{(N^{s_i}, s) \mid i \in A\}$, chosen uniformly at random, where $N^{s_i} = (A, E^{s_i})$ is given by*

$$E^{s_i}(kl) = \begin{cases} +, & \text{if } s_i(N) = (+, i, j) \text{ and } kl = ij, \\ -, & \text{if } s_i(N) = (-, i, j) \text{ and } kl = ij, \\ E(kl), & \text{otherwise.} \end{cases}$$

- (Utility) *The utility function $u = (u_1, \ldots, u_n)$, where u_i is the utility of player i, is given recursively by*

$$u_i(N, s) = val_i(N) + \delta \cdot \frac{1}{n} \cdot (\textstyle\sum_{j \in A} u_i(N^{s_j}, s) - c_j),$$

where $c_j = 1$ if $i = j$ and $N \neq N^{s_j}$, and $c_j = 0$ otherwise.

The recursive definition of utility does not immediately provide a practical way to compute $u_i(N, s)$. It is therefore useful to also have a direct characterization of $u_i(N, s)$. For this purpose, we use the concept of *timelines*. Given a strategy profile s, an s-*timeline* is an infinite sequence $l = \langle N_0, N_1, \ldots \rangle$ such that for every $t \in \mathbb{N}$, $N_{t+1} \in \{N_t^{s_i} \mid i \in A\}$. The utility of agent i in such a timeline is given by $u_i(l) = \sum_{t=0}^{\infty} \delta^t (val_i(N_t) - c)$, where $c = 1$ if i brought about a change from N_{t-1} to N_t and $c = 0$ otherwise. The utility $u_i(N, s)$ is then simply the expected value of $\{u_i(l) \mid l = \langle N, N_1, \ldots \rangle$ is an s-timeline$\}$.

For a given s-timeline $l = \langle N_0, N_1, \ldots \rangle$, if there is a natural number T such that $N_{t_1} = N_{t_2}$ for all $t_1, t_2 \geq T$, then we say l *finalizes* in N_T, or N_T is the *final* of l.

We write $N \rightsquigarrow_i N'$ if there is a strategy s_i for agent i such that $N' = N^{s_i}$, and we write $N \rightsquigarrow N'$ if there is at least one i such that $N \rightsquigarrow_i N'$.

As usual, we say a strategy profile is *Pareto optimal* (or simply, *optimal*) if there is no other strategy profile with which all players receive no less utility and at least one player gets a higher utility. A strategy profile is called a *subgame perfect Nash equilibrium* (or simply, an *equilibrium*), if no player could obtain a higher utility in any network by unilaterally changing its strategy.

2.3 Example

Consider the network $N(m)$ for a given $m \geq 2$ as depicted in Fig. 1c. In this network, most triads are balanced, but some remain unbalanced: the triads ikl and jkl are unbalanced for every $k \in \{k_1, \ldots, k_m\}$ and every $l \in \{l_1, \ldots, l_m\}$, since those triads are of the form $---$.

The agents could choose to pass, leaving the network in the state $N(m)$ forever. Alternatively, the agents could take actions that change the network. Taking such an action would incur a cost of change, however, so a rational agent will only do so in the expectation of a sufficiently high reward later. The main reward which all agents would like to obtain (although they may or may not be willing to pay the price for doing so) would be a balanced network.

There are many ways in which $N(m)$ can be changed to a balanced network. For example, all agents could decide to become friends with one another. That change would be very costly, however. Rational agents would instead aim for a balanced state that is easier to reach. A more feasible way to reach balance would be for the agents i and j to join the k-party or l-party, as shown in Figs. 2a and b.

Suppose that i joins the k-party. So eventually i will become friends with every agent k_x. Then at first, a friendship between i and some agent k_x must form. Without loss of generality, we can assume that this first friendship is with k_1, as shown in Fig. 2. Consider the effect this has on the valuation of the different agents. Triads ik_1k_y and ik_1j used to be of the form $+--$ but are now $++-$. So they have turned from balanced to unbalanced. Triads ik_1l_y, on the other hand, used to be $---$ and have become $+--$, so they have turned from unbalanced to balanced. All other triads are unaffected. In total, there are $m-1$ triads ik_1k_y, 1 triad ik_1k_y and m triads ik_1l_y. So the number of triads that become balanced and the number of triads that become unbalanced are both m.

The agents i and k_1 are part of all triads that change, so their valuation is unchanged. One of them does have to pay the cost of change, but they suffer no harm from the change in the network. Agents l_y are part of one triad that changes, and it turns balanced. So their valuation increases, without them having to take any action. They quite like this change. The agents j and k_y are less happy, however: they too are part of one triad that changes, but theirs turns unbalanced. So they lose out due to this new friendship.

Once this first friendship has been established, all other members of the k-clique have an incentive to follow k_1 and become friends with i as well: currently, $k_1 k_y i$ if of the type $++-$, but by becoming friends with i they can turn this into the balanced type $+++$. So the first friendship ik_1 is likely to be followed by a flood of new friendships between i and the members of the k-party. Every such new friendship will be welcomed by the l-party, by i and by all k_y that are already friends with i, since it makes their relations more balanced. For those k_y that are not yet friends with i, the situation turns even worse, however. Every time an agent k_x becomes friends with i, the triad ik_yk_x becomes unbalanced, depriving k_y of another 2 points of valuation. In particular, if k_m is the last agent to become friends with i then just before they do so their valuation is $2(m-1)$ lower than it was in $N(m)$. Eventually, however, the network reaches one of the balanced states depicted in Fig. 2, at which point all temporary losses are wiped away and replaced by the benefits of being part of a balanced network.

For highly impatient agents, paying the initial cost of change is not worth it, so remaining in $N(m)$ is the only rational option. If agents are more patient, however, aiming for balance may be the only rational choice. How patient agents have to be in order for remaining in $N(m)$ not to be an option depends on whether we are considering optimal strategy profiles or equilibria. The fact that the agents who are late to become friends with i (or j) suffer until balance is achieved means that remaining in $N(m)$ remains optimal until δ becomes very high. But the agents that experience a loss in valuation are not the ones that take action, it's the ones that have not yet taken action. So if the agents are even a little bit patient ($\delta = 0.5$ suffices, for example), the agents who decide to initiate the friendships will benefit by doing so, thereby making the strategy of remaining in $N(m)$ not an equilibrium.

2.4 Related Work

Our definition of balance is called *3-balance* in the classical literature (e.g., [3]), where the number 3 refers to the length of the cycles to be examined – 3-cycles for triangles. In general, *k-balance* of a network requires that all cycles of length up to k contain an even number of negative edges. There is also pressure of balance from longer cycles, but it is considered of less effect. This leads to a difference between viewing balance of networks as a *property* or a *process*. Taking the former view, as in the classical literature, all cycles of all lengths are examined before we can determine the balance of the whole network. The lesser effect of longer cycles is modeled by assigning a *weight* or *strength* to each length [3,23]. In the latter view as proposed in [17] and adopted in this paper, however, the balance of a network lies in the balance of its local parts. The balance of longer cycles is achieved gradually over time by the constraints of balance among shortest cycles (triads in the case of undirected graphs).

The *structure theorem* [3,13] states that a balanced network can be partitioned into two mutually antagonistic and self-solidary components. The structure theorem was later generalized in [5] to consider a weaker version of balance which corresponds to more than two partitions. This gives a different way of studying the tendency of balance: it can be viewed as a process of partitioning a network. This approach has been developed in [7,8,24].

In recent years the study of link formation has drawn much attention in various fields including social network analysis, economics, information and computer science. Some of these are empirical studies that investigate into, say, the formation of social networks or how technology is adopted in a network [4,28], and some are theoretical studies that focus on, say, the prediction, formal model, statistical and computational results of network formation [6,21,29,30]. This paper falls into theoretical side, and we focus on the formal model of a type of link formation from the viewpoint of game theory.

The study of structural balance theory has not been limited to a single field since the very beginning. It was initiated in Heider's work [15,16] in social psychology and reinvented by Harary et al. [3,11–13] using graph theory. Empirical studies on the impact of structural balance theory was carried out in the area of social network analysis (see, e.g., [25,26]). The trend to study and adopt the theory from new perspectives and in new fields has not come to an end. For example, the impact of structural balance on opinion formation has been evaluated in the framework of evolutionary games [20]. In our paper we also have structural balance and games in the same framework, but we focus more on the theoretical aspects of the structural balance of social networks.

Another area of related work is that of games on networks, a sub-discipline of game theory concerned with networks. See for example [9,19,22]. Balance games can be considered part of this field, but they differ significantly from the games that have been studied before. Other disciplines of game theory, such as coalition formation and evolutionary games (see, e.g., [31]), are also related to balance games but very different from a technical point of view.

3 Patient Players

We show that for sufficiently patient players, a Pareto optimal strategy profile finalizes in a balanced network with probability 1.

Lemma 4. *Let s be a strategy profile, N_0 a network and L the set of s-timeline starting in N_0 that do not finalize in balance. If L occurs in the game (N_0, s) with probability greater than 0, then there is a $\delta_{high} < 1$ such that for all $\delta > \delta_{high}$, s is not Pareto optimal.*

Proof. Every agent is part of $b := \frac{(n-1)\cdot(n-2)}{2}$ different triads. In a balanced network, all triads are balanced so every agent has a valuation of b. In every non-balanced network, every agent has a valuation of at most $b - 2$, since by Lemma 3 every agent is part of at least one unbalanced triad.

Let s be any strategy profile that finalizes in a balanced network with probability 1, and s' any strategy profile that does not. Then after some number of time steps, the expected valuation under s will be higher than under s'. For sufficiently patient agents, s therefore Pareto dominates s'.

Theorem 1. *For a given number of players, there exists a discount factor δ_{high} such that for every $\delta > \delta_{high}$ and every Pareto optimal strategy profile s the following hold:*

1. *every s-timeline that contains a balanced network finalizes in that network;*
2. *for every N, the game (N, s) reaches a balanced network with probability 1.*

Note that the bound δ_{high} depends on the number of agents. In fact, it can be seen that $\lim_{n\to\infty} \delta_{high} = 1$, so the required amount of patience approached 1 as the number of agents increases.

This can, for example, be seen from the network $N(m)$ depicted in Fig. 1c. In order for $N(m)$ to become balanced, the central two agents i and j need to join either the clique $k_1, \dots, k_m$ or the clique $l_1, \dots, l_m$. While i is in the process of joining a clique, those members of the clique that are not yet friends with i experience a loss in valuation equal to twice the number of agents that are already friends with i. This loss is temporary, but both its magnitude and its duration increase with the number of agents. The amount of patience needed for any "go to balance" strategy to beat the "everyone passes in $N(m)$" strategy for every agent therefore increase with m.

4 Impatient Players

Here we show that if the discount factor δ is sufficiently close to 0, then every subgame perfect Nash equilibrium finalizes in a stable state with probability 1.

Unlike the case for patient agents, where the bound depends on the number of agents, our bound δ_{low} for impatient agents is constant. More concretely, $\delta_{low} = \frac{1}{10}$ suffices.

In order to prove this bound, we first need a few lemmas. For most of these lemmas the existence of a bound is relatively easy to see, but finding a precise number for the bound requires a lot of tedious calculations. We therefore do not prove the precise bound, and only give a qualitative argument for the existence of a bound.

Lemma 5. *Let N_0 be a network, and let m be the maximum increase of valuation brought about by any action of agent i, i.e., $m = \max\{val_i(N_1) - val_i(N_0) \mid N_0 \rightsquigarrow_i N_1\}$. Then for any strategy profile s, any s-timeline $\langle N_0, N_1, N_2, \dots\rangle$ and any $t \in \mathbb{N}$ we have $val(N_t) \leq val(N_0) + (m + 2t)t$.*

Proof. Consider the same action carried out in N_0 and N_k. This action will make some triads balanced, while making others unbalanced. Since N_0 and N_k differ in at most k edges, the number of triads made balanced when performing the action in N_k is at most k higher than in N_0, and the number of triads made unbalanced is at most k lower.

Turning a triad balanced increases valuation by 2, turning it unbalanced decreases it by 2. So in N_k the action yields at most $2k + 2k$ more valuation than in N_0, where it yields at most m. So the increase in valuation from N_k to N_{k+1} is at most $4k$. It follows that $val(N_t) \leq val(N_0) + \sum_{k=0}^{t-1}(m + 4k) \leq val(N_0) + m \cdot t + \frac{4t}{2} \cdot t = val(N_0) + (m + 2t)t$.

It follows that for sufficiently small δ, agents will not take any action that would cause a loss of valuation to them.

Lemma 6. *Let $\delta < \frac{1}{10}$ and s a Nash equilibrium. Then at every game (N, s), none of the agents take any action that changes the network unless that action increases their valuation.*

Proof. Taking an action that changes the network will incur the cost of change. An action that changes the network but does not increase the agent's valuation therefore causes a short term loss in utility for that agent. A sufficiently impatient agent will never take such an action.

Lemma 5 gives an upper bound on the long term benefit of taking a short term loss. Since this bound does not depend on n, the bound δ_{low} below which agents are sufficiently impatient does not depend on n either.

Finally, if some agent has a valuation increasing move available, then such a move will be taken by at least one agent.

Lemma 7. *Let $\delta < \frac{1}{10}$ and s a Nash equilibrium. Then in every subgame (N, s), if any agent has an available action that will increase its valuation, then at least one agent takes an action that increases its valuation.*

Proof. Any action that increases valuation increases it by at least two, so the increase in valuation outweighs the cost of change, resulting in a short term increase in utility.

Theorem 2. *Let $\delta_{low} = \frac{1}{10}$. Then for any discount factor $\delta < \delta_{low}$ and any subgame perfect Nash equilibrium s, the following holds:*

1. *every s-timeline that contains a stable network finalizes in that stable network;*
2. *for every N, the subgame (N, s) reaches a stable network with probability 1.*

Proof. The first part of the theorem follows from Lemma 6. The second part follows from Lemmas 6 and 7.

5 Discussion

Accuracy. Balance theory predicts that social networks broadly tend towards balance, but that a fully balanced network is not always reached. This is also confirmed by empirical studies. The same general behavior is observed in balance games: rational agents will generally increase the amount of balance in the network, but under most circumstances a fully balanced outcome is not guaranteed.

Whether balance games accurately predict agents' behaviour on a more detailed level is not currently known, and remains an interesting question for further research.

Pareto Optimality for Low δ and Subgame Perfect Nash Equilibria for High δ. Our results are "asymmetric", in the sense that δ_{high} is related to optimality while δ_{low} is related to equilibria. We conjecture that this asymmetry is fundamental: we think that for arbitrarily high $\delta < 1$ there remain equilibria that do not finalize in balanced networks and that for arbitrarily low $\delta > 0$ there remain Pareto optimal strategy profiles that do not finalize in stable networks. Unfortunately, the strategy space for balance games is very large and hard to describe. So while we have reasons to believe that there are no lower bound for optimality and upper bound for equilibria, we have not yet managed to find the counterexamples that prove this to be the case.

Cost of Change. Changing a relation takes some amount of effort, so it should be associated with some cost $c > 0$. Furthermore, agents seem willing to incur this cost in order to make their relations more balanced. This suggests that the increase in valuation caused by the increase in balance is higher than the cost of change, so $c < 2$. We therefore consider values of c outside the interval $(0, 2)$ to be implausible. Still, for the sake of completeness we explain how out results change for any $c \in [0, \infty)$.

The bound δ_{high} is not qualitatively affected by the cost of change: for every $c \in [0, \infty)$, there is still a bound δ_{high} above which every optimal solution finalizes in balance with probability 1 and δ_{high} approaches 1 as n approaches infinity.

For any $c \in (0, 2)$, the bound δ_{low} is also qualitatively unaffected. The exact value of the bound may change, but a bound δ_{low} still exists and $\lim_{n\to\infty} \delta_{low} > 0$.

The first statement of Theorem 2 still applies: every equilibrium timeline that contains a stable network finalizes in that network. But the second part of Theorem 2 does not hold for $c \in (2, \infty)$. If $c > 2$ and δ is sufficiently low then some timelines finalize before reaching a stable network.

This leaves the two cases $c = 0$ and $c = 2$. If $c = 0$, then no bound δ_{low} exists: for every $\delta \in (0, 1)$ there are equilibria where agents move out of a locally optimal stable state and eventually reach a globally optimal balanced state. Finally, for $c = 2$, there is a bound δ_{low}, but in that case we do not know whether $\lim_{n\to\infty} \delta_{low} = 0$.

Complete Graphs. We assumed all edges to be either positive or negative, unlike some works on social balance we do not consider neutral relations. This is because for networks with neutral edges we do not consider there to be sufficient data to accurately determine the agents' preferences.

6 Conclusion

In this paper we viewed structural balance of a social network as a result of its agents playing a *balance game*. When the agents are patient, their Pareto optimal strategies result in a *balanced* network as the game proceeds. When, on the other hand, the agents are impatient, their subgame perfect Nash equilibrium strategies result in a *stable* network. By a framework accommodating both the concepts of balance and stability, our work bridged the classical literature on social balance [3] and its recent development using a logical approach [17].

There is still work that remains to be done. In particular, while we have shown that bounds δ_{high} and δ_{low} exist, we have not yet found tight bounds. Furthermore, as mentioned in Sect. 5, we conjecture that an equilibrium for patient agents may not finalize in balance and that an optimal profile for impatient agents may not finalize in balance. A proof (or, for that matter, a disproof) of these conjectures would be interesting. It would also be good to know more about the behaviour of agents that are neither as patient as to guarantee balance nor so impatient to guarantee stability.

Additionally, there are a number of further questions related to generalizations of the balance game. The balance game could, for example, be generalized to different kinds of networks. These include incomplete networks (were agents i and j may be neither friends nor foes), weighted networks (where some friendships/enmities are stronger than others) and directed networks (where i's relation towards j may be different from j's relation towards i).

It should also be interesting to allow different kinds of agents. Some agents might be more patient than others, or have a higher tolerance for unbalance. The framework of Boolean games [10,14] seems to be appropriate for modelling the diversity of agents in their goals.

Another way to increase diversity is in the strategies of agents. By going further to formalizing the dynamics of balance games in the framework of temporal logic, in particular, alternating-time temporal logic [1,2], we can get a better characterization of the time evolution and the flexibility of modeling agent's strategies in a formal and unified manner. We leave, however, all these for future work.

References

1. Alur, R., Henzinger, T.A., Kupferman, O.: Alternating-time temporal logic. In: Proceedings of FOCS 1997, pp. 100–109. IEEE (1997)
2. Alur, R., Henzinger, T.A., Kupferman, O.: Alternating-time temporal logic. J. ACM **49**(5), 672–713 (2002)

3. Cartwright, D., Harary, F.: Structure balance: a generalization of Heider's theory. Psychol. Rev. **63**(5), 277–293 (1956)
4. Conley, T.G., Udry, C.R.: Learning about a new technology: pineapple in ghana. Am. Econ. Rev. **100**(1), 35–69 (2010)
5. Davis, J.A.: Clustering and structural balance in graphs. Hum. Relat. **20**(2), 181–187 (1967)
6. Dev, P.: Networks of information exchange: are link formation decisions strategic? Econ. Lett. **162**, 86–92 (2018)
7. Doreian, P., Mrvar, A.: A partitioning approach to structural balance. Soc. Netw. **18**(2), 149–168 (1996)
8. Doreian, P., Mrvar, A.: Partitioning signed social networks. Soc. Netw. **31**, 1–11 (2009)
9. Goyal, S.: Connections: An Introduction to the Economics of Networks. Princeton (2007)
10. Gutierrez, J., Harrenstein, P., Wooldridge, M.: Iterated boolean games. Inf. Comput. **242**, 53–79 (2015)
11. Harary, F.: On the notion of balance of a signed graph. Mich. Math. J. **2**(2), 143–146 (1953)
12. Harary, F.: On local balance and n-balance in signed graphs. Mich. Math. J. **3**(1), 37–41 (1955)
13. Harary, F., Norman, R.Z., Cartwright, D.: Structural Models: An Introduction to the Theory of Directed Graphs. Wiley, New York (1965)
14. Harrenstein, P., van der Hoek, W., Meyer, J.J., Witteveen, C.: Boolean games. In: Proceedings of the 8th Conference on Theoretical Aspects of Rationality and Knowledge, pp. 287–298 (2001)
15. Heider, F.: Social perception and phenomenal causality. Psychol. Rev. **51**(6), 358–374 (1944)
16. Heider, F.: Attitudes and cognitive organization. J. Psychol. **21**(1), 107–112 (1946)
17. van der Hoek, W., Kuijer, L.B., Wáng, Y.N.: A logic of allies and enemies. In: 13th Conference on Logic and the Foundations of Game and Decision Theory (LOFT), Italy (2018)
18. Hummon, N.P., Doreian, P.: Some dynamics of social balance processes: bringing Heider back into balance theory. Soc. Netw. **25**(1), 17–49 (2003)
19. Jackson, M.O., Zenou, Y.: Games on networks. In: Handbook of Game Theory, vol. 4, pp. 95–164 (2015)
20. Li, W., Li, P., Wang, H., Fan, P.: Evolutionary game of opinion dynamics under impact of structural balance. In: Proceedings of the 3rd International Conference on System Science, Engineering Design and Manufacturing Informatization, pp. 208–211. IEEE (2012)
21. Liben-Nowell, D., Kleinberg, J.: The link-prediction problem for social networks. J. Am. Soc. Inf. Sci. Technol. **58**(7), 1019–1031 (2007)
22. de Martí, J., Zenou, Y.: Social networks. In: Handbook of the Philosophy of Social Sciences, pp. 339–361. SAGE Publications, London (2011)
23. Morrissette, J.O.: An experimental study of the theory of structural balance. Hum. Relat. **11**(3), 239–254 (1958)
24. Mrvar, A., Doreian, P.: Partitioning signed two-mode networks. J. Math. Sociol. **33**, 196–221 (2009)
25. Newcomb, T.M.: Acquaintance Process. Holt, Rinehart & Winston, New York (1961)
26. Newcomb, T.M.: Reciprocity of interpersonal attraction: a nonconfirmation of a plausible hypothesis. Soc. Psychol. Q. **42**(4), 299–306 (1979)

27. Sampson, S.F.: A Novitiate in a Period of Change: An Experimental and Case Study of Social Relationships. Ph.D. thesis, Cornell University (1968)
28. Santos, P., Barrett, C.B.: Identity, interest and information search in a dynamic rural economy. World Dev. **38**(12), 1788–1796 (2010)
29. Smets, S., Velázquez-Quesada, F.R.: How to make friends: a logical approach to social group creation. In: Baltag, A., Seligman, J., Yamada, T. (eds.) LORI 2017. LNCS, vol. 10455, pp. 377–390. Springer, Heidelberg (2017). https://doi.org/10.1007/978-3-662-55665-8_26
30. Stattner, E.: Involvement of node attributes in the link formation process into a telecommunication network. Soc. Netw. Anal. Min. **5**(1), 64 (2015)
31. Young, H.P., Zamir, S. (eds.): Handbook of Game Theory, vol. 4. North-Holland (2015)

Epistemic Logic with Partial Dependency Operator

Xinyu Wang(✉)

School of Electronics Engineering and Computer Science,
Peking University, Beijing, China
xinyuwang1998@pku.edu.cn

Abstract. In this paper, we introduce *partial* dependency modality $\mathcal{D}$ into epistemic logic so as to reason about *partial* dependency relationship in Kripke models. The resulted dependence epistemic logic possesses decent expressivity and beautiful properties. Several interesting examples are provided, which highlight this logic's practical usage. The logic's bisimulation is then discussed, and we give a sound and strongly complete axiomatization for a sub-language of the logic.

Keywords: Epistemic logic · Knowing value · Partial dependency

1 Introduction

Following some previous fundamental work on "knowing value" [2,4,10,15,16], recent years have seen an abundance of interest in this novel kind of non-standard epistemic logic. There has been epistemic logic with functional dependency operator [3], which can help us reason about knowing that the value of certain variable is functionally decided by some other variables. For instance, the agent knows that $y = x^2$, so he knows that y functionally depends on x even if without knowing the exact values of x or y.

Nevertheless, the real world is never so ideal as a simple parabola. As a matter of fact, in a lot of practical cases, the value of a dependent variable y is usually influenced by thousands of independent factors as x_1, x_2, ... in a quite complicated way, such that it is virtually impossible to obtain a detailed function to precisely determine the value of y. Therefore, in both scientific and social study, the method of control variable gets widely used. We often set the values of all the other variables rigid, only change the value of an independent variable x and observe the change of the dependent variable y. If the value of y varies with the value of x, then we conclude that y *partially* depends on x. In this paper, we introduce modality $\mathcal{D}$ in order to express this kind of *partial* dependency relationship.

There have also been dependence and independence logics dealing with dependency relationship between variables [6,8,9,13], and we will discuss our logic's connection to them in Remark 2. A similar definition for dependency

P. Blackburn et al. (Eds.): LORI 2019, LNCS 11813, pp. 385–398, 2019.
https://doi.org/10.1007/978-3-662-60292-8_28

relationship also appears in Halpern's recent book, pp. 14–19. [11] However, the start point of our work is epistemic logic as well as the Kripke model, and we would like to incorporate *partial* dependency relationship between variables into the agent's knowledge so that we shall obtain an epistemic logic of "knowing dependency", which is hence named as dependence epistemic logic. This dependence epistemic logic proves to possess further affluent expressivity as well as rather straightforward properties.

In the Kripke model for our dependence epistemic logic, besides a usual $\sim_i$ S5 equivalence relation representing the agent's knowledge, i.e., all the possible worlds that the agent cannot distinguish, there also exists another $\approx$ S5 equivalence relation representing the physical probability, i.e., all the possible worlds that share the same set of physical laws with the current world. Generally speaking, these two equivalence relations do not have to have any correlation, and thus in the language, the former is characterized by an S5 modality $\mathcal{K}$, while the latter is characterized by another independent S5 modality $\mathcal{A}$. This kind of framework is first introduced by another recent work [14], and so readers who get confused with the conception of two independent equivalence relations in the model are *strongly* recommended to refer to that paper.

Then the *partial* dependency relationship is valuated in the $\approx$ equivalence class, since dependency relationship between variables is in fact related to some universal physical law and thus concerns not only the current exact world but also all the other worlds that are physically potentially possible. Actually, we introduce two different modalities $\mathcal{D}_g$ and $\mathcal{D}_l$ to characterize *partial* dependency relationship. Their respective semantics is both based on the discussion in the beginning about what modality $\mathcal{D}$ should be like, except for that, the former $\mathcal{D}_g$ is valuated globally in a whole $\approx$ equivalence class, while the latter $\mathcal{D}_l$ fixes one reference point as the current exact world and so is valuated locally. Readers will soon become clear about what $\mathcal{D}_g$ and $\mathcal{D}_l$ mean respectively through the following Sect. 2 on preliminaries including the language, model and semantics, and the correlation between these two modalities also gets discussed in Remark 1. Examples in Sect. 3 illustrate that $\mathcal{D}_g$ is helpful in analyzing universal physical laws while $\mathcal{D}_l$ is useful in expressing counterfactual assumptions, in surprising accordance with our very intuition as well as commonsense, so the practicality of $\mathcal{D}_g$ and $\mathcal{D}_l$ counts to why we introduce both modalities.

The rest of the paper is organized as follows. We lay out the basics of the language and the semantics in Sect. 2. Several interesting examples are illustrated in Sect. 3. A bisimulation notion for this dependence epistemic logic then gets thoroughly discussed in Sect. 4, followed by a sound and strongly complete axiomatization for a sub-language in Sect. 5. We finally conclude this paper and propose future research directions in Sect. 6.

2 Preliminaries

Definition 1 (Language EDL). *For a fixed countable set of propositions* $\mathbb{P}$, *and a fixed countable set of variables* $\mathbb{V}$, *the language* **EDL** *of dependence* *epistemic logic is defined recursively as:*

$$\varphi ::= \top \mid p \mid \neg\varphi \mid (\varphi \wedge \varphi) \mid \mathcal{K}\varphi \mid \mathcal{A}\varphi \mid \mathcal{D}_g(X,Y) \mid \mathcal{D}_l(X,Y)$$

where $p \in \mathbb{P}$, *and* X *as well as* Y *are finite subsets of* $\mathbb{V}$. $\mathcal{D}_g(X,Y)$ *reads as* Y *depends on* X *globally, while* $\mathcal{D}_l(X,Y)$ *reads as* Y *depends on* X *locally. We define* $\bot$, $\vee$ *and* $\to$ *as usual.*

Important Notation. In the following parts of this paper, when some property applies to both $\mathcal{D}_g$ and $\mathcal{D}_l$, we will simply omit the subscript and write down only one theorem, lemma, axiom, etc. concerning $\mathcal{D}$ for convenience, and the omitting is also similar for other notations derived from $\mathcal{D}$.

If $X = \{x\}$, we will also denote $\mathcal{D}(\{x\}, Y)$ as $\mathcal{D}(x, Y)$ for simplicity, and likely for Y if $Y = \{y\}$.

Definition 2 (Model). *A dependence epistemic model* $\mathcal{M}$ *is* $\langle S, T, V, U, \sim_i, \approx\rangle$*:*

- S *is a set of possible worlds.*
- $T\colon S \times \mathbb{P} \to \{0,1\}$.
- $V \supseteq \mathbb{V}$ *is a countable set of variable objects.*
- $U\colon S \times V \to \mathbb{N}$.
- $\sim_i$ *is an equivalence relation over* S.
- $\approx$ *is an equivalence relation over* S.

As the convention in first-order logic, while $\mathbb{V}$ *in the language are names for variables,* V *in the model interpret each name with a concrete object and also may consist of other variable objects whose names are not included in the language. Since the language* ***EDL*** *excludes the equal sign* $=$, *every name in* $\mathbb{V}$ *can be managed to be interpreted differently in* V, *so we simply let* $V \supseteq \mathbb{V}$ *and do not make explicit distinctions between names and objects in the following without causing any confusion. Then* U *is the function that assigns each variable on each possible world with a (countably possible) value, which is supposed to be uniformly numbered by* $\mathbb{N}$ *for convenience.*

Sometimes we apply another extra stipulation on the model in order to satisfy our practical needs: for any proposition $p \in \mathbb{P}$, *it may have its corresponding variable* $\overline{p} \in \mathbb{V}$. *If so, we then stipulate that* $\forall s \in S$, $U(s,\overline{p}) = T(s,p)$. *The following Subsects. 3.1 and 3.3 present examples of this kind.*

Definition 3 (Semantics). *We define that* $\forall s,t \in S$, $\forall$ *subset* $X \subseteq V$, $X_s = X_t$ *iff* $\forall x \in X, U(s,x) = U(t,x)$, *while of course,* $X_s \neq X_t$ *iff* $\exists x \in X, U(s,x) \neq U(t,x)$. *A pointed model* $\mathcal{M}, s$ *is a dependence epistemic model* $\mathcal{M}$ *with a possible world* $s \in S$.

$$
\begin{array}{lll}
\mathcal{M}, s \vDash \top & \iff & \textit{always} \\
\mathcal{M}, s \vDash p & \iff & T(s,p)=1 \\
\mathcal{M}, s \vDash \neg\varphi & \iff & \textit{not } \mathcal{M}, s \vDash \varphi \\
\mathcal{M}, s \vDash (\varphi \wedge \psi) & \iff & \mathcal{M}, s \vDash \varphi \textit{ and } \mathcal{M}, s \vDash \psi \\
\mathcal{M}, s \vDash \mathcal{K}\varphi & \iff & \forall t \in S,\ t \sim_i s,\ \mathcal{M}, t \vDash \varphi \\
\mathcal{M}, s \vDash \mathcal{A}\varphi & \iff & \forall t \in S,\ t \approx s,\ \mathcal{M}, t \vDash \varphi \\
\mathcal{M}, s \vDash \mathcal{D}_g(X,Y) & \iff & \exists u,\ v \in S,\ u \approx v \approx s, \\
 & & (V \backslash (X \cup Y))_u = (V \backslash (X \cup Y))_v,\ X_u \neq X_v,\ Y_u \neq Y_v \\
\mathcal{M}, s \vDash \mathcal{D}_l(X,Y) & \iff & \exists t \in S,\ t \approx s, \\
 & & (V \backslash (X \cup Y))_t = (V \backslash (X \cup Y))_s,\ X_t \neq X_s,\ Y_t \neq Y_s
\end{array}
$$

When it is not that $\mathcal{M}, s \vDash \varphi$, we denote it as $\mathcal{M}, s \nvDash \varphi$.

Remark 1 (Expressivity of $\mathcal{D}_g$ and $\mathcal{D}_l$). We are able to perceive through Definition 3 that $\mathcal{D}_g$ is actually definable using $\neg$, $\mathcal{A}$ and $\mathcal{D}_l$, demonstrated as the following:

$$\mathcal{D}_g(X,Y) \leftrightarrow \neg\mathcal{A}\neg\mathcal{D}_l(X,Y)$$

In fact, $\mathcal{D}_l$ is strictly more expressive than $\mathcal{D}_g$, which will become clear to readers through our discussion for bisimulation in Sect. 4. Nevertheless, due to $\mathcal{D}_g$'s simplicity and usefulness, we will take the language with $\mathcal{D}_g$ but without $\mathcal{D}_l$ as a sub-language of **EDL**.

Definition 4 (Language EDG). *For a fixed countable set of propositions $\mathbb{P}$, and a fixed countable set of variables $\mathbb{V}$, the language* ***EDG*** *is defined recursively as:*

$$\varphi ::= \top \mid p \mid \neg\varphi \mid (\varphi \wedge \varphi) \mid \mathcal{K}\varphi \mid \mathcal{A}\varphi \mid \mathcal{D}_g(X,Y)$$

where $p \in \mathbb{P}$, and X as well as Y are finite subsets of $\mathbb{V}$.

The model and semantics are the same.

Remark 2 (Connection to Independence Logic). If the total set of variables V is finite and explicitly known, then modality $\mathcal{D}_g$ can be expressed in inclusion logic, a sub-language of independence logic [5], as the following:[1]

$$\mathcal{D}_g(X,Y) \iff \exists \overrightarrow{w_1}\overrightarrow{x_1}\overrightarrow{y_1} \exists \overrightarrow{w_2}\overrightarrow{x_2}\overrightarrow{y_2} (\overrightarrow{w_1}\overrightarrow{x_1}\overrightarrow{y_1} \subseteq (V \backslash (X \cup Y))XY \wedge \overrightarrow{w_2}\overrightarrow{x_2}\overrightarrow{y_2} \subseteq (V \backslash (X \cup Y))XY \wedge \overrightarrow{w_1} = \overrightarrow{w_2} \wedge \neg\overrightarrow{x_1} = \overrightarrow{x_2} \wedge \neg\overrightarrow{y_1} = \overrightarrow{y_2})$$

However, this form puts too many restrictions and becomes too lengthy, while we actually want the total set V to be clear from our language so that we can reason with simple and compact logic. In fact, the team model on which independence logic is based is quite different from the Kripke possible world model [12], both in technique and in philosophical explanation, and hence they are very unlike logics. While independence logic, inherited from first order logic, always reasons globally, epistemic logic, rooted from modal logic, usually reasons locally, which is demonstrated by this obvious fact that local modality $\mathcal{D}_l$ can surely not be defined in independence logic.

[1] As for the notation, we prefer to use X and Y instead of $\overrightarrow{x}$ or $\overrightarrow{y}$. Anyway, their respective meanings in this specific context should be clear to readers.

3 Examples

3.1 An Open Door

Let p denote that the door of the room is open now, q denote that the agent possesses the key of the door, and r denote that the agent is able to enter the room. Let us suppose that the agent has perfect knowledge, so $\sim_i$ relation is only reflexive. Then we have:[2]

$$\begin{array}{ccccccc}
s: p,q,r & & p,\neg q,r & & \neg p,q,r & & \neg p,\neg q,\neg r \\
\overline{p}=1 & \text{—}\approx\text{—} & \overline{p}=1 & \text{—}\approx\text{—} & \overline{p}=0 & \text{—}\approx\text{—} & \overline{p}=0 \\
\overline{q}=1 & & \overline{q}=0 & & \overline{q}=1 & & \overline{q}=0 \\
\overline{r}=1 & & \overline{r}=1 & & \overline{r}=1 & & \overline{r}=0
\end{array}$$

It is not difficult to observe that $\mathcal{M}, s \vDash \mathcal{KD}_g(\overline{p}, \overline{r})$ and $\mathcal{M}, s \vDash \mathcal{K}\neg\mathcal{D}_l(\overline{p}, \overline{r})$. The former says that the agent knows whether he is able to enter the room is somewhat related to whether the door is open now – if he did not possess the key. And the latter says that under the present situation, since the agent does possess the key, he surely knows that if this precondition is kept unchanged, then he was still able to open the door to enter the room even if the door was now closed. Namely, whether he is able to enter the room does not depend on whether the door is open now, which provides us with a fancy way to express counterfactual assumptions.

3.2 Error-Included Experiment

Suppose we are carrying out an experiment, and we know from theory that there are two independent variables x and y which may influence the value of the dependent variable z, where the value of x is well under control but y represents some random experimental error, and so of course, we cannot control or even measure the value of y. The only thing we know about y is that it will be either 1 or 2 during every experiment.

Now we have done this experiment twice. When $x = 1$, $z = 1$. When $x = 2$, $z = 2$. By combining all kinds of possibilities, we can have the model as:

$$\begin{array}{ccccccc}
x=1 & & x=1 & & x=1 & & x=1 \\
y=1 & \text{—}i\text{—} & y=1 & \text{—}i\text{—} & y=2 & \text{—}i\text{—} & y=2 \\
z=1 & & z=1 & & z=1 & & z=1 \\
\updownarrow\approx & & \updownarrow\approx & & \updownarrow\approx & & \updownarrow\approx \\
x=2 & & x=2 & & x=2 & & x=2 \\
y=1 & \text{—}i\text{—} & y=2 & \text{—}i\text{—} & y=1 & \text{—}i\text{—} & y=2 \\
z=2 & & z=2 & & z=2 & & z=2
\end{array}$$

Can we be confident that z depends on x? Certainly not, because the change of z may be brought about by the change of y. As a matter of fact, on every possible world s there is $\mathcal{M}, s \nvDash \mathcal{KD}_g(x, z)$.

[2] When drawing all these figures in this paper, for brevity we will omit some relation lines which can be deduced from S5 equivalence class requirements.

However, if we have further done the third experiment, and when $x = 3$, $z = 3$. Now can we be confident that z depends on x? Indeed we can. This fact can be easily observed through the following huge model, where $\mathcal{M}, s \vDash \mathcal{KD}_g(x, z)$ on every possible world s:

$$\begin{array}{ccccccccccccccc}
\begin{matrix}x=1\\y=1\\z=1\end{matrix} & \text{-}i\text{-} & \begin{matrix}x=1\\y=1\\z=1\end{matrix} & \text{-}i\text{-} & \begin{matrix}x=1\\y=1\\z=1\end{matrix} & \text{-}i\text{-} & \begin{matrix}x=1\\y=1\\z=1\end{matrix} & \text{-}i\text{-} & \begin{matrix}x=1\\y=2\\z=1\end{matrix} & \text{-}i\text{-} & \begin{matrix}x=1\\y=2\\z=1\end{matrix} & \text{-}i\text{-} & \begin{matrix}x=1\\y=2\\z=1\end{matrix} & \text{-}i\text{-} & \begin{matrix}x=1\\y=2\\z=1\end{matrix} \\
\mid\approx & & \mid\approx & & \mid\approx & & \mid\approx & & \mid\approx & & \mid\approx & & \mid\approx & & \mid\approx \\
\begin{matrix}x=2\\y=1\\z=2\end{matrix} & \text{-}i\text{-} & \begin{matrix}x=2\\y=1\\z=2\end{matrix} & \text{-}i\text{-} & \begin{matrix}x=2\\y=2\\z=2\end{matrix} & \text{-}i\text{-} & \begin{matrix}x=2\\y=2\\z=2\end{matrix} & \text{-}i\text{-} & \begin{matrix}x=2\\y=1\\z=2\end{matrix} & \text{-}i\text{-} & \begin{matrix}x=2\\y=1\\z=2\end{matrix} & \text{-}i\text{-} & \begin{matrix}x=2\\y=2\\z=2\end{matrix} & \text{-}i\text{-} & \begin{matrix}x=2\\y=2\\z=2\end{matrix} \\
\mid\approx & & \mid\approx & & \mid\approx & & \mid\approx & & \mid\approx & & \mid\approx & & \mid\approx & & \mid\approx \\
\begin{matrix}x=3\\y=1\\z=3\end{matrix} & \text{-}i\text{-} & \begin{matrix}x=3\\y=2\\z=3\end{matrix} & \text{-}i\text{-} & \begin{matrix}x=3\\y=1\\z=3\end{matrix} & \text{-}i\text{-} & \begin{matrix}x=3\\y=2\\z=3\end{matrix} & \text{-}i\text{-} & \begin{matrix}x=3\\y=1\\z=3\end{matrix} & \text{-}i\text{-} & \begin{matrix}x=3\\y=2\\z=3\end{matrix} & \text{-}i\text{-} & \begin{matrix}x=3\\y=1\\z=3\end{matrix} & \text{-}i\text{-} & \begin{matrix}x=3\\y=2\\z=3\end{matrix}
\end{array}$$

Whatever values y may be in the three experiments, there must be at least two experiments in which y is the same, so we can only explain the difference between z in these two experiments as caused by the difference between the value of x. This scenario clearly explains why in all the natural science experiments, despite the universal existence of errors, we can still manage to obtain useful conclusions concerning our interested variables, by multiple experiments with relatively large data range.

3.3 Judging a Case

We have seen that global modality $\mathcal{D}_g$ can help us analyze complicated experimental results, while local modality $\mathcal{D}_l$ is very helpful in expressing counterfactual assumptions. And there are still trickier things worth examining. Until now, we have only proposed examples including modality $\mathcal{D}$ affecting solely on singletons. It may seem by intuitive guess that $\mathcal{D}(\{a, b\}, c)$ tells very similar thing as $\mathcal{D}(a, c) \vee \mathcal{D}(b, c)$. Nevertheless, these two expressions are not exactly the same, and in fact, they may result in quite opposite epistemic consequences, as demonstrated by the following scenario.

Unfortunately, Charles got killed in a tragedy (c), which was related to Alan having done something (a) and/or Bob having done something (b). Firstly, let us suppose that either a or b could happen so as to cause c, and only one of them could have happened to be c's indeed cause. However, on the current world s we are yet not sure whether a or b actually happened to be the exact cause of c. This can be modeled as the following:

$$\begin{array}{ccccc}
s : a, \neg b, c & & \neg a, b, c & & \neg a, \neg b, \neg c \\
\begin{matrix}\overline{a}=1\\\overline{b}=0\\\overline{c}=1\end{matrix} & \xrightarrow{\;i,\approx\;} & \begin{matrix}\overline{a}=0\\\overline{b}=1\\\overline{c}=1\end{matrix} & \xrightarrow{\;\approx\;} & \begin{matrix}\overline{a}=0\\\overline{b}=0\\\overline{c}=0\end{matrix}
\end{array}$$

It is not difficult to observe that $\mathcal{M}, s \vDash \mathcal{K}\mathcal{D}_l(\{\overline{a}, \overline{b}\}, \overline{c}) \wedge \mathcal{K}(\mathcal{D}_l(\overline{a}, \overline{c}) \vee \mathcal{D}_l(\overline{b}, \overline{c}))$. This is to say, it is within our knowledge that not only the whole group event $\{a, b\}$ is related to c, but also either a or b itself is alone related to c, namely, their influences on c can be separated in concept. Hence, unless we obtain further evidence to pin down our knowledge in order to determine whether Alan or Bob was the real criminal, by presumption of innocence neither of them can be sentenced guilty for Charles' death.

Now let us turn to a second phenomenon, where b's happening was a direct consequence of a's happening. For instance, let b denote that Bob killed Charles, and a denote that Alan compelled Bob to kill Charles, either by threatening that he would have killed Bob otherwise or by Alan's mind control over Bob through magic or science fiction. In other words, we restrict ourselves to only consider possible worlds on which $a \to b$ holds in our Kripke model. Under this circumstance, we can model our knowledge as the following:

$$
\begin{array}{ccccc}
s : a, b, c & & \neg a, b, c & & \neg a, \neg b, \neg c \\
\overline{a} = 1 & \overset{\approx}{\text{———}} & \overline{a} = 0 & \overset{\approx}{\text{———}} & \overline{a} = 0 \\
\overline{b} = 1 & & \overline{b} = 1 & & \overline{b} = 0 \\
\overline{c} = 1 & & \overline{c} = 1 & & \overline{c} = 0
\end{array}
$$

At present, even physically speaking b should be the only direct cause of c, which is demonstrated by $\mathcal{A}(b \leftrightarrow c)$ holding throughout the model, to our little surprise $\mathcal{K}\mathcal{D}_l(\overline{b}, \overline{c})$ does not hold on the current world s. As a matter of fact, we have $\mathcal{M}, s \vDash \mathcal{K}\mathcal{D}_l(\{\overline{a}, \overline{b}\}, \overline{c}) \wedge \mathcal{K}(\neg\mathcal{D}_l(\overline{a}, \overline{c}) \wedge \neg\mathcal{D}_l(\overline{b}, \overline{c}))$, a direct contrast against the former scene. This time we not only know that c locally depends on $\{a, b\}$ as a whole, but also know that this dependency relationship should be viewed as an entirety instead of conceptually separable, and therefore, both Alan and Bob should be responsible for Charles' death. Further considering that $\mathcal{K}\mathcal{A}(a \to b)$ holds on s, a legal and rational sentence ought to be that Alan is the principal criminal while Bob is the coerced criminal, which precisely captures the meanings of all the formulae mentioned above.

4 Bisimulation

Definition 5 ($\Delta(u, v)$)**.** *For any two possible worlds $u, v \in S$, we define:*

$$
\Delta(u, v) = \begin{cases} \{x \mid x \in \mathbb{V}, U(u, x) \neq U(v, x)\}, & \textit{if } (V \backslash \mathbb{V})_u = (V \backslash \mathbb{V})_v \\ \emptyset, & \textit{otherwise} \end{cases}
$$

Definition 6 (Evidence). *For any three sets W, X and Y, W is called an evidence of $\langle X, Y \rangle$, iff $W \cap X \neq \emptyset$, $W \cap Y \neq \emptyset$, and $W \subseteq X \cup Y$.*

Compared with the original semantics defined in Definition 3, we manage to rewrite part of it in an equivalent form as the following:

Lemma 1 (Evidence Lemma I).

$$\mathcal{M}, s \vDash \mathcal{D}_g(X,Y) \iff \exists u,v \in S,\ u \approx v \approx s,\ \Delta(u,v) \textit{ is an evidence of } \langle X,Y \rangle$$
$$\mathcal{M}, s \vDash \mathcal{D}_l(X,Y) \iff \exists t \in S,\ t \approx s,\ \Delta(t,s) \textit{ is an evidence of } \langle X,Y \rangle$$

Proof. Directly from the semantics defined in Definition 3. ∎

Definition 7 ($\mathfrak{P}(s)$)**.** *For any possible world $s \in S$, we define:*

$$\mathfrak{P}_g(s) = \{\textit{nonempty finite set } \Delta(u,v) \mid u,v \in S, u \approx v \approx s\}$$
$$\mathfrak{P}_l(s) = \{\textit{nonempty finite set } \Delta(t,s) \mid t \in S, t \approx s\}$$

It is obvious that $\forall s \in S$, $\mathfrak{P}_l(s) \subseteq \mathfrak{P}_g(s) \subseteq \{$*nonempty finite set* $W \mid W \subseteq \mathbb{V}\}$.

We again manage to rewrite part of the semantics in another equivalent form as the following, making use of the newly defined $\mathfrak{P}(s)$:

Lemma 2 (Evidence Lemma II).

$$\mathcal{M}, s \vDash \mathcal{D}_g(X,Y) \iff \exists W \in \mathfrak{P}_g(s), W \textit{ is an evidence of } \langle X,Y \rangle$$
$$\mathcal{M}, s \vDash \mathcal{D}_l(X,Y) \iff \exists W \in \mathfrak{P}_l(s), W \textit{ is an evidence of } \langle X,Y \rangle$$

Proof. By Lemma 1. ∎

Definition 8 (Generative). $\forall s \in S$, *any nonempty finite set $W \subseteq \mathbb{V}$ is called generative from $\mathfrak{P}(s)$, iff for any two finite sets $X, Y \subseteq \mathbb{V}$, such that W is an evidence of $\langle X,Y \rangle$, there exists $W' \in \mathfrak{P}(s)$, such that W' is also an evidence of $\langle X,Y \rangle$.*

Theorem 1 (Equivalence Theorem I). *For any two pointed models $\mathcal{M}, s$ and $\mathcal{M}', s'$, they satisfy exactly the same $\mathcal{D}(X,Y)$ formulae for any two finite sets $X, Y \subseteq \mathbb{V}$ iff:*

- *Zig:* $\forall W \in \mathfrak{P}(s)$, *$W$ is generative from $\mathfrak{P}(s')$.*
- *Zag:* $\forall W \in \mathfrak{P}(s')$, *$W$ is generative from $\mathfrak{P}(s)$.*

Proof. For the direction from left to right, we first concentrate on the Zig condition. If there exists $W \in \mathfrak{P}(s)$, such that W is not generative from $\mathfrak{P}(s')$, then by definition, there exist two finite sets $X, Y \subseteq \mathbb{V}$, such that W is an evidence of $\langle X,Y \rangle$, but there does not exist $W' \in \mathfrak{P}(s')$, such that W' is an evidence of $\langle X,Y \rangle$. By Lemma 2, this is equivalent to that $\mathcal{M}, s \vDash \mathcal{D}(X,Y)$ but $\mathcal{M}', s' \nvDash \mathcal{D}(X,Y)$, a contradiction. The Zag condition follows by symmetry.

The other direction can also be verified similarly and easily. ∎

Definition 9 ($\mathfrak{G}(s)$)**.** *For any possible world $s \in S$, we define:*

$$\mathfrak{G}(s) = \{W \mid W \textit{ is generative from } \mathfrak{P}(s)\}$$

It is obvious that $\forall s \in S$, $\mathfrak{G}_l(s) \subseteq \mathfrak{G}_g(s) \subseteq \{$*nonempty finite set* $W \mid W \subseteq \mathbb{V}\}$.

Theorem 2 (Equivalence Theorem II). *For any two pointed models $\mathcal{M}, s$ and $\mathcal{M}', s'$, they satisfy exactly the same $\mathcal{D}(X,Y)$ formulae for any two finite sets $X, Y \subseteq \mathbb{V}$ iff $\mathfrak{G}(s) = \mathfrak{G}(s')$.*

Proof. Similar to the proof of Theorem 1. ■

Actually, the set $\mathfrak{G}(s)$ is the existent and the only greatest generative set from the original $\mathfrak{P}(s)$ while keeping satisfying the same formulae for modality $\mathcal{D}$. Therefore, it is worthwhile investigating what characteristics $\mathfrak{G}(s)$ possesses, since it precisely determines the modal property of the pointed model $\mathcal{M}, s$. In the following theorem, we manage to express the generative condition for a nonempty finite set W from $\mathfrak{P}(s)$ in several different equivalent forms.

Theorem 3 (Generative Theorem). *$\forall s \in S$, for any nonempty finite set $W \subseteq \mathbb{V}$, we define $\Sigma(s, W) = \{W' \mid W' \in \mathfrak{P}(s), W' \subseteq W\}$, then:*

$$\begin{aligned}
& W \textit{ is generative from } \mathfrak{P}(s) \\
\iff & \textstyle\bigcup \Sigma(s,W) = W,\ \forall Z \subset W \textit{ such that } Z \neq \emptyset, \\
& \exists W' \in \Sigma(s,W) \textit{ such that } W' \cap Z \neq \emptyset \wedge W' \cap (W \backslash Z) \neq \emptyset \\
\iff & \textstyle\bigcup \Sigma(s,W) = W,\ \forall \Gamma \subset \Sigma(s,W) \textit{ such that } \Gamma \neq \emptyset, \\
& (\textstyle\bigcup \Gamma) \cap (\bigcup(\Sigma(s,W) \backslash \Gamma)) \neq \emptyset \\
\iff & \textstyle\bigcup \Sigma(s,W) = W,\ \forall W_1', W_2' \in \Sigma(s,W),\ \textit{define } \mathcal{R}W_1'W_2' \textit{ iff } W_1' \cap W_2' \neq \emptyset, \\
& \textit{then } \forall W_1', W_2' \in \Sigma(s,W),\ W_1' \textit{ connects to } W_2' \textit{ by a chain of } \mathcal{R} \textit{ relations}
\end{aligned}$$

Proof. Let us concentrate on the following crucial lemma, from which the proof of this theorem follows not difficultly. ■

Lemma 3 (Generative Lemma). *$\forall s \in S$, for any nonempty finite set $W \subseteq \mathbb{V}$, W is generative from $\mathfrak{P}(s)$ iff:*

- *if $|W| = 1$, then $W \in \mathfrak{P}(s)$.*
- *if $|W| \geqslant 2$, then $\forall Z \subset W$ such that $Z \neq \emptyset$, $\exists W' \in \mathfrak{P}(s)$ such that W' is an evidence of $\langle Z, W \backslash Z \rangle$.*

Proof. The direction from left to right is immediate. For the direction from right to left, we only have to make use of one simple fact about evidence:

- If W is an evidence of $\langle X, Y \rangle$ and $X \subseteq X'$, then W is an evidence of $\langle X', Y \rangle$.

which, as a matter of fact, can be correspondingly written into a sound axiom regarding modality $\mathcal{D}$:

$$\mathcal{D}(X,Y) \rightarrow \mathcal{D}(X',Y), \text{ given } X \subseteq X' \text{ (Weakening Rule)}$$

Full axiomatization will later be discussed in the following Sect. 5. ■

The last equivalent condition in Theorem 3 is to say, we can construct an undirected graph over $\mathfrak{P}(s)$ by its elements' intersection relation, and all the generative sets are exactly union of some connected nonempty finite subgraph. This provides us with a clear picture and an intuitive understanding about where every generative set comes from and what $\mathfrak{G}(s)$ looks like. Hence given $\mathfrak{P}(s)$, there is an explicit algorithm to calculate all the generative nonempty finite sets $W \subseteq \mathbb{V}$ so as to obtain $\mathfrak{G}(s)$.

Finally, taking into account all the modalities including $\mathcal{K}$, $\mathcal{A}$, $\mathcal{D}_g$ and $\mathcal{D}_l$, we are able to define the full bisimulation relation between two models $\mathcal{M}$ and $\mathcal{M}'$:

Definition 10 (Bisimulation). *A nonempty binary relation $B \subseteq S \times S'$ is called a bisimulation between two models $\mathcal{M}$ and $\mathcal{M}'$ iff:*

- *If sBs', then $\forall p \in \mathbb{P}$, $T(s,p) = T(s',p)$.*
- *If sBs', then $\mathfrak{G}_g(s) = \mathfrak{G}_g(s')$.*
- *If sBs', then $\mathfrak{G}_l(s) = \mathfrak{G}_l(s')$.*
- *Zig for $\mathcal{K}$: if sBs' and $s \sim_i t$, then $\exists t' \in S'$ such that tBt' and $s' \sim_i t'$.*
- *Zig for $\mathcal{A}$: if sBs' and $s \approx t$, then $\exists t' \in S'$ such that tBt' and $s' \approx t'$.*
- *Zag for $\mathcal{K}$: if sBs' and $s' \sim_i t'$, then $\exists t \in S$ such that tBt' and $s \sim_i t$.*
- *Zag for $\mathcal{A}$: if sBs' and $s' \approx t'$, then $\exists t \in S$ such that tBt' and $s \approx t$.*

When B is a bisimulation between two models $\mathcal{M}$ and $\mathcal{M}'$, we write $B : \mathcal{M} \,\underline{\leftrightarrow}\, \mathcal{M}'$. Furthermore if sBs', we write $B : \mathcal{M}, s \,\underline{\leftrightarrow}\, \mathcal{M}', s'$. If there is a bisimulation B such that $B : \mathcal{M}, s \,\underline{\leftrightarrow}\, \mathcal{M}', s'$, we write $\mathcal{M}, s \,\underline{\leftrightarrow}\, \mathcal{M}', s'$.

*We write $\mathcal{M}, s \leftrightsquigarrow \mathcal{M}', s'$, when for any **EDL**-formula φ, $\mathcal{M}, s \vDash \varphi$ iff $\mathcal{M}', s' \vDash \varphi$.*

Theorem 4 (Hennessy-Milner Theorem). *For any two m-saturated models $\mathcal{M}$ and $\mathcal{M}'$, $\forall s \in S$, $\forall s' \in S'$, $\mathcal{M}, s \,\underline{\leftrightarrow}\, \mathcal{M}', s'$ iff $\mathcal{M}, s \leftrightsquigarrow \mathcal{M}', s'$.*

Proof. See [1]. The definition of m-saturated models also appears as Definition 2.53 in that book. It is only the cases for modalities $\mathcal{D}_g$ and $\mathcal{D}_l$ that are added, which just follow from Theorem 2. ■

5 Axiomatization

We only provide a sound and strongly complete axiomatization for language **EDG**. Nevertheless, the same as the assumed routine in this paper, axioms without subscripts attached to $\mathcal{D}$ are sound with respect to both $\mathcal{D}_g$ and $\mathcal{D}_l$.

To start with, we may notice some obviously sound axioms to characterize the properties of modality $\mathcal{D}$:

1. $\mathcal{D}(\emptyset, X) \leftrightarrow \bot$ (Empty Set Rule)
2. $\mathcal{D}(X,Y) \leftrightarrow \mathcal{D}(Y,X)$ (Symmetry Rule)
3. $\mathcal{D}(X,Y) \rightarrow \mathcal{D}(X',Y)$, given $X \subseteq X'$ (Weakening Rule)
4. $\mathcal{D}(X,Y) \leftrightarrow \mathcal{D}(X \backslash Y, Y) \vee \mathcal{D}(X \cap Y, Y)$ (Separation Rule)

Although these naïve axioms indeed look very similar to those in independence logic [7], pitifully in our dependence epistemic logic, they alone are away from being complete. The good news is that, we can instead find some conciser axioms, which entirely grasp the full properties of modality $\mathcal{D}$ itself, and from which all the above sound axioms can surely be deduced.

For brevity, let us first define an auxiliary notation:

Definition 11 ($\mathcal{Q}(W)$). *For any nonempty finite set $W \subseteq \mathbb{V}$, we define:*

$$\mathcal{Q}(W) ::= \begin{cases} \mathcal{D}(W, W), & |W| = 1 \\ \bigwedge_{Z \subset W, Z \neq \emptyset} \mathcal{D}(Z, W \backslash Z), & |W| \geqslant 2 \end{cases}$$

Recall Lemma 3, readers should be aware that this $\mathcal{Q}(W)$ precisely depicts the minimum necessary $\mathcal{D}(X, Y)$ formulae, such that W is an evidence of $\langle X, Y\rangle$. Taking advantage of this notation, we can write down rather concise sound axioms about modality $\mathcal{D}$ so as to obtain a complete axiomatization, as the following Q and E Axioms for $\mathcal{D}$ in Theorem 5:

Theorem 5 (Axiomatization). *The following proof system is sound and strongly complete with respect to language* **EDG**.

TAUT	*all instances of tautologies*
MP	*from* φ *and* $\varphi \to \psi$ *infer* ψ
NEC for $\mathcal{K}$	*from* φ *infer* $\mathcal{K}\varphi$
DIST for $\mathcal{K}$	$\mathcal{K}(\varphi \to \psi) \to (\mathcal{K}\varphi \to \mathcal{K}\psi)$
T for $\mathcal{K}$	$\mathcal{K}\varphi \to \varphi$
4 for $\mathcal{K}$	$\mathcal{K}\varphi \to \mathcal{K}\mathcal{K}\varphi$
5 for $\mathcal{K}$	$\neg\mathcal{K}\varphi \to \mathcal{K}\neg\mathcal{K}\varphi$
NEC for $\mathcal{A}$	*from* φ *infer* $\mathcal{A}\varphi$
DIST for $\mathcal{A}$	$\mathcal{A}(\varphi \to \psi) \to (\mathcal{A}\varphi \to \mathcal{A}\psi)$
T for $\mathcal{A}$	$\mathcal{A}\varphi \to \varphi$
4 for $\mathcal{A}$	$\mathcal{A}\varphi \to \mathcal{A}\mathcal{A}\varphi$
5 for $\mathcal{A}$	$\neg\mathcal{A}\varphi \to \mathcal{A}\neg\mathcal{A}\varphi$
Q for $\mathcal{D}$	$\mathcal{D}(X, Y) \leftrightarrow \bigvee_{X' \subseteq X, Y' \subseteq Y, X', Y' \neq \emptyset} \mathcal{Q}(X' \cup Y')$, *given* $X, Y \neq \emptyset$
E for $\mathcal{D}$	$\mathcal{D}(\emptyset, X) \leftrightarrow \mathcal{D}(X, \emptyset) \leftrightarrow \bot$
4 for $\mathcal{D}_g$	$\mathcal{D}_g(X, Y) \to \mathcal{A}\mathcal{D}_g(X, Y)$

Proof. We only show completeness. The proof is almost routine, so we concentrate on how the canonical model is built and on the Truth Lemma for modality $\mathcal{D}_g$. Notice that the Axiom of Choice has to be made use of in the proof. ■

Definition 12 (Canonical Model). *For a fixed language with a set of propositions $\mathbb{P}$ and a set of variables $\mathbb{V}$, we first expand this language to $\mathbb{P}^C$ and $\mathbb{V}^C$, such that $\mathbb{P}^C = \mathbb{P}$, $\mathbb{V}^C \supseteq \mathbb{V}$, and that $\mathbb{V}^C$ is countably infinite. Obviously, if an MCS is satisfied in the canonical model of the expanded language, its restriction down to the original language will also be satisfied in the same model.*

The canonical dependence epistemic model $\mathcal{M}^C$ is $\langle S^C, T^C, V^C, U^C, \sim_i^C, \approx^C\rangle$:

- *S^C is the set of all MCSs.*
- *$T^C : S^C \times \mathbb{P}^C \to \{0, 1\}$. $\forall s \in S^C$, $\forall p \in \mathbb{P}^C$, $T^C(s, p) = 1$ iff $p \in s$.*
- *$V^C = \mathbb{V}^C$.*
- *$\sim_i^C$ is an equivalence relation over S^C. $\forall s, t \in S^C$, $s \sim_i^C t$ iff $\{\mathcal{K}\varphi \mid \mathcal{K}\varphi \in s\} = \{\mathcal{K}\varphi \mid \mathcal{K}\varphi \in t\}$.*

- $\approx^C$ *is an equivalence relation over* S^C. $\forall s, t \in S^C$, $s \approx^C t$ *iff* $\{\mathcal{A}\varphi \mid \mathcal{A}\varphi \in s\} = \{\mathcal{A}\varphi \mid \mathcal{A}\varphi \in t\}$.
- $U^C : S^C \times V^C \to \mathbb{N}$. *For each fixed* $\approx^C$ *equivalence class* $S_\approx \subseteq S^C$, *we assign* V^C*'s values on every possible world* $s \in S_\approx$ *as the following procedure:*

By the 4 Axiom for $\mathcal{D}_g$ in Theorem 5, it is easy to see that if $s \approx^C t$, then $\{\mathcal{D}_g(X,Y) \mid \mathcal{D}_g(X,Y) \in s\} = \{\mathcal{D}_g(X,Y) \mid \mathcal{D}_g(X,Y) \in t\}$. So suppose arbitrary $s \in S_\approx$, $W_\approx = \{\text{nonempty finite set } W \mid W \subset \mathbb{V}^C, \mathcal{Q}_g(W) \in s\}$ is a well defined set, regardless of which possible world s we choose from $S_\approx$.

Claim. $W_\approx$ is countable. Therefore, we can suppose a well order $<_W \cong \omega$ on it.

We define a constant function $f_0 : V^C \to \mathbb{N}$, $\forall x \in V^C$, $f_0(x) = 0$.

Lemma 4 (Canonical Assignment). *For every* $W \in W_\approx$, *we can simultaneously find two corresponding functions* $f_1^W : V^C \to \mathbb{N}$ *and* $f_2^W : V^C \to \mathbb{N}$ *such that:*

- $\{f_1^W(x) \neq f_2^W(x) \mid x \in V^C\} = W$;
- *if* $W_1, W_2 \in W_\approx$, $W_1 \neq W_2$, *then* $\{f_i^{W_1}(x) \neq f_j^{W_2}(x) \mid x \in V^C\}$ *is countably infinite,* $i, j \in \{1, 2\}$;
- $\{f_i^W(x) \neq f_0(x) \mid x \in V^C\}$ *is countably infinite,* $i \in \{1, 2\}$.

Proof. Noticing that there are countably infinite variables in V^C which can be assigned to countably infinite values, while $W_\approx$ is also countable and all the sets $W \in W_\approx$ are finite, we are sure that these requirements can be satisfied. For example, we manage to designate f_1^W and f_2^W for every $W \in W_\approx$ one by one, along the well order $<_W$. Since every W is finite, to satisfy the first requirement, the ranges of f_1^W and f_2^W can be controlled to be both finite. For the second requirement, if $W_1 <_W W_2$, we let the ranges of $f_i^{W_2}$ and $f_j^{W_1}$ not intersect. For the third requirement, we let 0 not be in f_i^W's range. ■

We collect all these functions as $F_\approx = \{f_i^W \mid W \in W_\approx, i \in \{1, 2\}\} \cup \{f_0\}$.

Claim. $F_\approx$ is countable. Therefore, we can suppose a well order $<_F$ on it.

Then by the Well-ordering Theorem, we can also suppose a well order $<_S$ on $S_\approx$. By correlating these two well orders $<_F$ and $<_S$, we can use function $f \in F_\approx$ to assign V^C's values on possible world $s \in S_\approx$, such that $\forall x \in V^C$, $U^C(s, x) = f(x)$. As any two well orders can be compared, during this correlating procedure, one and only one of the following three conditions will occur:

- If $<_F \cong <_S$, done.
- If we first run out of functions from $F_\approx$, then we use f_0 to assign V^C's values for all the other left possible worlds in $S_\approx$.
- If we first run out of possible worlds from $S_\approx$, then we arbitrarily choose one possible world $s \in S_\approx$, and copy it many times so as to match all the other left functions in $F_\approx$. All these copies of s, along with the original one, of course share the same T^C, and are in the same $\sim_i^C$ and $\approx^C$ equivalence classes. Obviously, this copy will not cause any unpleasant consequences.

Lemma 5 (Truth Lemma for Modality $\mathcal{D}_g$). *$\forall s \in S^C$, $\forall$ finite subsets $X, Y \subset \mathbb{V}^C$, $\mathcal{D}_g(X,Y) \in s \iff \mathcal{M}^C, s \vDash \mathcal{D}_g(X,Y)$.*

Proof. The cases when $X = \emptyset$ or $Y = \emptyset$ follow immediately from the E Axiom for $\mathcal{D}$ in Theorem 5, so we concentrate on the situations when $X \neq \emptyset$ and $Y \neq \emptyset$. By Lemma 1, $\mathcal{M}^C, s \vDash \mathcal{D}_g(X,Y) \iff \exists u, v \in S^C$, $u \approx^C v \approx^C s$, such that $\Delta(u,v)$ is an evidence of $\langle X, Y\rangle$.

For the direction from right to left, from the above assignment procedure of U^C in the canonical model, we have $\mathcal{Q}_g(\Delta(u,v)) \in s$. Since $\Delta(u,v)$ is an evidence of $\langle X, Y\rangle$, by making use of the Weakening Rule it is not difficult to reason that $\mathcal{D}_g(X,Y) \in s$.

For the direction from left to right, considering the Q Axiom for $\mathcal{D}$ in Theorem 5, at least one of the $\mathcal{Q}_g(X' \cup Y')$ in the big disjunction is in s, and thus from the above assignment procedure of U^C in the canonical model, $\exists u, v \in S^C$, $u \approx^C v \approx^C s$, such that $\Delta(u,v) = X' \cup Y'$. Since $X' \subseteq X, Y' \subseteq Y, X', Y' \neq \emptyset$, obviously $X' \cup Y'$ is just an evidence of $\langle X, Y\rangle$ and hence $\mathcal{M}^C, s \vDash \mathcal{D}_g(X,Y)$. ∎

6 Conclusions and Future Work

In this paper, we come up with dependence epistemic logic in order to reason about *partial* dependency relationship between variables under an epistemic scenario. Several interesting examples are proposed, which demonstrate our language's affluent expressivity and practical usage. Besides that, the essential properties of the logic are straightforward to understand, and hence we further discuss its bisimulation relation and manage to provide a sound and strongly complete axiomatization system for the simpler sub-language **EDG**.

Nevertheless, there still remains much work to be done in the future. The axiomatization of the full language **EDL** is yet unknown. It will also be helpful to elaborate on other computational properties of this logic, such as decidability. Besides, as we only deal with the presence of a single agent in this paper, extending this dependence epistemic logic to cases with multiple agents may result in more interesting results. Moreover, it seems to be an exciting idea to add other modalities into this framework so that we will be able to reason about knowing dependency, knowing value, knowing how as well as many other epistemic assertions all together.

Acknowledgements. The author would like to thank Yanjing Wang for proposing the $\approx$ equivalence relation as well as a lot of other useful comments and suggestions on this paper.

The author would like to thank Fan Yang for coming up with the formula in Remark 2.

The author would also like to thank all the teachers and students who have been taking part in the epistemic logic course, from discussions with whom a lot of inspirations have been stimulated.

References

1. Blackburn, P., de Rijke, M., Venema, Y.: Modal Logic. Cambridge University Press, New York (2001)
2. Ding, Y.: Axiomatization and complexity of modal logic with knowing-what operator on model class k (2016). http://arxiv.org/abs/1609.07684
3. Ding, Y.: Epistemic logic with functional dependency operator. Stud. Log. **9**(4), 55–84 (2016)
4. van Eijck, J., Gattinger, M., Wang, Y.: Knowing values and public inspection. In: Ghosh, S., Prasad, S. (eds.) ICLA 2017. LNCS, vol. 10119, pp. 77–90. Springer, Heidelberg (2017). https://doi.org/10.1007/978-3-662-54069-5_7
5. Galliani, P.: Inclusion and exclusion dependencies in team semantics - on some logics of imperfect information. Ann. Pure Appl. Log. **163**(1), 68–84 (2012)
6. Galliani, P.: Dependence logic. In: Stanford Encyclopedia of Philosophy (2017). https://plato.stanford.edu/entries/logic-dependence/
7. Galliani, P., Hella, L.: Inclusion logic and fixed point logic. In: Computer Science Logic 2013 (CSL 2013). Leibniz International Proceedings in Informatics (LIPIcs), Schloss Dagstuhl–Leibniz-Zentrum fuer Informatik, Dagstuhl, Germany, vol. 23, pp. 281–295 (2013)
8. Galliani, P., Väänänen, J.: On dependence logic. In: Baltag, A., Smets, S. (eds.) Johan van Benthem on Logic and Information Dynamics. OCL, vol. 5, pp. 101–119. Springer, Cham (2014). https://doi.org/10.1007/978-3-319-06025-5_4
9. Grädel, E., Väänänen, J.: Dependence and independence. Stud. Logica **101**(2), 399–410 (2013)
10. Gu, T., Wang, Y.: "Knowing value" logic as a normal modal logic. In: Proceedings of Advances in Modal Logic 2016, vol. 11, pp. 362–381. College Publications (2016)
11. Halpern, J.Y.: Actual Causality. The MIT Press, Cambridge (2016)
12. Hodges, W.: Compositional semantics for a language of imperfect information. Log. J. IGPL **5**(4), 539–563 (1997)
13. Väänänen, J.: Dependence Logic: A New Approach to Independency Friendly Logic. Cambridge University Press, Cambridge (2007)
14. Wang, X., Wang, Y.: Knowledge-now and knowledge-all. In: Blackburn, P., Lorini, E., Guo, M. (eds.) LORI 2019. LNCS, vol. 11813, pp. 399–412. Springer, Heidelberg (2019)
15. Wang, Y., Fan, J.: Knowing that, knowing what, and public communication: public announcement logic with Kv operators. In: Proceedings of International Joint Conference on Artificial Intelligence 2013, pp. 1147–1154. AAAI Press (2013)
16. Wang, Y., Fan, J.: Conditionally knowing what. In: Proceedings of Advances in Modal Logic 2014, vol. 10, pp. 569–587. College Publications (2014)

Knowledge-Now and Knowledge-All

Xinyu Wang[1] and Yanjing Wang[2](✉)

[1] School of Electronics Engineering and Computer Science, Peking University, Beijing, China
xinyuwang1998@pku.edu.cn
[2] Department of Philosophy, Peking University, Beijing, China
y.wang@pku.edu.cn

Abstract. In this paper, we propose a logical framework extending the standard epistemic logic with a new knowledge operator $\mathcal{G}_i$ which captures the knowledge about (physically) necessary facts, e.g., scientific knowledge. Semantically, the truth of $\mathcal{G}_i\varphi$ depends on not only the epistemically indistinguishable worlds from the current real world but also the relevant (physically) possible worlds which are clearly distinguishable. Essentially, $\mathcal{G}_i$ is a bundle of the standard epistemic modality and a necessity-like modality. We axiomatize the corresponding epistemic logic completely in single- and multi-agent cases with interesting interaction axioms between the two epistemic operators.

Keywords: Knowledge-now · Knowledge-all · Epistemic logic · Axiomatization

1 Introduction

The semantics of standard epistemic logic is based on the idea that knowing that φ *iff* you can rule out the epistemic alternatives (of the current state of the world) on which φ is false [5]. In the Kripke semantics for epistemic logic, knowing that φ amounts to that φ holds on all the epistemic alternatives which you cannot distinguish from the current state. This semantics works well with knowledge about the *current state*, to which we shall refer as *knowledge-now* in this paper.[1]

Nevertheless, both in everyday life and in scientific studies, we are interested in not only knowledge-now but also *knowledge-all*, the law-like propositions which are true on *all* the relevant possible states even though they are not epistemic alternatives of the current one. For example, as a piece of common sense (physical) knowledge, we know that if it rains then the ground exposed to the rain gets wet (φ). This knowledge is about not only the states which you cannot distinguish from the real one, but also those physically possible states which are clearly distinguishable from the real one for you. In fact, as φ is not a material

[1] Although the term knowledge-*now* seems to have a temporal flavor, the emphasis is not about the temporal issues, as it will become more clear later on.

P. Blackburn et al. (Eds.): LORI 2019, LNCS 11813, pp. 399–412, 2019.
https://doi.org/10.1007/978-3-662-60292-8_29

implication, just knowing that the ground is wet *right now* does not necessarily entail that you know this common sense physical law.

Knowledge-all also plays an important role when we move to knowledge expressions beyond *knowing that* [9]. For example, *knowing how* to cook a Japanese dish is about not only the current state but also whether you can do it on all the possible states where the relevant ingredients and equipments are available [8]. *Knowing why* thunder often follows lightning is not merely about the current state, either [10]. As another example, to know whether variable X depends on Y, it is clearly not enough just to check the current values of X and Y [1,2].

In this paper, we propose a simple framework to treat both knowledge-now and knowledge-all by including two epistemic modalities $\mathcal{K}_i$ and $\mathcal{G}_i$ for these two kinds of knowledge, respectively. We hope this framework can serve as a playground for further studies of more concrete knowledge-all in various settings.

According to the above-mentioned intuition, that φ is a piece of knowledge-all means that knowing that φ holds on *all* relevant possible states, which can be viewed as that knowing that φ is *necessary* over the relevant state space ($\mathcal{K}_i\Box\varphi$). This brings us to the scattered discussions in the literature about the logic of knowledge and necessity (e.g., [4]). A well-known issue in this setting is the *knowability paradox* by Fitch [3]: if we assume that all truths are knowable ($\varphi \to \Diamond\mathcal{K}_i\varphi$), then based on some innocent assumptions about $\Diamond$ and $\mathcal{K}_i$ we can derive $\varphi \to \mathcal{K}_i\varphi$ by using the Moore sentences. See [6] for further discussions about logic of knowledge, necessity and *a priori* knowledge. In this paper, we take a more pragmatic view on necessity beyond the typical philosophical reading, and use the symbol $\mathcal{A}$ instead of $\Box$ ($\mathcal{A}$ for *all*), which can be about all physically possible states or only a few possible states specified by a context. We then define knowledge-all formula $\mathcal{G}_i\varphi$ as $\mathcal{K}_i\mathcal{A}\varphi$.

The modality $\mathcal{G}_i$ may look simple at the first glance, as it might simply share most (if not all) properties of $\mathcal{K}_i$. However, it turns out that $\mathcal{G}_i$ is *not* exactly as $\mathcal{K}_i$, and the interaction between the two operators is highly non-trivial, especially in the multi-agent setting. For now, note that the introspection axioms for $\mathcal{G}_i$ are not valid intuitively, e.g., if φ is known by i as a general law ($\mathcal{G}_i\varphi$), it does not follow that knowing this general law is itself a known general law too ($\mathcal{G}_i\mathcal{G}_i\varphi$). We will also see the crucial role played by the interaction axiom $\mathcal{K}_i\varphi \to \mathcal{G}_i\neg\mathcal{G}_j\neg\varphi$ as well as its variant in the completeness proof. These features can be best observed in axioms when we only have $\mathcal{K}_i$ and $\mathcal{G}_i$ as primitive modalities in the language without $\mathcal{A}$. Our main results are complete axiomatizations of the logic of both knowledge-now and knowledge-all in single- and multi-agent cases, demonstrating the striking difference between the two cases which does not show up in standard epistemic logic.

The rest of the paper is structured as follows: we lay out the basics of the language and the semantics in Sect. 2, propose axiomatizations for both the single-agent and the multi-agent cases and prove their completeness in Sect. 3, and conclude with future directions in Sect. 4.

2 Preliminaries

We first introduce a background language with both $\mathcal{K}_i$ and $\mathcal{A}$.

Definition 1 (Language ELA). *Given a countable non-empty set $\boldsymbol{P}$ of basic proposition letters and a countable non-empty set $\boldsymbol{I}$ of agents, the formulae of the epistemic language with $\mathcal{A}$ operator are constructed as follows:*

$$\varphi ::= p \mid \neg\varphi \mid (\varphi \wedge \varphi) \mid \mathcal{K}_i\varphi \mid \mathcal{A}\varphi$$

where $i \in \boldsymbol{I}$, $p \in \boldsymbol{P}$. We define $\vee$ and $\rightarrow$ as usual, and define $\mathcal{G}_i\varphi$ as the abbreviation of $\mathcal{K}_i\mathcal{A}\varphi$. $\widehat{\mathcal{K}}_i, \widehat{\mathcal{A}}, \widehat{\mathcal{G}}_i$ are the abbreviations of $\neg\mathcal{K}_i\neg, \neg\mathcal{A}\neg, \neg\mathcal{G}_i\neg$, respectively.

When $\boldsymbol{I}$ is a singleton, without loss of generality we always denote $\boldsymbol{I} = \{i\}$, and we further omit this i in the subscripts of $\mathcal{K}_i$ and $\mathcal{G}_i$ so as to simply write $\mathcal{K}$ and $\mathcal{G}$ when no confusion occurs.

Most philosophers would agree that the necessity operator is an S5 modality, and therefore, an equivalence relation is a good candidate to interpret $\mathcal{A}$ in the model. Moreover, note that the equivalence relation induces a partition over the space of all the potential states. For each state w, its relevant state space is simply the equivalence class of w. If we interpret $\mathcal{A}$ as physically necessary then it is also intuitive to see that any single state should not be in two state spaces, since each state should already contain all the general physical laws for all the possible states in the same "physical universe". Finally, $\mathcal{A}$ should not be interpreted as a usual *universal modality* either, for the agent may have uncertainty about the state space in terms of the general (physical) laws.

Hence the language is interpreted in the following model:

Definition 2 (Model). *An epistemic model for* **ELA** *is $\langle S, \{\sim_i \mid i \in \boldsymbol{I}\}, V, \approx\rangle$ where:*

- *S is a non-empty set of epistemically possible worlds.*
- *For every $i \in \boldsymbol{I}$, $\sim_i$ is an equivalence relation over S.*
- *$V : S \rightarrow \mathcal{P}(\boldsymbol{P})$ is a valuation function.*
- *$\approx$ is an equivalence relation over S representing the (physical) possibilities.*

Let $[s]_i$ and $[s]_{\mathcal{A}}$ be the equivalence classes generated by s with respect to $\sim_i$ and $\approx$, respectively. Let $[s]^i_{\mathcal{A}}$ be $\bigcup\limits_{t\in[s]_i} [t]_{\mathcal{A}}$. Intuitively $[s]^i_{\mathcal{A}}$ is the set of worlds which are reachable by the sequential composition of $\sim_i$ and $\approx$ ($\approx \circ \sim_i$). Note that $\approx \circ \sim_i$ is **not** *necessarily an equivalence relation, although both $\sim_i$ and $\approx$ are.*

Important Notation. When $\mathcal{M}$ is an epistemic model, we use M in italics to represent the underlying set of possible worlds S of $\mathcal{M}$.

Definition 3 (Semantics). *A pointed model $\mathcal{M}, s$ is a model $\mathcal{M}$ with a possible world $s \in M$.*

$$
\begin{array}{|lcl|}
\hline
\mathcal{M}, s \vDash p & \iff & p \in V(s) \\
\mathcal{M}, s \vDash \neg\varphi & \iff & \mathcal{M}, s \nvDash \varphi \\
\mathcal{M}, s \vDash \varphi \wedge \psi & \iff & \mathcal{M}, s \vDash \varphi \textit{ and } \mathcal{M}, s \vDash \psi \\
\mathcal{M}, s \vDash \mathcal{K}_i\varphi & \iff & \mathcal{M}, t \vDash \varphi \textit{ for all } t \in [s]_i \\
\mathcal{M}, s \vDash \mathcal{A}\varphi & \iff & \mathcal{M}, t \vDash \varphi \textit{ for all } t \in [s]_{\mathcal{A}} \\
\hline
\end{array}
$$

Given the semantics, it is not hard to see that

$$\boxed{\mathcal{M}, s \vDash \mathcal{G}_i\varphi \iff \mathcal{M}, t \vDash \varphi \text{ for all } t \in [s]^i_{\mathcal{A}}}$$

Example 1. Consider the following model $\mathcal{M}$ on the left (reflexive arrows omitted) where g means that God exists, and the pointed model $\mathcal{M}, s$ represents a situation where God does not exist but i is unsure about it. Depending on whether God exists or not, some physical laws p, q may hold or not. It is easy to verify that $\mathcal{M}, s \vDash \mathcal{A}(\neg g \wedge p) \wedge \neg\mathcal{K}_i\neg g \wedge \mathcal{G}_i(p \vee q) \wedge \widehat{\mathcal{G}}_i(g \wedge p \wedge q)$. □

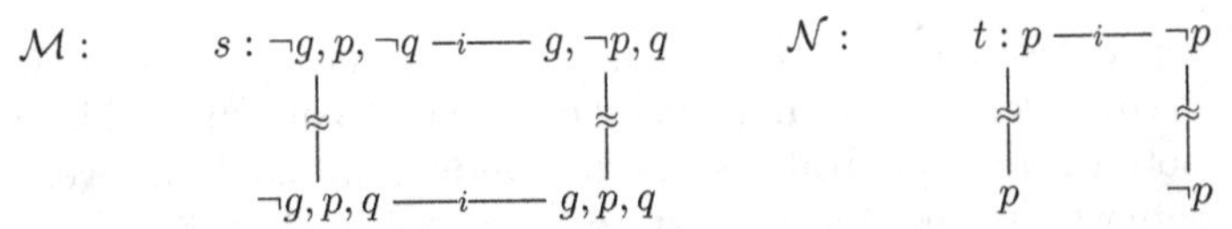

Remark 1. Note that $\mathcal{A}\varphi \to \mathcal{K}_i\varphi$ is not valid as the above example shows. In fact, in this work we do not assume *any* properties between $\sim_i$ and $\approx$. Actually neither KA2AK: $\mathcal{K}_i\mathcal{A}\varphi \to \mathcal{A}\mathcal{K}_i\varphi$ nor AK2KA: $\mathcal{A}\mathcal{K}_i\varphi \to \mathcal{K}_i\mathcal{A}\varphi$ is intuitively valid in general. For example, consider the above model $\mathcal{N}$ on the right (reflexive arrows omitted) and it is easy to verify that $\mathcal{N}, t \vDash \mathcal{A}\mathcal{K}_i\widehat{\mathcal{K}}_ip \wedge \neg\mathcal{K}_i\mathcal{A}\widehat{\mathcal{K}}_ip$.

Since both $\mathcal{A}$ and $\mathcal{K}_i$ are just normal S5 modalities and we do not assume any interactions between them, it is obvious that the following proof system is sound and strongly complete:

Definition 4 (System SELA).

Axioms			
TAUT	*all axioms of propositional logic*		
DISTK	$\mathcal{K}_i\varphi \wedge \mathcal{K}_i(\varphi \to \psi) \to \mathcal{K}_i\psi$	DISTA	$\mathcal{A}\varphi \wedge \mathcal{A}(\varphi \to \psi) \to \mathcal{A}\psi$
T	$\mathcal{K}_i\varphi \to \varphi$	TA	$\mathcal{A}\varphi \to \varphi$
4	$\mathcal{K}_i\varphi \to \mathcal{K}_i\mathcal{K}_i\varphi$	4A	$\mathcal{A}\varphi \to \mathcal{A}\mathcal{A}\varphi$
5	$\neg\mathcal{K}_i\varphi \to \mathcal{K}_i\neg\mathcal{K}_i\varphi$	5A	$\neg\mathcal{A}\varphi \to \mathcal{A}\neg\mathcal{A}\varphi$
Rules			
MP	$\dfrac{\varphi, \varphi \to \psi}{\psi}$		
NECK	$\dfrac{\varphi}{\mathcal{K}_i\varphi}$	NECA	$\dfrac{\varphi}{\mathcal{A}\varphi}$

We define the proofs in this system as usual. As an example, we show how to derive the following valid formula, which will play an important role later. Intuitively, if i knows that φ, then i knows that on *all* the states that i can think of, everyone at least considers φ possible on *some* state. It also helps to verify it semantically based on the properties of $\sim_i$ and $\approx$ to see its merit.

Proposition 1. *$\mathcal{K}_i\varphi \to \mathcal{G}_i\widehat{\mathcal{G}}_j\varphi$ is provable in SELA and thus it is valid.*

Proof. By TA we have $\varphi \to \widehat{\mathcal{A}}\varphi$. By 5A we have $\widehat{\mathcal{A}}\varphi \to \mathcal{A}\widehat{\mathcal{A}}\varphi$, so $\varphi \to \mathcal{A}\widehat{\mathcal{A}}\varphi$. By T we have $\widehat{\mathcal{A}}\varphi \to \widehat{\mathcal{K}}_j\widehat{\mathcal{A}}\varphi$, so by NECA and DISTA we have $\mathcal{A}\widehat{\mathcal{A}}\varphi \to \mathcal{A}\widehat{\mathcal{K}}_j\widehat{\mathcal{A}}\varphi$, hence follows $\varphi \to \mathcal{A}\widehat{\mathcal{K}}_j\widehat{\mathcal{A}}\varphi$. Then by NECK and DISTK we have $\mathcal{K}_i\varphi \to \mathcal{K}_i\mathcal{A}\widehat{\mathcal{K}}_j\widehat{\mathcal{A}}\varphi$, namely $\mathcal{K}_i\varphi \to \mathcal{G}_i\widehat{\mathcal{G}}_j\varphi$. ∎

However, this simple proof system does not reveal the interaction between knowledge-now $\mathcal{K}_i\varphi$ and knowledge-all $\mathcal{G}_i\varphi$ explicitly. Therefore, in the rest of the paper, we will consider the following language taking $\mathcal{G}_i$ as the primitive modalities without using $\mathcal{A}$.

Definition 5 (Language ELG). *Given a countable non-empty set $\boldsymbol{P}$ of basic proposition letters and a countable non-empty set $\boldsymbol{I}$ of agents, the formulae of the epistemic language with $\mathcal{G}_i$ operator are constructed as follows:*

$$\varphi ::= p \mid \neg\varphi \mid (\varphi \wedge \varphi) \mid \mathcal{K}_i\varphi \mid \mathcal{G}_i\varphi$$

where $i \in \boldsymbol{I}$, $p \in \boldsymbol{P}$.

The models and semantics are as before. Intuitively, in **ELG** we only allow a particular combination of $\mathcal{K}_i$ and $\mathcal{A}$. We can actually show that **ELG** is strictly less expressive than **ELA** by using a suitable bisimulation notion, but due to the limit of space we leave the detailed discussion to a future occasion.

In the next section, we will see that the logic of **ELG** brings us some interesting new axioms, which make the logic technically non-trivial to axiomatize.

3 Axiomatization

In this section we axiomatize the logics of knowledge-now and knowledge-all in the single-agent and the multi-agent cases. It is interesting to discover that the logic of the multi-agent case is more complex than that of the single-agent case: more agents bring about highly non-trivial interaction between $\mathcal{K}_i$ and $\mathcal{G}_i$.

3.1 Single-Agent Case

We first consider the single-agent case when both $\mathcal{K}$ and $\mathcal{G}$ are not indexed.

Definition 6 (System SKNKA).

Axioms			
TAUT	*all axioms of propositional logic*		
DISTK	$\mathcal{K}\varphi \wedge \mathcal{K}(\varphi \to \psi) \to \mathcal{K}\psi$	DISTG	$\mathcal{G}\varphi \wedge \mathcal{G}(\varphi \to \psi) \to \mathcal{G}\psi$
T	$\mathcal{K}\varphi \to \varphi$	GK	$\mathcal{G}\varphi \to \mathcal{K}\varphi$
4	$\mathcal{K}\varphi \to \mathcal{K}\mathcal{K}\varphi$	4G	$\mathcal{G}\varphi \to \mathcal{K}\mathcal{G}\varphi$
5	$\neg\mathcal{K}\varphi \to \mathcal{K}\neg\mathcal{K}\varphi$	5G	$\neg\mathcal{G}\varphi \to \mathcal{K}\neg\mathcal{G}\varphi$
		BG	$\mathcal{K}\varphi \to \mathcal{G}\widehat{\mathcal{G}}\varphi$
Rules			
MP	$\dfrac{\varphi, \varphi \to \psi}{\psi}$		
NECK	$\dfrac{\varphi}{\mathcal{K}\varphi}$	NECG	$\dfrac{\varphi}{\mathcal{G}\varphi}$

Remark 2. Note that for the intuitive reading, we also include some axioms not independent from the others. For example, 4 is derivable from T and 5, while 5G is derivable from T, 5 and 4G. Also note that the following $\mathcal{G}$-counterparts for Axioms 4, 5 and B in standard modal logic are *not* valid:

$$\mathcal{G}\varphi \to \mathcal{G}\mathcal{G}\varphi, \neg\mathcal{G}\varphi \to \mathcal{G}\neg\mathcal{G}\varphi, \varphi \to \mathcal{G}\widehat{\mathcal{G}}\varphi.$$

Theorem 1. *SKNKA is sound.*

Proof. The nontrivial part is BG, whose validity follows from Proposition 1. ■

As an example, we show how to derive the following valid formula, whose variant will play an important role in axiomatization for the multi-agent case.

Proposition 2. *The following formula is provable in SKNKA:*

$$\mathcal{K}(\bigvee_{1\leqslant a\leqslant n} \varphi_a \vee \bigvee_{1\leqslant b\leqslant m} \mathcal{G}\psi_b) \to \mathcal{G}(\bigvee_{1\leqslant a\leqslant n} \widehat{\mathcal{G}}\varphi_a \vee \bigvee_{1\leqslant b\leqslant m} \psi_b),$$
$$m, n \in \omega, m > 0 \text{ or } n > 0$$

Proof. The degenerated cases when $n = 0$ or $m = 0$ are relatively simple to show based on the normality of $\mathcal{K}$ and $\mathcal{G}$, and BG. Here we prove the case when $m > 0$ *and* $n > 0$. On the one hand, for each $1 \leqslant b \leqslant m$, by 5G we have $\widehat{\mathcal{K}}\mathcal{G}\psi_b \to \mathcal{G}\psi_b$, hence follows $\widehat{\mathcal{K}}\mathcal{G}\psi_b \to \mathcal{G}(\bigvee_{1\leqslant a\leqslant n} \widehat{\mathcal{G}}\varphi_a \vee \bigvee_{1\leqslant b\leqslant m} \psi_b)$ (i) by weakening the consequent. On the other hand, by BG we have $\mathcal{K}(\bigvee_{1\leqslant a\leqslant n} \varphi_a) \to \mathcal{G}\widehat{\mathcal{G}}(\bigvee_{1\leqslant a\leqslant n} \varphi_a)$, which, by the normality of $\mathcal{G}$, is equivalent to $\mathcal{K}(\bigvee_{1\leqslant a\leqslant n} \varphi_a) \to \mathcal{G}(\bigvee_{1\leqslant a\leqslant n} \widehat{\mathcal{G}}\varphi_a)$, hence follows $\mathcal{K}(\bigvee_{1\leqslant a\leqslant n} \varphi_a) \to \mathcal{G}(\bigvee_{1\leqslant a\leqslant n} \widehat{\mathcal{G}}\varphi_a \vee \bigvee_{1\leqslant b\leqslant m} \psi_b)$ (ii) by weakening the consequent.

Therefore, by (i) and (ii), in SKNKA we can prove $\mathcal{K}(\bigvee_{1\leqslant a\leqslant n} \varphi_a) \vee \bigvee_{1\leqslant b\leqslant m} \widehat{\mathcal{K}}\mathcal{G}\psi_b \to \mathcal{G}(\bigvee_{1\leqslant a\leqslant n} \widehat{\mathcal{G}}\varphi_a \vee \bigvee_{1\leqslant b\leqslant m} \psi_b)$. Also note that the following formula $\mathcal{K}(\bigvee_{1\leqslant a\leqslant n} \varphi_a \vee \bigvee_{1\leqslant b\leqslant m} \mathcal{G}\psi_b) \to \mathcal{K}(\bigvee_{1\leqslant a\leqslant n} \varphi_a) \vee \bigvee_{1\leqslant b\leqslant m} \widehat{\mathcal{K}}\mathcal{G}\psi_b$ can be proved by using the normality of $\mathcal{K}$. By combining the above two formulae we finally obtain $\mathcal{K}(\bigvee_{1\leqslant a\leqslant n} \varphi_a \vee \bigvee_{1\leqslant b\leqslant m} \mathcal{G}\psi_b) \to \mathcal{G}(\bigvee_{1\leqslant a\leqslant n} \widehat{\mathcal{G}}\varphi_a \vee \bigvee_{1\leqslant b\leqslant m} \psi_b)$. ■

To show the completeness of SKNKA is quite non-trivial. First note that to build a canonical model, it is insufficient just to use the set of maximal consistent sets as the set of all the possible worlds: for there are cases where in order to satisfy a set of consistent formulae, in the model we must need different worlds satisfying exactly the same **ELG**-formulae, which suggests that we must need *copies* of possible worlds in the canonical model, on which the same maximal consistent set is attached. The following example demonstrates such a case. Recall that we use M in italics to represent the underlying set of possible worlds S of a model $\mathcal{M}$ (similarly for other letters).

Example 2. Let $\Gamma = \{\varphi \to \mathcal{K}\varphi \mid \varphi \in \mathbf{ELG}\} \cup \{\widehat{\mathcal{G}}\mathcal{G}p, \widehat{\mathcal{G}}\neg p\}$. Γ is consistent, as it has a model $\mathcal{M}$ below (reflexive arrows are omitted) for which we can ravel the $\approx \circ \sim_i$ relation on the right to evaluate **ELG**-formulae more conveniently (reflexive arrows of $\approx \circ \sim_i$ are also omitted). Recall that i is the single agent.

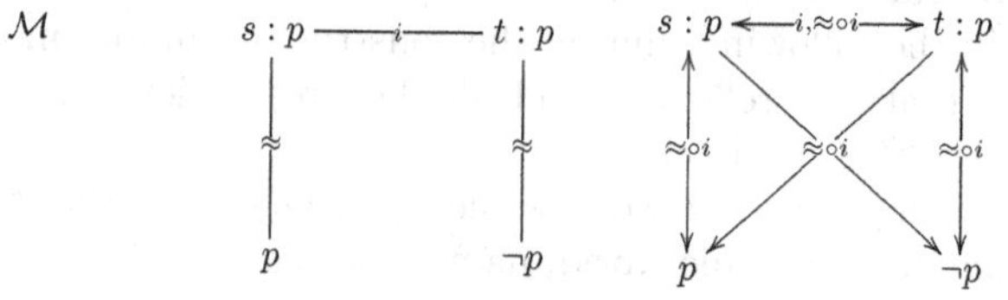

By a simple induction, it can be shown that s and t satisfy exactly the same **ELG**-formulae and therefore it is not hard to show that $\mathcal{M}, s \vDash \Gamma$. However, the apparent duplicate t is necessary here to make Γ true. In fact, we can show that if Γ is satisfiable at some $\mathcal{N}, u$ then we need a different world which satisfies exactly the same **ELG** formulae as u. Suppose $\mathcal{N}, u \vDash \Gamma$ then for all $v \sim_i u$, v satisfies exactly the same formulae as u as $\{\varphi \to \mathcal{K}\varphi \mid \varphi \in \mathbf{ELG}\} \subseteq \Gamma$. Suppose there is no such v except u itself then it is not hard to show that $\{\widehat{\mathcal{G}}\mathcal{G}p, \widehat{\mathcal{G}}\neg p\}$ cannot be satisfied at u due to the fact that $\approx$ is transitive and $\sim_i$ is reflexive. □

Fixing any single maximal consistent set (MCS) σ, and starting from a single possible world s, we build the canonical model $\mathcal{M}(s)$ for σ in a step-by-step method. There is a function $\Theta(t)$ that maps each possible world $t \in M(s)$ to an MCS, and $\Theta(s) = \sigma$. For clarity we always use lowercase English letters such as r, s, t, u and v to represent possible worlds, and lowercase Greek letters such as σ, μ and ν to represent MCSs. The core idea of our construction can be illustrated as the following graph:

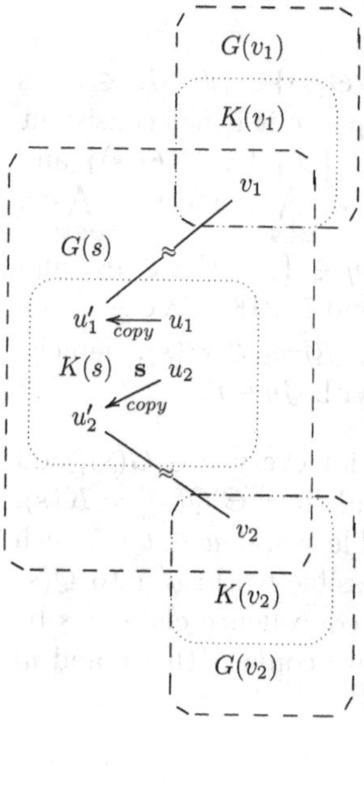

We start from a single possible world s with the MCS σ, and find all the $\sim_i$-connected MCSs (the set $K(s)$) and all the $\approx \circ \sim_i$-connected MCSs (the set $G^+(s)$) via the usual canonical relation for $\mathcal{K}$ and $\mathcal{G}$ respectively. Note that $\sigma \in K(s)$ by T, and that $K(s) \subseteq G^+(s)$ by GK. Let $G(s) = G^+(s) \setminus K(s)$. Then we will recover the $\approx$ relation, by connecting via $\approx$ relation the corresponding possible world v of each $\nu \in G(s)$ to a copy u' of some possible world u corresponding to an MCS $\mu \in K(s)$ (we will show this is always possible). Therefore each possible world in the model can only be connected to *at most one* another via $\approx$ relation. We call the resulting model $\mathcal{N}(s)$, and let $\mathcal{M}_0 = \mathcal{N}(s)$. Then we repeat this construction for every possible world v corresponding to an MCS $\nu \in G(s)$ to form many $\mathcal{N}(v)$, and union all these $\mathcal{N}(v)$ as well as $\mathcal{M}_0$ to form $\mathcal{M}_1$, and so on.

Formally, the construction works as follows. As for the function $\Theta(t)$, in our step-by-step method, we always add fresh possible worlds into the model without modifying existing ones, so we should only pay attention to designate the value of $\Theta(t)$ whenever we add a fresh possible world t into the model, and then in any model $\mathcal{M}$ the total function $\Theta(t)$ comes out by union naturally. Also, without saying explicitly in the following, during the construction in any model $\mathcal{M}$, $\sim_i$ and $\approx$ relations are always reflexive, and all the propositions are canonically evaluated in $\mathcal{M}$, i.e., $\forall t \in M$, $V(t) = \mathbf{P} \cap \Theta(t)$.

Starting from a single possible world s such that $\Theta(s) = \sigma$, the initial model $\mathcal{O}(s)$ consists of only one possible world, namely s. Let

$$K(s) = \{\mu \mid \mu \text{ is an MCS}, \forall \mathcal{K}\varphi \in \Theta(s), \varphi \in \mu\}$$

By T of course $\Theta(s) \in K(s)$. The model $\mathcal{Q}(s)$ extending $\mathcal{O}(s)$ by adding the $\sim_i$-neighbors: for every $\mu \in K(s)$ and $\mu \neq \Theta(s)$, add one possible world u into $\mathcal{O}(s)$ such that $\Theta(u) = \mu$, and finally connect all these possible worlds (including s) by $\sim_i$ relation. In the following, we will build the $\approx$ relation. Firstly, let

$$G(s) = \{\nu \mid \nu \text{ is an MCS}, \nu \notin K(s), \forall \mathcal{G}\varphi \in \Theta(s), \varphi \in \nu\}$$

Proposition 3. *$K(s) \cup G(s) = \{\chi \mid \chi$ is an MCS, $\forall \mathcal{G}\varphi \in \Theta(s), \varphi \in \chi\}$.*

Proof. $\forall \mu \in K(s)$, $\forall \mathcal{G}\varphi \in \Theta(s)$, by GK we have $\mathcal{K}\varphi \in \Theta(s)$, so $\varphi \in \mu$. ∎

For the construction to go on we first prove a Lemma.

Lemma 1 (Backward Lemma). *$\forall \nu \in G(s)$, $\exists \mu \in K(s)$, such that $\forall \mathcal{G}\psi \in \nu$, $\psi \in \mu$.*

Proof. We can construct μ by showing that $\{\varphi \mid \mathcal{K}\varphi \in \Theta(s)\} \cup \{\psi \mid \mathcal{G}\psi \in \nu\}$ is consistent for all $\nu \in G(s)$. Towards a contradiction suppose it is not consistent for some $\nu \in G(s)$ then there are two finite sets $\Phi \subseteq \{\varphi \mid \mathcal{K}\varphi \in \Theta(s)\}$ and $\Psi \subseteq \{\psi \mid \mathcal{G}\psi \in \nu\}$ such that $\Phi \cup \Psi$ is inconsistent. Let $\zeta = \bigwedge_{\varphi \in \Phi} \varphi$ and $\eta = \bigwedge_{\psi \in \Psi} \psi$, it is easy to reason that $\zeta \in \{\varphi \mid \mathcal{K}\varphi \in \Theta(s)\}$ and $\eta \in \{\psi \mid \mathcal{G}\psi \in \nu\}$, and that $\{\zeta, \eta\}$ is inconsistent, hence $\vdash \zeta \to \neg\eta$. By NECK and DISTK, $\vdash \mathcal{K}\zeta \to \mathcal{K}\neg\eta$, and since $\mathcal{K}\zeta \in \Theta(s)$, $\mathcal{K}\neg\eta \in \Theta(s)$, and thus by BG, $\mathcal{G}\widehat{\mathcal{G}}\neg\eta \in \Theta(s)$, namely, $\mathcal{G}\neg\mathcal{G}\eta \in \Theta(s)$. Therefore, $\neg\mathcal{G}\eta \in \nu$, which contradicts with $\mathcal{G}\eta \in \nu$. ∎

The model $\mathcal{N}(s)$ extending $\mathcal{Q}(s)$ is built as such: for every $\nu \in G(s)$, add one possible world v into $\mathcal{Q}(s)$ such that $\Theta(v) = \nu$, and since $\Theta[Q(s)] = K(s)$, by the above Backward Lemma, we can find a possible world $u \in Q(s)$ such that $\forall \mathcal{G}\varphi \in \nu$, $\varphi \in \Theta(u)$, and then we add a *fresh* possible world u' into $\mathcal{Q}(s)$ such that $\Theta(u') = \Theta(u)$, also add this u' into the $\sim_i$ equivalence class of s by connecting all the necessary $\sim_i$ relations, and finally we connect this v and u' by $\approx$ relation, and make $\approx$ reflexive.

The canonical model $\mathcal{M}(s)$ is built by induction. Let $\mathcal{M}_0 = \mathcal{N}(s)$ and $H_0 = \{t \mid t \in N(s), \Theta(t) \in G(s)\} \subseteq M_0$. Suppose we have already built a partial canonical model $\mathcal{M}_l$ and a set of possible worlds $H_l \subseteq M_l$, where $l \in \omega$, we move on to build $\mathcal{M}_{l+1}$ extending $\mathcal{M}_l$ and $H_{l+1} \subseteq M_{l+1}$, as follows. For every $v \in H_l$, we view v itself as a starting point like s and construct its corresponding $K(v)$, $\mathcal{Q}(v)$, $G(v)$ and $\mathcal{N}(v)$. Except for v itself, which already exists in $\mathcal{M}_l$, all the other possible worlds in $\mathcal{N}(v)$ are introduced fresh in order to avoid interference. Then $\mathcal{M}_{l+1} = \mathcal{M}_l \cup \bigcup_{v \in H_l} \mathcal{N}(v)$ and $H_{l+1} = \bigcup_{v \in H_l} \{t \mid t \in N(v), \Theta(t) \in G(v)\}$. Note that this construction is monotonic. Finally let $\mathcal{M}(s) = \bigcup_{l \in \omega} \mathcal{M}_l$.

Proposition 4. *The canonical model $\mathcal{M}(s)$ is indeed an epistemic model.*

Proof. (Sketch) Our construction of $\mathcal{N}(v)$ guarantees that $\sim_i$ is an equivalence relation for the freshly added worlds $v \in H_l$ at each step l. Also note that those fresh worlds are not connected by $\sim_i$ to the existing worlds.

To show that $\approx$ is an equivalence relation over $M(s)$, note that our construction makes sure that $\approx$ is always reflexive and symmetric and moreover every world is $\approx$-connected to *at most* two worlds, including itself.[2] Therefore the transitivity also holds trivially. Again, it is crucial that the freshly added worlds do not interfere with the old ones. ∎

To show strong completeness in this canonical model $\mathcal{M}(s)$, the only non-trivial part concerns the Truth Lemma for modality $\mathcal{G}$.

Lemma 2 (Truth Lemma). *For any $t \in M(s)$, any $\varphi \in$ **ELG**: $\mathcal{M}(s), t \vDash \varphi \iff \varphi \in \Theta(t)$.*

Proof. We only show $\mathcal{M}(s), t \vDash \mathcal{G}\varphi \iff \mathcal{G}\varphi \in \Theta(t)$ based on the induction hypothesis. Other cases are routine as in standard modal logic.

According to our construction, any possible world $t \in M(s)$ falls into exactly one of the following three cases.

First, t is s. Then $\Theta[[t]^i_{\mathcal{A}}] = K(t) \cup G(t)$, and by Proposition 3, $\Theta[[t]^i_{\mathcal{A}}] = \{\chi \mid \chi \text{ is an MCS}, \forall \mathcal{G}\varphi \in \Theta(t), \varphi \in \chi\}$.

Second, $t \in H_l$, where $l \in \omega$. Then there exists $w \in M(s)$ such that $t \in N(w) - Q(w)$, and $\Theta[[t]^i_{\mathcal{A}}] = K(t) \cup G(t) \cup \{\Theta(u')\}$, where $u' \in N(w) - Q(w)$, $u' \neq t$, and t is connected with u' by $\approx$ relation. By Proposition 3, $K(t) \cup G(t) = \{\chi \mid \chi \text{ is an MCS}, \forall \mathcal{G}\varphi \in \Theta(t), \varphi \in \chi\}$, and by our construction, $\forall \mathcal{G}\varphi \in \Theta(t)$, $\varphi \in \Theta(u')$, so $\Theta(u') \in K(t) \cup G(t)$ and $\Theta[[t]^i_{\mathcal{A}}] = \{\chi \mid \chi \text{ is an MCS}, \forall \mathcal{G}\varphi \in \Theta(t), \varphi \in \chi\}$.

Third, otherwise. Then by our construction, there exists $v \in M(s)$ such that t and v are connected by $\sim_i$ relation and that v is s or $v \in H_l$, where $l \in \omega$. By 4G we have $\forall \mathcal{G}\varphi \in$ **ELG**, $\mathcal{G}\varphi \in \Theta(v) \iff \mathcal{G}\varphi \in \Theta(t)$, so $\Theta[[t]^i_{\mathcal{A}}] = \Theta[[v]^i_{\mathcal{A}}] = \{\chi \mid \chi \text{ is an MCS}, \forall \mathcal{G}\varphi \in \Theta(v), \varphi \in \chi\} = \{\chi \mid \chi \text{ is an MCS}, \forall \mathcal{G}\varphi \in \Theta(t), \varphi \in \chi\}$.

In sum, we have $\forall t \in M(s)$, $\Theta[[t]^i_{\mathcal{A}}] = \{\chi \mid \chi \text{ is an MCS}, \forall \mathcal{G}\varphi \in \Theta(t), \varphi \in \chi\}$. As $\mathcal{G}$ is a normal modality, the Truth Lemma follows by routine arguments. ∎

[2] Such a simple construction works essentially because $\mathcal{A}_i$ can only occur in the bundle $\mathcal{G}_i = \mathcal{K}_i\mathcal{A}_i$ in **ELG**. We cannot express the $\approx$ possibilities directly.

Based on the above Truth Lemma, strong completeness follows immediately.

Theorem 2. *SKNKA is strongly complete over single-agent epistemic models.*

3.2 Multi-agent Case

As shown in Proposition 1, when there present multiple agents, the following stronger version of BG is still valid:

$$\mathcal{K}_i\varphi \to \mathcal{G}_i\widehat{\mathcal{G}}_j\varphi$$

which enables us to go from one agent's knowledge-now to its own knowledge-all about another agent's knowledge-all. We also have another valid formula KG to go from one agent's knowledge-now about another agent's knowledge-all to its own knowledge-all, shown by the following:

$$\mathcal{K}_i\mathcal{G}_j\varphi \to \mathcal{G}_i\varphi$$

However, just adding the above two valid formulae into SKNKA is not enough. Indeed we need a further stronger axiom KGP:

$$\mathcal{K}_i(\bigvee_{1\leqslant a\leqslant n}\varphi_a \vee \bigvee_{1\leqslant b\leqslant m}\mathcal{G}_{j_b}\psi_b) \to \mathcal{G}_i(\bigvee_{1\leqslant a\leqslant n}\widehat{\mathcal{G}}_{k_a}\varphi_a \vee \bigvee_{1\leqslant b\leqslant m}\psi_b),$$

$$m, n \in \omega, m > 0 \text{ or } n > 0, i, j_b, k_a \in \mathbf{I}$$

Theorem 3. KGP *is valid.*

Proof. Fix a pointed model $\mathcal{M}, s$ and suppose that the antecedent is true. Then $\forall t \in [s]_i$, at least one of the branches in the disjunction $\bigvee_{1\leqslant a\leqslant n}\varphi_a \vee \bigvee_{1\leqslant b\leqslant m}\mathcal{G}_{j_b}\psi_b$ is true on $\mathcal{M}, t$.

Suppose φ_a is true on $\mathcal{M}, t$. Then $\forall r \in [t]_\mathcal{A}$, for arbitrary (fixed) agent $k_a \in \mathbf{I}$, $\widehat{\mathcal{G}}_{k_a}\varphi_a$ is true on $\mathcal{M}, r$.

Suppose $\mathcal{G}_{j_b}\psi_b$ is true on $\mathcal{M}, t$, where $j_b \in \mathbf{I}$ is an arbitrary (fixed) agent. Then $\forall r \in [t]_\mathcal{A}$, ψ_b is true on $\mathcal{M}, r$.

Therefore, $\forall t \in [s]_i$ and $\forall r \in [t]_\mathcal{A}$, $\bigvee_{1\leqslant a\leqslant n}\widehat{\mathcal{G}}_{k_a}\varphi_a \vee \bigvee_{1\leqslant b\leqslant m}\psi_b$ is true on $\mathcal{M}, r$. Since $[s]^i_\mathcal{A} = \bigcup_{t\in[s]_i}[t]_\mathcal{A}$, this is to say that $\forall r \in [s]^i_\mathcal{A}$, $\bigvee_{1\leqslant a\leqslant n}\widehat{\mathcal{G}}_{k_a}\varphi_a \vee \bigvee_{1\leqslant b\leqslant m}\psi_b$ is true on $\mathcal{M}, r$, so the consequent is true on $\mathcal{M}, s$. ∎

Note that if $\mathcal{K}_i$ were distributive over $\vee$ then we could easily prove KGP from stronger BG and KG. However, as $\mathcal{K}_i$ is a normal modality, this is not the case.

Although the single-agent version of KGP is provable in SKNKA as Proposition 2 shows, the similar proof method for the multi-agent case does not work, e.g., the axiom 5G in the multi-agent setting cannot play the same role as before in the proof of Proposition 2. We conjecture that KGP is not provable from the stronger BG and KG based on SKNKA. Therefore, we replace the weaker axiom BG by KGP, resulting in the following sound proof system MSKNKA for the multi-agent case:

Definition 7 (System MSKNKA).

Axioms			
TAUT	*all axioms of propositional logic*		
DISTK	$\mathcal{K}_i\varphi \wedge \mathcal{K}_i(\varphi \to \psi) \to \mathcal{K}_i\psi$	DISTG	$\mathcal{G}_i\varphi \wedge \mathcal{G}_i(\varphi \to \psi) \to \mathcal{G}_i\psi$
T	$\mathcal{K}_i\varphi \to \varphi$	GK	$\mathcal{G}_i\varphi \to \mathcal{K}_i\varphi$
4	$\mathcal{K}_i\varphi \to \mathcal{K}_i\mathcal{K}_i\varphi$	4G	$\mathcal{G}_i\varphi \to \mathcal{K}_i\mathcal{G}_i\varphi$
5	$\neg\mathcal{K}_i\varphi \to \mathcal{K}_i\neg\mathcal{K}_i\varphi$	5G	$\neg\mathcal{G}_i\varphi \to \mathcal{K}_i\neg\mathcal{G}_i\varphi$
KGP	$\mathcal{K}_i(\bigvee_{1\leqslant a\leqslant n} \varphi_a \vee \bigvee_{1\leqslant b\leqslant m} \mathcal{G}_{j_b}\psi_b) \to \mathcal{G}_i(\bigvee_{1\leqslant a\leqslant n} \widehat{\mathcal{G}}_{k_a}\varphi_a \vee \bigvee_{1\leqslant b\leqslant m} \psi_b),$ $m, n \in \omega, m > 0$ *or* $n > 0, i, j_b, k_a \in \boldsymbol{I}$		
Rules			
MP	$\dfrac{\varphi, \varphi \to \psi}{\psi}$		
NECK	$\dfrac{\varphi}{\mathcal{K}_i\varphi}$	NECG	$\dfrac{\varphi}{\mathcal{G}_i\varphi}$

We then prove the strong completeness. As in the single-agent case, fixing any single maximal consistent set (MCS) σ, and starting from a single possible world s, we build the canonical model $\mathcal{D}(s)$ for σ and a corresponding function $\Theta(t)$ for every $t \in D(s)$, such that $\Theta(s) = \sigma$. The construction of the canonical model $\mathcal{D}(s)$ in the multi-agent case is based on the construction of $\mathcal{M}(s)$ in the single-agent case, the core idea being illustrated as the following graph:

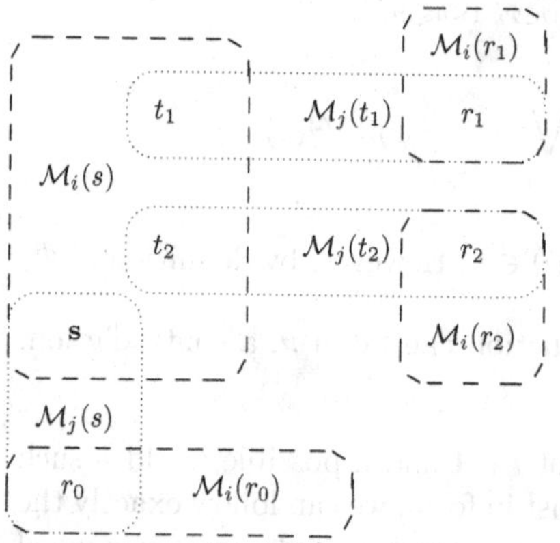

For example, suppose there are only two agents i and j. We start from a single possible world s with some MCS σ, and go through nearly the same process (only replacing the Backward Lemma with a stronger version proved below) as in the single-agent case to construct the single-agent canonical model $\mathcal{M}_i(s)$ with respect to agent i, while making sure $\sim_j$ relation is only reflexive. Let $\mathcal{D}_1 = \mathcal{M}_i(s)$. Then in the second step, for every possible world $t \in D_1$, we construct their single-agent canonical model $\mathcal{M}_j(t)$ with respect to agent j by introducing fresh possible worlds except t, and union all these $\mathcal{M}_j(t)$ as well as $\mathcal{D}_1$ to form $\mathcal{D}_2$. In the third step, we repeat this construction for every possible world $t \in D_2 \setminus D_1$ again with respect to agent i, and in the fourth step with respect to j, and so on.

Formally, the construction works as follows. To begin with, we prove a stronger version of the Backward Lemma.

Lemma 3 (Stronger Backward Lemma). *Fix an agent $i \in \boldsymbol{I}$ and a possible world s with the value of $\Theta(s)$ having been designated. Similarly to the single-agent case, we define $K_i(s) = \{\mu \mid \mu$ is an MCS, $\forall \mathcal{K}_i\varphi \in \Theta(s), \varphi \in \mu\}$ and $G_i(s) = \{\nu \mid \nu$ is an MCS, $\nu \notin K_i(s), \forall \mathcal{G}_i\varphi \in \Theta(s), \varphi \in \nu\}$. Then $\forall \nu \in G_i(s)$, $\exists \mu \in K_i(s)$, such that $\forall k \in \boldsymbol{I}$, $\forall \mathcal{G}_k\varphi \in \nu$, $\varphi \in \mu$, and that $\forall j \in \boldsymbol{I}$, $\forall \mathcal{G}_j\psi \in \mu$, $\psi \in \nu$.*

Proof. First note that $\forall j \in \mathbf{I}$, $\forall \mathcal{G}_j\psi \in \mu$, $\psi \in \nu \iff \forall j \in \mathbf{I}$, $\forall \psi \in \nu$, $\widehat{\mathcal{G}}_j\psi \in \mu$. Given $\nu \in G_i(s)$, we can construct μ by showing that $\{\varphi \mid \mathcal{G}_k\varphi \in \nu, k \in \mathbf{I}\} \cup \{\widehat{\mathcal{G}}_j\psi \mid \psi \in \nu, j \in \mathbf{I}\} \cup \{\xi \mid \mathcal{K}_i\xi \in \Theta(s)\}$ is consistent. Suppose not, then there exist two finite sets of agents $\{k_a \mid k_a \in \mathbf{I}, 1 \leqslant a \leqslant n\}$ and $\{j_b \mid j_b \in \mathbf{I}, 1 \leqslant b \leqslant m\}$, where $m, n \in \omega$, and for every agent its corresponding finite set $\Phi_a \subseteq \{\varphi \mid \mathcal{G}_{k_a}\varphi \in \nu\}$ or $\Psi_b \subseteq \{\widehat{\mathcal{G}}_{j_b}\psi \mid \psi \in \nu\}$, and yet another finite set $\Xi \subseteq \{\xi \mid \mathcal{K}_i\xi \in \Theta(s)\}$, such that $\Xi \cup \bigcup_{1\leqslant a\leqslant n} \Phi_a \cup \bigcup_{1\leqslant b\leqslant m} \Psi_b$ is inconsistent. Let $\zeta = \bigwedge_{\xi\in\Xi} \xi$, and it is easy to see that $\zeta \in \{\xi \mid \mathcal{K}_i\xi \in \Theta(s)\}$ and that $\{\zeta\} \cup \bigcup_{1\leqslant a\leqslant n} \Phi_a \cup \bigcup_{1\leqslant b\leqslant m} \Psi_b$ is inconsistent, hence $\vdash \zeta \to \neg(\bigwedge_{\varphi\in\Phi_a,1\leqslant a\leqslant n} \varphi \wedge \bigwedge_{\widehat{\mathcal{G}}_{j_b}\psi\in\Psi_b,1\leqslant b\leqslant m} \widehat{\mathcal{G}}_{j_b}\psi)$, namely, $\vdash \zeta \to (\bigvee_{\varphi\in\Phi_a,1\leqslant a\leqslant n} \neg\varphi \vee \bigvee_{\widehat{\mathcal{G}}_{j_b}\psi\in\Psi_b,1\leqslant b\leqslant m} \mathcal{G}_{j_b}\neg\psi)$. Thus by NECK and DISTK, $\vdash \mathcal{K}_i\zeta \to \mathcal{K}_i(\bigvee_{\varphi\in\Phi_a,1\leqslant a\leqslant n} \neg\varphi \vee \bigvee_{\widehat{\mathcal{G}}_{j_b}\psi\in\Psi_b,1\leqslant b\leqslant m} \mathcal{G}_{j_b}\neg\psi)$, and by KGP,

$$\vdash \mathcal{K}_i\zeta \to \mathcal{G}_i(\bigvee_{\varphi\in\Phi_a,1\leqslant a\leqslant n} \widehat{\mathcal{G}}_{k_a}\neg\varphi \vee \bigvee_{\widehat{\mathcal{G}}_{j_b}\psi\in\Psi_b,1\leqslant b\leqslant m} \neg\psi),$$

namely, $\vdash \mathcal{K}_i\zeta \to \mathcal{G}_i(\bigvee_{\varphi\in\Phi_a,1\leqslant a\leqslant n} \neg\mathcal{G}_{k_a}\varphi \vee \bigvee_{\widehat{\mathcal{G}}_{j_b}\psi\in\Psi_b,1\leqslant b\leqslant m} \neg\psi)$. As $\mathcal{K}_i\zeta \in \Theta(s)$, we have

$$\mathcal{G}_i(\bigvee_{\varphi\in\Phi_a,1\leqslant a\leqslant n} \neg\mathcal{G}_{k_a}\varphi \vee \bigvee_{\widehat{\mathcal{G}}_{j_b}\psi\in\Psi_b,1\leqslant b\leqslant m} \neg\psi) \in \Theta(s),$$

so $(\bigvee_{\varphi\in\Phi_a,1\leqslant a\leqslant n} \neg\mathcal{G}_{k_a}\varphi \vee \bigvee_{\widehat{\mathcal{G}}_{j_b}\psi\in\Psi_b,1\leqslant b\leqslant m} \neg\psi) \in \nu$. However, by definition of Φ_a and Ψ_b none of these branches in the conjunction can be in ν, a contradiction. ■

Now suppose that we have fixed an agent $i \in \mathbf{I}$ and a possible world v such that $\Theta(v)$ is defined. Then starting from v, just in form, we can follow exactly the same procedure as in the single-agent case to construct a single-agent canonical model $\mathcal{M}(v)$ with respect to agent i, by changing every appearance of $\mathcal{K}$ and $\mathcal{G}$ into $\mathcal{K}_i$ and $\mathcal{G}_i$, and also changing the use of the original Backward Lemma when constructing $\mathcal{N}(t)$ for any possible world t into the stronger version above. For clarification we denote this single-agent canonical model as $\mathcal{M}_i(v)$, with a subscript showing that it is built with respect to agent i. In $\mathcal{M}_i(v)$ of course, for any $j \in \mathbf{I}$ and $j \neq i$, the $\sim_j$ relation is only reflexive. After the construction of $\mathcal{M}_i(v)$, we say that an operation with respect to agent i has been done on all the possible worlds $t \in M_i(v)$; by default no operation with respect to any agent has been done on a possible world, of course.

Having finished the preparation above, we then construct a (fixed) countably infinite sequence of agents as $i_0, i_1, \ldots, i_l, \ldots$, in the order of which we build the canonical model $\mathcal{D}(s)$.

Since **I** is countable, we can have a countable sequence of agents such that each agent appears countably infinite times.

Proposition 5. *A countably infinite sequence* $i_0, i_1, \ldots, i_l, \ldots$ *can be constructed, which is isomorphic to* ω *in order, such that* $\forall l \in \omega$, $i_l \in \boldsymbol{I}$, *and that* $\forall j \in \boldsymbol{I}$, $|\{l \mid l \in \omega, i_l = j\}| = \aleph_0$.

The canonical model $\mathcal{D}(s)$ is built by induction. Let $\mathcal{D}_0 = \mathcal{O}(s)$ and no operation with respect to any agent has been done on s. Suppose we have already built a partial canonical model $\mathcal{D}_l$, where $l \in \omega$, we move on to build $\mathcal{D}_{l+1}$ extending $\mathcal{D}_l$, as the follows. Let $F_l = \{v \mid v \in D_l$, no operation with respect to agent i_l has been done on $v\} \subseteq D_l$. For every $v \in F_l$, we view v itself as a starting point like $\mathcal{O}(v)$ and construct its single-agent canonical model $\mathcal{M}_{i_l}(v)$ with respect to agent i_l. Except for v itself, which already exists in $\mathcal{D}_l$, all the other possible worlds in $\mathcal{M}_{i_l}(v)$ are freshly introduced in order to avoid interference. Then $\mathcal{D}_{l+1} = \mathcal{D}_l \cup \bigcup_{v \in F_l} \mathcal{M}_{i_l}(v)$. Finally let $\mathcal{D}(s) = \bigcup_{l \in \omega} \mathcal{D}_l$.

Proposition 6. *The canonical model* $\mathcal{D}(s)$ *is indeed an epistemic model.*

Proof. (Sketch) To show that $\forall i \in \mathbf{I}$, $\sim_i$ is an equivalence relation over $D(s)$, we claim that $\forall i \in \mathbf{I}$, $\forall t \in D(s)$, an operation with respect to agent i has been done on t once and only once. This statement is ensured by our construction of the sequence i_l and the canonical model $\mathcal{D}(s)$, because $\forall i \in \mathbf{I}$, i appears in the sequence i_l for countably infinite times.

To show that $\approx$ is an equivalence relation over $D(s)$, we claim a stronger statement, that in every $\approx$ equivalence class there are at most two possible worlds. This statement holds just as the single-agent case. ■

To show strong completeness in this canonical model $\mathcal{D}(s)$, the only non-trivial part concerns the Truth Lemma for modality $\mathcal{G}_i$. By symmetry among the agents we fix an arbitrary $i \in \mathbf{I}$.

Lemma 4 (Truth Lemma). *For any* $t \in D(s)$, *any* $\varphi \in \mathbf{ELG}$: $\mathcal{D}(s), t \vDash \varphi \iff \varphi \in \Theta(t)$.

Proof. Similarly to the single-agent case, we only show $\mathcal{D}(s), t \vDash \mathcal{G}_i\varphi \iff \mathcal{G}_i\varphi \in \Theta(t)$. As before, by using the conditions in the Stronger Backward Lemma, we have $\forall t \in D(s)$, $\Theta[[t]^i_{\mathcal{A}}] = \{\chi \mid \chi$ is an MCS, $\forall \mathcal{G}_i\varphi \in \Theta(t), \varphi \in \chi\}$. As $\mathcal{G}_i$ is a normal modality, the Truth Lemma follows by routine arguments. ■

Based on the above Truth Lemma strong completeness follows immediately.

Theorem 4. MSKNKA *is strongly complete over multi-agent epistemic models.*

4 Conclusions and Future Work

In this paper, we have proposed a framework to deal with both knowledge-now and knowledge-all together. The logic, being conceptually simple and clear, actually possesses quite complicated properties. We obtain sound and strongly complete axiomatizations for both the single-agent and the multi-agent cases, where the latter case is significantly more complex. Note that the decidability of the satisfiability problem of our logic follows immediately from the fact that **ELG** is essentially a fragment of **ELA** whose logic a version of multi-modal S5.

Under this framework, there is an abundance of work worth investigating in the future. On the one hand, we can further examine different properties of this logic with respect to various frame classes, such as belief-all (KD45) frames and confluent frames. On the other hand, as mentioned in the introduction, adding all sorts of knowledge operators beyond "knowing that" into this framework may result in novel logics with more affluent expressivity as well as more interesting characteristics. A first attempt to handle knowing-value and *partial* dependency in our setting is carried out in [7]. Finally, the dynamics of knowledge in this setting can be non-trivial, e.g., how to update the model with both $\sim_i$ and $\approx$ w.r.t. a public announcement $[\varphi]$? The simple-minded solution of deleting all the $\neg\varphi$ worlds would ruin the other physically possible worlds. We also leave this to a future occasion.

Acknowledgement. The first author thanks Shengyang Zhong for the help in making the canonical model construction more precise. The authors are grateful to the anonymous reviewers for their insightful comments.

References

1. Baltag, A.: To know is to know the value of a variable. Adv. Modal Logic **11**, 135–155 (2016)
2. Ding, Y.: Epistemic logic with functional dependency operator. Stud. Logic **9**(4), 55–84 (2016)
3. Fitch, F.: A logical analysis of some value concepts. J. Symb. Logic **28**(2), 135–142 (1963)
4. Hart, W., McGinn, C.: Knowledge and necessity. J. Philos. Logic **5**(2), 205–208 (1976)
5. Hintikka, J.: Knowledge and Belief: An Introduction to the Logic of the Two Notions. Cornell University Press, Ithaca (1962)
6. Rescher, N.: Epistemic logic: a survey of the logic of knowledge. University of Pittsburgh Press (2005)
7. Wang, X.: Epistemic logic with partial dependency operator. In: Blackburn, P., Lorini, E., Guo, M. (eds.) LORI 2019, Proceedingsof LORI-VII, LNCS, vol. 11813, pp. 385–398 (2019)
8. Wang, Y.: A logic of goal-directed knowing how. Synthese **195**(10), 4419–4439 (2018)
9. Wang, Y.: Beyond knowing that: a new generation of epistemic logics. In: van Ditmarsch, H., Sandu, G. (eds.) Jaakko Hintikka on Knowledge and Game-Theoretical Semantics. OCL, vol. 12, pp. 499–533. Springer, Cham (2018). https://doi.org/10.1007/978-3-319-62864-6_21
10. Xu, C., Wang, Y., Studer, T.: A logic of knowing why. Synthese (2019, forthcoming)

A Logic of Knowing How with Skippable Plans

Xun Wang(✉)

Department of Philosophy, Peking University, Beijing, China
wangxun123@pku.edu.cn

Abstract. The paper expands upon the work by Wang [16], who proposes a single-agent modal logic framework for reasoning about "knowing how". This paper proposes a more flexible semantics to the knowing-how operator. According to this semantics, an agent knows how to achieve φ given ψ if there exists a finite linear plan such that it will end up with some φ-state from any ψ-state by executing the plan, either fully or skipping some non-executable steps. We give a sound and complete axiomatization of this logic. Finally we introduce a suitable notion of bisimulation for this logic.

Keywords: Knowing how · Epistemic logic · Skippable plans

1 Introduction

A large body of automated planning concerns the realization of strategies or action sequences for autonomous agents to achieve some goals, i.e., to obtain goal-directed knowledge-how. Several formalizations of knowing how were suggested before under different ways (see [2,5]). The attempt to develop logics of knowing how begins from [9–11]. The situation calculus as one of the cornerstone formalisms of AI suggests possible interpretations of what it means for a computer program to be able to achieve a state of affairs [14]. Coalition Logic (CL) [13] formalises the ability of groups of agents to achieve certain goals in strategic games. The framework of Alternating-time Temporal Logic (ATL) [1] is concerned with reasoning about the abilities of agents in game-like systems. In the above-mentioned works, knowledge-how is usually expressed in powerful quantified logical language involving quantifiers or various complicated modalities. Recently logicians attempted to formalize some knowledge expressions (such as knowing what, knowing how and so on) as a single modality, similar to the way of epistemic logic dealing with knowing-that (see [3,6–8,15–17]).

In particular, Wang [16] proposes a single-agent modal logic framework for reasoning about goal-directed knowing how, which includes formula $\mathcal{K}h(\psi, \varphi)$ expressing that the agent knows how to achieve φ given the precondition ψ. This conditionalization reconciles our intuitive acquaintance about knowing how. More specifically, when you claim that you know how to go to Hong Kong, what you actually claim is that you can get to Hong Kong under some implicit

P. Blackburn et al. (Eds.): LORI 2019, LNCS 11813, pp. 413–424, 2019.
https://doi.org/10.1007/978-3-662-60292-8_30

preconditions, such as there are flights from Beijing to Hong Kong and you can afford a flight ticket. In [16], the models are labelled transition systems which represent the agent's abilities. $\mathcal{K}h(\psi, \varphi)$ holds globally in a labelled transition system if there is a strongly executable plan such that from each ψ-state the plan can always be successfully executed to reach some φ-state. For example, $\mathcal{K}h(p, q)$ holds in the model depicted by Fig. 1[1], since ru is the plan which can always work to reach a q-state from any p-state.

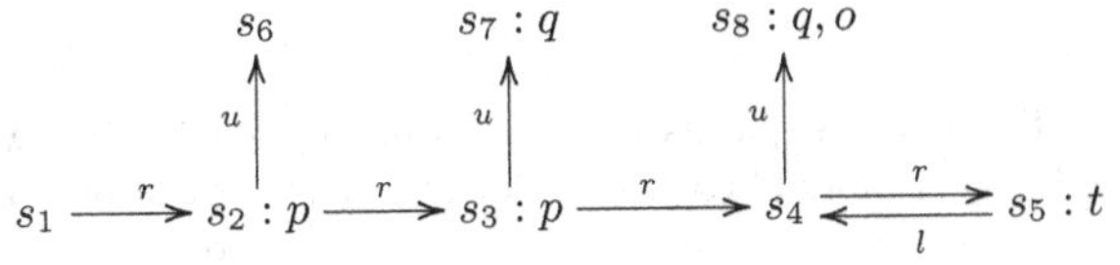

Fig. 1. An example

However, the demands that a strongly executable plan asks may be too strong, in the sense that the plan can always be fully executed. In this model, $\mathcal{K}h(p, t)$ does not hold since there does not exist a plan which can be successfully executed to reach s_5 from s_2 and s_3, but we usually think that "we know how to reach a t-state from p-states" because we can get there by moving right at most three times. Li [6] proposes a weaker semantics to the knowing-how operator, where an agent knows how to achieve φ given ψ if it has a weak conformant plan by which it can always end up with a φ-state when the execution of the plan terminates, either successfully or not. According to Li's semantics, $\mathcal{K}h(p, t)$ holds since the plan rrr can guarantee the agent ending up with s_5 when the execution of the plan starting from s_2 or s_3 terminates. But Li's weaker interpretation of knowing-how results in a weaker logic where the composition axiom $(\mathcal{K}h(p, r) \wedge \mathcal{K}h(r, q) \rightarrow \mathcal{K}h(p, q))$ in Wang's paper no longer holds. For example, $\mathcal{K}h(t, o)$ also holds in the above model under Li's interpretation because of the plan lu, but $\mathcal{K}h(p, o)$ does not hold for the execution of the plan $rrrlu$ starting from s_3 will terminate at s_5 but not s_8. However it is counterintuitive that you know how to achieve r given p and achieve q given r, but do not know how to achieve q given p.

Intuitively, we think a reasonable agent should also know how to get to some o-state if it knows it is at some p-state but does not know exactly where it is, since it can keep moving r until it cannot move r any more and then move l and u. However, $\mathcal{K}h(p, o)$ does not hold under the before two kinds of semantics, since the execution of the plan $rrrlu$ starting from s_3 will fail to continue at s_5. Usually, it is appropriate to say "we know how to achieve φ given ψ" only if we will always end up with a φ-state by executing the plan, either fully or skipping some non-executable steps in the plan. Let us go back to the example of $\mathcal{K}h(p, o)$. The plan $rrrlu$ is neither a strongly executable plan nor a weak

[1] It is a variant of the example in [16].

conformant plan. But this plan will still guarantee us reaching the o-state s_8 in the sense that we can end up with s_8 from s_2 by fully executing the plan, and can also end up with s_8 from s_3 by executing the plan but skipping the third r-moving which is not executable.

In this paper, the execution of a plan means executing the executable steps and skipping the non-executable steps. We interpret $\mathcal{K}h(\psi, \varphi)$ as that there is such a plan achieving φ given ψ. Intuitively, such a plan is enough for us to know how to achieve φ given ψ. Our interpretation is weaker than the interpretation in [16] but the composition axiom is still valid under our weaker semantics. Compared to the interpretation in [6,16], our interpretation is more realistic.

The rest of the paper is organized as follows. We introduce the language, the semantics, and a sound and complete proof system in Sect. 2. In Sect. 3, we introduce a notion of bisimulation for our logic. In the last section, we point out some future directions.

2 The Logic

Definition 1 (Language). *Given a set of propositional letters* $\boldsymbol{P}$*, the language* $\boldsymbol{L}_{\boldsymbol{Khs}}$ *is defined by the following BNF where* $p \in \boldsymbol{P}$*:*

$$\varphi ::= \top \mid p \mid \neg\varphi \mid (\varphi \wedge \varphi) \mid \mathcal{K}hs(\varphi, \varphi).$$

$\mathcal{K}hs(\psi, \varphi)$ expresses that the agent knows how to achieve φ given ψ. We use the standard abbreviations $\bot, \varphi \vee \psi$ and $\varphi \to \psi$, and define $\mathcal{U}\varphi$ as the abbreviation of $\mathcal{K}hs(\neg\varphi, \bot)$. $\mathcal{U}$ is intended to be a universal modality, and it will become clearer after we define the semantics.

Definition 2 (Models). *A model is essentially a labelled transition system* $(\mathcal{S}, \Sigma, \mathcal{R}, \mathcal{V})$ *where:*

- $\mathcal{S}$ *is a non-empty set of states,*
- Σ *is a non-empty set of actions,*
- $\mathcal{R} : \Sigma \to 2^{\mathcal{S} \times \mathcal{S}}$ *is a collection of transitions labelled by actions in* Σ*,*
- $\mathcal{V} : \mathcal{S} \to 2^{\boldsymbol{P}}$ *is a valuation function.*

We write $s \xrightarrow{a} t$ *and say* t *is an* a*-successor of* s*, if* $(s, t) \in \mathcal{R}(a)$*.*

Note that the actions in Σ can be non-deterministic.

Definition 3 (Arrival States). *Given a model* $(\mathcal{S}, \Sigma, \mathcal{R}, \mathcal{V})$*, a state* $w \in \mathcal{S}$ *and an action* $a \in \Sigma$*,* $\text{ArrSta}(w, a)$ *is the set of states at which executing* a *on* w *might arrive. Formally, it is defined as:*

$$\text{ArrSta}(w, a) = \begin{cases} \{w\}, & \textit{if } w \textit{ has no } a\textit{-successor} \\ \{t \in \mathcal{S} \mid w \xrightarrow{a} t\}, & \textit{otherwise} \end{cases}$$

We write $w \xrightarrow{a}_s t$ *if* $t \in \text{ArrSta}(w, a)$. *More generally, given a state* $w \in \mathcal{S}$ *and an action sequence* $\sigma = a_1 \cdots a_n \in \Sigma^*$, $\text{ArrSta}(w, \sigma)$ *is the set of states at which executing* σ *on* w *might arrive. Formally, it is defined as:*

$$\text{ArrSta}(w, \sigma) = \{t \mid \exists t_1 \cdots t_{n-1} : w \xrightarrow{a_1}_s t_1 \xrightarrow{a_2}_s \cdots t_{n-1} \xrightarrow{a_n}_s t\}.$$

We write $w \xrightarrow{\sigma}_s t$ *if* $t \in \text{ArrSta}(w, \sigma)$. *In particular,* σ *can be the empty sequence* ϵ *(when* $n = 0$*), and we set* $\text{ArrSta}(w, \epsilon) = \{w\}$ *for any* $w \in \mathcal{S}$, *i.e.,* $w \xrightarrow{\epsilon}_s w$.

Note that $\text{ArrSta}(w, a) \neq \emptyset$, and it follows that $\text{ArrSta}(w, \sigma) \neq \emptyset$.

Definition 4 (Semantic). *Suppose* s *is a state in a model* $\mathcal{M} = (\mathcal{S}, \Sigma, \mathcal{R}, \mathcal{V})$. *The satisfaction relation* $\vDash$ *is defined as follows:*

$\mathcal{M}, s \vDash \top$		*always*
$\mathcal{M}, s \vDash p$	$\Longleftrightarrow$	$p \in \mathcal{V}(s)$
$\mathcal{M}, s \vDash \neg\varphi$	$\Longleftrightarrow$	$\mathcal{M}, s \nvDash \varphi$
$\mathcal{M}, s \vDash \varphi \wedge \psi$	$\Longleftrightarrow$	$\mathcal{M}, s \vDash \varphi$ *and* $\mathcal{M}, s \vDash \psi$
$\mathcal{M}, s \vDash \mathcal{K}hs(\psi, \varphi)$	$\Longleftrightarrow$	*there is* $\sigma \in \Sigma^*$ *such that for each* $w \in [\![\psi]\!]^{\mathcal{M}}$ *and each* $t \in \text{ArrSta}(w, \sigma)$ *we have* $\mathcal{M}, t \vDash \varphi$

where $[\![\psi]\!]^{\mathcal{M}} = \{s \in \mathcal{S} \mid \mathcal{M}, s \vDash \psi\}$.

Note that the semantics of $\mathcal{K}hs$-formulas ignores the current state s. It is not hard to see that a formula of the form $\mathcal{K}hs(\psi, \varphi)$ is globally true or false. Now we can check that the operator $\mathcal{U}$ defined by $\mathcal{K}hs$ is actually a universal modality:

$\mathcal{M}, s \vDash \mathcal{U}\varphi \Longleftrightarrow \mathcal{M}, s \vDash \mathcal{K}hs(\neg\varphi, \bot)$
$\Longleftrightarrow$ there is $\sigma \in \Sigma^*$ such that for each $w \in [\![\neg\varphi]\!]^{\mathcal{M}}$ and each $t \in \text{ArrSta}(w, \sigma)$: $\mathcal{M}, t \vDash \bot$
$\Longleftrightarrow$ there is $\sigma \in \Sigma^*$ such that for each $w \in [\![\neg\varphi]\!]^{\mathcal{M}}$: there is no t such that $t \in \text{ArrSta}(w, \sigma)$
$\Longleftrightarrow$ there is $\sigma \in \Sigma^*$ such that there is no w such that $w \in [\![\neg\varphi]\!]^{\mathcal{M}}$
$\Longleftrightarrow \mathcal{M}, w \vDash \varphi$ for all $w \in \mathcal{S}$

Definition 5 (System $\mathbb{SKHS}$). *The proof system* $\mathbb{SKHS}$[2] *is defined as follows:*

Axioms		**Rules**	
TAUT	*all axioms of propositional logic*	MP	$\dfrac{\varphi,\ \varphi \to \psi}{\psi}$
DISTU	$\mathcal{U}p \wedge \mathcal{U}(p \to q) \to \mathcal{U}q$	NECU	$\dfrac{\varphi}{\mathcal{U}\varphi}$
COMPKh	$\mathcal{K}hs(p, r) \wedge \mathcal{K}hs(r, q) \to \mathcal{K}hs(p, q)$	SUB	$\dfrac{\varphi(p)}{\varphi[(\psi/p)]}$
EMP	$\mathcal{U}(p \to q) \to \mathcal{K}hs(p, q)$		
TU	$\mathcal{U}p \to p$		
4KU	$\mathcal{K}hs(p, q) \to \mathcal{U}\mathcal{K}hs(p, q)$		
5KU	$\neg\mathcal{K}hs(p, q) \to \mathcal{U}\neg\mathcal{K}hs(p, q)$		

[2] $\mathbb{SKHS}$ is exactly the same proof system as in [16].

We can derive the following in $\mathbb{SKHS}$. In the rest of the paper we often use it.

Proposition 1. $\vdash \mathcal{U}\psi \wedge \mathcal{U}\varphi \rightarrow \mathcal{U}(\psi \wedge \phi)$ [3]

Theorem 1. $\mathbb{SKHS}$ *is sound w.r.t. the class of all models.*

Proof. The only non-trivial case is COMPKh. Note that if there is an action sequence σ leading you from any p-state to some r-state, and there is a sequence η from any r-state to some q-state, then $\sigma\eta$ will make sure that you end up with q-states from any p-state.

But $\sigma\eta$ is not necessarily a weak conformant plan for achieving q-states from any p-state, since the execution of σ may interrupt and then the execution of $\sigma\eta$ would terminate at r-states but not necessarily q-states. This is the reason why COMPKh does not hold in [6].

Here are some notions before we prove the completeness. Given a set of $\mathbf{L_{Khs}}$ formulas Δ, let $\Delta|_{\mathcal{K}hs}$ and $\Delta|_{\neg\mathcal{K}hs}$ be the collections of its positive and negative $\mathcal{K}hs$ formulas:

$$\Delta|_{\mathcal{K}hs} = \{\theta \mid \theta = \mathcal{K}hs(\psi,\varphi) \in \Delta\},$$

$$\Delta|_{\neg\mathcal{K}hs} = \{\theta \mid \theta = \neg\mathcal{K}hs(\psi,\varphi) \in \Delta\}.$$

Next we built a separate canonical model for each maximal consistent set Γ, for it is not possible to satisfy all $\mathcal{K}hs$ formulas simultaneously in a single model since $\mathcal{K}hs$ formulas are globally true or false.

Definition 6 (Canonical Models). *Given a maximal consistent set Γ w.r.t.* $\mathbb{SKHS}$, *the canonical model for Γ is $\mathcal{M}^c_\Gamma = \langle \mathcal{S}^c_\Gamma, \Sigma_\Gamma, \mathcal{R}^c, \mathcal{V}^c\rangle$ where:*

- $\mathcal{S}^c_\Gamma = \{\Delta \mid \Delta$ *is a maximal consistent set w.r.t.* $\mathbb{SKHS}$ *and* $\Gamma|_{\mathcal{K}hs} = \Delta|_{\mathcal{K}hs}\}$,
- $\Sigma_\Gamma = \{\langle\psi,\varphi\rangle \mid \mathcal{K}hs(\psi,\varphi) \in \Gamma\}$,
- $\Delta \xrightarrow{\langle\psi,\varphi\rangle}_c \Theta$ *iff* 1. $\mathcal{K}hs(\psi,\varphi) \in \Gamma$, $\psi \in \Delta$, $\varphi \in \Theta$, *or*
  2. $\mathcal{K}hs(\psi,\varphi) \in \Gamma$, $\neg\psi \in \Delta$, $\Delta = \Theta$, *or*
  3. $\mathcal{K}hs(\psi,\varphi) \in \Gamma$, $\neg\psi \in \Delta$, $\psi \in \Theta$,
- $p \in \mathcal{V}^c(\Delta)$ *iff* $p \in \Delta$.

We say that $\Delta \in \mathcal{S}^c_\Gamma$ is a φ-state if $\varphi \in \Delta$.

Let us compare our canonical models to Wang's [16], where $\Delta \xrightarrow{\langle\psi,\varphi\rangle}_c \Theta$ iff $\mathcal{K}h(\psi,\varphi) \in \Gamma$, $\psi \in \Delta$ and $\varphi \in \Theta$. Each state has a successor for each action in our canonical models,[4] but not in [16]. For example, $\Delta \xrightarrow{\langle\bot,\bot\rangle}_c \Delta$ in our canonical models since $\mathcal{K}hs(\bot,\bot) \in \Gamma$ and $\top \in \Delta$, but Δ has no $\langle\bot,\bot\rangle$-successor in [16] since $\bot \notin \Delta$.

Every state in $\mathcal{S}^c_\Gamma$ has the same collection of Γ's $\mathcal{K}hs$ formulas. It follows immediately that:

---

[3] Please refer to the proof of Proposition 2 in [16].
[4] It is exactly the Proposition 5.

Proposition 2. *For any* Δ, Δ' *in* $\mathcal{S}^c_\Gamma$, $\Delta|_{\mathcal{K}hs} = \Delta'|_{\mathcal{K}hs}$.

Now we give an important proposition to be used later.

Proposition 3. *If* $\varphi \in \Delta$ *for all* $\Delta \in \mathcal{S}^c_\Gamma$*, then* $\mathcal{U}\varphi \in \Delta$ *for all* $\Delta \in \mathcal{S}^c_\Gamma$.[5]

Now we prove the following two key propositions for the truth lemma. The first one reflects our intuition that if we know how to achieve φ from ψ and we are at a ψ-state, then there must be a φ-state where we could arrive. Moreover, the second one reflects the intuition that the states where we arrive after executing the plan for achieving φ must be φ-states.

Proposition 4. *For any* $\mathcal{K}hs(\psi, \varphi) \in \Gamma$*, any* $\Delta \in \mathcal{S}^c_\Gamma$*, if* $\psi \in \Delta$ *then there exists* $\Delta' \in \mathcal{S}^c_\Gamma$ *such that* $\varphi \in \Delta'$.

Proof. For suppose not. Then for all $\Delta' \in \mathcal{S}^c_\Gamma$, we have that $\neg\varphi \in \Delta'$, i.e., $\varphi \to \bot \in \Delta'$. By Proposition 3, $\mathcal{U}(\varphi \to \bot) \in \Delta$. It follows by EMP that $\mathcal{K}hs(\varphi, \bot) \in \Delta$. By COMPKh, $\mathcal{K}hs(\psi, \bot) \in \Delta$, i.e., $\mathcal{U}\neg\psi \in \Delta$. By TU, $\neg\psi \in \Delta$, contradicting that $\psi \in \Delta$. Therefore, there exists $\Delta' \in \mathcal{S}^c_\Gamma$ such that $\varphi \in \Delta'$.

Proposition 5. *For any* $\langle\psi, \varphi\rangle \in \Sigma_\Gamma$ *and any* $\Delta \in \mathcal{S}^c_\Gamma$*,* Δ *has a* $\langle\psi, \varphi\rangle$*-successor. Moreover, if* $\psi \in \Delta$ *then* $\text{ArrSta}(\Delta, \langle\psi, \varphi\rangle) = \{\Pi \in \mathcal{S}^c_\Gamma \mid \varphi \in \Pi\} \neq \emptyset$.

Proof. There are two cases:

Case 1. $\psi \notin \Delta$. It follows by the construction of $\mathcal{R}^c$ that $\Delta \xrightarrow{\langle\psi,\varphi\rangle}_c \Delta$. Thus, Δ has a $\langle\psi, \varphi\rangle$-successor.

Case 2. $\psi \in \Delta$. Then by Proposition 4, there exists $\Delta' \in \mathcal{S}^c_\Gamma$ such that $\varphi \in \Delta'$. It follows by the construction of $\mathcal{R}^c$ that $\Delta \xrightarrow{\langle\psi,\varphi\rangle}_c \Delta'$. Thus Δ has a $\langle\psi, \varphi\rangle$-successor. Moreover, $\text{ArrSta}(\Delta, \langle\psi, \varphi\rangle) = \{\Pi \in \mathcal{S}^c_\Gamma \mid \Delta \xrightarrow{\langle\psi,\varphi\rangle}_c \Pi\} = \{\Pi \in \mathcal{S}^c_\Gamma \mid \varphi \in \Pi\}$. Since $\varphi \in \Delta'$, $\{\Pi \in \mathcal{S}^c_\Gamma \mid \varphi \in \Pi\} \neq \emptyset$.

Now we prove that if there exists a plan for achieving φ from ψ, then there exists a one-step plan for achieving φ from ψ.

Proposition 6. *If there is* $\sigma = \langle\psi_1, \varphi_1\rangle \cdots \langle\psi_n, \varphi_n\rangle \in \Sigma^*_\Gamma$ $(n \geq 0)$ *such that for each* ψ*-state* $\Delta \in \mathcal{S}^c_\Gamma$ *and each state* $\Theta \in \text{ArrSta}(\Delta, \sigma)$ *we have* $\varphi \in \Theta$*, then there is* $\langle\psi', \varphi'\rangle \in \Sigma_\Gamma$ *such that for each* ψ*-state* $\Delta \in \mathcal{S}^c_\Gamma$ *and each state* $\Theta \in \text{ArrSta}(\Delta, \langle\psi', \varphi'\rangle)$ *we have* $\varphi \in \Theta$.

Proof. Suppose that there is $\sigma = \langle\psi_1, \varphi_1\rangle \cdots \langle\psi_n, \varphi_n\rangle \in \Sigma^*_\Gamma$ such that for each ψ-state $\Delta \in \mathcal{S}^c_\Gamma$ and each state $\Theta \in \text{ArrSta}(\Delta, \sigma)$ we have $\varphi \in \Theta$. We have that $\top$ is in every state in $\mathcal{S}^c_\Gamma$. By Proposition 3, $\mathcal{U}(\top) \in \Gamma$, i.e., $\mathcal{K}hs(\bot, \bot) \in \Gamma$. Thus $\langle\bot, \bot\rangle \in \Sigma_\Gamma$. If there does not exist a ψ-state in $\mathcal{S}^c_\Gamma$, then $\langle\psi', \varphi'\rangle = \langle\bot, \bot\rangle$ satisfies the conditions. In the following we suppose that there exists some ψ-state in $\mathcal{S}^c_\Gamma$. Next we prove the proposition by induction on the length n of σ. Assume that $\Delta \in \mathcal{S}^c_\Gamma$ is a ψ-state.

[5] Please refer to the proof of Proposition 6 in [16].

$n = 0$. Since $\text{ArrSta}(\Delta, \epsilon) = \text{ArrSta}(\Delta, \langle\bot,\bot\rangle) = \{\Delta\}$, $\langle\psi',\varphi'\rangle = \langle\bot,\bot\rangle$ satisfies the conditions.

$n = 1$. $\langle\psi',\varphi'\rangle = \langle\psi_1,\varphi_1\rangle$ satisfies the conditions.

$n \geq 2$. There are two cases:

Case 1. There exists a ψ-state $\Delta' \in \mathcal{S}^c_\Gamma$ such that $\neg\psi_1 \in \Delta'$. Either $\psi_1 \in \Delta$ or $\neg\psi_1 \in \Delta$, then by the construction of $\mathcal{R}^c$, either $\Delta' \xrightarrow{\langle\psi_1,\varphi_1\rangle}_c \Delta$ or $\Delta \xrightarrow{\langle\psi_1,\varphi_1\rangle}_c \Delta$. It follows that $\text{ArrSta}(\Delta, \langle\psi_2,\varphi_2\rangle\cdots\langle\psi_n,\varphi_n\rangle) \subseteq \text{ArrSta}(\Delta',\sigma)$, or $\text{ArrSta}(\Delta, \langle\psi_2,\varphi_2\rangle\cdots\langle\psi_n,\varphi_n\rangle) \subseteq \text{ArrSta}(\Delta,\sigma)$. Let $\tau_1 = \langle\psi_2,\varphi_2\rangle\cdots\langle\psi_n,\varphi_n\rangle$. Then τ_1 is an action sequence of length $n-1$ such that for each ψ-state $\Delta \in \mathcal{S}^c_\Gamma$ and each state $\Theta \in \text{ArrSta}(\Delta,\tau_1)$ we have $\varphi \in \Theta$.

Case 2. Every ψ-state in $\mathcal{S}^c_\Gamma$ is a ψ_1-state. Then ψ-state Δ is also a ψ_1-state. There are two cases:

Case 2.1. There is a φ_1-state $\Theta' \in \mathcal{S}^c_\Gamma$ such that $\neg\psi_2 \in \Theta'$. By Proposition 5, $\text{ArrSta}(\Delta, \langle\psi_1,\varphi_1\rangle) = \{\Pi \mid \varphi_1 \in \Pi\}$. Analogous to Case 1, $\langle\psi_3,\varphi_3\rangle\cdots\langle\psi_n,\varphi_n\rangle$ is an action sequence such that for each φ_1-state Υ and each state $\Theta \in \text{ArrSta}(\Upsilon, \langle\psi_3,\varphi_3\rangle\cdots\langle\psi_n,\varphi_n\rangle)$ we have $\varphi \in \Theta$. Let $\tau_{2_1} = \langle\psi_1,\varphi_1\rangle\langle\psi_3,\varphi_3\rangle\cdots\langle\psi_n,\varphi_n\rangle$. Thus τ_{2_1} is an action sequence of length $n-1$ such that for each ψ-state $\Delta \in \mathcal{S}^c_\Gamma$ and each state $\Theta \in \text{ArrSta}(\Delta,\tau_{2_1})$ we have $\varphi \in \Theta$.

Case 2.2. Every φ_1-state in $\mathcal{S}^c_\Gamma$ is a ψ_2-state, i.e., $\neg\varphi_1$ or ψ_2 is in each state in $\mathcal{S}^c_\Gamma$. Each state in $\mathcal{S}^c_\Gamma$ is a maximal consistent set, so $\varphi_1 \to \psi_2$ is in each state in $\mathcal{S}^c_\Gamma$. By Proposition 3, $\mathcal{U}(\varphi_1 \to \psi_2) \in \Gamma$. By EMP, $\mathcal{K}hs(\varphi_1,\psi_2) \in \Gamma$. We have that $\mathcal{K}hs(\psi_1,\varphi_1) \in \Gamma$ and $\mathcal{K}hs(\psi_2,\varphi_2) \in \Gamma$. By using COMPKh twice, $\mathcal{K}hs(\psi_1,\varphi_2) \in \Gamma$, i.e., $\langle\psi_1,\varphi_2\rangle \in \Sigma_\Gamma$.

Now we show that $\text{ArrSta}(\Delta, \langle\psi_1,\varphi_1\rangle\langle\psi_2,\varphi_2\rangle) = \text{ArrSta}(\Delta, \langle\psi_1,\varphi_2\rangle)$. By Proposition 5, $\text{ArrSta}(\Delta, \langle\psi_1,\varphi_1\rangle) = \{\Pi \mid \varphi_1 \in \Pi\}$ and $\text{ArrSta}(\Delta, \langle\psi_1,\varphi_2\rangle) = \{\Pi \mid \varphi_2 \in \Pi\}$. For any $\Lambda \in \text{ArrSta}(\Delta, \langle\psi_1,\varphi_1\rangle)$, $\varphi_1 \in \Lambda$, then $\psi_2 \in \Lambda$ since every φ_1-state is a ψ_2-state. Then for any $\Lambda \in \text{ArrSta}(\Delta, \langle\psi_1,\varphi_1\rangle)$, $\text{ArrSta}(\Lambda, \langle\psi_2,\varphi_2\rangle) = \{\Pi \mid \varphi_2 \in \Pi\}$. It follows that $\text{ArrSta}(\Delta, \langle\psi_1,\varphi_1\rangle\langle\psi_2,\varphi_2\rangle) = \{\Pi \mid \varphi_2 \in \Pi\}$. So $\text{ArrSta}(\Delta, \langle\psi_1,\varphi_1\rangle\langle\psi_2,\varphi_2\rangle) = \text{ArrSta}(\Delta, \langle\psi_1,\varphi_2\rangle)$.

Then, $\text{ArrSta}(\Delta, \langle\psi_1,\varphi_1\rangle\cdots\langle\psi_n,\varphi_n\rangle) = \text{ArrSta}(\Delta, \langle\psi_1,\varphi_2\rangle\langle\psi_3,\varphi_3\rangle\cdots\langle\psi_n, \varphi_n\rangle)$. Let $\tau_{2_2} = \langle\psi_1,\varphi_2\rangle\langle\psi_3,\varphi_3\rangle\cdots\langle\psi_n,\varphi_n\rangle$. Then τ_{2_2} is an action sequence of length $n-1$ such that for each ψ-state $\Delta \in \mathcal{S}^c_\Gamma$ and each state $\Theta \in \text{ArrSta}(\Delta,\tau_{2_2})$ we have $\varphi \in \Theta$.

In all cases, there exists $\tau \in \Sigma^*_\Gamma$ of length $n-1$ such that for each ψ-state $\Delta \in \mathcal{S}^c_\Gamma$ and each state $\Theta \in \text{ArrSta}(\Delta,\sigma)$ we have $\varphi \in \Theta$. The induction hypothesis is that if there is $\sigma \in \Sigma^*$ of length $n-1$ such that for each ψ-state $\Delta \in \mathcal{S}^c_\Gamma$ and each state $\Theta \in \text{ArrSta}(\Delta,\sigma)$ we have $\varphi \in \Theta$, then there exists $\langle\psi',\varphi'\rangle \in \Sigma_\Gamma$ such that for each ψ-state $\Delta \in \mathcal{S}^c_\Gamma$ and each state $\Theta \in \text{ArrSta}(\Delta, \langle\psi',\varphi'\rangle)$ we have $\varphi \in \Theta$. It follows by IH that there exists $\langle\psi',\varphi'\rangle \in \Sigma_\Gamma$ such that for each ψ-state $\Delta \in \mathcal{S}^c_\Gamma$ and each state $\Theta \in \text{ArrSta}(\Delta, \langle\psi',\varphi'\rangle)$ we have $\varphi \in \Theta$. The induction is complete.

Now we are ready to prove the truth lemma.

Lemma 1 (Truth Lemma). *For any formula* $\varphi \in \boldsymbol{L_{Khs}}$, $\mathcal{M}^{c}_{\Gamma}, \Delta \models \varphi$ *iff* $\varphi \in \Delta$.

Proof. Boolean cases are trivial. We only focus on the case of $\mathcal{K}hs(\psi, \varphi)$.

For the left to right direction, suppose that $\mathcal{M}^{c}_{\Gamma}, \Delta \models \mathcal{K}hs(\psi, \varphi)$. Then there exists $\sigma \in \Sigma^{*}_{\Gamma}$ such that for each $\mathcal{M}^{c}_{\Gamma}, \Delta' \models \psi$: if $\Delta' \xrightarrow{\sigma}_{s} \Delta''$ then $\mathcal{M}^{c}_{\Gamma}, \Delta'' \models \varphi$. By IH, there exists $\sigma \in \Sigma^{*}_{\Gamma}$ such that for each ψ-state Δ': if $\Delta' \xrightarrow{\sigma}_{s} \Delta''$ then $\varphi \in \Delta''$. By Proposition 6, there exists $\langle \psi', \varphi' \rangle \in \Sigma_{\Gamma}$ such that for each ψ-state Δ': if $\Delta' \xrightarrow{\langle \psi', \varphi' \rangle}_{s} \Delta''$ then $\varphi \in \Delta''$. (▲)

If there is no ψ-state, then by Proposition 3, $\mathcal{U}\neg\psi \in \Delta$, i.e., $\mathcal{K}hs(\psi, \bot) \in \Delta$. Since $\bot \rightarrow \varphi$ is a tautology, $\mathcal{U}(\bot \rightarrow \varphi) \in \Delta$ by NECU. Then by EMP and COMPKh, $\mathcal{K}hs(\psi, \varphi) \in \Delta$. In the following we suppose that there exists some ψ-state. Then there are two cases:

Case 1. There exists a ψ-state Λ such that $\neg\psi' \in \Lambda$. For each ψ-state Δ', either $\psi' \in \Delta'$ or $\neg\psi' \in \Delta'$, it follows that either $\Lambda \xrightarrow{\langle \psi', \varphi' \rangle}_{c} \Delta'$ or $\Delta' \xrightarrow{\langle \psi', \varphi' \rangle}_{c} \Delta'$. So each ψ-state is reached via $\langle \psi', \varphi' \rangle$ from some ψ-state. By (▲), every ψ-state is a φ-state. Similarly to the case 2.2 in the proof of Proposition 6, $\mathcal{K}hs(\psi, \varphi) \in \Gamma$. Thus $\mathcal{K}hs(\psi, \varphi) \in \Delta$.

Case 2. Every ψ-state is a ψ'-state. Similarly, we have that $\mathcal{K}hs(\psi, \psi') \in \Delta$. By Proposition 5, each φ'-state is reached via $\langle \psi', \varphi' \rangle$ from some ψ-state. It follows by (▲) that each φ'-state is a φ-state. Similarly, $\mathcal{K}hs(\varphi', \varphi) \in \Delta$. And we have that $\mathcal{K}hs(\psi', \varphi') \in \Delta$. Then by using COMPKh twice, $\mathcal{K}hs(\psi, \varphi) \in \Delta$.

In both cases, we have that $\mathcal{K}hs(\psi, \varphi) \in \Delta$. Thus the left to right direction holds.

For the other direction, suppose that $\mathcal{K}hs(\psi, \varphi) \in \Delta$, i.e., $\mathcal{K}hs(\psi, \varphi) \in \Gamma$. Then $\langle \psi, \varphi \rangle \in \Sigma_{\Gamma}$. Next we show that $\mathcal{M}^{c}_{\Gamma}, \Delta \models \mathcal{K}hs(\psi, \varphi)$. Suppose $\mathcal{M}^{c}_{\Gamma}, \Delta \models \psi$. By IH, $\psi \in \Delta$. By Proposition 5, $\Delta \xrightarrow{\langle \psi, \varphi \rangle}_{s} \Theta$ iff $\varphi \in \Theta$. By IH, $\varphi \in \Theta$ iff $\mathcal{M}^{c}_{\Gamma}, \Theta \models \varphi$. Thus, $\mathcal{M}^{c}_{\Gamma}, \Delta \models \mathcal{K}hs(\psi, \varphi)$. The right to left direction also holds.

The completeness of SKHS follows immediately.

Theorem 2. *SKHS is strongly complete w.r.t. the class of all models.*

Moreover, SKHS is decidable.

Theorem 3. *If* φ *is satisfiable then it is satisfiable on a finite model. Indeed, it is satisfiable on a finite model containing at most* 2^k *states, where* k *is the number of subformulas of* φ. *It follows that SKHS is decidable.*[6]

[6] We can prove it via the filtration method. Consider the filtrations of canonical models through the subformula closed set generated by φ. The filtration model is indeed a bounded small model for φ. Here we do not show the details of the proof, since the proposition is a corollary of Proposition 8 in [16], for SKHS is same as the proof system in [16] and both systems are sound and complete.

3 Bisimulation

Now we introduce a suitable notion of bisimulation for our logic, and study the model theoretical aspects of our logic.[7]

Here are some notions before we introduce the bisimulation. We write $U \xrightarrow{\sigma}_s V$ whenever $V = \bigcup_{u_i \in U} \text{ArrSta}(u_i, \sigma)$. We write $U \rightarrow_s V$ whenever there is a σ such that $U \xrightarrow{\sigma}_s V$.

Definition 7 ($\mathbf{L_{Khs}}$-equivalence). *Let $\mathcal{M}, w$ and $\mathcal{M}', w'$ be two pointed models. We say $\mathcal{M}, w$ and $\mathcal{M}', w'$ are $\mathbf{L_{Khs}}$-equivalent and write $\mathcal{M}, w \equiv_{\mathbf{L_{Khs}}} \mathcal{M}', w'$, if and only if for all $\varphi \in \mathbf{L_{Khs}}$, $\mathcal{M}, w \vDash \varphi$ iff $\mathcal{M}', w' \vDash \varphi$.*

Definition 8 (Definability). *Let $\mathcal{M} = \langle \mathcal{S}, \Sigma, \mathcal{R}, \mathcal{V} \rangle$ be a model. A set $U \subseteq \mathcal{S}$ is $\mathbf{L_{Khs}}$-definable in $\mathcal{M}$ iff there exists $\varphi \in \mathbf{L_{Khs}}$ such that $U = [\![\varphi]\!]^{\mathcal{M}}$.*

Here we have a notion of bisimulation for $\mathbf{L_{Khs}}$.

Definition 9 (Bisimulation). *Let $\mathcal{M} = \langle \mathcal{S}, \Sigma, \mathcal{R}, \mathcal{V} \rangle$ and $\mathcal{M}' = \langle \mathcal{S}', \Sigma', \mathcal{R}', \mathcal{V}' \rangle$ be two models. A non-empty relation $Z \subseteq \mathcal{S} \times \mathcal{S}'$ is called an $\mathbf{L_{Khs}}$-bisimulation between $\mathcal{M}$ and $\mathcal{M}'$ iff wZw' implies:*
Atom: $\mathcal{V}(w) = \mathcal{V}(w')$.
Khs-**Zig**: *for any $\mathbf{L_{Khs}}$-definable $U \subseteq \mathcal{S}$, if $U \rightarrow_s V$ for some $V \subseteq \mathcal{S}$, then there is $V' \subseteq \mathcal{S}'$ such that (i) $Z[U] \rightarrow_s V'$ and (ii) for each $v' \in V'$ there is $v \in V$ such that vZv'.*
Khs-**Zag**: *for any $\mathbf{L_{Khs}}$-definable $U' \subseteq \mathcal{S}'$, if $U' \rightarrow_s V'$ for some $V' \subseteq \mathcal{S}'$, then there is $V \subseteq \mathcal{S}$ such that (i) $Z^{-1}[U'] \rightarrow_s V$ and (ii) for each $v \in V$ there is $v' \in V'$ such that vZv'.*
A-Zig: *for all v in $\mathcal{S}$ there is v' in $\mathcal{S}'$ such that vZv'.*
A-Zag: *for all v' in $\mathcal{S}'$ there is v in $\mathcal{S}$ such that vZv'.*
where $Z[U] = \{w' \mid wZw'$ for some $w \in U\}$ and $Z^{-1}[U'] = \{w \mid wZw'$ for some $w' \in U'\}$. We write $\mathcal{M}, w \leftrightarroweq_{\mathbf{L_{Khs}}} \mathcal{M}', w'$ when there is an $\mathbf{L_{Khs}}$-bisimulation Z between $\mathcal{M}$ and $\mathcal{M}'$ such that wZw'.

Now we show that $\mathbf{L_{Khs}}$-bisimularity implies $\mathbf{L_{Khs}}$-equivalence.

Theorem 4. *Let $\mathcal{M} = \langle \mathcal{S}, \Sigma, \mathcal{R}, \mathcal{V} \rangle$ and $\mathcal{M}' = \langle \mathcal{S}', \Sigma', \mathcal{R}', \mathcal{V}' \rangle$ be two models, $w \in \mathcal{S}$ and $w' \in \mathcal{S}'$. If $\mathcal{M}, w \leftrightarroweq_{\mathbf{L_{Khs}}} \mathcal{M}', w'$, then $\mathcal{M}, w \equiv_{\mathbf{L_{Khs}}} \mathcal{M}', w'$.*

Proof. Suppose that $\mathcal{M}, w \leftrightarroweq_{\mathbf{L_{Khs}}} \mathcal{M}', w'$, i.e., there is an $\mathbf{L_{Khs}}$-bisimulation $Z \subseteq (\mathcal{S} \times \mathcal{S}')$ such that wZw'. The proof is by structural induction on $\mathbf{L_{Khs}}$-formulas. Boolean cases are trivial. We only prove the case for $\mathcal{K}hs(\psi, \varphi)$.

Suppose that $\mathcal{M}, w \vDash \mathcal{K}hs(\psi, \varphi)$. Then there exists $\sigma \in \Sigma^*$ such that $[\![\psi]\!]^{\mathcal{M}} \xrightarrow{\sigma}_s V$ and $V \subseteq [\![\varphi]\!]^{\mathcal{M}}$. We first claim that $Z[[\![\psi]\!]^{\mathcal{M}}] = [\![\psi]\!]^{\mathcal{M}'}$.
($\subseteq$) Let $v' \in Z[[\![\psi]\!]^{\mathcal{M}}]$. Then there exists $v \in [\![\psi]\!]^{\mathcal{M}}$ such that vZv'. By IH, $v' \in [\![\psi]\!]^{\mathcal{M}'}$.

[7] This section borrows ideas from [4].

($\supseteq$) Let $v' \in [\![\psi]\!]^{\mathcal{M}'}$. By A-Zag there is $v \in \mathcal{S}$ such that vZv'. By IH, $v \in [\![\psi]\!]^{\mathcal{M}}$. Then, $v' \in Z[[\![\psi]\!]^{\mathcal{M}}]$.

Now we show that $\mathcal{M}', w' \vDash \mathcal{K}hs(\psi, \varphi)$. From Khs-Zig and the fact that $[\![\psi]\!]^{\mathcal{M}} \rightarrow_s V$, there is $V' \subseteq \mathcal{S}'$ such that (i) $Z[[\![\psi]\!]^{\mathcal{M}}] \rightarrow_s V'$ (i.e., $[\![\psi]\!]^{\mathcal{M}'} \rightarrow_s V'$), and (ii) for every $v' \in V'$ there is $v \in V$ such that vZv'. From (ii) and the fact that $V \subseteq [\![\varphi]\!]^{\mathcal{M}}$, for every $v' \in V'$ there is $v \in [\![\varphi]\!]^{\mathcal{M}}$ such that vZv'. By IH, $v' \in [\![\varphi]\!]^{\mathcal{M}'}$. Thus, $[\![\psi]\!]^{\mathcal{M}'} \rightarrow_s V'$ and $V' \subseteq [\![\varphi]\!]^{\mathcal{M}'}$. Hence, $\mathcal{M}', w' \vDash \mathcal{K}hs(\psi, \varphi)$.

For the other direction use Khs-Zag and A-Zig.

Now we prove the other direction. Here we focus on finite models rather than image-finite models. This is because the global modality is definable in $\mathbf{L_{Khs}}$, and thus a finite domain is required in order to ensure the image-finiteness property.

Theorem 5. *Let $\mathcal{M} = \langle \mathcal{S}, \Sigma, \mathcal{R}, \mathcal{V} \rangle$ and $\mathcal{M}' = \langle \mathcal{S}', \Sigma', \mathcal{R}', \mathcal{V}' \rangle$ be two finite models, $w \in \mathcal{S}$ and $w' \in \mathcal{S}'$. If $\mathcal{M}, w \equiv_{\mathrm{L_{Khs}}} \mathcal{M}', w'$, then $\mathcal{M}, w \leftrightarrows_{\mathrm{L_{Khs}}} \mathcal{M}', w'$.*

Proof. Let $Z = \{(v, v') \in (\mathcal{S} \times \mathcal{S}') \mid \mathcal{M}, v \equiv_{\mathrm{L_{Khs}}} \mathcal{M}', v'\}$. Next we show that Z is an $\mathbf{L_{Khs}}$-bisimulation. Let (w, w') be an element in Z.
(Atom): It is trivial.
(A-Zig): Let v be an element in $\mathcal{S}$. Towards a contradiction, suppose that there is no $v' \in V'$ such that vZv'. Let $\mathcal{S}' = \{v_1', \cdots, v_n'\}$, $n \in \mathbb{N}$. Then, for each $v_i' \in \mathcal{S}'$ there is a formula θ_i such that $\mathcal{M}, v \vDash \theta_i$ but $\mathcal{M}, v_i' \nvDash \theta_i$. Let $\theta = \theta_1 \wedge \cdots \wedge \theta_n$. Then $\mathcal{M}, v \vDash \theta$ but $\mathcal{M}, v_i' \nvDash \theta$ for each $v_i' \in \mathcal{S}'$. It follows that $\mathcal{M}, w \vDash \neg\mathcal{U}\neg\theta$ but $\mathcal{M}, w \nvDash \neg\mathcal{U}\neg\theta$, contradicting wZw'.
(A-Zag): Analogous to the A-Zig case.
(Khs-Zig): Suppose that $\psi \in \mathbf{L_{Khs}}$ and $[\![\psi]\!]^{\mathcal{M}} \rightarrow_s V$ for some $V \subseteq \mathcal{S}$. It suffices to find a $V' \subseteq \mathcal{S}'$ such that $Z[[\![\psi]\!]^{\mathcal{M}}] \rightarrow_s V'$ and for each $v' \in V'$ there is $v \in V$ such that vZv'. We first prove that $Z[[\![\psi]\!]^{\mathcal{M}}] = [\![\psi]\!]^{\mathcal{M}'}$.
($\subseteq$) Let $v' \in Z[[\![\psi]\!]^{\mathcal{M}}]$. Then there exists $v \in [\![\psi]\!]^{\mathcal{M}}$ such that vZv'. By Z's definition, $v' \in [\![\psi]\!]^{\mathcal{M}'}$.
($\supseteq$) Let $v' \in [\![\psi]\!]^{\mathcal{M}'}$. By A-Zag there exists $v \in \mathcal{S}$ such that vZv'. By Z's definition, $v \in [\![\psi]\!]^{\mathcal{M}}$. Then, $v' \in Z[[\![\psi]\!]^{\mathcal{M}}]$.

Then we just need to find an appropriate V' for $[\![\psi]\!]^{\mathcal{M}'}$. Note that if $[\![\psi]\!]^{\mathcal{M}}$ is empty, then $[\![\psi]\!]^{\mathcal{M}'} = Z[[\![\psi]\!]^{\mathcal{M}}]$ is empty too and we can just let $V' = \emptyset$. In the following we assume that $[\![\psi]\!]^{\mathcal{M}} \neq \emptyset$. Then $[\![\psi]\!]^{\mathcal{M}'} = Z[[\![\psi]\!]^{\mathcal{M}}] \neq \emptyset$ by A-Zig. Thus $V \neq \emptyset$ and $V' \neq \emptyset$ by the definition of arrival states. Towards a contradiction, suppose that for each $V' \subseteq \mathcal{S}'$ such that $[\![\psi]\!]^{\mathcal{M}'} \rightarrow_s V'$, there is $v_{V'} \in V'$ such that there is no $v \in V$ such that $vZv_{V'}$. Then for each $v \in V$ we have a formula $\varphi^v_{V'}$ such that $\mathcal{M}, v \vDash \varphi^v_{V'}$ but $\mathcal{M}', v_{V'} \nvDash \varphi^v_{V'}$. Since the models are finite, we can define $\theta_{V'} = \bigvee_{v \in V} \varphi^v_{V'}$ and $\theta = \bigwedge_{\{V' \mid [\![\psi]\!]^{\mathcal{M}'} \rightarrow_s V'\}} \theta_{V'}$. Then $\mathcal{M}, v \vDash \theta$ for all $v \in V$, but there is $v_{V'} \in V'$ such that $\mathcal{M}', v_{V'} \nvDash \theta$ for each V' such that $[\![\psi]\!]^{\mathcal{M}'} \rightarrow_s V'$. Since $\mathcal{K}hs$-formulas are global, $\mathcal{M}, w \vDash \mathcal{K}hs(\psi, \theta)$ but

$\mathcal{M}', w' \nvDash \mathcal{K}hs(\psi, \theta)$. Contradiction.
(Khs-Zag): Analogous to the Khs-Zig case.

4 Conclusion

In this paper, we interpret the knowing how formula $\mathcal{K}hs(\psi, \varphi)$ as that the agent has a plan for achieving φ given ψ. The plan for achieving φ given ψ is a finite linear action sequence such that the execution of the action sequence at each ψ-state will end up with some φ-state. We do not demand that every action is executable. Instead, if one action is not executable at the current state, then skip it and continue to the next action. We give a sound and complete axiomatic system $\mathbb{SKHS}$ which is same as the system in [16]. Borrowing the idea from the completeness proof in [16], we revise the transitions of the canonical model, which makes that there is a one-step plan if you know how to achieve φ given ψ. Actually every action is executable at every state in our canonical models, therefore our canonical models also apply to the completeness proof in [16], but not vice versa. Moreover, $\mathbb{SKHS}$ is decidable. We also study the model theoretical aspects of our logic by introducing a suitable notion of bisimulation for the logic. We prove that bisimularity implies modal equivalence, and over finite models, modal equivalence implies bisimulation.

As we point out, $\mathbb{SKHS}$ is same as the proof system in [16]. Different interpretations of knowing-how lead to the same logic, which shows that this axiomatization of knowing-how just illustrate our daily understanding of knowing how. For future research, we can explore other semantics of the knowing-how operator which result in the same logic as ours.

An important next step is to consider the multi-agent version of knowing how. Distributed knowing-how based on single-step plans has been studied (see [12]). We can consider distributed knowledge-how based on multistep plans, where a group of agents may achieve a lot more together. For example, if you know how to get from A to B and I know how to get from B to C, then we together could know how to get from A to C.

Acknowledgments. The author thanks Yanjing Wang for giving the author the ideas of the skippable plan and the bisimulation, and his helpful comments to make the paper more readable. The author thanks the three anonymous reviewers for their insightful comments on the early version of the paper.

References

1. Alur, R., Henzinger, T.A., Kupferman, O.: Alternating-time temporal logic. J. ACM **49**(5), 672–713 (2002). https://doi.org/10.1145/585265.585270
2. Ågotnes, T., Goranko, V., Jamroga, W., Wooldrige, M.: Knowledge and ability. In: van Ditmarsch, H., Halperen, J., van der Hoek, H., Kooi, B. (eds.) Handbook of Epistemic Logic, pp. 543–589. College Publications, London (2015)
3. Fervari, R., Herzig, A., Li, Y., Wang, Y.: Strategically knowing how. In: Proceedings of IJCAI, pp. 1031–1038 (2017). https://doi.org/10.24963/ijcai.2017/143

4. Fervari, R., Velázquez-Quesada, F.R., Wang, Y.: Bisimulation for knowing how logics. Manuscript, Presented at SR 2017 (2017)
5. Gochet, P.: An open question in the logic of knowing how. In: Open Problems in Epistemlogy. The Philosophical Society of Finland (2013)
6. Li, Y.: Stopping means achieving: a wearker logic of knowing how. Stud. Log. **9**(4), 34–54 (2016)
7. Li, Y.: Knowing what to do: a logical approach to planning and knowing how. Ph.D. thesis, University of Groningen (2017)
8. Li, Y., Wang, Y.: Achieving while maintaining: a logic of knowing how with intermediate constraints. In: Ghosh, S., Prasad, S. (eds.) ICLA 2017. LNCS, vol. 10119, pp. 154–167. Springer, Heidelberg (2017). https://doi.org/10.1007/978-3-662-54069-5_12
9. McCarthy, J., Hayes, J.P.: Some philosophical problems from the standpoint of artificial intelligence. In: Meltzer, B., Michie, D. (eds.) Machine Intelligence, pp. 463–502. Edinburgh University Press, Edinburgh (1969)
10. McCarthy, J.: First-order theories of individual concepts and propositions. Mach. Intell. **9**, 129–147 (1979)
11. Moore, R.C.: A formal theory of knowledge and action. In: Hobbs, J.R., Moore, R.C. (eds.) Formal Theories of the Commonsense World. Ablex Publishing Corporation, Norwood (1985)
12. Naumov, P., Tao, J.: Together we know how to achieve: an epistemic logic of know-how. Artif. Intell. **262**, 279–300 (2018). https://doi.org/10.1016/j.artint.2018.06.007
13. Pauly, M.: A modal logic for coalitional power in games. J. Log. Comput. **12**(1), 149–166 (2002). https://doi.org/10.1093/logcom/12.1.149
14. Raymond, R.: Knowledge in Action. The MIT Press, Cambridge (2001)
15. Wang, Y.: A logic of knowing how. In: van der Hoek, W., Holliday, W.H., Wang, W. (eds.) LORI 2015. LNCS, vol. 9394, pp. 392–405. Springer, Heidelberg (2015). https://doi.org/10.1007/978-3-662-48561-3_32
16. Wang, Y.: A logic of goal-directed knowing how. Synthese **195**(10), 4419–4439 (2018). https://doi.org/10.1007/s11229-016-1272-0
17. Wang, Y.: Beyond knowing that: a new generation of epistemic logics. In: van Ditmarsch, H., Sandu, G. (eds.) Jaakko Hintikka on Knowledge and Game-Theoretical Semantics. OCL, vol. 12, pp. 499–533. Springer, Cham (2018). https://doi.org/10.1007/978-3-319-62864-6_21

A Dynamic Hybrid Logic for Followership

Zuojun Xiong(✉) and Meiyun Guo

Institute of Logic and Intelligence, Southwest University, Chongqing, China
{zuojunxiong,guomy007}@swu.edu.cn

Abstract. This paper formalises the followership in networks that agents following or unfollowing each other dynamically. The semantics is based on the basic hybrid logic and we extend the logic with a propositional action modality $[a \uparrow \theta]$ for the changes of followership. The main contribution of this paper is the completeness result. Moreover, we have proved that all pure axiomatic extensions have completeness and discussed some possible future works and extensions, in particular, some features of the extended action modality $[a \uparrow \varphi]$, like *repetition regrets* are discussed.

1 Introduction

Logics in social networks get more and more attentions in recent years [5,11–13,18]. In [18,19], they characterised the action of messaging in networks by followership in standard modal logic. This paper focuses on actions of changing followership by propositions. In other words, we assume that agents could choose to follow a particular group of agents. In social networks, this can be understood as agents looking for some particular agents, i.e., searching on Twitter, Weibo, Facebook etc.[1] Like most dynamic logics [7,14,17], we use the notation $[a \uparrow \theta]$ standing for "agent a is following all the θ-agents". For "θ-agents", we mean that agents have the property of θ. The logic extends the basic hybrid logic with an action operator $[a \uparrow \theta]$ for changing followership, the completeness of the logic and pure axiomatic extensions have been contributed.

The logic can be categorized as a modal logic for *binary relation changes*. A feature of dynamic operators in the logic is keeping *states* stable but changing a particular group of *relations*. There are some literature contributed to that areas. In relation-changing modal logic [1], it introduced several different modalities that are able to *delete*, *add* or *swap* an edge between a pair of states. That logic is able to interpret some relation-changing games called *sabotage games* [3,15]. A difference with relation-changing modal logic is that we only have one kind of modality for the interpretation of adding, deleting, and swapping relations. Also, based on hybrid logic and the semantic of "states as agents", relations of agents are syntactically presented in a simple way.

[1] Our action here is a little bit in difference, agents will follow and only all the agents satisfied by the claimed property. Action itself does not preserving the old relations.

P. Blackburn et al. (Eds.): LORI 2019, LNCS 11813, pp. 425–439, 2019.
https://doi.org/10.1007/978-3-662-60292-8_31

In [2], they introduced some data structure modifiers for the relation changing, it can be used for reasoning about changes in graph. They have discussed two different actions for changing relations, $[a + (\varphi, \psi)]$ and $[a - (\varphi, \psi)]$. $[a + (\varphi, \psi)]$ means adding an accessible arrow from φ-states to ψ-states for agent a, and $[a - (\varphi, \psi)]$ for removing. Semantically, our logic are different, we simply add arrows from *agent* a to all ψ-agents, and remove arrows from *agent* a to all $\neg\psi$-agents by action $[a \uparrow \psi]$. It is easier and simpler for the interpretation of relation changes, if we only want to focus on the structural dynamics among a group of agents.

Moreover, extensions on the logic of relation changes have been studied in literature like logics of *public assignments* (see, e.g., [6,10]), of *public announcements* (see, e.g., [8]), and of *preference modification* [16]. In particular, *arrow update logic* (see, e.g., [9,10]) focuses on the study of epistemic access elimination, and can be used for the reasoning of multi-agent belief changes. The logic allows a statement eliminating the access to all epistemic possibilities in which the statement does not satisfied. Technically, it's similar as the logic we introduced in this paper, but arrow update logic is an extension of the basic epistemic logic and our logic is an extension of hybrid modal logic. The main difference between arrow updates and the action of followership is that arrow updates remove edges according to a pre and a post-condition, and action of followership can remove arbitrary arrows in named models (with hybrid modal logic semantics).

Finally, not like the logics we mentioned above, our logic is based on hybrid modal logic with a simple extension of modality. Dynamic modalities in our logic are not reducible, and there is no reduction axioms for dynamic operators.

As this is a preliminary logic for the action of followership. We give our actions of following a number of idealising assumptions:

Universality	Executing an action is universally no difference.
Privacy	The outcome of an action only influents the actor.
Invariance	Actions does not changing any propositional sentences.
Sincerity	After an action, the actor will follow and only follow those agents satisfied by the claimed propositions.

These assumptions allows us to focus on the core concept of network structure dynamics, and will be taken as axiom schemata in Subsect. 3.1.

The structure of the paper goes as follows. We will give the language and semantics of the logic of followership in the coming section, and discuss some properties of our semantics. Then we offer the axiomatisation and discuss related issues, the completeness proof will be offered in Sect. 3. Finally, we conclude in Sect. 4 and elaborate on some property of extensions, like *repetition regrets.*

2 Language and Semantics

Fix `Agt` to be a non-empty names set of agents, called set of *nominals*, and Let `Prop` be a non-empty set of *propositional letters*. The language of Dynamic

Hybrid Logic for Followership (DHLF) is defined as follows.

$$\theta ::= a \mid p \mid \neg\theta \mid (\theta \wedge \theta)$$

$$\varphi ::= \theta \mid \neg\varphi \mid (\varphi \wedge \varphi) \mid \Diamond\varphi \mid @_a\varphi \mid [a \uparrow \theta]\varphi$$

where $a \in$Agt, $p \in$Prop.

For brevity, PROP stands for the propositional language of generated by θ.[2] θ will stand for propositional formulas in the paper by default.

We define a *Network Model* as follows.

Definition 1 (Network Model). *A network model $M = (W, R, G, V)$ is a tuple where*

- *W is a non-empty set of agents;*
- *$R \subseteq W \times W$ is a following relation;*
- *G : Agt $\rightarrow W$ assigns each nominal (name) an element of W, called a name assignment;*
- *V : Prop $\rightarrow \wp(W)$ assigns each propositional letter a subset of W, called a valuation.*

The pair $F = (W, R)$ is called a Kripke frame, the network model M is said to be based on the Kripke frame F.

For brevity, we will say *model* and *frame* instead of *network model* and *network frame* respectively when no confusion may occur, and similar for *models* and *frames*. We will say *nominal(s)* instead of *agent's name(s)*, and *agent* if we do not need to mention its name.

Since this is a logic about followership, it is allowed that agent can simply follow everyone in a network. This is to say that $[a \uparrow \theta]$ means 'a chooses to only follow all the agents satisfying by θ'. For formal details, we refer to the following definition.

Definition 2 (Interpretation). *Let $M = (W, R, G, V)$ be a network model. For any $\varphi \in$DHLF, the relation $M, w \models \varphi$ is defined by induction, where $w \in W$.*

$$\begin{array}{lll}
M, w \models a & \textit{iff} & G(a) = w \\
M, w \models p & \textit{iff} & w \in V(p) \\
M, w \models \neg\varphi & \textit{iff} & M, w \not\models \varphi \\
M, w \models (\psi_1 \wedge \psi_2) & \textit{iff} & M, w \models \psi_1 \textit{ and } M, w \models \psi_2 \\
M, w \models \Diamond\psi & \textit{iff} & \textit{there exists } u \in W, wRu \textit{ and } M, u \models \psi \\
M, w \models @_a\psi & \textit{iff} & M, G(a) \models \psi \\
M, w \models [a \uparrow \theta]\psi & \textit{iff} & [\![a \uparrow \theta]\!]M, w \models \psi
\end{array}$$

where $[\![a \uparrow \theta]\!]M = (W, R^{[\frac{a}{\theta}]}, G, V)$ is a network model with $R^{[\frac{a}{\theta}]}$ defined as follows.

[2] In the later section, when we talk about the extended languages, the set of Agt will then be substituted by the extended set as well.

– *For any* $w, u \in W$, $wR^{[^a_\theta]}u$ *iff* $\begin{cases} wRu & \text{if } G(a) \neq w \\ M, u \models \theta & \text{otherwise} \end{cases}$

The dual operators $\Box := \neg\Diamond\neg$ and common boolean operators $\vee, \rightarrow, \leftrightarrow$ are interpreted as usual. Also, $\top$ is indicating a *tautology* and $\bot := \neg\top$ is for a *contradiction.* Moreover, we have the usual inductive definition over $\models$ by convention.

Observing Definition 2, we have a general inductive definition for a sequence of action modalities. Before we present the details, here we introduce some shortenings.

Definition 3 (Sequences). *Let* $a_i \in$Agt *and* $\theta_i \in$PROP *for any* $i \in \mathbb{N}$. $\overrightarrow{a}$ *be a finite sequence of* $a_1\theta_1, a_2\theta_2, \ldots, a_n\theta_n$. *We write the converse of* $\overrightarrow{a}$ *as* $\overleftarrow{a}$: $a_n\theta_n, \ldots, a_2\theta_2, a_1\theta_1$. *A finite sequence of actions modalities* $[\overrightarrow{a}]$ [3] *is defined by* $[a_1 \uparrow \theta_1][a_2 \uparrow \theta_2] \ldots [a_n \uparrow \theta_n]$. $[\![\overrightarrow{a}]\!] := [^{a_1, \ldots, a_n}_{\theta_1, \ldots, \theta_n}]$ *and* $[\![\overleftarrow{a}]\!] := [^{a_n, \ldots, a_1}_{\theta_n, \cdots, \theta_1}]$ *indicate a sequence of action updates.*

Now, we will just say "sequence(s)" instead of "finite sequence(s)", and $\overrightarrow{a}$ stands for arbitrary well defined sequence in general. The following is the generalisation of model updating by any sequences.

Proposition 1. *Let* $M = (W, R, G, V)$ *be a network model and* $w \in W$ *be a state. We have a general semantics for any finite sequence of action modalities.*

$$M, w \models [\overrightarrow{a}]\psi \quad \textit{iff} \quad [\![\overleftarrow{a}]\!]M, w \models \psi$$

where network model $[\![\overleftarrow{a}]\!]M = (W, R^{[\![\overleftarrow{a}]\!]}, G, V)$, *and* $R^{[\![\overleftarrow{a}]\!]}$ *is defined as follows.*

– *For any* $w, u \in W$ *and the sequence* $\overleftarrow{a} = a_n\theta_n, \overleftarrow{a'}$,

$$wR^{[\![\overleftarrow{a}]\!]}u \textit{ iff } \begin{cases} wR^{[\![\overleftarrow{a'}]\!]}u & \text{if } G(a_n) \neq w \\ M, u \models \theta_n & \text{otherwise} \end{cases}$$

Proof. Induction on the length of $\overrightarrow{a}$, and using Definition 2.

Before we move to the next section, we elaborate on some features of our action operator here. Actions are not partial: $\not\models [a \uparrow \theta]\Diamond\varphi \rightarrow [a \uparrow \theta]\Box\varphi$. Let $\theta = p \vee (b \wedge \neg p)$ and $\varphi = p$, we can define a model such that $M, w \models [a \uparrow \theta]\Diamond\varphi$ but $M, w \not\models [a \uparrow \theta]\Box\varphi$. That means the operator can interpret *exceptions* of followership. Even our action operator is *memoryless,*[4] we can use *disjunct-forms* to preserve *historical actions.*

Moreover, we can define some *agents' abilities* in social networks. For example, pure formula like $[i \uparrow k]@_i\Diamond\Diamond j$ reflects *"the updated distance of followership between i and j is at most in 2"*; $(@_j\neg\Diamond k \wedge @_k\Diamond j) \rightarrow [k \uparrow \neg j]@_k\neg\Diamond j$ means *"the reason of why unfollowing, as they are unfollowers"*, etc. We will return for more discussion in Sect. 4.

3 The converse of $[\overrightarrow{a}]$ is $[\overleftarrow{a}]$ which is defined by $[a_n \uparrow \theta_n] \ldots [a_2 \uparrow \theta_2][a_1 \uparrow \theta_1]$.

4 It means that agents are dropping p-property, if they update with q-property.

3 Axiomatisation

The axiomatisation is a conservative extension on the hybrid logic as in [4, Section 7.3]. We will first offer an axiomatisation of our logic and then discuss the issues of *names* in Subsect. 3.1. By offering a *named model* in Subsect. 3.2, the completeness results are proved in Subsect. 3.3.

Let $\varphi \in \texttt{Lan}$ be a formula and $\texttt{L}$ be an axiomatisation. φ is provable in $\texttt{L}$, notated $\vdash_{\texttt{L}} \varphi$, if there exists a finite sequence $\psi_0, \ldots, \psi_n \in \texttt{Lan}$, such that $\psi_0 \wedge \ldots \wedge \psi_n$ and $(\psi_0 \wedge \ldots \wedge \psi_n) \rightarrow \varphi$ are provable in $\texttt{L}$. Let $\Gamma \subseteq \texttt{Lan}$. Γ is consistent in $\texttt{L}$ if $\Gamma \nvdash_{\texttt{L}} \perp$: for any finite sequence, $\chi_0, \ldots, \chi_n \in \Gamma$, we have no $\vdash_{\texttt{L}} (\chi_0 \wedge \ldots \wedge \chi_n) \rightarrow \perp$, written $\nvdash_{\texttt{L}} (\chi_0 \wedge \ldots \wedge \chi_n) \rightarrow \perp$; Γ is inconsistent in $\texttt{L}$ if Γ is not consistent in $\texttt{L}$.

For brevity, when the axiomatisation $\texttt{L}$ is fixed, we will just write $\vdash \varphi$ instead of $\vdash_{\texttt{L}} \varphi$. Furthermore, we say that φ is provable instead of saying that φ is provable in $\texttt{L}$. The set of axiom schemata for hybrid logic in DHLF is defined as in Fig. 1, and the set of inference rules is given in Definition 4.

PROP	all propositional tautology instances	K	$\Box(\varphi \rightarrow \psi) \rightarrow (\Box\varphi \rightarrow \Box\psi)$
K@	$@_a(\varphi \rightarrow \psi) \rightarrow (@_a\varphi \rightarrow @_a\psi)$	sdual	$@_a\varphi \leftrightarrow \neg @_a \neg\varphi$
intro	$(a \wedge \varphi) \rightarrow @_a\varphi$	nom	$(@_a b \wedge @_b\varphi) \rightarrow @_a\varphi$
refx	$@_a a$	symc	$@_a b \leftrightarrow @_b a$
agree	$@_a @_b\varphi \leftrightarrow @_b\varphi$	back	$\Diamond @_a\varphi \rightarrow @_a\varphi$

Fig. 1. Axiom schemata for hybrid logic, where $a, b \in$Agt, $\varphi, \psi \in$DHLF.

[]K	$[a\uparrow\theta](\psi \rightarrow \varphi) \rightarrow ([a\uparrow\theta]\psi \rightarrow [a\uparrow\theta]\varphi)$	[]dual	$[a\uparrow\theta]\psi \leftrightarrow \neg[a\uparrow\theta]\neg\psi$
[]dia	$(\neg a \wedge [a\uparrow\theta]\Diamond\psi) \leftrightarrow (\neg a \wedge \Diamond[a\uparrow\theta]\psi)$	[]ignr	$[a\uparrow\theta]\chi \leftrightarrow \chi$
[]@	$[a\uparrow\theta]@_b\varphi \leftrightarrow @_b[a\uparrow\theta]\varphi$	[]sinc	$@_b\theta \leftrightarrow @_a[a\uparrow\theta]\Diamond b$

Fig. 2. Axioms schemata for actions, where $a, b \in$Agt, $\theta, \chi \in$PROP, $\varphi, \psi \in$DHLF.

Definition 4 (Inference rules). *For any $\varphi \in$DHLF, $\theta \in$PROP, and $a, b \in$Agt, we have*

- MP*: if φ and $\varphi \rightarrow \psi$ are both provable, then so is ψ.*
- Nec*: if φ is provable, then so is $\Box\varphi$.*
- Nec@*: if φ is provable, then so is $@_a\varphi$.*
- Nec[]*: if φ is provable, then so is $[a\uparrow\theta]\varphi$.*

3.1 States Without Names

In this subsection, we offer the interpretation and soundness proofs of axiomatisation, and discuss the problems of "canonical model" in our logic like the "unique nominal property" and the "sufficiency of maximal consistent sets".

Definition 5 (DHLF). *The logic* **DHLF** *is the smallest set containing all axiom schemata in Figs. 1 and 2, and inference rules in Definition 4. Let* $\Gamma \subseteq \mathtt{DHLF}$ *and* Γ *is maximal consistent in* **DHLF**, *written* `mcs` Γ, *which means:*

- Γ *is consistent in* **DHLF**: $\Gamma \nvdash \bot$;
- Γ *is maximal: for any* $\Gamma' \subseteq \mathtt{DHLF}$, *if* $\Gamma \subset \Gamma'$ *then* $\Gamma' \vdash \bot$.

Observing that Γ is maximal can also be interpreted as 'for any well-formed formula φ, either $\varphi \in \Gamma$ or $\neg\varphi \in \Gamma$'.

Proposition 2. *The following are some derivable formulas.*

`[]or`	$[a \uparrow p](q \vee r) \leftrightarrow ([a \uparrow p]q \vee [a \uparrow p]r)$	`[]falsum`	$[a \uparrow p]\bot \leftrightarrow \bot$
`[]and`	$[a \uparrow p](q \wedge r) \leftrightarrow ([a \uparrow p]q \wedge [a \uparrow p]r)$	`bridge`	$(\Diamond a \wedge @_a p) \to \Diamond p$

Proof. The proof of theorems with action operators can be showed simply from the semantics, we show `bridge` here only. We show the form $\vdash (\Diamond a \wedge \Box\psi) \to @_a\psi$ only using Fig. 1. By `PROP`, $\vdash (\Diamond a \wedge \Box\psi) \to \Diamond(a \wedge \psi)$. Apply `PROP` on `intro`, $\vdash \Diamond(a \wedge \psi) \to \Diamond @_a\psi$, we then have $\vdash \Diamond(a \wedge \psi) \to @_a\psi$ with `back` and `PROP`. With $\vdash (\Diamond a \wedge \Box\psi) \to \Diamond(a \wedge \psi)$, we have $\vdash (\Diamond a \wedge \Box\psi) \to @_a\psi$, the derivation of `bridge` then is followed.

Proposition 3 (Soundness). *All axiom schemata are valid, and rules preserve validity in* **DHLF**.

Proof. We show the soundness of `[]dia` and `[]sinc` the rest are omitted. Let M, w be an arbitrary pointed model. For `[]dia`: $M, w \models \neg a \wedge [a \uparrow \theta]\Diamond\psi$; iff $G(a) \neq w$ and $[\![a \uparrow \theta]\!]M, w \models \Diamond\psi$; iff $G(a) \neq w$ and for some $v \in W$, $wR^{[\frac{a}{\theta}]}v$ and $[\![a \uparrow \theta]\!]M, v \models \psi$; iff $G(a) \neq w$ and for some $v \in W$, wRv and $M, v \models [a \uparrow \theta]\psi$; iff $M, w \models \neg a \wedge \Diamond[a \uparrow \theta]\psi$.

For `[]sinc`: We prove it by two directions: ($\Rightarrow$) Let $M, w \models @_b\theta$. (1) There is a $v \in W$ such that $G(b) = v$ and $M, v \models \theta$. Also $M, w \models @_a @_b\theta$ by `agree`, and $M, u \models @_b\theta$ for some $G(a) = u$ by semantics. Then we have $uR^{[\frac{a}{\theta}]}v$, and $[\![a \uparrow \theta]\!]M, u \models \Diamond b$ from (1). That's $M, u \models [a \uparrow \theta]\Diamond b$ and $M, w \models @_a[a \uparrow \theta]\Diamond b$ since $G(a) = u$. ($\Leftarrow$) Let $M, w \models @_a[a \uparrow \theta]\Diamond b$, then $M, u \models [a \uparrow \theta]\Diamond b$ for some $G(a) = u$, and $[\![a \uparrow \theta]\!]M, u \models \Diamond b$. There exists $v \in W$, such that $uR^{[\frac{a}{\theta}]}v$ and $[\![a \uparrow \theta]\!]M, v \models b$. We then have $uR^{[\frac{a}{\theta}]}v$ and $G(b) = v$ by semantics. Since $G(a) = u$, then by semantics, we have $M, v \models \theta$ from $uR^{[\frac{a}{\theta}]}v$. That's $M, v \models b \wedge \theta$ since $G(b) = v$. Then by `intro`, $M, v \models @_b\theta$, and $M, w \models @_b\theta$ by semantics.

Recall axiom schemata in Fig. 2 and idealising assumptions of the introduction, we find that assumptions are turned into axiom schemata. Like `[]sinc` stands for *Sincerity*,[5] $[]@$ stands for *Universality*, $[]\mathtt{dia}$ stands for *Privacy*, $[]\mathtt{ignr}$ stands for *Invariance*. We will continue the discussion of semantics in Sect. 4.

Now we are moving for elaborating on the properties of maximal consistent sets in our logic. For brevity, the following definition introduced some necessary terminology.

Definition 6. *Let Φ be a set of formulas, $a \in$Agt, $\theta \in$PROP and $\varphi \in$DHLF. We define that $[a \uparrow \theta]\Phi = \{\psi \mid [a \uparrow \theta]\psi \in \Phi\}$, called $[a \uparrow \theta]\Phi$ is upgraded by $[a \uparrow \theta]$. Similar, we defined $\Phi_i = \{\psi \mid @_i\psi \in \Phi\}$, it's yielded by Φ. Generally, for $\overleftarrow{e}$, let $\overleftarrow{c} = (c_n\theta_n, \overleftarrow{e})$, $[\overleftarrow{c}]\Phi$ is upgraded by $[c_n \uparrow \theta_n]$.*

According to Definition 6, we have the following lemma for upgraded mcss.

Lemma 1 (mcs upgrading). *Let Δ be an mcs and $\overleftarrow{a}$ be any sequence, $[\overleftarrow{a}]\Delta$ is an mcs.*

Proof. Proof by induction on $length(\overleftarrow{a})$. When $length(\overleftarrow{a}) = 0$, it is trivial. We show the case that $length(\overleftarrow{a}) = 1$. Let $\varphi \in$DHLF, and Δ is an mcs. We prove that $[a \uparrow \theta]\Delta$ is an mcs.

- Assume that $[a \uparrow \theta]\Delta$ is not maximal, then there is a Δ' such that $[a \uparrow \theta]\Delta \subset \Delta'$ and $\Delta' \not\vdash \bot$. Then there exists a $\chi \in \Delta'$ but $\chi \notin [a \uparrow \theta]\Delta$. By Definition 6, we have $[a \uparrow \theta]\chi \notin \Delta$. For the property of mcs, we have $\neg[a \uparrow \theta]\chi \in \Delta$, by $[]\mathtt{dual}$, $[a \uparrow \theta]\neg\chi \in \Delta$. By Definition 6, $\neg\chi \in [a \uparrow \theta]\Delta$, that's $\neg\chi \in \Delta'$, then $\Delta' \vdash \bot$, a contradiction.
- Assume that $[a \uparrow \theta]\Delta$ is not consistent. There exists $\psi_1, \ldots, \psi_m \in [a \uparrow \theta]\Delta$ such that $\vdash (\psi_1 \wedge \cdots \wedge \psi_m) \rightarrow \bot$, that's $\vdash \neg(\psi_1 \wedge \cdots \wedge \psi_m)$. By $\mathtt{Nec}[]$, $\vdash [a \uparrow \theta]\neg(\psi_1 \wedge \cdots \wedge \psi_m)$, and we have $[a \uparrow \theta]\neg(\psi_1 \wedge \cdots \wedge \psi_m) \in \Delta$ by mcs property, thus $\neg[a \uparrow \theta](\psi_1 \wedge \cdots \wedge \psi_m) \in \Delta$ by $[]\mathtt{dual}$, $[a \uparrow \theta](\psi_1 \wedge \cdots \wedge \psi_m) \notin \Delta$ by consistency. Also, we have $[a \uparrow \theta]\psi_1, \ldots, [a \uparrow \theta]\psi_m \in \Delta$ from Definition 6 and $\psi_1, \ldots, \psi_m \in [a \uparrow \theta]\Delta$. Then $[a \uparrow \theta]\psi_1 \wedge \ldots \wedge [a \uparrow \theta]\psi_m \in \Delta$ by mcs property, thus $[a \uparrow \theta](\psi_1 \wedge \ldots \wedge \psi_m) \in \Delta$ by $[]\mathtt{and}$, a contradiction to the consistency of Δ.

As θ, φ are arbitrary, then we have proved that for $length([\overleftarrow{a}]) = 1$. As $[\overleftarrow{a}]\Delta$ is an mcs, then the cases for $length([\overleftarrow{a}]) > 1$ can easily be proved by induction.

For any mcs Δ, if there is a nominal $j \in \Delta$, then we call Δ a *named* mcs. Recall Definition 6, we know that if Δ is an mcs and Δ_i is yielded by Δ, then $i \in \Delta_i$ from $@_i i \in \Delta$ (by refx). Δ_i is *named* (by i), and have the following lemma.

Lemma 2. *For any named mcs Δ_i, we have $[\overleftarrow{a}]\Delta_i$ is a named mcs and $i \in [\overleftarrow{a}]\Delta_i$.*

[5] $@_b\theta \rightarrow @_a[a \uparrow \theta]\Diamond b$ indicates that if b satisfies θ, then b will be followed by a after the executing $[a \uparrow \theta]$. $@_a[a \uparrow \theta]\Diamond b \rightarrow @_b\theta$ says that since agent a is following b after executing $[a \uparrow \theta]$, then agent b satisfies θ.

Proof. Let Δ_i be a named mcs, by Lemma 1, we have $[\overleftarrow{a}]\Delta_i$ as an mcs. We then show that $[\overleftarrow{a}]\Delta_i$ is named by i. Induction on $length([\overleftarrow{a}])$, we have $\vdash [\overleftarrow{a}]i \leftrightarrow i$ from [] ignr. Therefore $i \in [\overleftarrow{a}]\Delta_i$ from $i \in \Delta_i$ and Definition 6.

The named mcs Δ_i and upgraded mcs $[i \uparrow \theta]\Delta_i$ are not necessarily to be equal, i.e., we can have $\neg\Diamond j \in \Delta_i$, but $\Diamond j \in [i \uparrow j]\Delta_i$. Therefore, the set of all mcss (including named mcss and upgraded mcss) can not be the "agents set" for our network models, because the *unique property* of nominals is lost.

Therefore, to get a "proper model", we need a "proper set of mcss". We will use the "yield-idea" in Definition 6 for the solution. As we know that a named model (agents are all named) has nice properties for *pure formulas* (formulas contains no propositional variables, only nominals), we want to build a *named model* for our completeness proof.

Then the other problem is that we can not guarantee that all mcss are *named*. If we are able to make our "proper model" named (all mcss in the model are named), we then can prove the completeness of pure axiomatic extensions (with pure formulas as axioms or rules) by using a similar lemma as in [4, Lemma 7.22].

3.2 Axiomatisation for Named Model

We will introduce two important rules, Name and Paste, for defining named models. To get the proof, firstly, we extend our nominals set Agt with Agt′, and then update DHLF to DHLF′. When we refer to the lemmas or propositions before, we insist that using npemcs instead of mcs if it was talking about mcs.

Definition 7 (Name and Paste). *Let $a \neq b$ be two nominals, and both not occurring in φ or ψ, we have the following rules:*
Name *if $a \rightarrow \varphi$ is provable, then so is φ.*
Paste *$(@_a\Diamond b \wedge @_b\varphi) \rightarrow \psi$ is provable, then so is $@_a\Diamond\varphi \rightarrow \psi$.*

The soundness of the above rules are trivial from the basic hybrid logic. Name rule helps us to name each mcss, and Paste rule helps us to prove the Existence lemma. Now let **DHLF** + Name + Paste be the logic obtained by adding Name and Paste rules to **DHLF**.

We say that a **DHLF** + Name + Paste-mcs Δ is *pasted*, iff $@_i\Diamond\varphi \in \Delta$ implies that for some nominal j, $@_i\Diamond j \wedge @_j\varphi \in \Delta$. Then we will show the extended Lindenbaum's lemma and define named models.

Lemma 3 (Extended Lindenbaum lemma). *Let* Agt′ *be a countably infinite collection of nominals disjoint from* Agt, *and* DHLF′ *is the language generated by adding* Agt′ *to* DHLF. *Then every* **DHLF** + Name + Paste-*consistent set of formulas in* DHLF *can be extended to a named and pasted* **DHLF**+Name+Paste-mcs.

Proof. The proof is similar as the proof in [4, Lemma 7.25]. Name rule has been used for giving a fresh name from Agt′ to a consistent set. Paste rule has contributed for the set's paste-property (guaranteeing that we have sufficient named sets), we can then use nominals as Henkin constants.

Now we can move to the definition of "named models". For brevity and clarity, from now on, we use emcs stands for "**DHLF** + Name + Paste-mcs", and npemcs for "named and pasted emcs".

Lemma 4 (Nominals). *Let Γ be an* npemcs*. For every nominal i, we define $\Delta_i = \{\varphi \mid @_i\varphi \in \Gamma\}$. Then we have:*

1. *For each nominal i, Δ_i is an* npemcs *that contains i.*
2. *For all nominals i and j, if $i \in \Delta_j$, then $\Delta_i = \Delta_j$.*
3. *For all nominals i and j, $@_i\varphi \in \Delta_j$ iff $@_i\varphi \in \Gamma$.*
4. *If d is a name for Γ, then $\Gamma = \Delta_d$.*

Proof. We refer the proof to [4, Lemma 7.24].

Corollary 1 (Upgraded nominals). *Let Γ be an* npemcs*, and $\overrightarrow{a}$ be a sequence, we define $[\overleftarrow{a}]\Gamma = \{\varphi \mid [\overleftarrow{a}]\varphi \in \Gamma\}$. Then we have:*

1. *$[\overleftarrow{a}]\Delta_i = \{\varphi \mid @_i\varphi \in [\overleftarrow{a}]\Gamma\}$, for $\Delta_i = \{\varphi \mid @_i\varphi \in \Gamma\}$.*
2. *For each nominal i, $[\overleftarrow{a}]\Delta_i$ is an* npemcs *that contains i.*
3. *For all nominals i and j, if $i \in [\overleftarrow{a}]\Delta_j$, then $[\overleftarrow{a}]\Delta_i = [\overleftarrow{a}]\Delta_j$.*
4. *For all nominals i and j, $@_i\psi \in [\overleftarrow{a}]\Delta_j$ iff $@_i\psi \in [\overleftarrow{a}]\Gamma$.*
5. *If d is a name for Γ, then $[\overleftarrow{a}]\Gamma = [\overleftarrow{a}]\Delta_d$.*

Proof. (2), (3), and (5) are easy by using Lemmas 1, 2 and 4. We just show (1) and (4). For (1), by induction on the sequence of $\overrightarrow{a}$ with []@, we can see that (i) $[\overleftarrow{a}]@_i\varphi \leftrightarrow @_i[\overleftarrow{a}]\varphi$ is a theorem. By Definition 6, we know that $[\overleftarrow{a}]\Delta_i = \{\varphi \mid [\overleftarrow{a}]\varphi \in \Delta_i\}$. As $\Delta_i = \{\varphi \mid @_i\varphi \in \Gamma\}$, then $[\overleftarrow{a}]\Delta_i = \{\varphi \mid @_i[\overleftarrow{a}]\varphi \in \Gamma\}$. Apply (i), $[\overleftarrow{a}]\Delta_i = \{\varphi \mid [\overleftarrow{a}]@_i\varphi \in \Gamma\}$, then by Definition 6, $[\overleftarrow{a}]\Delta_i = \{\varphi \mid @_i\varphi \in [\overleftarrow{a}]\Gamma\}$, we know that $[\overleftarrow{a}]\Delta_i$ is yielded by $[\overleftarrow{a}]\Gamma$.

For (4), $@_i\psi \in [a \uparrow \theta]\Delta_j$; iff $[a \uparrow \theta]@_i\psi \in \Delta_j$ by Definition 6; iff $@_i[a \uparrow \theta]\psi \in \Delta_j$ by []@; iff $@_i[a \uparrow \theta]\psi \in \Gamma$ by Lemma 4; iff $[a \uparrow \theta]@_i\psi \in \Gamma$ by []@; iff $@_i\psi \in [a \uparrow \theta]\Gamma$ by Definition 6.

3.3 Completeness for DHLF

Definition 8 (Named model). *Let Γ be an* npemcs*. The named model $\mathcal{M}^\Gamma$ yielded by Γ is defined as follows. For any $\Delta, \Delta' \in \mathcal{W}^\Gamma$, $a \in$*Agt′ *and $p \in$*Prop*:*

- *$\mathcal{W}^\Gamma$ is the set of all named set yielded by Γ;*
- *$\Delta\mathcal{R}^\Gamma\Delta'$ iff $\varphi \in \Delta'$ implies $\Diamond\varphi \in \Delta$;*
- *$\mathcal{G}^\Gamma(a) = \Delta$ iff $a \in \Delta$;*
- *$\mathcal{V}^\Gamma(p) = \{\Delta \mid p \in \Delta\}$.*

Meanwhile, the named model $\mathcal{M}^\Gamma$ updated by sequence $\overrightarrow{a}$ is $[\![\overleftarrow{a}]\!]\mathcal{M}^\Gamma$. We define $\mathcal{M}^{[\overleftarrow{a}]\Gamma}$ to be the named model yielded by $[\overleftarrow{a}]\Gamma$. In particular, $[\![\overleftarrow{a}]\!]\mathcal{M}^\Gamma = \mathcal{M}^{[\overleftarrow{a}]\Gamma} = \mathcal{M}^\Gamma$ if $length(\overleftarrow{a}) = 0$.

Now we show the Existence lemma as follows.

Lemma 5 (Existence lemma). *Let Γ be an* `npemcs`. *$\mathcal{M}^\Gamma = (\mathcal{W}^\Gamma, \mathcal{R}^\Gamma, \mathcal{G}^\Gamma, \mathcal{V}^\Gamma)$ be the named model yielded by Γ. If $\Delta \in \mathcal{W}^\Gamma$ and $\Diamond\varphi \in \Delta$, then there exists $\Delta' \in \mathcal{W}^\Gamma$ such that $\Delta \mathcal{R}^\Gamma \Delta'$ and $\varphi \in \Delta'$.*

Proof. Let $\Delta \in \mathcal{W}^\Gamma$ and $\Diamond\varphi \in \Delta$, then there exists a nominal i such that $\Delta = \Delta_i$. From $\Diamond\varphi \in \Delta_i$, we have $@_i\Diamond\varphi \in \Gamma$. Since Γ is pasted, we have $@_i\Diamond j \wedge @_j\varphi$ for some j. So we have $\Diamond j \in \Delta_i$ and $\varphi \in \Delta_j$. If we can show that $\Delta_i \mathcal{R}^\Gamma \Delta_j$, Δ_j will be a suitable choice of Δ'. Let $\psi \in \Delta_j$, then we have $@_j\psi \in \Gamma$. Then by (3) of Lemma 4, we have $@_j\psi \in \Delta_i$. Also we have $\Diamond j \in \Delta_i$, by theorem $(\Diamond j \wedge @_j\psi) \to \Diamond\psi$ (`bridge`), therefore, $\Diamond\psi \in \Delta_i$. $\Delta_i \mathcal{R}^\Gamma \Delta_j$ is proved, and we let $\Delta' = \Delta_j$.

Corollary 2. *Let Γ be an* `npemcs`. *$\mathcal{M}^\Gamma = (\mathcal{W}^\Gamma, \mathcal{R}^\Gamma, \mathcal{G}^\Gamma, \mathcal{V}^\Gamma)$ be the named model yielded by Γ. For any $\Delta_i, \Delta_j \in \mathcal{W}^\Gamma$, we have that $\Delta_i \mathcal{R}^\Gamma \Delta_j$ iff $\Diamond j \in \Delta_i$.*

Proof. The left-to-right direction is trivial, since $j \in \Delta_j$. For the other direction. Assume that $\Diamond j \in \Delta_i$ but not $\Delta_i \mathcal{R}^\Gamma \Delta_j$. Then there exists $\varphi \in \Delta_j$ but $\Diamond\varphi \notin \Delta_i$. That's $\neg\Diamond\varphi \in \Delta_i$, $\Box\neg\varphi \in \Delta_i$. Therefore, $\Diamond j \wedge \Box\neg\varphi \in \Delta_i$, by `PROP`, $\Diamond(j \wedge \neg\varphi) \in \Delta_i$. By theorem of `bridge` (contraposed case), $@_j\neg\varphi \in \Delta_i$, that's $@_j\neg\varphi \in \Delta_j$ and $\neg\varphi \in \Delta_j$, a contradiction. Then we must have $\Delta_i \mathcal{R}^\Gamma \Delta_j$.

As we have action operators for changing models, we show that $[\![\overleftarrow{a}]\!]\mathcal{M}^\Gamma$ and $\mathcal{M}^{[\overleftarrow{a}]\Gamma}$ are modal equivalent in our logic. To achieve this goal, we show the following proposition first. The Truth lemma result of propositional cases (propositional language `PROP` generated from `Agt'`$\cup$`Prop`) will be used: (Atom)[6] for any $\theta \in$`PROP`, $\mathcal{M}^\Gamma, \Delta_i \models \theta$ iff $\theta \in \Delta_i$.

Proposition 4. *Let Γ be an* `npemcs` *and $[a \uparrow \theta]\Gamma$ is upgraded by $[a \uparrow \theta]$. $\mathcal{M}^\Gamma$ and $\mathcal{M}^{[a\uparrow\theta]\Gamma}$ be two named models yielded by Γ and $[a \uparrow \theta]\Gamma$ respectively. We have for any $\Delta_i, \Delta_j \in \mathcal{W}^\Gamma$: $[a \uparrow \theta]\Delta_i \mathcal{R}^{[a\uparrow\theta]\Gamma}[a \uparrow \theta]\Delta_j$ iff $\Delta_i \mathcal{R}^{\Gamma^{[\frac{a}{\theta}]}}\Delta_j$.*

Proof. We prove by discussion on $\mathcal{G}^\Gamma(a)$:

- $\mathcal{G}^\Gamma(a) = \Delta_i$, we show that $[a \uparrow \theta]\Delta_i \mathcal{R}^{[a\uparrow\theta]\Gamma}[a \uparrow \theta]\Delta_j$ iff $\Delta_i \mathcal{R}^{\Gamma^{[\frac{a}{\theta}]}}\Delta_j$ as follows.
  $[a \uparrow \theta]\Delta_i \mathcal{R}^{[a\uparrow\theta]\Gamma}[a \uparrow \theta]\Delta_j$ iff $j \in [a \uparrow \theta]\Delta_j$ and $\Diamond j \in [a \uparrow \theta]\Delta_i$ by Corollary 2
  iff $[a \uparrow \theta]j \in \Delta_j$ and $[a \uparrow \theta]\Diamond j \in \Delta_i$
  iff $[a \uparrow \theta]j \in \Delta_j$ and $@_a[a \uparrow \theta]\Diamond j \in \Delta_i$ by $a \in \Delta_i$
  iff $j \in \Delta_j$ and $@_j\theta \in \Delta_i$ by `[]ignr` and `[]sinc`
  iff $@_j\theta \in \Delta_j$ by Lemma 4
  iff $\mathcal{M}^\Gamma, \Delta_j \models \theta$ by (Atom), since $\theta \in$`PROP`
- $\mathcal{G}^\Gamma(a) \neq \Delta_i$, we show that $[a \uparrow \theta]\Delta_i \mathcal{R}^{[a\uparrow\theta]\Gamma}[a \uparrow \theta]\Delta_j$ iff $\Delta_i \mathcal{R}^\Gamma \Delta_j$ as follows.
  $[a \uparrow \theta]\Delta_i \mathcal{R}^{[a\uparrow\theta]\Gamma}[a \uparrow \theta]\Delta_j$ iff $j \in [a \uparrow \theta]\Delta_j$ and $\Diamond j \in [a \uparrow \theta]\Delta_i$ by Corollary 2
  iff $[a \uparrow \theta]j \in \Delta_j$ and $[a \uparrow \theta]\Diamond j \in \Delta_i$
  iff $[a \uparrow \theta]j \in \Delta_j$ and $\Diamond[a \uparrow \theta]j \in \Delta_i$
  by `[]dia`, and $\neg a \in \Delta_i$, as $\mathcal{G}^\Gamma(a) \neq \Delta_i$
  iff $j \in \Delta_j$ and $\Diamond j \in \Delta_i$ by `[]ignr`
  iff $\Delta_i \mathcal{R}^\Gamma \Delta_j$ by Corollary 2

6 As the propositional case is trivial, and it's also a technical reason why we need to restrict our actions into propositional formulas.

From Proposition 4, by induction on the $length(\overrightarrow{a})$, we can easily prove the following corollary.

Corollary 3. *Let $\overrightarrow{c}, a\theta$ be a sequence, and Γ be an* `npemcs`. *$\mathcal{M}^{\Gamma}$ and $\mathcal{M}^{[a\uparrow\theta][\overleftarrow{c}]\Gamma}$ be two named models yielded by Γ and $[a \uparrow \theta][\overleftarrow{c}]\Gamma$ respectively. For any $\Delta_i, \Delta_j \in \mathcal{W}^{\Gamma}$:*

$$[a \uparrow \theta][\overleftarrow{c}]\Delta_i \mathcal{R}^{[a\uparrow\theta][\overleftarrow{c}]\Gamma}[a \uparrow \theta][\overleftarrow{c}]\Delta_j \textit{ iff } \Delta_i \mathcal{R}^{\Gamma[\![\overrightarrow{c}, a\theta]\!]}\Delta_j.$$

Now we can show the correspondence theorem for $[\![\overleftarrow{a}]\!]\mathcal{M}^{\Gamma}$ and $\mathcal{M}^{[\overleftarrow{a}]\Gamma}$ as follows.

Theorem 1. *Let Γ be an* `npemcs`, *and $\overrightarrow{a}$ be a sequence. $\mathcal{M}^{\Gamma}$ and $\mathcal{M}^{[\overleftarrow{a}]\Gamma}$ are named models yielded by Γ and $[\overleftarrow{a}]\Gamma$ respectively. We have for any φ and $\Delta_i \in \mathcal{W}^{\Gamma}$: $[\![\overleftarrow{a}]\!]\mathcal{M}^{\Gamma}, \Delta_i \models \varphi$ iff $\mathcal{M}^{[\overleftarrow{a}]\Gamma}, [\overleftarrow{a}]\Delta_i \models \varphi$.*

Proof. As we just showed in (1) of Corollary 1 that for any Δ_i yielded by Γ, $[\overleftarrow{a}]\Delta_i$ is yielded by $[\overleftarrow{a}]\Gamma$ as well. Then we know that for any nominal i,

(i) $\Delta_i \in \mathcal{V}^{\Gamma}(p)$ iff $[\overleftarrow{a}]\Delta_i \in \mathcal{V}^{[\overleftarrow{a}]\Gamma}(p)$, and
(ii) $\mathcal{G}^{\Gamma}(i) = \Delta_i$ iff $\mathcal{G}^{[\overleftarrow{a}]\Gamma}(i) = [\overleftarrow{a}]\Delta_i$

by Definition 8 and theorems $[\overleftarrow{a}]p \leftrightarrow p, [\overleftarrow{a}]i \leftrightarrow i$ from `[]ignr`. We prove by induction on the complexity of φ.

- $\varphi = p$. The following are equivalent:

$[\![\overleftarrow{a}]\!]\mathcal{M}^{\Gamma}, \Delta \models p$	
$\mathcal{M}^{\Gamma}, \Delta \models [\overleftarrow{a}]p$	Definition 2
$\mathcal{M}^{\Gamma}, \Delta \models p$	theorem of `[]ignr`
$\Delta \in \mathcal{V}^{\Gamma}(p)$	Definition 2
$[\overleftarrow{a}]\Delta \in \mathcal{V}^{[\overleftarrow{a}]\Gamma}(p)$	(i)
$\mathcal{M}^{[\overleftarrow{a}]\Gamma}, [\overleftarrow{a}]\Delta \models p$	Definition 2

- $\varphi = i$. Similar as above.
- $\varphi = \neg\psi$. Omitted.
- $\varphi = (\psi \wedge \psi')$. Omitted.
- $\varphi = \Diamond\psi$. The following are equivalent:

$[\![\overleftarrow{a}]\!]\mathcal{M}^{\Gamma}, \Delta \models \Diamond\psi$	
$[\![\overleftarrow{a}]\!]\mathcal{M}^{\Gamma}, \Delta' \models \psi$ and $\Delta\mathcal{R}^{\Gamma[\![\overrightarrow{a}]\!]}\Delta'$ for some Δ'	Definition 2
$[\![\overleftarrow{a}]\!]\mathcal{M}^{\Gamma}, \Delta' \models \psi$ and $[\overleftarrow{a}]\Delta\mathcal{R}^{[\overleftarrow{a}]\Gamma}[\overleftarrow{a}]\Delta'$	Corollary 3.
$\mathcal{M}^{[\overleftarrow{a}]\Gamma}, [\overleftarrow{a}]\Delta' \models \psi$ and $[\overleftarrow{a}]\Delta\mathcal{R}^{[\overleftarrow{a}]\Gamma}[\overleftarrow{a}]\Delta'$	I.H.
$\mathcal{M}^{[\overleftarrow{a}]\Gamma}, [\overleftarrow{a}]\Delta \models \Diamond\psi$	Definition 2

- $\varphi = @_b\psi$. The following are equivalent:

$[\![\overleftarrow{a}]\!]\mathcal{M}^{\Gamma}, \Delta \models @_b\psi$	
$[\![\overleftarrow{a}]\!]\mathcal{M}^{\Gamma}, \Delta' \models \psi$	where $\mathcal{G}^{\Gamma}(b) = \Delta'$
$\mathcal{M}^{[\overleftarrow{a}]\Gamma}, [\overleftarrow{a}]\Delta' \models \psi$	I.H.
$\mathcal{M}^{[\overleftarrow{a}]\Gamma}, [\overleftarrow{a}]\Delta \models @_b\psi$	Definition 2, as $\mathcal{G}^{[\overleftarrow{a}]\Gamma}(b) = [\overleftarrow{a}]\Delta'$

– $\varphi = [b \uparrow \chi]\psi$. The following are equivalent:

$$\begin{array}{ll}
[\![\overleftarrow{a}]\!]\mathcal{M}^{\Gamma}, \Delta \models [b \uparrow \chi]\psi & \\
[\![b \uparrow \chi]\!][\![\overleftarrow{a}]\!]\mathcal{M}^{\Gamma}, \Delta \models \psi & \text{Definition 2} \\
\mathcal{M}^{[b \uparrow \chi][\overleftarrow{a}]\Gamma}, [b \uparrow \chi][\overleftarrow{a}]\Delta \models \psi & \text{I.H. for sequence } \overrightarrow{a}, b\chi \\
[\![b \uparrow \chi]\!]\mathcal{M}^{[\overleftarrow{a}]\Gamma}, [\overleftarrow{a}]\Delta \models \psi & \text{I.H. for sequence } b\chi \\
\mathcal{M}^{[\overleftarrow{a}]\Gamma}, [\overleftarrow{a}]\Delta \models [b \uparrow \chi]\psi & \text{Definition 2}
\end{array}$$

Lemma 6 (Truth lemma). *Let Γ be an* `npemcs`*, and $\mathcal{M}^{\Gamma} = (\mathcal{W}^{\Gamma}, \mathcal{R}^{\Gamma}, \mathcal{G}^{\Gamma}, \mathcal{V}^{\Gamma})$ be the named model yielded by Γ. We have for any $\Delta \in \mathcal{W}^{\Gamma}$ and $\psi \in$*DHLF*,*

$$\mathcal{M}^{\Gamma}, \Delta \models \psi \quad \textit{iff} \quad \psi \in \Delta.$$

Proof. Induction on the complexity of ψ. Cases for $\psi = p, i, \neg\varphi, (\varphi \wedge \varphi')$ are easy, we just show the rest here.

– $\psi = \Diamond\varphi$. The following are equivalent:

$$\begin{array}{ll}
\mathcal{M}^{\Gamma}, \Delta \models \Diamond\varphi & \\
\mathcal{M}^{\Gamma}, \Delta' \models \chi \text{ and } \Delta\mathcal{R}^{\Gamma}\Delta' & \text{by Definition 2} \\
\chi \in \Delta' \text{ and } \Delta\mathcal{R}^{\Gamma}\Delta' & \text{by I.H.} \\
\Diamond\varphi \in \Delta & \text{by Definition 8}
\end{array}$$

– $\psi = @_a\varphi$. The following are equivalent:

$$\begin{array}{ll}
\mathcal{M}^{\Gamma}, \Delta \models @_a\varphi & \\
\mathcal{M}^{\Gamma}, \Delta' \models \chi \text{ and } \mathcal{G}^{\Gamma}(a) = \Delta' & \text{by Definition 2} \\
\varphi \in \Delta' \text{ and } \mathcal{G}^{\Gamma}(a) = \Delta' & \text{by I.H.} \\
@_a\varphi \in \Delta' & \text{by } \texttt{intro} \\
@_a\varphi \in \Delta & \text{by (3) of Lemma 4}
\end{array}$$

– $\psi = [a \uparrow \theta]\varphi$. The following are equivalent:

$$\begin{array}{ll}
\mathcal{M}^{\Gamma}, \Delta \models [a \uparrow \theta]\varphi & \\
[\![a \uparrow \theta]\!]\mathcal{M}^{\Gamma}, \Delta \models \varphi & \text{by Definition 2} \\
\mathcal{M}^{[a \uparrow \theta]\Gamma}, [a \uparrow \theta]\Delta \models \varphi & \text{by Theorem 1} \\
\varphi \in [a \uparrow \theta]\Delta & \text{by I.H., Definition 8} \\
[a \uparrow \theta]\varphi \in \Delta & \text{by Definition 6}
\end{array}$$

Theorem 2 (Completeness). *(1) Every named and pasted consistent set of formulas in* DHLF *is satisfiable in a countable named model. (2) Moreover, if Π is a set of pure formulas in* DHLF*, and* **Ex** *is adding all the formulas in Π as extra axioms to* **DHLF**+`Name`+`Paste`*, then every* **Ex***-consistent set of formulas is satisfiable in a countable named model based on a frame which validates every formula in Π.*

Proof. For (1): Proof by contraposition. Assume that Δ is not a consistent set of formulas in DHLF: $\Delta \nvdash \varphi$. Then $\Delta \cup \{\neg\varphi\}$ is consistent. We know that $\Delta \cup \{\neg\varphi\}$ is a set of formulas generated from DHLF, then there exists an **npemcs**, called Γ, from Lemma 3 such that $\Delta \cup \{\neg\varphi\} \subseteq \Gamma$. By Definition 8, we generate the named model yielded by Γ, $\mathcal{M}^\Gamma$ (The model is countable, since every the state is named by some nominals in Agt$\cup$ Agt$'$, and Agt$\cup$ Agt$'$ is countable). Then $\mathcal{M}^\Gamma, \Gamma \models \neg\varphi$ by Lemma 6, that's $\mathcal{M}^\Gamma, \Gamma \not\models \varphi$, and $\not\models \varphi$. The proof of (1) is completed.

For (2): Given an **Ex**-consistent set of formulas Ω, using Lemma 3, we can extend it to a named and pasted **Ex-mcs** Ω^+. Clearly we have $\mathcal{M}^{\Omega^+}, \Omega^+ \models \Omega$. Since for pure formulas φ and it's *pure instance* φ' (φ' is obtained from φ by uniformly substituting nominals for nominals), we have for any $M = (F, G, V)$:

$$M \models \varphi \text{ iff } F \models \varphi'.$$

Also we know that $\mathcal{M}^{\Omega^+}, \Omega^+ \models \Pi$ from the property that every formulas in Π belongs to every **Ex**-mcs. Therefore, the frame underlying $\mathcal{M}^{\Omega^+}$ validates Π.

4 Conclusion

We have offered an axiomatisation of DHLF and proved its completeness. Moreover, the completeness of pure axiomatic extensions is offered. With pure formulas, we can define many properties *not definable* in the basic modal logic, like *irreflexivity:* $i \to \neg\Diamond i$, *asymmetry:* $i \to \neg\Diamond i$, *intransitivity:* $\Diamond\Diamond i \to \neg\Diamond i$, etc.

In [16], they introduced a similar method of changing relation for preference, called *upgrading preference*, but our semantics are different, our actions are marked by agents, and then it's possible to talk about multi-agents' choices. Also our semantics are based on hybrid logic, we automatically have the hybrid logic property.

Moreover, if we release the action operators from the restriction of propositional formulas, and allow agents to following by arbitrary φ-property as in public announcement logic [17], we will have more interesting features of actions. Like $[a \uparrow @_b\theta]$ will add or delete all "arrows" from a to all the other agents, by the property of @-formulas. If b has the θ-property, then adding; deleting, otherwise.

With this extension, repeating same actions could be different. Under the same semantics, action $[a \uparrow \neg\Diamond b]$ means that agent a chooses to follow all the agents that not a follower of b, and let's consider Fig. 3.[7] We call it *repetition regrets*, since agent will regret with some "arrows" till the structure is *stable* under that action-repetition.

The completeness of this extension so far is not clear, as the proof of Proposition 4 has required the propositional Truth lemma property. It will be our next job to solve the problem and offer the axiomatisation of this extension.

---

[7] Due to the property of $\Diamond$, it is not possible to interpret the action that "agent a chooses to follow all agents that are followed by b", actions like $[a \uparrow (i \wedge @_b\Diamond i)]$ only means that "agent a chooses to follow agent i and b is following i".

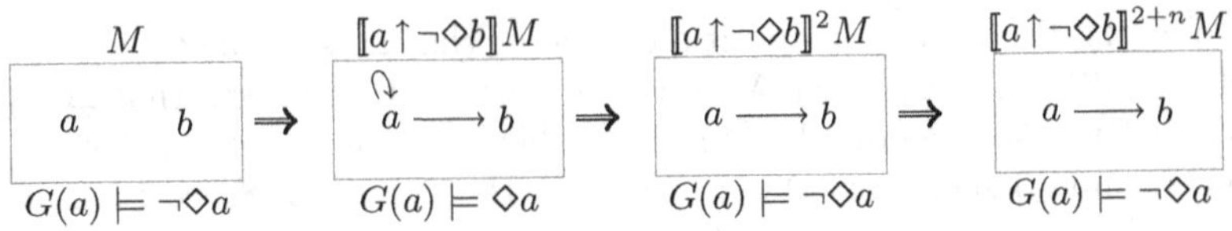

Fig. 3. a, b are nominals, numbers indicating times for executing $[a \uparrow \neg\Diamond b]$.

Another extension is extending the language with knowledge and messaging operators like $[a : \theta]$ as in [18, 19] to interpret the information flow on a particular networks. For instance, $[a \uparrow b][b : \theta]K_a\chi$ means that "If agent a chooses to follow b, then a will know χ if b messages θ". Semantically, we need introduce epistemic relations in our models to talk about knowledge, one strategy is to consider the epistemic indistinguishability of network models. Saying that "agent c does not know that "b is following a" in a model" means that the epistemic relation of c is connecting to models where b is not always following a, action models in dynamic epistemic logic (see, e.g., [17, Chapter 6]) would be a good tool to look at. We leave all the mentioned works to the future.

Acknowledgment. We thank anonymous reviewers of LORI-2019 for many helpful comments, and in particular, for the contribution of related works. The first author is supported by the Fundamental Research Funds for the Central Universities under research no. SWU1809669, and by the Key Research Funds for the Key Liberal Science Research Base of Chongqing under research no. 18SK045.

References

1. Areces, C., Fervari, R., Hoffmann, G.: Relation-changing modal operators. Log. J. IGPL **23**(4), 601–627 (2015)
2. Aucher, G., Balbiani, P., Cerro, L.F.D., Herzig, A.: Global and local graph modifiers. Electron. Notes Theor. Comput. Sci. **231**, 293–307 (2009)
3. Aucher, G., van Benthem, J., Grossi, D.: Sabotage modal logic: some model and proof theoretic aspects. In: van der Hoek, W., Holliday, W.H., Wang, W. (eds.) LORI 2015. LNCS, vol. 9394, pp. 1–13. Springer, Heidelberg (2015). https://doi.org/10.1007/978-3-662-48561-3_1
4. Blackburn, P., de Rijke, M., Venema, Y.: Modal Logic. Cambridge University Press, Cambridge (2001)
5. Christoff, Z., Hansen, J.U.: A logic for diffusion in social networks. J. Appl. Log. **13**(1), 48–77 (2015)
6. Renardel De Lavalette, G.R.: Changing modalities. J. Log. Comput. **14**(2), 251–275 (2004)
7. Harel, D., Tiuryn, J., Kozen, D.: Dynamic Logic. MIT Press, Cambridge (2000)
8. Kooi, B.: Expressivity and completeness for public update logics via reduction axioms. J. Appl. Non-Class. Log. **17**(2), 231–253 (2007)
9. Kooi, B., Renne, B.: Arrow update logic. Rev. Symbolic Log. **4**(4), 536–559 (2011)
10. Kooi, B., Renne, B.: Generalized arrow update logic. In: Proceedings of the 13th Conference on Theoretical Aspects of Rationality and Knowledge, TARK XIII, pp. 205–211. ACM, New York (2011)

11. Liu, F., Seligman, J., Girard, P.: Logical dynamics of belief change in the community. Synthese **191**(11), 2403–2431 (2014)
12. Seligman, J., Liu, F., Girard, P.: Logic in the community. In: Banerjee, M., Seth, A. (eds.) ICLA 2011. LNCS (LNAI), vol. 6521, pp. 178–188. Springer, Heidelberg (2011). https://doi.org/10.1007/978-3-642-18026-2_15
13. Seligman, J., Liu, F., Girard, P.: Facebook and the epistemic logic of friendship. In: Schipper, B.C. (ed.) Proceedings of the 14th Conference on Theoretical Aspects of Rationality and Knowledge, pp. 229–238, Chennai (2013)
14. van Benthem, J.: Dynamic logic for belief revision. J. Appl. Non-Class. Log. **17**, 129–156 (2007)
15. Benthem, J.: An essay on sabotage and obstruction. In: Hutter, D., Stephan, W. (eds.) Mechanizing Mathematical Reasoning. LNCS (LNAI), vol. 2605, pp. 268–276. Springer, Heidelberg (2005). https://doi.org/10.1007/978-3-540-32254-2_16
16. van Benthem, J., Liu, F.: The dynamics of preference upgrade. J. Appl. Non-Class. Log. **17**(2), 157–182 (2007)
17. van Ditmarsch, H., van der Hoek, W., Kooi, B.: Dynamic Epistemic Logic. Springer, Dordrecht (2007). https://doi.org/10.1007/978-1-4020-5839-4
18. Xiong, Z.: On the logic of multicast messaging and balance in social networks. Ph.D. thesis, University of Bergen (2017). ISBN 978-82-308-3928-7
19. Xiong, Z., Ågotnes, T., Seligman, J., Zhu, R.: Towards a logic of tweeting. In: Baltag, A., Seligman, J., Yamada, T. (eds.) LORI 2017. LNCS, vol. 10455, pp. 49–64. Springer, Heidelberg (2017). https://doi.org/10.1007/978-3-662-55665-8_4

Author Index

Printed in the United States
By Bookmasters